Student's Solutions Manual

CONCEPTS AND APPLICATIONS OF
Intermediate Algebra

SECOND EDITION

Bittinger / Keedy / Ellenbogen

Judith A. Penna

ADDISON-WESLEY PUBLISHING COMPANY
Reading, Massachusetts • Menlo Park, California • New York
Don Mills, Ontario • Wokingham, England • Amsterdam • Bonn
Sydney • Singapore • Tokyo • Madrid • San Juan

Reproduced by Addison-Wesley from camera-ready copy supplied by the author.

ISBN 0-201-17317-4

Reprinted with corrections February, 1992.

3 4 5 6 7 8 9 10 MU 95949392

TABLE OF CONTENTS

Special thanks are extended to Patsy Hammond for her
excellent typing. Her skill, patience, and efficiency
made the author's work much easier.

Exercise Set 1.1

1. Four less than some number
   Let n represent the number. Then we have
      n - 4.

2. n + 6, or 6 + n

3. Twice a number
   Let x represent the number. Then we have
      2x.

4. 5n

5. Thirty-two percent of some number
   Let n represent the number. Then we have
      32%n, or 0.32n.

6. 0.47n

7. Seven more than half of a number
   Let x represent the number. Then we have
      $\frac{1}{2}x + 7$, or $7 + \frac{1}{2}x$, or $\frac{x}{2} + 7$, or $7 + \frac{x}{2}$.

8. 2x - 8

9. Four less than nineteen percent of some number
   Let t represent the number. Then we have
      0.19t - 4.

10. 0.82x + 3

11. Five more than the difference of two numbers
    Let x represent the first number and y the
    second. Then we have
       x - y + 5, or 5 + x - y.

12. x + y - 6

13. Four less than the product of two numbers
    Let x and y represent the numbers. Then we have
       xy - 4, or yx - 4.

14. xy + 10

15. One more than thirty-five percent of some number
    Let t represent the number. Then we have
       0.35t + 1, or 1 + 0.35t.

16. 0.08x - 2

17. Substitute and carry out the multiplication and
    addition:
       2x + y = 2·3 + 7
             = 6 + 7
             = 13

18. 16

19. Substitute and multiply:
       7abc = 7·2·1·3
            = 42

20. 20

21. Substitute and carry out the operations indicated:
       8mn - p = 8·1·2 - 9
              = 16 - 9
              = 7

22. 47

23. Substitute and carry out the operations indicated:
       5ab ÷ c = 5·4·2 ÷ 10
              = 40 ÷ 10
              = 4

24. 6

25. Substitute and carry out the operations indicated:
       2ab - a = 2·5·3 - 5
              = 30 - 5
              = 25

26. 21

27. Substitute and carry out the operations indicated:
       pqr ÷ q = 2·3·2 ÷ 3
              = 12 ÷ 3
              = 4

28. 3

29. We substitute 3 for x and see whether we get a
    true sentence.

    $$\begin{array}{c|c} x - 2 = 1 \\ \hline 3 - 2 & 1 \\ 1 & \end{array}$$

    We get a true sentence, so 1 is a solution.

30. 4 is a solution.

31. We substitute 7 for a and see whether we get a
    true sentence.

    $$\begin{array}{c|c} 4 + a = 13 \\ \hline 4 + 7 & 13 \\ 11 & \end{array}$$

    We do not get a true sentence, so 7 is not a
    solution.

32. 6 is not a solution.

33. We substitute 6 for y and see whether we get a true sentence.

$$\frac{13 - y = 7}{13 - 6 \;\big|\; 7}$$
$$7 \;\big|$$

We get a true sentence, so 6 is a solution.

34. 7 is not a solution.

35. We substitute 5 for x and see whether we get a true sentence.

$$\frac{2x + 3 = 13}{2 \cdot 5 + 3 \;\big|\; 13}$$
$$10 + 3 \;\big|$$
$$13 \;\big|$$

We get a true sentence, so 5 is a solution.

36. 4 is a solution.

37. We substitute 0.4 for n and see whether we get a true sentence.

$$\frac{8n - 1.7 = 2.5}{8(0.4) - 1.7 \;\big|\; 2.5}$$
$$3.2 - 1.7 \;\big|$$
$$1.5 \;\big|$$

We do not get a true sentence, so 0.4 is not a solution.

38. 0.5 is a solution.

39. We substitute $\frac{17}{3}$ for x and see whether we get a true sentence.

$$\frac{3x - 4 = 13}{3\left[\frac{17}{3}\right] - 4 \;\big|\; 13}$$
$$17 - 4 \;\big|$$
$$13 \;\big|$$

We get a true sentence, so $\frac{17}{3}$ is a solution.

40. $\frac{9}{4}$ is a solution.

41. List the letters in the set: {a, e, i, o, u}, or {a, e, i, o, u, y}

42. {Sunday, Monday, Tuesday, Wednesday, Thursday, Friday, Saturday}

43. List the numbers in the set: {2, 4, 6, 8, . . .}

44. {1, 3, 5, 7, . . .}

45. List the numbers in the set: {5, 10, 15, 20, . . .}

46. {3, 6, 9, 12, . . .}

47. Specify the conditions under which a number is in the set: {x⎪x is an odd number between 10 and 30}

48. {x⎪x is a multiple of 4 between 22 and 45}

49. Specify the conditions under which a number is in the set: {x⎪x is a whole number less than 5}

50. {x⎪x is an integer less than 3 and greater than -4}

51. Specify the conditions under which a number is in the set: {n⎪n is a multiple of 5 between 7 and 79}

52. {x⎪x is an even number between 9 and 99}

53. Since $\frac{2}{5}$ means $2 \div 5$, we divide:

$$\begin{array}{r} 0.4 \\ 5\overline{)2.0} \\ 2.0 \end{array}$$

Decimal notation for $\frac{2}{5}$ is 0.4.

54. 0.6

55. We divide:

$$\begin{array}{r} 1.6\,6. . . \\ 3\overline{)5.0\;0} \\ 3 \\ \overline{2\;0} \\ 1\;8 \\ \overline{2\;0} \\ 1\;8 \\ \overline{2} \end{array}$$

We have a repeating decimal. Decimal notation for $\frac{5}{3}$ is $1.6\overline{6}$.

56. $1.1\overline{1}$

57. We divide:

$$\begin{array}{r} 0.1\;2\;5 \\ 8\overline{)1.0\;0\;0} \\ 8 \\ \overline{2\;0} \\ 1\;6 \\ \overline{4\;0} \\ 4\;0 \end{array}$$

Decimal notation for $\frac{1}{8}$ is 0.125.

58. 0.875

59. We divide:

$$\begin{array}{r} 0.8\;1\;8\;1. . . \\ 11\overline{)9.0\;0\;0\;0} \\ 8\;8 \\ \overline{2\;0} \\ 1\;1 \\ \overline{9\;0} \\ 8\;8 \\ \overline{2\;0} \\ 1\;1 \\ \overline{9} \end{array}$$

Decimal notation for $\frac{9}{11}$ is $0.\overline{81}$.

60. $0.\overline{63}$

61. We divide:

```
       1.4
   5 )7.0
     5
     2 0
     2 0
```

Decimal notation for $\frac{7}{5}$ is 1.4.

62. 2.25

63. We divide:

```
           0.1 2 3 . . .
   333 )4 1.0 0 0 0
        3 3 3
          7 7 0
          6 6 6
          1 0 4 0
            9 9 9
            4 1 0
```

Decimal notation for $\frac{41}{333}$ is $0.\overline{123}$.

64. $0.\overline{231}$

65. We divide:

```
         1.2 8 5 7 1 4 . . .
   7 )9.0 0 0 0 0 0
     7
     2 0
     1 4
       6 0
       5 6
         4 0
         3 5
           5 0
           4 9
             1 0
              7
              3 0
              2 8
               2
```

Decimal notation for $\frac{9}{7}$ is $1.\overline{285714}$.

66. $1.\overline{142857}$

67. $2.7 = 2\frac{7}{10} = \frac{27}{10}$

The last decimal place is tenths, so the denominator is 10.

68. $\frac{301}{100}$

69. $13.91 = 13\frac{91}{100} = \frac{1391}{100}$

The denominator is 100, because the last decimal place is hundredths.

70. $\frac{325}{10}$

71. $0.024 = \frac{24}{1000}$

The denominator is 1000, because the last decimal place is thousandths.

72. $\frac{103}{1000}$

73. $32.525 = 32\frac{525}{1000} = \frac{32,525}{1000}$

74. $\frac{46,375}{1000}$

75. $0.0275 = \frac{275}{10,000}$

The denominator is 10,000, because the last decimal place is ten thousandths.

76. $\frac{625}{10,000}$

77. $3.001 = 3\frac{1}{1000} = \frac{3001}{1000}$

78. $\frac{70,001}{10,000}$

79. $7.05 = 7\frac{5}{100} = \frac{705}{100}$

80. $\frac{305}{100}$

81. Three times the sum of the two numbers

Let x and y represent the numbers. The sum of the two numbers is x + y. Then three times their sum is 3(x + y).

82. $\frac{1}{2}(x - y)$

83. The quotient of the difference of two numbers and their sum

Let n and m represent the numbers. Their difference is n − m, and their sum is n + m. The quotient desired is then $\frac{n - m}{n + m}$.

84. (x + y)(x − y)

85. The only whole number that is not also a natural number is 0. Using roster notation to name the set, we have {0}.

86. {−1, −2, −3, . . .}

87. Observe that 2(x + 3) = 2x + 6, so the equation is an identity (is true for all real numbers). Using set-builder notation to name the set, we have {x∣x is a real number}.

88. {x∣x is a real number}

Exercise Set 1.2

---

1. |−7| = 7   −7 is 7 units from 0.

2. 9

3. |9| = 9   9 is 9 units from 0.

4. 12

5. |-6.2| = 6.2　　　-6.2 is 6.2 units from 0.

6. 7.9

7. |0| = 0　　　0 is 0 units from itself.

8. $3\frac{3}{4}$

9. $\left|1\frac{7}{8}\right| = 1\frac{7}{8}$　　　$1\frac{7}{8}$ is $1\frac{7}{8}$ units from 0.

10. 0.91

11. |-4.21| = 4.21　　　-4.21 is 4.21 units from 0.

12. 5.309

13. $-9 \leqslant -1$　　　-9 is less than or equal to -1, a true statement since -9 is to the left of -1 on the number line.

14. -1 is less than or equal to -5; false

15. -7 > 1　　　-7 is greater than 1, a false statement since -7 is to the left of 1 on the number line.

16. 7 is greater than or equal to -2; true

17. $3 \geqslant -5$　　　3 is greater than or equal to -5, a true statement since -5 is to the left of 3.

18. 9 is less than or equal to 9; true

19. -9 < -4　　　-9 is less than -4, a true statement since -9 is to the left of -4

20. 7 is greater than or equal to -8; true

21. $-4 \geqslant -4$　　　-4 is greater than or equal to -4. Since -4 = -4 is true, $-4 \geqslant -4$ is true.

22. 2 is less than 2; false

23. -5 < -5　　　-5 is less than -5, a false statement since -5 does not lie to the left of itself.

24. -2 is greater than -12; true

25. 5 + 12

Two positive numbers: Add the numbers, getting 17. The answer is positive, 17.

26. 16

27. -4 + (-7)

Two negative numbers: Add the absolute values, getting 11. The answer is negative, -11.

28. -11

29. -5.9 + 2.7

A negative and a positive number: The absolute values are 5.9 and 2.7. Subtract 2.7 from 5.9 to get 3.2. The larger absolute value came from the negative number, so the answer is negative, -3.2.

30. 5.4

31. $\frac{2}{7} + \left(-\frac{3}{5}\right) = \frac{10}{35} + \left(-\frac{21}{35}\right)$

A positive and a negative number. The absolute values are $\frac{10}{35}$ and $\frac{21}{35}$. Subtract $\frac{10}{35}$ from $\frac{21}{35}$ to get $\frac{11}{35}$. The larger absolute value came from the negative number, so the answer is negative, $-\frac{11}{35}$.

32. $-\frac{1}{40}$

33. -4.9 + (-3.6)

Two negative numbers: Add the absolute values, getting 8.5. The answer is negative, -8.5.

34. -9.6

35. $-\frac{1}{9} + \frac{2}{3} = -\frac{1}{9} + \frac{6}{9}$

A negative and a positive number. The absolute values are $\frac{1}{9}$ and $\frac{6}{9}$. Subtract $\frac{1}{9}$ from $\frac{6}{9}$ to get $\frac{5}{9}$. The larger absolute value came from the positive number, so the answer is positive, $\frac{5}{9}$.

36. $\frac{3}{10}$

37. 0 + (-4.5)

One number is zero: The sum is the other number, -4.5.

38. -3.19

39. -7.24 + 7.24

A negative and a positive number: The numbers have the same absolute value, 7.24, so the answer is 0.

40. 0

41. 15.9 + (-22.3)

A positive and a negative number: The absolute values are 15.9 and 22.3. Subtract 15.9 from 22.3 to get 6.4. The larger absolute value came from the negative number, so the answer is negative, -6.4.

42. -6.6

43. The additive inverse of 7.29 is -7.29, because -7.29 + 7.29 = 0.

44. -5.43

45. The additive inverse of -4.8 is 4.8, because
    -4.8 + 4.8 = 0.

46. 8.1

47. The additive inverse of 0 is 0, because
    0 + 0 = 0.

48. $2\frac{3}{4}$

49. The additive inverse of $-6\frac{1}{3}$ is $6\frac{1}{3}$, because
    $-6\frac{1}{3} + 6\frac{1}{3} = 0$.

50. $-4\frac{1}{5}$

51. If x = 7, then -x = -7. (The opposite of 7 is
    -7.)

52. -3

53. If x = -2.7, then -x = -(-2.7) = 2.7. (The
    opposite of -2.7 is 2.7.)

54. 1.9

55. If x = 1.79, then -x = -1.79. (The opposite of
    1.79 is -1.79.)

56. -3.14

57. If x = 0, then -x = 0. (The opposite of 0 is 0.)

58. 1

59. If x = -0.03, then -x = -(-0.03) = 0.03. (The
    opposite of -0.03 is 0.03.)

60. 1.09

61. 9 - 7 = 9 + (-7)   Change the sign and add.
          = 2

62. 5

63. 4 - 9 = 4 + (-9)   Change the sign and add.
          = -5

64. -7

65. -6 - (-10) = -6 + 10   Change the sign and add
               = 4

66. 6

67. -4 - 13 = -4 + (-13) = -17

68. -15

69. 2.7 - 5.8 = 2.7 + (-5.8) = -3.1

70. -0.5

71. $-\frac{3}{5} - \frac{1}{2} = -\frac{3}{5} + \left(-\frac{1}{2}\right)$

    $= -\frac{6}{10} + \left(-\frac{5}{10}\right)$ Finding a common denominator

    $= -\frac{11}{10}$

72. $-\frac{13}{15}$

73. -3.9 - (-6.8) = -3.9 + 6.8 = 2.9

74. -1.1

75. 0 - (-7.9) = 0 + 7.9 = 7.9

76. -5.3

77. (-4)7
    Two numbers with unlike signs: Multiply their
    absolute values, getting 28. The answer is
    negative, -28.

78. -45

79. (-3)(-8)
    Two numbers with the same sign: Multiply their
    absolute values, getting 24. The answer is
    positive, 24.

80. 56

81. (4.2)(-5)
    Two numbers with unlike signs: Multiply their
    absolute values, getting 21. The answer is
    negative, -21.

82. -28

83. (-7.2)(1)
    Two numbers with unlike signs: Multiply their
    absolute values, getting 7.2. The answer is
    negative, -7.2.

84. 5.9

85. 15.2 × 0 = 0

86. 0

87. (-3.2) × (-1.7)
    Two numbers with the same sign: Multiply their
    absolute values, getting 5.44. The answer is
    positive, 5.44.

88. 8.17

89. $\frac{-10}{-2}$

     Two numbers with the same sign: Divide their absolute values, getting 5. The answer is positive, 5.

90. 5

91. $\frac{-100}{20}$

     Two numbers with unlike signs: Divide their absolute values, getting 5. The answer is negative, -5.

92. -10

93. $\frac{73}{-1}$

     Two numbers with unlike signs: Divide their absolute values, getting 73. The answer is negative, -73.

94. -62

95. $\frac{0}{-7} = 0$

96. 0

97. $\frac{-42}{-6}$

     Two numbers with the same sign: Divide their absolute values, getting 7. The answer is positive, 7.

98. 8

99. The reciprocal of 5 is $\frac{1}{5}$, because $5 \cdot \frac{1}{5} = 1$.

100. $\frac{1}{3}$

101. The reciprocal of -9 is $\frac{1}{-9}$, or $-\frac{1}{9}$, because $-9\left(-\frac{1}{9}\right) = 1$.

102. $-\frac{1}{7}$

103. The reciprocal of $\frac{2}{3}$ is $\frac{3}{2}$, because $\frac{2}{3} \cdot \frac{3}{2} = 1$.

104. $\frac{7}{4}$

105. The reciprocal of $-\frac{3}{11}$ is $-\frac{11}{3}$, because $-\frac{3}{11}\left(-\frac{11}{3}\right) = 1$.

106. $-\frac{3}{7}$

107. $\frac{2}{3} \div \frac{4}{5} = \frac{2}{3} \cdot \frac{5}{4}$ Multiplying by the reciprocal of 4/5

      $= \frac{10}{12}$, or $\frac{5}{6}$

108. $\frac{5}{21}$

109. $\left(-\frac{3}{5}\right) \div \frac{1}{2} = -\frac{3}{5} \cdot \frac{2}{1}$ Multiplying by the reciprocal of 1/2

      $= -\frac{6}{5}$

110. $-\frac{12}{7}$

111. $-\frac{2}{9} \div \left(-\frac{3}{4}\right) = -\frac{2}{9} \cdot \left(-\frac{4}{3}\right) = \frac{8}{27}$

112. $-\frac{7}{22}$

113. $-\frac{3}{8} \div 1 = -\frac{3}{8} \cdot \frac{1}{1} = -\frac{3}{8}$

114. $\frac{2}{7}$

115. $\frac{7}{3} \div (-1) = \frac{7}{3}\left(-\frac{1}{1}\right) = -\frac{7}{3}$

116. $-\frac{5}{4}$

117. Since $35 = 7 \cdot 5$ we use $\frac{5}{5}$ as a name for 1:

      $\frac{3x}{7} = \frac{3x}{7} \cdot \frac{5}{5} = \frac{3x \cdot 5}{35} = \frac{15x}{35}$

118. $\frac{16a}{40}$

119. Since $32 = 8 \cdot 4$, we use $\frac{4}{4}$ as a name for 1:

      $-\frac{3a}{8} = -\frac{3a}{8} \cdot \frac{4}{4} = -\frac{3a \cdot 4}{32} = -\frac{12a}{32}$

120. $-\frac{44x}{33}$

121. $\frac{50x}{5} = \frac{10x \cdot 5}{1 \cdot 5}$ Factor numerator and denominator after identifying a common factor of 5

      $= \frac{10x}{1} \cdot \frac{5}{5}$ Factor the fractional expression, with one factor delete of 1

      $= 10x$ Leave off the factor of 1

122. 25x

123. $\frac{56a}{7} = \frac{8a \cdot 7}{1 \cdot 7} = \frac{8a}{1} \cdot \frac{7}{7} = 8a$

124. 8a

125. 5y + 2x

      = 2x + 5y   Commutative law of addition

126. x + 4y

127. 2x - 3y = 2x + (-3y)

             = -3y + 2x   Commutative law of addition

128. -4x + 5y

129. $\frac{x}{2} \cdot \frac{y}{3}$

     $= \frac{y}{3} \cdot \frac{x}{2}$ Commutative law of multiplication

130. $\frac{y}{4} \cdot 3$

131. $2 \cdot (8x)$

     $= (2 \cdot 8)x$ Associative law of multiplication

132. $(x \cdot 3)y$

133. $x + (2y + 5)$

     $= (x + 2y) + 5$ Associative law of addition

134. $3y + (4 + 10)$

135. $3(a + 1)$

     $= 3 \cdot a + 3 \cdot 1$ Using the distributive law

     $= 3a + 3$

136. $8x + 8$

137. $4(x - y)$

     $= 4 \cdot x - 4 \cdot y$ Using the distributive law

     $= 4x - 4y$

138. $9a - 9b$

139. $-5(2a + 3b)$

     $= -5 \cdot 2a + (-5) \cdot 3b$

     $= -10a - 15b$

140. $-6c - 10d$

141. $2a(b - c + d)$

     $= 2a \cdot b - 2a \cdot c + 2a \cdot d$

     $= 2ab - 2ac + 2ad$

142. $5xy - 5xz + 5xw$

143. $2\pi r(h + 1)$

     $= 2\pi r \cdot h + 2\pi r \cdot 1$

     $= 2\pi rh + 2\pi r$

144. $P + Prt$

145. $\frac{1}{2}h(a + b)$

     $= \frac{1}{2}h \cdot a + \frac{1}{2}h \cdot b$

     $= \frac{1}{2}ha + \frac{1}{2}hb$

146. $\pi r + \pi rs$

147. $8x + 8y$

     $= 8 \cdot x + 8 \cdot y$

     $= 8(x + y)$

148. $7(a + b)$

149. $9p - 9$

     $= 9 \cdot p - 9 \cdot 1$

     $= 9(p - 1)$

150. $12(x - 1)$

151. $7x - 21$

     $= 7 \cdot x - 7 \cdot 3$

     $= 7(x - 3)$

152. $6(y - 6)$

153. $xy + x$

     $= x \cdot y + x \cdot 1$

     $= x(y + 1)$

154. $a(b + 1)$

155. $2x - 2y + 2z$

     $= 2 \cdot x - 2 \cdot y + 2 \cdot z$

     $= 2(x - y + z)$

156. $3(x + y - z)$

157. $3x + 6y - 3$

     $= 3 \cdot x + 3 \cdot 2y - 3 \cdot 1$

     $= 3(x + 2y - 1)$

158. $4(a + 2b - 1)$

159. $4a + 5a$

     $= (4 + 5)a$

     $= 9a$

160. $12x$

161. $8b - 11b$

     $= (8 - 11)b$

     $= -3b$

162. $-3c$

163. $14y + y$

     $= 14y + 1y$

     $= (14 + 1)y$

     $= 15y$

164. $14x$

165. $12a - a$

     $= 12a - 1a$

     $= (12 - 1)a$

     $= 11a$

166. $14x$

167. $t - 9t$

  $= 1t - 9t$

  $= (1 - 9)t$

  $= -8t$

168. $-5x$

169. $5x - 3x + 8x$

  $= (5 - 3 + 8)x$

  $= 10x$

170. $-6x$

171. $5x - 8y + 3x$

  $= (5 + 3)x - 8y$

  $= 8x - 8y$

172. $13a - 10b$

173. $7c + 8d - 5c + 2d$

  $= (7 - 5)c + (8 + 2)d$

  $= 2c + 10d$

174. $7a + 9b$

175. $4x - 7 + 18x + 25$

  $= (4 + 18)x + (-7 + 25)$

  $= 22x + 18$

176. $9p + 12$

177. $13x + 14y - 11x - 47y$

  $= (13 - 11)x + (14 - 47)y$

  $= 2x - 33y$

178. $5a - 21b$

179. $a - (2a + 5)$

  $= a - 2a - 5$

  $= -a - 5$

180. $-4x - 9$

181. $4m - (3m - 1)$

  $= 4m - 3m + 1$

  $= m + 1$

182. $a + 3$

183. $3d - 7 - (5 - 2d)$

  $= 3d - 7 - 5 + 2d$

  $= 5d - 12$

184. $13x - 16$

185. Five less than seventy percent of a number

  Let x represent the number. Then we have $0.7x - 5$.

186. $\frac{1}{2}x + 2$

187. Substitute and carry out the operations indicated:

  $7xy - z = 7 \cdot 2 \cdot 1 - 20 = 14 - 20 = -6$

188. 23

189. $\frac{2x + 10}{3x + 15} = \frac{2(x + 5)}{3(x + 5)} = \frac{2}{3} \cdot \frac{x + 5}{x + 5} = \frac{2}{3}$

190. $\frac{2}{3}$

191. $\frac{6x - 12}{3x + 6} = \frac{3 \cdot 2(x - 2)}{3(x + 2)} = \frac{3}{3} \cdot \frac{2(x - 2)}{x + 2} = \frac{2(x - 2)}{x + 2}$

192. $\frac{x - 2}{2x + 3}$

193. $(0.0079x - 0.000843y) - (-0.00793x - 0.000853y)$

  $= 0.0079x - 0.000843y + 0.00793x + 0.000853y$

  $= 0.01583x + 0.00001y$

194. $38,520x - 10,102y$

195. Remove the innermost grouping symbols first, and work outward.

  $-\{-[-(-9)]\} = -\{-[9]\} = -\{-9\} = 9$

196. $-a + b$

197. $-|-[-4]| = -|4| = -4$

198. 3

Exercise Set 1.3

1. $3x = 12$ and $2x = 8$

  When x is replaced by 4, both equations are true, and for any other replacement, both equations are false. Thus the equations are equivalent.

2. Equivalent

3. $2x - 1 = -7$ and $x = -3$

  When x is replaced by -3, both equations are true, and for any other replacement, both equations are false. Thus the equations are equivalent.

4. Equivalent

5. $x + 5 = 11$ and $3x = 18$

  When x is replaced by 6, both equations are true, and for any other replacement, both equations are false. Thus the equations are equivalent.

6. Not equivalent

7. 13 - x = 4 and 2x = 20

When x is replaced by 9, the first equation is true, but the second equation is false. Thus the equations are not equivalent.

8. Equivalent

9. 5x = 2x and $\frac{4}{x} = 3$

When x is replaced by 0, the first equation is true, but the second equation is not defined. Thus the equations are not equivalent.

10. Not equivalent

11. $x - 5.2 = 9.4$

$x - 5.2 + 5.2 = 9.4 + 5.2$ Addition principle; adding 5.2

$x + 0 = 9.4 + 5.2$ Law of additive inverses

$x = 14.6$

Check: $\dfrac{x - 5.2 = 9.4}{\begin{array}{c|c} 14.6 - 5.2 & 9.4 \\ 9.4 & \end{array}}$

The solution is 14.6.

12. 6.9

13. $9y = 72$

$\frac{1}{9} \cdot 9y = \frac{1}{9} \cdot 72$ Multiplication principle; multiplying by $\frac{1}{9}$, the reciprocal of 9

$1 \cdot y = 8$

$y = 8$

Check: $\dfrac{9y = 72}{\begin{array}{c|c} 9 \cdot 8 & 72 \\ 72 & \end{array}}$

The solution is 8.

14. 9

15. $4x - 12 = 60$

$4x - 12 + 12 = 60 + 12$

$4x = 72$

$\frac{1}{4} \cdot 4x = \frac{1}{4} \cdot 72$

$1 \cdot x = \frac{72}{4}$

$x = 18$

Check:

$\dfrac{4x - 12 = 60}{\begin{array}{c|c} 4 \cdot 18 - 12 & 60 \\ 72 - 12 & \\ 60 & \end{array}}$

The solution is 18.

16. 19

17. $5y + 3 = 28$

$5y + 3 + (-3) = 28 + (-3)$

$5y = 25$

$\frac{1}{5} \cdot 5y = \frac{1}{5} \cdot 25$

$1 \cdot y = \frac{25}{5}$

$y = 5$

Check:

$\dfrac{5y + 3 = 28}{\begin{array}{c|c} 5 \cdot 5 + 3 & 28 \\ 25 + 3 & \\ 28 & \end{array}}$

The solution is 5.

18. 9

19. $2y - 11 = 37$

$2y - 11 + 11 = 37 + 11$

$2y = 48$

$\frac{1}{2} \cdot 2y = \frac{1}{2} \cdot 48$

$1 \cdot y = \frac{48}{2}$

$y = 24$

Check:

$\dfrac{2y - 11 = 37}{\begin{array}{c|c} 2 \cdot 24 - 11 & 37 \\ 48 - 11 & \\ 37 & \end{array}}$

The solution is 24.

20. 14

21. $-4x - 7 = -35$

$-4x - 7 + 7 = -35 + 7$

$-4x = -28$

$-\frac{1}{4} \cdot (-4x) = -\frac{1}{4} \cdot (-28)$

$1 \cdot x = \frac{28}{4}$

$x = 7$

Check:

$\dfrac{-4x - 7 = -35}{\begin{array}{c|c} -4 \cdot 7 - 7 & -35 \\ -28 - 7 & \\ -35 & \end{array}}$

The solution is 7.

22. 11

23. $5x + 2x = 56$

$7x = 56$

$\frac{1}{7} \cdot 7x = \frac{1}{7} \cdot 56$

$x = 8$

The solution is 8.

Check:

$\dfrac{5x + 2x = 56}{\begin{array}{c|c} 5 \cdot 8 + 2 \cdot 8 & 56 \\ 40 + 16 & \\ 56 & \end{array}}$

24. 12

25. $9y - 7y = 42$

$2y = 42$

$\frac{1}{2} \cdot 2y = \frac{1}{2} \cdot 42$

$y = 21$

The solution is 21.

Check:

$\dfrac{9y - 7y = 42}{\begin{array}{c|c} 9 \cdot 21 - 7 \cdot 21 & 42 \\ 189 - 147 & \\ 42 & \end{array}}$

26. 13

27. $-6y - 10y = -32$

          $-16y = -32$

  $-\dfrac{1}{16} \cdot (-16y) = -\dfrac{1}{16} \cdot (-32)$

            $y = 2$

The solution is 2.

Check:

$-6y - 10y = -32$	
$-6 \cdot 2 - 10 \cdot 2$	-32
$-12 - 20$	
-32	

28. -2

29. $7y - 1 = 23 - 5y$

  $7y - 1 + 5y = 23 - 5y + 5y$

     $12y - 1 = 23$

  $12y - 1 + 1 = 23 + 1$

        $12y = 24$

  $\dfrac{1}{12} \cdot 12y = \dfrac{1}{12} \cdot 24$

          $y = 2$

The solution is 2.

Check:

$7y - 1 = 23 - 5y$	
$7 \cdot 2 - 1$	$23 - 5 \cdot 2$
$14 - 1$	$23 - 10$
13	13

30. -6

31. $5 - 4a = a - 13$

  $5 - 4a + 4a = a - 13 + 4a$

        $5 = 5a - 13$

    $5 + 13 = 5a - 13 + 13$

       $18 = 5a$

   $\dfrac{1}{5} \cdot 18 = \dfrac{1}{5} \cdot 5a$

      $\dfrac{18}{5} = a$

The solution is $\dfrac{18}{5}$.

Check:

$5 - 4a = a - 13$	
$5 - 4 \cdot \dfrac{18}{5}$	$\dfrac{18}{5} - 13$
$\dfrac{25}{5} - \dfrac{72}{5}$	$\dfrac{18}{5} - \dfrac{65}{5}$
$-\dfrac{47}{5}$	$-\dfrac{47}{5}$

32. 4

33. $3m - 7 = -7 - 4m - m$

       $3m - 7 = -7 - 5m$

  $3m - 7 + 5m = -7 - 5m + 5m$

       $8m - 7 = -7$

  $8m - 7 + 7 = -7 + 7$

         $8m = 0$

    $\dfrac{1}{8} \cdot 8m = \dfrac{1}{8} \cdot 0$

          $m = 0$

Check:

$3m - 7 = -7 - 4m - m$	
$3 \cdot 0 - 7$	$-7 - 4 \cdot 0 - 0$
$0 - 7$	$-7 - 0 - 0$
-7	-7

The solution is 0.

34. 0

35. $5r - 2 + 3r = 2r + 6 - 4r$

        $8r - 2 = 6 - 2r$

    $8r - 2 + 2r = 6 - 2r + 2r$

      $10r - 2 = 6$

  $10r - 2 + 2 = 6 + 2$

        $10r = 8$

  $\dfrac{1}{10} \cdot 10r = \dfrac{1}{10} \cdot 8$

         $r = \dfrac{8}{10}$

         $r = \dfrac{4}{5}$

Check:

$5r - 2 + 3r = 2r + 6 - 4r$	
$5 \cdot \dfrac{4}{5} - 2 + 3 \cdot \dfrac{4}{5}$	$2 \cdot \dfrac{4}{5} + 6 - 4 \cdot \dfrac{4}{5}$
$\dfrac{20}{5} - \dfrac{10}{5} + \dfrac{12}{5}$	$\dfrac{8}{5} + \dfrac{30}{5} - \dfrac{16}{5}$
$\dfrac{22}{5}$	$\dfrac{22}{5}$

The solution is $\dfrac{4}{5}$.

36. -8

37. $\dfrac{1}{4} + \dfrac{3}{8}y = \dfrac{3}{4}$

  $\dfrac{1}{4} + \dfrac{3}{8}y - \dfrac{1}{4} = \dfrac{3}{4} - \dfrac{1}{4}$

        $\dfrac{3}{8}y = \dfrac{1}{2}$

  $\dfrac{8}{3} \cdot \dfrac{3}{8}y = \dfrac{8}{3} \cdot \dfrac{1}{2}$

         $y = \dfrac{4}{3}$

The solution is $\dfrac{4}{3}$.

Check:

$\dfrac{1}{4} + \dfrac{3}{8}y = \dfrac{3}{4}$	
$\dfrac{1}{4} + \dfrac{3}{8} \cdot \dfrac{4}{3}$	$\dfrac{3}{4}$
$\dfrac{1}{4} + \dfrac{1}{2}$	
$\dfrac{3}{4}$	

38. 2

39. $-\dfrac{5}{2}x + \dfrac{1}{2} = -18$

  $-\dfrac{5}{2}x + \dfrac{1}{2} - \dfrac{1}{2} = -18 - \dfrac{1}{2}$

      $-\dfrac{5}{2}x = -\dfrac{37}{2}$

  $-\dfrac{2}{5}\left(-\dfrac{5}{2}x\right) = -\dfrac{2}{5}\left(-\dfrac{37}{2}\right)$

        $x = \dfrac{37}{5}$

The solution is $\dfrac{37}{5}$.

Check:

$-\dfrac{5}{2}x + \dfrac{1}{2} = -18$	
$-\dfrac{5}{2} \cdot \dfrac{37}{5} + \dfrac{1}{2}$	-18
$-\dfrac{37}{2} + \dfrac{1}{2}$	
$-\dfrac{36}{2}$	
-18	

40. $\dfrac{49}{9}$

41. $0.8t - 0.3t = 6.5$

        $0.5t = 6.5$

    $\dfrac{1}{0.5}(0.5t) = \dfrac{1}{0.5}(6.5)$

           $t = \dfrac{6.5}{0.5}$

           $t = 13$

Check:

$0.8t - 0.3t$	6.5
$0.8(13) - 0.3(13)$	6.5
$10.4 - 3.9$	
	6.5

The solution is 13.

42. -5.02

43. $-2(x + 3) - 5(x - 4)$

  $= -2x - 6 - 5x + 20$

  $= -7x + 14$

44. $-15y - 45$

45. $5x - 7(2x - 3)$

  $= 5x - 14x + 21$

  $= -9x + 21$

46. $-12y + 24$

47. $9a - [7 - 5(7a - 3)]$

  $= 9a - [7 - 35a + 15]$

  $= 9a - [22 - 35a]$

  $= 9a - 22 + 35a$

  $= 44a - 22$

48. $47b - 51$

49. $5\{-2 + 3[4 - 2(3 + 5)]\}$

  $= 5\{-2 + 3[4 - 2 \cdot 8]\}$

  $= 5\{-2 + 3[4 - 16]\}$

  $= 5\{-2 + 3[-12]\}$

  $= 5\{-2 - 36\}$

  $= 5\{-38\}$

  $= -190$

50. -1449

51. $2y + \{7[3(2y - 5) - (8y + 7)] + 9\}$

  $= 2y + \{7[6y - 15 - 8y - 7] + 9\}$

  $= 2y + \{7[-2y - 22] + 9\}$

  $= 2y + \{-14y - 154 + 9\}$

  $= 2y + \{-14y - 145\}$

  $= 2y - 14y - 145$

  $= -12y - 145$

52. $-11b + 217$

53. $2(x + 6) = 8x$

   $2x + 12 = 8x$

      $12 = 6x$

       $2 = x$

Check:

$2(x + 6)$	$8x$
$2(2 + 6)$	$8 \cdot 2$
$2 \cdot 8$	16
16	

The solution is 2.

54. 3

55. $80 = 10(3t + 2)$

   $80 = 30t + 20$

   $60 = 30t$

    $2 = t$

Check:

80	$10(3t + 2)$
80	$10(3 \cdot 2 + 2)$
	$10(6 + 2)$
	$10 \cdot 8$
	80

The solution is 2.

56. 1

57. $180(n - 2) = 900$

   $180n - 360 = 900$

       $180n = 1260$

         $n = 7$

Check:

$180(n - 2)$	900
$180(7 - 2)$	900
$180 \cdot 5$	
900	

The solution is 7.

58. 7

59. $5y - (2y - 10) = 25$

   $5y - 2y + 10 = 25$

      $3y + 10 = 25$

         $3y = 15$

          $y = 5$

Check:

$5y - (2y - 10)$	25
$5 \cdot 5 - (2 \cdot 5 - 10)$	25
$25 - (10 - 10)$	
$25 - 0$	
25	

The solution is 5.

60. 7

61. $0.7(3x + 6) = 1.1 - (x + 2)$

   $2.1x + 4.2 = 1.1 - x - 2$

   $2.1x + 4.2 = -x - 0.9$

   $3.1x + 4.2 = -0.9$

       $3.1x = -5.1$

          $x = -\dfrac{5.1}{3.1}$

          $x = -\dfrac{51}{31}$

The number $-\dfrac{51}{31}$ checks and is the solution.

62. $\dfrac{39}{14}$

63. $\frac{1}{8}(16y + 8) - 17 = -\frac{1}{4}(8y - 16)$

$2y + 1 - 17 = -2y + 4$

$2y - 16 = -2y + 4$

$4y - 16 = 4$

$4y = 20$

$y = 5$

The number 5 checks and is the solution.

64. 6

65. $a + (a - 3) = (a + 2) - (a + 1)$

$a + a - 3 = a + 2 - a - 1$

$2a - 3 = 1$

$2a = 4$

$a = 2$

Check:

$a + (a - 3)$	$(a + 2) - (a + 1)$
$2 + (2 - 3)$	$(2 + 2) - (2 + 1)$
$2 - 1$	$4 - 3$
1	1

The solution is 2.

66. -7.4

67. $5[2 + 3(x - 1)] = 4$

$5[2 + 3x - 3] = 4$

$5[-1 + 3x] = 4$

$-5 + 15x = 4$

$15x = 9$

$\frac{1}{15} \cdot 15x = \frac{1}{15} \cdot 9$

$x = \frac{3}{5}$

Check:

$5[2 + 3(x - 1)] = 4$	
$5\left[2 + 3\left(\frac{3}{5} - 1\right)\right]$	4
$5\left[2 + 3\left(-\frac{2}{5}\right)\right]$	
$5\left[2 - \frac{6}{5}\right]$	
$5\left[\frac{4}{5}\right]$	
4	

The solution is $\frac{3}{5}$.

68. -9

69. $5 + 2(x - 3) = 2[5 - 4(x + 2)]$

$5 + 2x - 6 = 2[5 - 4x - 8]$

$2x - 1 = 2[-4x - 3]$

$2x - 1 = -8x - 6$

$2x = -8x - 5$

$10x = -5$

$\frac{1}{10} \cdot 10x = \frac{1}{10}(-5)$

$x = -\frac{1}{2}$

Check:

$5 + 2(x - 3) = 2[5 - 4(x + 2)]$	
$5 + 2\left[-\frac{1}{2} - 3\right]$	$2\left[5 - 4\left[-\frac{1}{2} + 2\right]\right]$
$5 + 2\left[-\frac{7}{2}\right]$	$2\left[5 - 4\left[\frac{3}{2}\right]\right]$
$5 - 7$	$2[5 - 6]$
-2	$2[-1]$
	-2

The solution is $-\frac{1}{2}$.

70. $\frac{23}{8}$

71. $2\{9 - 3[2x - 4(x + 1)]\} = 5(2x + 8)$

$2\{9 - 3[2x - 4x - 4]\} = 10x + 40$

$2\{9 - 3[-2x - 4]\} = 10x + 40$

$2\{9 + 6x + 12\} = 10x + 40$

$2\{21 + 6x\} = 10x + 40$

$42 + 12x = 10x + 40$

$12x = 10x - 2$

$2x = -2$

$x = -1$

Check:

$2\{9 - 3[2x - 4(x + 1)]\} = 5(2x + 8)$	
$2\{9 - 3[2(-1) - 4(-1 + 1)]\}$	$5(2 \cdot -1 + 8)$
$2\{9 - 3[-2 - 4 \cdot 0]\}$	$5(-2 + 8)$
$2\{9 - 3[-2 + 0]\}$	$5(6)$
$2\{9 - 3[-2]\}$	30
$2\{9 + 6\}$	
$2\{15\}$	
30	

The solution is -1.

72. $\frac{1}{7}$

73. $4x - 2x - 2 = 2x$

$2x - 2 = 2x$

$-2x + 2x - 2 = -2x + 2x$

$-2 = 0$

Since the original equation is equivalent to the false equation -2 = 0, there is no solution.

74. All real numbers

75.
$$2 + 9x = 3(3x + 1) - 1$$
$$2 + 9x = 9x + 3 - 1$$
$$2 + 9x = 9x + 2$$
$$2 + 9x - 9x = 9x + 2 - 9x$$
$$2 = 2$$

The original equation is equivalent to the equation 2 = 2 which is true for all real numbers. Thus the solution set is the set of all real numbers.

76. No solution

77.
$$-8x + 5 = 14 - 8x$$
$$-8x + 5 + 8x = 14 - 8x + 8x$$
$$5 = 14$$

Since the original equation is equivalent to the false equation 5 = 14, there is no solution.

78. All real numbers

79.
$$2\{9 - 3[-2x - 4]\} = 12x + 42$$
$$2\{9 + 6x + 12\} = 12x + 42$$
$$2\{21 + 6x\} = 12x + 42$$
$$42 + 12x = 12x + 42$$
$$42 + 12x - 12x = 12x + 42 - 12x$$
$$42 = 42$$

The original equation is equivalent to the equation 42 = 42, which is true for all real numbers. Thus the solution set is the set of all real numbers.

80. All real numbers

81. Roster notation: List the numbers in the set.

{1, 2, 3, 4, 5, 6, 7, 8, 9}

Set-builder notation: Specify the conditions under which a number is in the set.

{x | x is a positive integer less than 10}

82. {-8, -7, -6, -5, -4, -3, -2, -1};

{x | x is a negative integer greater than -9}

83. -9.4 + 7.2 = -2.2

(Find the difference in absolute values, and make the result negative.)

84. -3.4

85. -7 - (-5.3) = -7 + 5.3 = -1.7

(Change the sign of the number being subtracted and add.)

86. -5.3

87. (-9)(-6) = 54

(The product of two numbers with the same sign is positive.)

88. 36

89. (-12) ÷ (-3) = 4

(The quotient of two numbers with the same sign is positive.)

90. 3

91.
$$0.0008x = 0.0000056$$
$$x = \frac{0.0000056}{0.0008}$$
$$x = 0.007$$

92. 0.0279282

93.
$$4.23x - 17.898 = -1.65x - 42.454$$
$$5.88x - 17.898 = -42.454$$
$$5.88x = -24.556$$
$$x = -\frac{24.556}{5.88}$$
$$x = -4.1762$$

94. 0.2140224

95.
$$x - \{3x - [2x - (5x - (7x - 1))]\} = x + 7$$
$$x - \{3x - [2x - (5x - 7x + 1)]\} = x + 7$$
$$x - \{3x - [2x - (-2x + 1)]\} = x + 7$$
$$x - \{3x - [2x + 2x - 1]\} = x + 7$$
$$x - \{3x - [4x - 1]\} = x + 7$$
$$x - \{3x - 4x + 1\} = x + 7$$
$$x - \{-x + 1\} = x + 7$$
$$x + x - 1 = x + 7$$
$$2x - 1 = x + 7$$
$$x - 1 = 7$$
$$x = 8$$

The check is left to the student. The solution is 8.

96. 4

97.
$$7x - 2\{3x + 4[x - 3(x - 2(x + 1))]\} = 14$$
$$7x - 2\{3x + 4[x - 3(x - 2x - 2)]\} = 14$$
$$7x - 2\{3x + 4[x - 3(-x - 2)]\} = 14$$
$$7x - 2\{3x + 4[x + 3x + 6]\} = 14$$
$$7x - 2\{3x + 4[4x + 6]\} = 14$$
$$7x - 2\{3x + 16x + 24\} = 14$$
$$7x - 2\{19x + 24\} = 14$$
$$7x - 38x - 48 = 14$$
$$-31x - 48 = 14$$
$$-31x - 48 + 48 = 14 + 48$$
$$-31x = 62$$
$$-\frac{1}{31}(-31x) = -\frac{1}{31} \cdot 62$$
$$x = -2$$

The check is left to the student. The solution is -2.

<u>98.</u> $-\frac{188}{145}$

<u>99.</u> $17 - 3\{5 + 2[x - 2]\} + 4\{x - 3(x + 7)\} =$
$$9\{x + 3[2 + 3(4 - x)]\}$$
$$17 - 3\{5 + 2x - 4\} + 4\{x - 3x - 21\} =$$
$$9\{x + 3[2 + 12 - 3x]\}$$
$$17 - 3\{1 + 2x\} + 4\{-2x - 21\} = 9\{x + 3[14 - 3x]\}$$
$$17 - 3 - 6x - 8x - 84 = 9\{x + 42 - 9x\}$$
$$-14x - 70 = 9\{-8x + 42\}$$
$$-14x - 70 = -72x + 378$$
$$58x - 70 = 378$$
$$58x = 448$$
$$x = \frac{448}{58}, \text{ or } \frac{224}{29}$$

The check is left to the student. The solution is $\frac{224}{29}$.

<u>100.</u> $\frac{19}{46}$

Exercise Set 1.4

<u>1.</u> <u>Familiarize:</u> There are two numbers involved, and we want to find both of them. We can let x represent the first number and note that the second number is 9 more than the first. Also, the sum of the numbers is 81.

<u>Translate:</u> The second number can be named x + 9. We translate to an equation:

First Number plus Second Number is 81.
$$x \quad + \quad x + 9 \quad = 81$$

<u>2.</u> Let x and x + 11 be the numbers; $x + (x + 11) = 95$

<u>3.</u> <u>Familiarize:</u> Let t represent the time required, and list the relevant information in a table.

Swimmer's speed in still water	5 mph
Speed of current	3.2 mph
Swimmer's speed up river	(5 − 3.2) mph
Distance to be traveled	2.7 mi
Time required	t

<u>Translate:</u> Speed × Time = Distance
$$(5 - 3.2) \quad \times \quad t \quad \quad 2.7$$

<u>4.</u> Let t be the swimmer's time; $(4 - 1.5)t = 3.75$

<u>5.</u> <u>Familiarize:</u> Let t represent the time required, and list the relevant information in a table.

Boat's speed in still water	12 km/h
Speed of current	3 km/h
Boat's speed down river	(12 + 3) km/h
Distance to be traveled	35 km
Time required	t

<u>Translate:</u> Speed × Time = Distance
$$(12 + 3) \quad \times \quad t \quad = \quad 35$$

<u>6.</u> Let t be the boat's time; $(14 + 7)t = 56$

<u>7.</u> <u>Familiarize:</u> There are three angle measures involved, and we want to find all three. We can let x represent the smallest angle measure and note that the second is one more than x and the third is one more than the second, or two more than x. We also note that the sum of the three angle measures must be 180°.

<u>Translate:</u> The three angles measured are x, x + 1, and x + 2. We translate to an equation:

First plus second plus third is 180°.
$$x \quad + \quad (x + 1) \quad + \quad (x + 2) \quad = \quad 180$$

<u>8.</u> Let x, x + 2, and x + 4 be the angle measures; $x + (x + 2) + (x + 4) = 180$

<u>9.</u> <u>Familiarize:</u> If we let x represent the shorter length, then the other length is 4 more than x. The sum of the two lengths must be 12 ft.

<u>Translate:</u> The two lengths are x and x + 4. Translate to an equation:

Shorter plus longer is 12 ft.
$$x \quad + \quad (x + 4) \quad = \quad 12$$

<u>10.</u> Let x represent the longer length;
$$x + \frac{2}{3}x = 10$$

<u>11.</u> <u>Familiarize:</u> Let x represent the measure of the second angle. Then the first angle is three times x, and the third is 12° less than twice x. The sum of the three angle measures is 180°.

<u>Translate:</u> The first angle is 3x, the second is x, and the third is 2x − 12. Translate to an equation:

First plus second plus third is 180°.
$$3x \quad + \quad x \quad + (2x - 12) \quad = \quad 180$$

<u>12.</u> Let x represent the measure of the second angle; $4x + x + (2x + 5) = 180$

13. <u>Familiarize</u>: Note that each odd integer is two more than the one preceding it. If we let n represent the first odd integer, then the second is 2 more than the first and the third is 2 more than the second, or 4 more than the first. We are told that the sum of the first, twice the second, and three times the third is 70.

    <u>Translate</u>: The three odd integers are n, n + 2, and n + 4. Translate to an equation.

First	plus	two times second	plus	three times third	is	70.
n	+	2(n + 2)	+	3(n + 4)	=	70

14. Let x represent the first number;

    $2x + 3(x + 2) = 76$

15. <u>Familiarize</u>: Recall that the perimeter of each square is four times the length of a side. If s represents the length of a side of the smaller square, then (s + 2) represents the length of a side of the larger square. The sum of the two perimeters must be 100 cm.

    <u>Translate</u>:

Perimeter of smaller square	plus	perimeter of larger square	is	100 cm.
4s	+	4(s + 2)	=	100

16. Let x represent the length of one piece;

    $$\left(\frac{x}{4}\right)^2 = \left(\frac{100 - x}{4}\right)^2 + 144$$

17. <u>Familiarize</u>: If we let x represent the first number, then the second is six less than 3 times x and the third is two more than $\frac{2}{3}$ of the second. The sum of the three numbers is 172.

    <u>Translate</u>:

First	plus	second	plus	third	is	172.
x	+	3x - 6	+	$\left[\frac{2}{3}\left[3x - 6\right] + 2\right]$	=	172

18. Let x represent the price of the least expensive set;

    $(x + 20) + (x + 6 \cdot 20) = x + 12 \cdot 20$

19. <u>Familiarize</u>: After the next test there will be six test scores. The average of the six scores is their sum divided by 6. We let x represent the next test score.

    <u>Translate</u>:

    Average of the six scores is 88.

    $$\frac{93 + 89 + 72 + 80 + 96 + x}{6} = 88$$

20. Let x represent the percent of total change;

    $1.2(1.3)(0.8) - 1 = x$

21. <u>Familiarize and translate</u>: Let x represent the unknown number.

Three times	some number	is	12.3.
3 ·	x	=	12.3

    <u>Carry out</u>: $3x = 12.3$
                    $x = 4.1$

    <u>Check</u>: The product of 3 and 4.1 is 12.3, so the answer checks.

    <u>State</u>: The other number is 4.1.

22. 37.5

23. <u>Familiarize and translate</u>: Let x represent the number.

Some number	less	12.1	is	38.2.
x	-	12.1	=	38.2

    <u>Carry out</u>: $x - 12.1 = 38.2$
                       $x = 50.3$

    <u>Check</u>: 50.3 less 12.1 is 38.2, so the answer checks.

    <u>State</u>: The number is 50.3.

24. 156.7

25. <u>Familiarize and translate</u>: Let y represent the unknown number.

128	is	0.4 of	what number?
128	=	0.4·	y

    <u>Carry out</u>: $128 = 0.4y$
                   $320 = y$

    <u>Check</u>: 0.4 of 320 is 128, so the answer checks.

    <u>State</u>: The number is 320.

26. 1368

27. <u>Familiarize</u>: If we let x represent the smaller number, then the larger is x + 12. The sum of the numbers is 114. We want to find the larger number.

    <u>Translate</u>:

Smaller number	plus	larger number	is	114.
x	+	(x + 12)	=	114

    <u>Carry out</u>: $x + x + 12 = 114$
                $2x + 12 = 114$
                    $2x = 102$
                      $x = 51$

    If x = 51, then x + 12 = 51 + 12, or 63.

    <u>Check</u>: The larger number, 63, is 12 more than the smaller number, 51. Also, 51 + 63 = 114. The numbers check.

    <u>State</u>: The larger number is 63.

28. 13.5

29. Familiarize: If we let x represent the first number, then the second number is 2x. The sum of the numbers is 495. We want to find both numbers.

    Translate:

    First plus second is 495.

    $$x + 2x = 495$$

    Carry out: x + 2x = 495

    $$3x = 495$$
    $$x = 165$$

    If x = 165, then 2x = 2·165, or 330.

    Check: 330 is two times 165. Also, 165 + 330 = 495. The numbers check.

    State: The numbers are 165 and 330.

30. $78\frac{2}{3}$, $393\frac{1}{3}$

31. The Familiarize and Translate steps were completed in Exercise 4.

    Carry out: (4 - 1.5)t = 3.75

    $$2.5t = 3.75$$
    $$t = 1.5$$

    Check: The person's speed up river is (4 - 1.5), or 2.5 mph. In 1.5 hr the person will travel 2.5(1.5), or 3.75 mi. The answer checks.

    State: It will take the person 1.5 hr to swim 3.75 mi up river.

32. $1\frac{1}{2}$ hr

33. The Familiarize and Translate steps were completed in Exercise 14.

    Carry out: 2x + 3(x + 2) = 76

    $$2x + 3x + 6 = 76$$
    $$5x + 6 = 76$$
    $$5x = 70$$
    $$x = 14$$

    If x = 14, the x + 2 = 14 + 2, or 16.

    Check: 14 and 16 are consecutive even integers. Two times 14 plus three times 16 is 2·14 + 3·16, or 28 + 48, or 76. The numbers check.

    State: The numbers are 14 and 16.

34. 9, 11, 13

35. The Familiarize and Translate steps were completed in Exercise 10.

    Carry out: $x + \frac{2}{3}x = 10$

    $$\frac{5}{3}x = 10$$
    $$x = 6$$

    If x = 6, $\frac{2}{3}x = \frac{2}{3} \cdot 6$, or 4.

    Check: 4 is $\frac{2}{3}$ of 6. Also, 4 + 6 = 10. The numbers check.

    State: The wire should be cut into a 6 m piece and a 4 m piece.

36. 4 ft, 8 ft

37. The Familiarize and Translate steps were completed in Exercise 11.

    Carry out: 3x + x + 2x - 12 = 180

    $$6x - 12 = 180$$
    $$6x = 192$$
    $$x = 32$$

    If x = 32, 3x = 3·32, or 96, and 2x - 12 = 2·32 - 12, or 52.

    Check: The first angle, 96°, is three times the second angle, 32°. The third angle, 52°, is 12° less than twice the second angle, 32°. Also, 96° + 32° + 52° = 180°. The numbers check.

    State: The angle measures are 96°, 32°, and 52°.

38. 46 cm, 54 cm

39.
    $$3[2x - (5 + 4x)] = 5 - 7x$$
    $$3[2x - 5 - 4x] = 5 - 7x$$
    $$3[-2x - 5] = 5 - 7x$$
    $$-6x - 15 = 5 - 7x$$
    $$x - 15 = 5$$
    $$x = 20$$

    The solution is 20.

40. -19

41.
    $$2(3 - 5t) = 3(4 - 3t) - t$$
    $$6 - 10t = 12 - 9t - t$$
    $$6 - 10t = 12 - 10t$$
    $$6 = 12$$

    There is no solution.

42. All real numbers

43.
    $$4 + 2(a - 7) = 2(a - 5)$$
    $$4 + 2a - 14 = 2a - 10$$
    $$-10 + 2a = 2a - 10$$
    $$-10 = -10$$

    The solution is all real numbers.

44. No solution

45. Familiarize: If we let x represent the height, then the three sides are x + 1, x + 2, and x + 3, with x + 3 representing the base. The perimeter is 42 in.

    Translate: The perimeter is 42 in.

    $$(x + 1) + (x + 2) + (x + 3) = 42$$

    Carry out: x + 1 + x + 2 + x + 3 = 42

    $$3x + 6 = 42$$
    $$3x = 36$$
    $$x = 12$$

    If x = 12, then x + 1 = 12 + 1, or 13; x + 2 = 12 + 2, or 14; and x + 3 = 12 + 3, or 15.

45. (continued)

Check: 12, 13, 14, and 15 are four consecutive integers. Also, 13 + 14 + 15 = 42. The numbers check. The height is 12 in., and the base is 15 in.

Now we find the area of the triangle.

$$\text{Area} = \frac{1}{2} \cdot \text{base} \cdot \text{height}$$

$$= \frac{1}{2} \cdot 15 \text{ in.} \cdot 12 \text{ in.}$$

$$= 90 \text{ sq in.}$$

State: The area of the triangle is 90 sq in.

46. $\frac{100n}{100 - n}\%$

47. Familiarize: Let p represent the population at the start of 1986. The population at the end of 1986 is 108% of p, or 1.08p. At the end of 1987 the population is 110% of the population at the end of 1986, or 1.1(1.08p). At the end of 1988 the population is 111% of the population at the end of 1987, or 1.11(1.1)(1.08p).

Translate:

Population at the end of 1988 is 1,582,416.

$$1.11(1.1)(1.08p) = 1,582,416$$

Carry out: $1.11(1.1)(1.08p) = 1,582,416$

$$1.31868p = 1,582,416$$

$$p = 1,200,000$$

Check: If the population was 1,200,000 at the beginning of 1986, then at the end of 1986 the population would have been 1.08(1,200,000), or 1,296,000. At the end of 1987 the population would have been 1.1(1,296,000), or 1,425,600, and at the end of 1988 if would have been 1.11(1,425,600), or 1,582,416. The answer checks.

State: The population at the start of 1986 was 1,200,000.

48. 10 points

Exercise Set 1.5

---

1. $3^2 \cdot 3^5 = 3^{2+5} = 3^7$

2. 2^{11}

3. $5^6 \cdot 5^3 = 5^{6+3} = 5^9$

4. 6^8

5. $a^3 \cdot a^0 = a^{3+0} = a^3$

6. x^5

7. $5x^4 \cdot 3x^2 = 5 \cdot 3 \cdot x^4 \cdot x^2 = 15x^{4+2} = 15x^6$

8. $8a^{10}$

9. $(-3m^4)(-7m^9) = (-3)(-7)m^4 \cdot m^9 = 21m^{4+9} = 21m^{13}$

10. $-14a^9$

11. $(x^3y^4)(x^7y^6z^0) = (x^3x^7)(y^4y^6)(z^0) = x^{3+7}y^{4+6} \cdot 1 = x^{10}y^{10}$

12. $m^{10}n^{12}$

13. $\frac{a^9}{a^3} = a^{9-3} = a^6$

14. x^9

15. $\frac{8x^7}{4x^4} = \frac{8}{4} \cdot x^{7-4} = 2x^3$

16. $4a^{16}$

17. $\frac{m^7n^9}{m^2n^5} = m^{7-2} \cdot n^{9-5} = m^5n^4$

18. m^8n^3

19. $\frac{35x^8y^5}{7x^2y} = \frac{35}{7} \cdot x^{8-2} \cdot y^{5-1} = 5x^6y^4$

20. $9x^6y^6$

21. $\frac{-49a^5b^{12}}{7a^2b^2} = \frac{-49}{7} \cdot a^{5-2} \cdot b^{12-2} = -7a^3b^{10}$

22. $-6ab^3$

23. $6^{-3} = \frac{1}{6^3}$

24. $\frac{1}{8^4}$

25. $9^{-5} = \frac{1}{9^5}$

26. $\frac{1}{16^2}$

27. $(-11)^{-1} = \frac{1}{(-11)^1}$, or $-\frac{1}{11}$

28. $\frac{1}{(-4)^3}$

29. $(5x)^{-3} = \frac{1}{(5x)^3}$

30. $\frac{1}{(4xy)^5}$

31. $x^2y^{-3} = x^2 \cdot \frac{1}{y^3} = \frac{x^2}{y^3}$

32. $\frac{2a^2}{b^5}$

33. $x^2y^{-2} = x^2 \cdot \frac{1}{y^2} = \frac{x^2}{y^2}$

34. $\frac{a^2c^4}{b^3d^5}$

35. $\frac{x^3}{y^{-2}} = x^3 \cdot \frac{1}{y-2} = x^3 y^2$

36. $xy^4 z^3$

37. $\frac{y^{-5}}{x^2} = \frac{1}{x^2} \cdot y^{-5} = \frac{1}{x^2} \cdot \frac{1}{y^5} = \frac{1}{x^2 y^5}$

38. $\frac{1}{3x^5 z^4}$

39. $\frac{y^{-5}}{x^{-3}} = \frac{1}{x^{-3}} \cdot y^{-5} = x^3 \cdot \frac{1}{y^5} = \frac{x^3}{y^5}$

40. $\frac{y^7}{(7x)^4}$

41. $\frac{x^{-2}y^7}{z^{-4}} = x^{-2} \cdot y^7 \cdot \frac{1}{z^{-4}} = \frac{1}{x^2} \cdot y^7 \cdot z^4 = \frac{y^7 z^4}{x^2}$

42. $\frac{x^2 y^4}{z^3}$

43. $\frac{1}{3^4} = 3^{-4}$

44. 9^{-2}

45. $\frac{1}{(-16)^2} = (-16)^{-2}$

46. $(-8)^{-6}$

47. $6^4 = \frac{1}{6^{-4}}$

48. $\frac{1}{8^{-5}}$

49. $6x^2 = 6 \cdot \frac{1}{x^{-2}} = \frac{6}{x^{-2}}$

50. $\frac{-4}{y^{-5}}$

51. $\frac{1}{(5y)^3} = (5y)^{-3}$

52. $(5x)^{-5}$

53. $\frac{1}{3y^4} = \frac{1}{3} \cdot \frac{1}{y^4} = \frac{1}{3} \cdot y^{-4} = \frac{y^{-4}}{3}$

54. $\frac{b^{-3}}{4}$

55. $8^{-6} \cdot 8^2 = 8^{-6+2} = 8^{-4}$

56. 9^{-2}

57. $8^{-2} \cdot 8^{-4} = 8^{-2+(-4)} = 8^{-6}$

58. 9^{-7}

59. $b^2 \cdot b^{-5} = b^{2+(-5)} = b^{-3}$

60. a

61. $a^{-3} \cdot a^4 \cdot a^2 = a^{-3+4+2} = a^3$

62. 1

63. $(2x)^3 (3x)^2 = 2^3 \cdot x^3 \cdot 3^2 \cdot x^2 = 8 \cdot 9 \cdot x^3 \cdot x^2 = 72x^{3+2}$
    $= 72x^5$

64. $648y^5$

65. $(14m^2 n^3)(-2m^3 n^2) = 14 \cdot (-2) \cdot m^2 \cdot m^3 \cdot n^3 \cdot n^2$
    $= -28m^{2+3} n^{3+2}$
    $= -28m^5 n^5$

66. $-18x^7 y$

67. $(-2x^{-3})(7x^{-8}) = -2 \cdot 7 \cdot x^{-3} \cdot x^{-8} = -14x^{-3+(-8)}$
    $= -14x^{-11}$

68. $-24x^{-12} y$

69. $5^{a+1} \cdot 5^{2a-1} = 5^{a+1+2a-1} = 5^{3a}$

70. 7^{2a+b+1}

71. $(3x^a y^b)(2x^{a-1} y^{3-b}) = 3 \cdot 2 \cdot x^a \cdot x^{a-1} \cdot y^b \cdot y^{3-b}$
    $= 6x^{a+a-1} y^{b+3-b}$
    $= 6x^{2a-1} y^3$

72. $12x^{4a+3} y^3$

73. $\frac{4^3}{4^{-2}} = 4^{3-(-2)} = 4^{3+2} = 4^5$

74. 5^{11}

75. $\frac{10^{-3}}{10^6} = 10^{-3-6} = 10^{-9}$

76. 12^{-12}

77. $\frac{9^{-4}}{9^{-6}} = 9^{-4-(-6)} = 9^{-4+6} = 9^2$, or 81

78. 2^{-2}, or $\frac{1}{2^2}$

79. $\frac{a^3}{a^{-2}} = a^{3-(-2)} = a^{3+2} = a^5$

80. y^9

81. $\frac{9a^2}{(-3a)^2} = \frac{9a^2}{(-3)^2 a^2} = \frac{9a^2}{9a^2} = 1$

82. $-3ab^2$

83. $\frac{-24x^6 y^7}{18x^{-3} y^9} = \frac{-24}{18} x^{6-(-3)} y^{7-9} = -\frac{4}{3} x^9 y^{-2}$

84. $-\frac{7}{4} a^{-4} b^2$

85. $\frac{-18x^{-2} y^3}{-12x^{-5} y^5} = \frac{-18}{-12} x^{-2-(-5)} y^{3-5} = \frac{3}{2} x^3 y^{-2}$

86. $\dfrac{7}{9}a^{16}b^5$

87. $\dfrac{10^{2a}}{10^a} = 10^{2a-a} = 10^a$

88. 11^{-2b}

89. $(4^3)^2 = 4^{3 \cdot 2} = 4^6$

90. 5^{20}

91. $(8^4)^{-3} = 8^{4(-3)} = 8^{-12}$

92. 9^{-12}

93. $(6^{-4})^{-3} = 6^{-4(-3)} = 6^{12}$

94. 7^{40}

95. $(3x^2y^2)^3 = 3^3(x^2)^3(y^2)^3 = 27x^6y^6$

96. $32a^{15}b^{20}$

97. $(-2x^3y^{-4})^{-2} = (-2)^{-2}(x^3)^{-2}(y^{-4})^{-2} = (-2)^{-2}x^{-6}y^8$,

    or $\dfrac{1}{4}x^{-6}y^8$

98. $-\dfrac{1}{27}a^{-6}b^{15}$

99. $(-6a^{-2}b^3c)^{-2} = (-6)^{-2}(a^{-2})^{-2}(b^3)^{-2}c^{-2}$

    $= (-6)^{-2}a^4b^{-6}c^{-2}$,   or $\dfrac{1}{36}a^4b^{-6}c^{-2}$

100. $\dfrac{1}{(-8)^4}x^{16}y^{-20}z^{-8}$

101. $\left[\dfrac{4^{-3}}{3^4}\right]^3 = \dfrac{(4^{-3})^3}{(3^4)^3} = \dfrac{4^{-9}}{3^{12}}$,   or $\dfrac{1}{4^9 \cdot 3^{12}}$

102. $\dfrac{1}{5^6 \cdot 4^9}$

103. $\left[\dfrac{2x^3y^{-2}}{3y^{-3}}\right]^3 = \dfrac{(2x^3y^{-2})^3}{(3y^{-3})^3} = \dfrac{2^3(x^3)^3(y^{-2})^3}{3^3(y^{-3})^3} = \dfrac{8x^9y^{-6}}{27y^{-9}}$

     $= \dfrac{8x^9y^3}{27}$

104. $\dfrac{625}{256}x^{-20}y^{24}$,   or $\dfrac{625y^{24}}{256x^{20}}$

105. $(7^a)^{2b} = 7^{a \cdot 2b} = 7^{2ab}$

106. 3^{5xy}

107. $\left[\dfrac{30x^5y^{-7}}{6x^{-2}y^{-6}}\right]^0 = 1$ (Any nonzero real number raised to the zero power is 1.)

108. $\dfrac{9}{a^4b^{10}}$

109. $\left[\dfrac{5x^0y^{-7}}{2x^{-2}y^4}\right]^{-2} = \dfrac{(5x^0y^{-7})^{-2}}{(2x^{-2}y^4)^{-2}} = \dfrac{5^{-2}x^0y^{14}}{2^{-2}x^4y^{-8}} = $

     $\dfrac{2^2y^{22}}{5^2x^4}$,   or $\dfrac{4y^{22}}{25x^4}$

110. 1

111. $5 + 2 \cdot 3^2 = 5 + 2 \cdot 9$ Evaluating the exponential expression

     $= 5 + 18$ Multiplying

     $= 23$ Adding

112. -3

113. $12 - (9 - 3 \cdot 2^3) = 12 - (9 - 3 \cdot 8)$ Working within

     $= 12 - (9 - 24)$ the parentheses

     $= 12 - (-15)$ first

     $= 12 + 15$

     $= 27$

114. -3

115. $\dfrac{5 \cdot 2 - 4^2}{27 - 2^4} = \dfrac{5 \cdot 2 - 16}{27 - 16} = \dfrac{10 - 16}{11} = \dfrac{-6}{11}$,   or $-\dfrac{6}{11}$

116. $-\dfrac{4}{17}$

117. $\dfrac{3^4 - (5 - 3)^4}{1 - 2^3} = \dfrac{3^4 - 2^4}{1 - 8} = \dfrac{81 - 16}{-7} = \dfrac{65}{-7}$,   or $-\dfrac{65}{7}$

118. $\dfrac{55}{2}$

119. $5^3 - [2(4^2 - 3^2 - 6)]^3 = 5^3 - [2(16 - 9 - 6)]^3 = $
     $5^3 - [2 \cdot 1]^3 = 5^3 - 2^3 = 125 - 8 = 117$

120. 13

121. $5*(6 + 2) \wedge 2/(5 - 1) = 5*8 \wedge 2/4$ Carrying out operations in parentheses first

     $= 5*64/4$ Evaluating the exponential expression

     $= 320/4$ Multiplying

     $= 80$ Dividing

122. 32

123. $3 \wedge 4*5 - 7 + 3 \wedge (2 + 1) = 3 \wedge 4*5 - 7 + 3 \wedge 3 = $
     $81*5 - 7 + 27 = 405 - 7 + 27 = 425$

124. 125

125. $4*(3 - (2 + 4 \wedge 2) + 5) \wedge 2 = $
     $4*(3 - (2 + 16) + 5) \wedge 2 = 4*(3 - 18 + 5) \wedge 2 = $
     $4*(-10) \wedge 2 = 4*100 = 400$

126. 0

127. Refer to the text for the relationship between BASIC notation and algebraic notation.

     $7*X \wedge 2 - 3/4 = 7x^2 - \dfrac{3}{4}$

128. $5x^3 + \dfrac{4}{5}$

129. $[(5 - X) \wedge 2*4]/2*X = \dfrac{(5-x)^2 4}{2} \cdot x$

130. $\dfrac{7(x-5)^2}{3} \cdot x$

131. Refer to the text for the relationship between algebraic notation and BASIC notation.

$x^3 - 9x = X \wedge 3 - 9*X$

132. $M \wedge 5 - 7*M$

133. $\dfrac{7a^3 + a^2}{5 - 2a} = (7*A \wedge 3 + A \wedge 2)/(5 - 2*A)$

134. $(9*N \wedge 2 - N \wedge 3)/(7 + 2*N)$

135. $\dfrac{(3n+2)^2 - 5}{4+n} = ((3*N + 2) \wedge 2 - 5)/(4 + N)$

136. $(9 - (7*X+ 1) \wedge 3)/(X - 5)$

137. $\dfrac{9 + 3 \cdot 4 + 1}{2}$

⑨ ⊞ ③ ⊠ ④ ⊞ ① ⊜ ⊟ ② ⊜

138. ⑦ ⊟ ② ⊠ ③ ⊞ ⑤ ⊜ ⊟ ④ ⊜

139. $8 - 2 \cdot 3 + \dfrac{4}{15}$

⑧ ⊟ ② ⊠ ③ ⊞ ④ ⊟ ① ⑤ ⊜

140. ① ② ⊞ ③ ⊠ ② ⊟ ⑦ ⊟ ① ② ⊜

141. $\dfrac{9 - 2^8 + 4}{3}$

⑨ ⊟ ② yˣ ⑧ ⊞ ④ ⊜ ⊟ ③ ⊜

142. ① ⑤ ⊟ ③ yˣ ⑦ ⊞ ⑧ ⊜ ⊟
① ④ ⊜

143. $\left(7 - \dfrac{2}{3}\right)^4 - \dfrac{5}{8}$

⑦ ⊟ ② ⊟ ③ ⊜ yˣ ④ ⊟ ⑤ ⊟ ⑧ ⊜

144. ⑨ ⊟ ④ ⊟ ⑦ ⊜ yˣ ⑤ ⊞ ③ ⊟
① ① ⊜

145. $x + (2xy - z)^2 = 3 + (2 \cdot 3(-4) - (-20))^2 =$
$3 + (-24 + 20)^2 = 3 + (-4)^2 = 3 + 16 = 19$

146. -8

147. Familiarize and Translate: Let x represent the first integer, x + 2 the second, and x + 4 the third.

First plus second plus third is 183.
$x + (x + 2) + (x + 4) = 183$
Carry out: $x + x + 2 + x + 4 = 183$
$3x + 6 = 183$
$3x = 177$
$x = 59$

If x = 59, then x + 2 = 59 + 2, or 61, and x + 4 = 59 + 4, or 63.

Check: The check is left to the student.

State: The integers are 59, 61, and 63.

148. 36, 38, 40

149. $\dfrac{9a^{x-2}}{3a^{2x+2}} = \dfrac{9}{3} \cdot a^{x-2-(2x+2)} = 3a^{x-2-2x-2} = 3a^{-x-4}$

150. $-3x^{2a-1}$

151. $\dfrac{45x^{2a+4}y^{b+1}}{-9x^{a+3}y^{2+b}} = \dfrac{45}{-9} \cdot x^{2a+4-(a+3)}y^{b+1-(2+b)} =$
$-5x^{2a+4-a-3}y^{b+1-2-b} = -5x^{a+1}y^{-1}$

152. $-4x^{10}y^8$

153. $(3^{a+2})a = 3^{(a+2)(a)} = 3^{a^2+2a}$

154. 12^{6b-2ab}

155. $\dfrac{4x^{2a+3}y^{2b-1}}{2x^{a+1}y^{b+1}} = \dfrac{4}{2} \cdot x^{2a+3-(a+1)}y^{2b-1-(b+1)} =$
$2x^{2a+3-a-1}y^{2b-1-b-1} = 2x^{a+2}y^{b-2}$

156. $-5x^{2b}y^{-2a}$

157. $\dfrac{(2^{-2})a \cdot (2^b)^{-a}}{(2^{-2})^{-b}(2^b)^{-2a}} = \dfrac{2^{-2a} \cdot 2^{-ab}}{2^{2b} \cdot 2^{-2ab}} = \dfrac{2^{-2a-ab}}{2^{2b-2ab}} =$
$2^{-2a-ab-(2b-2ab)} = 2^{-2a-ab-2b+2ab} = 2^{-2a-2b+ab}$

158. 8^{-2abc}

159. $\left[\dfrac{(-3x^{-2a}y^{5b})^{-2}}{(2x^{4a}y^{-8b})^{-3}}\right]^2 = \dfrac{(-3x^{-2a}y^{5b})^{-4}}{(2x^{4a}y^{-8b})^{-6}} =$
$\dfrac{(-3)^{-4}x^{8a}y^{-20b}}{2^{-6}x^{-24a}y^{48b}} = \dfrac{2^6}{(-3)^4}x^{32a}y^{-68b} =$
$\dfrac{64}{81}x^{32a}y^{-68b}$, or $\dfrac{64x^{32a}}{81y^{68b}}$

160. $\dfrac{a^{-14ac}}{b^{27ac}}$

Exercise Set 1.6

1. $A = \ell w$

$\dfrac{1}{\ell} \cdot A = \dfrac{1}{\ell} \cdot \ell w$ Multiplying by $\dfrac{1}{\ell}$

$\dfrac{A}{\ell} = w$ Simplifying

2. $a = \dfrac{F}{m}$

3. $W = EI$

$\dfrac{1}{E} \cdot W = \dfrac{1}{E} \cdot EI$ Multiplying by $\dfrac{1}{E}$

$\dfrac{W}{E} = I$ Simplifying

4. $E = \dfrac{W}{I}$

5. $d = rt$

$d \cdot \dfrac{1}{t} = rt \cdot \dfrac{1}{t}$

$\dfrac{d}{t} = r$

6. $t = \dfrac{d}{r}$

7. $V = \ell wh$

$V \cdot \dfrac{1}{wh} = \ell wh \cdot \dfrac{1}{wh}$

$\dfrac{V}{wh} = \ell$

8. $r = \dfrac{I}{Pt}$

9. $E = mc^2$

$E \cdot \dfrac{1}{c^2} = mc^2 \cdot \dfrac{1}{c^2}$

$\dfrac{E}{c^2} = m$

10. $c^2 = \dfrac{E}{m}$

11. $P = 2\ell + 2w$

$P - 2w = 2\ell + 2w - 2w$

$P - 2w = 2\ell$

$\dfrac{1}{2}(P - 2w) = \dfrac{1}{2} \cdot 2\ell$

$\dfrac{P - 2w}{2} = \ell$

12. $w = \dfrac{P - 2\ell}{2}$

13. $c^2 = a^2 + b^2$

$c^2 - b^2 = a^2 + b^2 - b^2$

$c^2 - b^2 = a^2$

14. $b^2 = c^2 - a^2$

15. $A = \pi r^2$

$\dfrac{1}{\pi} \cdot A = \dfrac{1}{\pi} \cdot \pi r^2$

$\dfrac{A}{\pi} = r^2$

16. $\pi = \dfrac{A}{r^2}$

17. $W = \dfrac{11}{2}(h - 40)$

$\dfrac{2}{11} \cdot W = \dfrac{2}{11} \cdot \dfrac{11}{2}(h - 40)$

$\dfrac{2}{11}W = h - 40$

$\dfrac{2}{11}W + 40 = h - 40 + 40$

$\dfrac{2}{11}W + 40 = h$

18. $F = \dfrac{9}{5}C + 32$

19. $V = \dfrac{4}{3}\pi r^3$

$V = \dfrac{4\pi}{3}r^3$

$\dfrac{3}{4\pi} \cdot V = \dfrac{3}{4\pi} \cdot \dfrac{4\pi}{3}r^3$

$\dfrac{3V}{4\pi} = r^3$

20. $\pi = \dfrac{3V}{4r^3}$

21. $A = \dfrac{h}{2}(b_1 + b_2)$

$2 \cdot A = 2 \cdot \dfrac{h}{2}(b_1 + b_2)$

$2A = h(b_1 + b_2)$

$2A \cdot \dfrac{1}{b_1 + b_2} = h(b_1 + b_2) \cdot \dfrac{1}{b_1 + b_2}$

$\dfrac{2A}{b_1 + b_2} = h$

22. $b_2 = \dfrac{2A}{h} - b_1$

23. $F = \dfrac{mv^2}{r}$

$F = m \cdot \dfrac{v^2}{r}$

$F \cdot \dfrac{r}{v^2} = m \cdot \dfrac{v^2}{r} \cdot \dfrac{r}{v^2}$

$\dfrac{Fr}{v^2} = m$

24. $v^2 = \dfrac{rF}{m}$

25. $A = \dfrac{q_1 + q_2 + q_3}{n}$

$n \cdot A = n \cdot \dfrac{q_1 + q_2 + q_3}{n}$ Clearing the fraction

$nA = q_1 + q_2 + q_3$

$nA \cdot \dfrac{1}{A} = (q_1 + q_2 + q_3) \cdot \dfrac{1}{A}$

$n = \dfrac{q_1 + q_2 + q_3}{A}$

26. $d = \dfrac{s + t}{r}$

27.
$$v = \frac{d_2 - d_1}{t}$$

$$t \cdot v = t \cdot \frac{d_2 - d_1}{t}$$

$$tv = d_2 - d_1$$

$$tv \cdot \frac{1}{v} = (d_2 - d_1) \cdot \frac{1}{v}$$

$$t = \frac{d_2 - d_1}{v}$$

28. $m = \dfrac{s_2 - s_1}{v}$

29.
$$v = \frac{d_2 - d_1}{t}$$

$$t \cdot v = t \cdot \frac{d_2 - d_1}{t}$$

$$tv = d_2 - d_1$$

$$tv - d_2 = d_2 - d_1 - d_2$$

$$tv - d_2 = -d_1$$

$$-1 \cdot (tv - d_2) = -1 \cdot (-d_1)$$

$$-tv + d_2 = d_1,$$

$$\text{or } d_2 - tv = d_1$$

30. $s_1 = s_2 - vm$

31.
$$r = m + mnp$$

$$r = m(1 + np) \qquad \text{Factoring}$$

$$r \cdot \frac{1}{1 + np} = m(1 + np) \cdot \frac{1}{1 + np}$$

$$\frac{r}{1 + np} = m$$

32. $x = \dfrac{p}{1 - yz}$

33.
$$y = ab - ac^2$$

$$y = a(b - c^2)$$

$$y \cdot \frac{1}{b - c^2} = a(b - c^2) \cdot \frac{1}{b - c^2}$$

$$\frac{y}{b - c^2} = a$$

34. $m = \dfrac{d}{n - p^3}$

35. Translate. The formula for the area of a parallelogram with base b and height h is

$$A = bh.$$

Since we want to find the length of the base, we solve the formula for b.

$$A = bh$$

$$A \cdot \frac{1}{h} = bh \cdot \frac{1}{h}$$

$$\frac{A}{h} = b$$

We now have a formula that says to find b, we divide A by h.

35. (continued)

Carry out. Substitute 72 for A and 6 for h in the formula and calculate:

$$\frac{72}{6} = b$$

$$12 = b$$

We leave the check to the student. The base of the parallelogram is 12 cm.

36. 6 cm

37. Translate. The simple interest formula is I = Prt, where I is the amount of interest, P is the principal, r is the interest rate, and t is the time in years. What we want to know is the amount to invest, so we solve the formula for P.

$$I = Prt$$

$$\frac{I}{rt} = P$$

We now have a formula that gives P.

Carry out. We substitute $110 for I, 7% (or 0.07) for r and 1 for t.

$$P = \frac{I}{rt}$$

$$P = \frac{110}{0.07 \cdot 1}$$

$$P = 1571.43 \qquad \text{Rounding to the nearest cent}$$

The check is left to the student. The solution is that you will have to invest $1571.43.

38. 6.4%

39. Translate. The formula for the area of a trapezoid is $A = \frac{1}{2}h(b_1 + b_2)$, where A is the area, h is the height, and b_1 and b_2 are the bases. The unknown dimension in this problem is h, so we solve the formula for h.

$$A = \frac{1}{2}h(b_1 + b_2)$$

$$2A = h(b_1 + b_2)$$

$$\frac{2A}{b_1 + b_2} = h$$

We now have a formula that gives h.

Carry out. We substitute 90 for A, 8 for b_1, and 12 for b_2.

$$h = \frac{2A}{b_1 + b_2}$$

$$h = \frac{2 \cdot 90}{8 + 12}$$

$$h = \frac{180}{20}$$

$$h = 9$$

The check is left to the student. The unknown dimension is 9 ft.

40. 25 ft

41. Translate. The formula A = P + Prt tells us how much a principal P, in dollars, will be worth when invested at simple interest at a rate r for t years. We want to find the length of the investment period, so we solve the formula for t.

$$A = P + Prt$$

$$A - P = Prt$$

$$\frac{A - P}{Pr} = t$$

Carry out. Substitute 2608 for A, 1600 for P and 9%(or 0.09) for r.

$$\frac{A - P}{Pr} = t$$

$$\frac{2608 - 1600}{1600(0.09)} = t$$

$$\frac{1008}{144} = t$$

$$7 = t$$

The check is left to the student. It will take 7 years for the investment to be worth $2608.

42. 6 years

43. Translate. We solve the formula $D = \frac{m}{V}$ for m. (See Example 5.)

$$D = \frac{m}{V}$$

$$DV = m$$

Carry out. Substitute 7.5 for D and 61.5 for V.

$$DV = m$$

$$7.5(61.5) = m$$

$$461.25 = m$$

The check is left to the student. The anchor's mass is 461.25 g.

44. 358.85205 g

45. Familiarize and Translate. Recall that "%" means "× 0.01." Let n represent the percent.

What percent of 5800 is 4176?

n   × 0.01   × 5800 = 4176

Carry out. n × 0.01 × 5800 = 4176

$$58n = 4176$$

$$n = \frac{4176}{58}$$

$$n = 72$$

The check is left to the student.

State. 4176 is 72% of 5800.

46. -8a + 13b

47. -72.5 - (-14.06) = -72.5 + 14.06 = -58.44

48. 3

49.
$$s = v_1 t + \frac{1}{2}at^2$$

$$s - v_1 t = \frac{1}{2}at^2$$

$$2(s - v_1 t) = at^2$$

$$\frac{2(s - v_1 t)}{t^2} = a,$$

or $\dfrac{2s - 2v_1 t}{t^2} = a$

50. $\ell = \dfrac{A - w^2}{4w}$

51.
$$\frac{P_1 V_1}{T_1} = \frac{P_2 V_2}{T_2}$$

$$\frac{T_1}{P_1} \cdot \frac{P_1 V_1}{T_1} = \frac{T_1}{P_1} \cdot \frac{P_2 V_2}{T_2}$$

$$V_1 = \frac{T_1 P_2 V_2}{P_1 T_2}$$

52. $T_2 = \dfrac{T_1 P_2 V_2}{P_1 V_1}$

53.
$$x = \frac{a}{b + c}$$

$$x(b + c) = a$$

$$xb + xc = a$$

$$xc = a - xb$$

$$c = \frac{a - xb}{x}, \quad \text{or} \quad \frac{a}{x} - b$$

54. $d = \dfrac{me^2}{f}$

55.
$$\frac{a}{b} = \frac{c}{d}$$

$$\frac{a}{b} \cdot b = \frac{c}{d} \cdot b$$

$$a = \frac{cb}{d}$$

$$a \cdot \frac{1}{c} = \frac{cb}{d} \cdot \frac{1}{c}$$

$$\frac{a}{c} = \frac{b}{d}$$

56. $f = \dfrac{me^2}{d}$

57.
$$ab = c - bd$$

$$ab + bd = c$$

$$b(a + d) = c$$

$$b = \frac{c}{a + d}$$

58. $m = \dfrac{nr + np}{n - p}$

59. From Example 2 we know that $t = \frac{I}{Pr}$.

Carry out. Substitute 6 for I, 200 for P, and 12% (or 0.12) for r.

$$t = \frac{I}{Pr}$$

$$t = \frac{6}{200(0.12)}$$

$$t = \frac{6}{24}$$

$$t = \frac{1}{4}$$

The check is left to the student. It would take $\frac{1}{4}$ year, or 3 months $\left[\frac{1}{4} \times 12 \text{ months}\right]$.

60. 11%

61. Take the square root on both sides.

62. Take the cube root on both sides.

63. We will use the formulas for density and for the volume of a right circular cylinder.

Translate. Solving the formula $D = \frac{m}{V}$ for V, we get $V = \frac{m}{D}$. (See Example 5.) Also, the volume of a right circular cylinder with radius r and height h is given by $V = \pi r^2 h$, so we have $\pi r^2 h = \frac{m}{D}$. Solve for h:

$$h = \frac{m}{\pi r^2 D}$$

Note also that the radius of a penny is $\frac{1.85}{2}$, or 0.925.

Carry out. Substitue 8.93 for D, 177.6 for m, 0.925 for r, and 3.14 for π:

$$h = \frac{m}{\pi r^2 D}$$

$$h = \frac{177.6}{3.14(0.925)^2(8.93)}$$

$$h \approx 7.4$$

The check is left to the student. The roll of pennies is about 7.4 cm tall.

64. 610.55 cm

Exercise Set 1.7

1. $47{,}000{,}000{,}000 = 47{,}000{,}000{,}000 \times (10^{-10} \times 10^{10})$
   $= (47{,}000{,}000{,}000 \times 10^{-10}) \times 10^{10}$
   $= 4.7 \times 10^{10}$

2. 2.6×10^{12}

3. $863{,}000{,}000{,}000{,}000{,}000$
   $= 863{,}000{,}000{,}000{,}000{,}000 \times (10^{-17} \times 10^{17})$
   $= (863{,}000{,}000{,}000{,}000{,}000 \times 10^{-17}) \times 10^{17}$
   $= 8.63 \times 10^{17}$

4. 9.57×10^{17}

5. $0.000000016 = 0.000000016 \times (10^8 \times 10^{-8})$
   $= (0.000000016 \times 10^8) \times 10^{-8}$
   $= 1.6 \times 10^{-8}$

6. 2.63×10^{-7}

7. $0.00000000007 = 0.00000000007 \times (10^{11} \times 10^{-11})$
   $= (0.00000000007 \times 10^{11}) \times 10^{-11}$
   $= 7 \times 10^{-11}$

8. 9×10^{-11}

9. $407{,}000{,}000{,}000 = 407{,}000{,}000{,}000 \times (10^{-11} \times 10^{11})$
   $= (407{,}000{,}000{,}000 \times 10^{-11}) \times 10^{11}$
   $= 4.07 \times 10^{11}$

10. 3.09×10^{12}

11. $0.000000603 = 0.000000603 \times (10^7 \times 10^{-7})$
    $= (0.000000603 \times 10^7) \times 10^{-7}$
    $= 6.03 \times 10^{-7}$

12. 8.02×10^{-9}

13. $492{,}700{,}000{,}000 = 492{,}700{,}000{,}000 \times (10^{-11} \times 10^{11})$
    $= (492{,}700{,}000{,}000 \times 10^{-11}) \times 10^{11}$
    $= 4.927 \times 10^{11}$

14. 9.534×10^{11}

15. $4 \times 10^{-4} = 0.0004$ Moving decimal point 4 places to the left

16. 0.00005

17. $6.73 \times 10^8 = 673{,}000{,}000$ Moving decimal point 8 places to the right

18. $92{,}400{,}000$

19. $8.923 \times 10^{-10} = 0.0000000008923$ Moving decimal point 10 places to the left

20. 0.07034

21. $9.03 \times 10^{10} = 90{,}300{,}000{,}000$ Moving decimal point 10 places to the right

22. $1{,}010{,}000{,}000{,}000$

23. $4.037 \times 10^{-8} = 0.00000004037$ Moving decimal point 8 places to the left

24. 0.000000003007

25. $8.007 \times 10^{12} = 8,007,000,000,000$ Moving decimal point 12 places to the right

26. $90,010,000,000$

27. $(2.3 \times 10^6)(4.2 \times 10^{-11})$
    $= (2.3 \times 4.2)(10^6 \times 10^{-11})$
    $= 9.66 \times 10^{-5}$

28. 3.38×10^{-4}

29. $(2.34 \times 10^{-8})(5.7 \times 10^{-4})$
    $= (2.34 \times 5.7)(10^{-8} \times 10^{-4})$
    $= 13.338 \times 10^{-12}$
    $= (1.3338 \times 10^1) \times 10^{-12}$
    $= 1.3338 \times (10^1 \times 10^{-12})$
    $= 1.3338 \times 10^{-11}$

30. 2.6732×10^{-11}

31. $(3.2 \times 10^6)(2.6 \times 10^4) = (3.2 \times 2.6)(10^6 \times 10^4)$
    $= 8.32 \times 10^{10}$

32. 3.1411×10^{16}

33. $(3.01 \times 10^{-5})(6.5 \times 10^7)$
    $= (3.01 \times 6.5)(10^{-5} \times 10^7)$
    $= 19.565 \times 10^2$
    $= (1.9565 \times 10^1) \times 10^2$
    $= 1.9565 \times (10^1 \times 10^2)$
    $= 1.9565 \times 10^3$

34. 3.1416×10^{-4}

35. $(5.01 \times 10^{-7})(3.02 \times 10^{-6})$
    $= (5.01 \times 3.02)(10^{-7} \times 10^{-6})$
    $= 15.1302 \times 10^{-13}$
    $= (1.51302 \times 10^1) \times 10^{-13}$
    $= 1.51302 \times (10^1 \times 10^{-13})$
    $= 1.51302 \times 10^{-12}$

36. 6.34304×10^{-15}

37. $\dfrac{8.5 \times 10^8}{3.4 \times 10^5} = \dfrac{8.5}{3.4} \times \dfrac{10^8}{10^5}$
    $= 2.5 \times 10^3$

38. 1.5×10^3

39. $\dfrac{4.0 \times 10^{-6}}{8.0 \times 10^{-3}} = \dfrac{4.0}{8.0} \times \dfrac{10^{-6}}{10^{-3}}$
    $= 0.5 \times 10^{-3} = (5 \times 10^{-1}) \times 10^{-3}$
    $= 5 \times (10^{-1} \times 10^{-3}) = 5 \times 10^{-4}$

40. 3×10^{-5}

41. $\dfrac{12.6 \times 10^8}{4.2 \times 10^{-3}} = \dfrac{12.6}{4.2} \times \dfrac{10^8}{10^{-3}}$
    $= 3 \times 10^{11}$

42. 4×10^{-16}

43. $\dfrac{2.42 \times 10^5}{1.21 \times 10^{-5}} = \dfrac{2.42}{1.21} \times \dfrac{10^5}{10^{-5}}$
    $= 2 \times 10^{10}$

44. 3×10^{-22}

45. $\dfrac{4.7 \times 10^{-9}}{2.35 \times 10^7} = \dfrac{4.7}{2.35} \times \dfrac{10^{-9}}{10^7}$
    $= 2 \times 10^{-16}$

46. 2×10^{26}

47. $\dfrac{1.05 \times 10^{-6}}{4.2 \times 10^{-7}} = \dfrac{1.05}{4.2} \times \dfrac{10^{-6}}{10^{-7}}$
    $= 0.25 \times 10^1$
    $= (2.5 \times 10^{-1}) \times 10^1$
    $= 2.5 \times (10^{-1} \times 10^1)$
    $= 2.5 \times 10^0$
    $= 2.5 \times 1$
    $= 2.5$

48. 1.25×10^{-4}

49. $\dfrac{(6.1 \times 10^4)(7.2 \times 10^{-6})}{9.8 \times 10^{-4}} = \dfrac{6.1 \times 7.2}{9.8} \times \dfrac{10^4 \times 10^{-6}}{10^{-4}}$
    $\approx \dfrac{6 \times 7}{10} \times \dfrac{10^{-2}}{10^{-4}}$
    $\approx \dfrac{42}{10} \times 10^2$
    $\approx 4.2 \times 10^2$

50. 1.6×10^{-17}

51. $\dfrac{780,000,000 \times 0.00071}{0.000005} = \dfrac{(7.8 \times 10^8) \times (7.1 \times 10^{-4})}{5 \times 10^{-6}}$
    $= \dfrac{7.8 \times 7.1}{5} \times \dfrac{10^8 \times 10^{-4}}{10^{-6}}$
    $\approx \dfrac{8 \times 7}{5} \times \dfrac{10^4}{10^{-6}}$
    $\approx \dfrac{56}{5} \times 10^{10}$
    $\approx 11 \times 10^{10}$
    $\approx (1.1 \times 10^1) \times 10^{10}$
    $\approx 1.1 \times (10^1 \times 10^{10})$
    $\approx 1.1 \times 10^{11}$

52. 2.7×10

53.
$$\frac{43,000,000 \times 0.095}{63,000} = \frac{(4.3 \times 10^7) \times (9.5 \times 10^{-2})}{6.3 \times 10^4}$$

$$= \frac{4.3 \times 9.5}{6.3} \times \frac{10^7 \times 10^{-2}}{10^4}$$

$$\approx \frac{4 \times 10}{6} \times \frac{10^5}{10^4}$$

$$\approx \frac{40}{6} \times 10$$

$$\approx 6.7 \times 10$$

54. 9.3×10^2

55.
$$\frac{40,000 \times 0.29}{0.057 \times 160,000} = \frac{(4 \times 10^4) \times (2.9 \times 10^{-1})}{(5.7 \times 10^{-2}) \times (1.6 \times 10^5)}$$

$$= \frac{4 \times 2.9}{5.7 \times 1.6} \times \frac{10^4 \times 10^{-1}}{10^{-2} \times 10^5}$$

$$\approx \frac{4 \times 3}{6 \times 2} \times \frac{10^3}{10^3}$$

$$\approx \frac{12}{12} \times 1$$

$$\approx 1$$

56. 3.2

57.
$$\frac{0.000012 \times 98,000,000}{19,000,000,000} = \frac{(1.2 \times 10^{-5}) \times (9.8 \times 10^7)}{1.9 \times 10^{10}}$$

$$= \frac{1.2 \times 9.8}{1.9} \times \frac{10^{-5} \times 10^7}{10^{10}}$$

$$\approx \frac{1 \times 10}{2} \times \frac{10^2}{10^{10}}$$

$$\approx \frac{10}{2} \times 10^{-8}$$

$$\approx 5 \times 10^{-8}$$

58. 1×10^{-5}, or 10^{-5}

59.
$$\frac{0.0000038 \times 102,000,000}{36,000 \times 0.0000123}$$

$$= \frac{(3.8 \times 10^{-6}) \times (1.02 \times 10^8)}{(3.6 \times 10^4) \times (1.23 \times 10^{-5})}$$

$$= \frac{3.8 \times 1.02}{3.6 \times 1.23} \times \frac{10^{-6} \times 10^8}{10^4 \times 10^{-5}}$$

$$\approx \frac{4 \times 1}{4 \times 1} \times \frac{10^2}{10^{-1}}$$

$$\approx \frac{4}{4} \times 10^3$$

$$\approx 1 \times 10^3, \text{ or } 10^3$$

60. 1.75×10^3

61. Familiarize. We will let y represent the number of light years from one end of the galaxy to the other (that is, the number of light years in the diameter of the galaxy). Recall from Example 1 that 1 light year = 5.88×10^{12} mi.

Translate. Note that the diameter of the galaxy is $(5.88 \times 10^{12})y$ mi. We are also told that this distance is 5.88×10^{17} mi. Since these quantities represent the same number, we write the equation

$$(5.88 \times 10^{12})y = 5.88 \times 10^{17}.$$

61. (continued)

Carry out. Solve the equation:
$$(5.88 \times 10^{12})y = 5.88 \times 10^{17}$$

$$\frac{1}{5.88 \times 10^{12}}(5.88 \times 10^{12})y = \frac{1}{5.88 \times 10^{12}} \times 5.88 \times 10^{17}$$

$$y = \frac{5.88 \times 10^{17}}{5.88 \times 10^{12}}$$

$$y = \frac{5.88}{5.88} \times \frac{10^{17}}{10^{12}}$$

$$y = 1 \times 10^5, \text{ or } 10^5$$

The check is left to the student.

State. It is 10^5, or 100,000, light years from one end of the Milky Way galaxy to the other.

62. 8 light years

63. Familiarize. We are told that 1 Angstrom = 10^{-10} m, one parsec ≈ 3.26 light years, and 1 light year = 9.46×10^{15} m. Let a represent the number of Angstroms in one parsec.

Translate. The length of one parsec is $a \times 10^{-10}$ m. It can also be expressed as 3.26 light years, or $3.26 \times 9.46 \times 10^{15}$ m. Since these quantities represent the same number, we can write the equation

$$a \times 10^{-10} = 3.26 \times 9.46 \times 10^{15}.$$

Carry out. Solve the equation:
$$a \times 10^{-10} = 3.26 \times 9.46 \times 10^{15}$$

$$a \times 10^{-10} \times \frac{1}{10^{-10}} = 3.26 \times 9.46 \times 10^{15} \times \frac{1}{10^{-10}}$$

$$a = \frac{3.26 \times 9.46 \times 10^{15}}{10^{-10}}$$

$$= (3.26 \times 9.46) \times \frac{10^{15}}{10^{-10}}$$

$$= 30.8396 \times 10^{25}$$

$$= (3.08396 \times 10) \times 10^{25}$$

$$= 3.08396 \times (10 \times 10^{25})$$

$$\approx 3.084 \times 10^{26}$$

The check is left to the student.

State. There are about 3.084×10^{26} Angstroms in one parsec.

64. 3.084×10^{13} km

65. Familiarize. We have a very long cylinder. Its length is the average distance from the earth to the sun, 1.5×10^{11} m, and the diameter of its base is 3Å. We will use the formula for the volume of a cylinder, $V = \pi r^2 h$. (See Example 11.)

Translate. We will express all distances in Angstroms.

Height (length): 1.5×10^{11} m $= \frac{1.5 \times 10^{11}}{10^{-10}}$ Å, or

$$1.5 \times 10^{21} \text{ Å}$$

Diameter: 3Å

The radius is half the diameter:

Radius: $\frac{1}{2} \times 3$Å $= 1.5$ Å

65. (continued)

Now substitute into the formula (using 3.14 for π):

$$V = \pi r^2 h$$
$$V = 3.14 \times 1.5^2 \times 1.5 \times 10^{21}$$

Carry out. Do the calculations.

$$V = 3.14 \times 1.5^2 \times 1.5 \times 10^{21}$$
$$= 10.5975 \times 10^{21}$$
$$\approx 1.06 \times 10^{22}$$

Check. In this case, about all we can do is recheck the translation and the calculations.

State. The volume of the sunbeam is about 1.06×10^{22} cu Å.

66. 2.944×10^{22} cu Å

67. Familiarize. When the roll of plastic is unrolled, its length, width, and height (thickness) are 30 m, 1 m, and 0.8 mm, respectively. We will use the formula for the volume of a rectangular solid, $V = \ell wh$, where ℓ is length, w is width, and h is height.

Translate. We will express all dimensions in meters.

Length: 30 m

Width: 1 m

Height: 0.8 mm = 0.8×10^{-3} m, or 8×10^{-4} m

Substitute into the formula:

$$V = \ell wh$$
$$= 30 \times 1 \times 8 \times 10^{-4}$$

Carry out. Do the calculations.

$$V = 30 \times 1 \times 8 \times 10^{-4}$$
$$= 240 \times 10^{-4}$$
$$= 2.4 \times 10^{-2}$$

Check. Recheck the translation and the calculations.

State. The volume of plastic in a roll is 2.4×10^{-2} m³, or 0.0024 m³.

68. 1.475×10^{12} mi

69. Familiarize. We can use a circle whose radius is the average distance of the earth from the sun to approximate the earth's orbit about the sun. The distance the earth travels in a yearly orbit about the sun is given by the circumference of that circle. Recall that the formula for the circumference of a circle is $C = 2\pi r$, where r is the radius.

Translate. Substitute 3.14 for π and 9.3×10^7 for r in the formula.

$$C = 2\pi r$$
$$= 2 \times 3.14 \times 9.3 \times 10^7$$

Carry out. Do the calculations.

$$C = 2 \times 3.14 \times 9.3 \times 10^7$$
$$= 58.404 \times 10^7$$
$$\approx 5.84 \times 10^8$$

69. (continued)

Check. Recheck the translation and the calculations.

State. The earth travels approximately 5.84×10^8 mi in a yearly orbit about the sun.

70. $67,000

71. $-\dfrac{5}{6} - \left(-\dfrac{3}{4}\right) = -\dfrac{5}{6} + \dfrac{3}{4} = -\dfrac{10}{12} + \dfrac{9}{12} = -\dfrac{1}{12}$

72. 30.96

73. $-2(x - 3) - 3(4 - x) = -2x + 6 - 12 + 3x$
$$= (-2x + 3x) + (6 - 12)$$
$$= x - 6$$

74. $\dfrac{7}{4}$

75. The larger number is the one in which the power of ten has the larger exponent. Since -90 is larger than -91, $8 \cdot 10^{-90}$ is larger than $9 \cdot 10^{-91}$.

$$8 \cdot 10^{-90} - 9 \cdot 10^{-91} = 10^{-90}(8 - 9 \cdot 10^{-1})$$
$$= 10^{-90}(8 - 0.9)$$
$$= 7.1 \times 10^{-90}$$

76. $\dfrac{1}{3}$

77. $(4096)^{0.05}(4096)^{0.2} = 4096^{0.25}$
$$= 4096^{\frac{1}{4}}$$
$$= \sqrt[4]{4096}$$
$$= 8$$

78. 7

79. Solve $a = 4 \cdot 3^b$ for 3^b.

$$a = 4 \cdot 3^b$$
$$\frac{a}{4} = 3^b$$

Substitute a/4 for 3^b in $c = 2 - 3^{-b}$.

$$c = 2 - 3^{-b}$$
$$c = 2 - \frac{1}{3^b}$$
$$c = 2 - \frac{1}{a/4} = 2 - \frac{4}{a} = \frac{2a - 4}{a}$$

80. $\dfrac{1}{y^4}$

81. $\dfrac{3^{q+3} - 3^2(3^q)}{3(3^{q+4})} = \dfrac{3^3(3^q) - 3^2(3^q)}{3(3^q)(3^4)}$
$$= \frac{3^2 \cdot 3^q(3 - 1)}{3^2 \cdot 3^q(3^3)}$$
$$= \frac{2}{27}$$

82. $\dfrac{3^{12S+4}}{625}$

83. <u>Familiarize</u>. Observe that there are 2^{n-1} grains of sand on the nth square of the chessboard. Let g represent this quantity. Recall that a chessboard has 64 squares. Note also that $2^{10} \approx 10^3$.

    <u>Translate</u>. We write the equation

    $g = 2^{n-1}$.

    To find the number of grains of sand on the last (or 64th) square, substitute 64 for n: $g = 2^{64-1}$

    <u>Carry out</u>. Do the calculations, expressing the result in scientific notation.

    $g = 2^{64-1} = 2^{63} = 2^3(2^{10})^6$

    $\approx 2^3(10^3)^6 \approx 8 \times 10^{18}$

    <u>Check</u>. Recheck the translation and the calculations.

    <u>State</u>. Approximately 8×10^{18} grains of sand are required for the last square.

Exercise Set 2.1

1.

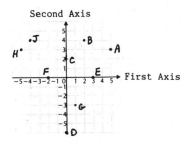

A(5,3) is 5 units right and 3 units up.
B(2,4) is 2 units right and 4 units up.
C(0,2) is 0 units left or right and 2 units up.
D(0,-6) is 0 units left or right and 6 units down.
E(3,0) is 3 units right and 0 units up or down.
F(-2,0) is 2 units left and 0 units up or down.
G(1,-3) is 1 unit right and 3 units down.
H(-5,3) is 5 units left and 3 units up.
J(-4,4) is 4 units left and 4 units up.

2.

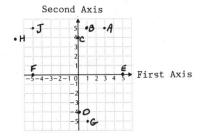

3.

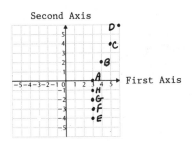

A(3,0) is 3 units right and 0 units up or down.
B(4,2) is 4 units right and 2 units up.
C(5,4) is 5 units right and 4 units up.
D(6,6) is 6 units right and 6 units up.
E(3,-4) is 3 units right and 4 units down.
F(3,-3) is 3 units right and 3 units down.
G(3,-2) is 3 units right and 2 units down.
H(3,-1) is 3 units right and 1 unit down.

4.

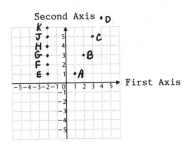

5.

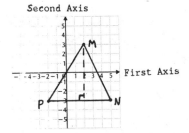

A triangle is formed. The area of a triangle is
found by using the formula $A = \frac{1}{2}bh$. In this
triangle the base and height are 7 units and 6
units, respectively.

$A = \frac{1}{2}bh = \frac{1}{2} \cdot 7 \cdot 6 = \frac{42}{2} = 21$ square units

6.

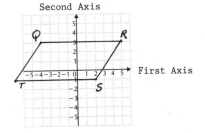

A parallelogram is formed. The area is 36 square
units.

7. Both coordinates are negative, so the point
(-3,-5) is in quadrant III.

8. I

9. The first coordinate is negative and the second
positive, so the point (-6,1) is in quadrant II.

10. IV

11. Both coordinates are positive, so the point
$\left(3,\frac{1}{2}\right)$ is in quadrant I.

12. III

13. The first coordinate is positive and the second
negative, so the point (7,-0.2) is in quadrant
IV.

14. II

15.
$$\begin{array}{c|l} \multicolumn{2}{l}{y = 2x - 3} \\ \hline -1 & 2 \cdot 1 - 3 \quad \text{Substituting 1 for x and -1 for y} \\ & \quad -1 \qquad \text{(alphabetical order of variables)} \end{array}$$
Since -1 = -1 is true, (1,-1) <u>is</u> a solution of
y = 2x - 3.

16. Yes

17.
$$\frac{3s + t = 4}{\begin{array}{c|c} 3\cdot3 + 4 & 4 \\ 9 + 4 & \\ 13 & \end{array}}$$
Substituting 3 for s and 4 for t
(alphabetical order of variables)

Since 13 = 4 is false, (3,4) <u>is</u> <u>not</u> a solution of
3s + t = 4.

18. No

19.
$$\frac{4x - y = 7}{\begin{array}{c|c} 4\cdot3 - 5 & 7 \\ 12 - 5 & \\ 7 & \end{array}}$$
Substituting 3 for x and 5 for y
(alphabetical order of variables)

Since 7 = 7 is true, (3,5) <u>is</u> a solution of
4x - y = 7.

20. Yes

21.
$$\frac{2a + 5b = 3}{\begin{array}{c|c} 2\cdot0 + 5\cdot\frac{3}{5} & 3 \\ 0 + 3 & \\ 3 & \end{array}}$$
Substituting 0 for a and $\frac{3}{5}$
for b
(alphabetical order of variables)

Since 3 = 3 is true, $\left[0,\frac{3}{5}\right]$ <u>is</u> a solution of
2a + 5b = 3.

22. Yes

23.
$$\frac{4r + 3s = 5}{\begin{array}{c|c} 4\cdot2 + 3\cdot(-1) & 5 \\ 8 - 3 & \\ 5 & \end{array}}$$
Substituting 2 for r and -1
for s
(alphabetical order of
variables)

Since 5 = 5 is true, (2,-1) <u>is</u> a solution of
4r + 3s = 5.

24. Yes

25.
$$\frac{3x - 2y = -4}{\begin{array}{c|c} 3\cdot3 - 2\cdot2 & -4 \\ 9 - 4 & \\ 5 & \end{array}}$$
Substituting 3 for x and 2 for y
(alphabetical order of variables)

Since 5 = -4 is false, (3,2) <u>is</u> <u>not</u> a solution of
3x - 2y = -4.

26. No

27.
$$\frac{y = 3x^2}{\begin{array}{c|c} 3 & 3(-1)^2 \\ & 3\cdot1 \\ & 3 \end{array}}$$
Substitute -1 for x and 3 for y
(alphabetical order of variables)

Since 3 = 3 is true, (-1,3) <u>is</u> a solution of
y = 3x².

28. No

29.
$$\frac{5s^2 - t = 7}{\begin{array}{c|c} 5(2)^2 - 3 & 7 \\ 5\cdot4 - 3 & \\ 20 - 3 & \\ 17 & \end{array}}$$
Substitute 2 for s and 3 for t
(alphabetical order of variables)

Since 17 = 7 is false, (2,3) <u>is</u> <u>not</u> a solution of
5s² - t = 7.

30. Yes

31. y = -2x

To find an ordered pair, we choose any number for
x and then determine y by substitution.

When x = 0, y = -2·0 = 0.
When x = 3, y = -2·3 = -6.
When x = -2, y = -2·(-2) = 4.

x	y	(x,y)
0	0	(0,0)
3	-6	(3,-6)
-2	4	(-2,4)

Plot these points, draw the line they determine,
and label the graph y = -2x.

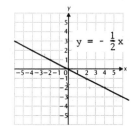

32.

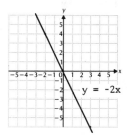

33. y = x + 3

To find an ordered pair we choose any number for x
and then determine y. For example, if we choose
1 for x, then y = 1 + 3, or 4. We find several
ordered pairs, plot them, and draw the line.

x	y	(x,y)
1	4	(1,4)
2	5	(2,5)
-1	2	(-1,2)
-3	0	(-3,0)

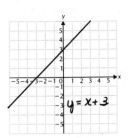

34.

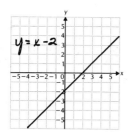

35. y = 3x - 2

To find an ordered pair, we choose any number
for x and then determine y. For example, if
x = 2, then y = 3·2 - 2 = 6 - 2 = 4. We find
several ordered pairs, plot them, and draw the
line.

x	y	(x,y)
2	4	(2,4)
0	-2	(0,-2)
-1	-5	(-1,-5)
1	1	(1,1)

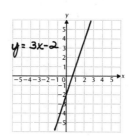

36.

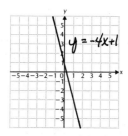

37. y = -2x + 3

To find an ordered pair, we choose any number for
x and then determine y. For example, if x = 1,
then y = -2·1 + 3 = -2 + 3 = 1. We find several
ordered pairs, plot them, and draw the line.

x	y
1	1
3	-3
-1	5
0	3

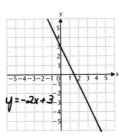

38.

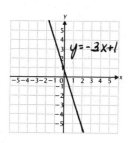

39. y = $\frac{2}{3}$x + 1

To find an ordered pair, we choose any number for
x and then determine y. For example, if x = 3,
then y = $\frac{2}{3}$ · 3 + 1 = 2 + 1 = 3. We find several
ordered pairs, plot them, and draw the line.

x	y
3	3
0	1
-3	-1

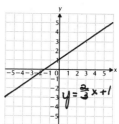

40.

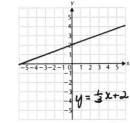

41. y = - $\frac{3}{2}$x + 1

To find an ordered pair, we choose any number for
x and then determine y. For example, if x = 2,
then y = - $\frac{3}{2}$ · 2 + 1 = -3 + 1 = -2. We find
several ordered pairs, plot them, and draw the
line.

x	y
2	-2
4	-5
0	1
-2	4

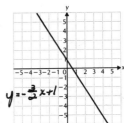

42.

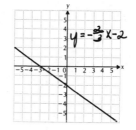

31

43. $y = \frac{3}{4}x + 1$

To find an ordered pair, we choose any number for x and then determine y. For example, if x = 4, $y = \frac{3}{4} \cdot 4 + 1 = 3 + 1 = 4$. We find several ordered pairs, plot them and draw the line.

x	y
4	4
0	1
-4	-2

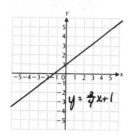

47. $y = x^2 - 2$

To find an ordered pair, we choose any number for x and then determine y. For example, if x = 2, $y = 2^2 - 2 = 4 - 2 = 2$. We find several ordered pairs, plot them, and connect them with a smooth curve.

x	y
2	2
1	-1
0	-2
-1	-1
-2	2

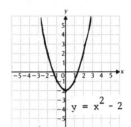

44.

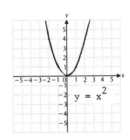

48.

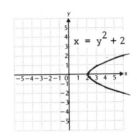

45. $y = -x^2$

To find an orderd pair, we choose any number for x and then determine y. For example, if x = 2, then $y = -(2)^2 = -4$. We find several ordered pairs, plot them, and connect them with a smooth curve.

x	y
2	-4
1	-1
0	0
-1	-1
-2	-4

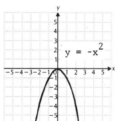

49. $y = 3 - x^2$

To find an ordered pair, we choose any number for x and then determine y. For example, if x = 2, $y = 3 - 2^2 = 3 - 4 = -1$. We find several ordered pairs, plot them, and connect them with a smooth curve.

x	y
2	-1
1	2
0	3
-1	2
-2	-1

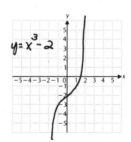

46.

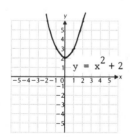

50.

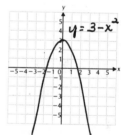

51. $y = -\frac{1}{x}$

We select x-values and find the corresponding y-values. The table lists some ordered pairs.

x	y
4	$-\frac{1}{4}$
2	$-\frac{1}{2}$
1	-1
$\frac{1}{2}$	-2
$-\frac{1}{2}$	2
-1	1
-2	$\frac{1}{2}$
-4	$\frac{1}{4}$

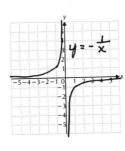

We plot these points. Note that we cannot use 0 as a first-coordinate, since -1/0 is undefined. Thus the graph has two branches, one on either side of the y-axis.

52.

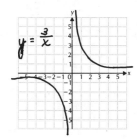

53. $y = |x| + 2$

We select x-values and find the corresponding y-values. The table lists some ordered pairs.

x	y
3	5
1	3
0	2
-1	3
-3	5

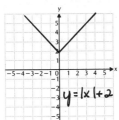

We plot these points. Note that the graph is V-shaped, centered at (0,2).

54.

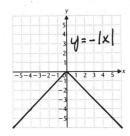

55. <u>Familiarize</u>. The formula for the area of a triangle with base b and height h is $A = \frac{1}{2}bh$.

<u>Translate</u>. Substitute 156 for A and 12 for b in the formula.

$$A = \frac{1}{2}bh$$

$$156 = \frac{1}{2} \cdot 12 \cdot h$$

<u>Carry out</u>. Solve the equation.

$$156 = \frac{1}{2} \cdot 12 \cdot h$$

$$156 = 6h$$

$$26 = h$$

<u>Check</u>. Left to the student.

<u>State</u>. The triangle should be 26 ft tall.

56. 11%

57. $-3.9 - (-2.5) = -3.9 + 2.5 = -1.4$

58. 5

59.

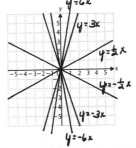

The number in front of the x tells the direction of the slant and how steep the slant is.

60. (-1,-2)

61. Substitute $-\frac{1}{3}$ for x and $\frac{1}{4}$ for y in each equation.

a)

$$\frac{-\frac{3}{2}x - 3y = -\frac{1}{4}}{}$$

$-\frac{3}{2}\left(-\frac{1}{3}\right) - 3\left(\frac{1}{4}\right)$	$-\frac{1}{4}$
$\frac{1}{2} - \frac{3}{4}$	
$-\frac{1}{4}$	

Since $-\frac{1}{4} = -\frac{1}{4}$ is true, $\left(-\frac{1}{3}, \frac{1}{4}\right)$ <u>is</u> a solution.

b)

$$\frac{8y - 15x = \frac{7}{2}}{}$$

$8\left(\frac{1}{4}\right) - 15\left(-\frac{1}{3}\right)$	$\frac{7}{2}$
$2 + 5$	
7	

Since $7 = \frac{7}{2}$ is false, $\left(-\frac{1}{3}, \frac{1}{4}\right)$ <u>is not</u> a solution.

61. (continued)

c)

$$\frac{0.16y = -0.09x + 0.1}{\begin{array}{c|c} 0.16\left(\frac{1}{4}\right) & -0.09\left(-\frac{1}{3}\right) + 0.1 \\ 0.04 & 0.03 + 0.1 \\ & 0.13 \end{array}}$$

Since 0.04 = 0.13 is false, $\left(-\frac{1}{3}, \frac{1}{4}\right)$ is not a solution.

d)

$$\frac{2(-y + 2) - \frac{1}{4}(3x - 1) = 4}{\begin{array}{c|c} 2\left[-\frac{1}{4} + 2\right] - \frac{1}{4}\left[3\left[-\frac{1}{3}\right] - 1\right] & 4 \\ 2\left[\frac{7}{4}\right] - \frac{1}{4}(-2) & \\ \frac{14}{4} + \frac{2}{4} & \\ \frac{16}{4} & \\ 4 & \end{array}}$$

Since 4 = 4 is true, $\left(-\frac{1}{3}, \frac{1}{4}\right)$ is a solution.

62.

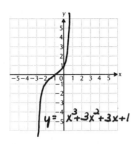

$y = x^3 + 3x^2 + 3x + 1$

63. $y = x^3 - 6x^2 + 12x - 8$

We select x-values between 0 and 4 and use a calculator to find the corresponding y-values. Plot the ordered pairs, and connect them with a smooth curve.

x	y
0	-8
0.25	-5.359375
0.5	-3.375
1	-1
1.5	-0.125
2	0
2.5	0.125
3	1
3.5	3.375
4	8

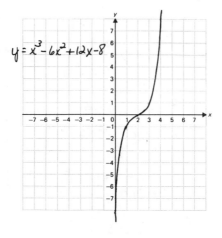

$y = x^3 - 6x^2 + 12x - 8$

64.

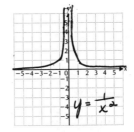

$y = \frac{1}{x^2}$

65. $y = -\frac{1}{x^2}$

Choose x-values from -3 to 3, and use a calculator to find the corresponding y-values. [Note that we cannot choose 0 as a first coordinate, since $-\frac{1}{0^2}$, or $-\frac{1}{0}$, is not defined.] Plot the points, and draw the graph. Note that it has two branches, one on either side of the y-axis.

x	y
-3	$-\frac{1}{9}$
-2	$-\frac{1}{4}$
-1	-1
-0.5	-4
-0.25	-16
0.25	-16
0.5	-4
1	-1
2	$-\frac{1}{4}$
3	$-\frac{1}{9}$

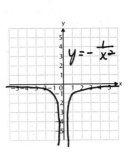

$y = -\frac{1}{x^2}$

66.

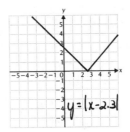

$y = |x - 2.3|$

67. $y = \frac{1}{x - 2}$

Choose x-values from -1 to 5, and use a calculator to find the corresponding y-values. (Note that we cannot choose 2 as a first coordinate since $\frac{1}{2 - 2}$, or $\frac{1}{0}$, is not defined.) Plot the points, and draw the graph. Note that it has two branches, one on either side of a vertical line through (2,0).

67. (continued)

x	y
-1	$-0.\overline{3}$
-0.5	-0.4
0	-0.5
1	-1
1.9	-10
2.1	10
2.5	2
3	1
4	0.5
5	$0.\overline{3}$

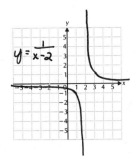

68.

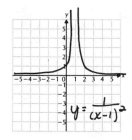

69.

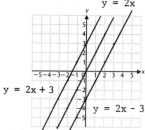

If the sign is negative, the graph is moved down. If the sign is positive, the graph is moved up. The three lines are parallel.

70. (-1,-4) and (4,1)

71.

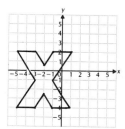

72.

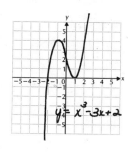

73. Use a compute software package or a graphing calculator to graph the equation.

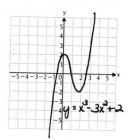

74.

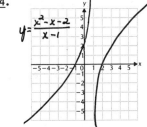

75. Use a computer software package or a graphing calculator to graph the equation.

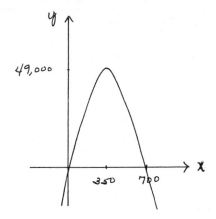

Exercise Set 2.2

---

1. The correspondence is not a function, because a member of the domain (a) corresponds to more than one member of the range.

2. Yes

3. The correspondence is a function, because for each member of the domain there corresponds just one member of the range.

4. Yes

5. The correspondence is a function, because for each member of the domain there corresponds just one member of the range.

6. No

7. The correspondence is not a function, because a member of the domain (Viola) corresponds to more than one member of the range.

8. No

9. The correspondence is a function, because for each member of the domain there corresponds just one member of the range.

10. Yes

11. This correspondence is a function, because each class member has only one seat number.

12. Function

13. This correspondence is a function, because each geometric figure has only one number for its area.

14. Function

15. The correspondence is not a function, since it is reasonable to assume that at least one person in the town has more than one aunt.

    The correspondence is a relation, since it is reasonable to assume that each person in the town has at least one aunt.

16. A relation but not a function

17. $g(x) = x + 1$
    a) $g(0) = 0 + 1 = 1$
    b) $g(-4) = -4 + 1 = -3$
    c) $g(-7) = -7 + 1 = -6$
    d) $g(8) = 8 + 1 = 9$
    e) $g(a + 2) = a + 2 + 1 = a + 3$

18. a) 0,   b) 4,   c) -7,   d) -8,   e) a - 5

19. $f(n) = 5n^2 + 4$
    a) $f(0) = 5(0)^2 + 4 = 0 + 4 = 4$
    b) $f(-1) = 5(-1)^2 + 4 = 5 + 4 = 9$
    c) $f(3) = 5(3)^2 + 4 = 45 + 4 = 49$
    d) $f(t) = 5(t)^2 + 4 = 5t^2 + 4$
    e) $f(2a) = 5(2a)^2 + 4 = 5 \cdot 4a^2 + 4 = 20a^2 + 4$

20. a) -2, b) 1, c) 25, d) $3t^2 - 2$, e) $12a^2 - 2$

21. $g(r) = 3r^2 + 2r - 1$
    a) $g(2) = 3(2)^2 + 2(2) - 1 = 12 + 4 - 1 = 15$
    b) $g(3) = 3(3)^2 + 2(3) - 1 = 27 + 6 - 1 = 32$
    c) $g(-3) = 3(-3)^2 + 2(-3) - 1 = 27 - 6 - 1 = 20$
    d) $g(1) = 3(1)^2 + 2(1) - 1 = 3 + 2 - 1 = 4$
    e) $g(3r) = 3(3r)^2 + 2(3r) - 1 = 3 \cdot 9r^2 + 6r - 1 =$
         $27r^2 + 6r - 1$

22. a) 35, b) 2, c) 7, d) 20, e) $36r^2 - 3r + 2$

23. $f(x) = \dfrac{x - 3}{2x - 5}$

    a) $f(0) = \dfrac{0 - 3}{2 \cdot 0 - 5} = \dfrac{-3}{0 - 5} = \dfrac{-3}{-5} = \dfrac{3}{5}$

    b) $f(4) = \dfrac{4 - 3}{2 \cdot 4 - 5} = \dfrac{1}{8 - 5} = \dfrac{1}{3}$

    c) $f(-1) = \dfrac{-1 - 3}{2(-1) - 5} = \dfrac{-4}{-2 - 5} = \dfrac{-4}{-7} = \dfrac{4}{7}$

    d) $f(3) = \dfrac{3 - 3}{2 \cdot 3 - 5} = \dfrac{0}{6 - 5} = \dfrac{0}{1} = 0$

    e) $f(x + 2) = \dfrac{x + 2 - 3}{2(x + 2) - 5} = \dfrac{x - 1}{2x + 4 - 5} = \dfrac{x - 1}{2x - 1}$

24. a) $\dfrac{26}{25}$, b) $\dfrac{2}{9}$, c) undefined, d) $-\dfrac{7}{3}$, e) $\dfrac{3x + 5}{2x + 11}$

25. $g(x) = -2x - 4$ $h(x) = 3x^2$
    $g(5) = -2 \cdot 5 - 4$ $h(-2) = 3(-2)^2$
       $= -10 - 4$ $= 3 \cdot 4$
       $= -14$ $= 12$
    $g(5) + h(-2) = -14 + 12 = -2$

26. 192

27. $g(x) = -2x - 4$ $h(x) = 3x^2$
    $g(-3) = -2(-3) - 4$ $h(12) = 3 \cdot 12^2$
       $= 6 - 4$ $= 3 \cdot 144$
       $= 2$ $= 432$
    $2g(-3) - 5h(12) = 2 \cdot 2 - 5 \cdot 432 = 4 - 2160 = -2156$

28. -8

29. $A(s) = s^2 \dfrac{\sqrt{3}}{4}$

    $A(4) = 4^2 \dfrac{\sqrt{3}}{4} = 4\sqrt{3}$

    The area is $4\sqrt{3}$ cm².

30. $9\sqrt{3}$ in²

31. $V(r) = 4\pi r^2$
    $V(3) = 4\pi(3)^2 = 36\pi$
    The area is 36π in² ≈ 113.04 in².

32. 100π cm² ≈ 314 cm²

33. $F(C) = \dfrac{9}{5}C + 32$

    $F(-10) = \dfrac{9}{5}(-10) + 32 = -18 + 32 = 14$

    The equivalent temperature is 14° F.

34. 41° F

35. $H(x) = 2.75x + 71.48$
    $H(40) = 2.75(40) + 71.48 = 181.48$
    The predicted height is 181.48 cm.

36. 189.73 cm

37. From the table, we know that the % charged is 30%, so we have

   Cost = 30% × $78.50 = 0.3 × $78.50 = $23.55

38. $57.92

39. Plot and connect the points, using body weight as the first coordinate and the corresponding number of drinks as the second coordinate.

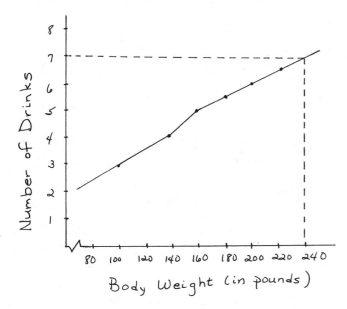

To estimate the number of drinks that a 240-lb person would have to drink to be considered intoxicated, first find the number 240 on the horizontal axis. Then move along a vertical line above it to the graph, and from there move to the left to the vertical axis. Read the approximate function value there. The estimated number of drinks is 7.

40. 2.5 drinks

41.

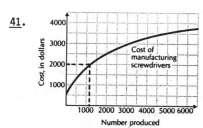

Locate 1200 on the horizontal axis. Move along a vertical line above it to the curve, and from there move to the left to the vertical axis. Read the function value there. The cost of producing 1200 screwdrivers is approximately $2000.

42. $3500

43. Locate 225 on the horizontal axis, move along a vertical line above it to the curve, and from there move to the left to the vertical axis. Read the function value there. The rate is approximately 75.

44. 125

45. Plot and connect the points, using the year as the first coordinate and the population as the second.

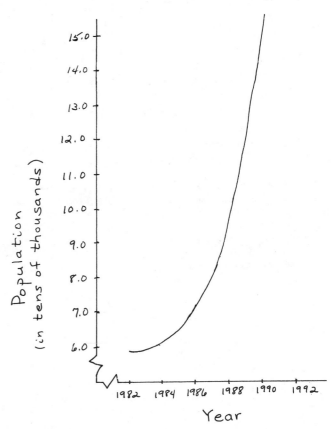

46. 150,000

47. Locate 1985 on the horizontal axis, move up a vertical line above it to the curve, and from there move to the left to the vertical axis. Read the function value there. In 1985 the population was about 6.4 ten thousands, or 64,000.

48. 80,000

49. We can use the vertical line test:

   If it is possible for a vertical line to intersect a graph more than once, the graph is not the graph of a function.

   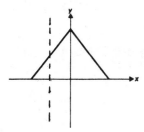

   Visualize moving this vertical line across the graph. Ask yourself the question:

   Will this line ever intersect the graph more than once?

   If the answer is yes, the graph <u>is not</u> a graph of a function. If the answer is no, the graph <u>is</u> a graph of a function.

   In this problem the vertical line will not intersect the graph more than once. Thus, the graph is a graph of a function.

50. No

51. We can use the vertical line test:

   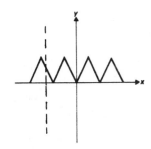

   Visualize moving this vertical line across the graph. The vertical line will not intersect the graph more than once. Thus, the graph is a graph of a function.

52. No

53. We can use the vertical line test:

   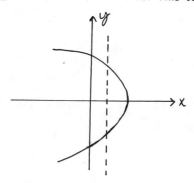

   It is possible for a vertical line to intersect the graph more than once. Thus this is not the graph of a function.

54. Yes

55. Use the vertical line test. It is not possible for a vertical line to intersect the graph more than once. Thus this is a graph of a function.

56. No

57. Use the vertical line test. It is possible for a vertical line to intersect the graph more than once. Thus this is not a graph of a function.

58. Yes

59. <u>Familiarize</u>. If x represents the first integer, then x + 2 represents the second integer, and x + 4 represents the third.

   <u>Translate</u>. We write an equation.

First integer	plus	two times the second	plus	three times the third	is	124.
x	+	2(x + 2)	+	3(x + 4)	=	124

   <u>Carry out</u>. Solve the equation.

$$x + 2(x + 2) + 3(x + 4) = 124$$
$$x + 2x + 4 + 3x + 12 = 124$$
$$6x + 16 = 124$$
$$6x = 108$$
$$x = 18$$

   If x = 18, then x + 2 = 18 + 2, or 20, and x + 4 = 18 + 4, or 22.

   <u>Check</u>. 18, 20, and 22 are consecutive even integers. Also, 18 + 2·20 + 3·22 = 18 + 40 + 66 = 124. The numbers check.

   <u>State</u>. The integers are 18, 20, and 22.

60. 20.175% increase

61.
$$S = 2\ell h + 2\ell w + 2wh$$
$$S - 2wh = 2\ell h + 2\ell w$$
$$S - 2wh = \ell(2h + 2w)$$
$$\frac{S - 2wh}{2h + 2w} = \ell$$

62. $\frac{5}{3}$

63.
$$f(x) = 4.3x^2 - 1.4x$$

a) $f(1.034) = 4.3(1.034)^2 - 1.4(1.034)$
$= 4.3(1.069156) - 1.4(1.034)$
$= 4.5973708 - 1.4476$
$= 3.1497708$

b) $f(-3.441) = 4.3(-3.441)^2 - 1.4(-3.441)$
$= 4.3(11.840481) - 1.4(-3.441)$
$= 50.9140683 + 4.8174$
$= 55.7314683$

c) $f(27.35) = 4.3(27.35)^2 - 1.4(27.35)$
$= 4.3(748.0225) - 1.4(27.35)$
$= 3216.49675 - 38.29$
$= 3178.20675$

d) $f(-16.31) = 4.3(-16.31)^2 - 1.4(-16.31)$
$= 4.3(266.0161) - 1.4(-16.31)$
$= 1143.86923 + 22.834$
$= 1166.70323$

64. a) 11,394.477, b) -582,136.93,
c) 3.5018862, d) -554.4995

65. The correspondence is a function because for each ordered pair (a,b), a ≠ b, there is only one number which represents the larger of a and b.

66. Yes

67. We know that (-1,-7) and (3,8) are both solutions of g(x) = mx + b. Substituting, we have
$$-7 = m(-1) + b, \quad or \quad -7 = -m + b,$$
and $8 = m(3) + b, \quad or \quad 8 = 3m + b.$

Solve the first equation for b and substitute that expression into the second equation.

$-7 = -m + b$	First equation
$m - 7 = b$	Solving for b
$8 = 3m + b$	Second equation
$8 = 3m + (m - 7)$	Substituting
$8 = 3m + m - 7$	
$8 = 4m - 7$	
$15 = 4m$	
$\frac{15}{4} = m$	

We know that $m - 7 = b$, so $\frac{15}{4} - 7 = b$, or $-\frac{13}{4} = b$.

Then we have $m = \frac{15}{4}$, $b = -\frac{13}{4}$. We can express g(x) as $g(x) = \frac{15}{4}x - \frac{13}{4}$.

68. 35

69. The correspondence in the chart is a function, because for each member of the domain, or activity, there is just one member of the range.

70. 18 mm

71. Locate the highest point on the graph. For this graph, there are two equally high points that are higher than all the other points of the graph. Then move down a vertical line through each point to the horizontal axis and read the corresponding times.

The times are 2 min, 40 sec and 5 min, 40 sec.

72. 15 mm

73. From Exercise 71 we know that the first largest contraction occurred at 2 min, 40 sec, and the second occurred at 5 min, 40 sec. Find the time between the contractions:

```
  5 min  40 sec
- 2 min  40 sec
  3 min
```

Then the frequency of the largest contractions is 1 contraction every 3 min, or $\frac{1}{3}$ contraction per min.

74.
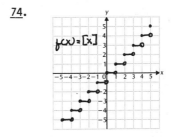

Exercise Set 2.3

1. y = 4x + 5
The slope is 4, and the y-intercept is (0,5), or 5.

2. Slope is 5; y-intercept is (0,3), or 3.

3. f(x) = -2x - 6
The slope is -2, and the y-intercept is (0,-6), or -6.

4. Slope is -5; y-intercept is (0,7), or 7.

5. $y = -\frac{3}{8}x - 0.2$

   $y = \quad mx + b$

   The slope is $-\frac{3}{8}$, and the y-intercept is (0,-0.2), or -0.2.

6. Slope is $\frac{15}{7}$; y-intercept is (0,2.2), or 2.2.

7. $g(x) = 0.5x - 9$

   $g(x) = \quad m x + b$

   The slope is 0.5, and the y-intercept is (0,-9), or -9.

8. Slope is -3.1; y-intercept is (0,5), or 5.

9. $y = 7$

   Think of this as $y = 0 \cdot x + 7$.

   $y = m x + b$

   The slope is 0, and the y-intercept is (0,7), or 7.

10. Slope is 0; y-intercept is (0,-2), or -2.

11. $f(x) = 3.7$

    Think of this as $f(x) = 0 \cdot x + 3.7$.

    $f(x) = m x + b$

    The slope is 0, and the y-intercept is (0,3.7), or 3.7.

12. Slope is 0; y-intercept is (0,5.2), or 5.2.

13. Use the slope-intercept equation, $y = mx + b$, with $m = \frac{2}{3}$ and $b = -7$.

    $y = mx + b$

    $y = \frac{2}{3}x + (-7)$

    $y = \frac{2}{3}x - 7$

14. $y = -\frac{3}{4}x + 5$

15. Use the slope-intercept equation, $y = mx + b$, with $m = -4$ and $b = 2$.

    $y = mx + b$

    $y = -4x + 2$

16. $y = 2x - 1$

17. Use the slope-intercept equation, $y = mx + b$, with $m = -\frac{7}{9}$ and $b = 3$.

    $y = mx + b$

    $y = -\frac{7}{9}x + 3$

18. $y = -\frac{4}{11}x + 9$

19. Use the slope-intercept equation, $y = mx + b$, with $m = 5$ and $b = \frac{1}{2}$.

    $y = mx + b$

    $y = 5x + \frac{1}{2}$

20. $y = 6x + \frac{2}{3}$

21. Use the slope-intercept equation, $y = mx + b$, with $m = 0.7$ and $b = 3.8$.

    $y = mx + b$

    $y = 0.7x + 3.8$

22. $y = 1.7x - 4.3$

23. $y = \frac{5}{2}x + 1$

    Slope is $\frac{5}{2}$; y-intercept (0,1).

    From the y-intercept, we go up 5 units and to the right 2 units. This gives us the point (2,6). We can now draw the graph.

    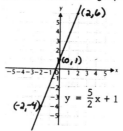

    As a check, we can rename the slope and find another point.

    $$\frac{5}{2} = \frac{5}{2} \cdot \frac{-1}{-1} = \frac{-5}{-2}$$

    From the y-intercept, we go down 5 units and to the left 2 units. This gives us the point (-2,-4). Plot this point to see if it is on the line.

24. Slope is $\frac{2}{5}$; y-intercept is (0,4).

    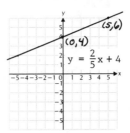

25. $y = -\frac{5}{2}x + 4$

Slope is $-\frac{5}{2}$, or $\frac{-5}{2}$; y-intercept is $(0,4)$.

From the y-intercept, we go <u>down</u> 5 units and to the <u>right</u> 2 units. This gives us the point $(2,-1)$. We can now draw the graph.

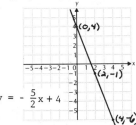

As a check, we can rename the slope and find another point.

$$\frac{-5}{2} = \frac{-5}{2} \cdot \frac{2}{2} = \frac{-10}{4}$$

From the y-intercept, we go <u>down</u> 10 units and to the <u>right</u> 4 units. This gives us the point $(4,-6)$. Plot this point to see if it is on the line.

26. Slope is $-\frac{2}{5}$; y-intercept is $(0,3)$.

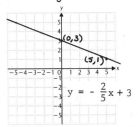

27. $y = 2x - 5$

Slope is 2, or $\frac{2}{1}$; y-intercept is $(0,-5)$.

From the y-intercept, we go <u>up</u> 2 units and to the <u>right</u> 1 unit. This gives us the point $(1,-3)$. We can now draw the graph.

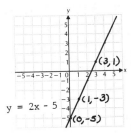

As a check, we can rename the slope and find another point.

$$2 = \frac{2}{1} \cdot \frac{3}{3} = \frac{6}{3}$$

From the y-intercept, we go <u>up</u> 6 units and to the <u>right</u> 3 units. This gives us the point $(3,1)$. Plot this point to see if it is on the line.

28. Slope is -2; y-intercept is $(0,4)$.

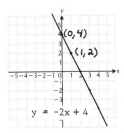

29. $y = \frac{1}{3}x + 6$

Slope is $\frac{1}{3}$; y-intercept is $(0,6)$.

From the y-intercept, we go <u>up</u> 1 unit and to the <u>right</u> 3 units. This gives us the point $(3,7)$. We can now draw the graph.

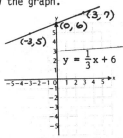

As a check, we can rename the slope and find another point.

$$\frac{1}{3} = \frac{1}{3} \cdot \frac{-1}{-1} = \frac{-1}{-3}$$

From the y-intercept, we go <u>down</u> 1 unit and to the <u>left</u> 3 units. This gives us the point $(-3,5)$. Plot this point to see if it is on the line.

30. Slope is -3; y-intercept is $(0,6)$.

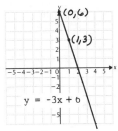

31. $y = -0.25x + 2$

Slope is -0.25, or $\frac{-1}{4}$; y-intercept is $(0,2)$.

From the y-intercept, we go <u>down</u> 1 unit and to the <u>right</u> 4 units. This gives us the point $(4,1)$. We can now draw the graph.

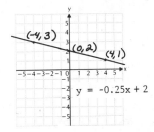

31. (continued)

As a check, we can rename the slope and find another point.

$$\frac{-1}{4} = \frac{-1}{4} \cdot \frac{-1}{-1} = \frac{1}{-4}$$

From the y-intercept, we go <u>up</u> 1 unit and to the <u>left</u> 4 units. This gives us the point (-4,3). Plot this point to see if it is on the line.

32. Slope is 1.5, or $\frac{3}{2}$; y-intercept is (0,-3).

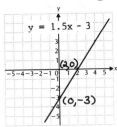

33. $y = \frac{4}{5}x - 2$

Slope is $\frac{4}{5}$; y-intercept is (0,-2).

From the y-intercept, we go <u>up</u> 4 units and to the <u>right</u> 5 units. This gives us the point (5,2). We can now draw the graph.

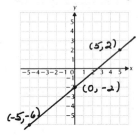

As a check, we choose some other value for x, say -5, and determine y:

$$y = \frac{4}{5}(-5) - 2 = -4 - 2 = -6$$

We plot the point (-5,-6) and see if it is on the line.

34. Slope is $-\frac{5}{4}$; y-intercept is (0,1).

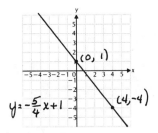

35. $f(x) = \frac{5}{4}x - 2$

Slope is $\frac{5}{4}$; y-intercept is (0,-2).

From the y-intercept, we go <u>up</u> 5 units and to the <u>right</u> 4 units. This gives us the point (4,3). We can now draw the graph.

35. (continued)

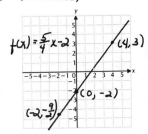

As a check, we choose some other value for x, say -2 and determine f(x):

$$f(x) = \frac{5}{4}(-2) - 2 = -\frac{5}{2} - 2 = -\frac{9}{2}$$

We plot the point $\left(-2, -\frac{9}{2}\right)$ and see if it is on the line.

36. Slope is $\frac{4}{3}$; y-intercept is (0,2).

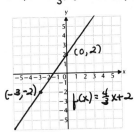

37. $f(x) = 4$

Think of this as $f(x) = 0 \cdot x + 4$.

Slope is 0; y-intercept is (0,4).

From the y-intercept we do not go up or down. We can go right or left any non-zero number of units, since $0/a = 0$, $a \neq 0$. Let's choose to go right 3 units. This gives us the point (3,4). We can now draw the graph.

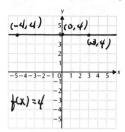

As a check, we choose some other value for x, say -4, and determine f(x):

$$f(x) = 0(-4) + 4 = 0 + 4 = 4.$$

We plot the point (-4,4) and see if it is on the line.

38. Slope is 0; y-intercept is (0,-1).

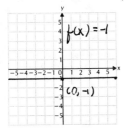

39. a) C(m) = 0.25m + 2

Slope is 0.25, or $\frac{1}{4}$; y-intercept is (0,2).

From the y-intercept we go up 1 unit and to the right 4 units. This gives us the point (4,3). We can now draw the graph. (We do not plot points for m < 0, since it does not make sense to talk about a taxi ride of a negative number of miles.)

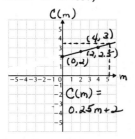

As a check, we note that the pair (2,2.5) also satisfies the equation.

b) To estimate the cost of a $5\frac{1}{2}$ - mile taxi ride, we determine the coordinate on the vertical (cost) axis that appears to be paired with the coordinate $5\frac{1}{2}$ on the horizontal (miles) axis. We do so by drawing a vertical line segment from $5\frac{1}{2}$ up to the line and then a horizontal segment from the line over to the cost axis. (See the graph in part a).) We estimate that the cost is approximately $3.50.

40. a) C(m)

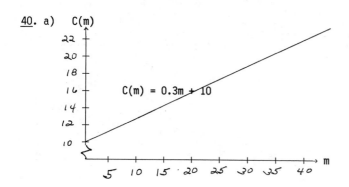

b) $22

41. a) D(t) = $\frac{1}{5}$t + 20

Slope is $\frac{1}{5}$; y-intercept is (0,20).

From the y-intercept go up 1 unit and to the right 5 units. This gives us the point (5,21). We can now draw the graph.

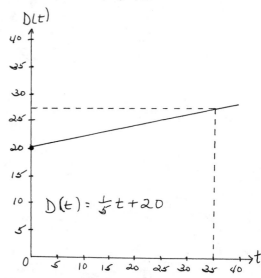

As a check, we note that the point (25,25) also satisfies the equation.

b) To predict the demand for gas in 1995 (35 years after 1960), draw a vertical line segment from 35 on the horizontal (time) axis up to the line and then a horizontal segment from the line over to the vertical (demand) axis. (See the graph in part a).) We estimate that the demand will be about 27 quadrillion joules.

42. a) N(t)

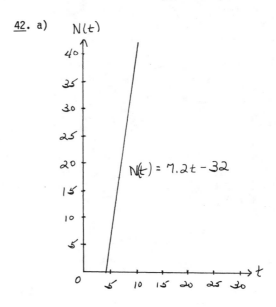

b) 4 chirps per minute

43. a)   C(m) = 0.80m + 1

Slope is 0.80, or $\frac{4}{5}$; y-intercept is (0,1).

From the y-intercept go <u>up</u> 4 units and to the <u>right</u> 5 units. This gives us the point (5,5). We can now draw the graph.

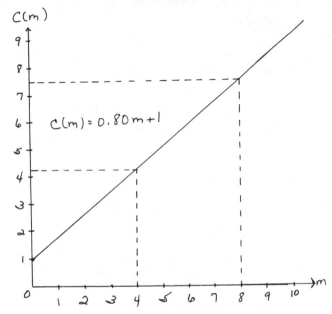

As a check, we note that the point (10,9) also satisfies the equation.

b)   Draw a vertical line segment from 4 on the horizontal (minutes) axis up to the line and then a horizontal line over to the vertical (cost) axis. (See the graph in part a).) We predict that the cost of a 4-min call is about $4.25.

c)   Draw a horizontal line segment from $7.40 (or 7.4) on the vertical (cost) axis over to the line and then down to the horizontal (minutes) axis. (See the graph in part a).) We determine that a phone call about 8 min long can be made for $7.40.

44. a)
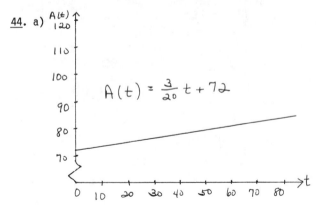

b)   About 80 yr

c)   1990

45.   We first solve for y.

3y = 9

 y = 3

Since x is missing, any number for x will do. Thus all ordered pairs (x,3) are solutions. A few solutions are listed below. Plot these and draw the line.

x	y
-2	3
0	3
3	3

(y must be 3; x can be any number.)

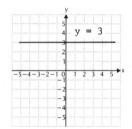

The graph is a line parallel to the x-axis with y-intercept (0,3).

46.

47.   We first solve for x.

3x = -15

 x = -5

Since y is missing, any number for y will do. Thus all ordered pairs (-5,y) are solutions. A few solutions are listed below. Plot these and draw the line.

x	y
-5	-3
-5	1
-5	4

(x must be -5; y can be any number.)

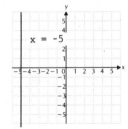

The graph is a line parallel to the y-axis with x-intercept (-5,0).

48.

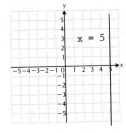

49. We first solve for g(x).

$4g(x) + 3x = 12 + 3x$

$\qquad 4g(x) = 12$

$\qquad\quad g(x) = 3$

Since x is missing, any number for x will do. Thus all ordered pairs (x,3) are solutions. A few solutions are listed below. Plot these and draw the line.

x	g(x)
-4	3
-1	3
0	3

(g(x) must be 3; x can be any number.)

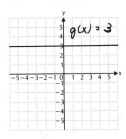

The graph is a line parallel to the x-axis with y-intercept (0,3).

50.

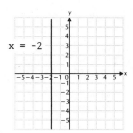

51. We first solve for y.

$6y + 3x = -2(4 - 3y)$

$6y + 3x = -8 + 6y$

$\qquad 3x = -8$

$\qquad\quad x = -\dfrac{8}{3}$

Since y is missing, any number for y will do. Thus all orderd pairs $\left(-\dfrac{8}{3}, y\right)$ are solutions. A few solutions are listed below. Plot these and draw the line.

51. (continued)

x	y
$-\dfrac{8}{3}$	-3
$-\dfrac{8}{3}$	1
$-\dfrac{8}{3}$	4

$\left[$ x must be $-\dfrac{8}{3}$; y can be any number $\right]$

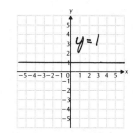

The graph is a line parallel to the y-axis with x-intercept $\left[-\dfrac{8}{3}, 0\right]$.

52.

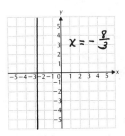

53. Try to put the equation in standard form, Ax + By = C, with A and B not both 0.

$3x + 5f(x) + 15 = 0$

$3x + 5y\quad + 15 = 0$ \quad Replace f(x) with y

$\qquad 3x + 5y = -15$

The equation can be put in standard form with A = 3, B = 5, and C = -15. It is linear.

Now convert standard form to a slope-intercept equation, y = mx + b.

$3x + 5y = -15$

$\qquad 5y = -3x - 15$

$\dfrac{1}{5} \cdot 5y = \dfrac{1}{5}(-3x - 15)$

$\qquad y = -\dfrac{3}{5}x - 3$

The slope is $-\dfrac{3}{5}$.

54. Linear; slope is $\dfrac{5}{3}$

45

55. Try to put the equation in standard form, $Ax + By = C$, with A and B not both 0.

$$3x - 12 = 0$$
$$3x = 12$$
$$x = 4$$

The equation can be put in standard form with $A = 1$, $B = 0$, and $C = 4$. It is linear. (It is the equation of a vertical line.)

56. Linear; slope is 0.

57. Try to put the equation in standard form.

$$2x + 4g(x) = 19$$
$$2x + 4y = 19 \quad \text{Replace } g(x) \text{ with } y$$

The equation can be put in standard form with $A = 2$, $B = 4$, and $C = 19$. It is linear.

Convert standard form to a slope-intercept equation.

$$2x + 4y = 19$$
$$4y = -2x + 19$$
$$\frac{1}{4} \cdot 4y = \frac{1}{4}(-2x + 19)$$
$$y = -\frac{1}{2}x + \frac{19}{4}$$

The slope is $-\frac{1}{2}$.

58. Not linear

59. $5x - 4xy = 12$

The equation is not linear, because it has an xy-term.

60. Not linear

61. Try to put the equation in standard form.

$$\frac{3y}{4x} = 5y + 2$$
$$4x \cdot \frac{3y}{4x} = 4x(5y + 2)$$
$$3y = 20xy + 8x$$
$$-8x + 3y - 20xy = 0$$

The equation is not linear, because it has an xy-term.

62. Not linear

63. Try to put the equation in standard form.

$$f(x) = x^3$$
$$y = x^3 \quad \text{Replace } f(x) \text{ with } y$$
$$-x^3 + y = 0$$

The equation is not linear, because it has an x^3-term.

64. Not linear

65. <u>Familiarize.</u> Let x represent the measure of the first angle. Then 2x represents the measure of the second angle and x + 44 represents the third angle's measure. Recall that the sum of the measures of the angles of a triangle is 180°.

<u>Translate.</u> Write an equation.

1st angle plus 2nd angle plus 3rd angle is 180°.
$$x \quad + \quad 2x \quad + \quad (x + 44) \quad = 180$$

<u>Carry out.</u>

$$x + 2x + x + 44 = 180$$
$$4x + 44 = 180$$
$$4x = 136$$
$$x = 34$$

If $x = 34$, then $2x = 2 \cdot 34$, or 68, and $x + 44 = 34 + 44$, or 78.

<u>Check.</u> Left to the student.

<u>State.</u> The measures of the angles are 34°, 68°, and 78°.

66. 12 ft, 20.8 ft

67.
$$9\{2x - 3[5x + 2(-3x + y^0 - 2)]\}$$
$$= 9\{2x - 3[5x + 2(-3x + 1 - 2)]\} \quad (y^0 = 1)$$
$$= 9\{2x - 3[5x + 2(-3x - 1)]\}$$
$$= 9\{2x - 3[5x - 6x - 2]\}$$
$$= 9\{2x - 3[-x - 2]\}$$
$$= 9\{2x + 3x + 6\}$$
$$= 9\{5x + 6\}$$
$$= 45x + 54$$

68. $-\frac{27}{14}$

69. We first solve for y.

$$ay = -5x + b$$
$$y = -\frac{5}{a}x + \frac{b}{a}$$

The slope is $-\frac{5}{a}$, and the y-intercept is $\left[0, \frac{b}{a}\right]$.

70. Slope is $\frac{b}{4}$; y-intercept is $\left[0, \frac{9a + 2}{4}\right]$.

71. We first solve for y.

$$ax + by = c$$
$$by = -ax + c$$
$$y = -\frac{a}{b}x + \frac{c}{b}$$

The slope is $-\frac{a}{b}$, and the y-intercept is $\left[0, \frac{c}{b}\right]$.

72. Slope is $-\frac{a}{a + b}$; y-intercept is $\left[0, \frac{c}{a + b}\right]$.

73. $ax + 3y = b - c$

The equation is in standard form with $A = a$, $B = 3$, and $C = b - c$. It is linear.

74. Linear

75. Try to put the equation in standard form.
$$a^2x = by + 5$$
$$a^2x - by = 5$$
The equation is in standard form with $A = a^2$, $B = -b$, and $C = 5$. It is linear.

76. Not linear

77. $\frac{x}{a} - by = 17$

The equation is in standard form with $A = \frac{1}{a}$, $B = -b$, and $C = 17$. It is linear.

78. Not linear

79. We know that the y-intercept is (0,3.1). We also know that to get another point on the graph, (2,7.8), we go up 4.7 units (7.8 - 3.1 = 4.7) and to the right 2 units (2 - 0 = 2). Then the slope of the line is $\frac{4.7}{2}$, or $\frac{47}{20}$.

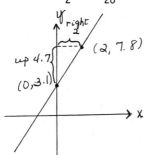

To write an equation for the function, use the slope-intercept equation with $m = \frac{47}{20}$ and b = 3.1:
$$y = mx + b$$
$$y = \frac{47}{20}x + 3.1$$

80. $y = -\frac{5}{3}x + \frac{14}{3}$

Exercise Set 2.4

1. Graph x - 2 = y.
To find the y-intercept, let x = 0.
$$x - 2 = y$$
$$0 - 2 = y, \quad \text{or} \quad -2 = y$$
The y-intercept is (0,-2).
To find the x-intercept, let y = 0.
$$x - 2 = y$$
$$x - 2 = 0, \quad \text{or} \quad x = 2$$
The x-intercept is (2,0).
Plot these points and draw the line. A third point could be used as a check.

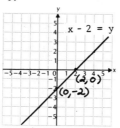

2.

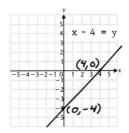

3. Graph 3x - 1 = y.
To find the y-intercept, let x = 0.
$$3x - 1 = y$$
$$3 \cdot 0 - 1 = y, \quad \text{or} \quad -1 = y$$
The y-intercept is (0,-1).

To find the x-intercept, let y = 0.
$$3x - 1 = y$$
$$3x - 1 = 0$$
$$3x = 1$$
$$x = \frac{1}{3}$$

The x-intercept is $\left(\frac{1}{3}, 0\right)$.

Plot these points and draw the line. A third point could be used as a check.

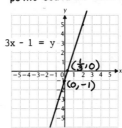

4.

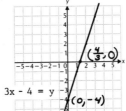

5. Graph 5x - 4y = 20.

    To find the y-intercept, let x = 0.

    5x - 4y = 20

    5·0 - 4y = 20

    -4y = 20

    y = -5

    The y-intercept is (0,-5).

    To find the x-intercept, let y = 0.

    5x - 4y = 20

    5x - 4·0 = 20

    5x = 20

    x = 4

    The x-intercept is (4,0).

    Plot these points and draw the line. A third point could be used as a check.

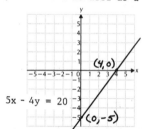

6.

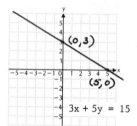

7. Graph y = -5 - 5x.

    To find the y-intercept, let x = 0.

    y = -5 - 5x

    y = -5 - 5·0

    y = -5

    (0,-5) is the y-intercept.

    To find the x-intercept, let y = 0.

    y = -5 - 5x

    0 = -5 - 5x

    5x = -5

    x = -1

    (-1,0) is the x-intercept.

    Plot these points and draw the line. A third point could be used as a check.

7. (continued)

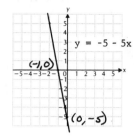

8.

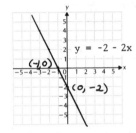

9. Graph 5y = -15 + 3x.

    To find the y-intercept, let x = 0.

    5y = -15 + 3x

    5y = -15 + 3·0

    5y = -15

    y = -3

    (0,-3) is the y-intercept.

    To find the x-intercept, let y = 0.

    5y = -15 + 3x

    5·0 = -15 + 3x

    15 = 3x

    5 = x

    (5,0) is the x-intercept.

    Plot these points and draw the line. A third point could be used as a check.

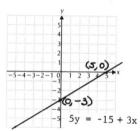

10.

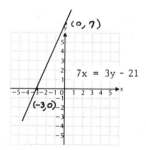

11. 6x − 7 + 3y = 9x − 2y + 8

    −3x + 5y = 15

Graph −3x + 5y = 15

To find the y-intercept, let x = 0.

      −3x + 5y = 15

      −3·0 + 5y = 15

          5y = 15

           y = 3

(0,3) is the y-intercept.

To find the x-intercept, let y = 0.

      −3x + 5y = 15

      −3x + 5·0 = 15

        −3x = 15

           x = −5

(−5,0) is the x-intercept.

Plot these points and draw the line. A third point could be used as a check.

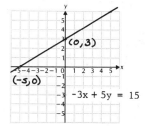

12.

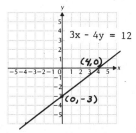

13. 1.4y − 3.5x = −9.8

      14y − 35x = −98   Multiplying by 10

       2y − 5x = −14   Multiplying by $\frac{1}{7}$

Graph 2y − 5x = −14.

To find the y-intercept, let x = 0.

      2y − 5x = −14

      2y − 5·0 = −14

          2y = −14

           y = −7

(0,−7) is the y-intercept.

To find the x-intercept, let y = 0.

      2y − 5x = −14

      2·0 − 5x = −14

        −5x = −14

           x = $\frac{14}{5}$

$\left[\frac{14}{5}, 0\right]$ is the x-intercept.

13. (continued)

Plot these points and draw the line. A third point could be used as a check.

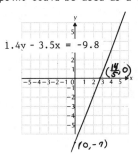

14.

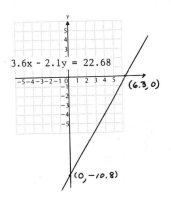

15. Graph 5x + 2y = 7

To find the y-intercept, let x = 0.

      5x + 2y = 7

      5·0 + 2y = 7

          2y = 7

           y = $\frac{7}{2}$

$\left[0, \frac{7}{2}\right]$ is the y-intercept.

To find the x-intercept, let y = 0.

      5x + 2y = 7

      5x + 2·0 = 7

        5x = 7

           x = $\frac{7}{5}$

$\left[\frac{7}{5}, 0\right]$ is the x-intercept.

Plot these points and draw the line. A third point could be used as a check.

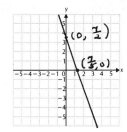

16.

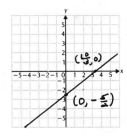

17. Let $(6,9) = (x_1,y_1)$ and $(4,5) = (x_2,y_2)$.

Slope $= \dfrac{y_2 - y_1}{x_2 - x_1} = \dfrac{5 - 9}{4 - 6} = \dfrac{-4}{-2} = 2$

18. $\dfrac{4}{3}$

19. Let $(3,8) = (x_1,y_1)$ and $(9,-4) = (x_2,y_2)$.

Slope $= \dfrac{y_2 - y_1}{x_2 - x_1} = \dfrac{-4 - 8}{9 - 3} = \dfrac{-12}{6} = -2$

20. $\dfrac{3}{26}$

21. Let $(-8,-7) = (x_1,y_1)$ and $(-9,-12) = (x_2,y_2)$.

Slope $= \dfrac{y_2 - y_1}{x_2 - x_1} = \dfrac{-12 - (-7)}{-9 - (-8)} = \dfrac{-5}{-1} = 5$

22. $-\dfrac{3}{4}$

23. Let $(-16.3,12.4) = (x_1,y_1)$ and $(-5.2,8.7) = (x_2,y_2)$.

Slope $= \dfrac{y_2 - y_1}{x_2 - x_1} = \dfrac{8.7 - 12.4}{-5.2 - (-16.3)} = \dfrac{-3.7}{11.1} = -\dfrac{37}{111} = -\dfrac{1}{3}$

24. $\dfrac{98}{269}$

25. Let $(3.2,-12.8) = (x_1,y_1)$ and $(3.2,2.4) = (x_2,y_2)$.

Slope $= \dfrac{y_2 - y_1}{x_2 - x_1} = \dfrac{2.4 - (-12.8)}{3.2 - 3.2} = \dfrac{15.2}{0}$

Since we cannot divide by 0, the slope is undefined.

26. Slope is undefined.

27. Let $(7,3.4) = (x_1,y_1)$ and $(-1,3.4) = (x_2,y_2)$.

Slope $= \dfrac{y_2 - y_1}{x_2 - x_1} = \dfrac{3.4 - 3.4}{-1 - 7} = \dfrac{0}{-8} = 0$

28. 0

29. Let $(0,9.1) = (x_1,y_1)$ and $(9.1,0) = (x_2,y_2)$.

Slope $= \dfrac{y_2 - y_1}{x_2 - x_1} = \dfrac{0 - 9.1}{9.1 - 0} = \dfrac{-9.1}{9.1} = -1$

30. 1

31. <u>Familiarize.</u> We might familiarize ourselves with this problem in one of two ways. First, we might use the formula $d = r \cdot t$. (See Example 5.) Second, we might use a graph to give a "picture" of the problem.

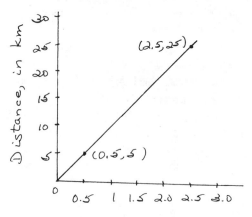

<u>Translate.</u> Using the first approach, we could substitute either 0.5 for t and 5 for d or 2.5 for t and 25 for d. We have the equation

$5 = r(0.5)$ or $25 = r(2.5)$.

Using the second approach, we could write the equation

rate = change in distance/change in time or

$r = (25 - 5)/(2.5 - 0.5)$.

<u>Carry out.</u> Using the first approach, we solve

$5 = r(0.5)$ or $25 = r(2.5)$

to obtain r = 10 km/hr.

Using the second approach, we solve

$r = \dfrac{25 - 5}{2.5 - 0.5} = \dfrac{20}{2} = 10$ km/hr.

<u>Check.</u> If the rate is 10 km/hr, in 0.5 hr the marathoner will have run 0.5(10) = 5 km, and in 2.5 hr, 2.5(10) = 25 km. The answer checks.

<u>State.</u> The marathoner's speed is 10 km/hr.

32. 12 km/hr

33. <u>Familiarize.</u> We can use a graph to give a "picture" of the problem.

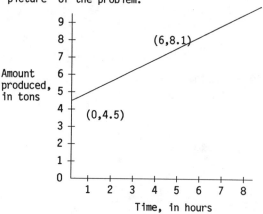

33. (continued)

Translate. We can write an equation:

rate of production = change in amount/change in time

or $r = (8.1 - 4.5)/(6 - 0)$

Carry out. Solve the equation.

$$r = \frac{8.1 - 4.5}{6 - 0} = \frac{3.6}{6} = 0.6 \text{ ton/hr}$$

Check. If the rate of production is 0.6 ton/hr, in 6 hr then 0.6(6) = 3.6 tons are produced. If 4.5 tons had already been produced at the beginning of the run, then the total amount after 6 hours is 4.5 + 3.6 = 8.1 tons. The answer checks.

State. The rate of production is 0.6 ton/hr.

34. $\frac{5}{96}$ per hour

35. Familiarize. We can use a graph to give a "picture" of the problem. Express both times in minutes.

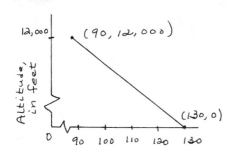

Translate. We can write an equation:

rate = change in altitude/change in time

or $r = (0 - 12{,}000)/(130 - 90)$

Carry out. Solve the equation.

$$r = \frac{0 - 12{,}000}{130 - 90} = \frac{-12{,}000}{40} = -300 \text{ ft/min}$$

The negative result indicates that the plane is descending. We can say the rate is -300 ft/min or the rate of descent is 300 ft/min.

Check. If the rate of descent is 300 ft/min, then in 40 min $\left[2 \text{ hr } 10 \text{ min} - 1\frac{1}{2} \text{ hr}\right]$ the plane descends 300(40) = 12,000 ft. The answer checks.

State. The rate of descent is 300 ft/min.

36. $1166\frac{2}{3}$ ft/hr

37. Familiarize. We can use a graph to give a "picture" of the problem.

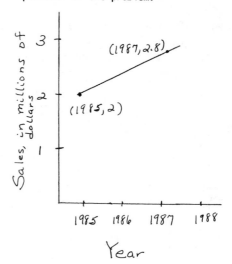

Translate. We can write an equation:

rate = change in sales/change in time

or $r = (2.8 - 2.0)/(1987 - 1985)$

Carry out. Solve the equation.

$$r = \frac{2.8 - 2.0}{1987 - 1985} = \frac{0.8}{2} = \$0.4 \text{ million/yr}$$

Check. If the rate at which sales are increasing is $0.4 million/yr, then in 2 yr the sales increase 0.4(2) = $0.8 million. The answer checks.

State. Sales are increasing at the rate of $0.4 million/yr.

38. 1.367 million/yr

39. $3x = 12 + y$

$3x - 12 = y$

    $y = 3x - 12$ $(y = mx + b)$

The slope is 3.

40. $\frac{3}{5}$

41. $5x - 6 = 15$

    $5x = 21$

    $x = \frac{21}{5}$

When y is missing, the line is parallel to the y-axis. The line is vertical, and the slope is undefined.

42. Undefined

43. $5y = 6$

    $y = \frac{6}{5}$

When x is missing, the line is parallel to the x-axis. The line is horizontal and has a slope of 0.

44. 0

45. y − 6 = 14

      y = 20

When x is missing, the line is parallel to the x-axis. The line is horizontal and has a slope of 0.

46. 0

47. 12 − 4x = 9 + x

      3 = 5x

      $\frac{3}{5}$ = x

When y is missing, the line is parallel to the y-axis. The line is vertical, and the slope is undefined.

48. Undefined

49. 2y − 4 = 35 + x

      2y = x + 39

      y = $\frac{1}{2}$x + $\frac{39}{2}$ (y = mx + b)

The slope is $\frac{1}{2}$.

50. −2

51. 3y + x = 3y + 2

      x = 2

When y is missing, the line is parallel to the y-axis. The line is vertical, and the slope is undefined.

52. Undefined

53. 3y − 2x = 5 + 9y − 2x

      3y = 5 + 9y

      0 = 5 + 6y

      −5 = 6y

      −$\frac{5}{6}$ = y

When x is missing, the line is parallel to the x-axis. The line is horizontal and has a slope of 0.

54. 0

55. Familiarize. The cost of the taxi ride is $1 the first $\frac{1}{2}$ mile and $1.20 per mile $\left[30¢ \text{ per } \frac{1}{4} \text{ mile} = \$1.20 \text{ per mile}\right]$ for the mileage beyond the first $\frac{1}{2}$ mile. We let x represent the number of miles from Johnson Street to Elm Street. Then x − $\frac{1}{2}$ represents the number of miles beyond the first $\frac{1}{2}$ mile.

55. (continued)

Translate. Write an equation.

Cost for first $\frac{1}{2}$ mile	+	Cost per additional mile	·	Mileage beyond first $\frac{1}{2}$ mile	=	Total cost
$1	+	$1.20	·	$\left(x - \frac{1}{2}\right)$	=	$5.20

Carry out. We solve the equation.

$1 + 1.20\left(x - \frac{1}{2}\right) = 5.20$

$1 + 1.2x - 0.6 = 5.20$

$1.2x + 0.4 = 5.20$

$1.2x = 4.80$

$x = \frac{4.80}{1.2}$

$x = 4$

Check. The first $\frac{1}{2}$ mile costs $1. The remaining $3\frac{1}{2}$ miles $\left[4 - \frac{1}{2} = 3\frac{1}{2}\right]$ cost $1.20 per mile which is $1.20 × 3\frac{1}{2}$, or $4.20. The total cost is 1 + 4.20, or $5.20. The value checks.

State. It is 4 miles from Johnson Street to Elm Street.

56. $3.708 × 10^9$

57. $f = \frac{F(c - v_0)}{c - v_S}$

$f(c - v_S) = F(c - v_0)$

$\frac{f(c - v_S)}{c - v_0} = F$

58. $y = mx - mx_1 + y_1$

59. Observe that for each point (x,y), y = −$\frac{1}{25}$x. To find other points of the line, choose any number for x (other than −100 and 0, which are already given) and find the corresponding y-value. When x = 50, y = −$\frac{1}{25}$(50), or −2, giving the point (50,−2). When x = 25, y = −$\frac{1}{25}$(25), or −1, giving the point (25,−1). When x = −25, y = −$\frac{1}{25}$(−25), or 1, giving the point (−25,1). When x = −50, y = −$\frac{1}{25}$(−50), or 2, giving the point (−50,2). There are many more correct answers.

60. 4x − 5y = 20

61. $y = mx + b$ $(m \neq 0)$

To find the x-intercept, let $y = 0$.

$$y = mx + b$$
$$0 = mx + b$$
$$-b = mx$$
$$-\frac{b}{m} = x$$

Thus, the x-intercept is $\left(-\frac{b}{m}, 0\right)$

62. $\frac{5}{8}$

63. a) Let $(5b, -6c) = (x_1, y_1)$ and $(b, -c) = (x_2, y_2)$.

$$\text{Slope} = \frac{y_2 - y_1}{x_2 - x_1} = \frac{-c - (-6c)}{b - 5b} = \frac{5c}{-4b} = -\frac{5c}{4b}$$

b) Let $(b, d) = (x_1, y_1)$ and $(b, d + e) = (x_2, y_2)$.

$$\text{Slope} = \frac{y_2 - y_1}{x_2 - x_1} = \frac{(d + e) - d}{b - b} = \frac{e}{0}$$

Since we cannot divide by 0, the slope is undefined.

c) Let $(c + f, a + d) = (x_1, y_1)$ and $(c - f, -a - d) = (x_2, y_2)$

$$\text{Slope} = \frac{y_2 - y_1}{x_2 - x_1} = \frac{(-a - d) - (a + d)}{(c - f) - (c + f)}$$

$$= \frac{-a - d - a - d}{c - f - c - f}$$

$$= \frac{-2a - 2d}{-2f}$$

$$= \frac{-2(a + d)}{-2f}$$

$$= \frac{a + d}{f}$$

64. The slope of equation A is two times the slope of equation B.

65. We first make a drawing.

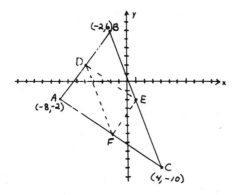

Let D = midpoint of $\overline{AB}$

E = midpoint of $\overline{BC}$

F = midpoint of $\overline{AC}$

First we use the midpoint formula to calculate the coordinates of the midpoints.

65. (continued)

$$\left(\frac{x_1 + x_2}{2}, \frac{y_1 + y_2}{2}\right)$$ (Midpoint formula)

$$D = \left(\frac{-8 + (-2)}{2}, \frac{-2 + 6}{2}\right) = (-5, 2)$$

$$E = \left(\frac{-2 + 4}{2}, \frac{6 + (-10)}{2}\right) = (1, -2)$$

$$F = \left(\frac{-8 + 4}{2}, \frac{-2 + (-10)}{2}\right) = (-2, -6)$$

Next we find the slopes of the sides of triangle DEF.

Slope of $\overline{DE} = \frac{-2 - 2}{1 - (-5)} = \frac{-4}{6} = -\frac{2}{3}$

Slope of $\overline{EF} = \frac{-6 - (-2)}{-2 - 1} = \frac{-4}{-3} = \frac{4}{3}$

Slope of $\overline{DF} = \frac{-6 - 2}{-2 - (-5)} = \frac{-8}{3} = -\frac{8}{3}$

Exercise Set 2.5

1. $y - y_1 = m(x - x_1)$ Point-slope equation

   $y - 2 = 4(x - 3)$ Substituting 4 for m, 3 for x_1, and 2 for y_1

2. $y - 4 = 5(x - 5)$

3. $y - y_1 = m(x - x_1)$ Point-slope equation

   $y - 7 = -2(x - 4)$ Substituting -2 for m, 4 for x_1, and 7 for y_1

4. $y - 3 = -3(x - 7)$

5. $y - y_1 = m(x - x_1)$ Point-slope equation

   $y - (-4) = 3[x - (-2)]$ Substituting 3 for m, -2 for x_1, and -4 for y_1

   $y + 4 = 3(x + 2)$

6. $y + 7 = x + 5$

7. $y - y_1 = m(x - x_1)$ Point-slope equation

   $y - 0 = -2(x - 8)$ Substituting -2 for m, 8 for x_1, and 0 for y_1

   $y = -2(x - 8)$

8. $y = -3(x + 2)$

9. $y - y_1 = m(x - x_1)$ Point-slope equation

   $y - (-7) = 0(x - 0)$ Substituting 0 for m, 0 for x_1, and -7 for y_1

   $y + 7 = 0$

10. $y - 4 = 0$

11. $y - y_1 = m(x - x_1)$ Point-slope equation

    $y - (-1) = \frac{3}{4}(x - 5)$ Substituting $\frac{3}{4}$ for m, 5 for x_1, and -1 for y_1

    $y + 1 = \frac{3}{4}(x - 5)$

12. $y - 7 = \frac{2}{5}(x + 3)$

13.

$y - y_1 = m(x - x_1)$	Point-slope equation
$y - (-3) = 5(x - 2)$	Substituting 5 for m, 2 for x_1, and -3 for y_1
$y + 3 = 5x - 10$	Simplifying
$y = 5x - 13$	Subtracting 3 on both sides

14. $y = -4x + 1$

15.

$y - y_1 = m(x - x_1)$	Point-slope equation
$y - (-7) = -\frac{2}{3}(x - 4)$	Substituting $-\frac{2}{3}$ for m, 4 for x_1, and -7 for y_1
$y + 7 = -\frac{2}{3}x + \frac{8}{3}$	Simplifying
$y = -\frac{2}{3}x - \frac{13}{3}$	Subtracting 7 on both sides

16. $y = \frac{3}{7}x + \frac{13}{7}$

17.

$y - y_1 = m(x - x_1)$	Point-slope equation
$y - (-4) = -0.6[x - (-3)]$	Substituting -0.6 for m, -3 for x_1, and -4 for y_1
$y + 4 = -0.6(x + 3)$	
$y + 4 = -0.6x - 1.8$	
$y = -0.6x - 5.8$	

18. $y = -3.1x + 13.5$

19. First find the slope of the line:

$$m = \frac{6 - 4}{5 - 1} = \frac{2}{4} = \frac{1}{2}$$

Use the point-slope equation with $m = \frac{1}{2}$ and $(1,4) = (x_1,y_1)$. (We could let $(5,6) = (x_1,y_1)$ instead and obtain an equivalent equation.)

$$y - 4 = \frac{1}{2}(x - 1)$$

$$y - 4 = \frac{1}{2}x - \frac{1}{2}$$

$$y = \frac{1}{2}x + \frac{7}{2}$$

$$f(x) = \frac{1}{2}x + \frac{7}{2} \quad \text{Using function notation}$$

20. $f(x) = -\frac{5}{2}x + 11$

21. First find the slope of the line:

$$m = \frac{2 - (-1)}{2 - (-1)} = \frac{2 + 1}{2 + 1} = \frac{3}{3} = 1$$

Use the point slope equation with $m = 1$ and $(-1,-1) = (x_1,y_1)$.

$$y - (-1) = 1[(x - (-1)]$$

$$y + 1 = x + 1$$

$$y = x$$

$$f(x) = x \quad \text{Using function notation}$$

22. $f(x) = x$

23. First find the slope of the line:

$$m = \frac{5 - 0}{0 - (-2)} = \frac{5}{2}$$

Use the point slope equation with $m = \frac{5}{2}$ and $(-2,0) = (x_1,y_1)$.

$$y - 0 = \frac{5}{2}[x - (-2)]$$

$$y = \frac{5}{2}(x + 2)$$

$$y = \frac{5}{2}x + 5$$

$$f(x) = \frac{5}{2}x + 5$$

24. $f(x) = \frac{1}{2}x - 3$

25. First find the slope of the line:

$$m = \frac{3 - 5}{-5 - 3} = \frac{-2}{-8} = \frac{1}{4}$$

Use the point-slope equation with $m = \frac{1}{4}$ and $(3,5) = (x_1,y_1)$.

$$y - 5 = \frac{1}{4}(x - 3)$$

$$y - 5 = \frac{1}{4}x - \frac{3}{4}$$

$$y = \frac{1}{4}x + \frac{17}{4}$$

$$f(x) = \frac{1}{4}x + \frac{17}{4}$$

26. $f(x) = \frac{1}{5}x + \frac{26}{5}$

27. First find the slope of the line:

$$m = \frac{2 - 0}{5 - 0} = \frac{2}{5}$$

Use the point-slope equation with $m = \frac{2}{5}$ and $(0,0) = (x_1,y_1)$.

$$y - 0 = \frac{2}{5}(x - 0)$$

$$y = \frac{2}{5}x$$

$$f(x) = \frac{2}{5}x$$

28. $f(x) = \frac{3}{7}x$

29. First find the slope of the line:

$$m = \frac{-1 - (-7)}{-2 - (-4)} = \frac{-1 + 7}{-2 + 4} = \frac{6}{2} = 3$$

Use the point-slope equation with $m = 3$ and $(-4, -7) = (x_1, y_1)$.

$$y - (-7) = 3[x - (-4)]$$
$$y + 7 = 3(x + 4)$$
$$y + 7 = 3x + 12$$
$$y = 3x + 5$$
$$f(x) = 3x + 5$$

30. $f(x) = \frac{3}{2}x$

31. a) We form pairs of the type (t, R) where t is the number of years since 1930 and R is the record. We have two pairs, $(0, 46.8)$ and $(40, 43.8)$. These are two points on the graph of the linear function we are seeking. We use the point-slope form to write an equation relating R to t:

$$m = \frac{43.8 - 46.8}{40 - 0} = \frac{-3}{40} = -0.075$$
$$R - 46.8 = -0.075(t - 0)$$
$$R - 46.8 = -0.075t$$
$$R = -0.075t + 46.8$$
$$R(t) = -0.075t + 46.8 \quad \text{Using function notation}$$

b) To predict the record in 1990, we find $R(160)$:

$$R(60) = -0.075(60) + 46.8$$
$$= -4.5 + 46.8$$
$$= 42.3$$

We predict that the record will be 42.3 seconds in 1990.

To predict the record in 2000, we find $R(70)$:

$$R(70) = -0.075(70) + 46.8$$
$$= -5.25 + 46.8$$
$$= 41.55$$

We predict that the record will be 41.55 seconds in 2000.

c) Substitute 40 for $R(t)$ and solve for t:

$$40 = -0.075t + 46.8$$
$$-6.8 = -0.075t$$
$$91 \approx t$$

The record will be 40 seconds about 91 years after 1930, or in 2021.

32. a) $R(t) = -0.075t + 3.85$

b) 3.45 minutes, 3.41 minutes

c) One third of the way through 2003

33. a) Form pairs of the type (x, A), where x is the price of coffee, per pound, in dollars, and A is the amount sold, in millions of pounds. We have two pairs, $(4, 6.5)$ and $(5, 4.0)$. These are two points on the graph of the linear function we are seeking. We use the point-slope form to write an equation relating A to x:

$$m = \frac{4.0 - 6.5}{5 - 4} = \frac{-2.5}{1} = -2.5$$
$$A - 6.5 = -2.5(x - 4)$$
$$A - 6.5 = -2.5x + 10$$
$$A = -2.5x + 16.5$$
$$A(x) = -2.5x + 16.5 \quad \text{Using function notation}$$

b) Find $A(2)$:

$$A(2) = -2.5(2) + 16.5$$
$$= -5 + 16.5$$
$$= 11.5$$

The demand would be 11.5 million pounds.

34. a) $E(t) = \frac{3}{20}t + 65$

b) 70.7 years, 71.75 years

35. a) Form pairs of the type (x, A), where x is the price of coffee, per pound, in dollars, and A is the amount supplied, in millions of pounds. We have two pairs, $(4, 5.0)$ and $(5, 7.0)$. These are two points on the graph of the linear function we are seeking. We use the point-slope form to write an equation relating A to x:

$$m = \frac{7.0 - 5.0}{5 - 4} = \frac{2.0}{1} = 2$$
$$A - 5.0 = 2(x - 4)$$
$$A - 5.0 = 2x - 8$$
$$A = 2x - 3$$
$$A(x) = 2x - 3 \quad \text{Using function notation}$$

b) Find $A(2)$:

$$A(2) = 2 \cdot 2 - 3$$
$$= 4 - 3$$
$$= 1$$

Suppliers are willing to sell 1 million pounds.

36. a) $T(x) = 10x + 20$

b) $190°$ C

37. We form pairs of the type (d,P), where d is the depth, in ft, and P is the pressure, in atmospheres. We have two pairs, (100,4) and (200,7). These are two points on the graph of the linear function we are seeking. We use the point-slope form to write an equation relating P to d:

$$m = \frac{7 - 4}{200 - 100} = \frac{3}{100}$$

$$P - 4 = \frac{3}{100}(d - 100)$$

$$P - 4 = \frac{3}{100}d - 3$$

$$P = \frac{3}{100}d + 1$$

$$P(d) = \frac{3}{100}d + 1$$

   b) Find P(690):

$$P(690) = \frac{3}{100}(690) + 1$$

$$= 20.7 + 1$$

$$= 21.7$$

   The pressure at a depth of 690 ft is 21.7 atm.

38. 28,914 items

39. Familiarize. Form pairs of the type (t,C), where t is the number of years since 1980 and C is the number of country radio stations. We have two pairs, (0,1534) and (6,2275). These are two points on the graph of the linear function we are seeking.

   Translate. We use the point-slope form to write an equation relating C to t:

$$m = \frac{2275 - 1534}{6 - 0} = \frac{741}{6} = 123.5$$

$$C - 1534 = 123.5(t - 0)$$

$$C - 1534 = 123.5t$$

$$C = 123.5t + 1534$$

$$C(t) = 123.5t + 1534$$

   Carry out. Find C(14):

$$C(14) = 123.5(14) + 1534$$

$$= 1729 + 1534$$

$$= 3263$$

   Check. Recheck the calculations, or graph the line containing the points (0,1534) and (6,2275) and see if the point (14,3263) appears to be on the line also.

   State. There will be 3263 country radio stations in 1994.

40. $151,000; $201,000

41. We first solve for y and determine the slope of each line.

$$x + 6 = y$$

$$y = x + 6 \quad \text{Reversing the order}$$

The slope of y = x + 6 is 1.

$$y - x = -2$$

$$y = x - 2$$

The slope of y = x - 2 is 1.

The slopes are the same; the lines are parallel.

42. Parallel

43. We first solve for y and determine the slope of each line.

$$y + 3 = 5x$$

$$y = 5x - 3$$

The slope of y = 5x - 3 is 5.

$$3x - y = -2$$

$$3x + 2 = y$$

$$y = 3x + 2 \quad \text{Reversing the order}$$

The slope of y = 3x + 2 is 3.

The slopes are <u>not</u> the same; the lines are <u>not</u> parallel.

44. Not parallel

45. We determine the slope of each line.

The slope of y = 3x + 9 is 3.

$$2y = 6x - 2$$

$$y = 3x - 1$$

The slope of y = 3x - 1 is 3.

The slopes are the same; the lines are parallel.

46. Parallel

47. First solve the equation for y and determine the slope of the given line.

$$x + 2y = 6 \qquad \text{Given line}$$

$$2y = -x + 6$$

$$y = -\frac{1}{2}x + 3$$

The slope of the given line is $-\frac{1}{2}$.

The slope of every line parallel to the given line must also be $-\frac{1}{2}$. We find the equation of the line with slope $-\frac{1}{2}$ and containing the point (3,7).

$$y - y_1 = m(x - x_1) \qquad \text{Point-slope equation}$$

$$y - 7 = -\frac{1}{2}(x - 3) \qquad \text{Substituting}$$

$$y - 7 = -\frac{1}{2}x + \frac{3}{2}$$

$$y = \frac{1}{2}x + \frac{17}{2}$$

48. y = 3x + 3

49. First solve the equation for y and determine the slope of the given line.

$$5x - 7y = 8 \qquad \text{Given line}$$
$$5x - 8 = 7y$$
$$\frac{5}{7}x - \frac{8}{7} = y$$
$$y = \frac{5}{7}x - \frac{8}{7}$$

The slope of the given line is $\frac{5}{7}$.

The slope of every line parallel to the given line must also be $\frac{5}{7}$. We find the equation of the line with slope $\frac{5}{7}$ and containing the point $(2,-1)$.

$$y - y_1 = m(x - x_1) \qquad \text{Point-slope equation}$$
$$y - (-1) = \frac{5}{7}(x - 2) \qquad \text{Substituting}$$
$$y + 1 = \frac{5}{7}x - \frac{10}{7}$$
$$y = \frac{5}{7}x - \frac{17}{7}$$

50. $y = -2x - 13$

51. First solve the equation for y and determine the slope of the given line.

$$3x - 9y = 2 \qquad \text{Given line}$$
$$3x - 2 = 9y$$
$$\frac{1}{3}x - \frac{2}{9} = y$$

The slope of the given line is $\frac{1}{3}$.

The slope of every line parallel to the given line must also be $\frac{1}{3}$. We find the equation of the line with slope $\frac{1}{3}$ and containing the point $(-6,2)$.

$$y - y_1 = m(x - x_1) \qquad \text{Point-slope equation}$$
$$y - 2 = \frac{1}{3}[x - (-6)] \qquad \text{Substituting}$$
$$y - 2 = \frac{1}{3}(x + 6)$$
$$y - 2 = \frac{1}{3}x + 2$$
$$y = \frac{1}{3}x + 4$$

52. $y = -\frac{5}{2}x - \frac{35}{2}$

53. First solve the equation for y and determine the slope of the given line.

$$3x + 2y = -7 \qquad \text{Given line}$$
$$2y = -3x - 7$$
$$y = -\frac{3}{2}x - \frac{7}{2}$$

The slope of the given line is $-\frac{3}{2}$.

53. (continued)

The slope of every line parallel to the given line must also be $-\frac{3}{2}$. We find the equation of the line with slope $-\frac{3}{2}$ and containing the point $(-3,-2)$.

$$y - y_1 = m(x - x_1) \qquad \text{Point-slope equation}$$
$$y - (-2) = -\frac{3}{2}[x - (-3)] \qquad \text{Substituting}$$
$$y + 2 = -\frac{3}{2}(x + 3)$$
$$y + 2 = -\frac{3}{2}x - \frac{9}{2}$$
$$y = -\frac{3}{2}x - \frac{13}{2}$$

54. $y = \frac{6}{5}x + \frac{39}{5}$

55. We determine the slope of each line.
The slope of $y = 4x - 5$ is 4.

$$4y = 8 - x$$
$$4y = -x + 8$$
$$y = -\frac{1}{4}x + 2$$

The slope of $4y = 8 - x$ is $-\frac{1}{4}$.

The product of their slopes is $4\left(-\frac{1}{4}\right)$, or -1; the lines are perpendicular.

56. Not perpendicular

57. We determine the slope of each line.

$$x + 2y = 5$$
$$2y = -x + 5$$
$$y = -\frac{1}{2}x + \frac{5}{2}$$

The slope of $x + 2y = 5$ is $-\frac{1}{2}$.

$$2x + 4y = 8$$
$$4y = -2x + 8$$
$$y = -\frac{1}{2}x + 2.$$

The slope of $2x + 4y = 8$ is $-\frac{1}{2}$.

The product of their slopes is $\left(-\frac{1}{2}\right)\left(-\frac{1}{2}\right)$, or $\frac{1}{4}$; the lines are _not_ perpendicular. For the lines to be perpendicular, the product must be -1.

58. Perpendicular

59. First solve the equation for y and determine the slope of the given line.

$$2x + y = -3 \qquad \text{Given line}$$
$$y = -2x - 3$$

The slope of the given line is -2.

To find the slope of a perpendicular line, take the reciprocal of -2 and change the sign. The slope is $\frac{1}{2}$.

We find the equation of the line with slope $\frac{1}{2}$ containing the point (2,5).

$$y - y_1 = m(x - x_1) \qquad \text{Point-slope equation}$$
$$y - 5 = \frac{1}{2}(x - 2) \qquad \text{Substituting}$$
$$y - 5 = \frac{1}{2}x - 1$$
$$y = \frac{1}{2}x + 4$$

60. $y = -3x + 12$

61. First solve the equation for y and determine the slope of the given line.

$$3x + 4y = 5 \qquad \text{Given line}$$
$$4y = -3x + 5$$
$$y = -\frac{3}{4}x + \frac{5}{4}$$

The slope of the given line is $-\frac{3}{4}$.

To find the slope of the perpendicular line, take the reciprocal of $-\frac{3}{4}$ and change the sign. The slope is $\frac{4}{3}$.

We find the equation of the line with slope $\frac{4}{3}$ and containing the point (3,-2).

$$y - y_1 = m(x - x_1) \qquad \text{Point-slope equation}$$
$$y - (-2) = \frac{4}{3}(x - 3) \qquad \text{Substituting}$$
$$y + 2 = \frac{4}{3}x - 4$$
$$y = \frac{4}{3}x - 6$$

62. $y = -\frac{2}{5}x - \frac{31}{5}$

63. First solve the equation for y and determine the slope of the given line.

$$2x + 5y = 7 \qquad \text{Given line}$$
$$5y = -2x + 7$$
$$y = -\frac{2}{5}x + \frac{7}{5}$$

The slope of the given line is $-\frac{2}{5}$.

To find the slope of the perpendicular line, take the reciprocal of $-\frac{2}{5}$ and change the sign. The slope is $\frac{5}{2}$.

63. (continued)

We find the equation of the line with slope $\frac{5}{2}$ and containing the point (0,9).

$$y - y_1 = m(x - x_1) \qquad \text{Point-slope equation}$$
$$y - 9 = \frac{5}{2}(x - 0) \qquad \text{Substituting}$$
$$y - 9 = \frac{5}{2}x$$
$$y = \frac{5}{2}x + 9$$

64. $y = -2x - 10$

65. First solve the equation for y and find the slope of the given line.

$$3x - 5y = 6$$
$$-5y = -3x + 6$$
$$y = \frac{3}{5}x - \frac{6}{5}$$

The slope of the given line is $\frac{3}{5}$. To find the slope of the perpendicular line, take the reciprocal of $\frac{3}{5}$ and change the sign. The slope is $-\frac{5}{3}$.

We find the equation of the line with slope $-\frac{5}{3}$ and containing the point (-4,-7).

$$y - y_1 = m(x - x_1) \qquad \text{Point-slope equation}$$
$$y - (-7) = -\frac{5}{3}[x - (-4)]$$
$$y + 7 = -\frac{5}{3}(x + 4)$$
$$y + 7 = -\frac{5}{3}x - \frac{20}{3}$$
$$y = -\frac{5}{3}x - \frac{41}{3}$$

66. $y = -\frac{2}{7}x + \frac{27}{7}$

67. Familiarize. We find the sales tax by multiplying the listed price by the tax rate. We let x represent the price and 5%x represent the sales tax.

Translate. Listed price + Sales tax = Total price

$$\qquad x \qquad + \qquad 5\%x \qquad = \qquad \$36.75$$

Carry out. We solve the equation.

$$x + 0.05x = 36.75$$
$$1.05x = 36.75$$
$$x = \frac{36.75}{1.05}$$
$$x = 35$$

Check. If the listed price is \$35, then the sales tax is 5%·\$35, or \$1.75. The total price is 35 + 1.75, or \$36.75. The value checks.

State. The price of the radio before the tax is \$35.

68. 69 points

69. <u>Familiarize.</u> Let n represent the number.

    <u>Translate.</u>

    15% of what number is 12.4?

    15% ×   n   = 12.4

    <u>Carry out.</u> Solve the equation.

    $15\%n = 12.4$

    $0.15n = 12.4$

    $15n = 1240$ Multiplying by 100 to clear of decimals

    $n = \frac{1240}{15}$

    $n = \frac{248}{3}$, or $82\frac{2}{3}$

    <u>Check.</u> Left to the student.

    <u>State.</u> 15% of $82\frac{2}{3}$ is 12.4.

70. $-\frac{13}{12}$

71. <u>Familiarize.</u> Form pairs of the type (t,L), where t is the temperature, in °C, and L IS the length of the pipe in cm. We have two pairs, (18,100) and (20,100.00356). These are two points on the graph of the equation we are seeking.

    <u>Translate.</u> Use the point-slope form to write an equation relating L to t:

    $$m = \frac{100.00356 - 100}{20 - 18} = \frac{0.00356}{2} = 0.00178$$

    $L - 100 = 0.00178(t - 18)$

    $L - 100 = 0.00178t - 0.03204$

    $L = 0.00178t + 99.96796$

    $L(t) = 0.00178t + 99.96796$

    <u>Carry out.</u> First find L(40):

    $L(40) = 0.00178(40) + 99.96796$

    $= 0.0712 + 99.96796$

    $= 100.03916$

    Then find L(10):

    $L(0) = 0.00178(0) + 99.96796$

    $L(0) = 99.96796$

    <u>Check.</u> Recheck the calculations, or graph the line containing the points (18,100) and (20,100.00356) and see if the points (40,100.03916) and (0,99.96796) appear to be on the line also.

    <u>State.</u> At 40° C the pipe is 100.03916 cm long, and at 0° C the pipe is 99.96796 cm long.

72. $1300

73. a) Let x = the gross sales for the month, and let E = the amount earned.

    Plan A:

    Amount earned is $600 plus 4% of gross sales

    E   = 600 + 4% ·   x

    $E = 600 + 0.4x$

    Plan B:

    Amount earned is $700 if gross sales are no more than $10,000

    $E = 700$, $x \leqslant 10,000$

    and Amount earned is $700 plus 6% of gross sales over $10,000

    $E = 700 + 6\% \times (x - 10,000)$, $x \geqslant 10,000$,

    or $E = 700 + 0.06(x - 10,000)$, $x > 10,000$

    Summarizing Plan B we have:

    $$E = \begin{cases} 700, \text{ if } x \leqslant 10,000 \\ 700 + 0.06(x - 10,000), \text{ if } x > 10,000 \end{cases}$$

    b) For $x \leqslant 10,000$, we solve the inequality

    $600 + 0.04x < 700$

    $0.04x < 100$

    $x < 2500$

    Thus for gross sales less than $2500 plan B is preferable.

    For $x > 10,000$, we solve the inequality

    $600 + 0.04x < 700 + 0.06(x - 10,000)$

    $600 + 0.04x < 700 + 0.06x - 600$

    $600 + 0.04x < 100 + 0.06x$

    $500 < 0.02x$

    $25,000 < x$

    Thus for gross sales greater than $25,000 plan B is also preferable.

    Thus plan B is preferable for $x < 2500$ or $x > 25,000$.

74. 21.1° C

75. a) We have two pairs, (-1,3) and (2,4). Use the point-slope form:

    $$m = \frac{4 - 3}{2 - (-1)} = \frac{1}{2 + 1} = \frac{1}{3}$$

    $y - 3 = \frac{1}{3}[x - (-1)]$

    $y - 3 = \frac{1}{3}(x + 1)$

    $y - 3 = \frac{1}{3}x + \frac{1}{3}$

    $y = \frac{1}{3}x + \frac{10}{3}$

    $f(x) = \frac{1}{3}x + \frac{10}{3}$ Using function notation

    b) $f(x) = \frac{1}{3}(3) + \frac{10}{3} = 1 + \frac{10}{3} = \frac{13}{3}$

75. (continued)

c) $f(a) = \frac{1}{3}(a) + \frac{10}{3} = \frac{a}{3} + \frac{10}{3}$

If $f(a) = 100$, we have

$$100 = \frac{a}{3} + \frac{10}{3}$$
$$300 = a + 10 \quad \text{Multiplying by 3}$$
$$290 = a$$

76. a) $g(x) = x - 8$,   b) $g(-2) = -10$,   c) $a = 83$

77. Find the slope of $5y - kx = 7$:

$$5y - kx = 7$$
$$5y = kx + 7$$
$$y = \frac{k}{5}x + \frac{7}{5}$$

The slope is $\frac{k}{5}$.

Find the slope of the line containing the points $(7,-3)$ and $(-2,5)$:

$$m = \frac{5 - (-3)}{-2 - 7} = \frac{5 + 3}{-9} = -\frac{8}{9}$$

If the lines are parallel, their slopes must be equal:

$$\frac{k}{5} = -\frac{8}{9}$$
$$k = -\frac{40}{9}$$

78. $k = 7$

Exercise Set 2.6

1. Since $f(1) = -3 \cdot 1 + 1 = -2$ and $g(1) = 1^2 - 2 = 3$, we have $f(1) + g(1) = -2 + 3 = 1$.

2. 1

3. Since $f(-1) = -3(-1) + 1 = 4$ and $g(-1) = (-1)^2 + 2 = 3$ we have $f(-1) + g(-1) = 4 + 3 = 7$.

4. 13

5. Since $f(-7) = -3(-7) + 1 = 22$ and $g(-7) = (-7)^2 + 2 = 51$, we have $f(-7) - g(-7) = 22 - 51 = -29$.

6. -11

7. Since $f(5) = -3(5) + 1 = -14$ and $g(5) = 5^2 + 2 = 27$, we have $f(5) - g(5) = -14 - 27 = -41$.

8. -29

9. Since $f(2) = -3 \cdot 2 + 1 = -5$ and $g(2) = 2^2 + 2 = 6$, we have $f(2) \cdot g(2) = -5 \cdot 6 = -30$.

10. -88

11. Since $f(-3) = -3(-3) + 1 = 10$ and $g(-3) = (-3)^2 + 2 = 11$, we have $f(-3) \cdot g(-3) = 10 \cdot 11 = 110$.

12. 234

13. Since $f(0) = -3 \cdot 0 + 1 = 1$ and $g(0) = 0^2 + 2 = 2$, we have $f(0)/g(0) = 1/2$.

14. $-\frac{2}{3}$

15. Using our work in Exercise 11, we have $f(-3)/g(-3) = 10/11$.

16. $\frac{13}{18}$

17. $(F + G)(x) = F(x) + G(x)$
$$= x^2 - 3 + 4 - x$$
$$= x^2 - x + 1 \quad \text{Combining like terms}$$
$(F + G)(-3) = (-3)^2 - (-3) + 1 = 9 + 3 + 1 = 13$

18. 7

19. Using our work in Exercise 17, we have $(F + G)(x) = x^2 - x + 1$.

20. $a^2 - a + 1$

21. $(F - G)(x) = F(x) - G(x)$
$$= x^2 - 3 - (4 - x)$$
$$= x^2 - 3 - 4 + x$$
$$= x^2 + x - 7$$
$(F - G)(-4) = (-4)^2 + (-4) - 7 = 16 - 4 - 7 = 5$

22. 13

23. $(F \cdot G)(x) = F(x) \cdot G(x)$
$$= (x^2 - 3)(4 - x)$$
$$= 4x^2 - x^3 - 12 + 3x$$
$(F \cdot G)(2) = 4 \cdot 2^2 - 2^3 - 12 + 3 \cdot 2 =$
$$16 - 8 - 12 + 6 = 2$$

24. 6

25. Using our work in Exercise 23, we have:
$(F \cdot G)(-3) = 4(-3)^2 - (-3)^3 - 12 + 3(-3)$
$$= 4 \cdot 9 - (-27) - 12 - 9$$
$$= 36 + 27 - 12 - 9$$
$$= 42$$

26. 104

<u>27.</u> $(F/G)(x) = F(x)/G(x)$

$$= \frac{x^2 - 3}{4 - x}$$

$$(F/G)(0) = \frac{0^2 - 3}{4 - 0} = -\frac{3}{4}$$

<u>28.</u> $-\frac{2}{3}$

<u>29.</u> Using our work in Exercise 27, we have

$$(F/G)(-2) = \frac{(-2)^2 - 3}{4 - (-2)} = \frac{4 - 3}{4 + 2} = \frac{1}{6}.$$

<u>30.</u> $-\frac{2}{5}$

<u>31.</u> The domain of f and of g is all real numbers. Thus, Domain of f + g = Domain of f - g = Domain of f·g = {x|x is a real number}.

<u>32.</u> {x|x is a real number}

<u>33.</u> Because division by 0 is undefined, we have

Domain of f = {x|x is a real number and x ≠ 2}, and

Domain of g = {x|x is a real number}.

Thus, Domain of f + g = Domain of f - g = Domain of f·g = {x|x is a real number and x ≠ 2}.

<u>34.</u> {x|x is a real number and x ≠ 4}

<u>35.</u> Because division by 0 is undefined, we have

Domain of f = {x|x is a real number and x ≠ 0}, and

Domain of g = {x|x is a real number}.

Thus, Domain of f + g = Domain of f - g = Domain of f·g = {x|x is a real number and x ≠ 0}.

<u>36.</u> {x|x is a real number and x ≠ 0}

<u>37.</u> Because division by 0 is undefined, we have

Domain of f = {x|x is a real number and x ≠ 1}, and

Domain of g = {x|x is a real number}.

Thus, Domain of f + g = Domain of f - g = Domain of f·g = {x|x is a real number and x ≠ 1}.

<u>38.</u> {x|x is a real number and x ≠ 5}

<u>39.</u> Because division by 0 is undefined, we have

Domain of f = {x|x is a real number and x ≠ 2}, and

Domain of g = {x|x is a real number and x ≠ 4}.

Thus, Domain of f + g = Domain of f - g = Domain of f·g = {x|x is a real number and x ≠ 2 and x ≠ 4}.

<u>40.</u> {x|x is a real number and x ≠ 3 and x ≠ 2}

<u>41.</u> Because division by 0 is undefined, we have

Domain of f = {x|x is a real number and x ≠ -2}, and

Domain of g = {x|x is a real number and x ≠ 4}.

Thus, Domain of f + g = Domain of f - g = Domain of f·g = {x|x is a real number and x ≠ -2 and x ≠ 4}.

<u>42.</u> {x|x is a real number and x ≠ 3 and x ≠ 5}

<u>43.</u> Domain of f = Domain of g = {x|x is a real number}.

Since g(x) = 0 when x - 3 = 0, we have g(x) = 0 when x = 3. We conclude that Domain of f/g = {x|x is a real number and x ≠ 3}.

<u>44.</u> {x|x is a real number and x ≠ 5}

<u>45.</u> Domain of f = Domain of g = {x|x is a real number}.

Since g(x) = 0 when 2x - 8 = 0, we have g(x) = 0 when x = 4. We conclude that Domain of f/g = {x|x is a real number and x ≠ 4}.

<u>46.</u> {x|x is a real number and x ≠ 3}

<u>47.</u> Domain of f = {x|x is a real number and x ≠ 0}.

Domain of g = {x|x is a real number and x ≠ 4}.

Since g(x) is never 0, we conclude that Domain of f/g = {x|x is a real number and x ≠ 0 and x ≠ 4}.

<u>48.</u> {x|x is a real number and x ≠ 3 and x ≠ 0}

<u>49.</u> Domain of f = {x|x is a real number and x ≠ 4}.

Domain of g = {x|x is a real number}.

Since g(x) = 0 when 5 - x = 0, we have g(x) = 0 when x = 5. We conclude that Domain of f/g = {x|x is a real number and x ≠ 4 and x ≠ 5}.

<u>50.</u> {x|x is a real number and x ≠ 2 and x ≠ 7}

<u>51.</u> Domain of f = {x|x is a real number and x ≠ -1}.

Domain of g = {x|x is a real number}.

Since g(x) = 0 when 2x + 5 = 0, we have g(x) = 0 when $x = -\frac{5}{2}$. We conclude that Domain of f/g = $\left\{ x \middle| x \text{ is a real number and } x \neq -1 \text{ and } x \neq -\frac{5}{2} \right\}$.

<u>52.</u> $\left\{ x \middle| x \text{ is a real number and } x \neq 2 \text{ and } x \neq -\frac{7}{3} \right\}$

<u>53.</u> Domain of f = {x|x is a real number and x ≠ 4} Domain of g = {x|x is a real number}. Since g(x) = 0 when $3x^2 = 0$, we have g(x) = 0 when x = 0. We conclude that Domain of f/g = {x|x is a real number and x ≠ 4 and x ≠ 0}.

<u>54.</u> {x|x is a real number and x ≠ 0 and x ≠ -3}

55. Domain of f = {x|x is a real number}.
Domain of g = {x|x is a real number and x ≠ 4}.
Since g(x) = 0 when $\frac{x-1}{x-4}$ = 0, we have g(x) = 0
when x = 1. We conclude that Domain of f/g =
{x|x is a real number and x ≠ 4 and x ≠ 1}.

56. {x|x is a realnumber and x ≠ -2 and x ≠ -3}

57. Domain of f = {x|x is a real number and x ≠ 2}.
Domain of g = {x|x is a real number and x ≠ 4}.
Since g(x) = 0 when $\frac{x-3}{x-4}$ = 0, we have g(x) = 0
when x = 3. We conclude that Domain of f/g =
{x|x is a real number and x ≠ 2, x ≠ 4, and
x ≠ 3}.

58. {x|x is a real number and x ≠ -3, x ≠ -2, and
x ≠ -1}

59. Domain of f = {x|x is a real number and x ≠ -2}.
Domain of g = {x|x is a real number and x ≠ -1}.
Since g(x) = 0 when $\frac{3x-5}{x+1}$ = 0, we have g(x) = 0
when x = $\frac{5}{3}$. We conclude that Domain of f/g =
$\left\{x \middle| x \text{ is a real number and } x \neq -2, x \neq -1, \text{ and } x \neq \frac{5}{3}\right\}$.

60. $\left\{x \middle| x \text{ is a real number and } x \neq \frac{1}{5}\right\}$

61. Domain of f = $\left\{x \middle| x \text{ is a real number and } x \neq \frac{5}{2}\right\}$.
Domain of g = $\left\{x \middle| x \text{ is a real number and } x \neq \frac{1}{3}\right\}$.
Since g(x) = 0 when $\frac{x}{3x-1}$ = 0, we have g(x) = 0
when x = 0. We conclude that Domain of f/g =
$\left\{x \middle| x \text{ is a real number and } x \neq \frac{5}{2}, x \neq \frac{1}{3}, \text{ and } x \neq 0\right\}$.

62. $\left\{x \middle| x \text{ is a real number and } x \neq 0 \text{ and } x \neq \frac{6}{5}\right\}$

63. 5x - 7 = 0
   5x = 7   Adding 7 on both sides
   x = $\frac{7}{5}$ Multiplying by $\frac{7}{5}$ on both sides

64. -37

65. <u>Familiarize</u>. The average score on n tests is the
sum of the n test scores divided by n. If the
average on 4 tests is 78.5, then the sum of the
4 scores is 4(78.5), or 314. Let s represent
the score required on the fifth test to raise
the average to 80.

<u>Translate.</u>
   Sum of 5 scores divided by 5 is 80.
   (314 + s)   ÷   5 = 80

<u>Carry out.</u> Solve the equation.
   $\frac{314+s}{5}$ = 80
   314 + s = 400
   s = 86

<u>Check.</u> If the sum of the 5 scores is 314 + 86,
or 400, then the average is 400/5, or 80. The
value checks.

<u>State.</u> A score of 86 is needed on the fifth test
to raise the average to 80.

66. 6.31 × 10⁻⁶

67. Domain of p = {x|x is a real number and x ≠ 1}
(since p is not defined for x = 1).
Domain of q = {x|x is a real number and x ≠ 2}.
Since q(x) = 0 when $\frac{x-3}{x-2}$ = 0, we have q(x) = 0
when x = 3. We conlude that Domain of p/q =
{x|x is a real number and x ≠ 1, x ≠ 2, and
x ≠ 3}.

68. $\left\{x \middle| x \text{ is a real number and } x \neq \frac{3}{2} \text{ and } -1 < x < 5\right\}$

69. The domain of each function is the set of first
coordinates for that function.
Domain of f = {-2,-1,0,1,2}, and
Domain of g = {-4,-3,-2,-1,0,1}.
Domain of f + g = Domain of f - g = Domain of
f·g = {-2,-1,0,1}.
Since g(-1) = 0, we conclude that Domain of f/g =
{-2,0,1}.

70. (f + g)(-2) = 5, (f·g)(0) = 15, (f/g)(1) = 2/3

71. Domain of F = {x|x is a real number and x ≠ 4}.
Domain of G = {x|x is a real number and x ≠ 3}.
Since G(x) = 0 when $\frac{x^2-4}{x-3}$ = 0, we have G(x) = 0
when x² - 4 = 0, or when x = 2 or x = -2. We
conclude that Domain of F/G = {x|x is a real
number and x ≠ 4, x ≠ 3, x ≠ 2, and x ≠ -2}.

72. $\left\{x \middle| x \text{ is a real number and } x \neq -\frac{5}{2}, x \neq 1, x \neq -1, \text{ and } x \neq -3\right\}$

73. Domain of f = {x│x is a real number and
    −5 ⩽ x ⩽ 5}.

    Domain of g = {x│x is a real number and
    −3 ⩽ x ⩽ 7}.

    Domain of f + g = Domain of f − g = Domain of
    f·g = {x│x is a real number and −3 ⩽ x ⩽ 5}.

    Since g(x) = 0 when x = −1 and when x = 3 we
    conclude that Domain of f/g =
    {x│x is a real number and −3 ⩽ x ⩽ 5 and
    x ≠ −1 and x ≠ 3}.

74. Answers may vary. $f(x) = \dfrac{1}{x + 2}$, $g(x) = \dfrac{1}{x + 5}$

75. There are many correct answers. Graph functions
    f and g whose domains include {x│−2 ⩽ x ⩽ 3}.
    In addition, g must have only one x-intercept in
    that interval. One result is the following:

    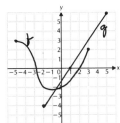

76. a) 67 million,   b) 50 million,   c) 50 million

Exercise Set 3.1

1. <u>Familiarize</u>. Let x = the first number and y = the second number.

   <u>Translate</u>.

   The sum of two numbers is -42.

   Rewording: The first number plus the second number is -42.

   $$x \quad + \quad y \quad = -42$$

   The first number minus the second number is 52.

   $$x \quad - \quad y \quad = 52$$

   We have a system of equations:

   $$x + y = -42$$
   $$x - y = 52$$

2. $x - y = 11$

   $3x + 2y = 123$

3. <u>Familiarize</u>. List the information in a table.

Kind of scarf	White	Printed	Total
Number sold	x	y	40
Price	$4.95	$7.95	
Amount taken in	4.95x	7.95y	282

   <u>Translate</u>: The "Number sold" row of the table gives us one equation:

   $$x + y = 40$$

   The "Amount taken in" row gives us a second equation:

   $$4.95x + 7.95y = 282$$

   We have a system of equations:

   $$x + y = 40$$
   $$4.95x + 7.95y = 282$$

   We can multiply the second equation on both sides by 100 to clear the decimals:

   $$x + y = 40$$
   $$495x + 795y = 28,200$$

4. $x + y = 45$

   $8.50x + 9.75y = 398.75$

5. <u>Familiarize</u>. The basketball court is a rectangle with perimeter 288 ft. Let ℓ = length and w = width. Recall that for a rectangle with length ℓ and width w, the perimeter P is given by $P = 2\ell + 2w$.

   <u>Translate</u>. The formula for perimeter gives us one equation:

   $$2\ell + 2w = 288$$

   The statement relating length and width gives us a second equation:

   Length is 44 ft longer than width

   $$\ell = 44 + w$$

   We have a system of equations:

   $$2\ell + 2w = 288$$
   $$\ell = 44 + w$$

6. $2w + 2\ell = 228$

   $w = \ell - 42$

7. <u>Familiarize</u>. Let x = the measure of one angle and y = the measure of the other angle. Recall that two angles are supplementary if the sum of their measures is 180°.

   <u>Translate</u>. The fact that the angles are supplementary gives us one equation.

   Rewording: The sum of the measures is 180°.

   $$x + y = 180$$

   The second statement gives us another equation:

   One angle is 3° less than twice the other.

   $$x = 2y - 3$$

   We have a system of equations:

   $$x + y = 180$$
   $$x = 2y - 3$$

8. $x + y = 90$

   $x + \frac{1}{2}y = 64$

9. <u>Familiarize</u>. List the information in a table.

Type of score	Field goal	Free throw	Total
Number scored	x	y	18
Points per score	2	1	
Points scored	2x	1·y, or y	30

   <u>Translate</u>. The "Number scored" row of the table gives us one equation:

   $$x + y = 18$$

   The "Points scored" row gives us a second equation:

   $$2x + y = 30$$

   We have a system of equations:

   $$x + y = 18$$
   $$2x + y = 30$$

10. $x + y = 250$

    $1.50x + 2y = 441$

11. <u>Familiarize</u>. Let x = number of games won and y = number of games tied. The total points earned in x wins is 2x; the total points earned in y ties is 1·y, or y.

    <u>Translate</u>.

    Points from wins plus points from ties is 60.

    $$2x \qquad + \qquad y \qquad = 60$$

    Number of wins is 9 more than number of ties.

    $$x \qquad = 9 + y$$

    We have a system of equations:

    $$2x + y = 60$$
    $$x = 9 + y$$

12. $x + y = 152$
    $x = 5 + 6y$

13. <u>Familiarize</u>. Let x = number of 30-sec commercials and y = number of 60-sec commercials. The total time used by x 30-sec commercials is 30x; the total time used by y 60-sec commercials is 60y. Also note that 10 min = 10 × 60, or 600 sec.

    <u>Translate</u>.

    Total number of commercials is 12.

    $$x + y \qquad = 12$$

    Total commercial time is 10 min, or 600 sec.

    $$30x + 60y \qquad = \qquad 600$$

    We have a system of equations:

    $$x + y = 12$$
    $$30x + 60y = 600$$

14. $x + y = 400$
    $20x + 30y = 11,000$

15. We use alphabetical order for the variables. We replace x by 1 and y by 2.

    $$\frac{4x - y = 2}{4 \cdot 1 - 2 \mid 2}$$
    $$4 - 2$$
    $$2 \mid$$

    $$\frac{10x - 3y = 4}{10 \cdot 1 - 3 \cdot 2 \mid 4}$$
    $$10 - 6$$
    $$4 \mid$$

    The pair (1,2) makes both equations true, so it <u>is</u> a solution of the system.

16. Yes

17. We use alphabetical order for the variables. We replace x by 2 and y by 5.

    $$\frac{y = 3x - 1}{5 \mid 3 \cdot 2 - 1}$$
    $$6 - 1$$
    $$5$$

    $$\frac{2x + y = 4}{2 \cdot 2 + 5 \mid 4}$$
    $$4 + 5$$
    $$9$$

    The pair (2,5) is not a solution of 2x + y = 4. Therefore it <u>is not</u> a solution of the system of equations.

18. No

19. We replace x by 1 and y by 5.

    $$\frac{x + y = 6}{1 + 5 \mid 6}$$
    $$6 \mid$$

    $$\frac{y = 2x + 3}{5 \mid 2 \cdot 1 + 3}$$
    $$2 + 3$$
    $$5$$

    The pair (1,5) makes both equations true, so it <u>is</u> a solution of the system.

20. Yes

21. We replace a by 2 and b by -7.

    $$\frac{3a + b = -1}{3 \cdot 2 + (-7) \mid -1}$$
    $$6 - 7$$
    $$-1 \mid$$

    $$\frac{2a - 3b = -8}{2 \cdot 2 - 3 \cdot (-7) \mid -8}$$
    $$4 + 21$$
    $$25 \mid$$

    The pair (2,-7) is not a solution of 2a - 3b = -8. Therefore it <u>is not</u> a solution of the system of equations.

22. No

23. We replace x by 3 and y by 1.

    $$\frac{3x + 4y = 13}{3 \cdot 3 + 4 \cdot 1 \mid 13}$$
    $$9 + 4$$
    $$13 \mid$$

    $$\frac{5x - 4y = 11}{5 \cdot 3 - 4 \cdot 1 \mid 11}$$
    $$15 - 4$$
    $$11$$

    The pair (3,1) makes both equations true, so it <u>is</u> a solution of the system.

24. No

25. We replace x by -2 and y by 3.

    $$\frac{6x - y = -15}{6(-2) - 3 \mid -15}$$
    $$-12 - 3$$
    $$-15 \mid$$

    $$\frac{y = 4x + 11}{3 \mid 4(-2) + 11}$$
    $$-8 + 11$$
    $$3$$

    The pair (-2,3) makes both equations true, so it <u>is</u> a solution of the system.

26. No

27. Graph both lines on the same set of axes.

    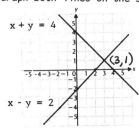

    The solution (point of intersection) seems to be the point (3,1).

    Check:

    $$\frac{x + y = 4}{3 + 1 \mid 4}$$
    $$4 \mid$$

    $$\frac{x - y = 2}{3 - 1 \mid 2}$$
    $$2 \mid$$

    The solution is (3,1).

<u>28.</u> (4,1)

<u>29.</u> Graph both lines on the same set of axes.

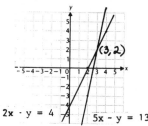

The solution (point of intersection) seems to be the point (3,2).

Check:

$2x - y = 4$		$5x - y = 13$	
$2 \cdot 3 - 2$	4	$5 \cdot 3 - 2$	13
$6 - 2$		$15 - 2$	
4		13	

The solution is (3,2).

<u>30.</u> (2,-1)

<u>31.</u> Graph both lines on the same set of axes.

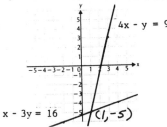

The solution seems to be the point (1,-5).

Check:

$4x - y = 9$		$x - 3y = 16$	
$4 \cdot 1 - (-5)$	9	$1 - 3(-5)$	16
$4 + 5$		$1 + 15$	
9		16	

The solution is (1,-5).

<u>32.</u> $\left[3, \frac{3}{2}\right]$

<u>33.</u> Graph both lines on the same set of axes.

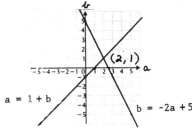

The solution seems to be the point (2,1).

<u>33.</u> (continued)

Check:

$a = 1 + b$		$b = -2a + 5$	
2	$1 + 1$	1	$-2 \cdot 2 + 5$
	2		$-4 + 5$
			1

The solution is (2,1).

<u>34.</u> (-3,-2)

<u>35.</u> Graph both lines on the same set of axes.

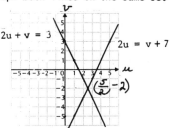

The solution seems to be $\left[\frac{5}{2}, -2\right]$.

Check:

$2u + v = 3$		$2u = v + 7$	
$2 \cdot \frac{5}{2} + (-2)$	3	$2 \cdot \frac{5}{2}$	$-2 + 7$
$5 - 2$		5	5
3			

The solution is $\left[\frac{5}{2}, -2\right]$.

<u>36.</u> (7,2)

<u>37.</u> Graph both lines on the same set of axes.

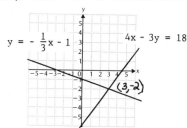

The ordered pair (3,-2) checks in both equations. It is the solution.

<u>38.</u> (4,0)

39. Graph both lines on the same set of axes.

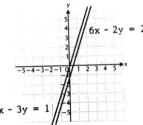

   The lines are parallel. There is <u>no</u> solution.

40. No solution

41. Graph both lines on the same set of axes.

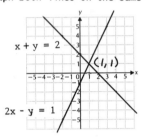

   The ordered pair (1,1) checks in both equations.
   It is the solution.

42. (-3,2)

43. Graph both lines on the same set of axes.

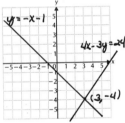

   The ordered pair (3,-4) checks in both equations.
   It is the solution.

44. $\left[-\frac{2}{3},\frac{7}{3}\right]$

45. Graph both lines on the same set of axes.

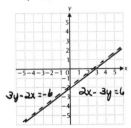

   The graphs are the same. Any solution of one
   equation is a solution of the other. Each
   equation has infinitely many solutions. The
   solution set is the set of all pairs (x,y) for
   which 2x - 3y = 6, or {(x,y)|2x - 3y = 6}. (In
   place of 2x - 3y = 6 we could have used
   3y - 2x = -6 since the two equations are
   equivalent.)

46. {(x,y)|y = 3 - x}

47. A system of equations is consistent if it has at
   least one solution. Of the systems under
   consideration, only the one in Exercise 39 has no
   solution. Therefore, all except the system in
   Exercise 39 are consistent.

48. All except 40

49. A system of two equations in two variables is
   dependent if it has infinitely many solutions.
   Only the system in Exercise 45 is dependent.

50. 46

51. 3x + 4 = x - 2
   2x + 4 = -2   Adding -x on both sides
   2x = -6   Adding -4 on both sides
   x = -3   Multiplying by $\frac{1}{2}$ on both sides

52. -35

53. 4x - 5x = 8x - 9 + 11x
   -x = 19x - 9   Collecting like terms
   -20x = -9   Adding -19x on both sides
   x = $\frac{9}{20}$ Multiplying by $-\frac{1}{20}$ on both sides

54. b = a - 4Q

55. Substitute 4 for x and -5 for y in the first
   equation:
   $$A(4) - 6(-5) = 13$$
   $$4A + 30 = 13$$
   $$4A = -17$$
   $$A = -\frac{17}{4}$$

   Substitute 4 for x and -5 for y in the second
   equation:
   $$4 - B(-5) = -8$$
   $$4 + 5B = -8$$
   $$5B = -12$$
   $$B = -\frac{12}{5}$$

   We have A = $-\frac{17}{4}$, B = $-\frac{12}{5}$.

56. Answers may vary.
   x + y = 6
   x - y = 4

57. There are many correct answers. One can be found
   by expressing the sum and difference of the two
   numbers:
   x + y = 9
   x - y = -3

58. Answers may vary.

$$x + y = 1$$
$$2x + 2y = 3$$

59. There are many correct answers. One can be found by writing an equation in two variables and then writing a constant multiple of that equation:

$$x + y = 1$$
$$2x + 2y = 2$$

60. a) Answers may vary. (4,-5); b) Infinitely many

61. There are many correct answers. Find one by expressing the sum of the two numbers:

$$x + y = -1$$

62. $x = 2y$

$x + 20 = 3y$

63. <u>Familiarize</u>. Let x = number of years Juan has taught and y = number of years Juanita has taught. Two years ago, Juan and Juanita had taught $x - 2$ and $y - 2$ years, respectively.

<u>Translate</u>.

Total years of service is 46.

$$x + y \qquad\quad = 46$$

Two years ago

Juan's years of teaching were 2.5 times Juanita's

$$x - 2 \qquad\quad = 2.5(y - 2)$$

We have a system of equations:

$$x + y = 46$$
$$x - 2 = 2.5(y - 2)$$

64. $2\ell + 2w = 156$

$\ell = 4(w - 6)$

65. <u>Familiarize</u>. Let w = number of wins and ℓ = number of losses. Before winning the last eight games in a row, the team had won $w - 8$ games.

<u>Translate</u>.

Total number of games is 104.

$$w + \ell \qquad\qquad = 104$$

Eight consecutive wins ago,

number of wins was 3 times number of losses.

$$w - 8 \qquad = 3\ell$$

We have a system of equations:

$$w + \ell = 104$$
$$w - 8 = 3\ell$$

66. $d + q = 58$

$q + 7 = 4d$

67. Graph both equations on the same set of axes.

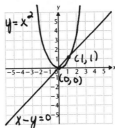

The points (0,0) and (1,1) appear to be solutions.

Check (0,0):

$x - y = 0$				$y = x^2$	
$0 - 0$	0			0	0^2
0					0

(0,0) is a solution.

Check (1,1):

$x - y = 0$				$y = x^2$	
$1 - 1$	0			1	1^2
0					1

(1,1) is a solution.

The solutions are (0,0) and (1,1).

68. (1,3), (-2,0)

69. Graph both equations on the same set of axes.

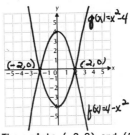

The points (-2,0) and (2,0) appear to be solutions.

Check (-2,0):

$f(x) = 4 - x^2$			$g(x) = x^2 - 4$	
0	$4 - (2)^2$		0	$(-2)^2 - 4$
	$4 - 4$			$4 - 4$
	0			0

(-2,0) is a solution.

Check (2,0):

$f(x) = 4 - x^2$			$g(x) = x^2 - 4$	
0	$4 - 2^2$		0	$2^2 - 4$
	$4 - 4$			$4 - 4$
	0			0

(2,0) is a solution.

The solutions are (-2,0) and (2,0).

70. (3,3), (-5,5)

Exercise Set 3.2

1. $3x + 5y = 3$ (1)

  $x = 8 - 4y$ (2)

We substitute $8 - 4y$ for x in the first equation and solve for y.

$3x + 5y = 3$ (1)

$3(8 - 4y) + 5y = 3$ Substituting

$24 - 12y + 5y = 3$

$24 - 7y = 3$

$-7y = -21$

$y = 3$

Next we substitute 3 for y in either equation of the original system and solve for x.

$x = 8 - 4y$ (2)

$x = 8 - 4 \cdot 3$ Substituting

$x = 8 - 12$

$x = -4$

We check the ordered pair $(-4, 3)$.

$$\frac{3x + 5y = 3}{3(-4) + 5 \cdot 3 \mid 3}$$
$$-12 + 15$$
$$3$$

$$\frac{x = 8 - 4y}{-4 \mid 8 - 4 \cdot 3}$$
$$8 - 12$$
$$-4$$

Since $(-4, 3)$ checks, it is the solution.

2. $(2, -3)$

3. $9x - 2y = 3$ (1)

  $3x - 6 = y$ (2)

We substitute $3x - 6$ for y in the first equation and solve for x.

$9x - 2y = 3$ (1)

$9x - 2(3x - 6) = 3$ Substituting

$9x - 6x + 12 = 3$

$3x + 12 = 3$

$3x = -9$

$x = -3$

Next we substitute -3 for x in either equation of the original system and solve for y.

$3x - 6 = y$ (2)

$3(-3) - 6 = y$ Substituting

$-9 - 6 = y$

$-15 = y$

We check the ordered pair $(-3, -15)$.

$$\frac{9x - 2y = 3}{9(-3) - 2(-15) \mid 3}$$
$$-27 + 30$$
$$3$$

$$\frac{3x - 6 = y}{3(-3) - 6 \mid -15}$$
$$-9 - 6$$
$$-15$$

Since $(-3, -15)$ checks, it is the solution.

4. $\left(\frac{21}{5}, \frac{12}{5}\right)$

5. $5m + n = 8$ (1)

  $3m - 4n = 14$ (2)

We solve the first equation for n.

$5m + n = 8$ (1)

$n = 8 - 5m$

We substitute $8 - 5m$ for n in the second equation and solve for m.

$3m - 4n = 14$ (2)

$3m - 4(8 - 5m) = 14$ Substituting

$3m - 32 + 20m = 14$

$23m - 32 = 14$

$23m = 46$

$m = 2$

Now we substitute 2 for m in either equation of the original system and solve for n.

$5m + n = 8$ (1)

$5 \cdot 2 + n = 8$ Substituting

$10 + n = 8$

$n = -2$

We check the ordered pair $(2, -2)$.

$$\frac{5m + n = 8}{5 \cdot 2 + (-2) \mid 8}$$
$$10 - 2$$
$$8$$

$$\frac{3m - 4n = 14}{3 \cdot 2 - 4(-2) \mid 14}$$
$$6 + 8$$
$$14$$

Since $(2, -2)$ checks, it is the solution.

6. $(2, -7)$

7. $4x + 12y = 4$ (1)

  $-5x + y = 11$ (2)

We solve the second equation for y.

$-5x + y = 11$ (2)

$y = 5x + 11$

We substitute $5x + 11$ for y in the first equation and solve for x.

$4x + 12y = 4$ (1)

$4x + 12(5x + 11) = 4$ Substituting

$4x + 60x + 132 = 4$

$64x + 132 = 4$

$64x = -128$

$x = -2$

Now substitute -2 for x in either equation of the original system and solve for y.

$-5x + y = 11$ (2)

$-5(-2) + y = 11$ Substituting

$10 + y = 11$

$y = 1$

We check the ordered pair $(-2, 1)$.

$$\frac{4x + 12y = 4}{4(-2) + 12 \cdot 1 \mid 4}$$
$$-8 + 12$$
$$4$$

$$\frac{-5x + y = 11}{-5(-2) + 1 \mid 11}$$
$$10 + 1$$
$$11$$

Since $(-2, 1)$ checks, it is the solution.

8. (4,-1)

9. $3x - y = 1$ (1)

$2x + 2y = 2$ (2)

We solve the first equation for y.

$3x - y = 1$ (1)

$3x - 1 = y$

We substitute $3x - 1$ for y in the second equation and solve for x.

$2x + 2y = 2$ (2)

$2x + 2(3x - 1) = 2$ Substituting

$2x + 6x - 2 = 2$

$8x - 2 = 2$

$8x = 4$

$x = \frac{1}{2}$

Next we substitute $\frac{1}{2}$ for x in either equation of the original system and solve for y.

$3x - y = 1$ (1)

$3 \cdot \frac{1}{2} - y = 1$ Substituting

$\frac{3}{2} - y = \frac{2}{2}$

$-y = -\frac{1}{2}$

$y = \frac{1}{2}$

We check the ordered pair $\left[\frac{1}{2},\frac{1}{2}\right]$.

$3x - y = 1$		$2x + 2y = 2$	
$3 \cdot \frac{1}{2} - \frac{1}{2}$	1	$2 \cdot \frac{1}{2} + 2 \cdot \frac{1}{2}$	2
$\frac{3}{2} - \frac{1}{2}$			
		$1 + 1$	
1			2

Since $\left[\frac{1}{2},\frac{1}{2}\right]$ checks, it is the solution.

10. (3,-2)

11. $3x - y = 7$ (1)

$2x + 2y = 5$ (2)

We solve the first equation for y.

$3x - y = 7$ (1)

$3x - 7 = y$

We substitute $3x - 7$ for y in the second equation and solve for x.

$2x + 2y = 5$ (2)

$2x + 2(3x - 7) = 5$ Substituting

$2x + 6x - 14 = 5$

$8x - 14 = 5$

$8x = 19$

$x = \frac{19}{8}$

11. (continued)

Now we substitute $\frac{19}{8}$ for x in either equation of the original system and solve for y.

$3x - y = 7$ (1)

$3 \cdot \frac{19}{8} - y = 7$ Substituting

$\frac{57}{8} - y = \frac{56}{8}$

$-y = -\frac{1}{8}$

$y = \frac{1}{8}$

The ordered pair $\left[\frac{19}{8},\frac{1}{8}\right]$ checks in both equations. It is the solution.

12. $\left[\frac{25}{23},\frac{11}{23}\right]$

13. $x + 2y = 6$ (1)

$x = 4 - 2y$ (2)

We substitute $4 - 2y$ for x in the first equation and solve for y.

$x + 2y = 6$ (1)

$(4 - 2y) + 2y = 6$ Substituting

$4 - 2y + 2y = 6$

$4 = 6$ Collecting like terms

We have a false equation. Therefore, there is no solution.

14. Ø

15. $x - 3 = y$ (1)

$2x - 2y = 6$ (2)

Substitute $x - 3$ for y in the second equation and solve for x.

$2x - 2y = 6$ (2)

$2x - 2(x - 3) = 6$ Substituting

$2x - 2x + 6 = 6$

$6 = 6$ Collecting like terms

We have an equation that is true for all numbers x and y. The system is dependent and has an infinite number of solutions. Using (1), we can express the solution set as $\{(x,y)|x - 3 = y\}$.

16. $\{(x,y)|3y = x - 2\}$

17. $x + 3y = 7$
      $\underline{-x + 4y = 7}$
      $0 + 7y = 14$ Adding
      $\quad\quad 7y = 14$
      $\quad\quad\quad y = 2$

Substitute 2 for y in one of the original equations and solve for x.

$\quad x + 3y = 7$
$\quad x + 3\cdot2 = 7$ Substituting
$\quad\quad x + 6 = 7$
$\quad\quad\quad\quad x = 1$

Check: For (1,2)

$$\begin{array}{c|c} x + 3y = 7 & -x + 4y = 7 \\ \hline 1 + 3\cdot2 \;\vert\; 7 & -1 + 4\cdot2 \;\vert\; 7 \\ 1 + 6 & -1 + 8 \\ 7 & 7 \end{array}$$

Since (1,2) checks, it is the solution.

18. (2,7)

19. $2x + y = 6$
     $\underline{x - y = 3}$
     $3x + 0 = 9$ Adding
     $\quad\quad 3x = 9$
     $\quad\quad\; x = 3$

Substitute 3 for x in one of the original equations and solve for y.

$\quad 2x + y = 6$
$\quad 2\cdot3 + y = 6$ Substituting
$\quad\quad 6 + y = 6$
$\quad\quad\quad\; y = 0$

We obtain (3,0). This checks, so it is the solution.

20. (10,2)

21. $9x + 3y = -3$
      $\underline{2x - 3y = -8}$
      $11x + 0 = -11$ Adding
      $\quad\quad 11x = -11$
      $\quad\quad\quad x = -1$

Substitute -1 for x in one of the original equations and solve for y.

$\quad 9x + 3y = -3$
$\quad 9(-1) + 3y = -3$ Substituting
$\quad\quad -9 + 3y = -3$
$\quad\quad\quad\quad 3y = 6$
$\quad\quad\quad\quad\; y = 2$

We obtain (-1,2). This checks, so it is the solution.

22. $\left(\frac{1}{2}, -5\right)$

23. $5x + 3y = 19$ (1)
     $2x - 5y = 11$ (2)

We multiply twice to make two terms become additive inverses.

From (1): $25x + 15y = 95$ Multiplying by 5
From (2): $\underline{6x - 15y = 33}$ Multiplying by 3
$\quad\quad\quad 31x \quad\quad = 128$ Adding
$$x = \frac{128}{31}$$

Substitute $\frac{128}{31}$ for x in one of the original equations and solve for y.

$\quad\quad 5x + 3y = 19$
$\quad 5 \cdot \frac{128}{31} + 3y = 19$ Substituting
$\quad\quad \frac{640}{31} + 3y = \frac{589}{31}$
$\quad\quad\quad\quad 3y = -\frac{51}{31}$
$\quad\quad \frac{1}{3} \cdot 3y = \frac{1}{3} \cdot \left(-\frac{51}{31}\right)$
$\quad\quad\quad\quad\; y = -\frac{17}{31}$

We obtain $\left(\frac{128}{31}, -\frac{17}{31}\right)$. This checks, so it is the solution.

24. $\left(\frac{10}{21}, \frac{11}{14}\right)$

25. $5r - 3s = 24$ (1)
     $3r + 5s = 28$ (2)

We multiply twice to make two terms become additive inverses.

From (1): $25r - 15s = 120$ Multiplying by 5
From (2): $\underline{9r + 15s = 84}$ Multiplying by 3
$\quad\quad\quad 34r \quad\quad = 204$ Adding
$\quad\quad\quad\quad\; r = 6$

Substitute 6 for r in one of the original equations and solve for s.

$\quad 3r + 5s = 28$
$\quad 3\cdot6 + 5s = 28$ Substituting
$\quad\quad 18 + 5s = 28$
$\quad\quad\quad\; 5s = 10$
$\quad\quad\quad\; s = 2$

We obtain (6,2). This checks, so it is the solution.

26. (1,3)

27. $0.3x - 0.2y = 4$

$0.2x + 0.3y = 1$

We first multiply each equation by 10 to clear decimals.

$3x - 2y = 40$ (1)

$2x + 3y = 10$ (2)

We use the multiplication principle with both equations of the resulting system.

From (1): $9x - 6y = 120$ Multiplying by 3

From (2): $\underline{4x + 6y = 20}$ Multiplying by 2

           $13x \quad\quad = 140$ Adding

                $x = \dfrac{140}{13}$

Substitute $\dfrac{140}{13}$ for x in one of the equations in which the decimals were cleared and solve for y.

        $2x + 3y = 10$

$2 \cdot \dfrac{140}{13} + 3y = 10$ Substituting

    $\dfrac{280}{13} + 3y = \dfrac{130}{13}$

           $3y = -\dfrac{150}{13}$

             $y = -\dfrac{50}{13}$

We obtain $\left(\dfrac{140}{13}, -\dfrac{50}{13}\right)$. This checks, so it is the solution.

28. $(2,3)$

29. $\dfrac{1}{2}x + \dfrac{1}{3}y = 4$

$\dfrac{1}{4}x + \dfrac{1}{3}y = 3$

We first multiply each equation by the LCM of the denominators to clear fractions.

$3x + 2y = 24$ Multiplying by 6

$3x + 4y = 36$ Multiplying by 12

We multiply by -1 on both sides of the first equation and then add.

$-3x - 2y = -24$ Multiplying by -1

$\underline{3x + 4y = 36}$

     $2y = 12$ Adding

      $y = 6$

Substitute 6 for y in one of the equations in which the fractions were cleared and solve for x.

    $3x + 2y = 24$

$3x + 2 \cdot 6 = 24$ Substituting

    $3x + 12 = 24$

        $3x = 12$

         $x = 4$

We obtain $(4,6)$. This checks, so it is the solution.

30. $(-21,21)$

31. $3p - 3q = 7$

$3p + 5q = 17$

We multiply by -1 on both sides of the first equation and then add.

$-3p + 3q = -7$ Multiplying by -1

$\underline{3p + 5q = 17}$

     $8q = 10$ Adding

       $q = \dfrac{10}{8}$

       $q = \dfrac{5}{4}$

Substitute $\dfrac{5}{4}$ for q in one of the original equations and solve for p.

      $3p + 5q = 17$

$3p + 5 \cdot \dfrac{5}{4} = 17$ Substituting

    $3p + \dfrac{25}{4} = \dfrac{68}{4}$

         $3p = \dfrac{43}{4}$

          $p = \dfrac{43}{12}$

We obtain $\left(\dfrac{43}{12}, \dfrac{5}{4}\right)$. This checks, so it is the solution.

32. $\left(\dfrac{8}{5}, \dfrac{1}{5}\right)$

33. $\dfrac{2}{5}x + \dfrac{1}{2}y = 2$

$\dfrac{1}{2}x - \dfrac{1}{6}y = 3$

We first multiply each equation by the LCM of the denominators to clear fractions.

$4x + 5y = 20$ Multiplying by 10

$3x - y = 18$ Multiplying by 6

We multiply by 5 on both sides of the second equation and then add.

   $4x + 5y = 20$

$\underline{15x - 5y = 90}$ Multiplying by 5

$19x \quad\quad = 110$ Adding

        $x = \dfrac{110}{19}$

Substitute $\dfrac{110}{19}$ for x in one of the equations in which the fractions were cleared and solve for y.

      $3x - y = 18$

$3\left(\dfrac{110}{19}\right) - y = 18$ Substituting

    $\dfrac{330}{19} - y = \dfrac{342}{19}$

          $-y = \dfrac{12}{19}$

           $y = -\dfrac{12}{19}$

We obtain $\left(\dfrac{110}{19}, -\dfrac{12}{19}\right)$. This checks, so it is the solution.

34. (12,15)

35. $2x + 3y = 1$

    $4x + 6y = 2$

    Multiply the first equation by -2 and then add.

    $-4x - 6y = -2$

    $\underline{4x + 6y = 2}$

    $0 = 0$ Adding

    We have an equation that is true for all numbers x and y. The system is dependent and has an infinite number of solutions. The solution set can be expressed using either equation. Using the first equation, we have $\{(x,y)|2x + 3y = 1\}$.

36. $\{(x,y)|3x - 2y = 1\}$

37. $2x - 4y = 5$

    $2x - 4y = 6$

    Multiply the first equation by -1 and then add.

    $-2x + 4y = -5$

    $\underline{2x - 4y = 6}$

    $0 = 1$

    We have a false equation. The system has no solution.

38. Ø

39. $5x - 9y = 7$

    $7y - 3x = -5$

    We first write the second equation in the form $Ax + By = C$.

    $5x - 9y = 7$

    $-3x + 7y = -5$

    We use the multiplication principle with both equations and then add.

    $15x - 27y = 21$ Multiplying by 3

    $\underline{-15x + 35y = -25}$ Multiplying by 5

    $8y = -4$ Adding

    $y = -\frac{1}{2}$

    Substitute $-\frac{1}{2}$ for y in one of the original equations and solve for x.

    $5x - 9y = 7$

    $5x - 9\left(-\frac{1}{2}\right) = 7$ Substituting

    $5x + \frac{9}{2} = \frac{14}{2}$

    $5x = \frac{5}{2}$

    $x = \frac{1}{2}$

    We obtain $\left(\frac{1}{2}, -\frac{1}{2}\right)$. This checks, so it is the solution.

40. (-2.-9)

41. $3(a - b) = 15$

    $4a = b + 1$

    We first write each equation in the form $Ax + By = C$.

    $3a - 3b = 15$

    $4a - b = 1$

    We multiply by -3 on both sides of the second equation and then add.

    $3a - 3b = 15$

    $\underline{-12a + 3b = -3}$ Multiplying by -3

    $-9a = 12$

    $a = -\frac{12}{9}$

    $a = -\frac{4}{3}$

    Substitute $-\frac{4}{3}$ for a in one of the original equations and solve for b.

    $4a - b = 1$

    $4\left(-\frac{4}{3}\right) - b = 1$

    $-\frac{16}{3} - b = \frac{3}{3}$

    $-b = \frac{19}{3}$

    $b = -\frac{19}{3}$

    We obtain $\left(-\frac{4}{3}, -\frac{19}{3}\right)$. This checks, so it is the solution.

42. (30,6)

43. $x - \frac{1}{10}y = 100$

    $y - \frac{1}{10}x = -100$

    We first write the second equation in the form $Ax + By = C$.

    $x - \frac{1}{10}y = 100$

    $-\frac{1}{10}x + y = -100$

    Next we multiply each equation by 10 to clear fractions.

    $10x - y = 1000$

    $-x + 10y = -1000$

    We multiply by 10 on both sides of the first equation and then add.

    $100x - 10y = 10,000$ Multiplying by 10

    $\underline{-x + 10y = -1000}$

    $99x = 9000$

    $x = \frac{9000}{99}$

    $x = \frac{1000}{11}$

    Substitute $\frac{1000}{11}$ for x in one of the equations in which the fractions were cleared and solve for y.

43. (continued)

$$10x - y = 1000$$

$$10\left[\frac{1000}{11}\right] - y = 1000 \qquad \text{Substituting}$$

$$\frac{10,000}{11} - y = \frac{11,000}{11}$$

$$-y = \frac{1000}{11}$$

$$y = -\frac{1000}{11}$$

We obtain $\left[\frac{1000}{11}, -\frac{1000}{11}\right]$. This checks, so it is the solution.

44. $(-20, 20)$

45. $0.05x + 0.25y = 22$

$0.15x + 0.05y = 24$

We first multiply each equation by 100 to clear decimals.

$5x + 25y = 2200$

$15x + 5y = 2400$

We multiply by -5 on both sides of the second equation and add.

$$\begin{array}{rl} 5x + 25y = & 2200 \\ -75x - 25y = & -12,000 \qquad \text{Multiplying by } -5 \\ \hline -70x = & -9800 \qquad \text{Adding} \\ x = & \dfrac{-9800}{-70} \\ x = & 140 \end{array}$$

Substitute 140 for x in one of the equations in which the decimals were cleared and solve for y.

$$5x + 25y = 2200$$

$$5 \cdot 140 + 25y = 2200 \qquad \text{Substituting}$$

$$700 + 25y = 2200$$

$$25y = 1500$$

$$y = 60$$

We obtain $(140, 60)$. This checks, so it is the solution.

46. $(10, 5)$

47. <u>Familiarize.</u> Recall the formula for simple interest, $I = Prt$, where I = interest, P = principal, r = interest rate, and t = time (in years).

<u>Translate.</u> Substitute \$17.60 for I, \$320 for P, and $\frac{1}{2}$ for t.

$$I = Prt$$

$$17.60 = 320r\left[\frac{1}{2}\right]$$

47. (continued)

<u>Carry out.</u> Solve the equation.

$$17.60 = 320r\left[\frac{1}{2}\right]$$

$$17.60 = 160r$$

$$\frac{17.60}{160} = r$$

$$0.11 = r$$

<u>Check.</u> $320(0.11)\left[\frac{1}{2}\right] = 17.60$. The answer checks.

<u>State.</u> The rate of interest is 0.11, or 11%.

48. $-15y - 39$

49. $3.5x - 2.1y = 106.2$

$4.1x + 16.7y = -106.28$

Since this is a calculator exercise, you may choose not to clear the decimals. We will do so here, however.

$35x - 21y = 1062 \qquad \text{Multiplying by 10}$

$410x + 1670y = -10,628 \qquad \text{Multiplying by 100}$

Multiply twice to make two terms become additive inverses.

$$\begin{array}{rl} 58,450x - 35,070y = & 1,773,540 \quad \text{Multiplying by 1670} \\ 8610x + 35,070y = & -223,188 \quad \text{Multiplying by 21} \\ \hline 67,060x = & 1,550,352 \quad \text{Adding} \\ x \approx & 23.118879 \end{array}$$

Substitute 23.118879 for x in one of the equations in which the decimals were cleared and solve for y.

$$35x - 21y = 1062$$

$$35(23.118879) - 21y = 1062 \qquad \text{Substituting}$$

$$809.160765 - 21y = 1062$$

$$-21y = 252.839235$$

$$y \approx -12.039964$$

The numbers check, so the solution is $(23.118879, -12.039964)$.

50. $\left[-\frac{32}{17}, \frac{38}{17}\right]$

51. $5x + 2y = a$

$x - y = b$

We multiply by 2 on both sides of the second equation and then add.

$$\begin{array}{rl} 5x + 2y = & a \\ 2x - 2y = & 2b \qquad \text{Multiplying by 2} \\ \hline 7x = & a + 2b \qquad \text{Adding} \\ x = & \dfrac{a + 2b}{7} \end{array}$$

Next we multiply by -5 on both sides of the second equation and then add.

51. (continued)

$$5x + 2y = a$$
$$\underline{-5x + 5y = -5b} \qquad \text{Multiplying by } -5$$
$$7y = a - 5b$$
$$y = \frac{a - 5b}{7}$$

We obtain $\left(\frac{a + 2b}{7}, \frac{a - 5b}{7}\right)$. This checks, so it is the solution.

52. (5,2)

53. (0,-3) and $\left(-\frac{3}{2}, 6\right)$ are two solutions of $px - qy = -1$.

Substitute 0 for x and -3 for y.
$$p \cdot 0 - q \cdot (-3) = -1$$
$$3q = -1$$
$$q = -\frac{1}{3}$$

Substitute $-\frac{3}{2}$ for x and 6 for y.
$$p \cdot \left(-\frac{3}{2}\right) - q \cdot 6 = -1$$
$$-\frac{3}{2}p - 6q = -1$$

Substitute $-\frac{1}{3}$ for q and solve for p.
$$-\frac{3}{2}x - 6 \cdot \left(-\frac{1}{3}\right) = -1$$
$$-\frac{3}{2}p + 2 = -1$$
$$-\frac{3}{2}p = -3$$
$$-\frac{2}{3} \cdot \left(-\frac{3}{2}p\right) = -\frac{2}{3} \cdot (-3)$$
$$p = 2$$

Thus, $p = 2$ and $q = -\frac{1}{3}$.

54. $m = -\frac{1}{2}, \ b = \frac{5}{2}$

55. (0,3) and (-2,3) are two solutions of $g(x) = ax^2 + c$.

Substitute 0 for x and 3 for g(x).
$$3 = a \cdot 0^2 + c$$
$$3 = a \cdot 0 + c$$
$$3 = c$$

Substitute -2 for x and 3 for g(x).
$$3 = a(-2)^2 + c$$
$$3 = 4a + c$$

Substitute 3 for c and solve for a.
$$3 = 4a + 3$$
$$0 = 4a$$
$$0 = a$$

Thus, $a = 0$ and $c = 3$.

56. $\left(-\frac{1}{4}, -\frac{1}{2}\right)$

57. $\dfrac{2}{x} + \dfrac{1}{y} = 0 \qquad$ or $\qquad 2 \cdot \dfrac{1}{x} + 1 \cdot \dfrac{1}{y} = 0$

$\dfrac{5}{x} + \dfrac{2}{y} = -5 \qquad\qquad 5 \cdot \dfrac{1}{x} + 2 \cdot \dfrac{1}{y} = -5$

Substitute u for $\dfrac{1}{x}$ and v for $\dfrac{1}{y}$.

$$2u + v = 0$$
$$5u + 2v = -5$$

We multiply by -2 on both sides of the first equation and then add.

$$-4u - 2v = 0 \qquad \text{Multiplying by } -2$$
$$\underline{5u + 2v = -5}$$
$$u = -5 \qquad \text{Adding}$$

Substitute -5 for u and solve for v.
$$2u + v = 0$$
$$2(-5) + v = 0 \qquad \text{Substituting}$$
$$-10 + v = 0$$
$$v = 10$$

If $u = -5$, then $\dfrac{1}{x} = -5$. Thus $x = -\dfrac{1}{5}$.

If $v = 10$, then $\dfrac{1}{y} = 10$. Thus $y = \dfrac{1}{10}$.

The solution is $\left(-\dfrac{1}{5}, \dfrac{1}{10}\right)$.

Exercise Set 3.3

1. The Familiarize and Translate steps were done in Exercise 1 of Exercise Set 3.1.

<u>Carry out.</u> We solve the system of equations
$$x + y = -42$$
$$x - y = 52$$
where x = the first number and y = the second number. We use the elimination method.

$$x + y = -42$$
$$\underline{x - y = 52}$$
$$2x = 10 \qquad \text{Adding}$$
$$x = 5$$

Substitute 5 for x in one of the equations and solve for y.
$$x + y = -42$$
$$5 + y = -42$$
$$y = -47$$

<u>Check.</u> The sum of the numbers is 5 + (-47), or -42. The difference is 5 - (-47), or 52. The numbers check.

<u>State.</u> The numbers are 5 and -47.

2. 29 and 18

<u>3</u>. The Familiarize and Translate steps were done in Exercise 3 of Exercise Set 3.1.

<u>Carry out</u>. We solve the system of equations

$$x + y = 40$$
$$495x + 795y = 28,200$$

where x = the number of white scarves sold and y = the number of printed scarves sold. We use the elimination method. We eliminate x by multiplying the first equation by -495 and then adding it to the second.

$$-495x - 495y = -19,800 \quad \text{Multiplying by -495}$$
$$\underline{495x + 795y = 28,200}$$
$$300y = 8400 \quad \text{Adding}$$
$$y = 28$$

Substitute 28 for y in one of the original equations and solve for x.

$$x + 28 = 40$$
$$x = 12$$

<u>Check</u>. The number of scarves sold is 12 + 28, or 40.

Money from white scarves = $4.95 × 12 = $ 59.40
Money from printed scarves = $7.95 × 28 = $222.60
Total = $282.00

The numbers check.

<u>State</u>. 12 white scarves and 28 printed scarves were sold.

<u>4</u>. $8.50 pens: 32; $9.75 pens: 13

<u>5</u>. The Familiarize and Translate steps were done in Exercise 5 of Exercise Set 3.1.

<u>Carry out</u>. We solve the system of equations

$$2\ell + 2w = 288 \quad (1)$$
$$\ell = 44 + w \quad (2)$$

where ℓ = length and w = width. We use substitution.

$$2(44 + w) + 2w = 288 \quad \text{Substituting } 44 + w \text{ for } \ell \text{ in (1)}$$
$$88 + 2w + 2w = 288$$
$$88 + 4w = 288$$
$$4w = 200$$
$$w = 50$$

$$\ell = 44 + 50 \quad \text{Substituting 50 for w in (2)}$$
$$\ell = 94$$

<u>Check</u>. The perimeter of a 94 ft by 50 ft rectangle is 2·94 + 2·50 = 188 + 100 = 288. Also, 94 ft is 44 ft longer than 50 ft. The numbers check.

<u>State</u>. The length is 94 ft, and the width is 50 ft.

<u>6</u>. Width = 36 ft, length = 78 ft

<u>7</u>. The Familiarize and Translate steps were done in Exercise 7 of Exercise Set 3.1.

<u>Carry out</u>. We solve the system of equations

$$x + y = 180 \quad (1)$$
$$x = 2y - 3 \quad (2)$$

where x = the measure of one angle and y = the measure of the other angle. We use substitution.

$$(2y - 3) + y = 180 \quad \text{Substituting } 2y - 3 \text{ for x in (1)}$$
$$3y - 3 = 180$$
$$3y = 183$$
$$y = 61$$

$$x + 61 = 180 \quad \text{Substituting 61 for y in (1)}$$
$$x = 119$$

<u>Check</u>. The sum of the angles is 61° + 119°, or 180°, so they are supplementary. Also, two times the 61° angle minus 3° is 119°, the other angle. The numbers check.

<u>State</u>. The measure of one angle is 61°, and the measure of the other is 119°.

<u>8</u>. 38° and 52°

<u>9</u>. The Familiarize and Translate steps were done in Exercise 9 of Exercise Set 3.1.

<u>Carry out</u>. We solve the system of equations

$$x + y = 18 \quad (1)$$
$$2x + y = 30 \quad (2)$$

where x = the number of field goals and y = the number of free throws. We use the elimination method.

$$-x - y = -18 \quad \text{Multiplying (1) by -1}$$
$$\underline{2x + y = 30}$$
$$x = 12$$

$$12 + y = 18 \quad \text{Substituting 12 for x in (1)}$$
$$y = 6$$

<u>Check</u>. The total number of times the player scored is 12 + 6, or 18.

Points from field goals: 12 × 2 = 24
Points from free throws: 6 × 1 = 6
Total 30

The numbers check.

<u>State</u>. The player made 12 field goals and 6 free throws.

<u>10</u>. Children's: 118; adult's: 132

11. The Familiarize and Translate steps were done in Exercise 11 of Exercise Set 3.1.

    Carry out. We solve the system of equations

    $$2x + y = 60 \qquad (1)$$
    $$x = 9 + y \qquad (2)$$

    where x = number of games won and y = number of games tied. We use substitution.

    $$2(9 + y) + y = 60 \qquad \text{Substituting } 9 + y \text{ for } x \text{ in (1)}$$
    $$18 + 2y + y = 60$$
    $$18 + 3y = 60$$
    $$3y = 42$$
    $$y = 14$$

    $$x = 9 + 14 \qquad \text{Substituting 14 for } y \text{ in (2)}$$
    $$x = 23$$

    Check. The number of wins, 23, is 9 more than the number of ties, 14.

    Points from wins: $23 \times 2 = 46$

    Points from ties: $14 \times 1 = \underline{14}$

                Total   60

    The numbers check.

    State. The team had 23 wins and 14 ties.

12. Coach: 131; first-class: 21

13. The Familiarize and Translate steps were done in Exercise 13 of Exercise Set 3.1.

    Carry out. We solve the system of equations

    $$x + y = 12 \qquad (1)$$
    $$30x + 60y = 600 \qquad (2)$$

    where x = the number of 30-sec commercials and y = the number of 60-sec commercials. We use the elimination method.

    $$-30x - 30y = -360 \qquad \text{Multiplying (1) by } -30$$
    $$\underline{30x + 60y = 600}$$
    $$30y = 240 \qquad \text{Adding}$$
    $$y = 8$$

    $$x + 8 = 12 \qquad \text{Substituting 8 for } y \text{ in (1)}$$
    $$x = 4$$

    Check. The total number of commercials is 4 + 8, or 12.

    Time for 30-sec commercials: $30 \times 4 = 120$ sec

    Time for 60-sec commercials: $60 \times 8 = \underline{480 \text{ sec}}$

                            600 sec,
    or 10 min

    The numbers check.

    State. There were 4 30-sec commercials and 8 60-sec commercials.

14. Lumber: 100; plywood: 300

15. Familiarize. Let x = the larger number and y = the smaller number.

    Translate.

    The difference of the numbers is 16.

    $$x - y \qquad\qquad = 16$$

    Three times the larger is nine times the smaller.

    $$3x \qquad = \qquad 9y$$

    We have a system of equations:

    $$x - y = 16 \qquad (1)$$
    $$3x = 9y \qquad (2)$$

    Carry out. Solve the system of equations. We use the substitution method.

    $$x = 3y \qquad \text{Solving (2) for } x$$
    $$3y - y = 16 \qquad \text{Substituting } 3y \text{ for } x \text{ in (1)}$$
    $$2y = 16$$
    $$y = 8$$

    $$x - 8 = 16 \qquad \text{Substituting 8 for } y \text{ in (1)}$$
    $$x = 24$$

    Check. The difference of the numbers is 24 - 8, or 16. Also, $3 \cdot 24 = 72 = 9 \cdot 8$. The numbers check.

    State. The larger number is 24 and the smaller is 8.

16. 20 and 30

17. Familiarize. We organize the information in a table.

    Let x = the number of pounds of soybean meal and y = the number of pounds of corn meal.

Type of meal	Pounds of meal	Percent of protein	Pounds protein in meal
Soybean	x	16%	0.16x
Corn	y	9%	0.09y
Mixture	350	12%	0.12×350 or 42

    Translate. The "Pounds of meal" column gives us one equation: $x + y = 350$

    The last column gives us a second equation: $0.16x + 0.09y = 42$

    After clearing decimals, we have this system:

    $$x + y = 350 \qquad (1)$$
    $$16x + 9y = 4200 \qquad (2)$$

    Carry out. Solve the system of equations.

    $$-9x - 9y = -3150 \qquad \text{Multiplying (1) by } -9$$
    $$\underline{16x + 9y = 4200}$$
    $$7x = 1050$$
    $$x = 150$$

    $$150 + y = 350 \qquad \text{Substituting 150 for } x \text{ in (1)}$$
    $$y = 200$$

    Check. The total number of pounds is 150 + 200, or 350. Also, 16% of 150 is 24, and 9% of 200 is 18. Their total is 42. The numbers check.

17. (continued)

State. 150 lb of soybean meal and 200 lb of corn meal should be mixed.

18. 4 L of 25% solution, 6 L of 50% solution

19. Familiarize. We can organize the information in a table. Let x = the number of liters of the drink containing 15% orange juice and y = the number of liters of the drink containing 5% orange juice.

Type of canned juice drink	Amount of drink	Percent of orange juice	Amount of orange juice in drink
15% juice	x	15%	0.15x
5% juice	y	5%	0.05y
Mixture	10	10%	0.1 × 10 or 1

Translate. The "Amount of drink" column gives us one equation: x + y = 10

The last column gives us a second equation: 0.15x + 0.05y = 1

After clearing decimals, we have this system:

$$x + y = 10 \quad (1)$$
$$15x + 5y = 100 \quad (2)$$

Carry out. Solve the system of equations

$$-5x - 5y = -50 \quad \text{Multiplying (1) by -5}$$
$$\underline{15x + 5y = 100}$$
$$10x \qquad = 50 \quad \text{Adding}$$
$$x = 5$$

$$5 + y = 10 \quad \text{Substituting 5 for x in (1)}$$
$$y = 5$$

Check. The total number of liters is 5 + 5, or 10. Also, 15% of 5 is 0.75, and 5% of 5 is 0.25. Their sum is 1. The numbers check.

State. 5 L of each drink should be used.

20. $12\frac{1}{2}$ L of A, $7\frac{1}{2}$ L of B

21. Familiarize. Let x = one investment and y = the other investment. We organize the information in a table.

	Prin-cipal	Rate	Time	Interest (I = Prt)
1st investment	x	14%	1 yr	0.14x
2nd investment	y	16%	1 yr	0.16y
Total	$8800			$1326

21. (continued)

Translate. The first column gives us one equation: x + y = 8800

The last column gives us a second equation: 0.14x + 0.16y = 1326

After clearing decimals, we have this system:

$$x + y = 8800 \quad (1)$$
$$14x + 16y = 132,600 \quad (2)$$

Carry out. We solve the system of equations.

$$-14x - 14y = -123,200 \quad \text{Multiplying (1) by -14}$$
$$\underline{14x + 16y = 132,600}$$
$$2y = 9400 \quad \text{Adding}$$
$$y = 4700$$

$$x + 4700 = 8800 \quad \text{Substituting 4700 for y in (1)}$$
$$x = 4100$$

Check. The sum of the investments is $4100 + $4700, or $8800. The amounts of interest earned are 14% of $4100, or $574, and 16% of $4700, or $752. The total interest earned is $574 + $752, or $1326. The values check.

State. $4100 is invested at 14% and $4700 is invested at 16%.

22. $6800 at 9%, $8200 at 10%

23. Familiarize. Let x = one investment and y = the other investment. Organize the information in a table.

	Prin-cipal	Rate	Time	Interest (I = Prt)
1st investment	x	12%	1 yr	0.12x
2nd investment	y	11%	1 yr	0.11y
Total	$1150			$133.75

Translate. The first column gives us one equation: x + y = 1150

The last column gives us a second equation: 0.12x + 0.11y = 133.75

After clearing decimals, we have this system:

$$x + y = 1150 \quad (1)$$
$$12x + 11y = 13,375 \quad (2)$$

Carry out. We solve the system of equations.

$$-11x - 11y = -12,650 \quad \text{Multiplying (1) by -11}$$
$$\underline{12x + 11y = 13,375}$$
$$x \qquad = 725 \quad \text{Adding}$$

$$725 + y = 1150 \quad \text{Substituting 725 for x in (1)}$$
$$y = 425$$

Check. The sum of the investments is $725 + $425, or $1150. The amounts of interest earned are 12% of $725, or $87 and 11% of $425, or $46.75. The total interest earned is $87 + $46.75, or $133.75. The values check.

23. (continued)

    State. $725 was invested at 12% and $425 was invested at 11%.

24. $12,500 at 10%, $14,500 at 12%

25. Familiarize. Let x = the number of white sweatshirts sold and y = the number of yellow sweatshirts sold. Organize the information in a table.

Kind of sweatshirt	White	Yellow	Total
Number sold	x	y	30
Price	$9.95	$10.50	
Amount taken in	9.95x	10.50y	310.60

→ x + y = 30

→ 9.95x + 10.50y = 310.60

    Translate. Using the "Number sold" and "Amount taken in" rows we have a system of equations:

$$x + y = 30$$
$$9.95x + 10.50y = 310.60$$

    After clearing decimals, we have

$$x + y = 30 \quad (1)$$
$$995x + 1050y = 31,060 \quad (2)$$

    Carry out. We solve the system of equations.

$$-995x - 995y = -29,850 \quad \text{Multiplying (1) by } -995$$
$$995x + 1050y = 31,060$$
$$55y = 1210 \quad \text{Adding}$$
$$y = 22$$

$$x + 22 = 30 \quad \text{Substituting 22 for y in (1)}$$
$$x = 8$$

    Check. The total number of sweatshirts sold was 8 + 22, or 30.

    Money from white: $ 9.95 × 8 = $ 79.60

    Money from yellow: $10.50 × 22 = $231.00

    Total = $310.60

    The numbers check.

    State. 8 white sweatshirts and 22 yellow sweatshirts were sold.

26. 84 adult, 33 children

27. Familiarize. Let x = Carlos' age now and y = Maria's age now. Four years ago, Carlos' age was x – 4 and Maria's age was y – 4.

    Translate.

Carlos' age now	is	8	more than	Maria's age now.
x	=	8	+	y

    Four years ago

    ───────────────────────

    Maria's age was $\frac{2}{3}$ of Carlos' age.

$$y - 4 = \frac{2}{3} \cdot (x - 4)$$

    We have a system of equations:

$$x = 8 + y \quad (1)$$
$$y - 4 = \frac{2}{3}(x - 4) \quad (2)$$

    Carry out. We use the substitution method.

$$y - 4 = \frac{2}{3}(8 + y - 4) \quad \text{Substituting } 8 + y \text{ for x in (2)}$$
$$y - 4 = \frac{2}{3}(4 + y)$$
$$3(y - 4) = 2(4 + y) \quad \text{Clearing the fraction}$$
$$3y - 12 = 8 + 2y$$
$$y - 12 = 8$$
$$y = 20$$

$$x = 8 + 20 \quad \text{Substituting 20 for y in (1)}$$
$$x = 28$$

    Check. Carlos, who is 28, is 8 years older than his sister Maria, who is 20. Four years ago, Carlos was 24 and Maria 16, and 16 is $\frac{2}{3}$ of 24. The numbers check.

    State. Now Carlos is 28 years old and Maria is 20 years old.

28. Paula is 32; Bob is 20.

29. Familiarize. We first make a drawing. We let ℓ = length and w = width.

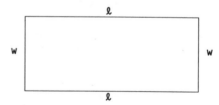

    The formula for perimeter is P = 2ℓ + 2w.

    Translate. The perimeter is 194 yd.

$$2\ell + 2w = 194$$

    The length is two more than four times the width.

ℓ	=	2	+	4	·	w

    We have a system of equations:

$$2\ell + 2w = 194 \quad (1)$$
$$\ell = 2 + 4w \quad (2)$$

<u>29.</u> (continued)

<u>Carry out.</u> We use the substitution method.

$$2(2 + 4w) + 2w = 194 \quad \text{Substituting } 2 + 4w$$
$$\text{for } \ell \text{ in (1)}$$
$$4 + 8w + 2w = 194$$
$$4 + 10w = 194$$
$$10w = 190$$
$$w = 19$$

$$\ell = 2 + 4 \cdot 19 \quad \text{Substituting 19 for w in (2)}$$
$$= 2 + 76$$
$$= 78$$

<u>Check.</u> The perimeter of a rectangle with width 19 yd and length 78 yd is $2 \cdot 78 + 2 \cdot 19 = 156 + 38 = 194$ yd. Also, 78 is 2 more than 4 times 19. The numbers check.

<u>State.</u> The width is 19 yd and the length is 78 yd.

<u>30.</u> Length = 76 m, width = 19 m

<u>31.</u> <u>Familiarize.</u> The amount of change is $20 - \$9.25$, or $10.75. Let q = the number of quarters and f = the number of fifty-cent pieces. Organize the information in a table.

Kind of coin	Quarter	Fifty-cent piece	Total	
Number	q	f	30	→ q + f = 30
Amount of change	0.25q	0.50f	$10.75	→ 0.25q + 0.50f = 10.75

<u>Translate.</u> The rows of the table give us two equations:

$$q + f = 30$$
$$0.25q + 0.50f = 10.75$$

After clearing decimals, we have

$$q + f = 30 \quad (1)$$
$$25q + 50f = 1075 \quad (2)$$

<u>Carry out.</u> Solve the system of equations.

$$-25q - 25f = -750 \quad \text{Multiplying (1) by } -25$$
$$\underline{25q + 50f = 1075}$$
$$25f = 325 \quad \text{Adding}$$
$$f = 13$$

$$q + 13 = 30 \quad \text{Substituting 13 for f in (1)}$$
$$q = 17$$

<u>Check.</u> The total number of coins is $17 + 13$, or 30. The value of 17 quarters is $4.25, and the value of 13 fifty-cent pieces is $6.50, so the total value of the coins is $4.25 + \$6.50$, or $10.75. The numbers check.

<u>State.</u> There are 17 quarters and 13 fifty-cent pieces.

<u>32.</u> $5 bills: 7, $1 bills: 15

<u>33.</u> <u>Familiarize.</u> We first make a drawing.

Slow train
d kilometers   75 km/h   (t + 2) hours

Fast train
d kilometers   125 km/h   t hours

From the drawing we see that the distances are the same. Now complete the chart.

$$d = r \cdot t$$

	Distance	Rate	Time	
Slow train	d	75	t + 2	→ d = 75(t + 2)
Fast train	d	125	t	→ d = 125t

<u>Translate.</u> Using d = rt in each row of the table, we get a system of equations:

$$d = 75(t + 2)$$
$$d = 125t$$

<u>Carry out.</u> We solve the system of equations.

$$125t = 75(t + 2) \quad \text{Using substitution}$$
$$125t = 75t + 150$$
$$50t = 150$$
$$t = 3$$

The time for the fast train should be 3 hr, and the time for the slow train $3 + 2$, or 5 hr.

Then $d = 125t = 125 \cdot 3 = 375$

<u>Check.</u> At 125 km/h, in 3 hr the fast train will travel $125 \cdot 3 = 375$ km. At 75 km/h, in 5 hr the slow train will travel $75 \cdot 5 = 375$ km. The numbers check.

<u>State.</u> The trains will meet 375 km from the station.

<u>34.</u> 3 hr

<u>35.</u> <u>Familiarize.</u> We first make a drawing.

Chicago   Indianapolis
110 km/h   t hours   t hours   90 km/h
|———————— 350 km ————————|

The sum of the distances is 350 km. The times are the same. We organize the information in a table.

$$d = r \cdot t$$

Motorcycle	Distance	Rate	Time	
From Chicago	d	110	t	→ d = 110t
From Indpls.	350 - d	90	t	→ 350 - d = 90t

We let t = the time, d = the distance from Chicago, and 350 - d = the distance from Indianapolis.

<u>Translate.</u> Using d = rt in each row of the table, we get a system of equations:

$$d = 110t \quad (1)$$
$$350 - d = 90t \quad (2)$$

35. (continued)

<u>Carry out.</u> We use the substitution method.

$$350 - 110t = 90t \quad \text{Substituting } 110t \text{ for } d$$
$$\text{in (2)}$$
$$350 = 200t$$
$$\frac{350}{200} = t$$
$$\frac{7}{4} = t$$

<u>Check.</u> The motorcycle from Chicago will travel $110 \cdot \frac{7}{4}$, or 192.5 km. The motorcycle from Indianapolis will travel $90 \cdot \frac{7}{4}$, or 157.5 km. The sum of the two distances is $192.5 + 157.5$, or 350 km. The value checks.

<u>State.</u> In $1\frac{3}{4}$ hours the motorcycles will meet.

36. 2 hr

37. <u>Familiarize.</u> We first make a drawing.

Downstream, 6 mph current

d mi, r + 6, 3 hr

Upstream, 6 mph current

d mi, r - 6, 5 hr

Let d = the distance and r = the speed of the boat in still water. Then when the boat travels downstream its speed is r + 6, and its speed upstream is r - 6. From the drawing we see that the distances are the same. Organize the information in a table.

	Distance	Rate	Time	
Downstream	d	r + 6	3	→ d = (r + 6)3
Upstream	d	r - 6	5	→ d = (r - 6)5

<u>Translate.</u> Using d = rt in each row of the table, we get a system of equations:

d = 3r + 18
d = 5r - 30

<u>Carry out.</u> Solve the system of equations.

$$3r + 18 = 5r - 30 \quad \text{Using substitution}$$
$$18 = 2r - 30$$
$$48 = 2r$$
$$24 = r$$

<u>Check.</u> When r = 24, r + 6 = 30, and the distance traveled in 3 hr is 30·3 = 90 mi. Also, r - 6 = 18, and the distance traveled in 5 hr is 18·5 = 90 mi. The value checks.

<u>State.</u> The speed of the boat in still water is 24 mph.

38. 14 km/h

39. The Familiarize and Translate steps were done in Exercise 62 of Exercise Set 3.1.

<u>Carry out.</u> We solve the system of equations

x = 2y   (1)
x + 20 = 3y   (2)

where x = Burl's age now and y = his son's age now.

$$2y + 20 = 3y \quad \text{Substituting } 2y \text{ for } x \text{ in (2)}$$
$$20 = y$$

$$x = 2 \cdot 20 \quad \text{Substituting 20 for } y \text{ in (1)}$$
$$x = 40$$

<u>Check.</u> Burl's age now, 40, is twice his son's age now, 20. Ten years ago Burl was 30 and his son was 10, and 30 = 3·10. The numbers check.

<u>State.</u> Now Burl is 40 and his son is 20.

40. Juan: 32 years; Juanita: 14 years

41. The Familiarize and Translate steps were done in Exercise 64 of Exercise Set 3.1.

<u>Carry out.</u> We solve the system of equations

$$2\ell + 2w = 156 \quad (1)$$
$$\ell = 4(w - 6) \quad (2)$$

where ℓ = length and w = width.

$$2 \cdot 4(w - 6) + 2w = 156 \quad \text{Substituting } 4(w - 6) \text{ for } \ell \text{ in (1)}$$
$$8w - 48 + 2w = 156$$
$$10w - 48 = 156$$
$$10w = 204$$
$$w = \frac{204}{10}, \text{ or } \frac{102}{5}$$

$$\ell = 4\left[\frac{102}{5} - 6\right] \quad \text{Substituting } \frac{102}{5} \text{ for } w \text{ in (2)}$$
$$\ell = 4\left[\frac{102}{5} - \frac{30}{5}\right]$$
$$\ell = 4\left[\frac{72}{5}\right]$$
$$\ell = \frac{288}{5}$$

<u>Check.</u> The perimeter of a rectangle with width $\frac{102}{5}$ in. and length $\frac{288}{5}$ in. is $2\left[\frac{288}{5}\right] + 2\left[\frac{102}{5}\right] = \frac{576}{5} + \frac{204}{5} = \frac{780}{5} = 156$ in. If 6 in. is cut off the width, the new width is $\frac{102}{5} - 6 = \frac{102}{5} - \frac{30}{5} = \frac{72}{5}$. The length, $\frac{288}{5}$, is $4\left[\frac{72}{5}\right]$. The numbers check.

<u>State.</u> The original piece of posterboard has width $\frac{102}{5}$ in. and length $\frac{288}{5}$ in.

42. 80 wins, 24 losses

43. The Familiarize and Translate steps were done in Exercise 66 of Exercise Set 3.1.

    Carry out. We solve the system of equations

    $$d + q = 58 \quad (1)$$
    $$q + 7 = 4d \quad (2)$$

    where d = the number of dimes and q = the number of quarters.

    $$d + q = 58$$
    $$\underline{4d - q = 7} \quad \text{Writing (2) as } Ax + By = C$$
    $$5d = 65 \quad \text{Adding}$$
    $$d = 13$$

    $$13 + q = 58 \quad \text{Substituting 13 for } d \text{ in (1)}$$
    $$q = 45$$

    Check. The total number of coins is 13 + 45, or 58. When John gets 7 more quarters, for a total of 52, he has 4 times as many quarters as dimes. The numbers check.

    State. John initially has 13 dimes and 45 quarters.

44. 4 km

45. Familiarize. Let x = the ten's digit and y = the unit's digit. Then the number is 10x + y. If the digits are interchanged, the new number is 10y + x.

    Translate.

    Ten's digit is 2 more than 3 times unit's digit.

    $$x = 2 + 3 \cdot y$$

    If the digits are interchanged,

    new number is half of given number minus 13.

    $$10y + x = \frac{1}{2} \cdot (10x + y) - 13$$

    The system of equations is

    $$x = 2 + 3y \quad (1)$$
    $$10y + x = \frac{1}{2}(10x + y) - 13 \quad (2)$$

    Carry out. We use the substitution method. Substitute 2 + 3y for x in (2).

    $$10y + (2 + 3y) = \frac{1}{2}[10(2 + 3y) + y] - 13$$
    $$13y + 2 = \frac{1}{2}[20 + 30y + y] - 13$$
    $$13y + 2 = \frac{1}{2}[20 + 31y] - 13$$
    $$13y + 2 = 10 + \frac{31}{2}y - 13$$
    $$13y + 2 = \frac{31}{2}y - 3$$
    $$5 = \frac{5}{2}y$$
    $$2 = y$$

    $$x = 2 + 3 \cdot 2 \quad \text{Substituting 2 for } y \text{ in (1)}$$
    $$x = 2 + 6$$
    $$x = 8$$

45. (continued)

    Check. If x = 8 and y = 2, the given number is 82 and the new number is 28. In the given number the ten's digit, 8, is two more than three times the unit's digit, 2. The new number is 13 less than one-half the given number: $28 = \frac{1}{2}(82) - 13$. The values check.

    State. The given integer is 82.

46. 180

47. Familiarize. Let n = the numerator and d = the denominator.

    Translate.

    The numerator is 12 more than the denominator.

    $$n = 12 + d$$

    The sum of the numerator and denominator is 5 more than 3 times the denominator.

    $$n + d = 5 + 3 \cdot d$$

    We have a system of equations:

    $$n = 12 + d \quad (1)$$
    $$n + d = 5 + 3d \quad (2)$$

    Carry out. We use the substitution method.

    $$(12 + d) + d = 5 + 3d \quad \text{Substituting 12 + d for n in (2)}$$
    $$12 + 2d = 5 + 3d$$
    $$12 = 5 + d$$
    $$7 = d$$

    $$n = 12 + 7 \quad \text{Substituting 7 for } d \text{ in (1)}$$
    $$n = 19$$

    Check. If the fraction is $\frac{19}{7}$, the numerator is 12 more than the denominator. Also, $3 \cdot 7 + 5 = 26$, the sum of the numerator and denominator. The numbers check.

    State. (Recall that we want the reciprocal of the fraction.) The reciprocal of the fraction is $\frac{7}{19}$.

48. $4\frac{4}{7}$ L

49. _Familiarize._ We first make a drawing. Let r_1 = speed of the first train and r_2 = the speed of the second train. If the first train leaves at 9 A.M. and the second at 10 A.M., we have:

```
                 Train 1              Train 2
                 r₁, 3 hr             r₂, 2 hr
Union   ─────────────────────→ ←──────────────── Central
Station ├                      │                ┤ Station
        ├────────── 216 km ──────────────┤
```

If the second train leaves at 9 A.M. and the first at 10:30 A.M. we have:

```
           Train 1          Train 2
           r₁, 3/2 hr       r₂, 3 hr
Union   ──────────→ ←────────────────── Central
Station ├          │                   ┤ Station
        ├────────── 216 km ──────┤
```

The total distance traveled in each case is 216 km and is equal to the sum of the distances traveled by each train.

Translate. We will use the formula $d = rt$. For each situation we have:

Total distance	is	Train 1's distance	plus	Train 2's distance.
216	=	$3r_1$	+	$2r_2$
and 216	=	$\frac{3}{2}r_1$	+	$3r_2$

Clearing the fraction, we have this system:

$$216 = 3r_1 + 2r_2 \qquad (1)$$
$$432 = 3r_1 + 6r_2 \qquad (2)$$

Carry out. Solve the system of equations.

$$-216 = -3r_1 - 2r_2 \qquad \text{Multiplying (1) by } -1$$
$$\underline{432 = 3r_1 + 6r_2}$$
$$216 = 4r_2$$
$$54 = r_2$$

$$216 = 3r_1 + 2(54) \qquad \text{Substituting 54 for } r_2 \text{ in (1)}$$
$$216 = 3r_1 + 108$$
$$108 = 3r_1$$
$$36 = r_1$$

Check. If Train 1 travels for 3 hr at 36 km/h and Train 2 travels for 2 hr at 54 km/h, the total distance traveled is $3 \cdot 36 + 2 \cdot 54 = 108 + 108 = 216$ km. If Train 1 travels for $\frac{3}{2}$ hr at 36 km/h and Train 2 travels for 3 hr at 54 km/h, the total distance traveled is $\frac{3}{2} \cdot 36 + 3 \cdot 54 = 54 + 162 = 216$ km. The numbers check.

State. The speed of the first train is 36km/h, and the speed of the second train is 54 km/h.

50. City: 261 miles;   highway: 204 miles

51. _Familiarize._ Let g = the number of girls and b = the number of boys. Then Phyllis has b brothers and g - 1 sisters, and Phil has b - 1 brothers and g sisters.

Translate.

Number of Phyllis' brothers	is	twice	number of her sisters.
b	=	2 ·	(g - 1)

Number of Phil's brothers	is same as	number of his sisters.
b - 1	=	g

We have a system of equations:

$$b = 2(g - 1) \qquad (1)$$
$$b - 1 = g \qquad (2)$$

Carry out. We use the substitution method.

$$b = 2[(b - 1) - 1] \qquad \text{Substituting } b - 1 \text{ for } g \text{ in (1)}$$
$$b = 2(b - 2)$$
$$b = 2b - 4$$
$$4 = b$$

$$4 - 1 = g \qquad \text{Substituting 4 for } b \text{ in (2)}$$
$$3 = g$$

Check. If there are 3 girls and 4 boys in the family, then Phyllis has 4 brothers and 2 sisters. She has twice as many brothers as sisters. Also, Phil has 3 brothers and 3 sisters, the same number of each. The numbers check.

State. There are 3 girls and 4 boys in the family.

52. Answers will vary.

Exercise Set 3.4

---

1. Substitute (1,-2,3) into the three equations, using alphabetical order.

$$\frac{x + y + z = 2}{1 + (-2) + 3 \mid 2}$$
$$2 \mid$$

$$\frac{x - 2y - z = 2}{1 - 2(-2) - 3 \mid 2}$$
$$1 + 4 - 3 \mid$$
$$2 \mid$$

$$\frac{3x + 2y + z = 2}{3 \cdot 1 + 2(-2) + 3 \mid 2}$$
$$3 - 4 + 3 \mid$$
$$2 \mid$$

The triple (1,-2,3) makes all three equations true, so it is a solution.

2. No

_3._ $x + y + z = 6,$ (1)

$2x - y + 3z = 9,$ (2)

$-x + 2y + 2z = 9$ (3)

Add equations (1) and (2) to eliminate y:

$x + y + z = 6$ (1)

$\underline{2x - y + 3z = 9}$ (2)

$3x \quad + 4z = 15$ (3)   Adding

Use a different pair of equations and eliminate y:

$4x - 2y + 6z = 18$ Multiplying (2) by 2

$\underline{-x + 2y + 2z = 9}$ (3)

$3x \quad + 8z = 27$ (5)

Now solve the system of equations (4) and (5).

$3x + 4z = 15$ (4)

$3x + 8z = 27$ (5)

$-3x - 4z = -15$ Multiplying (4) by -1

$\underline{3x + 8z = 27}$

$4z = 12$

$z = 3$

$3x + 4\cdot3 = 15$ Substituting 3 for z in (4)

$3x + 12 = 15$

$3x = 3$

$x = 1$

$1 + y + 3 = 6$ Substituting 1 for x and 3 for z in (1)

$y + 4 = 6$

$y = 2$

We obtain (1,2,3). This checks, so it is the solution.

_4._   (4,0,2)

_5._ $2x - y - 3z = -1,$ (1)

$2x - y + z = -9,$ (2)

$x + 2y - 4z = 17$ (3)

We start by eliminating z from two different pairs of equations.

$2x - y - 3z = -1$ (1)

$\underline{6x - 3y + 3z = -27}$ Multiplying (2) by 3

$8x - 4y \quad = -28$ (4)   Adding

$8x - 4y + 4z = -36$ Multiplying (2) by 4

$\underline{x + 2y - 4z = 17}$

$9x - 2y \quad = -19$ (5)   Adding

Now solve the system of equations (4) and (5).

$8x - 4y = -28$ (4)

$9x - 2y = -19$ (5)

$8x - 4y = -28$ (4)

$\underline{-18x + 4y = 38}$ Multiplying (5) by -2

$-10x \quad = 10$ Adding

$x = -1$

_5._   (continued)

$8(-1) - 4y = -28$ Substituting -1 for x in (4)

$-8 - 4y = -28$

$-4y = -20$

$y = 5$

$2(-1) - 5 + z = -9$

$-2 - 5 + z = -9$

$-7 + z = -9$

$z = -2$

We obtain (-1,5,-2). This checks, so it is the solution.

_6._   (2,-2,2)

_7._ $2x - 3y + z = 5,$ (1)

$x + 3y + 8z = 22,$ (2)

$3x - y + 2z = 12$ (3)

We start by eliminating y from two different pairs of equations.

$2x - 3y + z = 5$ (1)

$\underline{x + 3y + 8z = 22}$ (2)

$3x \quad + 9z = 27$ (4)   Adding

$x + 3y + 8z = 22$ (2)

$\underline{9x - 3y + 6z = 36}$ Multiplying (3) by 3

$10x \quad + 14z = 58$ (5)   Adding

Solve the system of equations (4) and (5).

$3x + 9z = 27$ (4)

$10x + 14z = 58$ (5)

$30x + 90z = 270$ Multiplying (4) by 10

$\underline{-30x - 42z = -174}$ Multiplying (5) by -3

$48z = 96$ Adding

$z = 2$

$3x + 9\cdot2 = 27$ Substituting 2 for z in (4)

$3x + 18 = 27$

$3x = 9$

$x = 3$

$2\cdot3 - 3y + 2 = 5$ Substituting 3 for x and 2 for z in (1)

$-3y + 8 = 5$

$-3y = -3$

$y = 1$

We obtain (3,1,2). This checks, so it is the solution.

_8._   (3,-2,1)

<u>9</u>. 3a - 2b + 7c = 13,   (1)
    a + 8b - 6c = -47,   (2)
    7a - 9b - 9c = -3   (3)

We start by eliminating a from two different pairs of equations.

$$3a - 2b + 7c = 13 \quad (1)$$
$$-3a - 24b + 18c = 141 \quad \text{Multiplying (2) by -3}$$
$$-26b + 25c = 154 \quad (4) \quad \text{Adding}$$

$$-7a - 56b + 42c = 329 \quad \text{Multiplying (2) by -7}$$
$$7a - 9b - 9c = -3 \quad (3)$$
$$-65b + 33c = 326 \quad (5) \quad \text{Adding}$$

Now solve the system of equations (4) and (5).

$$-26b + 25c = 154 \quad (4)$$
$$-65b + 33c = 326 \quad (5)$$

$$-130b + 125c = 770 \quad \text{Multiplying (4) by 5}$$
$$130b - 66c = -652 \quad \text{Multiplying (5) by -2}$$
$$59c = 118$$
$$c = 2$$

$$-26b + 25\cdot2 = 154 \quad \text{Substituting 2 for c in (4)}$$
$$-26b + 50 = 154$$
$$-26b = 104$$
$$b = -4$$

$$a + 8(-4) - 6(2) = -47 \quad \text{Substituting -4 for b and 2 for c in (2)}$$
$$a - 32 - 12 = -47$$
$$a - 44 = -47$$
$$a = -3$$

We obtain (-3,-4,2). This checks, so it is the solution.

<u>10</u>. (7,-3,-4)

<u>11</u>. 2x + 3y + z = 17,   (1)
    x - 3y + 2z = -8,   (2)
    5x - 2y + 3z = 5   (3)

We start by eliminating y from two different pairs of equations.

$$2x + 3y + z = 17 \quad (1)$$
$$x - 3y + 2z = -8 \quad (2)$$
$$3x + 3z = 9 \quad (4) \quad \text{Adding}$$

$$4x + 6y + 2z = 34 \quad \text{Multiplying (1) by 2}$$
$$15x - 6y + 9z = 15 \quad \text{Multiplying (3) by 3}$$
$$19x + 11z = 49 \quad (5) \quad \text{Adding}$$

Now solve the system of equations (4) and (5).

$$3x + 3z = 9 \quad (4)$$
$$19x + 11z = 49 \quad (5)$$

<u>11</u>. (continued)

$$33x + 33z = 99 \quad \text{Multiplying (4) by 11}$$
$$-57x - 33z = -147 \quad \text{Multiplying (5) by -3}$$
$$-24x = -48$$
$$x = 2$$

$$3\cdot2 + 3z = 9 \quad \text{Substituting 2 for x in (4)}$$
$$6 + 3z = 9$$
$$3z = 3$$
$$z = 1$$

$$2\cdot2 + 3y + 1 = 17 \quad \text{Substituting 2 for x and 1 for z in (1)}$$
$$3y + 5 = 17$$
$$3y = 12$$
$$y = 4$$

We obtain (2,4,1). This checks, so it is the solution.

<u>12</u>. (2,1,3)

<u>13</u>. 2x + y + z = -2,   (1)
    2x - y + 3z = 6,   (2)
    3x - 5y + 4z = 7   (3)

We start by eliminating y from two different pairs of equations.

$$2x + y + z = -2 \quad (1)$$
$$2x - y + 3z = 6 \quad (2)$$
$$4x + 4z = 4 \quad (4) \quad \text{Adding}$$

$$10x + 5y + 5z = -10 \quad \text{Multiplying (1) by 5}$$
$$3x - 5y + 4z = 7 \quad (3)$$
$$13x + 9z = -3 \quad (5) \quad \text{Adding}$$

Now solve the system of equations (4) and (5).

$$4x + 4z = 4 \quad (4)$$
$$13x + 9z = -3 \quad (5)$$

$$36x + 36z = 36 \quad \text{Multiplying (4) by 9}$$
$$-52x - 36z = 12 \quad \text{Multiplying (5) by -4}$$
$$-16x = 48 \quad \text{Adding}$$
$$x = -3$$

$$4(-3) + 4z = 4 \quad \text{Substituting -3 for x in (4)}$$
$$-12 + 4z = 4$$
$$4z = 16$$
$$z = 4$$

$$2(-3) + y + 4 = -2 \quad \text{Substituting -3 for x and 4 for z in (1)}$$
$$y - 2 = -2$$
$$y = 0$$

We obtain (-3,0,4). This checks, so it is the solution.

<u>14</u>. (2,-5,6)

15. $x - y + z = 4,$ (1)

$5x + 2y - 3z = 2,$ (2)

$3x - 7y + 4z = 8$ (3)

We start by eliminating z from two different pairs of equations.

$3x - 3y + 3z = 12$ Multiplying (1) by 3

$\underline{5x + 2y - 3z = 2}$ (2)

$8x - y \quad\quad = 14$ (4)   Adding

$-4x + 4y - 4z = -16$ Multiplying (1) by -4

$\underline{3x - 7y + 4z = 8}$ (3)

$-x - 3y \quad\quad = -8$ (5)   Adding

Now solve the system of equations (4) and (5).

$8x - y = 14$ (4)

$-x - 3y = -8$ (5)

$8x - y = 14$ (4)

$\underline{-8x - 24y = -64}$ Multiplying (5) by 8

$-25y = -50$

$y = 2$

$8x - 2 = 14$ Substituting 2 for y in (4)

$8x = 16$

$x = 2$

$2 - 2 + z = 4$ Substituting 2 for x and 2 for y in (1)

$z = 4$

We obtain (2,2,4). This checks, so it is the solution.

16. (-2,-1,4)

17. $4x - y - z = 4,$ (1)

$2x + y + z = -1,$ (2)

$6x - 3y - 2z = 3$ (3)

We start by eliminating y from two different pairs of equations.

$4x - y - z = 4$ (1)

$\underline{2x + y + z = -1}$ (2)

$6x \quad\quad = 3$ (4)   Adding

At this point we can either continue by eliminating y from a second pair of equations or we can solve (4) for x and substitute that value in a different pair of the original equations to obtain a system of two equations in two variables. We take the second option.

$6x = 3$ (4)

$x = \frac{1}{2}$

Substitute $\frac{1}{2}$ for x in (1):

$4\left(\frac{1}{2}\right) - y - z = 4$

$2 - y - z = 4$

$-y - z = 2$ (5)

17. (continued)

Substitute $\frac{1}{2}$ for x in (3):

$6\left(\frac{1}{2}\right) - 3y - 2z = 3$

$3 - 3y - 2z = 3$

$-3y - 2z = 0$ (6)

Solve the system of equations (5) and (6).

$2y + 2z = -4$ Multiplying (5) by -2

$\underline{-3y - 2z = 0}$ (6)

$-y \quad\quad = -4$

$y = 4$

$-4 - z = 2$ Substituting 4 for y in (5)

$-z = 6$

$z = -6$

We obtain $\left(\frac{1}{2},4,-6\right)$. This checks, so it is the solution.

18. (3,-5,8)

19. $2r + 3s + 12t = 4,$ (1)

$4r - 6s + 6t = 1,$ (2)

$r + s + t = 1$ (3)

We start by eliminating s from two different pairs of equations.

$4r + 6s + 24t = 8$ Multiplying (1) by 2

$\underline{4r - 6s + 6t = 1}$ (2)

$8r \quad\quad + 30t = 9$ (4)   Adding

$4r - 6s + 6t = 1$ (2)

$\underline{6r + 6s + 6t = 6}$ Multiplying (3) by 6

$10r \quad\quad + 12t = 7$ (5)   Adding

Solve the system of equations (4) and (5).

$40r + 150t = 45$ Multiplying (4) by 5

$\underline{-40r - 48t = -28}$ Multiplying (5) by -4

$102t = 17$

$t = \frac{17}{102}$

$t = \frac{1}{6}$

$8r + 30\left(\frac{1}{6}\right) = 9$ Substituting $\frac{1}{6}$ for t in (4)

$8r + 5 = 9$

$8r = 4$

$r = \frac{1}{2}$

$\frac{1}{2} + s + \frac{1}{6} = 1$ Substituting $\frac{1}{2}$ for r and $\frac{1}{6}$ for t in (3)

$s + \frac{2}{3} = 1$

$s = \frac{1}{3}$

We obtain $\left(\frac{1}{2},\frac{1}{3},\frac{1}{6}\right)$. This checks, so it is the solution.

20. $\left(\frac{3}{5},\frac{2}{3},-3\right)$

21. $4a + 9b \quad = 8,$ (1)

    $8a \quad + 6c = -1,$ (2)

    $\quad 6b + 6c = -1$ (3)

    We will use the elimination method. Note that there is no c in equation (1). We will use equations (2) and (3) to obtain another equation with no c terms.

    $\quad 8a \quad + 6c = -1$ (2)

    $\quad \underline{\; - 6b - 6c = \;\; 1} \quad$ Multiplying (3) by –1

    $\quad 8a - 6b \quad\;\; = 0$ (4)   Adding

    Now solve the system of equations (1) and (4).

    $\quad -8a - 18b = -16 \quad$ Multiplying (1) by –2

    $\quad \underline{\;\; 8a - \;\; 6b = \;\;\; 0}$

    $\quad\qquad -24b = -16$

    $\qquad\qquad b = \;\; \frac{2}{3}$

    $8a - 6\left(\frac{2}{3}\right) = 0 \quad$ Substituting $\frac{2}{3}$ for b in (4)

    $\qquad 8a - 4 = 0$

    $\qquad\quad\; 8a = 4$

    $\qquad\qquad a = \frac{1}{2}$

    $8\left(\frac{1}{2}\right) + 6c = \;\; -1 \quad$ Substituting $\frac{1}{2}$ for a in (2)

    $\qquad 4 + 6c = \;\; -1$

    $\qquad\quad\; 6c = \;\; -5$

    $\qquad\qquad c = -\frac{5}{6}$

    We obtain $\left(\frac{1}{2},\frac{2}{3},-\frac{5}{6}\right)$. This checks, so it is the solution.

22. $\left(4,\frac{1}{2},-\frac{1}{2}\right)$

23. $\quad x + y + z = 57,$ (1)

    $-2x + y \quad\;\; = 3,$ (2)

    $\quad x \quad - z = 6$ (3)

    We will use the substitution method. Solve equations (2) and (3) for y and z, respectively. Then substitute in equation (1) to solve for x.

    $\quad -2x + y = 3 \qquad$ Solving (2) for y

    $\qquad\qquad y = 2x + 3$

    $\quad x - z = 6 \qquad\quad$ Solving (3) for z

    $\qquad -z = -x + 6$

    $\qquad\; z = x - 6$

    $x + (2x + 3) + (x - 6) = 57 \quad$ Substituting in (1)

    $\qquad\qquad 4x - 3 = 57$

    $\qquad\qquad\quad\; 4x = 60$

    $\qquad\qquad\qquad x = 15$

    To find y, substitute 15 for x in y = 2x + 3:

    $\quad y = 2 \cdot 15 + 3 = 33$

23. (continued)

    To find z, substitute 15 for x in z = x – 6:

    $\quad z = 15 - 6 = 9$

    We obtain (15,33,9). This checks, so it is the solution.

24. (17,9,79)

25. $2a - 3b \quad\;\; = 2,$ (1)

    $7a \quad\;\; + 4c = \frac{3}{4},$ (2)

    $\quad -3b + 2c = 1$ (3)

    We will use the elimination method.

    $\quad 2a - 3b \quad\;\; = 2$ (1)

    $\quad \underline{\qquad 3b - 2c = -1} \quad$ Multiplying (3) by –1

    $\quad 2a \quad\;\; - 2c = \;\; 1$ (4)   Adding

    $\qquad 7a + 4c = \frac{3}{4}$ (2)

    $\qquad \underline{4a - 4c = 2} \quad$ Multiplying (4) by 2

    $\qquad 11a \quad\;\; = \frac{11}{4}$

    $\qquad\qquad a = \;\; \frac{1}{4}$

    $7\left(\frac{1}{4}\right) + 4c = \;\; \frac{3}{4} \quad$ Substituting $\frac{1}{4}$ for a in (2)

    $\qquad \frac{7}{4} + 4c = \;\; \frac{3}{4}$

    $\qquad\quad\; 4c = \;\; -1$

    $\qquad\qquad c = -\frac{1}{4}$

    $2\left(\frac{1}{4}\right) - 3b = \;\; 2 \quad$ Substituting $\frac{1}{4}$ for a in (1)

    $\qquad \frac{1}{2} - 3b = \;\; 2$

    $\qquad\quad\; -3b = \;\; \frac{3}{2}$

    $\qquad\qquad b = -\frac{1}{2}$

    We obtain $\left(\frac{1}{4},-\frac{1}{2},-\frac{1}{4}\right)$. This checks, so it is the solution.

26. (3,4,–1)

27. $x + y + z = 180,$ (1)

    $y = 2 + 3x,$ (2)

    $z = 80 + x$ (3)

    Substitute 2 + 3x for y and 80 + x for z in (1):

    $\quad x + (2 + 3x) + 80 + x = 180$

    $\qquad\qquad\quad 5x + 82 = 180$

    $\qquad\qquad\qquad\quad 5x = \;\; 98$

    $\qquad\qquad\qquad\qquad x = \;\; \frac{98}{5}$

27. (continued)

$y = 2 + 3\left[\frac{98}{5}\right]$ Substituting $\frac{98}{5}$ for x in (2)

$y = \frac{10}{5} + \frac{294}{5}$

$y = \frac{304}{5}$

$z = 80 + \frac{98}{5}$ Substituting $\frac{98}{5}$ for x in (3)

$z = \frac{498}{5}$

We obtain $\left(\frac{98}{5}, \frac{304}{5}, \frac{498}{5}\right)$. This checks, so it is the solution.

28. $(2, 5, -3)$

29.
$$x \quad\;\; + z = 0, \quad (1)$$
$$x + y + 2z = 3, \quad (2)$$
$$\quad\;\; y + z = 2 \quad (3)$$

The variable y is missing in equation (1). We use equations (2) and (3) to obtain another equation with no y term.

$$x + y + 2z = 3 \quad (2)$$
$$\underline{\;-y - z = -2\;}\quad \text{Multiplying (3) by -1}$$
$$x \quad\;\; + z = 1 \quad (4) \quad \text{Adding}$$

We then have the following system of equations:

$$x + z = 0 \quad (1)$$
$$x + z = 1 \quad (4)$$

Multiply (4) by -1 and add:

$$x + z = 0 \quad (1)$$
$$\underline{\;-x - z = -1\;}\quad \text{Multiplying (4) by -1}$$
$$\qquad\;\; 0 = -1 \quad \text{Adding}$$

We get a false equation. There is no solution. The solution set is $\emptyset$.

30. $\emptyset$

31.
$$x + y + z = 1, \quad (1)$$
$$-x + 2y + z = 2, \quad (2)$$
$$2x - y \quad\;\; = -1 \quad (3)$$

Use equations (1) and (2) to eliminate z:

$$x + y + z = 1 \quad (1)$$
$$\underline{\;x - 2y - z = -2\;}\quad \text{Multiplying 2 by -1}$$
$$2x - y \quad\;\; = -1 \quad (4) \quad \text{Adding}$$

Equations (3) and (4) are identical, so the system is dependent.

32. Dependent

33.
$$F = \frac{1}{2}t(c - d)$$
$$2F = t(c - d)$$
$$2F = tc - td$$
$$2F + td = tc$$
$$\frac{2F + td}{t} = c, \text{ or}$$
$$\frac{2F}{t} + d = c$$

34. $d = \dfrac{2F - tc}{-t}$, or $d = -\dfrac{2F}{t} + c$

35.
$$\frac{x + 2}{3} - \frac{y + 4}{2} + \frac{z + 1}{6} = 0,$$
$$\frac{x - 4}{3} + \frac{y + 1}{4} - \frac{z - 2}{2} = -1,$$
$$\frac{x + 1}{2} + \frac{y}{2} + \frac{z - 1}{4} = \frac{3}{4}$$

To clear fractions, we multiply each equation by the LCM of its denominators. The LCM's are 6, 12, and 4, respectively.

$$2(x + 2) - 3(y + 4) + (z + 1) = 0$$
$$2x + 4 - 3y - 12 + z + 1 = 0$$
$$2x - 3y + z = 7$$

$$4(x - 4) + 3(y + 1) - 6(z - 2) = -12$$
$$4x - 16 + 3y + 3 - 6z + 12 = -12$$
$$4x + 3y - 6z = -11$$

$$2(x + 1) + 2(y) + (z - 1) = 3$$
$$2x + 2 + 2y + z - 1 = 3$$
$$2x + 2y + z = 2$$

The resulting system is

$$2x - 3y + z = 7, \quad (1)$$
$$4x + 3y - 6z = -11, \quad (2)$$
$$2x + 2y + z = 2 \quad (3)$$

We start by eliminating z from two different pairs of equations.

$$12x - 18y + 6z = 42 \quad \text{Multiplying (1) by 6}$$
$$\underline{\;4x + 3y - 6z = -11\;}\quad (2)$$
$$16x - 15y \quad\;\; = 31 \quad (4) \quad \text{Adding}$$

$$2x - 3y + z = 7 \quad (1)$$
$$\underline{\;-2x - 2y - z = -2\;}\quad \text{Multiplying (3) by -1}$$
$$\qquad -5y \quad = 5 \quad (5) \quad \text{Adding}$$

Solve (5) for y: $-5y = 5$
$$y = -1$$

Substitute -1 for y in (4):
$$16x - 15(-1) = 31$$
$$16x + 15 = 31$$
$$16x = 16$$
$$x = 1$$

Substitute 1 for x and -1 for y in (1):
$$2 \cdot 1 - 3(-1) + z = 7$$
$$5 + z = 7$$
$$z = 2$$

35. (continued)

    We obtain $(1,-1,2)$. This checks, so it is the solution.

36. $(1,1,1)$

37. $w + x + y + z = 2,$ (1)

    $w + 2x + 2y + 4z = 1,$ (2)

    $w - x + y + z = 6,$ (3)

    $w - 3x - y + z = 2,$ (4)

    Start by eliminating w from three different pairs of equations.

    $$\begin{array}{ll} w + x + y + z = 2 & (1) \\ \underline{-w - 2x - 2y - 4z = -1} & \text{Multiplying (2) by } -1 \\ -x - y - 3z = 1 & (5)\quad\text{Adding} \end{array}$$

    $$\begin{array}{ll} w + x + y + z = 2 & (1) \\ \underline{-w + x - y - z = -6} & \text{Multiplying (3) by } -1 \\ 2x \qquad\quad = -4 & (6)\quad\text{Adding} \end{array}$$

    $$\begin{array}{ll} w + x + y + z = 2 & (1) \\ \underline{-w + 3x + y - z = -2} & \text{Multiplying (4) by } -1 \\ 4x + 2y \quad = 0 & (7)\quad\text{Adding} \end{array}$$

    We can solve (6) for x:

    $2x = -4$

    $x = -2$

    Substitute -2 for x in (7):

    $4(-2) + 2y = 0$

    $-8 + 2y = 0$

    $2y = 8$

    $y = 4$

    Substitute -2 for x and 4 for y in (5):

    $-(-2) - 4 - 3z = 1$

    $-2 - 3z = 1$

    $-3z = 3$

    $z = -1$

    Substitute -2 for x, 4 for y, and -1 for z in (1):

    $w - 2 + 4 - 1 = 2$

    $w + 1 = 2$

    $w = 1$

    We obtain $(1,-2,4,-1)$. This checks, so it is the solution.

38. $(-3,-1,0,4)$

39. $\dfrac{2}{x} - \dfrac{1}{y} - \dfrac{3}{z} = -1$

    $\dfrac{2}{x} - \dfrac{1}{y} + \dfrac{1}{z} = -9$

    $\dfrac{1}{x} + \dfrac{2}{y} - \dfrac{4}{z} = 17$

    Let u represent $\dfrac{1}{x}$, v represent $\dfrac{1}{y}$, and w represent $\dfrac{1}{z}$. Substituting, we have

    $2u - v - 3w = -1,$ (1)

    $2u - v + w = -9,$ (2)

    $u + 2v - 4w = 17.$ (3)

    We start by eliminating v from two different pairs of equations.

    $$\begin{array}{ll} 2u - v - 3w = -1 & (1) \\ \underline{-2u + v - w = 9} & \text{Multiplying (2) by } -1 \\ -4w = 8 & (4)\quad\text{Adding} \end{array}$$

    $$\begin{array}{ll} 4u - 2v - 6w = -2 & \text{Multiplying (1) by 2} \\ \underline{u + 2v - 4w = 17} & (3) \\ 5u \qquad - 10w = 15 & (5)\ \text{Adding} \end{array}$$

    We can solve (4) for w:

    $-4w = 8$

    $w = -2$

    Substitute -2 for w in (5):

    $5u - 10(-2) = 15$

    $5u + 20 = 15$

    $5u = -5$

    $u = -1$

    Substitute -1 for u and -2 for w in (1):

    $2(-1) - v - 3(-2) = -1$

    $-v + 4 = -1$

    $-v = -5$

    $v = 5$

    Solve for x, y, and z. We substitute -1 for u, 5 for v and -2 for w.

    $$u = \frac{1}{x} \qquad v = \frac{1}{y} \qquad w = \frac{1}{z}$$

    $$-1 = \frac{1}{x} \qquad 5 = \frac{1}{y} \qquad -2 = \frac{1}{z}$$

    $$x = -1 \qquad y = \frac{1}{5} \qquad z = -\frac{1}{2}$$

    We obtain $\left(-1, \frac{1}{5}, -\frac{1}{2}\right)$. This checks, so it is the solution.

40. $\left(-\frac{1}{2}, -1, -\frac{1}{3}\right)$

41. ax + by + cz = -11
     bx - cy + az = -19
     ax + cy - bz =   9

Since (2,3,-4) is a solution of the system, we substitute 2 for x, 3 for y, and -4 for z in each equation. The resulting system is

     2a + 3b - 4c = -11   (1)
     2b - 3c - 4a = -19   (2)
     2a + 3c + 4b =   9   (3)

We rewrite equations (2) and (3) with the variables in alphabetical order and then solve the system for a, b, and c.

     2a + 3b - 4c = -11   (1)
     -4a + 2b - 3c = -19   (2)
     2a + 4b + 3c =   9   (3)

Eliminate a from two different pairs of equations.

     4a + 6b - 8c = -22   Multiplying (1) by 2
     -4a + 2b - 3c = -19   (2)
     ─────────────────────
        8b - 11c = -41   (4)   Adding

     -4a + 2b - 3c = -19   (2)
     4a + 8b + 6c = 18   Multiplying (3) by 2
     ─────────────────────
       10b + 3c = -1   (5)   Adding

Now solve the system of equations (4) and (5).

      8b - 11c = -41   (4)
     10b + 3c = -1   (5)

      40b - 55c = -205   Multiplying (4) by 5
     -40b - 12c =   4   Multiplying (5) by -4
     ─────────────────────
          -67c = -201
             c =   3

     8b - 11·3 = -41   Substituting 3 for c in (4)
     8b - 33 = -41
          8b = -8
           b = -1

     2a + 3(-1) - 4(3) = -11   Substituting -1 for b and 3 for c in (1)
           2a - 15 = -11
              2a =   4
               a =   2

We have a = 2, b = -1, c = 3.

42. 3x + 4y + 2z = 12

43. z = b - mx - ny

Three solutions are (1,1,2), (3,2,-6), and $\left(\frac{3}{2},1,1\right)$. We substitute for x, y, and z and then solve for b, m, and n. (Note that we also clear the fraction in equation (3)).

     2 = b - m - n   or   b - m - n = 2   (1)
     -6 = b - 3m - 2n   b - 3m - 2n = -6 (2)
     1 = b - $\frac{3}{2}$m - n   2b - 3m - 2n = 2   (3)

43. (continued)

Eliminate b from two different pairs of equations.

      b - m - n = 2   (1)
     -b + 3m + 2n = 6   Multiplying (2) by -1
     ─────────────────────
         2m + n = 8   (4)   Adding

     -2b + 2m + 2n = -4   Multiplying (1) by -2
      2b - 3m - 2n = 2   (3)
     ─────────────────────
          -m   = -2   (5)   Adding

We can solve (5) for m:

     -m = -2
      m = 2

     2·2 + n = 8   Substituting 2 for m in (4)
       4 + n = 8
           n = 4

     b - 2 - 4 = 2   Substituting 2 for m and 4 for n in (1)
        b - 6 = 2
            b = 8

We have b = 8, m = 2, and n = 4. The equation is z = 8 - 2x - 4y.

Exercise Set 3.5

1. <u>Familiarize</u>. Let x = the first number, y = the second number, and z = the third number.

    <u>Translate</u>.

        The sum of three numbers is 105.
              x + y + z   = 105

        The third   is   ten times the   less   eleven.
        number   second number
          z   =   10y   -   11

        Twice the   is   seven more than   three times the
        first number   second number.
          2x   =   7 +   3y

    We now have a system of equations.

        x + y + z = 105   or   x +   y + z = 105
        z = 10y - 11   -10y + z = -11
        2x = 7 + 3y   2x - 3y   =   7

    <u>Carry out</u>. Solving the system we get (17,9,79).

    <u>Check</u>. The sum of the three numbers is 105. Ten times the second number is 90, and 11 less than 90 is 79, the third number. Three times the second number is 27, and 7 more than 27 is 34, which is twice the first number. The numbers check.

    <u>State</u>. The numbers are 17, 9, and 79.

2. 16, 19, 22

3. <u>Familiarize.</u> Let x = the first number, y = the second number, and z = the third number.

<u>Translate.</u>

The sum of the three numbers is 5.

$$x + y + z = 5$$

The first number minus the second plus the third is 1.

$$x - y + z = 1$$

The first number minus the third is 3 more than the second.

$$x - z = y + 3$$

We now have a system of equations.

x + y + z = 5 or x + y + z = 5
x - y + z = 1   x - y + z = 1
x - z = y + 3   x - y - z = 3

<u>Carry out.</u> Solving the system we get (4,2,-1).

<u>Check.</u> The sum of the numbers is 5. The first minus the second plus the third is 4 - 2 + (-1), or 1. The first minus the third is 5, which is three more than the second. The numbers check.

<u>State.</u> The numbers are 4, 2, and -1.

4. 8, 21, -3

5. <u>Familiarize.</u> We first make a drawing.

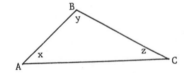

We let x, y, and z represent the measures of angles A, B, and C, respectively. The measures of the angles of a triangle add up to 180°.

<u>Translate.</u>

The sum of the measures is 180°.

$$x + y + z = 180$$

The measure of angle B is 2° more than three times the measure of angle A.

$$y = 3x + 2$$

The measure of angle C is 8° more than the measure of angle A.

$$z = x + 8$$

We now have a system of equations.

x + y + z = 180
y = 3x + 2
z = x + 8

<u>Carry out.</u> Solving the system we get (34,104,42).

<u>Check.</u> The sum of the numbers is 180, so that checks. Three times the measure of angle A is 3·34, or 102°, and 2° added to 102° is 104°, the measure of angle B. The measure of angle C, 42°, is 8° more than 34°, the measure of angle A. These values check.

<u>State.</u> Angles A, B, and C measure 34°, 104°, and 42°, respectively.

6. A = 30°, B = 90°, C = 60°

7. <u>Familiarize.</u> We first make a drawing.

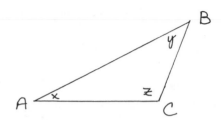

We let x, y, and z represent the measures of angles A, B, and C, respectively. The measures of the angles of a triangle add up to 180°.

<u>Translate.</u>

The sum of the measures is 180°.

$$x + y + z = 180$$

The measure of angle B is twice the measure of angle A.

$$y = 2 \cdot x$$

The measure of angle C is 80° more than the measure of angle A.

$$z = x + 80$$

We now have a system of equations.

x + y + z = 180
y = 2x
z = x + 80

<u>Carry out.</u> Solving the system we get (25,50,105).

<u>Check.</u> The sum of the numbers is 180, so that checks. The measure of angle B, 50°, is twice 25°, the measure of angle A. The measure of angle C, 105°, is 80° more than 25°, the measure of angle A. The values check.

<u>State.</u> Angles A, B, and C measure 25°, 50°, and 105°, respectively.

8. A = 32°, B = 96°, C = 52°

9. <u>Familiarize.</u> We let x, y, and z represent the amounts Gina took in on Thursday, Friday, and Saturday, respectively.

<u>Translate.</u>

The sum of the amounts for all three days is $66.

$$x + y + z = 66$$

Thursday's amount was $3 more than Friday's amount.

$$x = 3 + y$$

Saturday's amount was $6 more than Thursday's amount.

$$z = 6 + x$$

We now have a system of equations.

x + y + z = 66
x = 3 + y
z = 6 + x

9. (continued)

   Carry out. Solving the system we get (21,18,27).

   Check. The sum of the three numbers is 66. Thursday's amount, $21, is $3 more than $18, Friday's amount. Saturday's amount, $27, is $6 more than $21, Thursday's amount. The values check.

   State. Gina sold $21 on Thursday, $18 on Friday, and $27 on Saturday.

10. Fred: 195, Jane: 180, Mary: 200

11. Familiarize. Let s = the number of servings of steak, p = the number of baked potatoes, and b = the number of servings of broccoli. Then s servings of steak contain 300s calories, 20s g of protein, and no vitamin C. In p baked potatoes there are 100p calories, 5p g of protein, and 20p mg of vitamin C. And b servings of broccoli contain 50b calories, 5b g of protein, and 100b mg of vitamin C. The patient requires 800 calories, 55 g of protein, and 220 mg of vitamin C.

   Translate. Write equations for the total number of calories, the total amount of protein, and the total amount of vitamin C.

$$300s + 100p + 50b = 800 \quad \text{(calories)}$$
$$20s + 5p + 5b = 55 \quad \text{(protein)}$$
$$20p + 100b = 220 \quad \text{(vitamin C)}$$

   We now have a system of equations.

   Carry out. Solving the system we get (2,1,2).

   Check. Two servings of steak provide 600 calories, 40 g of protein, and no vitamin C. One baked potato provides 100 calories, 5 g of protein, and 20 mg of vitamin C. And 2 servings of broccoli provide 100 calories, 10 g of protein, and 200 mg of vitamin C. Together, then, they provide 800 calories, 55 g of protein, and 220 mg of vitamin C. The values check.

   State. The dietician should prepare 2 servings of steak, 1 baked potato, and 2 servings of broccoli.

12. Steak: $1\frac{1}{8}$, baked potato: $2\frac{3}{4}$, asparagus: $3\frac{3}{4}$

13. Familiarize. It helps to orgainze the information in a table.

Machines Working	A	B	C	A + B	B + C	A, B, & C
Weekly Production	x	y	z	3400	4200	5700

   We let x, y,and z represent the weekly productions of the individual machines.

   Translate. From the table, we obtain three equations.

$$x + y + z = 5700 \quad \text{(All three machines working)}$$
$$x + y = 3400 \quad \text{(A and B working)}$$
$$y + z = 4200 \quad \text{(B and C working)}$$

   Carry out. Solving the system we get (1500,1900,2300).

13. (continued)

   Check. The sum of the weekly productions of machines A, B & C is 1500 + 1900 + 2300, or 5700. The sum of the weekly productions of machines A and B is 1500 + 1900, or 3400. The sum of the weekly productions of machines B and C is 1900 + 2300, or 4200. The numbers check.

   State. In a week Machine A can polish 1500 lenses, Machine B can polish 1900 lenses, and Machine C can polish 2300 lenses.

14. A: 2200, B: 2500, C: 2700

15. Familiarize. It helps to organize the information in a table.

Pumps Working	A	B	C	A + B	A + C	A, B, & C
Gallons Per hour	x	y	z	2200	2400	3700

   We let x, y, and z represent the number of gallons per hour which can be pumped by pumps A, B, and C, respectively.

   Translate. From the table, we obtain three equations.

$$x + y + z = 3700 \quad \text{(All three pumps working)}$$
$$x + y = 2200 \quad \text{(A and B working)}$$
$$x + z = 2400 \quad \text{(A and C working)}$$

   Carry out. Solving the system we get (900,1300,1500).

   Check. The sum of the gallons per hour pumped when all three are pumping is 900 + 1300 + 1500, or 3700. The sum of the gallons per hour pumped when only pump A and pump B are pumping is 900 + 1300, or 2200. The sum of the gallons per hour pumped when only pump A and pump C are pumping is 900 + 1500, or 2400. The numbers check.

   State. The pumping capacities of pumps A, B, and C are respectively 900, 1300, and 1500 gallons per hour.

16. A: 10 linear ft, B: 12 linear ft, C: 15 linear ft

17.
$$3(5 - x) + 7 = 5(x + 3) - 9$$
$$15 - 3x + 7 = 5x + 15 - 9$$
$$-3x + 22 = 5x + 6$$
$$22 = 8x + 6$$
$$16 = 8x$$
$$2 = x$$

18. 8

19. $\dfrac{(a^2b^3)^5}{a^7b^{16}} = \dfrac{a^{10}b^{15}}{a^7b^{16}} = a^{10-7}b^{15-16} = a^3b^{-1}$, or $\dfrac{a^3}{b}$

20. $y = -\dfrac{3}{5}x - 7$

21. <u>Familiarize</u>. We let w, x, y, and z represent the ages of Tammy, Carmen, Dennis, and Mark, respectively.

<u>Translate</u>.

| Tammy's age | is | Carmen's age | plus | Dennis's age |
| w | = | x | + | y |

| Carmen's age | is | 2 | plus | Dennis's age | plus | Mark's age. |
| x | = | 2 | + | y | + | z |

| Dennis's age | is | four | times | Mark's age. |
| y | = | 4 | · | z |

The sum of all ages is 42.

$$w + x + y + z = 42$$

We now have a system of equations.

$$w = x + y$$
$$x = 2 + y + z$$
$$y = 4z$$
$$w + x + y + z = 42$$

<u>Carry out</u>. Solving the system we get (20,12,8,2).

<u>Check</u>. The sum of all four numbers is 42. Tammy's age, 20, is the sum of the ages of Carmen and Dennis, 12 + 8. Carmen's age, 12, is 2 more than the sum of the ages of Dennis and Mark, 8 + 2. Dennis's age, 8, is four times Mark's age, 2. The numbers check.

<u>State</u>. Tammy is 20 years old.

22. 464

23. <u>Familiarize</u>. We first make a drawing with additional labels.

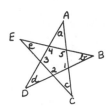

We let a, b, c, d, and e represent the angle measures at the tips of the star. We also label the interior angles of the pentagon 1, 2, 3, 4, and 5. We must recall the following geometric fact:

The sum of the measures of the interior angles of a polygon of n sides is given by (n - 2)180°.

Using this fact we know:

1. The sum of the angle measures of a triangle is (3 - 2)180°, or 180°.

2. The sum of the angle measures of a pentagon is (5 - 2)180°, or 3(180°).

23. (continued)

<u>Translate</u>. Using fact (1) listed above we obtain a system of 5 equations.

$$a + 1 + d = 180$$
$$b + 2 + e = 180$$
$$c + 3 + a = 180$$
$$d + 4 + b = 180$$
$$e + 5 + c = 180$$

<u>Carry out</u>. Adding we obtain

$$2a + 2b + 2c + 2d + 2e + 1 + 2 + 3 + 4 + 5 = 5(180)$$
$$2(a + b + c + d + e) + (1 + 2 + 3 + 4 + 5) = 5(180)$$

Using fact (2) listed above we substitute 3(180) for (1 + 2 + 3 + 4 + 5) and solve for (a + b + c + d + e).

$$2(a + b + c + d + e) + 3(180) = 5(180)$$
$$2(a + b + c + d + e) = 2(180)$$
$$a + b + c + d + e = 180$$

<u>Check</u>. We should repeat the above calculations.

<u>State</u>. The sum of the angle measures at the tips of the star is 180°.

24. Men: 5, women: 1, children: 94

25. We know the thousand's digit must be 1. The United States did not exist as a union until the 1700's, and we have not reached the year 2000.

1 x y z

We let x represent the hundred's digit, y the ten's digit, and z the one's digit. Since z is a multiple of 3, we know that z must be 3, 6, or 9. We have the following possibilities:

1 x y 3   1 x y 6   1 x y 9

We also know that z is one more than x.

When z = 3, x = 2.
When z = 6, x = 5.
When z = 9, x = 8.

Now the possibilities are:

1 2 y 3   1 5 y 6   1 8 y 9

Since the sum of the digits must be 24 and y is a multiple of 3 (3, 6, or 9), only 1 8 y 9 is a possibility.

$$1 + 8 + y + 9 = 24$$
$$y = 6$$

Thus, the year is 1869.

26. 35

27. <u>Familiarize</u>. We let x represent the number of adults' tickets sold, y represent the number of senior citizens' tickets sold, and z represent the number of children's tickets sold. We organize the information in a table.

Tickets	Number sold	Ticket price	Receipts
Adults	x	5.50	5.50x
Senior citizens	y	4.00	4.00y
Children	z	1.50	1.50z
		Total	$14,970

<u>Translate</u>.

Children's tickets + Senior citizens' tickets = 30 more than half the adults' tickets

$z + y = 30 + \frac{1}{2}x$

Senior citizens' tickets = 5 + four times the children's tickets

$y = 5 + 4z$

Total receipts = $14,970

$(5.50x + 4.00y + 1.50z) = 14,970$

We obtain a system of equations.

$z + y = 30 + \frac{1}{2}x$

$y = 5 + 4z$

$5.50x + 4.00y + 1.50z = 14,970$

<u>Carry out</u>. Solving the system we get $(2050,845,210)$.

<u>Check</u>. The number of children's and senior citizens' tickets sold was 210 + 845, or 1055. Half the number of adults' tickets sold was $\frac{1}{2}(2050)$, or 1025, and 30 more than 1025 is 1055. Four times the number of children's tickets sold was 4·210, or 840, and 5 more than 840 is 845. The total receipts were 5.50(2050) + 4.00(845) + 1.50(210), or $14,970. The numbers check.

<u>State</u>. 2050 adults' tickets, 845 senior citizens' tickets, and 210 children's tickets were sold.

28. a = 1, b = -5, c = 6

29. <u>Familiarize</u>. Since (-1,3), (2,5), and (1,-2) are all solutions of $f(x) = ax^2 + bx + c$, we can substitute to obtain a system of three equations.

<u>Translate</u>.

$3 = a - b + c$ Substituting (-1,3)

$5 = 4a + 2b + c$ Substituting (2,5)

$-2 = a + b + c$ Substituting (1,-2)

<u>Carry out</u>. Solving the system we get $\left(\frac{19}{6}, -\frac{5}{2}, -\frac{8}{3}\right)$.

29. (continued)

<u>Check</u>. Check to see if each pair makes the equation $f(x) = \frac{19}{6}x^2 - \frac{5}{2}x - \frac{8}{3}$ true. This is left to the student.

<u>State</u>. $a = \frac{19}{6}$, $b = -\frac{5}{2}$, $c = -\frac{8}{3}$, and the function is $f(x) = \frac{19}{6}x^2 - \frac{5}{2}x - \frac{8}{3}$.

30. First: $4.50, second: $10.00, third: $18.00

Exercise Set 3.6

1. $\begin{vmatrix} 2 & 7 \\ 1 & 5 \end{vmatrix} = 2 \cdot 5 - 1 \cdot 7 = 10 - 7 = 3$

2. -13

3. $\begin{vmatrix} 6 & -9 \\ 2 & 3 \end{vmatrix} = 6 \cdot 3 - 2 \cdot (-9) = 18 + 18 = 36$

4. 29

5. $\begin{vmatrix} 0 & 2 & 0 \\ 3 & -1 & 1 \\ 1 & -2 & 2 \end{vmatrix}$

$= 0 \begin{vmatrix} -1 & 1 \\ -2 & 2 \end{vmatrix} - 3 \begin{vmatrix} 2 & 0 \\ -2 & 2 \end{vmatrix} + 1 \begin{vmatrix} 2 & 0 \\ -1 & 1 \end{vmatrix}$

$= 0 - 3[2 \cdot 2 - (-2) \cdot 0] + 1[2 \cdot 1 - (-1) \cdot 0]$

$= 0 - 3 \cdot 4 + 1 \cdot 2$

$= 0 - 12 + 2$

$= -10$

6. 1

7. $\begin{vmatrix} -1 & -2 & -3 \\ 3 & 4 & 2 \\ 0 & 1 & 2 \end{vmatrix}$

$= -1 \begin{vmatrix} 4 & 2 \\ 1 & 2 \end{vmatrix} - 3 \begin{vmatrix} -2 & -3 \\ 1 & 2 \end{vmatrix} + 0 \begin{vmatrix} -2 & -3 \\ 4 & 2 \end{vmatrix}$

$= -1[4 \cdot 2 - 1 \cdot 2] - 3[-2 \cdot 2 - 1(-3)] + 0$

$= -1 \cdot 6 - 3 \cdot (-1) + 0$

$= -6 + 3 + 0$

$= -3$

8. 3

9. $\begin{vmatrix} 3 & 2 & -2 \\ -2 & 1 & 4 \\ -4 & -3 & 3 \end{vmatrix}$

$= 3 \begin{vmatrix} 1 & 4 \\ -3 & 3 \end{vmatrix} - (-2) \begin{vmatrix} 2 & -2 \\ -3 & 3 \end{vmatrix} + (-4) \begin{vmatrix} 2 & -2 \\ 1 & 4 \end{vmatrix}$

$= 3[1\cdot3 - (-3)\cdot4] + 2[2\cdot3 - (-3)(-2)] -$
$\hspace{6cm} 4[2\cdot4 - 1(-2)]$

$= 3\cdot15 + 2\cdot0 - 4\cdot10$

$= 45 + 0 - 40$

$= 5$

10. -6

11. $3x - 4y = 6$
    $5x + 9y = 10$

We compute D, D_x, and D_y.

$D = \begin{vmatrix} 3 & -4 \\ 5 & 9 \end{vmatrix} = 27 - (-20) = 47$

$D_x = \begin{vmatrix} 6 & -4 \\ 10 & 9 \end{vmatrix} = 54 - (-40) = 94$

$D_y = \begin{vmatrix} 3 & 6 \\ 5 & 10 \end{vmatrix} = 30 - 30 = 0$

Then,

$$x = \frac{D_x}{D} = \frac{94}{47} = 2$$

and

$$y = \frac{D_y}{D} = \frac{0}{47} = 0$$

The solution is $(2,0)$.

12. $(-3,2)$

13. $-2x + 4y = 3$
    $3x - 7y = 1$

We compute D, D_x, and D_y.

$D = \begin{vmatrix} -2 & 4 \\ 3 & -7 \end{vmatrix} = 14 - 12 = 2$

$D_x = \begin{vmatrix} 3 & 4 \\ 1 & -7 \end{vmatrix} = -21 - 4 = -25$

$D_y = \begin{vmatrix} -2 & 3 \\ 3 & 1 \end{vmatrix} = -2 - 9 = -11$

Then,

$$x = \frac{D_x}{D} = \frac{-25}{2} = -\frac{25}{2}$$

and

$$y = \frac{D_y}{D} = \frac{-11}{2} = -\frac{11}{2}$$

The solution is $\left(-\frac{25}{2}, -\frac{11}{2}\right)$.

14. $\left(\frac{9}{19}, \frac{51}{38}\right)$

15. $3x + 2y - z = 4$
    $3x - 2y + z = 5$
    $4x - 5y - z = -1$

We compute D, D_x, D_y, and D_z.

$D = \begin{vmatrix} 3 & 2 & -1 \\ 3 & -2 & 1 \\ 4 & -5 & -1 \end{vmatrix}$

$= 3 \begin{vmatrix} -2 & 1 \\ -5 & -1 \end{vmatrix} - 3 \begin{vmatrix} 2 & -1 \\ -5 & -1 \end{vmatrix} + 4 \begin{vmatrix} 2 & -1 \\ -2 & 1 \end{vmatrix}$

$= 3(7) - 3(-7) + 4(0)$

$= 21 + 21 + 0$

$= 42$

$D_x = \begin{vmatrix} 4 & 2 & -1 \\ 5 & -2 & 1 \\ -1 & -5 & -1 \end{vmatrix}$

$= 4 \begin{vmatrix} -2 & 1 \\ -5 & -1 \end{vmatrix} - 5 \begin{vmatrix} 2 & -1 \\ -5 & -1 \end{vmatrix} + (-1) \begin{vmatrix} 2 & -1 \\ -2 & 1 \end{vmatrix}$

$= 4(7) - 5(-7) - 1(0)$

$= 28 + 35 - 0$

$= 63$

$D_y = \begin{vmatrix} 3 & 4 & -1 \\ 3 & 5 & 1 \\ 4 & -1 & -1 \end{vmatrix}$

$= 3 \begin{vmatrix} 5 & 1 \\ -1 & -1 \end{vmatrix} - 3 \begin{vmatrix} 4 & -1 \\ -1 & -1 \end{vmatrix} + 4 \begin{vmatrix} 4 & -1 \\ 5 & 1 \end{vmatrix}$

$= 3(-4) - 3(-5) + 4(9)$

$= -12 + 15 + 36$

$= 39$

$D_z = \begin{vmatrix} 3 & 2 & 4 \\ 3 & -2 & 5 \\ 4 & -5 & -1 \end{vmatrix}$

$= 3 \begin{vmatrix} -2 & 5 \\ -5 & -1 \end{vmatrix} - 3 \begin{vmatrix} 2 & 4 \\ -5 & -1 \end{vmatrix} + 4 \begin{vmatrix} 2 & 4 \\ -2 & 5 \end{vmatrix}$

$= 3(27) - 3(18) + 4(18)$

$= 81 - 54 + 72$

$= 99$

Then,

$$x = \frac{D_x}{D} = \frac{63}{42} = \frac{3}{2},$$

$$y = \frac{D_y}{D} = \frac{39}{42} = \frac{13}{14},$$

and

$$z = \frac{D_z}{D} = \frac{99}{42} = \frac{33}{14}$$

The solution is $\left(\frac{3}{2}, \frac{13}{14}, \frac{33}{14}\right)$.

16. $\left(-1, -\dfrac{6}{7}, \dfrac{11}{7}\right)$

17. $2x - 3y + 5z = 27$
    $x + 2y - z = -4$
    $5x - y + 4z = 27$

We compute D, D_x, D_y, and D_z.

$$D = \begin{vmatrix} 2 & -3 & 5 \\ 1 & 2 & -1 \\ 5 & -1 & 4 \end{vmatrix}$$

$$= 2 \begin{vmatrix} 2 & -1 \\ -1 & 4 \end{vmatrix} - 1 \begin{vmatrix} -3 & 5 \\ -1 & 4 \end{vmatrix} + 5 \begin{vmatrix} -3 & 5 \\ 2 & -1 \end{vmatrix}$$

$$= 2(7) - 1(-7) + 5(-7)$$
$$= 14 + 7 - 35$$
$$= -14$$

$$D_x = \begin{vmatrix} 27 & -3 & 5 \\ -4 & 2 & -1 \\ 27 & -1 & 4 \end{vmatrix}$$

$$= 27 \begin{vmatrix} 2 & -1 \\ -1 & 4 \end{vmatrix} - (-4) \begin{vmatrix} -3 & 5 \\ -1 & 4 \end{vmatrix} + 27 \begin{vmatrix} -3 & 5 \\ 2 & -1 \end{vmatrix}$$

$$= 27(7) + 4(-7) + 27(-7)$$
$$= 189 - 28 - 189$$
$$= -28$$

$$D_y = \begin{vmatrix} 2 & 27 & 5 \\ 1 & -4 & -1 \\ 5 & 27 & 4 \end{vmatrix}$$

$$= 2 \begin{vmatrix} -4 & -1 \\ 27 & 4 \end{vmatrix} - 1 \begin{vmatrix} 27 & 5 \\ 27 & 4 \end{vmatrix} + 5 \begin{vmatrix} 27 & 5 \\ -4 & -1 \end{vmatrix}$$

$$= 2(11) - 1(-27) + 5(-7)$$
$$= 22 + 27 - 35$$
$$= 14$$

$$D_z = \begin{vmatrix} 2 & -3 & 27 \\ 1 & 2 & -4 \\ 5 & -1 & 27 \end{vmatrix}$$

$$= 2 \begin{vmatrix} 2 & -4 \\ -1 & 27 \end{vmatrix} - 1 \begin{vmatrix} -3 & 27 \\ -1 & 27 \end{vmatrix} + 5 \begin{vmatrix} -3 & 27 \\ 2 & -4 \end{vmatrix}$$

$$= 2(50) - 1(-54) + 5(-42)$$
$$= 100 + 54 - 210$$
$$= -56$$

Then,
$$x = \frac{D_x}{D} = \frac{-28}{-14} = 2,$$
$$y = \frac{D_y}{D} = \frac{14}{-14} = -1,$$
and
$$z = \frac{D_z}{D} = \frac{-56}{-14} = 4$$

The solution is $(2,-1,4)$.

18. $(-3,2,1)$

19. $r - 2s + 3t = 6$
    $2r - s - t = -3$
    $r + s + t = 6$

We compute D, D_r, D_s, and D_t.

$$D = \begin{vmatrix} 1 & -2 & 3 \\ 2 & -1 & -1 \\ 1 & 1 & 1 \end{vmatrix}$$

$$= 1 \begin{vmatrix} -1 & -1 \\ 1 & 1 \end{vmatrix} - 2 \begin{vmatrix} -2 & 3 \\ 1 & 1 \end{vmatrix} + 1 \begin{vmatrix} -2 & 3 \\ -1 & -1 \end{vmatrix}$$

$$= 1(0) - 2(-5) + 1(5)$$
$$= 0 + 10 + 5$$
$$= 15$$

$$D_r = \begin{vmatrix} 6 & -2 & 3 \\ -3 & -1 & -1 \\ 6 & 1 & 1 \end{vmatrix}$$

$$= 6 \begin{vmatrix} -1 & -1 \\ 1 & 1 \end{vmatrix} - (-3) \begin{vmatrix} -2 & 3 \\ 1 & 1 \end{vmatrix} + 6 \begin{vmatrix} -2 & 3 \\ -1 & -1 \end{vmatrix}$$

$$= 6(0) + 3(-5) + 6(5)$$
$$= 0 - 15 + 30$$
$$= 15$$

$$D_s = \begin{vmatrix} 1 & 6 & 3 \\ 2 & -3 & -1 \\ 1 & 6 & 1 \end{vmatrix}$$

$$= 1 \begin{vmatrix} -3 & -1 \\ 6 & 1 \end{vmatrix} - 2 \begin{vmatrix} 6 & 3 \\ 6 & 1 \end{vmatrix} + 1 \begin{vmatrix} 6 & 3 \\ -3 & -1 \end{vmatrix}$$

$$= 1(3) - 2(-12) + 1(3)$$
$$= 3 + 24 + 3$$
$$= 30$$

$$D_t = \begin{vmatrix} 1 & -2 & 6 \\ 2 & -1 & -3 \\ 1 & 1 & 6 \end{vmatrix}$$

$$= 1 \begin{vmatrix} -1 & -3 \\ 1 & 6 \end{vmatrix} - 2 \begin{vmatrix} -2 & 6 \\ 1 & 6 \end{vmatrix} + 1 \begin{vmatrix} -2 & 6 \\ -1 & -3 \end{vmatrix}$$

$$= 1(-3) - 2(-18) + 1(12)$$
$$= -3 + 36 + 12$$
$$= 45$$

Then,
$$r = \frac{D_r}{D} = \frac{15}{15} = 1,$$
$$s = \frac{D_s}{D} = \frac{30}{15} = 2,$$
and
$$t = \frac{D_t}{D} = \frac{45}{15} = 3$$

The solution is $(1,2,3)$.

20. (3,4,-1)

21. $0.5x - 2.34 + 2.4x = 7.8x - 9$
$$2.9x - 2.34 = 7.8x - 9$$
$$6.66 = 4.9x$$
$$\frac{6.66}{4.9} = x$$
$$\frac{666}{490} = x$$
$$\frac{333}{245} = x$$

22. -12

23. <u>Familiarize</u>. We first make a drawing.

 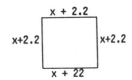

Let x represent the length of a side of the smaller square and x + 2.2 the length of a side of the larger square. The perimeter of the smaller square is 4x. The perimeter of the larger square is 4(x + 2.2).

<u>Translate</u>. The sum of the perimeters is 32.8 ft.
$$4x + 4(x + 2.2) = 32.8$$

<u>Carry out</u>. We solve the equation.
$$4x + 4x + 8.8 = 32.8$$
$$8x = 24$$
$$x = 3$$

<u>Check</u>. If x = 3 ft, then x + 2.2 = 5.2 ft. The perimeters are 4·3, or 12 ft, and 4(5.2), or 20.8 ft. The sum of the two perimeters is 12 + 20.8, or 32.8 ft. The values check.

<u>State</u>. The wire should be cut into two pieces measuring 12 ft and 20.8 ft.

24. 12

25. $\begin{vmatrix} 2 & x & -1 \\ -1 & 3 & 2 \\ -2 & 1 & 1 \end{vmatrix} = -12$

$2\begin{vmatrix} 3 & 2 \\ 1 & 1 \end{vmatrix} - (-1)\begin{vmatrix} x & -1 \\ 1 & 1 \end{vmatrix} + (-2)\begin{vmatrix} x & -1 \\ 3 & 2 \end{vmatrix} = -12$

$$2(1) + 1(x + 1) - 2(2x + 3) = -12$$
$$2 + x + 1 - 4x - 6 = -12$$
$$-3x - 3 = -12$$
$$-3x = -9$$
$$x = 3$$

26. 10

27. $\begin{vmatrix} x & y & 1 \\ x_1 & y_1 & 1 \\ x_2 & y_2 & 1 \end{vmatrix} = 0$

is equivalent to

$x\begin{vmatrix} y_1 & 1 \\ y_2 & 1 \end{vmatrix} - x_1\begin{vmatrix} y & 1 \\ y_2 & 1 \end{vmatrix} + x_2\begin{vmatrix} y & 1 \\ y_1 & 1 \end{vmatrix} = 0$

or
$$x(y_1 - y_2) - x_1(y - y_2) + x_2(y - y_1) = 0$$

or
$$xy_1 - xy_2 - x_1y + x_1y_2 + x_2y - x_2y_1 = 0.$$

An equation of the line through (x_1,y_1) and (x_2,y_2) is
$$y - y_1 = \frac{y_2 - y_1}{x_2 - x_1}(x - x_1)$$

which is equivalent to
$$(x_2 - x_1)(y - y_1) = (y_2 - y_1)(x - x_1)$$

or
$$x_2y - x_2y_1 - x_1y + x_1y_1 = y_2x - y_2x_1 - y_1x + y_1x_1$$

or
$$x_2y - x_2y_1 - x_1y - xy_2 + x_1y_2 + xy_1 = 0.$$

Exercise Set 3.7

1. $4x + 2y = 11,$
$$3x - y = 2$$
We first form the augmented matrix:
$$\begin{bmatrix} 4 & 2 & | & 11 \\ 3 & -1 & | & 2 \end{bmatrix}$$

Our goal is to transform the augmented matrix to one of the form $\begin{bmatrix} 1 & 0 & | & a \\ 0 & 1 & | & b \end{bmatrix}$.

$\begin{bmatrix} 4 & 2 & | & 11 \\ 12 & -4 & | & 8 \end{bmatrix}$ New Row 2 = 4(Row 2)

$\begin{bmatrix} 4 & 2 & | & 11 \\ 0 & -10 & | & -25 \end{bmatrix}$ New Row 2 = -3(Row 1) + (Row 2)

$\begin{bmatrix} 20 & 10 & | & 55 \\ 0 & -10 & | & -25 \end{bmatrix}$ New Row 1 = 5(Row 1)

$\begin{bmatrix} 20 & 0 & | & 30 \\ 0 & -10 & | & -25 \end{bmatrix}$ New Row 1 = (Row 2) + (Row 1)

$\begin{bmatrix} 1 & 0 & | & \frac{3}{2} \\ 0 & 1 & | & \frac{5}{2} \end{bmatrix}$ New Row 1 = $\frac{1}{20}$(Row 1)

New Row 2 = $-\frac{1}{10}$(Row 2)

1. (continued)

    We translate to the corresponding system of equations:
    $$x = \frac{3}{2}, \quad y = \frac{5}{2}$$

    The solution is $\left(\frac{3}{2},\frac{5}{2}\right)$.

2. $\left(-\frac{1}{3},-4\right)$

3. $x + 4y = 8,$
    $3x + 5y = 3$

    Write the augmented matrix and transform it to reduced row echelon form.

    $\begin{bmatrix} 1 & 4 & | & 8 \\ 3 & 5 & | & 3 \end{bmatrix}$ Augmented matrix

    $\begin{bmatrix} 1 & 4 & | & 8 \\ 0 & -7 & | & -21 \end{bmatrix}$ New Row 2 = -3(Row 1) + (Row 2)

    $\begin{bmatrix} 7 & 28 & | & 56 \\ 0 & -7 & | & -21 \end{bmatrix}$ New Row 1 = 7(Row 1)

    $\begin{bmatrix} 7 & 0 & | & -28 \\ 0 & -7 & | & -21 \end{bmatrix}$ New Row 1 = 4(Row 2) + (Row 1)

    $\begin{bmatrix} 1 & 0 & | & -4 \\ 0 & 1 & | & 3 \end{bmatrix}$ New Row 1 = $\frac{1}{7}$(Row 1)
    New Row 2 = $-\frac{1}{7}$(Row 2)

    Translate to the corresponding system of equations: $x = -4, \quad y = 3$
    The solution is (-4,3).

4. (-3,2)

5. $5x - 3y = -2,$
    $4x + 2y = 5$

    Write the augmented matrix and transform it to reduced row echelon form.

    $\begin{bmatrix} 5 & -3 & | & -2 \\ 4 & 2 & | & 5 \end{bmatrix}$ Augmented matrix

    $\begin{bmatrix} 5 & -3 & | & -2 \\ 20 & 10 & | & 25 \end{bmatrix}$ New Row 2 = 5(Row 2)

    $\begin{bmatrix} 5 & -3 & | & -2 \\ 0 & 22 & | & 33 \end{bmatrix}$ New Row 2 = -4(Row 1) + (Row 2)

    $\begin{bmatrix} 110 & -66 & | & -44 \\ 0 & 22 & | & 33 \end{bmatrix}$ New Row 1 = 22(Row 1)

5. (continued)

    $\begin{bmatrix} 110 & 0 & | & 55 \\ 0 & 22 & | & 33 \end{bmatrix}$ New Row 1 = 3(Row 2) + (Row 1)

    $\begin{bmatrix} 1 & 0 & | & \frac{1}{2} \\ 0 & 1 & | & \frac{3}{2} \end{bmatrix}$ New Row 1 = $\frac{1}{110}$(Row 1)
    New Row 2 = $\frac{1}{22}$(Row 2)

    Translate to the corresponding system of equations:
    $$x = \frac{1}{2}, \quad y = \frac{3}{2}$$

    The solution is $\left(\frac{1}{2},\frac{3}{2}\right)$.

6. $\left(-1,\frac{5}{2}\right)$

7. $2x + 3y = -7,$
    $-3x + 4y = 2$

    Write the augmented matrix and tranform it to reduced row echelon form:

    $\begin{bmatrix} 2 & 3 & | & -7 \\ -3 & 4 & | & 2 \end{bmatrix}$ Augmented matrix

    $\begin{bmatrix} 2 & 3 & | & -7 \\ -6 & 8 & | & 4 \end{bmatrix}$ New Row 2 = 2(Row 2)

    $\begin{bmatrix} 2 & 3 & | & -7 \\ 0 & 17 & | & -17 \end{bmatrix}$ New Row 2 = 3(Row 1) + (Row 2)

    $\begin{bmatrix} 34 & 51 & | & -119 \\ 0 & 17 & | & -17 \end{bmatrix}$ New Row 1 = 17(Row 1)

    $\begin{bmatrix} 34 & 0 & | & -68 \\ 0 & 17 & | & -17 \end{bmatrix}$ New Row 1 = -3(Row 2) + (Row 1)

    $\begin{bmatrix} 1 & 0 & | & -2 \\ 0 & 1 & | & -1 \end{bmatrix}$ New Row 1 = $\frac{1}{34}$(Row 1)
    New Row 2 = $\frac{1}{17}$(Row 2)

    Translate to the corresponding system of equations:
    $$x = -2, \quad y = -1$$
    The solution is (-2,-1).

8. (-127,100)

9. $4x - y - 3z = 1,$
    $8x + y - z = 5,$
    $2x + y + 2z = 5$

Our goal is to transform the augmented matrix to one of the form
$$\begin{bmatrix} 1 & 0 & 0 & | & a \\ 0 & 1 & 0 & | & b \\ 0 & 0 & 1 & | & c \end{bmatrix}.$$

$\begin{bmatrix} 4 & -1 & -3 & | & 1 \\ 8 & 1 & -1 & | & 5 \\ 2 & 1 & 2 & | & 5 \end{bmatrix}$ Augmented matrix

$\begin{bmatrix} 2 & 1 & 2 & | & 5 \\ 8 & 1 & -1 & | & 5 \\ 4 & -1 & -3 & | & 1 \end{bmatrix}$ Interchange Row 1 and Row 3

$\begin{bmatrix} 2 & 1 & 2 & | & 5 \\ 0 & -3 & -9 & | & -15 \\ 0 & -3 & -7 & | & -9 \end{bmatrix}$ New Row 2 = -4(Row 1) + (Row 2)
New Row 3 = -2(Row 1) + (Row 3)

$\begin{bmatrix} 6 & 3 & 6 & | & 15 \\ 0 & -3 & -9 & | & -15 \\ 0 & -3 & -7 & | & -9 \end{bmatrix}$ New Row 1 = 3(Row 1)

$\begin{bmatrix} 6 & 0 & -3 & | & 0 \\ 0 & -3 & -9 & | & -15 \\ 0 & 0 & 2 & | & 6 \end{bmatrix}$ New Row 1 = (Row 2) + (Row 1)
New Row 3 = -1(Row 2) + (Row 3)

$\begin{bmatrix} 12 & 0 & -6 & | & 0 \\ 0 & -6 & -18 & | & -30 \\ 0 & 0 & 2 & | & 6 \end{bmatrix}$ New Row 1 = 2(Row 1)
New Row 2 = 2(Row 2)

$\begin{bmatrix} 12 & 0 & 0 & | & 18 \\ 0 & -6 & 0 & | & 24 \\ 0 & 0 & 2 & | & 6 \end{bmatrix}$ New Row 1 = 3(Row 3) + (Row 1)
New Row 2 = 9(Row 3) + (Row 2)

$\begin{bmatrix} 1 & 0 & 0 & | & \frac{3}{2} \\ 0 & 1 & 0 & | & -4 \\ 0 & 0 & 1 & | & 3 \end{bmatrix}$ New Row 1 = $\frac{1}{12}$(Row 1)
New Row 2 = $-\frac{1}{6}$(Row 2)
New Row 3 = $\frac{1}{2}$(Row 3)

Translate to the corresponding system of equations:
$$x = \frac{3}{2}, \quad y = -4, \quad z = 3$$
The solution is $\left(\frac{3}{2}, -4, 3\right)$.

10. $\left(2, \frac{1}{2}, -2\right)$

11. $p + q + r = 1,$
     $p - 2q - 3r = 3,$
     $4p + 5q + 6r = 4$

Write the augmented matrix and transform it to reduced row echelon form:

$\begin{bmatrix} 1 & 1 & 1 & | & 1 \\ 1 & -2 & -3 & | & 3 \\ 4 & 5 & 6 & | & 4 \end{bmatrix}$ Augmented matrix

$\begin{bmatrix} 1 & 1 & 1 & | & 1 \\ 0 & -3 & -4 & | & 2 \\ 0 & 1 & 2 & | & 0 \end{bmatrix}$ New Row 2 = -1(Row 1) + (Row 2)
New Row 3 = -4(Row 1) + (Row 3)

$\begin{bmatrix} 3 & 3 & 3 & | & 3 \\ 0 & -3 & -4 & | & 2 \\ 0 & 3 & 6 & | & 0 \end{bmatrix}$ New Row 1 = 3(Row 1)
New Row 3 = 3(Row 3)

$\begin{bmatrix} 3 & 0 & -1 & | & 5 \\ 0 & -3 & -4 & | & 2 \\ 0 & 0 & 2 & | & 2 \end{bmatrix}$ New Row 1 = (Row 2) + (Row 1)
New Row 3 = (Row 2) + (Row 3)

$\begin{bmatrix} 6 & 0 & -2 & | & 10 \\ 0 & -3 & -4 & | & 2 \\ 0 & 0 & 2 & | & 2 \end{bmatrix}$ New Row 1 = 2(Row 1)

$\begin{bmatrix} 6 & 0 & 0 & | & 12 \\ 0 & -3 & 0 & | & 6 \\ 0 & 0 & 2 & | & 2 \end{bmatrix}$ New Row 1 = (Row 3) + (Row 1)
New Row 2 = 2(Row 3) + (Row 2)

$\begin{bmatrix} 1 & 0 & 0 & | & 2 \\ 0 & 1 & 0 & | & -2 \\ 0 & 0 & 1 & | & 1 \end{bmatrix}$ New Row 1 = $\frac{1}{6}$(Row 1)
New Row 2 = $-\frac{1}{3}$(Row 2)
New Row 3 = $\frac{1}{2}$(Row 3)

Translate to the corresponding system of equations:
$$x = 2, \quad y = -2, \quad z = 1$$
The solution is $(2, -2, 1)$.

12. $(-1, 2, -2)$

<u>13.</u> $x - y + 2z = 0,$
$\quad x - 2y + 3z = -1,$
$\quad 2x - 2y + z = -3$

Write the augmented matrix and transform it to reduced row echelon form:

$$\left[\begin{array}{rrr|r} 1 & -1 & 2 & 0 \\ 1 & -2 & 3 & -1 \\ 2 & -2 & 1 & -3 \end{array}\right] \text{ Augmented matrix}$$

$$\left[\begin{array}{rrr|r} 1 & -1 & 2 & 0 \\ 0 & -1 & 1 & -1 \\ 0 & 0 & -3 & -3 \end{array}\right]$$
New Row 2 = -1(Row 1) + (Row 2)
New Row 3 = -2(Row 1) + (Row 3)

$$\left[\begin{array}{rrr|r} 1 & 0 & 1 & 1 \\ 0 & -1 & 1 & -1 \\ 0 & 0 & -3 & -3 \end{array}\right]$$
New Row 1 = -1(Row 2) + (Row 1)

$$\left[\begin{array}{rrr|r} 3 & 0 & 3 & 3 \\ 0 & -3 & 3 & -3 \\ 0 & 0 & -3 & -3 \end{array}\right]$$
New Row 1 = 3(Row 1)
New Row 2 = 3(Row 2)

$$\left[\begin{array}{rrr|r} 3 & 0 & 0 & 0 \\ 0 & -3 & 0 & -6 \\ 0 & 0 & -3 & -3 \end{array}\right]$$
New Row 1 = (Row 3) + (Row 1)
New Row 2 = (Row 3) + (Row 2)

$$\left[\begin{array}{rrr|r} 1 & 0 & 0 & 0 \\ 0 & 1 & 0 & 2 \\ 0 & 0 & 1 & 1 \end{array}\right]$$
New Row 1 = $\frac{1}{3}$(Row 1)
New Row 2 = $-\frac{1}{3}$(Row 2)
New Row 3 = $-\frac{1}{3}$(Row 3)

Translate to the corresponding system of equations:
$\quad x = 0, \quad y = 2, \quad z = 1$
The solution is (0,2,1).

<u>14.</u> $\left(\frac{1}{2}, \frac{2}{3}, -\frac{5}{6}\right)$

<u>15.</u> $3p \quad + 2r = 11,$
$\quad\quad q - 7r = 4,$
$\quad p - 6q \quad = 1$

$$\left[\begin{array}{rrr|r} 3 & 0 & 2 & 11 \\ 0 & 1 & -7 & 4 \\ 1 & -6 & 0 & 1 \end{array}\right] \text{ Augmented matrix}$$

$$\left[\begin{array}{rrr|r} 1 & -6 & 0 & 1 \\ 0 & 1 & -7 & 4 \\ 3 & 0 & 2 & 11 \end{array}\right] \text{ Interchange Row 1 and Row 3}$$

<u>15.</u> (continued)

$$\left[\begin{array}{rrr|r} 1 & -6 & 0 & 1 \\ 0 & 1 & -7 & 4 \\ 0 & 18 & 2 & 8 \end{array}\right] \text{ New Row 3 = -3(Row 1) + (Row 3)}$$

$$\left[\begin{array}{rrr|r} 1 & 0 & -42 & 25 \\ 0 & 1 & -7 & 4 \\ 0 & 0 & 128 & -64 \end{array}\right]$$
New Row 1 = 6(Row 2) + (Row 1)

New Row 3 = -18(Row 2) + (Row 3)

$$\left[\begin{array}{rrr|r} 128 & 0 & -5376 & 3200 \\ 0 & 128 & -896 & 512 \\ 0 & 0 & 128 & -64 \end{array}\right]$$
New Row 1 = 128(Row 1)
New Row 2 = 128(Row 2)

$$\left[\begin{array}{rrr|r} 128 & 0 & 0 & 512 \\ 0 & 128 & 0 & 64 \\ 0 & 0 & 128 & -64 \end{array}\right]$$
New Row 1 = 42(Row 3) + (Row 1)
New Row 2 = 7(Row 3) + (Row 2)

$$\left[\begin{array}{rrr|r} 1 & 0 & 0 & 4 \\ 0 & 1 & 0 & \frac{1}{2} \\ 0 & 0 & 1 & -\frac{1}{2} \end{array}\right]$$
New Row 1 = $\frac{1}{128}$(Row 1)
New Row 2 = $\frac{1}{128}$(Row 2)
New Row 3 = $\frac{1}{128}$(Row 3)

Translate to the corresponding system of equations:
$\quad x = 4, \quad y = \frac{1}{2}, \quad z = -\frac{1}{2}$
The solution is $\left(4, \frac{1}{2}, -\frac{1}{2}\right)$.

<u>16.</u> (3,-1,4)

<u>17.</u> We will rewrite the equations with the variables in alphabetical order:
$\quad -2w + 2x + 2y - 2z = -10,$
$\quad\quad w + x + y + z = -5,$
$\quad 3w + x - y + 4z = -2,$
$\quad\quad w + 3x - 2y + 2z = -6$

$$\left[\begin{array}{rrrr|r} -2 & 2 & 2 & -2 & -10 \\ 1 & 1 & 1 & 1 & -5 \\ 3 & 1 & -1 & 4 & -2 \\ 1 & 3 & -2 & 2 & -6 \end{array}\right] \text{ Augmented Matrix}$$

$$\left[\begin{array}{rrrr|r} -1 & 1 & 1 & -1 & -5 \\ 1 & 1 & 1 & 1 & -5 \\ 3 & 1 & -1 & 4 & -2 \\ 1 & 3 & -2 & 2 & -6 \end{array}\right] \text{ New Row 1 = }\frac{1}{2}\text{(Row 1)}$$

17. (continued)

$$\begin{bmatrix} -1 & 1 & 1 & -1 & -5 \\ 0 & 2 & 2 & 0 & -10 \\ 0 & 4 & 2 & 1 & -17 \\ 0 & 4 & -1 & 1 & -11 \end{bmatrix}$$
New Row 2 = (Row 1) + (Row 2)
New Row 3 = 3(Row 1) + (Row 3)
New Row 4 = (Row 1) + (Row 4)

$$\begin{bmatrix} -2 & 2 & 2 & -2 & -10 \\ 0 & 2 & 2 & 0 & -10 \\ 0 & 4 & 2 & 1 & -17 \\ 0 & 4 & -1 & 1 & -11 \end{bmatrix}$$
New Row 1 = 2(Row 1)

$$\begin{bmatrix} -2 & 0 & 0 & -2 & 0 \\ 0 & 2 & 2 & 0 & -10 \\ 0 & 0 & -2 & 1 & 3 \\ 0 & 0 & -5 & 1 & 9 \end{bmatrix}$$
New Row 1 = -1(Row 2) + (Row 1)
New Row 3 = -2(Row 2) + (Row 3)
New Row 4 = -2(Row 2) + (Row 4)

$$\begin{bmatrix} -2 & 0 & 0 & -2 & 0 \\ 0 & 2 & 2 & 0 & -10 \\ 0 & 0 & -2 & 1 & 3 \\ 0 & 0 & -10 & 2 & 18 \end{bmatrix}$$
New Row 4 = 2(Row 4)

$$\begin{bmatrix} -2 & 0 & 0 & -2 & 0 \\ 0 & 2 & 0 & 1 & -7 \\ 0 & 0 & -2 & 1 & 3 \\ 0 & 0 & 0 & -3 & 3 \end{bmatrix}$$
New Row 2 = (Row 3) + (Row 2)
New Row 4 = -5(Row 3) + (Row 4)

$$\begin{bmatrix} -6 & 0 & 0 & -6 & 0 \\ 0 & 6 & 0 & 3 & -21 \\ 0 & 0 & -6 & 3 & 9 \\ 0 & 0 & 0 & -3 & 3 \end{bmatrix}$$
New Row 1 = 3(Row 1)
New Row 2 = 3(Row 2)
New Row 3 = 3(Row 3)

$$\begin{bmatrix} -6 & 0 & 0 & 0 & -6 \\ 0 & 6 & 0 & 0 & -18 \\ 0 & 0 & -6 & 0 & 12 \\ 0 & 0 & 0 & -3 & 3 \end{bmatrix}$$
New Row 1 = -2(Row 4) + (Row 1)
New Row 2 = (Row 4) + (Row 2)
New Row 3 = (Row 4) + (Row 3)

$$\begin{bmatrix} 1 & 0 & 0 & 0 & 1 \\ 0 & 1 & 0 & 0 & -3 \\ 0 & 0 & 1 & 0 & -2 \\ 0 & 0 & 0 & 1 & -1 \end{bmatrix}$$
New Row 1 = $-\frac{1}{6}$(Row 1)
New Row 2 = $\frac{1}{6}$(Row 2)
New Row 3 = $-\frac{1}{6}$(Row 3)
New Row 4 = $-\frac{1}{3}$(Row 4)

17. (continued)

Translate to the corresponding system of equations:

$$w = 1, \quad x = -3, \quad y = -2, \quad z = -1$$

The solution is $(1,-3,-2,-1)$.

18. $(7,4,5,6)$

19. Familiarize. Let d = the number of dimes and n = the number of nickels. The value of d dimes is $0.10d$, and the value of n nickels is $0.05n$.

Translate.

Total number of coins is 34.

$$d + n = 34$$

Total value of coins is $1.90.

$$0.10d + 0.05n = 1.90$$

After clearing decimals we have this system.

$$d + n = 34,$$
$$10d + 5n = 190$$

Carry out. Solve using matrices.

$$\begin{bmatrix} 1 & 1 & 34 \\ 10 & 5 & 190 \end{bmatrix}$$
Augmented matrix

$$\begin{bmatrix} 1 & 1 & 34 \\ 0 & -5 & -150 \end{bmatrix}$$
New Row 2 = -10(Row 1) + (Row 2)

$$\begin{bmatrix} 5 & 5 & 170 \\ 0 & -5 & -150 \end{bmatrix}$$
New Row 1 = 5(Row 1)

$$\begin{bmatrix} 5 & 0 & 20 \\ 0 & -5 & -150 \end{bmatrix}$$
New Row 1 = (Row 2) + (Row 1)

$$\begin{bmatrix} 1 & 0 & 4 \\ 0 & 1 & 30 \end{bmatrix}$$
New Row 1 = $\frac{1}{5}$(Row 1)
New Row 2 = $-\frac{1}{5}$(Row 2)

Translate to the corresponding system of equations:

$$d = 4, \quad n = 30$$

Check. The sum of the two numbers is 34. The total value is $0.10(4) + 0.05(30) = 0.40 + 1.50 = 1.90$. The numbers check.

State. There are 4 dimes and 30 nickels.

20. 21 dimes, 22 quarters

21. <u>Familiarize</u>. We let x, y, and z represent the number of nickels, dimes, and quarters, respectively. Then 0.05x, 0.10y, and 0.25z represent the monetary values of the coins.

<u>Translate</u>.

Total number of coins is 22.

$$x + y + z = 22$$

Total value of coins is $2.90.

$$0.05x + 0.10y + 0.25z = 2.90$$

or

$$5x + 10y + 25z = 290$$

The number of nickels is 6 more than the number of dimes.

$$x = 6 + y$$

We now have a system of equations.

$$x + y + z = 22,$$
$$5x + 10y + 25z = 290,$$
$$x - y = 6$$

<u>Carry out</u>. Solve using matrices.

$$\begin{bmatrix} 1 & 1 & 1 & 22 \\ 5 & 10 & 25 & 290 \\ 1 & -1 & 0 & 6 \end{bmatrix}$$ Augmented matrix

$$\begin{bmatrix} 1 & 1 & 1 & 22 \\ 0 & 5 & 20 & 180 \\ 0 & -2 & -1 & -16 \end{bmatrix}$$ New Row 2 = -5(Row 1) + (Row 2)
New Row 3 = -1(Row 1) + (Row 3)

$$\begin{bmatrix} 5 & 5 & 5 & 110 \\ 0 & 5 & 20 & 180 \\ 0 & -10 & -5 & -80 \end{bmatrix}$$ New Row 1 = 5(Row 1)

New Row 3 = 5(Row 3)

$$\begin{bmatrix} 5 & 0 & -15 & -70 \\ 0 & 5 & 20 & 180 \\ 0 & 0 & 35 & 280 \end{bmatrix}$$ New Row 1 = -1(Row 2) + (Row 1)

New Row 3 = 2(Row 2) + (Row 3)

$$\begin{bmatrix} 35 & 0 & -105 & -490 \\ 0 & 35 & 140 & 1260 \\ 0 & 0 & 35 & 280 \end{bmatrix}$$ New Row 1 = 7(Row 1)
New Row 2 = 7(Row 2)

$$\begin{bmatrix} 35 & 0 & 0 & 350 \\ 0 & 35 & 0 & 140 \\ 0 & 0 & 35 & 280 \end{bmatrix}$$ New Row 1 = 3(Row 3) + (Row 1)
New Row 2 = -4(Row 3) + (Row 2)

$$\begin{bmatrix} 1 & 0 & 0 & 10 \\ 0 & 1 & 0 & 4 \\ 0 & 0 & 1 & 8 \end{bmatrix}$$ New Row 1 = $\frac{1}{35}$(Row 1)
New Row 2 = $\frac{1}{35}$(Row 2)
New Row 3 = $\frac{1}{35}$(Row 3)

21. (continued)

Translate to the corresponding system of equations:

$$x = 10, \quad y = 4, \quad z = 8$$

<u>Check</u>. The sum of the numbers is 22. The total value is $0.05(10) + $0.10(4) + $0.25(8), or $0.50 + $0.40 + $2.00, or $2.90. The number of nickels, 10, is 6 more than 4, the number of dimes. The numbers check.

<u>State</u>. There are 10 nickels, 4 dimes, and 8 quarters.

22. 6 nickels, 5 dimes, 7 quarters

23. <u>Familiarize</u>. We let x represent the number of pounds of the $4.05 kind and y represent the number of pounds of the $2.70 kind of tobacco. We organize the information in a table.

Tobacco	Number of pounds	Price per pound	Value
$4.05 kind	x	$4.05	$4.05x
$2.70 kind	y	$2.70	$2.70y
Mixture	15	$3.15	$3.15 × 15 or $47.25

<u>Translate</u>.

Total number of pounds is 15.

$$x + y = 15$$

Total value of mixture is $47.25.

$$4.05x + 2.70y = 47.25$$

After clearing decimals we have this system:

$$x + y = 15,$$
$$405x + 270y = 4725$$

<u>Carry out</u>. Solve using matrices.

$$\begin{bmatrix} 1 & 1 & 15 \\ 405 & 270 & 4725 \end{bmatrix}$$ Augmented matrix

$$\begin{bmatrix} 1 & 1 & 15 \\ 0 & -135 & -1350 \end{bmatrix}$$ New Row 2 = -405(Row 1) + (Row 2)

$$\begin{bmatrix} 135 & 135 & 2025 \\ 0 & -135 & -1350 \end{bmatrix}$$ New Row 1 = 135(Row 1)

$$\begin{bmatrix} 135 & 0 & 675 \\ 0 & -135 & -1350 \end{bmatrix}$$ New Row 1 = (Row 2) + (Row 1)

$$\begin{bmatrix} 1 & 0 & 5 \\ 0 & 1 & 10 \end{bmatrix}$$ New Row 1 = $\frac{1}{135}$(Row 1)
New Row 2 = $-\frac{1}{135}$(Row 2)

23. (continued)

Translate to the corresponding system of equations:

$$x = 5, \quad y = 10$$

Check. The sum of the numbers is 15. The total value is $4.05(5) + $2.70(10), or $20.25 + $27.00, or $47.25. The numbers check.

State. 5 pounds of the $4.05 per lb kind and 10 pounds of the $2.70 per lb kind should be used.

24. Candy: 14 lb, Nuts: 6 lb

25. Familiarize. Let x, y, and z represent the amounts invested at 7%, 8%, and 9%, respectively. Organize the information in a table.

	Amount invested	Interest rate	Time	Interest ($I = Prt$)
	x	7%	1 yr	0.07x
	y	8%	1 yr	0.08y
	z	9%	1 yr	0.09z
Total	$2500			$212

Translate. The first column gives us one equation:

$$x + y + z = 2500$$

The last column gives a second equation:

$$0.07x + 0.08y + 0.09z = 212$$

Amount invested at 9%	is	$1100	more than	amount invested at 8%
z	=	$1100	+	y

After clearing decimals we have this system:

$$x + y + z = 2500$$
$$7x + 8y + 9z = 21{,}200$$
$$-y + z = 1100$$

Carry out. Solve using matrices.

$$\begin{bmatrix} 1 & 1 & 1 & | & 2500 \\ 7 & 8 & 9 & | & 21{,}200 \\ 0 & -1 & 1 & | & 1100 \end{bmatrix} \quad \text{Augmented matrix}$$

$$\begin{bmatrix} 1 & 1 & 1 & | & 2500 \\ 0 & 1 & 2 & | & 3700 \\ 0 & -1 & 1 & | & 1100 \end{bmatrix} \quad \begin{array}{l} \text{New Row 2} = -7(\text{Row 1}) + \\ (\text{Row 2}) \end{array}$$

$$\begin{bmatrix} 1 & 0 & -1 & | & -1200 \\ 0 & 1 & 2 & | & 3700 \\ 0 & 0 & 3 & | & 4800 \end{bmatrix} \quad \begin{array}{l} \text{New Row 1} = -1(\text{Row 2}) + \\ (\text{Row 1}) \\ \\ \text{New Row 3} = (\text{Row 2}) + \\ (\text{Row 3}) \end{array}$$

25. (continued)

$$\begin{bmatrix} 3 & 0 & -3 & | & -3600 \\ 0 & 3 & 6 & | & 11{,}100 \\ 0 & 0 & 3 & | & 4800 \end{bmatrix} \quad \begin{array}{l} \text{New Row 1} = 3(\text{Row 1}) \\ \\ \text{New Row 2} = 3(\text{Row 2}) \end{array}$$

$$\begin{bmatrix} 3 & 0 & 0 & | & 1200 \\ 0 & 3 & 0 & | & 1500 \\ 0 & 0 & 3 & | & 4800 \end{bmatrix} \quad \begin{array}{l} \text{New Row 1} = (\text{Row 3}) + (\text{Row 1}) \\ \\ \text{New Row 2} = -2(\text{Row 3}) + \\ (\text{Row 2}) \end{array}$$

$$\begin{bmatrix} 1 & 0 & 0 & | & 400 \\ 0 & 1 & 0 & | & 500 \\ 0 & 0 & 1 & | & 1600 \end{bmatrix} \quad \begin{array}{l} \text{New Row 1} = \frac{1}{3}(\text{Row 1}) \\ \\ \text{New Row 2} = \frac{1}{3}(\text{Row 2}) \\ \\ \text{New Row 3} = \frac{1}{3}(\text{Row 3}) \end{array}$$

Translate to the corresponding system of equations:

$$x = 400, \quad y = 500, \quad z = 1600$$

Check. The total investment is $400 + $500 + $1600, or $2500. The total interest is 0.07($400) + 0.08($500) + 0.09($1600) = $28 + $40 + $144 = $212. The amount invested at 9%, $1600, is $1100 more than the amount invested at 8%, $500. The numbers check.

State. $400 is invested at 7%, $500 is invested at 8%, and $1600 is invested at 9%.

26. $500 at 8%, $400 at 9%, $2300 at 10%

27.
$$0.1x - 12 = 3.6x - 2.34 - 4.9x$$
$$10x - 1200 = 360x - 234 - 490x \quad \begin{array}{l}\text{Multiplying by} \\ \text{100 to clear} \\ \text{decimals}\end{array}$$
$$10x - 1200 = -130x - 234$$
$$140x - 1200 = -234$$
$$140x = 966$$
$$x = \frac{966}{140}, \quad \text{or} \quad \frac{69}{10}, \quad \text{or} \quad 6.9$$

28. −20

29.
$$4(9 - x) - 6(8 - 3x) = 5(3x + 4)$$
$$36 - 4x - 48 + 18x = 15x + 20$$
$$-12 + 14x = 15x + 20$$
$$-12 = x + 20$$
$$-32 = x$$

30. $b = \dfrac{c}{5 + a}$

31. Elimination yields a row containing all 0's. (This corresponds to an equation that is true for all values of the variables.)

32. Elimination yields a row containing all 0's to the left of the vertical line and a nonzero number to the right of the vertical line.

33. <u>Familiarize</u>. Let w, x, y, and z represent the thousand's, hundred's, ten's, and one's digits, respectively.

<u>Translate</u>.

The sum of the digits is 10.

$$w + x + y + z \quad = \quad 10$$

Twice the sum of the thousand's and ten's digits is the sum of the hundred's and one's digits less one.

$$2(w + y) \quad = \quad x + z \quad - 1$$

The ten's digit is twice the thousand's digit.

$$y \quad = \quad 2 \cdot \quad w$$

The one's digit equals the sum of the thousand's and hundred's digits.

$$z \quad = \quad w + x$$

We have a system of equations which can be written as

$$w + x + \; y + z = 10,$$
$$2w - x + 2y - z = -1,$$
$$-2w \quad + \; y \quad = 0,$$
$$w + x \quad - z = 0.$$

<u>Carry out</u>. Solving the system we get (1,3,2,4).

<u>Check</u>. The sum of the digits is 10. Twice the sum of 1 and 2 is 6. This is one less than the sum of 3 and 4. The ten's digit, 2, is twice the thousand's digit, 1. The one's digit, 4, equals 1 + 3. The numbers check.

<u>State</u>. The number is 1324.

Exercise Set 3.8

1. $C(x) = 45x + 600,000 \qquad R(x) = 65x$

   a) $P(x) = R(x) - C(x)$
   $$= 65x - (45x + 600,000)$$
   $$= 65x - 45x - 600,000$$
   $$= 20x - 600,000$$

   b) We set $R(x) = C(x)$ and solve for x.
   $$65x = 45x + 600,000$$
   $$20x = 600,000$$
   $$x = 30,000$$
   Thus, 30,000 units must be produced and sold in order to break even.

2. a) $P(x) = 45x - 360,000$, b) 8000 units

3. $C(x) = 10x + 120,000 \qquad R(x) = 60x$

   a) $P(x) = R(x) - C(x)$
   $$= 60x - (10x + 120,000)$$
   $$= 60x - 10x - 120,000$$
   $$= 50x - 120,000$$

3. (continued)

   b) We set $R(x) = C(x)$ and solve for x.
   $$60x = 10x + 120,000$$
   $$50x = 120,000$$
   $$x = 2400$$
   Thus, 2400 units must be produced and sold in order to break even.

4. a) $P(x) = 55x - 49,500$, b) 900 units

5. $C(x) = 20x + 10,000 \qquad R(x) = 100x$

   a) $P(x) = R(x) - C(x)$
   $$= 100x - (20x + 10,000)$$
   $$= 100x - 20x - 10,000$$
   $$= 80x - 10,000$$

   b) We set $R(x) = C(x)$ and solve for x.
   $$100x = 20x + 10,000$$
   $$80x = 10,000$$
   $$x = 125$$
   Thus, 125 units must be produced and sold in order to break even.

6. a) $P(x) = 45x - 22,500$, b) 500 units

7. $C(x) = 15x + 75,000 \qquad R(x) = 55x$

   a) $P(x) = R(x) - C(x)$
   $$= 55x - (15x + 75,000)$$
   $$= 55x - 15x - 75,000$$
   $$= 40x - 75,000$$

   b) We set $R(x) = C(x)$ and solve for x.
   $$55x = 15x + 75,000$$
   $$40x = 75,000$$
   $$x = 1875$$
   To break even 1875 units must be produced and sold.

8. a) $P(x) = 18x - 16,000$ b) 889 units

9. $C(x) = 50x + 195,000 \qquad R(x) = 125x$

   a) $P(x) = R(x) - C(x)$
   $$= 125x - (50x + 195,000)$$
   $$= 125x - 50x - 195,000$$
   $$= 75x - 195,000$$

   b) We set $R(x) = C(x)$ and solve for x.
   $$125x = 50x + 195,000$$
   $$75x = 195,000$$
   $$x = 2600$$
   To break even 2600 units must be produced and sold.

10. $P(x) = 94x - 928,000$ b) 9873 units

11. $D(p) = 2000 - 60p$ $S(p) = 460 + 94p$

We set $D(p) = S(p)$ and solve.

$2000 - 60p = 460 + 94p$

$1540 = 154p$

$10 = p$

The equilibrium price is $10 per unit. To find the equilibrium quantity we substitute $10 into either $D(p)$ or $S(p)$.

$D(10) = 2000 - 60(10) = 2000 - 600 = 1400$

The equilibrium quantity is 1400 units.

The equilibrium point is ($10,1400).

12. ($50,500)

13. $D(p) = 760 - 13p$ $S(p) = 430 + 2p$

We set $D(p) = S(p)$ and solve.

$760 - 13p = 430 + 2p$

$330 = 15p$

$22 = p$

The equilibrium price is $22 per unit.

To find the equilibrium quantity we substitute $22 into either $D(p)$ or $S(p)$.

$S(22) = 430 + 2(22) = 430 + 44 = 474$

The equilibrium quantity is 474 units.

The equilibrium point is ($22,474).

14. ($10,370)

15. $D(p) = 7500 - 25p$ $S(p) = 6000 + 5p$

We set $D(p) = S(p)$ and solve.

$7500 - 25p = 6000 + 5p$

$1500 = 30p$

$50 = p$

The equilibrium price is $50 per unit.

To find the equilibrium quantity we substitute $50 into either $D(p)$ or $S(p)$.

$D(50) = 7500 - 25(50) = 7500 - 1250 = 6250$

The equilibrium quantity is 6250 units.

The equilibrium point is ($50,6250).

16. ($40,7600)

17. $D(p) = 1600 - 53p$ $S(p) = 320 + 75p$

We set $D(p) = S(p)$ and solve.

$1600 - 53p = 320 + 75p$

$1280 = 128p$

$10 = p$

The equilibrium price is $10 per unit.

To find the equilibrium quantity we substitute $10 into either $D(p)$ or $S(p)$.

$S(10) = 320 + 75(10) = 320 + 750 = 1070$

The equilibrium quantity is 1070 units.

The equilibrium point is ($10,1070).

18. ($36,4060)

19. $D(p) = 4750 - 62p$ $S(p) = 3350 + 38p$

We set $D(p) = S(p)$ and solve.

$4750 - 62p = 3350 + 38p$

$1400 = 100p$

$14 = p$

The equilibrium price is $14 per unit.

To find the equilibrium quantity we substitute $14 into either $D(p)$ or $S(p)$.

$D(14) = 4750 - 62(14) = 4750 - 868 = 3882$

The equilibrium quantity is 3882 units.

The equilibrium point is ($14,3882).

20. ($45.80,5366)

21. a) $C(x)$ = Fixed costs + Variable costs

$C(x) = 125,000 + 750x,$

where x is the number of computers produced.

b) Each computer sells for $1050. The total revenue is 1050 times the number of computers sold. We assume that all computers produced are sold.

$R(x) = 1050x$

c) $P(x) = R(x) - C(x)$

$P(x) = 1050x - (125,000 + 750x)$

$= 1050x - 125,000 - 750x$

$= 300x - 125,000$

d) $P(x) = 300x - 125,000$

$P(400) = 300(400) - 125,000$

$= 120,000 - 125,000$

$= -5000$

The company will realize a $5000 loss when 400 computers are produced and sold.

$P(700) = 300(700) - 125,000$

$= 210,000 - 125,000$

$= 85,000$

The company will realize a profit of $85,000 from the production and sale of 700 computers.

e) We set $R(x) = C(x)$ and solve for x.

$1050x = 125,000 + 750x$

$300x = 125,000$

$x \approx 417$

To break even 417 units must be produced and sold.

22. a) $C(x) = 22,500 + 40x$

b) $R(x) = 85x$

c) $P(x) = 45x - 22,500$

d) Profit of $112,500, loss of $4500

e) 500 units

23. a) C(x) = Fixed costs + Variable costs

    C(x) = 10,000 + 20x,

    where x is the number of sport coats produced.

 b) Each sport coat sells for $100. The total revenue is 100 times the number of coats sold. We assume that all coats produced are sold.

    R(x) = 100x

 c) P(x) = R(x) - C(x)

    P(x) = 100x - (10,000 + 20x)

    $\qquad$ = 100x - 10,000 - 20x

    $\qquad$ = 80x - 10,000

 d) $\quad$ P(x) = 80x - 10,000

    P(2000) = 80(2000) - 10,000

    $\qquad$ = 160,000 - 10,000

    $\qquad$ = 150,000

    The clothing firm will realize a $150,000 profit when 2000 sport coats are produced and sold.

    P(50) = 80(50) - 10,000

    $\qquad$ = 4000 - 10,000

    $\qquad$ = -6000

    The company will realize a $6000 loss when 50 coats are produced and sold.

 e) We set R(x) = C(x) and solve for x.

    100x = 10,000 + 20x

    $\ $ 80x = 10,000

    $\quad$ x = 125

    To break even 125 coats must be produced and sold.

24. a) C(x) = 16,400 + 6x

 b) R(x) = 18x

 c) P(x) = 12x - 16,400

 d) Profit of $19,600, loss of $4400

 e) 1367 units

25. $y - 3 = \frac{2}{5}(x - 1)$

    The equation of the line is in point-slope form. We see that the line has slope $\frac{2}{5}$ and contains the point (1,3). Plot (1,3). Then go up two units and right 5 units to find another point on the line, (5,6). A third point can be found as a check.

26. 5 and 6

27. 9x = 5x - {3(2x - 7) - 4}

    9x = 5x - {6x - 21 - 4}

    9x = 5x - {6x - 25}

    9x = 5x - 6x + 25

    9x = -x + 25

    10x = 25

    $\quad$ x = $\frac{25}{10}$

    $\quad$ x = $\frac{5}{2}$, or 2.5

28. $t = \frac{rw - v}{-s}$, or $\frac{v - rw}{s}$

29. The supply function contains the points ($2,100) and ($8,500). We find its equation:

    $m = \frac{500 - 100}{8 - 2} = \frac{400}{6} = \frac{200}{3}$

    $y - y_1 = m(x - x_1)$ $\quad$ Point-slope form

    $y - 100 = \frac{200}{3}(x - 2)$

    $y - 100 = \frac{200}{3}x - \frac{400}{3}$

    $\qquad y = \frac{200}{3}x - \frac{100}{3}$

    We can equivalently express supply S as a function of price p:

    $S(p) = \frac{200}{3}p - \frac{100}{3}$

    The demand function contains the points ($1,500) and ($9,100). We find its equation:

    $m = \frac{100 - 500}{9 - 1} = \frac{-400}{8} = -50$

    $y - y_1 = m(x - x_1)$

    $y - 500 = -50(x - 1)$

    $y - 500 = -50x + 50$

    $\qquad y = -50x + 550$

    We can equivalently express demand D as a function of price p:

    $D(p) = -50p + 550$

    To find the equilibrium price we set S(p) = D(p) and solve.

    $\frac{200}{3}p - \frac{100}{3} = -50p + 550$

    200p - 100 = -150p + 1650 $\quad$ Multiplying by 3 to clear fractions

    350p - 100 = 1650

    $\qquad$ 350p = 1750

    $\qquad\quad$ p = 5

    The equilibrium price is $5 per unit.

    To find the equilibrium quantity, we substitute $5 into either S(p) or D(p).

    $\quad$ D(5) = -50(5) + 550 = -250 + 550 = 300

    The equilibrium quantity is 300 units.

    The equilibrium point is ($5,300).

30. 308

Exercise Set 4.1

1. x - 2 ⩾ 6

   -4: We substitute and get -4 - 2 ⩾ 6, or -6 ⩾ 6,
   a false sentence. Therefore, -4 is not a
   solution.

   0: We substitute and get 0 - 2 ⩾ 6, or -2 ⩾ 6, a
   false sentence. Therefore, 0 is not a
   solution.

   4: We substitute and get 4 - 2 ⩾ 6, or 2 ⩾ 6, a
   false sentence. Therefore, 4 is not a
   solution.

   8: We substitute and get 8 - 2 ⩾ 6, or 6 ⩾ 6, a
   true sentence. Therefore, 8 is a solution.

2. Yes, yes, no, no

3. t - 8 > 2t - 3

   0: We substitute and get 0 - 8 > 2·0 - 3, or
   -8 > -3, a false sentence. Therefore, 0 is
   not a solution.

   -8: We substitute and get -8 - 8 > 2(-8) - 3,
   or -16 > -19, a true sentence. Therefore,
   -8 is a solution.

   -9: We substitute and get -9 - 8 > 2(-9) - 3,
   or -17 > -21, a true sentence. Therefore,
   -9 is a solution.

   -3: We substitute and get -3 - 8 > 2(-3) - 3,
   or -11 > -9, a false sentence. Therefore,
   -3 is not a solution.

4. No, yes, yes, no

5. x > 4

   Graph: The solutions consist of all real numbers
   greater than 4, so we shade all numbers to the
   right of 4 and use an open circle at 4 to
   indicate that it is not a solution.

   Set-builder notation: {x|x > 4}

   Interval notation: (4,∞)

6.

   {y|y < 5}, (-∞,5)

7. t ⩽ 6

   Graph: We shade all numbers to the left of 6 and
   use a solid circle at 6 to indicate that it is
   also a solution.

   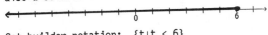

   Set-builder notation: {t|t ⩽ 6}

   Interval notation: (-∞,6]

8.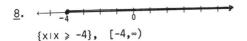

   {x|x ⩾ -4}, [-4,∞)

9. y < -3

   Graph: We shade all numbers to the left of -3
   and use an open circle at -3 to indicate that it
   is not a solution.

   Set-builder notation: {y|y < -3}

   Interval notation: (-∞,-3)

10. {t|t > -2}, (-2,∞)

11. x ⩾ -6

   Graph: We shade all numbers to the right of -6
   and use a solid circle at -6 to indicate that it
   is also a solution.

   Set-builder notation: {x|x ⩾ -6}

   Interval notation: [-6,∞)

12. {x|x ⩽ -5}, (-∞,-5]

13.   x + 8 > 3
   x + 8 + (-8) > 3 + (-8)   Adding -8
           x > -5
   The solution set is {x|x > -5}, or (-5,∞).

14. {x|x > -3}, or (-3,∞)

15.   y + 3 < 9
   y + 3 + (-3) < 9 + (-3)   Adding -3
           y < 6
   The solution set is {y|y < 6}, or (-∞,6).

16. {y|y < 6}, or (-∞,6)

17.   a + 9 ⩽ -12
   a + 9 + (-9) ⩽ -12 + (-9)   Adding -9
           a ⩽ -21
   The solution set is {a|a ⩽ -21}, or (-∞,-21].

18. {a|a ⩽ -20}, or (-∞,-20]

19.
$$t + 14 \geqslant 9$$
$$t + 14 + (-14) \geqslant 9 + (-14) \quad \text{Adding } -14$$
$$t \geqslant -5$$
The solution set is $\{t \mid t \geqslant -5\}$, or $[-5, \infty)$.

20. $\{x \mid x \leqslant 19\}$, or $(-\infty, 19]$

21.
$$y - 8 > -14$$
$$y - 8 + 8 > -14 + 8 \quad \text{Adding } 8$$
$$y > -6$$
The solution set is $\{y \mid y > -6\}$, or $(-6, \infty)$.

22. $\{y \mid y > -9\}$, or $(-9, \infty)$

23.
$$x - 11 \leqslant -2$$
$$x - 11 + 11 \leqslant -2 + 11 \quad \text{Adding } 11$$
$$x \leqslant 9$$
The solution set is $\{x \mid x \leqslant 9\}$, or $(-\infty, 9]$.

24. $\{y \mid y \leqslant 14\}$, or $(-\infty, 14]$

25.
$$8x \geqslant 24$$
$$\frac{1}{8} \cdot 8x \geqslant \frac{1}{8} \cdot 24 \quad \text{Multiplying by } \frac{1}{8}$$
$$x \geqslant 3$$
The solution set is $\{x \mid x \geqslant 3\}$, or $[3, \infty)$.

26. $\{t \mid t < -9\}$, or $(-\infty, -9)$

27.
$$0.3x < -18$$
$$\frac{1}{0.3}(0.3x) < \frac{1}{0.3}(-18) \quad \text{Multiplying by } \frac{1}{0.3}$$
$$x < -\frac{18}{0.3}$$
$$x < -60$$
The solution set is $\{x \mid x < -60\}$, or $(-\infty, -60)$.

28. $\{x \mid x < 50\}$, or $(-\infty, 50)$

29.
$$-9x \geqslant -8.1$$
$$-\frac{1}{9}(-9x) \leqslant -\frac{1}{9}(-8.1) \quad \text{Multiplying by } -\frac{1}{9} \text{ and reversing the inequality sign}$$
$$x \leqslant \frac{8.1}{9}$$
$$x \leqslant 0.9$$
The solution set is $\{x \mid x \leqslant 0.9\}$, or $(-\infty, 0.9]$.

30. $\{y \mid y \geqslant -0.4\}$, or $[-0.4, \infty)$

31.
$$-\frac{3}{4}x \geqslant -\frac{5}{8}$$
$$-\frac{4}{3}\left(-\frac{3}{4}x\right) \leqslant -\frac{4}{3}\left(-\frac{5}{8}\right) \quad \text{Multiplying by } -\frac{4}{3} \text{ and reversing the inequality sign}$$
$$x \leqslant \frac{20}{24}$$
$$x \leqslant \frac{5}{6}$$
The solution set is $\left\{x \mid x \leqslant \frac{5}{6}\right\}$, or $\left(-\infty, \frac{5}{6}\right]$.

32. $\left\{y \mid y \geqslant \frac{9}{10}\right\}$, or $\left[\frac{9}{10}, \infty\right]$

33.
$$2x + 7 < 19$$
$$2x + 7 + (-7) < 19 + (-7) \quad \text{Adding } -7$$
$$2x < 12$$
$$\frac{1}{2} \cdot 2x < \frac{1}{2} \cdot 12 \quad \text{Multiplying by } \frac{1}{2}$$
$$x < 6$$
The solution set is $\{x \mid x < 6\}$, or $(-\infty, 6)$.

34. $\{y \mid y > 3\}$, or $(3, \infty)$

35.
$$5y + 2y \leqslant -21$$
$$7y \leqslant -21 \quad \text{Collecting like terms}$$
$$\frac{1}{7}(7y) \leqslant \frac{1}{7}(-21) \quad \text{Multiplying by } \frac{1}{7}$$
$$y \leqslant -3$$
The solution set is $\{y \mid y \leqslant -3\}$, or $(-\infty, -3]$.

36. $\{x \mid x \leqslant 4\}$, or $(-\infty, 4]$

37.
$$2y - 7 < 5y - 9$$
$$-5y + 2y - 7 < -5y + 5y - 9 \quad \text{Adding } -5y$$
$$-3y - 7 < -9$$
$$-3y - 7 + 7 < -9 + 7 \quad \text{Adding } 7$$
$$-3y < -2$$
$$-\frac{1}{3}(-3y) > -\frac{1}{3}(-2) \quad \text{Multiplying by } -\frac{1}{3} \text{ and reversing the inequality sign}$$
$$y > \frac{2}{3}$$

The solution set is $\left\{y \middle| y > \frac{2}{3}\right\}$, or $\left[\frac{2}{3}, \infty\right)$.

38. $\left\{x \middle| x < -\frac{2}{5}\right\}$, or $\left[-\infty, -\frac{2}{5}\right]$

39.
$$0.4x + 5 \leqslant 1.2x - 4$$
$$-1.2x + 0.4x + 5 \leqslant -1.2x + 1.2x - 4 \quad \text{Adding } -1.2x$$
$$-0.8x + 5 \leqslant -4$$
$$-0.8x + 5 + (-5) \leqslant -4 + (-5) \quad \text{Adding } -5$$
$$-0.8x \leqslant -9$$
$$\left[-\frac{1}{0.8}\right](-0.8x) \geqslant \left[-\frac{1}{0.8}\right](-9)$$
$$\text{Multiplying by } -\frac{1}{0.8} \text{ and reversing the inequality sign}$$
$$x \geqslant \frac{9}{0.8}$$
$$x \geqslant 11.25$$

The solution set is $\{x \mid x \geqslant 11.25\}$, or $[11.25, \infty)$.

40. $\{y \mid y < 5\}$, or $(-\infty, 5)$

41.
$$3x - \frac{1}{8} \leqslant \frac{3}{8} + 2x$$
$$-2x + 3x - \frac{1}{8} \leqslant -2x + \frac{3}{8} + 2x \quad \text{Adding } -2x$$
$$x - \frac{1}{8} \leqslant \frac{3}{8}$$
$$x - \frac{1}{8} + \frac{1}{8} \leqslant \frac{3}{8} + \frac{1}{8} \quad \text{Adding } \frac{1}{8}$$
$$x \leqslant \frac{4}{8}$$
$$x \leqslant \frac{1}{2}$$

The solution set is $\left\{x \middle| x \leqslant \frac{1}{2}\right\}$, or $\left[-\infty, \frac{1}{2}\right]$.

42. $\{x \mid x \text{ is a real number}\}$, or $(-\infty, \infty)$

43.
$$4(3y - 2) \geqslant 9(2y + 5)$$
$$12y - 8 \geqslant 18y + 45$$
$$-6y - 8 \geqslant 45$$
$$-6y \geqslant 53$$
$$y \leqslant -\frac{53}{6}$$

The solution set is $\left\{y \middle| y \leqslant -\frac{53}{6}\right\}$, or $\left[-\infty, -\frac{53}{6}\right]$.
This can be expressed equivalently as
$\left\{y \middle| y \leqslant -8\frac{5}{6}\right\}$, or $\left[-\infty, -8\frac{5}{6}\right]$.

44. $\{m \mid m \leqslant 3.3\}$, or $(-\infty, 3.3]$

45.
$$3(2 - 5x) + 2x < 2(4 + 2x)$$
$$6 - 15x + 2x < 8 + 4x$$
$$6 - 13x < 8 + 4x$$
$$6 - 17x < 8$$
$$-17x < 2$$
$$x > -\frac{2}{17}$$

The solution set is $\left\{x \middle| x > -\frac{2}{17}\right\}$, or $\left[-\frac{2}{17}, \infty\right)$.

46. $\{y \mid y < 0.07\}$, or $(-\infty, 0.07)$

47.
$$5[3m - (m + 4)] > -2(m - 4)$$
$$5(3m - m - 4) > -2(m - 4)$$
$$5(2m - 4) > -2(m - 4)$$
$$10m - 20 > -2m + 8$$
$$12m - 20 > 8$$
$$12m > 28$$
$$m > \frac{28}{12}$$
$$m > \frac{7}{3}$$

The solution set is $\left\{m \middle| m > \frac{7}{3}\right\}$, or $\left[\frac{7}{3}, \infty\right)$.

48. $\left\{x \middle| x \leqslant -\frac{23}{2}\right\}$, or $\left[-\infty, -\frac{23}{2}\right]$

49.
$$3(r - 6) + 2 > 4(r + 2) - 21$$
$$3r - 18 + 2 > 4r + 8 - 21$$
$$3r - 16 > 4r - 13$$
$$-r - 16 > -13$$
$$-r > 3$$
$$r < -3$$

The solution set is $\{r \mid r < -3\}$, or $(-\infty, -3)$.

50. $\{t \mid t < -12\}$, or $(-\infty, -12)$

51. $19 - (2x + 3) \leqslant 2(x + 3) + x$

$19 - 2x - 3 \leqslant 2x + 6 + x$

$16 - 2x \leqslant 3x + 6$

$16 - 5x \leqslant 6$

$-5x \leqslant -10$

$x \geqslant 2$

The solution set is $\{x | x \geqslant 2\}$, or $[2, \infty)$.

52. $\{c | c \leqslant 1\}$, or $(-\infty, 1]$

53. $\frac{1}{4}(8y + 4) - 17 < -\frac{1}{2}(4y - 8)$

$2y + 1 - 17 < -2y + 4$

$2y - 16 < -2y + 4$

$4y - 16 < 4$

$4y < 20$

$y < 5$

The solution set is $\{y | y < 5\}$, or $(-\infty, 5)$.

54. $\{x | x > 6\}$, or $(6, \infty)$

55. $2[4 - 2(3 - x)] - 1 \geqslant 4[2(4x - 3) + 7] - 25$

$2[4 - 6 + 2x] - 1 \geqslant 4[8x - 6 + 7] - 25$

$2[-2 + 2x] - 1 \geqslant 4[8x + 1] - 25$

$-4 + 4x - 1 \geqslant 32x + 4 - 25$

$4x - 5 \geqslant 32x - 21$

$-28x - 5 \geqslant -21$

$-28x \geqslant -16$

$x \leqslant \frac{-16}{-28}$, or $\frac{4}{7}$

The solution set is $\left\{x | x \leqslant \frac{4}{7}\right\}$, or $\left[-\infty, \frac{4}{7}\right]$.

56. $\left\{t | t \geqslant -\frac{27}{19}\right\}$, or $\left[-\frac{27}{19}, \infty\right]$

57. $\frac{2}{3}(2x - 1) > 10$

$3 \cdot \frac{2}{3}(2x - 1) > 3 \cdot 10$ Multiplying by 3 to clear the fraction

$2(2x - 1) > 30$

$4x - 2 > 30$

$4x > 32$

$x > 8$

The solution set is $\{x | x > 8\}$, or $(8, \infty)$.

58. $\{x | x < 7\}$, or $(-\infty, 7)$

59. $\frac{3}{4}(3 + 2x) + 1 \geqslant 13$

$4\left[\frac{3}{4}(3 + 2x) + 1\right] \geqslant 4 \cdot 13$ Multiplying by 4 to clear the fraction

$3(3 + 2x) + 4 \geqslant 52$

$9 + 6x + 4 \geqslant 52$

$6x + 13 \geqslant 52$

$6x \geqslant 39$

$x \geqslant \frac{39}{6}$, or $\frac{13}{2}$

The solution set is $\left\{x | x \geqslant \frac{13}{2}\right\}$, or $\left[\frac{13}{2}, \infty\right]$.

60. $\left\{x | x \leqslant -\frac{405}{28}\right\}$, or $\left[-\infty, -\frac{405}{28}\right]$

61. $\frac{3}{4}\left[3x - \frac{1}{2}\right] - \frac{2}{3} < \frac{1}{3}$

$\frac{9x}{4} - \frac{3}{8} - \frac{2}{3} < \frac{1}{3}$

$24\left[\frac{9x}{4} - \frac{3}{8} - \frac{2}{3}\right] < 24 \cdot \frac{1}{3}$ Multiplying by 24 to clear fractions

$54x - 9 - 16 < 8$

$54x - 25 < 8$

$54x < 33$

$x < \frac{33}{54}$, or $\frac{11}{18}$

The solution set is $\left\{x | x < \frac{11}{18}\right\}$, or $\left[-\infty, \frac{11}{18}\right]$.

62. $\left\{x | x > -\frac{5}{32}\right\}$, or $\left[-\frac{5}{32}, \infty\right]$

63. $0.7(3x + 6) \geqslant 1.1 - (x + 2)$

$10[0.7(3x + 6)] \geqslant 10[1.1 - (x + 2)]$ Multiplying by 10 to clear decimals

$7(3x + 6) \geqslant 11 - 10(x + 2)$

$21x + 42 \geqslant 11 - 10x - 20$

$21x + 42 \geqslant -9 - 10x$

$31x + 42 \geqslant -9$

$31x \geqslant -51$

$x \geqslant -\frac{51}{31}$

The solution set is $\left\{x | x \geqslant -\frac{51}{31}\right\}$, or $\left[-\frac{51}{31}, \infty\right]$.

64. $\left\{x | x < \frac{39}{14}\right\}$, or $\left[-\infty, \frac{39}{14}\right]$

65. $a + (a - 3) \leqslant (a + 2) - (a + 1)$

$a + a - 3 \leqslant a + 2 - a - 1$

$2a - 3 \leqslant 1$

$2a \leqslant 4$

$a \leqslant 2$

The solution set is $\{a | a \leqslant 2\}$, or $(-\infty, 2]$.

66. $\left\{b | b < -\frac{37}{5}\right\}$, or $\left[-\infty, -\frac{37}{5}\right]$

67. 5y − 10 = 2x

x	y
0	2
−5	0
5	4

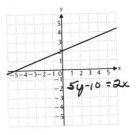

68. −2

69. |−16| = 16   −16 is 16 units from 0

70. −4

71. 3ax + 2x ⩾ 5ax − 4
$$2x - 2ax ⩾ -4$$
$$2x(1 - a) ⩾ -4$$
$$x(1 - a) ⩾ -2 \quad (a > 1,\ 1 - a < 0)$$
$$x ⩽ -\frac{2}{1 - a}, \text{ or } \frac{2}{a - 1}$$
The solution set is $\left\{ x \,\middle|\, x ⩽ \frac{2}{a - 1} \right\}$.

72. $\left\{ y \,\middle|\, y ⩾ -\frac{10}{b + 4} \right\}$

73. a(by − 2) ⩾ b(2y + 5)
$$aby - 2a ⩾ 2by + 5b$$
$$aby - 2by ⩾ 2a + 5b$$
$$y(ab - 2b) ⩾ 2a + 5b \quad (a > 2,\ ab - 2b > 0)$$
$$y ⩾ \frac{2a + 5b}{ab - 2b}, \text{ or } \frac{2a + 5b}{b(a - 2)}$$
The solution set is $\left\{ y \,\middle|\, y ⩾ \frac{2a + 5b}{b(a - 2)} \right\}$.

74. $\left\{ x \,\middle|\, x < \frac{4c + 3d}{6c - 2d} \right\}$

75. c(2 − 5x) + dx > m(4 + 2x)
$$2c - 5cx + dx > 4m + 2mx$$
$$-5cx + dx - 2mx > 4m - 2c$$
$$x(-5c + d - 2m) > 4m - 2c$$
$$x[d - (5c + 2m)] > 4m - 2c \quad (5c + 2m < d,$$
$$d - (5c + 2m) > 0)$$
$$x > \frac{4m - 2c}{d - (5c + 2m)}$$
The solution set is $\left\{ x \,\middle|\, x > \frac{4m - 2c}{d - (5c + 2m)} \right\}$.

76. $\left\{ x \,\middle|\, x < \frac{-3a + 2d}{c - (4a + 5d)} \right\}$

77. False. If a = 2, b = 3, c = 4, d = 5, then
2 − 4 = 3 − 5.

78. False, because −3 < −2, but 9 > 4.

79. Not equivalent.
x < 3 is defined at x = 0.
$x + \frac{1}{x} < 3 + \frac{1}{x}$ is undefined at x = 0.

80. Not equivalent, because 0·x < 0·3 is equivalent to
0 < 0 which is not true.

Exercise Set 4.2

1. A number <u>is</u> <u>less</u> <u>than</u> 8.
n < 8

2. n > −1.2

3. The price of a movie ticket <u>is greater than or
equal to</u> $6.
p ⩾ $6

4. p ⩽ 58

5. The price of compact disks <u>is at most</u> $17.95.
d ⩽ $17.95

6. s ⩾ $35,000

7. 24 minus 3 times a number <u>is less than</u> 16 plus
the number.
24 − 3x < 16 + x

8. −17 + 4x > x − 23

9. Fifteen times the sum of two numbers <u>is at least</u>
78.
15(a + b) ⩾ 78

10. 2xy + 27 ⩽ −3

11. <u>Familiarize.</u> We let x represent the number of
miles traveled in a day. Then the total rental
cost for a day is 30 + 0.20x.
<u>Translate.</u> The total cost for a day must be less
than or equal to $96. This translates to the
following inequality:
30 + 0.20x ⩽ 96
<u>Carry out.</u>
0.20x ⩽ 66   Adding −30
$x ⩽ \frac{66}{0.20}$ Multiplying by $\frac{1}{0.20}$
x ⩽ 330
<u>Check.</u> If you travel 330 miles, the total cost
is 30 + 0.20(330), or 30 + 66 = $96. Any mileage
less than 330 will also stay within the budget.

<u>State.</u> Mileage less than or equal to 330 miles
allows you to stay within budget.

12. More than 36.8 miles per day

13. <u>Familiarize</u>. List the information in a table. Let x represent the score on the fourth test.

Test	Score
Test 1	89
Test 2	92
Test 3	95
Test 4	x
Total	360 or more

<u>Translate</u>. We can easily get an inequality from the table.

$$88 + 92 + 95 + x \geqslant 360$$

<u>Carry out</u>.

$$276 + x \geqslant 360 \quad \text{Collecting like terms}$$
$$x \geqslant 84 \quad \text{Adding } -276$$

<u>Check</u>. If you get 84 on the fourth test, your total score will be 89 + 92 + 95 + 84, or 360. Any higher score will also give you an A.

<u>State</u>. A score of 84 or better will give you an A.

14. 77 or better

15. <u>Familiarize</u>. Organize the information in a table. Let x = the amount invested at 14%. Then 25,000 - x = the amount invested at 16%.

Amount invested	Rate of interest	Time	Interest (I = Prt)
x	14%	1 yr	0.14x
25,000 - x	16%	1 yr	0.16(25,000 - x)
Total $25,000			$3600 or more

<u>Translate</u>. Use the information in the table to write an inequality:

$$0.14x + 0.16(25,000 - x) \geqslant 3600$$

<u>Carry out</u>.

$$0.14x + 4000 - 0.16x \geqslant 3600$$
$$4000 - 0.02x \geqslant 3600 \quad \text{Collecting like terms}$$
$$-0.02x \geqslant -400 \quad \text{Adding } -4000$$
$$x \leqslant \frac{400}{0.02}$$
$$\quad \text{Multiplying by } -\frac{1}{0.02} \text{ and reversing the inequality sign}$$
$$x \leqslant 20,000$$

15. (continued)

<u>Check</u>. For x = $20,000, 14%(20,000) = 0.14(20,000), or $2800, and 16%(25,000 - 20,000) = 0.16(5000), or $800. The total interest earned is 2800 + 800, or $3600. We also calculate for some amount less than $20,000 and for some amount greater than $20,000. For x = $15,000, 14%(15,000) = 0.14(15,000), or $2100, and 16%(25,000 - 15,000) = 0.16(10,000), or $1600. The total interest earned is 2100 + 1600, or $3700. For x = $22,000, 14%(22,000) = 0.14(22,000), or $3080, and 16%(25,000 - 22,000) = 0.16(3000), or $480. The total interest earned is 3080 + 480, or $3560. For these values the inequality, x ⩽ 20,000, gives correct results.

<u>State</u>. To make at least $3600 interest per year, $20,000 is the most that can be invested at 14%.

16. $5000

17. <u>Familiarize</u>. We let x represent the ticket price. Then 300x represents the total receipts from the ticket sales assuming 300 people will attend. The first band will play for $250 + 50%(300x). The second band will play for $550.

<u>Translate</u>. For school profit to be greater when the first band plays, the amount the first band charges must be less than the amount the second band charges. We now have an inequality.

$$250 + 50\%(300x) < 550$$

<u>Carry out</u>.

$$250 + 0.5(300x) < 550$$
$$250 + 150x < 550$$
$$150x < 300 \quad \text{Adding } -250$$
$$x < 2 \quad \text{Multiplying by } \frac{1}{150}$$

<u>Check</u>. For x = $2, the total receipts are 300(2), or $600. The first band charges 250 + 50%(600), or $550. The second band also charges $550. The school profit is the same using either band. For x = $1.99, the total receipts are 300(1.99), or $597. The first band charges 250 + 50%(597), or $548.50. Using the first band, the school profit would be 597 - 548.50, or $48.50. Using the second band, the profit would be 597 - 550, or $47. Thus, the first band produces more profit. For x = $2.01, the total receipts are 300(2.01), or $603. The first band charges 250 + 50%(603), or $551.50. Using the first band, the school profit would be 603 - 551.50, or $51.50. Using the second band, the school profit would be 603 - 550, or $53. Thus, the second band produces more profit. For these values, the inequality x < 2 gives correct results.

<u>State</u>. The ticket price must be less than $2. The highest price, rounded to the nearest cent, less than $2 is $1.99.

18. More than 33

19. <u>Familiarize</u>. We make a table listing the information. We let x represent the total medical bill.

	Plan A	Plan B
You pay the first	100	250
Insurance pays	80%(x − 100)	90%(x − 250)
You also pay	20%(x − 100)	10%(x − 250)
Total you pay	100+20%(x−100)	250+10%(x−250)

<u>Translate</u>. We write an inequality stating that the amount you pay is less when you choose Plan B. This gives us an inequality.

$$250 + 10\%(x - 250) < 100 + 20\%(x - 100)$$

<u>Carry out</u>.

$$250 + 0.1x - 25 < 100 + 0.2x - 20$$
$$225 + 0.1x < 80 + 0.2x \quad \text{Collecting like terms}$$
$$145 < 0.1x \quad \text{Adding −80 and −0.1x}$$
$$1450 < x \quad \text{Multiplying by } \frac{1}{0.1}$$

or

$$x > 1450$$

<u>Check</u>. We calculate for x = $1450 and also for some amount greater than $1450 and some amount less than $1450.

Plan A:	Plan B:
100 + 20%(1450 − 100)	250 + 10%(1450 − 250)
100 + 0.2(1350)	250 + 0.1(1200)
100 + 270	250 + 120
$370	$370

When x = $1450, you pay the same with either plan.

Plan A:	Plan B:
100 + 20%(1500 − 100)	250 + 10%(1500 − 250)
100 + 0.2(1400)	250 + 0.1(1250)
100 + 280	250 + 125
$380	$375

When x = $1500, you pay less with Plan B.

Plan A:	Plan B:
100 + 20%(1400 − 100)	250 + 10%(1400 − 100)
100 + 0.2(1300)	250 + 0.1(1300)
100 + 260	250 + 130
$360	$380

When x = $1400, you pay more with Plan B.

For these values, the inequality x > $1450 gives correct results.

<u>State</u>. For Plan B to save you money, the total of the medical bills must be greater than $1450.

20. More than 8.75 pounds

21. <u>Familiarize</u>. We make a table of information.

Plan A: Monthly Income	Plan B: Monthly Income
$500 salary 4% of sales Total: 500 + 4% of sales	$750 salary 5% of sales over $8000 Total: 750 + 5% of sales over 8000

<u>Translate</u>. We write an inequality stating that the income from Plan B is greater than the income from Plan A. We let S represent gross sales. Then S − 8000 represents gross sales over 8000.

$$750 + 5\%(S - 8000) > 500 + 4\%S$$

<u>Carry out</u>.

$$750 + 0.05S - 400 > 500 + 0.04S$$
$$350 + 0.05S > 500 + 0.04S$$
$$0.01S > 150$$
$$S > \frac{150}{0.01}$$
$$S > 15,000$$

<u>Check</u>. We calculate for x = $15,000 and for some amount greater than $15,000 and some amount less than $15,000.

Plan A:	Plan B:
500 + 4%(15,000)	750 + 5%(15,000 − 8000)
500 + 0.04(15,000)	750 + 0.05(7000)
500 + 600	750 + 350
$1100	$1100

When x = $15,000, income from Plan A is equal to the income from Plan B.

Plan A:	Plan B:
500 + 4%(16,000)	750 + 5%(16,000 − 8000)
500 + 0.04(16,000)	750 + 0.05(8000)
500 + 640	750 + 400
$1140	$1150

When x = $16,000, income from Plan B is greater than the income from Plan A.

Plan A:	Plan B:
500 + 4%(14,000)	750 + 5%(14,000 − 8000)
500 + 0.04(14,000)	750 + 0.05(6000)
500 + 560	750 + 300
$1060	$1050

When x = $14,000, income from Plan B is less than the income from Plan A.

<u>State</u>. Plan B is better than Plan A when gross sales are greater than $15,000.

22. Less than $116,666.67 per year

23. <u>Familiarize</u>. We let n represent the number of hours it will take the mason to complete the job. Plan A will pay the mason 500 + 5n, while Plan B will pay the mason 8n.

    <u>Translate</u>. We write an inequality stating that the income from the job under Plan A is greater than the income from the job under Plan B.

    $$500 + 5n \geqslant 8n$$

    <u>Carry out</u>.

    $$500 \geqslant 3n$$

    $$\frac{500}{3} \geqslant n$$

    or

    $$n \leqslant 166\frac{2}{3}$$

    <u>Check</u>.

Plan A:	Plan B:
$500 + 5\left[\frac{500}{3}\right]$	$8\left[\frac{500}{3}\right]$
$\frac{1500}{3} + \frac{2500}{3}$	$\frac{4000}{3}$, or \$1333.33
$\frac{4000}{3}$, or \$1333.33	

    When n = $166\frac{2}{3}$ hours, Plan A = Plan B.

    We also calculate the income when n = 166 and when n = 167.

Plan A:	Plan B:
500 + 5(166)	8(166)
500 + 830	\$1328
\$1330	

    When n = 166, Plan A > Plan B.

Plan A:	Plan B:
500 + 5(167)	8(167)
500 + 835	\$1336
\$1335	

    When n = 167, Plan A < Plan B.

    <u>State</u>. Plan A is better than Plan B when n < $166\frac{2}{3}$ hours.

24. n > 85.7

25. <u>Familiarize</u>. We want to find the values of x for which f(x) > 36.

    <u>Translate</u>. 2(x + 10) > 36

    <u>Carry out</u>.

    $$2x + 20 > 36$$
    $$2x > 16$$
    $$x > 8$$

    <u>Check</u>. For x = 8, f(x) = f(8) = 2(8 + 10) = 2·18 = 36. Then any U.S. size larger than 8 will give a size larger than 36 in Italy.

    <u>State</u>. For U.S. dress sizes larger than 8, dress sizes in Italy will be larger than 36.

26. Greater than 132 ft

27. a) <u>Familiarize</u>. Find the values of F for which C is less than 1063°.

    <u>Translate</u>. $\frac{5}{9}$(F − 32) < 1063°.

    <u>Carry out</u>. 5(F − 32) < 9567
    $$5F - 160 < 9567$$
    $$5F < 9727$$
    $$F < 1945.4$$

    <u>Check</u>. When F = 1945.4, C = $\frac{5}{9}$(1945.4 − 32) = 1063. Then values of F less than 1945.4 will give values of C less than 1063.

    <u>State</u>. Gold is solid at temperatures less than 1945.4° F.

    b) <u>Familiarize</u>. Find the values of F for which C is less than 960.8.

    <u>Translate</u>. $\frac{5}{9}$(F − 32) < 960.8

    <u>Carry out</u>. 5(F − 32) < 8647.2
    $$5F - 160 < 8647.2$$
    $$5F < 8807.2$$
    $$F < 1761.44$$

    <u>Check</u>. When F = 1761.44, C = $\frac{5}{9}$(1761.44 − 32) = 960.8. Then values of F less than 1761.44 will give values of C less than 960.8.

    <u>State</u>. Silver is solid at temperatures less than 1761.44° F.

28. a) 44,000, 165,978, about 275,758;

    b) From 1987 on

29. a) <u>Familiarize</u>. Find the values of x for which R(x) < C(x).

    <u>Translate</u>. 26x < 90,000 + 15x

    <u>Carry out</u>.

    $$11x < 90,000$$
    $$x < 8181\frac{9}{11}$$

    <u>Check</u>. $R\left[8181\frac{9}{11}\right]$ = \$212,727.27 = $C\left[8181\frac{9}{11}\right]$. Calculate R(x) and C(x) for some x greater than $8181\frac{9}{11}$ and for some x less than $8181\frac{9}{11}$. Suppose x = 8200:

    $$R(x) = 26(8200) = 213,200 \quad \text{and}$$
    $$C(x) = 90,000 + 15(8200) = 213,000.$$

    In this case R(x) > C(x).
    Suppose x = 8000:

    $$R(x) = 26(8000) = 208,000 \quad \text{and}$$
    $$C(x) = 90,000 + 15(8000) = 210,000.$$

    In this case R(x) < C(x).

    Then for x < $8181\frac{9}{11}$, R(x) < C(x).

29. (continued)

    State. We will state the result in terms of integers, since the company cannot sell a fraction of a radio. For 8181 or fewer radios the company loses money.

    b) Our check in part a) shows that for $x > 8181\frac{9}{11}$, R(x) > C(x) and the company makes a profit. Again, we will state the result in terms of an integer. For 8182 or more radios the company makes money.

30. a) p < 10;

    b) p > 10

31. Familiarize. Find the values of t for which R < 10.3. Keep in mind that t represents the number of years after 1988.

    Translate. -0.03125t + 10.49 < 10.3

    Carry out. -0.03125t < -0.19

                        t > 6.08

    Check. When t = 6.08, R = -0.03125(6.08) + 10.49 = 10.3. Then for values of t greater than 6.08, R will be less than 10.3.

    State. In the years more than 6.08 years after 1988, or in the years after 1994, the record will be less than 10.3 sec. In terms of an inequality, we can say that the record will be less than 10.3 sec in the years y, y > 1994.

32. 2ℓ + 2w ⩽ 288

33. w wins yield 2w points, and t ties yield t points. Then for w wins and t ties the team gets a total of 2w + t points. This total must be at least 60 points, so 2w + t ⩾ 60.

34. 35c + 75a > 1000

Exercise Set 4.3

---

1. {5,6,7,8} ∩ {4,6,8,10}

   The numbers 6 and 8 are common to the two sets, so the intersection is {6,8}.

2. {10}

3. {2,4,6,8} ∩ {1,3,5}

   There are no numbers common to the two sets, so the intersection is the empty set, ∅.

4. ∅

5. {1,2,3,4} ∩ {1,2,3,4}

   The numbers 1, 2, 3, and 4 are common to the two sets, so the intersection is {1,2,3,4}.

6. ∅

7. 1 < x < 6; (1,6)

8. 0 ⩽ y ⩽ 3; [0,3]

9. -7 ⩽ y ⩽ -3; [-7,-3]

10. -9 ⩽ x < -5; [-9,5)

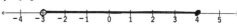

11. -4 ⩽ -x < 3

    4 ⩾ x > -3   Multiplying by -1

    (-3,4]

12. x > -8 and x < -3; (-8,-3)

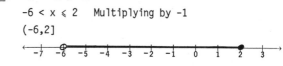

13. 6 > -x ⩾ -2

    -6 < x ⩽ 2   Multiplying by -1

    (-6,2]

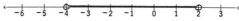

14. x > -4 and x < 2; (-4,2)

15. 5 > x ⩾ -2; [-2,5)

16. 3 > x ⩾ 0; [0,3)

17. x < 5 and x ⩾ 1; [1,5)

18. x ⩾ -2 and x < 2; [-2,2)

19.   -2 < x + 2 < 8

    -2 + (-2) < x + 2 + (-2) < 8 + (-2)   Adding -2

            -4 < x < 6

    The solution set is {x|-4 < x < 6}, or (-4,6).

20. {x|-2 < x ⩽ 5}, or (-2,5]

21. $$1 < 2y + 5 \leqslant 9$$
$$1 + (-5) < 2y + 5 + (-5) \leqslant 9 + (-5) \quad \text{Adding } -5$$
$$-4 < 2y \leqslant 4$$
$$\tfrac{1}{2} \cdot (-4) < \tfrac{1}{2} \cdot 2y \leqslant \tfrac{1}{2} \cdot 4 \quad \text{Multiplying by } \tfrac{1}{2}$$
$$-2 < y \leqslant 2$$
The solution set is $\{y | -2 < y \leqslant 2\}$, or $(-2,2]$.

22. $\{x | 0 \leqslant x \leqslant 1\}$, or $[0,1]$

23. $$-10 \leqslant 3x - 5 \leqslant -1$$
$$-10 + 5 \leqslant 3x - 5 + 5 \leqslant -1 + 5 \quad \text{Adding } 5$$
$$-5 \leqslant 3x \leqslant 4$$
$$\tfrac{1}{3} \cdot (-5) \leqslant \tfrac{1}{3} \cdot 3x \leqslant \tfrac{1}{3} \cdot 4 \quad \text{Multiplying by } \tfrac{1}{3}$$
$$-\tfrac{5}{3} \leqslant x \leqslant \tfrac{4}{3}$$
The solution set is $\left\{x \middle| -\tfrac{5}{3} \leqslant x \leqslant \tfrac{4}{3}\right\}$, or $\left[-\tfrac{5}{3}, \tfrac{4}{3}\right]$.

24. $\left\{x \middle| -\tfrac{7}{2} < x \leqslant \tfrac{11}{2}\right\}$, or $\left[-\tfrac{7}{2}, \tfrac{11}{2}\right]$

25. $$2 < x + 3 \leqslant 9$$
$$2 + (-3) < x + 3 + (-3) \leqslant 9 + (-3) \quad \text{Adding } -3$$
$$-1 < x \leqslant 6$$
The solution set is $\{x | -1 < x \leqslant 6\}$, or $(-1,6]$.

26. $\{x | -7 \leqslant x < 8\}$, or $[-7,8)$

27. $$-6 \leqslant 2x - 3 < 6$$
$$-6 + 3 \leqslant 2x - 3 + 3 < 6 + 3$$
$$-3 \leqslant 2x < 9$$
$$\tfrac{1}{2} \cdot (-3) \leqslant \tfrac{1}{2} \cdot 2x < \tfrac{1}{2} \cdot 9$$
$$-\tfrac{3}{2} \leqslant x < \tfrac{9}{2}$$
The solution set is $\left\{x \middle| -\tfrac{3}{2} \leqslant x < \tfrac{9}{2}\right\}$, or $\left[-\tfrac{3}{2}, \tfrac{9}{2}\right)$.

28. $\left\{m \middle| -\tfrac{11}{3} < m \leqslant -3\right\}$, or $\left[-\tfrac{11}{3}, -3\right]$

29. $$-\tfrac{1}{2} < \tfrac{1}{4}x - 3 \leqslant \tfrac{1}{2}$$
$$-\tfrac{1}{2} + 3 < \tfrac{1}{4}x - 3 + 3 \leqslant \tfrac{1}{2} + 3$$
$$\tfrac{5}{2} < \tfrac{1}{4}x \leqslant \tfrac{7}{2}$$
$$4 \cdot \tfrac{5}{2} < 4 \cdot \tfrac{1}{4}x \leqslant 4 \cdot \tfrac{7}{2}$$
$$10 < x \leqslant 14$$
The solution set is $\{x | 10 < x \leqslant 14\}$, or $(10,14]$.

30. $\left\{x \middle| \tfrac{40}{3} < x \leqslant \tfrac{56}{3}\right\}$, or $\left[\tfrac{40}{3}, \tfrac{56}{3}\right]$

31. $$-3 < \frac{2x - 5}{4} < 8$$
$$4(-3) < 4\left[\frac{2x - 5}{4}\right] < 4 \cdot 8$$
$$-12 < 2x - 5 < 32$$
$$-12 + 5 < 2x - 5 + 5 < 32 + 5$$
$$-7 < 2x < 37$$
$$\tfrac{1}{2}(-7) < \tfrac{1}{2} \cdot 2x < \tfrac{1}{2} \cdot 37$$
$$-\tfrac{7}{2} < x < \tfrac{37}{2}$$
The solution set is $\left\{x \middle| -\tfrac{7}{2} < x < \tfrac{37}{2}\right\}$, or $\left[-\tfrac{7}{2}, \tfrac{37}{2}\right]$.

32. $\left\{x \middle| -\tfrac{13}{3} \leqslant x \leqslant 9\right\}$, or $\left[-\tfrac{13}{3}, 9\right]$

33. $\{4,5,6,7,8\} \cup \{1,4,6,11\}$
The numbers in either or both sets are 1, 4, 5, 6, 7, 8, and 11, so the union is $\{1,4,5,6,7,8,11\}$.

34. $\{2,8,9,27\}$

35. $\{2,4,6,8\} \cup \{1,3,5\}$
The numbers in either or both sets are 1, 2, 3, 4, 5, 6, and 8, so the union is $\{1,2,3,4,5,6,8\}$.

36. $\{8,9,10\}$

37. $\{4,8,11\} \cup \emptyset$
The numbers in either or both sets are 4, 8, and 11, so the union is $\{4,8,11\}$.

38. $\emptyset$

39. $x < -1 \ \text{ or } \ x > 2$

40. $x < -2 \ \text{ or } \ x > 0$

41. $x \leqslant -3 \ \text{ or } \ x > 1$

42. $x \leqslant -1 \ \text{ or } \ x > 3$

43. $x < -8 \ \text{ or } \ x > -2$

44. $t \leqslant -10 \ \text{ or } \ t \geqslant -5$

45. $x + 7 < -2$ or $x + 7 > 2$
  $x + 7 + (-7) < -2 + (-7)$ or $x + 7 + (-7) > 2 + (-7)$
        $x < -9$ or $x > -5$
  The solution set is $\{x \mid x < -9 \text{ or } x > -5\}$, or
  $(-\infty, -9) \cup (-5, \infty)$.

46. $\{x \mid x < -13 \text{ or } x > -5\}$, or $(-\infty, -13) \cup (-5, \infty)$

47. $2x - 8 \leqslant -3$ or $x - 8 \geqslant 3$
  $2x - 8 + 8 \leqslant -3 + 8$ or $x - 8 + 8 \geqslant 3 + 8$
        $2x \leqslant 5$ or $x \geqslant 11$
  $\frac{1}{2} \cdot 2x \leqslant \frac{1}{2} \cdot 5$ or $x \geqslant 11$
        $x \leqslant \frac{5}{2}$ or $x \geqslant 11$

  The solution set is $\left\{x \mid x \leqslant \frac{5}{2} \text{ or } x \geqslant 11\right\}$, or
  $\left(-\infty, \frac{5}{2}\right] \cup [11, \infty)$.

48. $\{x \mid x \leqslant -9 \text{ or } x \geqslant 3\}$, or $(-\infty, -9] \cup [3, \infty)$

49. $3x - 9 < -5$ or $x - 9 > 6$
  $3x - 9 + 9 < -5 + 9$ or $x - 9 + 9 > 6 + 9$
        $3x < 4$ or $x > 15$
  $\frac{1}{3} \cdot 3x < \frac{1}{3} \cdot 4$ or $x > 15$
        $x < \frac{4}{3}$ or $x > 15$

  The solution set is $\left\{x \mid x < \frac{4}{3} \text{ or } x > 15\right\}$, or
  $\left(-\infty, \frac{4}{3}\right] \cup (15, \infty)$.

50. $\{x \mid x < -1 \text{ or } x > 16\}$, or $(-\infty, -1) \cup (16, \infty)$

51. $7 > -4x + 5$ or $10 \leqslant -4x + 5$
  $7 + (-5) > -4x + 5 + (-5)$ or $10 + (-5) \leqslant -4x + 5 + (-5)$
        $2 > -4x$ or $5 \leqslant -4x$
  $-\frac{1}{4} \cdot 2 < -\frac{1}{4}(-4x)$ or $-\frac{1}{4} \cdot 5 \geqslant -\frac{1}{4}(-4x)$
        $-\frac{1}{2} < x$ or $-\frac{5}{4} \geqslant x$

  The solution set is $\left\{x \mid x \leqslant -\frac{5}{4} \text{ or } x > -\frac{1}{2}\right\}$, or
  $\left(-\infty, -\frac{5}{4}\right] \cup \left(-\frac{1}{2}, \infty\right)$.

52. $\{x \mid x \text{ is a real number}\}$, or $(-\infty, \infty)$

53. $7 - x \leqslant -2$ or $7 - x > 2$
  $-7 + 7 - x \leqslant -7 - 2$ or $-7 + 7 - x > -7 + 2$
        $-x \leqslant -9$ or $-x > -5$
  $-1(-x) \geqslant -1(-9)$ or $-1(-x) < -1(-5)$
        $x \geqslant 9$ or $x < 5$
  The solution set is $\{x \mid x < 5 \text{ or } x \geqslant 9\}$, or
  $(-\infty, 5) \cup [9, \infty)$.

54. $\{x \mid x \leqslant 5 \text{ or } x > 13\}$, or $(-\infty, 5] \cup (13, \infty)$

55. $-2x - 2 < -6$ or $-2x - 2 > 6$
  $-2x - 2 + 2 < -6 + 2$ or $-2x - 2 + 2 > 6 + 2$
        $-2x < -4$ or $-2x > 8$
  $-\frac{1}{2}(-2x) > -\frac{1}{2}(-4)$ or $-\frac{1}{2}(-2x) < -\frac{1}{2} \cdot 8$
        $x > 2$ or $x < -4$
  The solution set is $\{x \mid x < -4 \text{ or } x > 2\}$, or
  $(-\infty, -4) \cup (2, \infty)$.

56. $\left\{m \mid m < -4 \text{ or } m > -\frac{2}{3}\right\}$, or $(-\infty, -4) \cup \left[-\frac{2}{3}, \infty\right)$

57. $\frac{2}{3}x - 14 < -\frac{5}{6}$ or $\frac{2}{3}x - 14 > \frac{5}{6}$
  $6\left[\frac{2}{3}x - 14\right] < 6\left[-\frac{5}{6}\right]$ or $6\left[\frac{2}{3}x - 14\right] > 6 \cdot \frac{5}{6}$
        $4x - 84 < -5$ or $4x - 84 > 5$
  $4x - 84 + 84 < -5 + 84$ or $4x - 84 + 84 > 5 + 84$
        $4x < 79$ or $4x > 89$
  $\frac{1}{4} \cdot 4x < \frac{1}{4} \cdot 79$ or $\frac{1}{4} \cdot 4x > \frac{1}{4} \cdot 89$
        $x < \frac{79}{4}$ or $x > \frac{89}{4}$

  The solution set is $\left\{x \mid x < \frac{79}{4} \text{ or } x > \frac{89}{4}\right\}$, or
  $\left(-\infty, \frac{79}{4}\right) \cup \left(\frac{89}{4}, \infty\right)$.

58. $\left\{x \mid x \leqslant -\frac{91}{100} \text{ or } x \geqslant \frac{79}{60}\right\}$, or $\left(-\infty, -\frac{91}{100}\right] \cup \left[\frac{79}{60}, \infty\right)$

59. $\frac{2x - 5}{6} \leqslant -3$ or $\frac{2x - 5}{6} \geqslant 4$
  $6\left[\frac{2x - 5}{6}\right] \leqslant 6(-3)$ or $6\left[\frac{2x - 5}{6}\right] \geqslant 6 \cdot 4$
        $2x - 5 \leqslant -18$ or $2x - 5 \geqslant 24$
  $2x - 5 + 5 \leqslant -18 + 5$ or $2x - 5 + 5 \geqslant 24 + 5$
        $2x \leqslant -13$ or $2x \geqslant 29$
  $\frac{1}{2} \cdot 2x \leqslant \frac{1}{2}(-13)$ or $\frac{1}{2} \cdot 2x \geqslant \frac{1}{2} \cdot 29$
        $x \leqslant -\frac{13}{2}$ or $x \geqslant \frac{29}{2}$

  The solution set is $\left\{x \mid x \leqslant -\frac{13}{2} \text{ or } x \geqslant \frac{29}{2}\right\}$, or
  $\left(-\infty, -\frac{13}{2}\right] \cup \left[\frac{29}{2}, \infty\right)$.

60. $\left\{x \mid x < -\frac{13}{3} \text{ or } x > 9\right\}$, or $\left(-\infty, -\frac{13}{3}\right] \cup (9, \infty)$

61. 2x - 3y = 7,   (1)
    3x + 2y = -10   (2)
    We will use the addition method.

    4x - 6y = 14   Multiplying (1) by 2
    9x + 6y = -30   Multiplying (2) by 3
    13x   = -16   Adding
         $x = -\frac{16}{13}$

    Substitute $-\frac{16}{13}$ for x in (1) and solve for y.

    $2\left(-\frac{16}{13}\right) - 3y = 7$

    $-\frac{32}{13} - 3y = 7$

    $\frac{32}{13} - \frac{32}{13} - 3y = \frac{32}{13} + \frac{91}{13}$

    $-3y = \frac{123}{13}$

    $y = -\frac{41}{13}$

    These numbers check, so the solution is
    $\left(-\frac{16}{13}, -\frac{41}{13}\right)$.

62. $x = -\frac{8}{3}$

63. 5(2x + 3) = 3(x - 4)
    10x + 15 = 3x - 12
    7x + 15 = -12
    7x = -27
    $x = -\frac{27}{7}$

64.

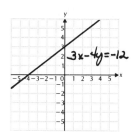

$3x - 4y = -12$

65. a) Substitute $\frac{5}{9}(F - 32)$ for C in the given
       inequality.
       $1063 \leqslant \frac{5}{9}(F - 32) < 2660$

       $9 \cdot 1063 \leqslant 9 \cdot \frac{5}{9}(F - 32) < 9 \cdot 2660$

       $9567 \leqslant 5(F - 32) < 23,940$
       $9567 \leqslant 5F - 160 < 23,940$
       $9727 \leqslant 5F < 24,100$
       $1945.4 \leqslant F < 4820$

       The inequality for Fahrenheit temperatures is
       1945.4° $\leqslant$ F < 4820°.

65. (continued)

    b) Substitute $\frac{5}{9}(F - 32)$ for C in the given
       inequality.
       $960.8 \leqslant \frac{5}{9}(F - 32) < 2180$

       $9(960.8) \leqslant 9 \cdot \frac{5}{9}(F - 32) < 9 \cdot 2180$

       $8647.2 \leqslant 5(F - 32) < 19,620$
       $8647.2 \leqslant 5F - 160 < 19,620$
       $8807.2 \leqslant 5F < 19,780$
       $1761.44 \leqslant F < 3956$

       The inequality for Fahrenheit temperatures is
       1761.44° $\leqslant$ F < 3956°.

66. All the numbers between -5 and 11

67. Solve 32 < f(x) < 46, or 32 < 2(x + 10) < 46.
       32 < 2(x + 10) < 46
       32 < 2x + 20 < 46
       12 < 2x < 26
        6 < x < 13

    For U.S. dress sizes between 6 and 13, dress
    sizes in Italy will be between 32 and 46.

68. 0 ft $\leqslant$ d $\leqslant$ 198 ft

69. Solve 50,000 $\leqslant$ N $\leqslant$ 250,000 where N = 12,197.8t +
    44,000. Keep in mind that t is the number of
    years since 1971.
       $50,000 \leqslant 12,197.8t + 44,000 \leqslant 250,000$
       $6000 \leqslant 12,197.8t \leqslant 206,000$
       $0.49 \leqslant t \leqslant 16.89$

    0.49 years after the end of 1971 is in 1972 and
    16.89 years after the end of 1971 is in 1988. We
    give the result in terms of _entire_ years that
    satisfy the inequality. Therefore, there will be
    at least 50,000 and at most 250,000 women in the
    active duty military force from 1973 to 1987.

70. 1957 $\leqslant$ y $\leqslant$ 1978

71. 4a - 2 $\leqslant$ a + 1 $\leqslant$ 3a + 4
    4a - 2 $\leqslant$ a + 1   and   a + 1 $\leqslant$ 3a + 4
       3a $\leqslant$ 3   and   -3 $\leqslant$ 2a
        a $\leqslant$ 1   and $-\frac{3}{2} \leqslant a$

    The solution set is $\left\{a \middle| -\frac{3}{2} \leqslant a \leqslant 1\right\}$, or $\left[-\frac{3}{2}, 1\right]$.

72. $\left\{m \middle| m < \frac{6}{5}\right\}$, or $\left(-\infty, \frac{6}{5}\right)$

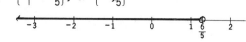

73. $x - 10 < 5x + 6 \leqslant x + 10$
$\quad -10 < 4x + 6 \leqslant 10$
$\quad -16 < 4x \leqslant 4$
$\quad -4 < x \leqslant 1$
The solution set is $\{x \mid -4 < x \leqslant 1\}$, or $(-4,1]$.

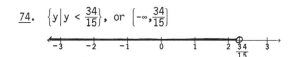

74. $\left\{y \mid y < \frac{34}{15}\right\}$, or $\left(-\infty, \frac{34}{15}\right)$

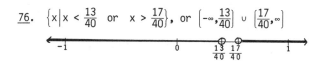

75. $-\frac{2}{15} \leqslant \frac{2}{3}x - \frac{2}{5} \leqslant \frac{2}{15}$
$\quad -\frac{2}{15} \leqslant \frac{2}{3}x - \frac{6}{15} \leqslant \frac{2}{15}$
$\quad \frac{4}{15} \leqslant \frac{2}{3}x \leqslant \frac{8}{15}$
$\quad \frac{3}{2} \cdot \frac{4}{15} \leqslant \frac{3}{2} \cdot \frac{2}{3}x \leqslant \frac{3}{2} \cdot \frac{8}{15}$
$\quad \frac{2}{5} \leqslant x \leqslant \frac{4}{5}$
The solution set is $\left\{x \mid \frac{2}{5} \leqslant x \leqslant \frac{4}{5}\right\}$, or $\left[\frac{2}{5}, \frac{4}{5}\right]$.

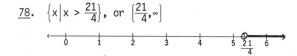

76. $\left\{x \mid x < \frac{13}{40} \text{ or } x > \frac{17}{40}\right\}$, or $\left(-\infty, \frac{13}{40}\right) \cup \left(\frac{17}{40}, \infty\right)$

77. $3x < 4 - 5x < 5 + 3x$
$\quad 0 < 4 - 8x < 5$
$\quad -4 < -8x < 1$
$\quad \frac{1}{2} > x > -\frac{1}{8}$
The solution set is $\left\{x \mid -\frac{1}{8} < x < \frac{1}{2}\right\}$, or $\left(-\frac{1}{8}, \frac{1}{2}\right)$.

78. $\left\{x \mid x > \frac{21}{4}\right\}$, or $\left(\frac{21}{4}, \infty\right)$

79. Given any two numbers b and c either $b < c$, $b = c$, or $b > c$. If $b > c$ is given, we know that $b \not< c$ and $b \neq c$. Thus $b \not\leqslant c$ is true.

80. True

81. Let $c = 6$ and $a = 10$. Then $c \neq a$, but $a \not< c$. The given statement is false.

82. False

83. Let $a = 5$, $c = 12$, and $b = 2$. Then $a < c$ and $b < c$, but $a \not< b$. The given statement is false.

84. False

85. $[4x - 2 < 8 \text{ or } 3(x - 1) < -2] \text{ and } -2 \leqslant 5x \leqslant 10$
$[4x - 2 < 8 \text{ or } 3x - 3 < -2] \text{ and } -\frac{2}{5} \leqslant x \leqslant 2$
$[4x < 10 \text{ or } 3x < 1] \text{ and } -\frac{2}{5} \leqslant x \leqslant 2$
$\left[x < \frac{5}{2} \text{ or } x < \frac{1}{3}\right] \text{ and } -\frac{2}{5} \leqslant x \leqslant 2$
$\left[x < \frac{5}{2}\right] \text{ and } -\frac{2}{5} \leqslant x \leqslant 2$
The solution set is $\left\{x \mid -\frac{2}{5} \leqslant x \leqslant 2\right\}$, or $\left[-\frac{2}{5}, 2\right]$.

86. $\emptyset$

87. $-3x + 2 < 15 < 4x - 3$
$\quad -3x + 2 < 15 \quad \text{and} \quad 15 < 4x - 3$
$\quad -3x < 13 \quad \text{and} \quad 18 < 4x$
$\quad x > -\frac{13}{3} \quad \text{and} \quad \frac{9}{2} < x$
The solution set is $\left\{x \mid x > \frac{9}{2}\right\}$, or $\left(\frac{9}{2}, \infty\right)$.

88. $\{x \mid x \leqslant -8.92 \text{ or } x > -4.8003\}$, or $(-\infty, -8.92] \cup (-4.8003, \infty)$

89. $-5.888 \leqslant x + 1.003 < 9.354$
$\quad -6.891 \leqslant x < 8.351$
The solution set is $\{x \mid -6.891 \leqslant x < 8.351\}$, or $[-6.891, 8.351)$.

90. $\left\{x \mid -\frac{1}{6} < x \leqslant \frac{37}{6}\right\}$, or $\left(-\frac{1}{6}, \frac{37}{6}\right]$

91. $0.9x > 1.8 - 0.1x$
$\quad 0.9x + 0.1x > 1.8 - 0.1x + 0.1x$
$\quad x > 1.8$
The solution set is $\{x \mid x > 1.8\}$, or $(1.8, \infty)$.

92. $\left\{x \mid x \leqslant \frac{1}{2}\right\}$, or $\left(-\infty, \frac{1}{2}\right]$

93. $2x + 3 \leqslant x - 6 \quad \text{or} \quad 3x - 2 \leqslant 4x + 5$
$2x - x \leqslant -6 - 3 \quad \text{or} \quad 3x - 4x \leqslant 5 + 2$
$x \leqslant -9 \quad \text{or} \quad -x \leqslant 7$
$x \leqslant -9 \quad \text{or} \quad x \geqslant -7$
The solution set is $\{x \mid x \leqslant -9 \text{ or } x \geqslant -7\}$, or $(-\infty, -9] \cup [-7, \infty)$.

94. $\{x \mid 10 < x \leqslant 18\}$, or $(10, 18]$

95. $2x - 3 < \frac{13}{4}x + 10 - 1.25x$

$2x - 3 < \frac{13}{4}x + 10 - \frac{5}{4}x$

$2x - 3 < 2x + 10$

$-3 < 10$

The inequality $-3 < 10$ is true for all real values of x. The solution set is {x|x is a real number}, or $(-\infty, \infty)$.

96. $\emptyset$

97. $x + \frac{1}{10} \leqslant -\frac{1}{10}$ or $x + \frac{1}{10} \geqslant \frac{1}{10}$

$x \leqslant -\frac{2}{10}$ or $x \geqslant 0$

$x \leqslant -\frac{1}{5}$ or $x \geqslant 0$

The solution set is $\left\{x \middle| x \leqslant -\frac{1}{5} \text{ or } x \geqslant 0\right\}$, or $\left(-\infty, -\frac{1}{5}\right] \cup [0, \infty)$.

Exercise Set 4.4

1. $|3x| = |3| \cdot |x| = 3|x|$

2. $17|x|$

3. $|9x^2| = |9| \cdot |x^2|$

   $= 9|x^2|$

   $= 9x^2$ x^2 is never negative

4. $6x^2$

5. $|-4x^2| = |-4| \cdot |x^2| = 4|x^2| = 4x^2$

6. $10x^2$

7. $|-8y| = |-8| \cdot |y| = 8|y|$

8. $13|y|$

9. $\left|\frac{-4}{x}\right| = \frac{|-4|}{|x|} = \frac{4}{|x|}$

10. $\frac{|y|}{7}$

11. $\left|\frac{x^2}{-y}\right| = \frac{|x^2|}{|-y|}$

    $= \frac{x^2}{|-y|}$

    $= \frac{x^2}{|y|}$ The absolute value of the additive inverse of a number is the same as the absolute value of the number.

12. $\frac{x^4}{|y|}$

13. $\left|\frac{-8x^2}{2x}\right| = |-4x| = |-4| \cdot |x| = 4|x|$

14. $\frac{3}{|y|}$

15. $|-8 - (-42)| = |34| = 34$

16. 27

17. $|26 - 15| = |11| = 11$

18. 36

19. $|-9 - 24| = |-33| = 33$

20. 19

21. $|-5 - 0| = |-5| = 5$

22. 23

23. $|x| = 3$

    $x = -3$ or $x = 3$ Using the absolute-value principle

    The solution set is {-3,3}.

24. {-5,5}

25. $|x| = -3$

    The absolute value of a number is always nonnegative. Therefore, the solution set is $\emptyset$.

26. $\emptyset$

27. $|p| = 0$

    The only number whose absolute value is 0 is 0. The solution set is {0}.

28. {-8.6,8.6}

29. $|t| = 5.5$

    $t = -5.5$ or $t = 5.5$ Using the absolute-value principle

    The solution set is {-5.5,5.5}.

30. {0}

31. $|x - 3| = 12$

    $x - 3 = -12$ or $x - 3 = 12$ Absolute-value principle

    $x = -9$ or $x = 15$

    The solution set is {-9,15}.

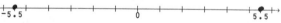

32. $\left\{-\frac{4}{3}, \frac{8}{3}\right\}$

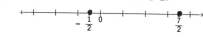
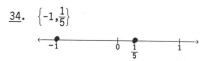

33. $|2x - 3| = 4$

$2x - 3 = -4$ or $2x - 3 = 4$ Absolute-value principle

$2x = -1$ or $2x = 7$

$x = -\frac{1}{2}$ or $x = \frac{7}{2}$

The solution set is $\left\{-\frac{1}{2}, \frac{7}{2}\right\}$.

34. $\left\{-1, \frac{1}{5}\right\}$

35. $|2y - 7| = 10$

$2y - 7 = -10$ or $2y - 7 = 10$

$2y = -3$ or $2y = 17$

$y = -\frac{3}{2}$ or $y = \frac{17}{2}$

The solution set is $\left\{-\frac{3}{2}, \frac{17}{2}\right\}$.

36. $\left\{-\frac{4}{3}, 4\right\}$

37. $|4x - 9| = 14$

$4x - 9 = -14$ or $4x - 9 = 14$

$4x = -5$ or $4x = 23$

$x = -\frac{5}{4}$ or $x = \frac{23}{4}$

The solution set $\left\{-\frac{5}{4}, \frac{23}{4}\right\}$.

38. $\left\{-\frac{5}{3}, \frac{19}{9}\right\}$

39. $|x| + 7 = 18$

$|x| + 7 - 7 = 18 - 7$ Adding -7

$|x| = 11$

$x = -11$ or $x = 11$ Absolute-value principle

The solution set is $\{-11, 11\}$.

40. $\{-8.3, 8.3\}$

41. $678 = 289 + |t|$

$389 = |t|$ Adding -289

$t = -389$ or $t = 389$ Absolute-value principle

The solution set is $\{-389, 389\}$.

42. $\{-433, 433\}$

43. $|5x| = 40$

$5x = -40$ or $5x = 40$

$x = -8$ or $x = 8$

The solution set is $\{-8, 8\}$.

44. $\{-9, 9\}$

45. $|3x| - 4 = 17$

$|3x| = 21$ Adding 4

$3x = -21$ or $3x = 21$

$x = -7$ or $x = 7$

The solution set is $\{-7, 7\}$.

46. $\{-4, 4\}$

47. $5|q| - 2 = 9$

$5|q| = 11$ Adding 2

$|q| = \frac{11}{5}$ Multiplying by $\frac{1}{5}$

$q = -\frac{11}{5}$ or $q = \frac{11}{5}$

The solution set is $\left\{-\frac{11}{5}, \frac{11}{5}\right\}$.

48. $\{-2, 2\}$

49. $\left|\frac{2x - 1}{3}\right| = 5$

$\frac{2x - 1}{3} = -5$ or $\frac{2x - 1}{3} = 5$

$2x - 1 = -15$ or $2x - 1 = 15$

$2x = -14$ or $2x = 16$

$x = -7$ or $x = 8$

The solution set is $\{-7, 8\}$.

50. $\left\{-\frac{38}{5}, \frac{46}{5}\right\}$

51. $|m + 5| + 9 = 16$

$|m + 5| = 7$ Adding -9

$m + 5 = -7$ or $m + 5 = 7$

$m = -12$ or $m = 2$

The solution set is $\{-12, 2\}$.

52. $\{6, 8\}$

53. $|g + 7| + 13 = 4$

$|g + 7| = -9$

The absolute value of a number is always nonnegative. The solution set is $\emptyset$.

54. $\{0, 7\}$

55. $\left|\dfrac{2x-1}{3}\right| = 1$

$\dfrac{2x-1}{3} = -1$ or $\dfrac{2x-1}{3} = 1$

$2x - 1 = -3$ or $2x - 1 = 3$

$2x = -2$ or $2x = 4$

$x = -1$ or $x = 2$

The solution set is $\{-1,2\}$.

56. $\left\{-\dfrac{8}{3},4\right\}$

57. $\left|\dfrac{2x-1}{0.0059}\right| = 1$

$\dfrac{2x-1}{0.0059} = -1$ or $\dfrac{2x-1}{0.0059} = 1$

$2x - 1 = -0.0059$ or $2x - 1 = 0.0059$

$2x = 0.9941$ or $2x = 1.0059$

$x = 0.49705$ or $x = 0.50295$

The solution set is $\{0.49705, 0.50295\}$.

58. $\left\{-1,\dfrac{7}{3}\right\}$

59. $\left|\dfrac{0.005x - 0.004}{0.0059}\right| = 0.0043$

$\dfrac{0.005x - 0.004}{0.0059} = -0.0043$

$0.005x - 0.004 = -0.00002537$

$0.005x = 0.00397463$

$x = 0.794926$

or

$\dfrac{0.005x - 0.004}{0.0059} = 0.0043$

$0.005x - 0.004 = 0.00002537$

$0.005x = 0.00402537$

$x = 0.805074$

The solution set is $\{0.794926, 0.805074\}$.

60. $\{0.1396, 0.1417\}$

61. $|3x - 4| = -2$

The absolute value of a number is always nonnegative. The solution set is $\emptyset$.

62. $\emptyset$

63. $\left|\dfrac{5}{9} + 3x\right| = \dfrac{1}{6}$

$\dfrac{5}{9} + 3x = -\dfrac{1}{6}$ or $\dfrac{5}{9} + 3x = \dfrac{1}{6}$

$3x = -\dfrac{13}{18}$ or $3x = -\dfrac{7}{18}$

$x = -\dfrac{13}{54}$ or $x = -\dfrac{7}{54}$

The solution set is $\left\{-\dfrac{13}{54}, -\dfrac{7}{54}\right\}$.

64. $\left\{-\dfrac{1}{30}, \dfrac{11}{30}\right\}$

65. $|3x + 4| = |x - 7|$

$3x + 4 = x - 7$ or $3x + 4 = -(x - 7)$

$2x + 4 = -7$ or $3x + 4 = -x + 7$

$2x = -11$ or $4x + 4 = 7$

$x = -\dfrac{11}{2}$ or $4x = 3$

$x = \dfrac{3}{4}$

The solution set is $\left\{-\dfrac{11}{2}, \dfrac{3}{4}\right\}$.

66. $\left\{11, \dfrac{5}{3}\right\}$

67. $|x + 5| = |x - 2|$

$x + 5 = x - 2$ or $x + 5 = -(x - 2)$

$5 = -2$ or $x + 5 = -x + 2$

False – yields no solution or $2x + 5 = 2$

$2x = -3$

$x = -\dfrac{3}{2}$

The solution set is $\left\{-\dfrac{3}{2}\right\}$.

68. $\left\{-\dfrac{1}{2}\right\}$

69. $|2a + 4| = |3a - 1|$

$2a + 4 = 3a - 1$ or $2a + 4 = -(3a - 1)$

$-a + 4 = -1$ or $2a + 4 = -3a + 1$

$-a = -5$ or $5a + 4 = 1$

$a = 5$ or $5a = -3$

$a = -\dfrac{3}{5}$

The solution set is $\left\{5, -\dfrac{3}{5}\right\}$.

70. $\left\{-4, -\dfrac{10}{9}\right\}$

71. $|y - 3| = |3 - y|$

$y - 3 = 3 - y$ or $y - 3 = -(3 - y)$

$2y - 3 = 3$ or $y - 3 = -3 + y$

$2y = 6$ or $-3 = -3$

$y = 3$ True for all real values of y

The solution set is $\{y \mid y \text{ is a real number}\}$.

72. $\{m \mid m \text{ is a real number}\}$

73. $|5 - p| = |p + 8|$

    $5 - p = p + 8$ or $5 - p = -(p + 8)$

    $5 - 2p = 8$ or $5 - p = -p - 8$

    $-2p = -3$ or $5 = -8$

       $p = \frac{3}{2}$ False

The solution set is $\left\{\frac{3}{2}\right\}$.

74. $\left\{-\frac{11}{2}\right\}$

75. $\left|\frac{2x - 3}{6}\right| = \left|\frac{4 - 5x}{8}\right|$

    $\frac{2x - 3}{6} = \frac{4 - 5x}{8}$ or $\frac{2x - 3}{6} = -\left(\frac{4 - 5x}{8}\right)$

  $24\left(\frac{2x - 3}{6}\right) = 24\left(\frac{4 - 5x}{8}\right)$ or $\frac{2x - 3}{6} = \frac{-4 + 5x}{8}$

    $8x - 12 = 12 - 15x$ or $24\left(\frac{2x - 3}{6}\right) = 24\left(\frac{-4 + 5x}{8}\right)$

    $23x - 12 = 12$ or $8x - 12 = -12 + 15x$

    $23x = 24$ or $-7x - 12 = -12$

      $x = \frac{24}{23}$ or $-7x = 0$

                            $x = 0$

The solution set is $\left\{\frac{24}{23}, 0\right\}$.

76. $\left\{-\frac{23}{31}, 47\right\}$

77. $\left|\frac{1}{2}x - 5\right| = \left|\frac{1}{4}x + 3\right|$

  $\frac{1}{2}x - 5 = \frac{1}{4}x + 3$ or $\frac{1}{2}x - 5 = -\left(\frac{1}{4}x + 3\right)$

  $\frac{1}{4}x - 5 = 3$ or $\frac{1}{2}x - 5 = -\frac{1}{4}x - 3$

  $\frac{1}{4}x = 8$ or $\frac{3}{4}x - 5 = -3$

   $x = 32$ or $\frac{3}{4}x = 2$

                          $x = \frac{8}{3}$

The solution set is $\left\{32, \frac{8}{3}\right\}$.

78. $\left\{-\frac{48}{37}, -\frac{144}{5}\right\}$

79. <u>Familiarize</u>. Let c = the number of coach seats and f = the number of first-class seats.

<u>Translate</u>.

Total number of seats is 153.

      $c + f$ = 153

Number of coach seats	is	5 more than	number of first-class seats.
c	$=$	5 $+$	f

We have a system of equations:

    $c + f = 153$,   (1)

       $c = 5 + f$ (2)

79. (continued)

<u>Carry out</u>. Solve using substitution. Substitute $5 + f$ for c in (1) and solve for f.

    $(5 + f) + f = 153$

        $5 + 2f = 153$

           $2f = 148$

            $f = 74$

Substitute 74 for f in (2) and solve for c.

    $c = 5 + 74 = 79$

<u>Check</u>. $74 + 79 = 153$, and 79 is 5 more than 74. The numbers check.

<u>State</u>. There are 74 first-class seats and 79 coach seats.

80. $\left\{\frac{1}{2}, \frac{15}{2}\right\}$

81. $3(x + 4) = -2(x - 2)$

    $3x + 12 = -2x + 4$

    $5x + 12 = 4$

        $5x = -8$

         $x = -\frac{8}{5}$

82. -8

83. $|3x^4| = |3| \cdot |x^4|$

    $= 3 \cdot |x^4|$

    $= 3x^4$ x^4 is never negative

84. $5y^6$

85. $|3x^3| = |3| \cdot |x^3| = 3 \cdot |x^2 \cdot x| = 3 \cdot |x^2| \cdot |x| = 3x^2|x|$

86. $5y^2|y|$

87. From the definition of absolute value, $|2x - 5| = 2x - 5$ only when $2x - 5 \geqslant 0$. Solve $2x - 5 \geqslant 0$.

    $2x - 5 \geqslant 0$

       $2x \geqslant 5$

        $x \geqslant \frac{5}{2}$

The solution set is $\left\{x \middle| x \geqslant \frac{5}{2}\right\}$.

88. $\{-31, -33\}$

89. $|x + 5| = x + 5$

From the definition of absolute value, $|x + 5| = x + 5$ only when $x + 5 \geqslant 0$, or $x \geqslant -5$. The solution set is $\{x | x \geqslant -5\}$.

90. $\{x | x \geqslant 1\}$

91. $|7x - 2| = x + 4$

From the definition of absolute value, we know
$x + 4 \geqslant 0$, or $x \geqslant -4$. So we have $x \geqslant -4$ and

$$7x - 2 = x + 4 \quad \text{or} \quad 7x - 2 = -(x + 4)$$
$$6x = 6 \quad \text{or} \quad 7x - 2 = -x - 4$$
$$x = 1 \quad \text{or} \quad 8x = -2$$
$$x = -\frac{1}{4}$$

The solution set is
$\left\{x \,\middle|\, x \geqslant -4 \text{ and } x = 1 \text{ or } x = -\frac{1}{4}\right\}$, or $\left\{1, -\frac{1}{4}\right\}$.

92. $\emptyset$

Exercise Set 4.5

1. $|x| = 3$

$x = -3$ or $x = 3$ Part (a)

The solution set is $\{-3, 3\}$.

2. $\{-5, 5\}$;

3. $|x| < 3$

$-3 < x < 3$ Part (b)

The solution set is $\{x \,|\, -3 < x < 3\}$, or $(-3, 3)$.

4. $\{x \,|\, -5 \leqslant x \leqslant 5\}$, or $[-5, 5]$;

5. $|x| \geqslant 2$

$x \leqslant -2$ or $x \geqslant 2$ Part (c)

The solution set is $\{x \,|\, x \leqslant -2 \text{ or } x \geqslant 2\}$, or
$(-\infty, -2] \cup [2, \infty)$.

6. $\{y \,|\, y < -8 \text{ or } y > 8\}$, or $(-\infty, -8) \cup (8, \infty)$

7. $|t| \geqslant 5.5$

$t \leqslant -5.5$ or $t \geqslant 5.5$ Part (c)

The solution set is $\{t \,|\, t \leqslant -5.5 \text{ or } t \geqslant 5.5\}$,
or $(-\infty, -5.5] \cup [5.5, \infty)$.

8. $\{m \,|\, m \neq 0\}$, or $(-\infty, 0) \cup (0, \infty)$

9. $|x - 3| < 1$

$-1 < x - 3 < 1$ Part (b)

$2 < x < 4$ Adding 3

The solution set is $\{x \,|\, 2 < x < 4\}$, or $(2, 4)$.

10. $\{x \,|\, -4 < x < 8\}$, or $(-4, 8)$

11. $|x + 2| \leqslant 5$

$-5 \leqslant x + 2 \leqslant 5$ Part (b)

$-7 \leqslant x \leqslant 3$ Adding -2

The solution set is $\{x \,|\, -7 \leqslant x \leqslant 3\}$, or $[-7, 3]$.

12. $\{x \,|\, -5 \leqslant x \leqslant -3\}$, or $[-5, -3]$

13. $|x - 3| > 1$

$x - 3 < -1$ or $x - 3 > 1$ Part (c)

$x < 2$ or $x > 4$ Adding 3

The solution set is $\{x \,|\, x < 2 \text{ or } x > 4\}$, or
$(-\infty, 2) \cup (4, \infty)$.

14. $\{x \,|\, x < -4 \text{ or } x > 8\}$, or $(-\infty, -4) \cup (8, \infty)$

15. $|2x - 3| \leqslant 4$

$-4 \leqslant 2x - 3 \leqslant 4$ Part (b)

$-1 \leqslant 2x \leqslant 7$ Adding 3

$-\frac{1}{2} \leqslant x \leqslant \frac{7}{2}$ Multiplying by $\frac{1}{2}$

The solution set is $\left\{x \,\middle|\, -\frac{1}{2} \leqslant x \leqslant \frac{7}{2}\right\}$, or $\left[-\frac{1}{2}, \frac{7}{2}\right]$.

16. $\left\{x \,\middle|\, -1 \leqslant x \leqslant \frac{1}{5}\right\}$, or $\left[-1, \frac{1}{5}\right]$

17. $|2y - 7| > 10$

$2y - 7 < -10$ or $2y - 7 > 10$ Part (c)

$2y < -3$ or $2y > 17$ Adding 7

$y < -\frac{3}{2}$ or $y > \frac{17}{2}$ Multiplying by $\frac{1}{2}$

The solution set is $\left\{y \,\middle|\, y < -\frac{3}{2} \text{ or } y > \frac{17}{2}\right\}$, or
$\left(-\infty, -\frac{3}{2}\right] \cup \left[\frac{17}{2}, \infty\right)$.

18. $\left\{y \,\middle|\, y < -\dfrac{4}{3} \text{ or } y > 4\right\}$, or $\left[-\infty, -\dfrac{4}{3}\right] \cup (4, \infty)$

19. $|4x - 9| \geqslant 14$

$4x - 9 \leqslant -14 \quad \text{or} \quad 4x - 9 \geqslant 14 \quad$ Part (c)

$\qquad 4x \leqslant -5 \quad \text{or} \qquad 4x \geqslant 23$

$\qquad x \leqslant -\dfrac{5}{4} \quad \text{or} \qquad x \geqslant \dfrac{23}{4}$

The solution set is $\left\{x \,\middle|\, x \leqslant -\dfrac{5}{4} \text{ or } x \geqslant \dfrac{23}{4}\right\}$, or

$\left[-\infty, -\dfrac{5}{4}\right] \cup \left[\dfrac{23}{4}, \infty\right]$.

20. $\left\{y \,\middle|\, y \leqslant -\dfrac{2}{9} \text{ or } y \geqslant \dfrac{4}{9}\right\}$, or $\left[-\infty, -\dfrac{2}{9}\right] \cup \left[\dfrac{4}{9}, \infty\right]$

21. $|y - 3| < 12$

$-12 < y - 3 < 12 \quad$ Part (b)

$-9 < y < 15 \qquad$ Adding 3

The solution set is $\{y \,|\, -9 < y < 15\}$, or $(-9, 15)$.

22. $\{p \,|\, -1 < p < 5\}$, or $(-1, 5)$

23. $|2x + 3| \leqslant 4$

$-4 \leqslant 2x + 3 \leqslant 4 \quad$ Part (b)

$-7 \leqslant 2x \leqslant 1 \qquad$ Adding -3

$-\dfrac{7}{2} \leqslant x \leqslant \dfrac{1}{2} \qquad$ Multiplying by $\dfrac{1}{2}$

The solution set is $\left\{x \,\middle|\, -\dfrac{7}{2} \leqslant x \leqslant \dfrac{1}{2}\right\}$, or $\left[-\dfrac{7}{2}, \dfrac{1}{2}\right]$.

24. $\left\{x \,\middle|\, -\dfrac{1}{5} \leqslant x \leqslant 1\right\}$, or $\left[-\dfrac{1}{5}, 1\right]$

25. $|4 - 3y| > 8$

$4 - 3y < -8 \quad \text{or} \quad 4 - 3y > 8 \quad$ Part (c)

$\quad -3y < -12 \quad \text{or} \qquad -3y > 4 \qquad$ Adding -4

$\qquad y > 4 \quad \text{or} \qquad y < -\dfrac{4}{3} \quad$ Multiplying by $-\dfrac{1}{3}$

The solution set is $\left\{y \,\middle|\, y < -\dfrac{4}{3} \text{ or } y > 4\right\}$, or

$\left[-\infty, -\dfrac{4}{3}\right] \cup (4, \infty)$.

26. $\{y \,|\, y < 1 \text{ or } y > 6\}$, or $(-\infty, 1) \cup (6, \infty)$

27. $|9 - 4x| \geqslant 14$

$9 - 4x \leqslant -14 \quad \text{or} \quad 9 - 4x \geqslant 14 \quad$ Part (c)

$\quad -4x \leqslant -23 \quad \text{or} \qquad -4x \geqslant 5 \qquad$ Adding -9

$\qquad x \geqslant \dfrac{23}{4} \quad \text{or} \qquad x \leqslant -\dfrac{5}{4} \quad$ Multiplying by $-\dfrac{1}{4}$

The solution set is $\left\{x \,\middle|\, x \leqslant -\dfrac{5}{4} \text{ or } x \geqslant \dfrac{23}{4}\right\}$, or

$\left[-\infty, -\dfrac{5}{4}\right] \cup \left[\dfrac{23}{4}, \infty\right]$.

28. $\left\{p \,\middle|\, p \leqslant -\dfrac{5}{3} \text{ or } p \geqslant \dfrac{19}{9}\right\}$, or $\left[-\infty, -\dfrac{5}{3}\right] \cup \left[\dfrac{19}{9}, \infty\right]$

29. $|3 - 4x| < 21$

$-21 < 3 - 4x < 21 \quad$ Part (b)

$-24 < -4x < 18 \qquad$ Adding -3

$6 > x > -\dfrac{9}{2} \qquad$ Multiplying by $-\dfrac{1}{4}$ and simplifying

The solution set is $\left\{x \,\middle|\, -\dfrac{9}{2} < x < 6\right\}$, or $\left[-\dfrac{9}{2}, 6\right]$.

30. $\left\{x \,\middle|\, -5 \leqslant x \leqslant \dfrac{25}{7}\right\}$, or $\left[-5, \dfrac{25}{7}\right]$

31. $\left|\dfrac{1}{2} + 3x\right| \geqslant 12$

$\dfrac{1}{2} + 3x \leqslant -12 \quad \text{or} \quad \dfrac{1}{2} + 3x \geqslant 12 \quad$ Part (c)

$\qquad 3x \leqslant -\dfrac{25}{2} \quad \text{or} \qquad 3x \geqslant \dfrac{23}{2} \quad$ Adding $-\dfrac{1}{2}$

$\qquad x \leqslant -\dfrac{25}{6} \quad \text{or} \qquad x \geqslant \dfrac{23}{6} \quad$ Multiplying by $\dfrac{1}{3}$

The solution set is $\left\{x \,\middle|\, x \leqslant -\dfrac{25}{6} \text{ or } x \geqslant \dfrac{23}{6}\right\}$, or

$\left[-\infty, -\dfrac{25}{6}\right] \cup \left[\dfrac{23}{6}, \infty\right]$.

32. $\{y \,|\, y < -72 \text{ or } y > 120\}$, or $(-\infty, -72) \cup (120, \infty)$

33. $\left|\dfrac{x-7}{3}\right| < 4$

    $-4 < \dfrac{x-7}{3} < 4$ Part (b)

    $-12 < x - 7 < 12$ Multiplying by 3

    $-5 < x < 19$ Adding 7

    The solution set is {x|-5 < x < 19}, or (-5,19).

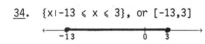

34. {x|-13 ≤ x ≤ 3}, or [-13,3]

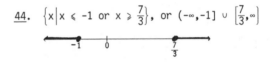

35. $\left|\dfrac{2-5x}{4}\right| \geqslant \dfrac{2}{3}$

    $\dfrac{2-5x}{4} \leqslant -\dfrac{2}{3}$ or $\dfrac{2-5x}{4} \geqslant \dfrac{2}{3}$ Part (c)

    $2 - 5x \leqslant -\dfrac{8}{3}$ or $2 - 5x \geqslant \dfrac{8}{3}$ Multiplying by 4

    $-5x \leqslant -\dfrac{14}{3}$ or $-5x \geqslant \dfrac{2}{3}$ Adding -2

    $x \geqslant \dfrac{14}{15}$ or $x \leqslant -\dfrac{2}{15}$ Multiplying by $-\dfrac{1}{5}$

    The solution set is $\left\{x \middle| x \leqslant -\dfrac{2}{15} \text{ or } x \geqslant \dfrac{14}{15}\right\}$, or $\left[-\infty,-\dfrac{2}{15}\right] \cup \left[\dfrac{14}{15},\infty\right)$.

36. $\left\{x \middle| x < -\dfrac{43}{24} \text{ or } x > \dfrac{9}{8}\right\}$, or $\left[-\infty,-\dfrac{43}{24}\right] \cup \left(\dfrac{9}{8},\infty\right]$

37. |m + 5| + 9 ≤ 16

    |m + 5| ≤ 7   Adding -9

    $-7 \leqslant m + 5 \leqslant 7$

    $-12 \leqslant m \leqslant 2$

    The solution set is {m|-12 ≤ m ≤ 2}, or [-12,2].

38. {t|t ≤ 6 or t ≥ 8}, or (-∞,6] ∪ [8,∞)

39. |g + 7| + 13 > 14

    |g + 7| > 1   Adding -13

    g + 7 < -1 or g + 7 > 1

    g < -8 or g > -6

    The solution set is {g|g < -8 or g > -6}, or (-∞,-8) ∪ (-6,∞).

40. {x|x < 0 or x > 7}, or (-∞,0) ∪ (7,∞)

41. $\left|\dfrac{2x-1}{3}\right| \leqslant 1$

    $-1 \leqslant \dfrac{2x-1}{3} \leqslant 1$

    $-3 \leqslant 2x - 1 \leqslant 3$

    $-2 \leqslant 2x \leqslant 4$

    $-1 \leqslant x \leqslant 2$

    The solution set is {x|-1 ≤ x ≤ 2}, or [-1,2].

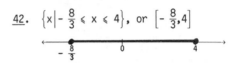

42. $\left\{x \middle| -\dfrac{8}{3} \leqslant x \leqslant 4\right\}$, or $\left[-\dfrac{8}{3},4\right]$

43. $\left|\dfrac{2x-1}{0.0059}\right| \leqslant 1$

    $-1 \leqslant \dfrac{2x-1}{0.0059} \leqslant 1$

    $-0.0059 \leqslant 2x - 1 \leqslant 0.0059$

    $0.9941 \leqslant 2x \leqslant 1.0059$

    $0.49705 \leqslant x \leqslant 0.50295$

    The solution set is {x|0.49705 ≤ x ≤ 0.50295}, or [0.49705,0.50295].

44. $\left\{x \middle| x \leqslant -1 \text{ or } x \geqslant \dfrac{7}{3}\right\}$, or $(-\infty,-1] \cup \left[\dfrac{7}{3},\infty\right)$

45. $\left|\dfrac{0.005x-0.004}{0.0059}\right| \leqslant 0.0043$

    $-0.0043 \leqslant \dfrac{0.005x-0.004}{0.0059} \leqslant 0.0043$

    $-0.00002537 \leqslant 0.005x - 0.004 \leqslant 0.00002537$

    $0.00397463 \leqslant 0.005x \leqslant 0.00402537$

    $0.794926 \leqslant x \leqslant 0.805074$

    The solution set is {x|0.794926 ≤ x ≤ 0.805074}, or [0.794926,0.805074].

46. {x|x ≤ 0.1396 or x ≥ 0.1417}, or (-∞,0.1396] ∪ [0.1417,∞)

47. <u>Familiarize</u>. Let ℓ represent the length and w represent the width. Recall that perimeter P = 2ℓ + 2w, and area A = ℓw.

<u>Translate</u>.

The perimeter is 628 m.

$$2ℓ + 2w = 628$$

The length is 6 m more than the width.

$$ℓ = 6 + w$$

We have a system of equations:

$$2ℓ + 2w = 628 \quad (1)$$
$$ℓ = 6 + w \quad (2)$$

<u>Carry out</u>. Use the substitution method to solve the system of equations and then compute the area. Substitute 6 + w for ℓ in (1).

$$2(6 + w) + 2w = 628$$
$$12 + 2w + 2w = 628$$
$$12 + 4w = 628$$
$$4w = 616$$
$$w = 154$$

Substitute 154 for w in (2).

$$ℓ = 6 + 154 = 160$$
$$A = ℓw = 160(154) = 24,640$$

<u>Check</u>. 2ℓ + 2w = 2·160 + 2·154 = 320 + 308 = 628. 160 is 6 more than 154. The numbers check.

<u>State</u>. The area is 24,640 m².

48. 132 adults, 118 children

49. |3x − 4| > −2

From the definition of absolute value we know that |3x − 4| ⩾ 0. Thus, |3x − 4| > −2 is true for all x. The solution set is the set of all real numbers, or {x|x is a real number}.

50. Ø

51. $\left|\dfrac{5}{9} + 3x\right| < -\dfrac{1}{6}$

From the definition of absolute value we know that $\left|\dfrac{5}{9} + 3x\right| ⩾ 0$. Thus $\left|\dfrac{5}{9} + 3x\right| < -\dfrac{1}{6}$ is false for all x. The solution is Ø.

52. {x|−33 < x < −31}, or (−33,−31)

53. |x + 5| > x

The inequality is true for all x < 0 (because absolute value must be nonnegative). The solution set in this case is {x|x < 0}. If x = 0, we have |0 + 5| > 0, which is true. The solution set in this case is {0}. If x > 0, we have the following:

$$x + 5 < -x \quad \text{or} \quad x + 5 > x$$
$$2x < -5 \quad \text{or} \quad 5 > 0$$
$$x < -\dfrac{5}{2}$$

53. (continued)

Although x > 0 and $x < -\dfrac{5}{2}$ yields no solution, x > 0 and 5 > 0 (true for all x) yield the solution set {x|x > 0} in this case. The solution set for the inequality is {x|x < 0} ∪ {0} ∪ {x|x > 0}, or {x|x is a real number}.

54. {x|−4 ⩽ x ⩽ −1 or 3 ⩽ x ⩽ 6}

55. |7x − 2| ⩽ x + 4

By the definition of absolute value, x + 4 ⩾ 0, or x ⩾ −4.

$$-(x + 4) ⩽ 7x - 2 ⩽ x + 4$$
$$-(x + 4) ⩽ 7x - 2 \quad \text{and} \quad 7x - 2 ⩽ x + 4$$
$$-x - 4 ⩽ 7x - 2 \quad \text{and} \quad 6x - 2 ⩽ 4$$
$$-4 ⩽ 8x - 2 \quad \text{and} \quad 6x ⩽ 6$$
$$-2 ⩽ 8x \quad \text{and} \quad x ⩽ 1$$
$$-\dfrac{1}{4} ⩽ x$$

The solution set is $\left\{x\middle|x ⩾ -4 \text{ and } -\dfrac{1}{4} ⩽ x ⩽ 1\right\}$, or $\left\{x\middle|-\dfrac{1}{4} ⩽ x ⩽ 1\right\}$, or $\left[-\dfrac{1}{4},1\right]$.

56. Ø

57. |x + 1| ⩽ |x − 3|

If x − 3 ⩾ 0, or x ⩾ 3, then |x − 3| = x − 3 and we have

$$-(x - 3) ⩽ x + 1 ⩽ x - 3 \quad \text{Part (b)}$$
$$-(x - 3) ⩽ x + 1 \text{ and } x + 1 ⩽ x - 3 \quad \text{Writing as}$$
$$\text{two inequalities}$$
$$-x + 3 ⩽ x + 1 \text{ and } \quad 1 ⩽ -3$$
$$-2x ⩽ -2$$
$$x ⩾ 1$$

Since 1 ⩽ −3 is false for all x, the solution set for this possibility is Ø.

If x − 3 < 0, or x < 3, then |x − 3| = −(x − 3) and we have

$$-[-(x - 3)] ⩽ x + 1 ⩽ -(x - 3) \quad \text{Part (b)}$$
$$-[-(x - 3)] ⩽ x + 1 \text{ and } x + 1 ⩽ -(x - 3) \quad \text{Writing}$$
$$\text{as two inequalities}$$
$$x - 3 ⩽ x + 1 \text{ and } x + 1 ⩽ -x + 3$$
$$-3 ⩽ 1 \quad \text{and} \quad 2x ⩽ 2$$
$$\text{and} \quad x ⩽ 1$$

The solution set for this possibility is {x|x < 3 and x ⩽ 1}, or {x|x ⩽ 1}.

The solution set of |x + 1| ⩽ |x − 3| is the union of the two solution sets above, or Ø ∪ {x|x ⩽ 1}, or {x|x ⩽ 1}, or (−∞,1].

58. |x| < 3

59. Using part (b), we find that −5 ⩽ y ⩽ 5 is equivalent to |y| ⩽ 5.

60. $|x| \geqslant 6$

61. Using part (c) we find that $p < -10$ or $p > 10$ is equivalent to $|p| > 10$.

62. $|x + 2| < 3$

63. $-1 \leqslant x \leqslant 7$

      $-1 - 3 \leqslant x - 3 \leqslant 7 - 3$ Adding -3

        $-4 \leqslant x - 3 \leqslant 4$

      $|x - 3| \leqslant 4$ Part (b)

64. $|x + 3| > 5$

65. $5 - \frac{1}{8} \leqslant p \leqslant 5 + \frac{1}{8}$

      $-\frac{1}{8} \leqslant p - 5 \leqslant \frac{1}{8}$ Adding -5

      $|p - 5| \leqslant \frac{1}{8}$ Part (b)

66. $\left\{ d \mid 5\frac{1}{2} \text{ ft} \leqslant d \leqslant 6\frac{1}{2} \text{ ft} \right\}$, or $\left[5\frac{1}{2} \text{ ft}, 6\frac{1}{2} \text{ ft} \right]$

Exercise Set 4.6

1. We replace x by -4 and y by 2.

    $$\frac{2x + y < -5}{2(-4) + 2 \mid -5}$$
    $$-8 + 2$$
    $$-6$$

    Since $-6 < -5$ is true, $(-4,2)$ is a solution.

2. Yes

3. We replace x by 8 and y by 14.

    $$\frac{2y - 3x > 5}{2 \cdot 14 - 3 \cdot 8 \mid 5}$$
    $$28 - 24$$
    $$4$$

    Since $4 > 5$ is false, $(8,14)$ is not a solution.

4. Yes

5. Graph: $y > 2x$

    We first graph the line $y = 2x$. We draw the line dashed since the inequality symbol is >. To determine which half-plane to shade, test a point not on the line. We try $(1,1)$ and substitute:

    $$\frac{y > 2x}{1 \mid 2 \cdot 1}$$
    $$2$$

    Since $1 > 2$ is false, $(1,1)$ is not a solution, nor are any points in the half-plane containing $(1,1)$. The points in the opposite half-plane are solutions, so we shade that half-plane and obtain the graph.

5. (continued)

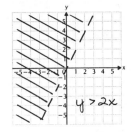

6.

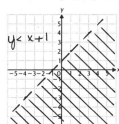

7. Graph: $y < x + 1$

    First graph the line $y = x + 1$. Draw it dashed since the inequality symbol is <. Test the point $(0,0)$ to determine if it is a solution.

    $$\frac{y < x + 1}{0 \mid 0 + 1}$$
    $$1$$

    Since $0 < 1$ is true, we shade the half-plane containing $(0,0)$ and obtain the graph.

8.

9. Graph: $y > x - 2$

    We first graph $y = x - 2$. Draw a dashed line since the inequality symbol is >. Test the point $(0,0)$ to determine if it is a solution.

    $$\frac{y > x - 2}{0 \mid 0 - 2}$$
    $$-2$$

    Since $0 > -2$ is true, we shade the half-plane containing $(0,0)$ and obtain the graph.

9. (continued)

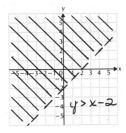
$y > x - 2$

10.

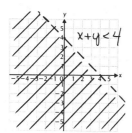
$y \geq x + 4$

11. Graph: x + y < 4

First graph x + y = 4. Draw the line dashed since the inequality symbol is <. Test the point (0,0) to determine if it is a solution.

$$\frac{x + y < 4}{0 + 0 \mid 4}$$
$$0 \mid$$

Since 0 < 4 is true, we shade the half-plane containing (0,0) and obtain the graph.

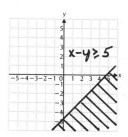
$x + y < 4$

12.

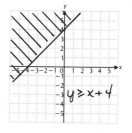
$x - y \geq 5$

13. Graph: 3x + 4y ≤ 12

We first graph 3x + 4y = 12. Draw the line solid since the inequality symbol is ≤. Test the point (0,0) to determine if it is a solution.

$$\frac{3x + 4y \leq 12}{3 \cdot 0 + 4 \cdot 0 \mid 12}$$
$$0 \mid$$

Since 0 ≤ 12 is true, we shade the half-plane containing (0,0) and obtain the graph.

13. (continued)

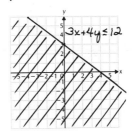
$3x + 4y \leq 12$

14.

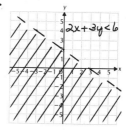
$2x + 3y < 6$

15. Graph: 2y - 3x > 6

We first graph 2y - 3x = 6. Draw the line dashed since the inequality symbol is >. Test the point (0,0) to determine if it is a solution.

$$\frac{2y - 3x > 6}{2 \cdot 0 - 3 \cdot 0 \mid 6}$$
$$0 \mid$$

Since 0 > 6 is false, we shade the half-plane that does not contain (0,0) and obtain the graph.

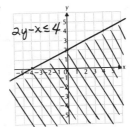
$2y - 3x > 6$

16.

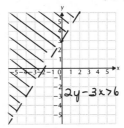
$2y - x \leq 4$

17. Graph: $3x - 2 \leqslant 5x + y$

                    $-2 \leqslant 2x + y$

We first graph $-2 = 2x + y$. Draw the line solid
since the inequality symbol is $\leqslant$. Test the point
(0,0) to determine if it is a solution.

$$\frac{-2 \leqslant 2x + y}{-2 \mid 2 \cdot 0 + 0}$$
$$0$$

Since $-2 \leqslant 0$ is true, we shade the half-plane
containing (0,0) and obtain the graph.

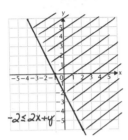

18.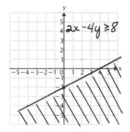

19. Graph: $x < -4$

We first graph $x = -4$. Draw the line dashed since
the inequality symbol is $<$. Test the point (0,0)
to determine if it is a solution.

$$\frac{x < -4}{0 \mid -4}$$

Since $0 < -4$ is false, we shade the half-plane
that does not contain (0,0) and obtain the graph.

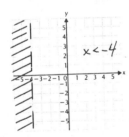

20.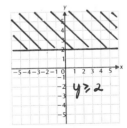

21. Graph: $y > -2$

We first graph $y = -2$. We draw the line dashed
since the inequality symbol is $>$. Test the point
(0,0) to determine if it is a solution.

$$\frac{y > -2}{0 \mid -2}$$

Since $0 > -2$ is true, we shade the half-plane
containing (0,0) and obtain the graph.

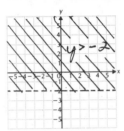

22.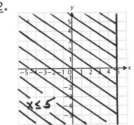

23. Graph: $-4 < y < -1$

This is a system of inequalities:

    $-4 < y,$

    $y < -1$

We graph the equation $-4 = y$ and see that the
graph of $-4 < y$ is the half-plane above the
line $-4 = y$. We also graph $y = -1$ and see that
the graph of $y < -1$ is the half-plane below the
line $y = -1$.

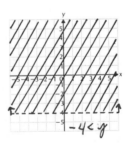 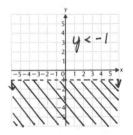

Finally, we shade the intersection of these
graphs.

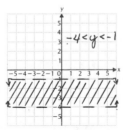

24.

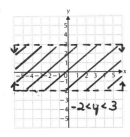

$-2 < y < 3$

25. Graph: $-3 \leqslant x \leqslant 3$

This is a system of inequalities:

$-3 \leqslant x$,

$x \leqslant 3$

Graph $-3 \leqslant x$ and $x \leqslant 3$.

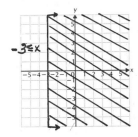 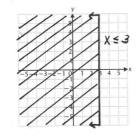

Then we shade the intersection of these graphs.

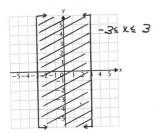

$-3 \leqslant x \leqslant 3$

26.

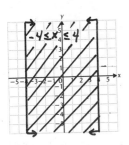

$-4 \leqslant x \leqslant 4$

27. Graph: $0 \leqslant x \leqslant 5$

This is a system of inequalities:

$0 \leqslant x$,

$x \leqslant 5$

Graph $0 \leqslant x$ and $x \leqslant 5$.

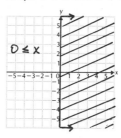 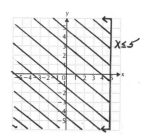

Then we shade the intersection of these graphs.

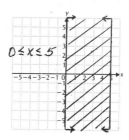

$0 \leqslant x \leqslant 5$

28.

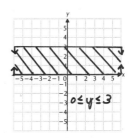

$0 \leqslant y \leqslant 3$

29. Graph: $y < x$,   (1)

$y > -x + 3$ (2)

We graph the lines $y = x$ and $y = -x + 3$, using dashed lines. We indicate the region for each inequality by the arrows at the ends of the lines. Note where the regions overlap and shade the region of solutions.

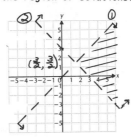

To find the vertex we solve the system of related equations:

$y = x$,

$y = -x + 3$

Solving, we obtain the vertex $\left(\frac{3}{2}, \frac{3}{2}\right)$.

30. y > x,   (1)
     y < -x + 1   (2)

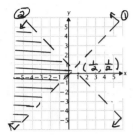

31. Graph: y ⩾ x,   (1)
             y ⩽ -x + 4   (2)

We graph the lines y = x and y = -x + 4, using
solid lines. We indicate the region for each
inequality by the arrows at the ends of lines.
Note where the regions overlap, and shade the
region of solutions.

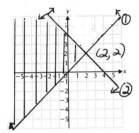

To find the vertex we solve the system of related
equations:

     y = x,
     y = -x + 4

Solving, we obtain the vertex (2,2).

32. y ⩾ x,   (1)
     y ⩽ -x + 2   (2)

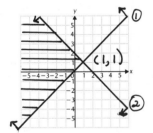

33. Graph: y ⩾ -2,   (1)
             x ⩾ 1   (2)

We graph the lines y = -2 and x = 1, using solid
lines. We indicate the region for each inequality
by arrows. Shade the region where they overlap.

33. (continued)

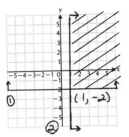

To find the vertex, we solve the system of related
equations:

     y = -2,
     x = 1

Solving, we obtain the vertex (1,-2).

34. y ⩽ -2,   (1)
     x ⩾ 2   (2)

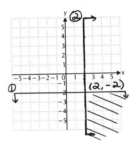

35. Graph: x ⩽ 3,   (1)
             y ⩾ -3x + 2   (2)

Graph the lines x = 3 and y = -3x + 2, using
solid lines. Indicate the region for each
inequality by arrows, and shade the region where
they overlap.

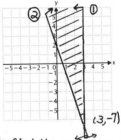

To find the vertex we solve the system of related
equations:

     x = 3,
     y = -3x + 2

Solving, we obtain the vertex (3,-7).

36. $x \geqslant -2$ (1)

     $y \leqslant -2x + 3$ (2)

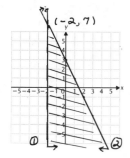

37. Graph: $y \geqslant -2$,   (1)

              $y \geqslant x + 3$ (2)

Graph the lines $y = -2$ and $y = x + 3$, using solid lines. Indicate the region for each inequality by arrows, and shade the region where they overlap.

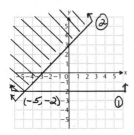

To find the vertex we solve the system of related equations:

     $y = -2$,

     $y = x + 3$

The vertex is $(-5, -2)$.

38. $y \leqslant 4$,   (1)

     $y \geqslant -x + 2$ (2)

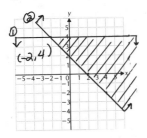

39. Graph: $x + y \leqslant 1$,   (1)

              $x - y \leqslant 2$ (2)

Graph the lines $x + y = 1$ and $x - y = 2$, using solid lines. Indicate the region for each inequality by arrows, and shade the region where they overlap.

39. (continued)

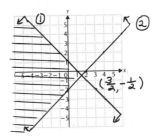

To find the vertex we solve the system of related equations:

     $x + y = 1$,

     $x - y = 2$

The vertex is $\left(\frac{3}{2}, -\frac{1}{2}\right)$.

40. $x + y \leqslant 3$,   (1)

     $x - y \leqslant 4$ (2)

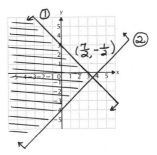

41. Graph: $y - 2x \geqslant 1$,   (1)

              $y - 2x \leqslant 3$ (2)

Graph the lines $y - 2x = 1$ and $y - 2x = 3$, using solid lines. Indicate the region for each inequality by arrows, and shade the region where they overlap.

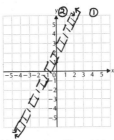

We can see from the graph that the lines are parallel. Hence there are no vertices.

<u>42.</u> y + 3x ⩾ 0,   (1)
y + 3x ⩽ 2   (2)

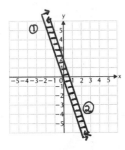

<u>43.</u> Graph: 2y - x ⩽ 2,   (1)
y - 3x ⩾ -1   (2)

Graph the lines 2y - x = 2 and y - 3x = -1, using solid lines. Indicate the region for each inequality by arrows, and shade the region where they overlap.

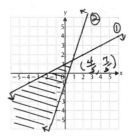

To find the vertex we solve the system of related equations:

2y - x = 2,
y - 3x = -1

The vertex is $\left(\frac{4}{5}, \frac{7}{5}\right)$.

<u>44.</u> y ⩽ 2x + 1,   (1)
y ⩾ -2x + 1,   (2)
x ⩽ 2   (3)

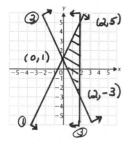

<u>45.</u> Graph: x - y ⩽ 2,   (1)
x + 2y ⩾ 8,   (2)
y ⩽ 4   (3)

Graph the lines x - y = 2, x + 2y = 8, and y = 4, using solid lines. Indicate the region for each inequality by arrows, and shade the region where they overlap.

<u>45.</u> (continued)

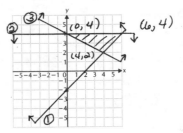

To find the vertices we solve three different systems of equations.

From (1) and (2) we have x - y = 2,
x + 2y = 8.

Solving, we obtain the vertex (4,2).

From (1) and (3) we have x - y = 2,
y = 4.

Solving, we obtain the vertex (6,4).

From (2) and (3) we have x + 2y = 8,
y = 4.

Solving, we obtain the vertex (0,4).

<u>46.</u> x + 2y ⩽ 12,   (1)
2x + y ⩽ 12   (2)
x ⩾ 0,   (3)
y ⩾ 0   (4)

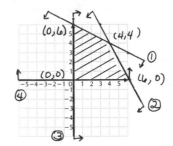

<u>47.</u> Graph: 4y - 3x ⩾ -12,   (1)
4y + 3x ⩾ -36,   (2)
y ⩽ 0,   (3)
x ⩽ 0   (4)

Graph the lines 4y - 3x = -12, 4y + 3x = -36, y = 0, and x = 0, using solid lines. Indicate the region for each inequality by arrows, and shade the region where they overlap.

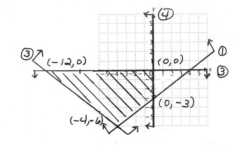

47. (continued)

To find the vertices we solve four different systems of equations.

From (1) and (2) we have 4y - 3x = -12,

4y + 3x = -36.

Solving we obtain the vertex (-4,-6).

From (1) and (4) we have 4y - 3x = -12,

x = 0.

Solving, we obtain the vertex (0,-3).

From (2) and (3) we have 4y + 3x = -36,

y = 0.

Solving, we obtain the vertex (-12,0).

From (3) and (4) we have y = 0,

x = 0.

Solving, we obtain the vertex (0,0).

48. 8x + 5y ⩽ 40,   (1)

x + 2y ⩽ 8,   (2)

x ⩾ 0,   (3)

y ⩾ 0   (4)

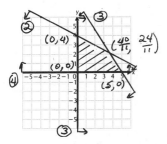

49. Graph: 3x + 4y ⩾ 12,   (1)

5x + 6y ⩽ 30,   (2)

1 ⩽ x ⩽ 3   (3)

Think of (3) as two inequalities:

1 ⩽ x,   (4)

x ⩽ 3   (5)

Graph the lines 3x + 4y = 12, 5x + 6y = 30, x = 1, and x = 3, using solid lines. Indicate the region for each inequality by arrows, and shade the region where they overlap.

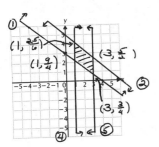

49. (continued)

To find the vertices we solve four different systems of equations.

From (1) and (4) we have 3x + 4y = 12,

x = 1.

Solving, we obtain the vertex $\left[1, \frac{9}{4}\right]$.

From (1) and (5) we have 3x + 4y = 12,

x = 3.

Solving, we obtain the vertex $\left[3, \frac{3}{4}\right]$.

From (2) and (4) we have 5x + 6y = 30,

x = 1.

Solving, we obtain the vertex $\left[1, \frac{25}{6}\right]$.

From (2) and (5) we have 5x + 6y = 30,

x = 3.

Solving, we obtain the vertex $\left[3, \frac{5}{2}\right]$.

50. y - x ⩾ 1,   (1)

y - x ⩽ 3,   (2)

2 ⩽ x ⩽ 5   (3)

Think of (3) as two inequalities:

2 ⩽ x,   (4)

x ⩽ 5   (5)

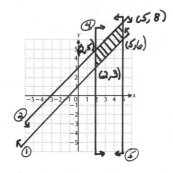

51. Familiarize. We first make a drawing. We let x represent the length of a side of the equilateral triangle. Then x - 5 represents the length of a side of the square.

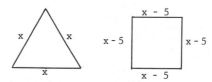

Translate.

Perimeter of triangle = Perimeter of square

3x   =   4(x - 5)

51. (continued)

    Carry out.

    $3x = 4(x - 5)$

    $3x = 4x - 20$

    $20 = x$

    Then $x - 5 = 20 - 5 = 15$.

    Check. If the length of a side of the triangle is 20 and the length of a side of the square is 15, the perimeter of the triangle is 3·20, or 60 and the perimeter of the square is 4·15, or 60. The values check.

    State. The length of a side of the square is 15, and the length of a side of the triangle is 20.

52. $\left(-\dfrac{44}{23}, \dfrac{13}{23}\right)$

53. $5(3x - 4) = -2(x + 5)$

    $15x - 20 = -2x - 10$

    $17x - 20 = -10$

    $17x = 10$

    $x = \dfrac{10}{17}$

54. $-\dfrac{14}{13}$

55. Graph: $x + y \geqslant 5$,   (1)

    $x + y \leqslant -3$ (2)

    Graph the lines $x + y = 5$ and $x + y = -3$, using solid lines, and indicate the region for each inequality by arrows. The regions do not overlap (the solution set is $\emptyset$), so we do not shade any portion of the graph.

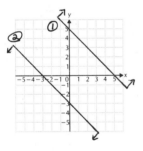

56. $x + y \leqslant 8$,   (1)

    $x + y \leqslant -2$ (2)

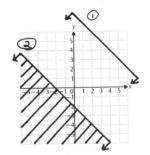

57. Graph: $x - 2y \leqslant 0$,   (1)

    $-2x + y \leqslant 2$,   (2)

    $x \leqslant 2$,   (3)

    $y \leqslant 2$,   (4)

    $x + y \leqslant 4$ (5)

    Graph the five inequalities above, and shade the region where they overlap.

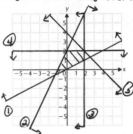

58. $x + y \geqslant 1$,   (1)

    $-x + y \geqslant 2$,   (2)

    $x \leqslant 4$,   (3)

    $y \geqslant 0$,   (4)

    $y \leqslant 4$,   (5)

    $x \leqslant 2$ (6)

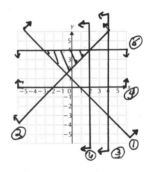

59. We graph the following system of inequalities:

    $0 \leqslant L \leqslant 94$,

    $0 \leqslant W \leqslant 50$

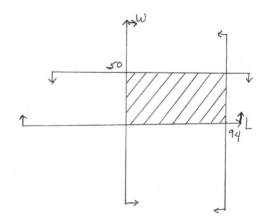

60. $2w + t \geqslant 60,$
     $\quad w \geqslant 0,$
     $\quad t \geqslant 0$

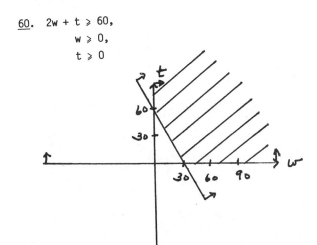

61. Let c = the number of children and a = the number of adults. Graph: $35c + 75a > 1000,$
     $\quad\quad c \geqslant 0,$
     $\quad\quad a \geqslant 0$

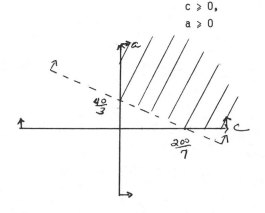

Exercise Set 4.7

---

1. Find the maximum and minimum values of $F(x,y) = 4x + 28y,$

    subject to

    $5x + 3y \leqslant 34,$ (1)
    $3x + 5y \leqslant 30,$ (2)
    $\quad\quad x \geqslant 0,$ (3)
    $\quad\quad y \geqslant 0.$ (4)

    Graph the system of inequalities and find the coordinates of the vertices.

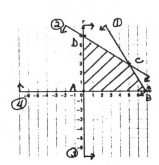

1. (continued)

    To find vertex A we solve the system
    $\quad x = 0$
    $\quad y = 0.$
    Vertex A is (0,0).
    To find vertex B we solve the system
    $\quad 5x + 3y = 34$
    $\quad\quad\quad y = 0.$
    Vertex B is $\left[\frac{34}{5}, 0\right].$
    To find vertex C we solve the system
    $\quad 5x + 3y = 34$
    $\quad 3x + 5y = 30.$
    Vertex C is (5,3).
    To find vertex D we solve the system
    $\quad 3x + 5y = 30$
    $\quad\quad\quad x = 0.$
    Vertex D is (0,6).
    Now compute the function values at the vertices.

Vertex (x,y)	Function $F(x,y) = 4x + 28y$	
A(0,0)	$4\cdot0 + 28\cdot0 = 0 + 0 = 0$	← Minimum
B$\left[\frac{34}{5},0\right]$	$4 \cdot \frac{34}{5} + 28\cdot0 = \frac{136}{5} + 0 = 27\frac{1}{5}$	
C(5,3)	$4\cdot5 + 28\cdot3 = 20 + 84 = 104$	
D(0,6)	$4\cdot0 + 28\cdot6 = 0 + 168 = 168$	← Maximum

    The maximum value of F is 168 when x = 0 and y = 6. The minimum value of F is 0 when x = 0 and y = 0.

2. Maximum 92.8 when x = 0 and y = 5.8; minimum 0 when x = 0 and y = 0

3. Find the maximum and minimum values of $P(x,y) = 16x - 2y + 40,$

    subject to

    $6x + 8y \leqslant 48,$ (1)
    $0 \leqslant y \leqslant 4,$ (2)
    $0 \leqslant x \leqslant 7.$ (3)
    Think of (2) as $0 \leqslant y,$ (4)
    $\quad\quad\quad\quad\quad y \leqslant 4.$ (5)
    Think of (3) as $0 \leqslant x,$ (6)
    $\quad\quad\quad\quad\quad x \leqslant 7.$ (7)

    Graph the system of inequalities and find the coordinates of the vertices.

3. (continued)

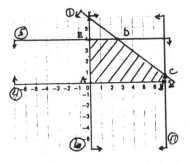

The vertices are $(0,0)$, $(7,0)$ $\left[7,\frac{3}{4}\right]$, $\left[\frac{8}{3},4\right]$, and $(0,4)$. Compute the function values at the vertices.

Vertex (x,y)	Function $P(x,y) = 16x - 2y + 40$	
A(0,0)	$16 \cdot 0 - 2 \cdot 0 + 40 =$ $0 - 0 + 40 = 40$	
B(7,0)	$16 \cdot 7 - 2 \cdot 0 + 40 =$ $112 - 0 + 40 = 152$	← Maximum
C$\left[7,\frac{3}{4}\right]$	$16 \cdot 7 - 2 \cdot \frac{3}{4} + 40 =$ $112 - \frac{3}{2} + 40 = 150\frac{1}{2}$	
D$\left[\frac{8}{3},4\right]$	$16 \cdot \frac{8}{3} - 2 \cdot 4 + 40 =$ $\frac{128}{3} - 8 + 40 = 74\frac{2}{3}$	
E(0,4)	$16 \cdot 0 - 2 \cdot 4 + 40 =$ $0 - 8 + 40 = 32$	← Minimum

The maximum is 152 when $x = 7$ and $y = 0$. The minimum is 32 when $x = 0$ and $y = 4$.

4. Maximum 124 when $x = 3$ and $y = 0$; minimum 40 when $x = 0$ and $y = 4$

5. <u>Familiarize.</u> We organize the information in a table. Let x represent the number of questions of type A, and let y represent the number of questions of type B.

Type	Number of points for each	Number answered	Total points for type
A	4	$5 \leqslant x \leqslant 10$	$4x$
B	7	$3 \leqslant y \leqslant 10$	$7y$
Total		$x + y \leqslant 18$	$4x + 7y$

No more than 18 questions can be answered.

This is the total score on the test.

5. (continued)

<u>Translate.</u> Suppose the total score on the test is T. Then $T(x,y) = 4x + 7y$.

Let us consider the domain of T. We know these facts (constraints) about x and y:

$$5 \leqslant x \leqslant 10,$$
$$3 \leqslant y \leqslant 10,$$
$$x + y \leqslant 18$$

<u>Carry out.</u> Graph the domain, find the coordinates of the vertices, and compute the function values at the vertices.

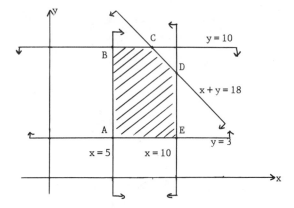

The vertices are A(0,0), B(5,10), C(8,10), D(10,8), and E(10,3).

Vertex (x,y)	Score $T(x,y) = 4x + 7y$
A(5,3)	$4 \cdot 5 + 7 \cdot 3 = 41$
B(5,10)	$4 \cdot 5 + 7 \cdot 10 = 90$
C(8,10)	$4 \cdot 8 + 7 \cdot 10 = 102$
D(10,8)	$4 \cdot 10 + 7 \cdot 8 = 96$
E(10,3)	$4 \cdot 10 + 7 \cdot 3 = 61$

The largest score in the table is 102, obtained when 8 questions of type A and 10 questions of type B are answered.

<u>Check.</u> Go over the algebra and arithmetic.

<u>State.</u> In order to maximize the score, 8 questions of type A and 10 questions of type B must be answered. The maximum score is 102.

6. 5 of A, 15 of B; maximum 425

7. **Familiarize.** We organize the information in a table. Let x = the amount invested in Bank X and y = the amount invested in Bank Y.

Bank	Amount invested	Interest rate	Interest earned
X	$2000 \leqslant x \leqslant 14{,}000$	6%	0.06x
Y	$0 \leqslant y \leqslant 15{,}000$	$6\frac{1}{2}\%$	0.065y
Total	$x + y \leqslant 22{,}000$		0.06x + 0.065y

No more than $22,000 will be invested.

This is the total interest earned.

Translate. Suppose the total interest earned is I. Then I(x,y) = 0.06x + 0.065y. We know these facts (constraints) about the domain of I:

$$2000 \leqslant x \leqslant 14{,}000,$$
$$0 \leqslant y \leqslant 15{,}000,$$
$$x + y \leqslant 22{,}000$$

Carry out. Graph the domain, find the coordinates of the vertices, and compute the function values at the vertices.

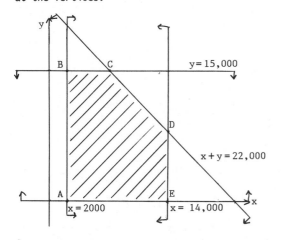

The vertices are A(2000,0), B(2000,15,000), C(7000,15,000), D(14,000,8000), and E(14,000,0).

Vertex (x,y)	Interest earned I(x,y) = 0.06x + 0.065y
A(2000,0)	0.06(2000) + 0.065(0) = 120
B(2000,15,000)	0.06(2000) + 0.065(15,000) = 1095
C(7000,15,000)	0.06(7000) + 0.065(15,000) = 1395
D(14,000,8000)	0.06(14,000) + 0.065(8000) = 1360
E(14,000,0)	0.06(14,000) + 0.065(0) = 840

The largest value of I in the table is 1395, obtained when x = 7000 and y = 15,000.

Check. Go over the algebra and arithmetic.

State. In order to maximize his income, the man should invest $7000 in Bank X and $15,000 in Bank Y. The maximum income is $1395.

8. $22,000 in corporate bonds, $18,000 in municipal bonds; maximum income is $3110

9. **Familiarize.** Let x represent the number of knit suits made per day and y represent the number of worsted suits made per day. Using the given information regarding cutting time, sewing time, and profit we set up a table.

Type of suit	Total cutting time	Total sewing time	Total profit
Knit	2x	4x	34x
Worsted	4y	2y	31y
Total	$2x + 4y \leqslant 20$	$4x + 2y \leqslant 16$	34x + 31y

Translate. Suppose P is the total profit. Then P(x,y) = 34x + 31y. We know these facts (constraints) about the domain of P:

$$2x + 4y \leqslant 20, \quad (1)$$
$$4x + 2y \leqslant 16, \quad (2)$$
$$x \geqslant 0, \quad (3)$$
$$y \geqslant 0 \quad (4)$$

(The number of suits made cannot be negative.)

Carry out. Graph the domain, find the coordinates of the vertices, and compute the function values at the vertices.

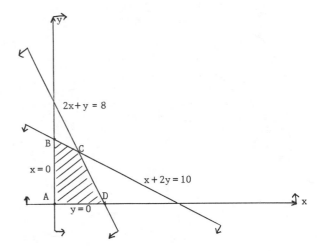

The vertices are A(0,0), B(0,5), C(2,4), and D(4,0).

Vertex (x,y)	Profit P(x,y) = 34x + 31y
A(0,0)	34·0 + 31·0 = 0
B(0,5)	34·0 + 31·5 = 155
C(2,4)	34·2 + 31·4 = 192
D(4,0)	34·4 + 31·0 = 136

The largest value of P in the table is 192, obtained when x = 2 and y = 4.

Check. Go over the algebra and arithmetic.

State. The tailoring firm should make 2 knit suits and 4 worsted suits per day. The maximum profit is $192.

10. 125 batches of SMELLO, 187.5 batches of ROPPO;
     maximum profit is $2520

12. 30 P-2's, 15 P-3's

11. Familiarize. Let x represent the number of P-1
     planes and y represent the number of P-2 planes.
     Organize the information in a table.

Plane	Number of planes	Passengers			Cost per mile
		First	Tourist	Economy	
P-1	x	40x	40x	120x	12,000x
P-2	y	80y	30y	40y	10,000y

Translate. Suppose C is the total cost per mile.
Then C(x,y) = 12,000x + 10,000y. We know these
facts (constraints) about the domain of C:

$$40x + 80y \geqslant 2000, \quad (1)$$
$$40x + 30y \geqslant 1500, \quad (2)$$
$$120x + 40y \geqslant 2400, \quad (3)$$
$$x \geqslant 0, \quad (4)$$
$$y \geqslant 0 \quad (5)$$

Carry out. Graph the domain, find the coordinates
of the vertices, and compute the function values
at the vertices.

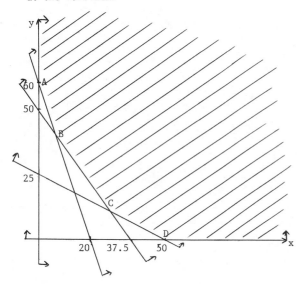

The vertices are A(0,60), B(6,42), C(30,10), and
D(50,0).

Vertex (x,y)	Cost C(x,y) = 12,000x + 10,000y
A(0,60)	12,000(0) + 10,000(60) = 600,000
B(6,42)	12,000(6) + 10,000(42) = 492,000
C(30,10)	12,000(30) + 10,000(10) = 460,000
D(50,0)	12,000(50) + 10,000(0) = 600,000

The smallest value of C in the table is 460,000,
obtained when x = 30 and y = 10.

Check. Go over the algebra and arithmetic.

State. In order to minimize the operating cost,
30 P-1 planes and 10 P-2 planes should be used.

Exercise Set 5.1

1. $-11x^4 - x^3 + x^2 + 3x - 9$

Term	$-11x^4$	$-x^3$	x^2	$3x$	-9
Degree	4	3	2	1	0
Degree of polynomial	4				

2. Degrees of terms: 3, 2, 1, 0; degree of polynomial: 3

3. $y^3 + 2y^7 + x^2y^4 - 8$

Term	y^3	$2y^7$	x^2y^4	-8
Degree	3	7	6	0
Degree of polynomial	7			

4. Degrees of terms: 2, 5, 7, 0; degree of polynomial: 7

5. $a^5 + 4a^2b^4 + 6ab + 4a - 3$

Term	a^5	$4a^2b^4$	$6ab$	$4a$	-3
Degree	5	6	2	1	0
Degree of polynomial	6				

6. Degrees of terms: 6, 8, 4, 2, 0; degree of polynomial: 8

7. $-4y^3 - 6y^2 + 7y + 23$; $-4y^3$; -4

8. $-18y^4 + 11y^3 + 6y^2 - 8y + 5$; $-18y^4$; -18

9. $3x^7 + 5x^2 - x + 12$; $3x^7$; 3

10. $-10x^4 + 7x^2 - 3x + 9$; $-10x^4$; -10

11. $-a^7 + 8a^5 + 5a^3 - 19a^2 + a$; $-a^7$; -1

12. $a^9 + 11a^4 + a^3 - 5a^2 - 7$; a^9; 1

13. $12 + 4x - 5x^2 + 3x^4$

14. $5 + 10x - 5x^2$

15. $3xy^3 + x^2y^2 - 9x^3y + 2x^4$

16. $-9xy + 5x^2y^2 + 8x^3y^2 - 5x^4$

17. $-7ab + 4ax - 7ax^2 + 4x^6$

18. $-12a + 5xy^8 + 4ax^3 - 3ax^5 + 5x^5$

19. $P(x) = 4x^2 - 3x + 2$

$P(4) = 4 \cdot 4^2 - 3 \cdot 4 + 2$
$\qquad = 64 - 12 + 2$
$\qquad = 54$

$P(0) = 4 \cdot 0^2 - 3 \cdot 0 + 2$
$\qquad = 0 - 0 + 2$
$\qquad = 2$

20. $Q(3) = -84$, $Q(-1) = 0$

21. $P(y) = 8y^3 - 12y - 5$

$P(-2) = 8(-2)^3 - 12(-2) - 5$
$\qquad = -64 + 24 - 5$
$\qquad = -45$

$P\left(\frac{1}{3}\right) = 8\left(\frac{1}{3}\right)^3 - 12 \cdot \frac{1}{3} - 5$

$\qquad = 8 \cdot \frac{1}{27} - 4 - 5$

$\qquad = \frac{8}{27} - 9$

$\qquad = \frac{8}{27} - \frac{243}{27}$

$\qquad = -\frac{235}{27}$, or $-8\frac{19}{27}$

22. $Q(-3) = -168$, $Q(0) = -9$

23. $P(x) = -4x^3 + 2x^2 - 7x + 5$

$P(a) = -4a^3 + 2a^2 - 7a + 5$

24. $Q(-b) = -6b^3 - 12b^2 + 8b - 3$

25. $P(x) = x^3 - 4x^2 + 3x - 7$

$P(-2a) = (-2a)^3 - 4(-2a)^2 + 3(-2a) - 7$
$\qquad = -8a^3 - 16a^2 - 6a - 7$

26. $Q(3c) = -189c^3 + 90c^2 + 6$

27. $P(x) = -3x^3 + 6x^2 - 4$

$P(5a) = -3(5a)^3 + 6(5a)^2 - 4$
$\qquad = -375a^3 + 150a^2 - 4$

28. $Q(-5b) = -750b^3 - 40b - 11$

29. Familiarize. We will find the numbers by substituting 8 and 20 for n in the polynomial $f(n) = \frac{1}{2}(n^2 - n)$.

Translate. The translation is the polynomial function.

Carry out. Substitute 8 for n and calculate:
$f(8) = \frac{1}{2}(8^2 - 8) = \frac{1}{2}(64 - 8) = \frac{1}{2}(56) = 28$

Substitute 20 for n and calculate:
$f(20) = \frac{1}{2}(20^2 - 20) = \frac{1}{2}(400 - 20) = \frac{1}{2}(380) = 190$

29. (continued)

     Check. Go over the calculations.

     State. When there are 8 teams, 28 games are played. When there are 20 teams, 190 games are played.

30. 132

31. Familiarize. We will find the surface area by substituting 4.7 for h, 1.2 for r, and 3.14 for π in the polynomial.

     Translate. The translation is the polynomial function.

     Carry out. Make the substitutions indicated above and calculate:

     $2\pi rh + 2\pi r^2 = 2(3.14)(1.2)(4.7) + 2(3.14)(1.2)^2 = 44.4624$

     Check. Go over the calculations.

     State. The surface area is 44.4624 in^2.

32. 56.52 in^2

33. Familiarize. We will find the answer by substituting for x in the polynomial function R(x).

     Translate. The translation is the polynomial function.

     Carry out. Substitute 200 for x and calculate:

     $R(200) = 280(200) - 0.4(200)^2 = 56,000 - 16,000 = 40,000$

     Check. Go over the calculation.

     State. The total revenue from the sale of 200 stereos is $40,000.

34. $45,000

35. Familiarize. We will find the answer by substituting for x in the polynomial function C(x).

     Translate. The translation is the polynomial function.

     Carry out. Substitute 200 for x and calculate:

     $C(200) = 7000 + 0.6(200)^2 = 7000 + 24,000 = 31,000$

     Check. Go over the calculation.

     State. The total cost of producing 200 stereos is $31,000.

36. $128,500

37. Familiarize. We will subtract the polynomial C(x) from the polynomial R(x).

     Translate. $P(x) = R(x) - C(x) = 280x - 0.4x^2 - (7000 + 0.6x^2)$

     Carry out. Do the subtraction.

     $P(x) = 280x - 0.4x^2 - (7000 + 0.6x^2)$

     $\quad = 280x - 0.4x^2 + (-7000 - 0.6x^2)$ Adding the opposite of the polynomial being subtracted

     $\quad = 280x - x^2 - 7000$

     Check. Go over the subtraction.

     State. $P(x) = 280x - x^2 - 7000$

38. $P(x) = 280x - 1.2x^2 - 8000$

39. $6x^2 - 7x^2 + 3x^2 = (6 - 7 + 3)x^2 = 2x^2$

40. $-4y^2$

41. $5x - 4y - 2x + 5y$

     $= (5 - 2)x + (-4 + 5)y$

     $= 3x + y$

42. $-2a - 6b$

43. $5a + 7 - 4 + 2a - 6a + 3$

     $= (5 + 2 - 6)a + (7 - 4 + 3)$

     $= a + 6$

44. $7x + 14$

45. $3a^2b + 4b^2 - 9a^2b - 6b^2$

     $= (3 - 9)a^2b + (4 - 6)b^2$

     $= -6a^2b - 2b^2$

46. $-8x^3 - 3x^2y^2$

47. $8x^2 - 3xy + 12y^2 + x^2 - y^2 + 5xy + 4y^2$

     $= (8 + 1)x^2 + (-3 + 5)xy + (12 - 1 + 4)y^2$

     $= 9x^2 + 2xy + 15y^2$

48. $11a^2 + 3ab - 3b^2$

49. $4x^2y - 3y + 2xy^2 - 5x^2y + 7y + 7xy^2$

     $= (4 - 5)x^2y + (-3 + 7)y + (2 + 7)xy^2$

     $= -x^2y + 4y + 9xy^2$

50. $-4xy^2 + 11xy + x^2y$

51. $(3x^2 + 5y^2 + 6) + (2x^2 - 3y^2 - 1)$

     $= (3 + 2)x^2 + (5 - 3)y^2 + (6 - 1)$

     $= 5x^2 + 2y^2 + 5$

52. $21y^2 + 3y + 4$

53. $(2a + 3b - c) + (4a - 2b + 2c)$
$= (2 + 4)a + (3 - 2)b + (-1 + 2)c$
$= 6a + b + c$

54. $14x + 8y - 6z$

55. $(a^2 - 3b^2 + 4c^2) + (-5a^2 + 2b^2 - c^2)$
$= (1 - 5)a^2 + (-3 + 2)b^2 + (4 - 1)c^2$
$= -4a^2 - b^2 + 3c^2$

56. $-5x^2 + 4y^2 - 11z^2$

57. $(x^2 + 2x - 3xy - 7) + (-3x^2 - x + 2xy + 6)$
$= (1 - 3)x^2 + (2 - 1)x + (-3 + 2)xy + (-7 + 6)$
$= -2x^2 + x - xy - 1$

58. $2a^2 + 3b - 4ab + 4$

59. $(7x^2y - 3xy^2 + 4xy) + (-2x^2y - xy^2 + xy)$
$= (7 - 2)x^2y + (-3 - 1)xy^2 + (4 + 1)xy$
$= 5x^2y - 4xy^2 + 5xy$

60. $20ab - 18ac - 3bc$

61. $(2r^2 + 12r - 11) + (6r^2 - 2r + 4) + (r^2 - r - 2)$
$= (2 + 6 + 1)r^2 + (12 - 2 - 1)r + (-11 + 4 - 2)$
$= 9r^2 + 9r - 9$

62. $-3x^2 - x - 3$

63. $(3 + 4a - 6a^2) + (-4 - 2a + 3a^2)$
$= (3 - 4) + (4 - 2)a + (-6 + 3)a^2$
$= -1 + 2a - 3a^2$

64. $-3x^2 - 5x - 27$

65. $\left[\dfrac{2}{3}xy + \dfrac{5}{6}xy^2 + 5.1x^2y\right] + \left[-\dfrac{4}{5}xy + \dfrac{3}{4}xy^2 - 3.4x^2y\right]$

$= \left[\dfrac{2}{3} - \dfrac{4}{5}\right]xy + \left[\dfrac{5}{6} + \dfrac{3}{4}\right]xy^2 + (5.1 - 3.4)x^2y$

$= \left[\dfrac{10}{15} - \dfrac{12}{15}\right]xy + \left[\dfrac{10}{12} + \dfrac{9}{12}\right]xy^2 + 1.7x^2y$

$= -\dfrac{2}{15}xy + \dfrac{19}{12}xy^2 + 1.7x^2y$

66. $-\dfrac{5}{24}xy - \dfrac{27}{20}x^3y^2 + 1.4y^3$

67. $5x^3 - 7x^2 + 3x - 6$
    a) $-(5x^3 - 7x^2 + 3x - 6)$ Writing an inverse
                                      sign in front
    b) $-5x^3 + 7x^2 - 3x + 6$.   Changing the sign of
                                        every term

68. $-(-8y^4 - 18y^3 + 4y - 9)$, $8y^4 + 18y^3 - 4y + 9$

69. $-12y^5 + 4ay^4 - 7by^2$
    a) $-(-12y^5 + 4ay^4 - 7by^2)$
    b) $12y^5 - 4ay^4 + 7by^2$.

70. $-(7ax^3y^2 - 8by^4 - 7abx - 12ay)$,
      $-7ax^3y^2 + 8by^4 + 7abx + 12ay$

71. $(8x - 4) - (-5x + 2)$
$= (8x - 4) + (5x - 2)$
$= 13x - 6$

72. $13y + 5$

73. $(-3x^2 + 2x + 9) - (x^2 + 5x - 4)$
$= (-3x^2 + 2x + 9) + (-x^2 - 5x + 4)$
$= -4x^2 - 3x + 13$

74. $-13y^2 + 2y + 11$

75. $(5a - 2b + c) - (3a + 2b - 2c)$
$= (5a - 2b + c) + (-3a - 2b + 2c)$
$= 2a - 4b + 3c$

76. $4x - 10y + 4z$

77. $(3x^2 - 2x - x^3) - (5x^2 - 8x - x^3)$
$= (3x^2 - 2x - x^3) + (-5x^2 + 8x + x^3)$
$= -2x^2 + 6x$

78. $5y^2 + 6y + 3y^3$

79. $(5a^2 + 4ab - 3b^2) - (9a^2 - 4ab + 2b^2)$
$= (5a^2 + 4ab - 3b^2) + (-9a^2 + 4ab - 2b^2)$
$= -4a^2 + 8ab - 5b^2$

80. $-3y^2 - 6yz - 12z^2$

81. $P(y) - Q(y) = 2 - 3y - (6y - 7)$
                  $= 2 - 3y + (-6y + 7)$
                  $= -9y + 9$

82. $10x^2 - 8x + 6$

83. $(6ab - 4a^2b + 6ab^2) - (3ab^2 - 10ab - 12a^2b)$
$= (6ab - 4a^2b + 6ab^2) + (-3ab^2 + 10ab + 12a^2b)$
$= 8a^2b + 16ab + 3ab^2$

84. $17xy + 5x^2y^2 - 7y^3$

85. $(0.09y^4 - 0.052y^3 + 0.93) -$
                  $(0.03y^4 - 0.084y^3 + 0.94y^2)$
$= (0.09y^4 - 0.052y^3 + 0.93) +$
                  $(-0.03y^4 + 0.084y^3 - 0.94y^2)$
$= 0.06y^4 + 0.032y^3 - 0.94y^2 + 0.93$

86. $0.44x^4 + 5.612x^3 - 0.04x^2 - 5.6x$

87. $\left[\dfrac{5}{8}x^4 - \dfrac{1}{4}x^2 - \dfrac{1}{2}\right] - \left[-\dfrac{3}{8}x^4 + \dfrac{3}{4}x^2 + \dfrac{1}{2}\right]$

$= \left[\dfrac{5}{8}x^4 - \dfrac{1}{4}x^2 - \dfrac{1}{2}\right] + \left[\dfrac{3}{8}x^4 - \dfrac{3}{4}x^2 - \dfrac{1}{2}\right]$

$= x^4 - x^2 - 1$

88. $\frac{29}{24}y^4 - \frac{5}{4}y^2 - 11.2y + \frac{8}{15}$

89. $3(y - 2) = 3y - 6$

90. $-2x - 92$

91. $2[P(x)] = 2(13x^5 - 22x^4 - 36x^3 + 40x^2 - 16x + 75)$
    $= 26x^5 - 44x^4 - 72x^3 + 80x^2 - 32x + 150$
    Use columns to add:
    $$26x^5 - 44x^4 - 72x^3 + 80x^2 - 32x + 150$$
    $$42x^5 - 37x^4 + 50x^3 - 28x^2 + 34x + 100$$
    $$68x^5 - 81x^4 - 22x^3 + 52x^2 + 2x + 250$$

92. $-3x^5 - 29x^4 - 158x^3 + 148x^2 - 82x + 125$

93. $2[Q(x)] = 2(42x^5 - 37x^4 + 50x^3 - 28x^2 + 34x + 100)$
    $= 84x^5 - 74x^4 + 100x^3 - 56x^2 + 68x + 200$

    $3[P(x)] = 3(13x^5 - 22x^4 - 36x^3 + 40x^2 - 16x + 75)$
    $= 39x^5 - 66x^4 - 108x^3 + 120x^2 - 48x + 225$
    Use columns to subtract:
    $$84x^5 - 74x^4 + 100x^3 - 56x^2 + 68x + 200$$
    $$-39x^5 + 66x^4 + 108x^3 - 120x^2 + 48x - 225$$
    $$45x^5 - 8x^4 + 208x^3 - 176x^2 + 116x - 25$$
    Adding the inverse of $3[P(x)]$

94. $178x^5 - 199x^4 + 6x^3 + 76x^2 + 38x + 600$

95. The area of the base is $x \cdot x$, or x^2.
    The area of each side is $x \cdot (x - 2)$.
    The total area of all four sides is $4x(x - 2)$.

    The surface area of this box can be expressed as a polynomial function.
    $S(x) = x^2 + 4x(x - 2)$
    $= x^2 + 4x^2 - 8x$
    $= 5x^2 - 8x$

96. $8x^{2a} + 7x^a + 7$

97. Writing in columns is helpful.
    Add:
    $$47x^{4a} + 3x^{3a} + 22x^{2a} + x^a + 1$$
    $$37x^{3a} + 8x^{2a} + 3$$
    $$47x^{4a} + 40x^{3a} + 30x^{2a} + x^a + 4$$

98. $x^{6a} - 5x^{5a} - 4a^{4a} + x^{3a} - 2x^{2a} + 8$

99. Writing in columns is helpful.
    Subtract:
    $$2x^5b + 4x^4b + 3x^3b \qquad\qquad + 8$$
    $$-(x^5b \qquad\quad + 2x^3b + 6x^2b + 9xb + 8)$$

    We rewrite as an addition problem.
    Add:
    $$2x^5b + 4x^4b + 3x^3b \qquad\qquad + 8$$
    $$-x^5b \qquad\quad - 2x^3b - 6x^2b - 9xb - 8$$
    $$x^5b + 4x^4b + x^3b - 6x^2b - 9xb$$

Exercise Set 5.2

1. $2y^2(5y) = (2 \cdot 5)(y^2 \cdot y) = 10y^3$

2. $-6x^3y$

3. $5x(-4x^2y) = 5(-4)(x \cdot x^2)y = -20x^3y$

4. $-6a^3b^4$

5. $2x^3y^2(-5x^2y^4) = 2(-5)(x^3 \cdot x^2)(y^2 \cdot y^4) = -10x^5y^6$

6. $-56a^3b^4c^6$

7. $2x(3 - x)$
   $= 2x \cdot 3 - 2x \cdot x$
   $= 6x - 2x^2$

8. $4a^3 - 20a^2$

9. $3ab(a + b)$
   $= 3ab \cdot a + 3ab \cdot b$
   $= 3a^2b + 3ab^2$

10. $4x^2y - 6xy^2$

11. $5cd(3c^2d - 5cd^2)$
    $= 5cd \cdot 3c^2d - 5cd \cdot 5cd^2$
    $= 15c^3d^2 - 25c^2d^3$

12. $2a^4 - 5a^5$

13. $(2x + 3)(3x - 4)$
    $= 6x^2 - 8x + 9x - 12$ FOIL
    $= 6x^2 + x - 12$

14. $8a^2 - 14ab + 3b^2$

15. $(s + 3t)(s - 3t)$
    $= s^2 - (3t)^2$ $(A + B)(A - B) = A^2 - B^2$
    $= s^2 - 9t^2$

16. $y^2 - 16$

17. $(x - y)(x - y)$
    $= x^2 - 2xy + y^2 \quad (A - B)^2 = A^2 - 2AB + B^2$

18. $a^2 + 4ab + 4b^2$

19. $(y + 8x)(2y - 7x)$
    $= 2y^2 - 7xy + 16xy - 56x^2 \quad$ FOIL
    $= 2y^2 + 9xy - 56x^2$

20. $x^2 - xy - 2y^2$

21. $(a^2 - 2b^2)(a^2 - 3b^2)$
    $= a^4 - 3a^2b^2 - 2a^2b^2 + 6b^4 \quad$ FOIL
    $= a^4 - 5a^2b^2 + 6b^4$

22. $6m^4 - 13m^2n^2 + 5n^4$

23. $(x - 4)(x^2 + 4x + 16)$
    $= (x - 4)(x^2) + (x - 4)(4x) + (x - 4)(16)$
    $\qquad\qquad\qquad\qquad\qquad$ Distributive law
    $= x(x^2) - 4(x^2) + x(4x) - 4(4x) + x(16) - 4(16)$
    $\qquad\qquad\qquad\qquad\qquad$ Distributive law
    $= x^3 - 4x^2 + 4x^2 - 16x + 16x - 64$
    $\qquad\qquad\qquad\qquad$ Multiplying monomials
    $= x^3 - 64 \qquad$ Collecting like terms

24. $y^3 + 27$

25. $(x + y)(x^2 - xy + y^2)$
    $= (x + y)x^2 + (x + y)(-xy) + (x + y)(y^2)$
    $= x(x^2) + y(x^2) + x(-xy) + y(-xy) + x(y^2) + y(y^2)$
    $= x^3 + x^2y - x^2y - xy^2 + xy^2 + y^3$
    $= x^3 + y^3$

26. $a^3 - b^3$

27.
$$
\begin{array}{l}
\quad\quad\quad a^2 + a - 1 \\
\quad\quad\quad \underline{a^2 + 4a - 5} \\
\quad\quad\quad -5a^2 - 5a + 5 \quad\text{Multiplying by } -5 \\
\quad 4a^3 + 4a^2 - 4a \quad\quad\text{Multiplying by } 4a \\
\underline{a^4 + a^3 - a^2} \quad\quad\quad\text{Multiplying by } a^2 \\
a^4 + 5a^3 - 2a^2 - 9a + 5 \quad\text{Adding}
\end{array}
$$

28. $x^4 - x^3 + x^2 - 3x + 2$

29.
$$
\begin{array}{l}
\quad\quad\quad 4a^2b - 2ab + 3b^2 \\
\quad\quad\quad \underline{ab - 2b + a} \\
\quad\quad\quad 4a^3b - 2a^2b + 3ab^2 \quad\quad\quad\quad\quad ① \\
-6b^3 \quad\quad\quad\quad + 4ab^2 - 8a^2b^2 \quad\quad ② \\
\underline{3ab^3 \quad\quad\quad\quad\quad\quad - 2a^2b^2 + 4a^3b^2} \quad ③ \\
3ab^3 - 6b^3 + 4a^3b - 2a^2b + 7ab^2 - 10a^2b^2 + 4a^3b^2 \quad ④
\end{array}
$$

① Multiplying by a
② Multiplying by $-2b$
③ Multiplying by ab
④ Adding

30. $2x^4 - 4x^3y - x^2y^2 + 3xy^3 - 2y^4$

31. $\left(x - \frac{1}{2}\right)\left(x - \frac{1}{4}\right)$
    $= x^2 - \frac{1}{4}x - \frac{1}{2}x + \frac{1}{8} \quad$ FOIL
    $= x^2 - \frac{1}{4}x - \frac{2}{4}x + \frac{1}{8}$
    $= x^2 - \frac{3}{4}x + \frac{1}{8}$

32. $b^2 - \frac{2}{3}b + \frac{1}{9}$

33. $(1.3x - 4y)(2.5x + 7y)$
    $= 3.25x^2 + 9.1xy - 10xy - 28y^2 \quad$ FOIL
    $= 3.25x^2 - 0.9xy - 28y^2$

34. $12a^2 + 399.928ab - 2.4b^2$

35. $P(x) \cdot Q(x) = (3x^2 - 5)(4x^2 - 7x + 2)$
    $= (3x^2 - 5)(4x^2) + (3x^2 - 5)(-7x) + (3x^2 - 5)(2)$
    $= 12x^4 - 20x^2 - 21x^3 + 35x + 6x^2 - 10$
    $= 12x^4 - 21x^3 - 14x^2 + 35x - 10$

36. $x^5 + x^3 + 2x^2 - x + 2$

37. $(a + 2)(a + 3)$
    $= a^2 + 3a + 2a + 6 \quad$ FOIL
    $= a^2 + 5a + 6$

38. $x^2 + 13x + 40$

39. $(y + 3)(y - 2)$
    $= y^2 - 2y + 3y - 6 \quad$ FOIL
    $= y^2 + y - 6$

40. $y^2 + 3y - 28$

41. $(x + 3)^2$
    $= x^2 + 2 \cdot x \cdot 3 + 3^2 \quad (A + B)^2 = A^2 + 2AB + B^2$
    $= x^2 + 6x + 9$

42. $y^2 - 14y + 49$

43. $(x - 2y)^2$
    $= x^2 - 2(x)(2y) + (2y)^2 \quad (A - B)^2 = A^2 - 2AB + B^2$
    $= x^2 - 4xy + 4y^2$

44. $4s^2 + 12st + 9t^2$

45. $\left(b - \frac{1}{3}\right)\left(b - \frac{1}{2}\right)$
    $= b^2 - \frac{1}{2}b - \frac{1}{3}b + \frac{1}{6} \quad$ FOIL
    $= b^2 - \frac{3}{6}b - \frac{2}{6}b + \frac{1}{6}$
    $= b^2 - \frac{5}{6}b + \frac{1}{6}$

46. $x^2 - \dfrac{13}{6}x + 1$

47. $(2x + 9)(x + 2)$
     $= 2x^2 + 4x + 9x + 18$ FOIL
     $= 2x^2 + 13x + 18$

48. $6b^2 - 11b - 10$

49. $(20a - 0.16b)^2$
     $= (20a)^2 - 2(20a)(0.16b) + (0.16b)^2$
                      $(A - B)^2 = A^2 - 2AB + B^2$
     $= 400a^2 - 6.4ab + 0.0256b^2$

50. $100p^4 + 46p^2q + 5.29q^2$

51. $(2x - 3y)(2x + y)$
     $= 4x^2 + 2xy - 6xy - 3y^2$ FOIL
     $= 4x^2 - 4xy - 3y^2$

52. $4a^2 - 8ab + 3b^2$

53. $\left(2a + \dfrac{1}{3}\right)^2$

     $= (2a)^2 + 2(2a)\left(\dfrac{1}{3}\right) + \left(\dfrac{1}{3}\right)^2$
                      $(A + B)^2 = A^2 + 2AB + B^2$
     $= 4a^2 + \dfrac{4}{3}a + \dfrac{1}{9}$

54. $9c^2 - 3c + \dfrac{1}{4}$

55. $(2x^3 - 3y^2)^2$
     $= (2x^3)^2 - 2(2x^3)(3y^2) + (3y^2)^2$
                      $(A - B)^2 = A^2 - 2AB + B^2$
     $= 4x^6 - 12x^3y^2 + 9y^4$

56. $9s^4 + 24s^2t^3 + 16t^6$

57. $(a^2b^2 + 1)^2$
     $= (a^2b^2)^2 + 2(a^2b^2)\cdot 1 + 1^2$
                      $(A + B)^2 = A^2 + 2AB + B^2$
     $= a^4b^4 + 2a^2b^2 + 1$

58. $x^4y^2 - 2x^3y^3 + x^2y^4$

59. $(20a^5 - 0.16b)^2$
     $= (20a^5)^2 - 2(20a^5)(0.16b) + (0.16b)^2$
                      $(A - B)^2 = A^2 - 2AB + B^2$
     $= 400a^{10} - 6.4a^5b + 0.0256b^2$

60. $100p^{10} + 46p^5q + 5.29q^2$

61. $P(x)\cdot P(x) = (3x - 4)(3x - 4) = (3x - 4)^2$
     $= (3x)^2 - 2(3x)(4) + (4)^2$
                      $(A - B)^2 = A^2 - 2AB + B^2$
     $= 9x^2 - 24x + 16$

62. $25x^4 - 110x^2 + 121$

63. $(c + 2)(c - 2) = c^2 - 2^2$
                      $(A + B)(A - B) = A^2 - B^2$
     $= c^2 - 4$

64. $x^2 - 9$

65. $(2a + 1)(2a - 1) = (2a)^2 - 1^2$
                      $(A + B)(A - B) = A^2 - B^2$
     $= 4a^2 - 1$

66. $9 - 4x^2$

67. $(3m - 2n)(3m + 2n) = (3m)^2 - (2n)^2$
                      $(A + B)(A - B) = A^2 - B^2$
     $= 9m^2 - 4n^2$

68. $9x^2 - 25y^2$

69. $(x^3 + yz)(x^3 - yz) = (x^3)^2 - (yz)^2$
                      $(A + B)(A - B) = A^2 - B^2$
     $= x^6 - y^2z^2$

70. $4a^6 - 25a^2b^2$

71. $(-mn + m^2)(mn + m^2) = (m^2 - mn)(m^2 + mn)$
     $= (m^2)^2 - (mn)^2$ $(A + B)(A - B) = A^2 - B^2$
     $= m^4 - m^2n^2$

72. $p^2q^2 - 2.56$

73. $\left(\dfrac{1}{2}p^4 - \dfrac{2}{3}q\right)\left(\dfrac{1}{2}p^4 + \dfrac{2}{3}q\right)$

     $= \left(\dfrac{1}{2}p^4\right)^2 - \left(\dfrac{2}{3}q\right)^2 = \dfrac{1}{4}p^8 - \dfrac{4}{9}q^2$

74. $\dfrac{9}{25}a^2b^2 - 16c^{10}$

75. $(x + 1)(x - 1)(x^2 + 1) = (x^2 - 1^2)(x^2 + 1)$
                      $= (x^2 - 1)(x^2 + 1)$
                      $= (x^2)^2 - 1^2$
                      $= x^4 - 1$

76. $y^4 - 16$

77. $(a - b)(a + b)(a^2 - b^2) = (a^2 - b^2)(a^2 - b^2)$
                      $= (a^2 - b^2)^2$
                      $= (a^2)^2 - 2(a^2)(b^2) +$
                       $(b^2)^2$
                      $= a^4 - 2a^2b^2 + b^4$

78. $16x^4 - 8x^2y^2 + y^4$

79. $(a + b + 1)(a + b - 1)$

$= [(a + b) + 1][(a + b) - 1]$

$= (a + b)^2 - 1^2$

$= a^2 + 2ab + b^2 - 1$

80. $m^2 + 2mn + n^2 - 4$

81. $(2x + 3y + 4)(2x + 3y - 4)$

$= [(2x + 3y) + 4][(2x + 3y) - 4]$

$= (2x + 3y)^2 - 4^2$

$= 4x^2 + 12xy + 9y^2 - 16$

82. $9a^2 - 12ab + 4b^2 - c^2$

83. $A = P(1 + i)^2$

$A = P(1 + 2i + i^2)$

$A = P + 2Pi + Pi^2$

84. $A = P + Pi + \dfrac{Pi^2}{4}$

85. a) Replace each occurrence of x by $t - 1$.

$f(t - 1) = 5(t - 1) + (t - 1)^2$

$= 5t - 5 + t^2 - 2t + 1$

$= t^2 + 3t - 4$

b) $f(a + h) - f(a)$

$= [5(a + h) + (a + h)^2] - [5a + a^2]$

$= 5a + 5h + a^2 + 2ah + h^2 - 5a - a^2$

$= 5h + 2ah + h^2$

86. a) $-p^2 + p + 6$

b) $3h - 2ah - h^2$

87. Familiarize. We let x represent the number of days worked during the week and y represent the number of days worked during the weekend. We organize the information in a table.

	Days worked	Pay per day	Total earned
During the week	x	$25	25x
During the weekend	y	$35	35y
Total	17		$485

Translate.

Total number of days worked is 17.

$x + y = 17$

Total earned is $485.

$25x + 35y = 485$

We now have a system of equations.

$x + y = 17,$

$25x + 35y = 485$

87. (continued)

Carry out. Solving the system we get (11,6).

Check. The total number of days is $11 + 6$, or 17. The total earned is $25(11) + 35(6)$, or $275 + 210$, or $485. The values check.

State. Jan worked 11 weekdays.

88. 56, 58, and 60

89. Familiarize. Let a, b, and c represent the daily production of machines A, B, and C, respectively.

Translate. Rewording, we have:

Production of A, B, and C is 222 per day.

$a + b + c = 222$

Production of A and B alone is 159.

$a + b = 159$

Production of B and C alone is 147.

$b + c = 147$

We have a system of equations.

$a + b + c = 222,$

$a + b = 159,$

$b + c = 147$

Carry out. Solving the system we get (75,84,63).

Check. The daily production of the three machines together is $75 + 84 + 63$, or 222. The daily production of A and B alone is $75 + 84$, or 159. The daily production of B and C alone is $84 + 63$, or 147. The numbers check.

State. The daily production of suitcases by machines A, B, and C is 75, 84, and 63, respectively.

90. Dimes: 5, nickels: 2, quarters: 6

91. $(6y)^2 \left[-\dfrac{1}{3}x^2y^3 \right]^3$

$= (36y^2)\left[-\dfrac{1}{27}x^6y^9 \right]$

$= -\dfrac{36}{27}x^6y^{11}$

$= -\dfrac{4}{3}x^6y^{11}$

92. $a^{16}n^2$

93. $(-r^6s^2)^3 \left[-\dfrac{r^2}{6} \right]^2 (9s^4)^2$

$= (-r^{18}s^6)\left(\dfrac{r^4}{36} \right)(81s^8)$

$= -\dfrac{81}{36}r^{22}s^{14}$

$= -\dfrac{9}{4}r^{22}s^{14}$

94. z^5n^5

95. $\left[-\frac{3}{8}xy^2\right]\left[-\frac{7}{9}x^3y\right]\left[-\frac{8}{7}x^2y\right]^2$

    $= \left[-\frac{3}{8}xy^2\right]\left[-\frac{7}{9}x^3y\right]\left[\frac{64}{49}x^4y^2\right]$

    $= \left[-\frac{3}{8}\right]\left[-\frac{7}{9}\right]\left[\frac{64}{49}\right](x^8y^5)$

    $= \frac{8}{21}x^8y^5$

96. $\frac{1}{4}a^7x b^2y+2$

97. $(-8s^3t)(2s^5 - 3s^3t^4 + st^7 - t^{10})$

    $= -8s^3t(2s^5) - (-8s^3t)3s^3t^4 +$
    $\qquad (-8s^3t)st^7 - (-8s^3t)t^{10}$

    $= -16s^8t + 24s^6t^5 - 8s^4t^8 + 8s^3t^{11}$

98. $y^{3n+3}z^{n+3} - 4y^4z^{3n}$

99. $[(2x - 1)^2 - 1]^2$

    $= [(4x^2 - 4x + 1) - 1]^2$

    $= [4x^2 - 4x]^2$

    $= (4x^2)^2 - 2(4x^2)(4x) + (4x)^2$

    $= 16x^4 - 32x^3 + 16x^2$

100. $x^3 + y^3 + 3y^2 + 3y + 1$

101. $[(a + b)(a - b)][5 - (a + b)][5 + (a + b)]$

    $= [a^2 - b^2][5^2 - (a + b)^2]$

    $= [a^2 - b^2][25 - (a^2 + 2ab + b^2)]$

    $= [a^2 - b^2][25 - a^2 - 2ab - b^2]$

    $= 25a^2 - a^4 - 2a^3b - a^2b^2 - 25b^2 + a^2b^2 + 2ab^3 + b^4$

    $= -a^4 - 2a^3b + 25a^2 + 2ab^3 - 25b^2 + b^4$

102. $y^{12} - 6y^{10} + 15y^8 - 20y^6 + 15y^4 - 6y^2 + 1$

103. $(r^2 + s^2)^2(r^2 + 2rs + s^2)(r^2 - 2rs + s^2)$

    $= (r^2 + s^2)^2(r + s)^2(r - s)^2$

    $= (r^2 + s^2)^2[(r + s)(r - s)]^2$

    $= (r^2 + s^2)^2(r^2 - s^2)^2$

    $= [(r^2 + s^2)(r^2 - s^2)]^2$

    $= (r^4 - s^4)^2$

    $= r^8 - 2r^4s^4 + s^8$

104. $9x^{10} - \frac{30}{11}x^5 + \frac{25}{121}$

105. $(a - b + c - d)(a + b + c + d)$

    $= [(a + c) - (b + d)][(a + c) + (b + d)]$

    $= (a + c)^2 - (b + d)^2$

    $= (a^2 + 2ac + c^2) - (b^2 + 2bd + d^2)$

    $= a^2 + 2ac + c^2 - b^2 - 2bd - d^2$

106. $\frac{4}{9}x^2 - \frac{1}{9}y^2 - \frac{2}{3}y - 1$

107. $[2(y - 3) - 6(x + 4)][5(y - 3) - 4(x + 4)]$

    $= [2y - 6 - 6x - 24][5y - 15 - 4x - 16]$

    $= (2y - 6x - 30)(5y - 4x - 31)$

$$
\begin{array}{r}
5y - 4x - 31 \\
2y - 6x - 30 \\
\hline
-150y + 120x + 930 \\
-30xy + 24x^2 \qquad + 186x \\
10y^2 - 8xy \qquad - 62y \\
\hline
10y^2 - 38xy + 24x^2 - 212y + 306x + 930
\end{array}
$$

    (Note: This exercise could also be done using FOIL.)

108. $x^3 - \frac{1}{343}$

109. $(4x^2 + 2xy + y^2)(4x^2 - 2xy + y^2)$

    $= [(4x^2 + y^2) + 2xy][(4x^2 + y^2) - 2xy]$

    $= (4x^2 + y^2)^2 - (2xy)^2$

    $= 16x^4 + 8x^2y^2 + y^4 - 4x^2y^2$

    $= 16x^4 + 4x^2y^2 + y^4$

110. $x^4 - 25x^2 + 144$

111. $(x^a + y^b)(x^a - y^b)(x^{2a} + y^{2b})$

    $= (x^{2a} - y^{2b})(x^{2a} + y^{2b})$

    $= x^{4a} - y^{4b}$

112. $\frac{1}{25}x^4 + \frac{4}{25}x^3 + \frac{16}{25}x^2 + \frac{24}{25}x + \frac{36}{25}$

113. $[a - (b - 1)][(b - 1)^2 + a(b - 1) + a^2]$

    $= (a - b + 1)(b^2 - 2b + 1 + ab - a + a^2)$

    $= ab^2 - 2ab + a + a^2b - a^2 + a^3 - b^3 + 2b^2 -$
    $\quad b - ab^2 + ab - a^2b + b^2 - 2b + 1 + ab - a + a^2$

    $= a^3 - b^3 + 3b^2 - 3b + 1$

114. $x^6 - 1$

115. $\left[\left[\frac{1}{3}x^3 - \frac{2}{3}y^2\right]\left[\frac{1}{3}x^3 + \frac{2}{3}y^2\right]\right]^2$

    $= \left[\left[\frac{1}{3}x^3\right]^2 - \left[\frac{2}{3}y^2\right]^2\right]^2$

    $= \left[\frac{1}{9}x^6 - \frac{4}{9}y^4\right]^2$

    $= \left[\frac{1}{9}x^6\right]^2 - 2\left[\frac{1}{9}x^6\right]\left[\frac{4}{9}y^4\right] + \left[\frac{4}{9}y^4\right]^2$

    $= \frac{1}{81}x^{12} - \frac{8}{81}x^6y^4 + \frac{16}{81}y^8$

116. $0.1x^8y^2 - 0.02x^5y^5 + 0.001x^2y^8$

117. $(x^{a-b})^{a+b} = x^{(a-b)(a+b)} = x^{a^2-b^2}$

118. $M^{x^2+2xy+y^2}$

Exercise Set 5.3

1. $4a^2 + 2a$
    $= 2a \cdot 2a + 2a \cdot 1$
    $= 2a(2a + 1)$

2. $3y(2y + 1)$

3. $y^2 - 5y$
    $= y \cdot y - 5 \cdot y$
    $= y(y - 5)$

4. $x(x + 9)$

5. $y^3 + 9y^2$
    $= y \cdot y^2 + 9 \cdot y^2$
    $= y^2(y + 9)$

6. $x^2(x + 8)$

7. $6x^2 - 3x^4$
    $= 3x^2 \cdot 2 - 3x^2 \cdot x^2$
    $= 3x^2(2 - x^2)$

8. $4y^2(2 + y^2)$

9. $4x^2y - 12xy^2$
    $= 4xy \cdot x - 4xy \cdot 3y$
    $= 4xy(x - 3y)$

10. $5x^2y^2(y + 3x)$

11. $3y^2 - 3y - 9$
    $= 3 \cdot y^2 - 3 \cdot y - 3 \cdot 3$
    $= 3(y^2 - y - 3)$

12. $5(x^2 - x + 3)$

13. $4ab - 6ac + 12ad$
    $= 2a \cdot 2b - 2a \cdot 3c + 2a \cdot 6d$
    $= 2a(2b - 3c + 6d)$

14. $2x(4y + 5z - 7w)$

15. $10a^4 + 15a^2 - 25a - 30$
    $= 5 \cdot 2a^4 + 5 \cdot 3a^2 - 5 \cdot 5a - 5 \cdot 6$
    $= 5(2a^4 + 3a^2 - 5a - 6)$

16. $4(3t^5 - 5t^4 + 2t^2 - 4)$

17. $-3x + 12 = -3(x - 4)$

18. $-5(x + 8)$

19. $-6y - 72 = -6(y + 12)$

20. $-8(t - 9)$

21. $-2x^2 + 4x - 12 = -2(x^2 - 2x + 6)$

22. $-2(x^2 - 6x - 20)$

23. $-3y^2 + 24x = -3(y^2 - 8x)$

24. $-7(x^2 + 8y)$

25. $-3y^3 + 12y^2 - 15y = -3y(y^2 - 4y + 5)$

26. $-4m(m^3 + 8m^2 - 16)$

27. $-x^2 + 3x - 7 = -(x^2 - 3x + 7)$

28. $-(p^3 + 4p^2 - 11)$

29. $-a^4 + 2a^3 - 13a = -a(a^3 - 2a^2 + 13)$

30. $-(m^3 + m^2 - m + 2)$

31. a) $h(0) = -16(0)^2 + 80(0) = 0$ ft
    $h(1) = -16(1)^2 + 80(1) = -16 + 80 = 64$ ft
    $h(2.5) = -16(2.5)^2 + 80(2.5) = -100 + 200 = 100$ ft
    $h(3) = -16(3)^2 + 80(3) = -144 + 240 = 96$ ft
    $h(5) = -16(5)^2 + 80(5) = -400 + 400 = 0$ ft

    b) $h(t) = -16t^2 + 80t$
    $h(t) = -16t(t - 5)$ Factoring out $-16t$

32. a) 0 ft, 80 ft, 140 ft, 144 ft, 80 ft

    b) $-16t(t - 6)$

33. $N(x) = \frac{1}{6}x^3 + \frac{1}{2}x^2 + \frac{1}{3}x$

    $N(x) = \frac{1}{6}x(x^2 + 3x + 2)$ Factoring out $\frac{1}{6}x$

34. $f(n) = \frac{1}{2}n(n - 1)$

35. $N(n) = n^2 - n$
    $N(n) = n(n - 1)$ Factoring out n

36. $2\pi r(h + r)$

37. $R(x) = 280x - 0.4x^2$
    $R(x) = 0.4x(700 - x)$

38. $C(x) = 0.6x(0.3 + x)$

39. $P(8) = \frac{1}{2}(8)^2 - \frac{3}{2}(8) = 32 - 12 = 20$

40. 54

41. $P(117) = \frac{1}{2}(117)^2 - \frac{3}{2}(117) = \frac{13,689}{2} - \frac{351}{2} =$
    $\frac{13,338}{2} = 6669$

42. $P(n) = \frac{1}{2}n(n - 3)$

43. $a(b - 2) + c(b - 2)$
$= (a + c)(b - 2)$

44. $(a - 2)(x^2 - 3)$

45. $(x - 2)(x + 5) + (x - 2)(x + 8)$
$= (x - 2)[(x + 5) + (x + 8)]$
$= (x - 2)(2x + 13)$

46. $2m(m - 4)$

47. $a^2(x - y) + a^2(x - y)$
$= 2a^2(x - y)$

48. $6x^2(x - 6)$

49. $ac + ad + bc + bd$
$= a(c + d) + b(c + d)$
$= (a + b)(c + d)$

50. $(x + w)(y + z)$

51. $b^3 - b^2 + 2b - 2$
$= b^2(b - 1) + 2(b - 1)$
$= (b^2 + 2)(b - 1)$

52. $(y^2 + 3)(y - 1)$

53. $a^3 - 3a^2 + 2a - 6$
$= a^2(a - 3) + 2(a - 3)$
$= (a^2 + 2)(a - 3)$

54. $(t^2 - 2)(t + 6)$

55. $24x^3 - 36x^2 + 72x$
$= 12x \cdot 2x^2 - 12x \cdot 3x + 12x \cdot 6$
$= 12x(2x^2 - 3x + 6)$

56. $3a(4a^3 - 7a^2 - 2)$

57. $x^6 + x^5 - x^3 + x^2$
$= x^2 \cdot x^4 + x^2 \cdot x^3 - x^2 \cdot x + x^2 \cdot 1$
$= x^2(x^4 + x^3 - x + 1)$

58. $y(y^3 - y^2 + y + 1)$

59. $2y^4 + 6y^2 + 5y^2 + 15$
$= 2y^2(y^2 + 3) + 5(y^2 + 3)$
$= (2y^2 + 5)(y^2 + 3)$

60. $(xy - 3)(2 + x)$

61. **Familiarize.** Let x, y, and z represent the number of nickels, dimes, and quarters, respectively. The value of x nickels is 0.05x, of y dimes is 0.10y, and of z quarters is 0.25z.

Translate.

The total value		is	$18.10.
0.05x + 0.10y + 0.25z		=	18.10

The number of quarters   is two more than twice   the number of dimes.
$$z = 2 + 2 \cdot y$$

The total number of coins is 120.
$$x + y + z = 120$$

We have a system of equations.
$0.05x + 0.10y + 0.25z = 18.10,$
$z = 2 + 2y,$
$x + y + z = 120$

Carry out. Solving the system we get (40,26,54).

Check. The total value of the coins is $0.05(40) + $0.10(26) + $0.25(54), or $18.10. The number of quarters, 54, is 2 more than twice the number of dimes, 26. The total number of coins is 40 + 26 + 54, or 120. The values check.

State. There are 40 nickels, 26 dimes, and 54 quarters.

62. $21°, 84°$

63. $P(x) = x^4 - 3x^3 - 5x^2 + 4x + 2$
$= x(x^3 - 3x^2 - 5x + 4) + 2$
$= x(x(x^2 - 3x - 5) + 4) + 2$
$= x(x(x(x - 3) - 5) + 4) + 2$

64. $x(x(x(x + 5) + 7) - 2) - 4$

65. $P(x) = 5x^5 + 0x^4 - 3x^3 + 4x^2 - 5x + 1$
$= x(5x^4 + 0x^3 - 3x^2 + 4x - 5) + 1$
$= x(x(5x^3 + 0x^2 - 3x + 4) - 5) + 1$
$= x(x(x(5x^2 + 0x - 3) + 4) - 5) + 1$
$= x(x(x(x(5x + 0) - 3) + 4) - 5) + 1$

66. $x(x(x(x(4x + 2) + 0) - 5) + 0) - 7$

67. $P(x) = 2x^4 - 3x^3 + 5x^2 + 6x - 4$
$= x(2x^3 - 3x^2 + 5x + 6) - 4$
$= x(x(2x^2 - 3x + 5) + 6) - 4$
$= x(x(x(2x - 3) + 5) + 6) - 4$
$P(5) = 5(5(5(2 \cdot 5 - 3) + 5) + 6) - 4$
$= 5(5(5 \cdot 7 + 5) + 6) - 4$
$= 5(5(5 \cdot 40 + 6) - 4$
$= 5(5 \cdot 40 + 6) - 4$
$= 5 \cdot 206 - 4$
$= 1030 - 4$
$= 1026$

67. (continued)

$$P(-2) = -2(-2(-2(2(-2) - 3) + 5) + 6) - 4$$
$$= -2(-2(-2 \cdot (-7) + 5) + 6) - 4$$
$$= -2(-2 \cdot 19 + 6) - 4$$
$$= -2 \cdot (-32) - 4$$
$$= 64 - 4$$
$$= 60$$

$$P(10) = 10(10(10(2 \cdot 10 - 3) + 5) + 6) - 4$$
$$= 10(10(10 \cdot 17 + 5) + 6) - 4$$
$$= 10(10 \cdot 175 + 6) - 4$$
$$= 10 \cdot 1756 - 4$$
$$= 17,560 - 4$$
$$= 17,556$$

68. $x(x(x(x(2x - 3) + 0) + 5) - 7) + 12;$
$Q(-4) = -2696, \ Q(7) = 26,619, \ Q(11) = 278,719$

69.
$$4y^{4a} + 12y^{2a} + 10y^{2a} + 30$$
$$= 2(2y^{4a} + 6y^{2a} + 5y^{2a} + 15)$$
$$= 2[2y^{2a}(y^{2a} + 3) + 5(y^{2a} + 3)]$$
$$= 2(2y^{2a} + 5)(y^{2a} + 3)$$

70. $(2x^{2p} + 5)(x^{2p} + 3)$

71.
$$4x^{a+b} + 7x^{a-b}$$
$$= 4 \cdot x^a \cdot x^b + 7 \cdot x^a \cdot x^{-b}$$
$$= x^a(4x^b + 7x^{-b})$$

72. $y^{a+b}(7y^a - 5 + 3y^b)$

Exercise Set 5.4

1. $x^2 + 9x + 20$

We look for two numbers whose product is 20 and whose sum is 9. Since both 20 and 9 are positive, we need only consider positive factors.

Pairs of Factors	Sums of Factors
1, 20	21
2, 10	12
4, 5	9

The numbers we need are 4 and 5. The factorization is $(x + 4)(x + 5)$.

2. $(x + 3)(x + 5)$

3. $t^2 - 8t + 15$

Since the constant term is positive and the coefficient of the middle term is negative, we look for a factorization of 15 in which both factors are negative. Their sum must be -8.

Pairs of Factors	Sums of Factors
-1, -15	-16
-3, -5	-8

The numbers we need are -3 and -5. The factorization is $(t - 3)(t - 5)$.

4. $(y - 3)(y - 9)$

5. $x^2 - 27 - 6x = x^2 - 6x - 27$

Since the constant term is negative, we look for a factorization of -27 in which one factor is positive and one factor is negative. Their sum must be -6, so the negative factor must have the larger absolute value. Thus we consider only pairs of factors in which the negative factor has the larger absolute value.

Pairs of Factors	Sums of Factors
-27, 1	-26
-9, 3	-6

The numbers we need are -9 and 3. The factorization is $(x - 9)(x + 3)$.

6. $(t - 5)(t + 3)$

7. $2y^2 - 16y + 32$

$= 2(y^2 - 8y + 16)$ Removing the common factor

We now factor $y^2 - 8y + 16$. We look for two numbers whose product is 16 and whose sum is -8. Since the constant term is positive and the coefficient of the middle term is negative, we look for a factorization of 16 in which both factors are negative.

Pairs of Factors	Sums of Factors
-1, -16	-17
-2, -8	-10
-4, -4	-8

The numbers we want are -4 and -4.

$y^2 - 8y + 16 = (y - 4)(y - 4)$

We must not forget to include the common factor 2.

$2y^2 - 16y + 32 = 2(y - 4)(y - 4)$.

8. $2(a - 5)(a - 5)$

9. $p^2 + 3p - 54$

Since the constant term is negative, we look for a factorization of -54 in which one factor is positive and one factor is negative. We consider only pairs of factors in which the positive factor has the larger absolute value, since the sum of the factors, 3, is positive.

Pairs of Factors	Sums of Factors
54, -1	53
27, -2	25
18, -3	15
9, -6	3

The numbers we need are 9 and -6. The factorization is $(p + 9)(p - 6)$.

10. $(m + 9)(m - 8)$

11. $14x + x^2 + 45 = x^2 + 14x + 45$

Since the constant term and the middle term are both positive, we look for a factorization of 45 in which both factors are positive. Their sum must be 14.

Pairs of Factors	Sums of Factors
45, 1	46
15, 3	18
9, 5	14

The numbers we need are 9 and 5. The factorization is $(x + 9)(x + 5)$.

12. $(y + 8)(y + 4)$

13. $y^2 + 2y - 63$

Since the constant term is negative, we look for a factorization of -63 in which one factor is positive and one factor is negative. We consider only pairs of factors in which the positive factor has the larger absolute value, since the sum of the factors, 2, is positive.

Pairs of Factors	Sums of Factors
63, -1	62
21, -3	18
9, -7	2

The numbers we need are 9 and -7. The factorization is $(y + 9)(y - 7)$.

14. $(p + 8)(p - 5)$

15. $t^2 - 11t + 28$

Since the constant term is positive and the coefficient of the middle term is negative, we look for a factorization of 28 in which both factors are negative. Their sum must be -11.

Pairs of Factors	Sums of Factors
-28, -1	-29
-14, -2	-16
-7, -4	-11

The numbers we need are -7 and -4. The factorization is $(t - 7)(t - 4)$.

16. $(y - 9)(y - 5)$

17. $3x + x^2 - 10 = x^2 + 3x - 10$

Since the constant term is negative, we look for a factorization of -10 in which one factor is positive and one factor is negative. We consider only pairs of factors in which the positive factor has the larger absolute value, since the sum of the factors, 3, is positive.

Pairs of Factors	Sums of Factors
10, -1	9
5, -2	3

The numbers we need are 5 and -2. The factorization is $(x + 5)(x - 2)$.

18. $(x + 3)(x - 2)$

19. $x^2 + 5x + 6$

We look for two numbers whose product is 6 and whose sum is 5. Since 6 and 5 are both positive, we need consider only positive factors.

Pairs of Factors	Sums of Factors
1, 6	7
2, 3	5

The numbers we need are 2 and 3. The factorization is $(x + 2)(x + 3)$.

20. $(y + 7)(y + 1)$

21. $56 + x - x^2 = -x^2 + x + 56 = -(x^2 - x - 56)$

We now factor $x^2 - x - 56$. Since the constant term is negative, we look for a factorization of -56 in which one factor is positive and one factor is negative. We consider only pairs of factors in which the negative factor has the larger absolute value, since the sum of the factors, -1, is negative.

Pairs of Factors	Sums of Factors
-56, 1	-55
-28, 2	-26
-14, 4	-10
-8, 7	-1

The numbers we need are -8 and 7. Thus, $x^2 - x - 56 = (x - 8)(x + 7)$. We must not forget to include the factor that was factored out earlier:

$56 + x - x^2 = -(x - 8)(x + 7)$, or

$(-x + 8)(x + 7)$, or $(8 - x)(7 + x)$

22. $(8 - y)(4 + y)$

23. $32y + 4y^2 - y^3$

There is a common factor, y. We also factor out -1 in order to make the leading coefficient positive.

$32y + 4y^2 - y^3 = -y(-32 - 4y + y^2)$

$= -y(y^2 - 4y - 32)$

Now we factor $y^2 - 4y - 32$. Since the constant term is negative, we look for a factorization of -32 in which one factor is positive and one factor is negative. We consider only pairs of factors in which the negative factor has the larger absolute value, since the sum of the factors, -4, is negative.

Pairs of Factors	Sums of Factors
-32, 1	-31
-16, 2	-14
-8, 4	-4

The numbers we need are -8 and 4. Thus, $y^2 - 4y - 32 = (y - 8)(y + 4)$. We must not forget to include the common factor:

$32y + 4y^2 - y^3 = -y(y - 8)(y + 4)$, or

$y(-y + 8)(y + 4)$, or $y(8 - y)(4 + y)$

24. $x(8 - x)(7 + x)$

25. $x^4 + 11x^2 - 80$

First make a substitution. We let $u = x^2$, so $u^2 = x^4$. Then we consider $u^2 + 11u - 80$. We look for pairs of factors of -80, one positive and one negative, such that the positive factor has the larger absolute value and the sum of the factors is 11.

Pairs of Factors	Sums of Factors
80, -1	79
40, -2	38
20, -4	16
16, -5	11
10, -8	2

The numbers we need are 16 and -5. Then $u^2 + 11u - 80 = (u + 16)(u - 5)$. Replacing u by x^2 we obtain the factorization of the original trinomial: $(x^2 + 16)(x^2 - 5)$

26. $(y^2 + 12)(y^2 - 7)$

27. $x^2 - 3x + 7$

There are no factors of 7 whose sum is -3. This trinomial is not factorable into binomials with integer coefficients.

28. Not factorable with integer coefficients

29. $x^2 + 12xy + 27y^2$

We look for numbers p and q such that $x^2 + 12xy + 27y^2 = (x + py)(x + qy)$. Our thinking is much the same as if we were factoring $x^2 + 12x + 27$. We look for factors of 27 whose sum is 12. Those factors are 9 and 3. Then

$$x^2 + 12xy + 27y^2 = (x + 9y)(x + 3y).$$

30. $(p - 8q)(p + 3q)$

31. $x^2 - 14x + 49$

Since the constant term is positive and the coefficient of the middle term is negative, we look for a factorization of 49 in which both factors are negative. Their sum must be -14.

Pairs of Factors	Sums of Factors
-49, -1	-50
-7, -7	-14

The numbers we need are -7 and -7. The factorization is $(x - 7)(x - 7)$.

32. $(y + 4)(y + 4)$

33. $x^4 + 50x^2 + 49$

Substitute u for x^2 (and hence u^2 for x^4). Consider $u^2 + 50u + 49$. We look for a pair of positive factors of 49 whose sum is 50.

Pairs of Factors	Sums of Factors
7, 7	14
1, 49	50

The numbers we need are 1 and 49. Then $u^2 + 50 + 49 = (u + 1)(u + 49)$. Replacing u by x^2 we have

$$x^4 + 50x^2 + 49 = (x^2 + 1)(x^2 + 49).$$

34. $(p^2 + 79)(p^2 + 1)$

35. $x^6 + 2x^3 - 63$

Substitute u for x^3 (and hence u^2 for x^6). Consider $u^2 + 2u - 63$. This is the same trinomial as in Exercise 13, so we know $u^2 + 2u - 63 = (u + 9)(u - 7)$. Replacing u by x^3 we have

$$x^6 + 2x^3 - 63 = (x^3 + 9)(x^3 - 7).$$

36. $(x^4 - 5)(x^4 - 2)$

37. $3x^2 - 16x - 12$

We will use the FOIL method.

a) There is no common factor (other than 1).

b) Factor the first term, $3x^2$. The factors are $3x$ and x. We have this possibility:

$$(3x \quad)(x \quad)$$

c) Factor the last term, -12. The factors are 12, -1 and -1, 12 and 6, -2 and -6, 2, and 4, -3 and -3, 4.

d) We look for factors in b) and c) such that the sum of their products is the middle term -16x. Since the constant term is negative, we know that one sign of the constant terms of the factors will be positive and the other will be negative. Try some possibilities and check by multiplying.

$$(3x - 3)(x + 4) = 3x^2 + 9x - 12$$

This gives a middle term with a positive coefficient. But, more significantly, the expression $3x - 3$ has a common factor of 3. But we know there is no common factor (other than 1) so we can eliminate $(3x - 3)$ and all other factors containing a common factor, such as $(3x + 6)$, $(3x - 12)$, and so on.

We try another possibility:

$$(3x + 2)(x - 6) = 3x^2 - 16x - 12$$

The factorization is $(3x + 2)(x - 6)$.

38. $(3x + 5)(2x - 5)$

39. $6x^3 - 15x - x^2 = 6x^3 - x^2 - 15x$

We will use the grouping method.

a) Look for a common factor. We factor out x:

$$x(6x^2 - x - 15)$$

b) Factor the trinomial $6x^2 - x - 15$. Multiply the leading coefficient, 6, and the constant, -15.

$$6(-15) = -90$$

c) Try to factor -90 so the sum of the factors is -1. We need only consider pairs of factors in which the negative term has the larger absolute value, since their sum is negative.

Pairs of Factors	Sums of Factors
-90, 1	-89
-45, 2	-43
-30, 3	-27
-16, 5	-11
-10, 9	-1

39. (continued)

d) We split the middle term, $-x$, as follows:

$-x = -10x + 9x$

e) Factor by grouping:

$$6x^2 - x - 15 = 6x^2 - 10x + 9x - 15$$
$$= 2x(3x - 5) + 3(3x - 5)$$
$$= (3x - 5)(2x + 3)$$

We must include the common factor to get a factorization of the original trinomial:

$$6x^3 - 15x - x^2 = x(3x - 5)(2x + 3)$$

40. $y(5y + 4)(2y - 3)$

41. $3a^2 - 10a + 8$

We will use the FOIL method.

a) There is no common factor (other than 1).

b) Factor the first term, $3a^2$. The factors are $3a$ and a. We have this possibility:
$(3a \quad)(a \quad)$

c) Factor the last term, 8. The factors are 8, 1 and -8, -1 and 4, 2 and -4, -2.

d) Look for factors in b) and c) such that the sum of the products is the middle term, $-10a$. Try some possibilities and check by multiplying. Trial and error leads us to the correct factorization, $(3a - 4)(a - 2)$.

42. $(4a - 1)(3a - 1)$

43. $35y^2 + 34y + 8$

We will use the grouping method.

a) There is no common factor (other than 1).

b) Multiply the leading coefficient, 35, and the constant, 8: $35(8) = 280$

c) Try to factor 280 so the sum of the factors is 34. We need only consider pairs of positive factors since 280 and 34 are both positive.

Pairs of Factors	Sums of Factors
280, 1	281
140, 2	142
70, 4	74
56, 5	61
40, 7	47
28, 10	38
20, 14	34

d) Split $34y$ as follows:

$34y = 20y + 14y$

e) Factor by grouping:

$$35y^2 + 34y + 8 = 35y^2 + 20y + 14y + 8$$
$$= 5y(7y + 4) + 2(7y + 4)$$
$$= (7y + 4)(5y + 2)$$

44. $(3a + 2)(3a + 4)$

45. $4t + 10t^2 - 6 = 10t^2 + 4t - 6$

We will use the FOIL method.

a) Factor out the common factor, 2:

$2(5t^2 + 2t - 3)$

b) Now we factor the trinomial $5t^2 + 2t - 3$.

Factor the first term, $5t^2$. The factors are $5t$ and t. We have this possibility:
$(5t \quad)(t \quad)$

c) Factor the last term, -3. The factors are 1, -3 and -1, 3.

d) Look for factors in b) and c) such that the sum of the products is the middle term, $2t$. Trial and error leads us to the correct factorization:
$5t^2 + 2t - 3 = (5t - 3)(t + 1)$

We must include the common factor to get a factorization of the original trinomial:

$$4t + 10t^2 - 6 = 2(5t - 3)(t + 1)$$

46. $2(5x + 3)(3x - 1)$

47. $8x^2 - 16 - 28x = 8x^2 - 28x - 16$

We will use the grouping method.

a) Factor out the common factor, 4:

$4(2x^2 - 7x - 4)$

b) Now we factor the trinomial $2x^2 - 7x - 4$. Multiply the leading coefficient, 2, and the constant, -4: $2(-4) = -8$

c) Factor -8 so the sum of the factors is -7. We need only consider pairs of factors in which the negative factor has the larger absolute value, since their sum is negative.

Pairs of Factors	Sums of Factors
-4, 2	-2
-8, 1	-7

d) Split $-7x$ as follows:

$-7x = -8x + x$

e) Factor by grouping:

$$2x^2 - 7x - 4 = 2x^2 - 8x + x - 4$$
$$= 2x(x - 4) + (x - 4)$$
$$= (x - 4)(2x + 1)$$

We must include the common factor to get a factorization of the original trinomial:

$$8x^2 - 16 - 28x = 4(x - 4)(2x + 1)$$

48. $6(3x - 4)(x + 1)$

49. $12x^3 - 31x^2 + 20x$

We will use the FOIL method.

a) Factor out the common factor, x:

$x(12x^2 - 31x + 20)$

b) We now factor the trinomial $12x^2 - 31x + 20$. Factor the first term, $12x^2$. The factors are 12x, x and 6x, 2x and 4x, 3x. We have these possibilities: (12x   )(x   ),
(6x   )(2x   ), (4x   )(3x   )

c) Factor the last term, 20. The factors are 20, 1 and -20, -1 and 10, 2 and -10, -2 and 5, 4 and -5, -4.

d) Look for factors in b) and c) such that the sum of the products is the middle term, -31x. Trial and error leads us to the correct factorization:
$12x^2 - 31x + 20 = (4x - 5)(3x - 4)$

We must include the common factor to get a factorization of the original trinomial:

$12x^3 - 31x^2 + 20x = x(4x - 5)(3x - 4)$

50. $x(5x + 2)(3x - 5)$

51. $14x^4 - 19x^3 - 3x^2$

We will use the grouping method.

a) Factor out the common factor, x^2:

$x^2(14x^2 - 19x - 3)$

b) Now we factor the trinomial $14x^2 - 19x - 3$. Multiply the leading coefficient, 14, and the constant, -3: $14(-3) = -42$

c) Factor -42 so the sum of the factors is -19. We need only consider pairs of factors in which the negative factor has the larger absolute value, since the sum is negative.

Pairs of Factors	Sums of Factors
-42, 1	-41
-21, 2	-19
-14, 3	-11
-7, 6	-1

d) Split -19x as follows:

$-19x = -21x + 2x$

e) Factor by grouping:

$14x^2 - 19x - 3 = 14x^2 - 21x + 2x - 3$
$= 7x(2x - 3) + 2x - 3$
$= (2x - 3)(7x + 1)$

We must include the common factor to get a factorization of the original trinomial:

$14x^4 - 19x^3 - 3x^2 = x^2(2x - 3)(7x + 1)$

52. $2x^2(5x - 2)(7x - 4)$

53. $3a^2 - a - 4$

We will use the FOIL method.

a) There is no common factor (other than 1).

b) Factor the first term, $3a^2$. The factors are 3a and a. We have this possibility:
(3a   )(a   )

53. (continued)

c) Factor the last term, -4. The factors are 4, -1 and -4, 1 and 2, -2.

d) Look for factors in b) and c) such that the sum of the products is the middle term, -a. Trial and error leads us to the correct factorization: $(3a - 4)(a + 1)$

54. $(6a + 5)(a - 2)$

55. $9x^2 + 15x + 4$

We will use the grouping method.

a) There is no common factor (other than 1).

b) Multiply the leading coefficient and the constant: $9(4) = 36$

c) Factor 36 so the sum of the factors is 15. We need only consider pairs of positive factors since 36 and 15 are both positive.

Pairs of Factors	Sums of Factors
36, 1	37
18, 2	20
12, 3	15
9, 4	13
6, 6	12

d) Split 15x as follows:

$15x = 12x + 3x$

e) Factor by grouping:

$9x^2 + 15x + 4 = 9x^2 + 12x + 3x + 4$
$= 3x(3x + 4) + 3x + 4$
$= (3x + 4)(3x + 1)$

56. $(3y - 2)(2y + 1)$

57. $3 + 35z - 12z^2 = -12z^2 + 35z + 3$

We will use the FOIL method.

a) Factor out -1 so the leading coefficient is positive: $-(12z^2 - 35z - 3)$

b) Now we factor the trinomial $12z^2 - 35z - 3$. Factor the first term, $12z^2$. The factors are 12z, z and 6z, 2z and 4z, 3z. We have these possibilities: (12z   )(z   ),
(6z   )(2z   ), (4z   )(3z   )

c) Factor the last term, -3. The factors are 3, -1 and -3, 1.

d) Look for factors in b) and c) such that the sum of the products is the middle term, -35z. Trial and error leads us to the correct factorization: (12z + 1)(z - 3)

We must include the common factor to get a factorization of the original trinomial:

$3 + 35z - 12z^2 = -(12z + 1)(z - 3)$, or $(12z + 1)(-z + 3)$, or $(1 + 12z)(3 - z)$

58. $(2 - 3a)(4 + 3a)$

59. $-4t^2 - 4t + 15$

We will use the grouping method.

a) Factor out -1 so the leading coefficient is positive: $-(4t^2 + 4t - 15)$

b) Now we factor the trinomial $4t^2 + 4t - 15$. Multiply the leading coefficient and the constant: $4(-15) = -60$

c) Factor -60 so the sum of the factors is 4. The desired factorization is $10(-6)$.

d) Split $4t$ as follows:
$$4t = 10t - 6t$$

e) Factor by grouping:
$$4t^2 + 4t - 15 = 4t^2 + 10t - 6t - 15$$
$$= 2t(2t + 5) - 3(2t + 5)$$
$$= (2t + 5)(2t - 3)$$

We must include the common factor to get a factorization of the original trinomial:
$$-4t^2 - 4t + 15 = -(2t + 5)(2t - 3)$$

60. $-(4a - 1)(3a - 1)$

61. $3x^3 - 5x^2 - 2x$

We will use the FOIL method.

a) Factor out the common factor, x:
$x(3x^2 - 5x - 2)$

b) Now we factor the trinomial $3x^2 - 5x - 2$. Factor the first term, $3x^2$. The factors are $3x$ and x. We have this possibility:
$(3x \quad)(x \quad)$

c) Factor the last term, -2. The factors are $2, -1$ and $-2, 1$.

d) Look for factors in b) and c) such that the sum of the products is the middle term, $-5x$. Trial and error leads us to the correct factorization: $(3x + 1)(x - 2)$

We must include the common factor to get a factorization of the original trinomial:
$$3x^3 - 5x^2 - 2x = x(3x + 1)(x - 2)$$

62. $y(6y - 5)(3y + 2)$

63. $24x^2 - 2 - 47x = 24x^2 - 47x - 2$

We will use the grouping method.

a) There is no common factor (other than 1).

b) Multiply the leading coefficient and the constant: $24(-2) = -48$

c) Factor -48 so the sum of the factors is -47. The desired factorization is $-48 \cdot 1$.

d) Split $-47x$ as follows:
$$-47x = -48x + x$$

63. (continued)

e) Factor by grouping:
$$24x^2 - 47x - 2 = 24x^2 - 48x + x - 2$$
$$= 24x(x - 2) + (x - 2)$$
$$= (x - 2)(24x + 1)$$

64. $(5y + 1)(3y - 10)$

65. $21x^2 + 37x + 12$

We will use the FOIL method.

a) There is no common factor (other than 1).

b) Factor the first term $21x^2$. The factors are $21x, x$ and $7x, 3x$. We have these possibilities: $(21x \quad)(x \quad)$ and $(7x \quad)(3x \quad)$.

c) Factor the last term, 12. The factors are $12, 1$ and $-12, -1$ and $6, 2$ and $-6, -2$ and $4, 3$ and $-4, -3$.

d) Look for factors in b) and c) such that the sum of the products is the middle term, $37x$. Trial and error leads us to the correct factorization: $(7x + 3)(3x + 4)$

66. $(5y + 4)(2y + 3)$

67. $40x^4 + 16x^2 - 12$

We will use the grouping method.

a) Factor out the common factor, 4.
$$4(10x^4 + 4x^2 - 3)$$

Now we will factor the trinomial $10x^4 + 4x^2 - 3$. Substitute u for x^2 (and u^2 for x^4), and factor $10u^2 + 4u - 3$.

b) Multiply the leading coefficient and the constant: $10(-3) = -30$

c) Factor -30 so the sum of the factors is 4. This cannot be done. The trinomial $10u^2 + 4u - 3$ cannot be factored into binomials with integer coefficients. We have
$$40x^4 + 16x^2 - 12 = 4(10x^4 + 4x^2 - 3)$$

68. $(6y^2 + 5)(4y^2 - 3)$

69. $12a^2 - 17ab + 6b^2$

We will use the FOIL method. (Our thinking is much the same as if we were factoring $12a^2 - 17a + 6$.)

a) There are no common factors (other than 1).

b) Factor the first term, $12a^2$. The factors are $12a, a$ and $6a, 2a$ and $4a, 3a$. We have these possibilities: $(12a \quad)(a \quad)$ and $(6a \quad)(2a \quad)$ and $(4a \quad)(3a \quad)$.

c) Factor the last term, $6b^2$. The factors are $6b, b$ and $-6b, -b$ and $3b, 2b$ and $-3b, -2b$.

d) Look for factors in b) and c) such that the sum of the products is the middle term, $-17ab$. Trial and error leads us to the correct factorization: $(4a - 3b)(3a - 2b)$

70. $(4p - 3q)(5p - 2q)$

71. $2x^2 + xy - 6y^2$

We will use the grouping method.

a) There is no common factor (other than 1).

b) Multiply the coefficients of the first and last terms: $2(-6) = -12$

c) Factor -12 so the sum of the factors is 1. The desired factorization is $4(-3)$.

d) Split xy as follows:
$$xy = 4xy - 3xy$$

e) Factor by grouping:
$$2x^2 + xy - 6y^2 = 2x^2 + 4xy - 3xy - 6y^2$$
$$= 2x(x + 2y) - 3y(x + 2y)$$
$$= (x + 2y)(2x - 3y)$$

72. $(4m + 3n)(2m - 3n)$

73. $6x^2 - 29xy + 28y^2$

We will use the FOIL method.

a) There is no common factor (other than 1).

b) Factor the first term, $6x^2$. The factors are $6x$, x and $3x$, $2x$. We have these possibilities: $(6x \quad)(x \quad)$ and $(3x \quad)(2x \quad)$.

c) Factor the last term, $28y^2$. The factors are $28y$, y and $-28y$, $-y$ and $14y$, $2y$ and $-14y$, $-2y$ and $7y$, $4y$ and $-7y$, $-4y$.

d) Look for factors in b) and c) such that the sum of the products is the middle term, $-29xy$. Trial and error leads us to the correct factorization: $(3x - 4y)(2x - 7y)$

74. $(2p + 3q)(5p - 4q)$

75. $9x^2 - 30xy + 25y^2$

We will use the grouping method.

a) There are no common factors (other than 1).

b) Multiply the coefficients of the first and last terms: $9(25) = 225$

c) Factor 225 so the sum of the factors is -30. The desired factorization is $-15(-15)$.

d) Split $-30xy$ as follows:
$$-30xy = -15xy - 15xy$$

e) Factor by grouping:
$$9x^2 - 30xy + 25y^2 = 9x^2 - 15xy - 15xy + 25y^2$$
$$= 3x(3x - 5y) - 5y(3x - 5y)$$
$$= (3x - 5y)(3x - 5y)$$

76. $(2p + 3q)(2p + 3q)$

77. $6x^6 + x^3 - 2$

We will use the FOIL method.

a) There is no common factor (other than 1). Substitute u for x^3 (and u^2 for x^6). We factor $6u^2 + u - 2$.

b) Factor the first term, $6u^2$. The factors are $6u$, u and $3u$, $2u$. We have these possibilities: $(6u \quad)(u \quad)$ and $(3u \quad)(2u \quad)$

c) Factor the last term, -2. The factors are 2, -1 and -2, 1.

d) Look for factors in b) and c) such that the sum of the products is the middle term, u. Trial and error leads us to the correct factorization: $(3u + 2)(2u - 1)$

Replacing u by x^3 we have
$$6x^6 + x^3 - 2 = (3x^3 + 2)(2x^3 - 1)$$

78. $(2p^4 + 5)(p^3 + 3)$

79. a) $h(0) = -16(0)^2 + 80(0) + 224 = 224$ ft
$h(1) = -16(1)^2 + 80(1) + 224 = 288$ ft
$h(3) = -16(3)^2 + 80(3) + 224 = 320$ ft
$h(4) = -16(4)^2 + 80(4) + 224 = 288$ ft
$h(6) = -16(6)^2 + 80(6) + 224 = 128$ ft

b) $h(t) = -16t^2 + 80t + 224$

We will use the grouping method.

A) Factor out -16 so the leading coefficient is positive: $-16(t^2 - 5t - 14)$

B) Factor the trinomial $t^2 - 5t - 14$. Multiply the leading coefficient and the constant: $1(-14) = -14$

C) Factor -14 so the sum of the factors is -5. The desired factorization is $-7 \cdot 2$.

D) Split $-5t$ as follows:
$$-5t = -7t + 2t$$

E) Factor by grouping:
$$t^2 - 5t - 14 = t^2 - 7t + 2t - 14$$
$$= t(t - 7) + 2(t - 7)$$
$$= (t - 7)(t + 2)$$

We must include the common factor to get a factorization of the original trinomial.
$$h(t) = -16(t - 7)(t + 2)$$

80. a) 880 ft, 960 ft, 1024 ft, 624 ft, 240 ft

b) $h(t) = -16(t - 11)(t + 5)$

81. $p^2q^2 + 7pq + 12$

    The factorization will be of the form
    (pq   )(pq   ). We look for factors of 12
    whose sum is 7. The factors we need are 4 and
    3. The factorization is $(pq + 4)(pq + 3)$.

82. $(2x^2y^3 + 5)(x^2y^3 - 4)$

83. $x^2 - \frac{4}{25} + \frac{3}{5}x = x^2 + \frac{3}{5}x - \frac{4}{25}$

    We look for factors of $-\frac{4}{25}$ whose sum is $\frac{3}{5}$.

    The factors are $\frac{4}{5}$ and $-\frac{1}{5}$. The factorization

    is $\left(x + \frac{4}{5}\right)\left(x - \frac{1}{5}\right)$.

84. $\left(y + \frac{4}{7}\right)\left(y - \frac{2}{7}\right)$

85. $y^2 + 0.4y - 0.05$

    We look for factors of -0.05 whose sum is 0.4.
    The factors are -0.1 and 0.5. The factorization
    is $(y - 0.1)(y + 0.5)$.

86. $(t + 0.9)(t - 0.3)$

87. $7a^2b^2 + 6 + 13ab = 7a^2b^2 + 13ab + 6$

    We will use the grouping method. There are no
    common factors (other than 1). Multiply the
    leading coefficient and the constant: $7(6) = 42$.
    Factor 42 so the sum of the factors is 13. The
    desired factorization is $6 \cdot 7$. Split the middle
    term and factor by grouping.

    $$7a^2b^2 + 13ab + 6 = 7a^2b^2 + 6ab + 7ab + 6$$
    $$= ab(7ab + 6) + 7ab + 6$$
    $$= (7ab + 6)(ab + 1)$$

88. $(9xy - 4)(xy + 1)$

89. $3x^2 + 12x - 495$

    Factor out the common factor, 3.

    $3(x^2 + 4x - 165)$

    Now factor $x^2 + 4x - 165$. Find factors of -165
    whose sum is 4. The factors are -11 and 15.
    Then $x^2 + 4x - 165 = (x - 11)(x + 15)$, and
    $3x^2 + 12x - 495 = 3(x - 11)(x + 15)$.

90. $15t(t - 7)(t + 3)$

91. $216x + 78x^2 + 6x^3 = 6x^3 + 78x^2 + 216x$

    Factor out the common factor, 6x.

    $6x(x^2 + 13x + 36)$

    Now factor $x^2 + 13x + 36$. Look for factors of
    36 whose sum is 13. The factors are 9 and 4.
    Then $x^2 + 13x + 36 = (x + 9)(x + 4)$, and
    $6x^3 + 78x^2 + 216x = 6x(x + 9)(x + 4)$.

92. $(x^a + 8)(x^a - 3)$

93. $x^2 + ax + bx + ab$
    $= x(x + a) + b(x + a)$ Factoring by grouping
    $= (x + a)(x + b)$

94. $(2x^a - 3)(2x^a + 1)$

95. $bdx^2 + adx + bcx + ac$
    $= dx(bx + a) + c(bx + a)$ Factoring by grouping
    $= (bx + a)(dx + c)$

96. $a^2(p^a + 2)(p^a - 1)$

97. $2ar^2 + 4asr + as^2 - asr$
    $= 2ar^2 + 3asr + as^2$ Combining like terms
    $= a(2r^2 + 3sr + s^2)$ Factoring out a

    Factor $2r^2 + 3sr + s^2$ using the grouping method.
    Multiply the first and last coefficients:
    $2 \cdot 1 = 2$. Factor 2 so the sum of the factors is
    3. The desired factorization is $2 \cdot 1$. Split the
    middle term and factor by grouping.

    $$2r^2 + 3sr + s^2 = 2r^2 + 2sr + sr + s^2$$
    $$= 2r(r + s) + s(r + s)$$
    $$= (2r + s)(r + s)$$

    We must include the common factor to get a
    factorization of the original expression. The
    factorization is $a(2r + s)(r + s)$.

98. $[(x + 3) - 7][(x + 3) + 5]$, or $(x - 4)(x + 8)$

99. $6(x - 7)^2 + 13(x - 7) - 5$

    Let $u = x - 7$, make the substitutions and factor
    using the grouping method. We have
    $6u^2 + 13u - 5$. Multiply the leading coefficient
    and the constant: $6(-5) = -30$. Factor -30 so
    the sum of the factors is 13. The desired
    factorization is $15(-2)$.

    Split the middle term and factor by grouping.

    $$6u^2 + 13u - 5 = 6u^2 + 15u - 2u - 5$$
    $$= 3u(2u + 5) - (2u + 5)$$
    $$= (2u + 5)(3u - 1)$$

    Replace u by $x - 7$. The factorization is
    $[2(x - 7) + 5][3(x - 7) - 1]$, or
    $(2x - 9)(3x - 22)$.

100. $76, -76, 28, -28, 20, -20$

101. $x^2 + qx - 32$

     All such q are the sums of the factors of -32.

Pairs of Factors	Sums of Factors
32, -1	31
-32, 1	-31
16, -2	14
-16, 2	-14
8, -4	4
-8, 4	-4

     q can be $31, -31, 14, -14, 4,$ or -4.

102. $(x - 365)$

Exercise Set 5.5

1. $y^2 - 6y + 9 = (y - 3)^2$ Find the square terms and write the quantities that were squared with a minus sign between them.

2. $(x - 4)^2$

3. $x^2 + 14x + 49 = (x + 7)^2$ Find the square terms and write the quantities that were squared with a plus sign between them.

4. $(x + 8)^2$

5. $x^2 + 1 + 2x = x^2 + 2x + 1$ Changing order
   $= (x + 1)^2$ Factoring the trinomial square

6. $(x - 1)^2$

7. $2a^2 + 8a + 8 = 2(a^2 + 4a + 4)$ Removing the common factor
   $= 2(a + 2)^2$ Factoring the trinomial square

8. $4(a - 2)^2$

9. $y^2 + 36 - 12y = y^2 - 12y + 36$ Changing order
   $= (y - 6)^2$ Factoring the trinomial square

10. $(y + 6)^2$

11. $-18y^2 + y^3 + 81y = y^3 - 18y^2 + 81y$ Changing order
    $= y(y^2 - 18y + 81)$ Removing the common factor
    $= y(y - 9)^2$ Factoring the trinomial square

12. $a(a + 12)^2$

13. $12a^2 + 36a + 27 = 3(4a^2 + 12a + 9)$ Removing the common factor
    $= 3(2a + 3)^2$ Factoring the trinomial square

14. $5(2y + 5)^2$

15. $2x^2 - 40x + 200 = 2(x^2 - 20x + 100)$
    $= 2(x - 10)^2$

16. $2(4x + 3)^2$

17. $1 - 8d + 16d^2 = (1 - 4d)^2$ Find the square terms and write the quantities that were squared with a minus sign between them.

18. $(5y - 8)^2$

19. $y^4 + 8y^2 + 16 = (y^2 + 4)^2$ Note that $y^4 = (y^2)^2$.

20. $(a^2 - 5)^2$

21. $0.25x^2 + 0.30x + 0.09 = (0.5x + 0.3)^2$ Find the square terms and write the quantities that were squared with a plus sign between them.

22. $(0.2x - 0.7)^2$

23. $p^2 - 2pq + q^2 = (p - q)^2$

24. $(m + n)^2$

25. $a^2 + 4ab + 4b^2 = (a + 2b)^2$

26. $(7p - q)^2$

27. $25a^2 - 30ab + 9b^2 = (5a - 3b)^2$

28. $(7p - 6q)^2$

29. $x^4 + 2x^2y^2 + y^4 = (x^2 + y^2)^2$ Note that $x^4 = (x^2)^2$ and $y^4 = (y^2)^2$.

30. $(p^4 + q^4)^2$

31. $x^2 - 16 = x^2 - 4^2 = (x + 4)(x - 4)$

32. $(y + 3)(y - 3)$

33. $p^2 - 49 = p^2 - 7^2 = (p + 7)(p - 7)$

34. $(m + 8)(m - 8)$

35. $p^2q^2 - 25 = (pq)^2 - 5^2 = (pq + 5)(pq - 5)$

36. $(ab + 9)(ab - 9)$

37. $6x^2 - 6y^2 = 6(x^2 - y^2)$ Removing the common factor
    $= 6(x + y)(x - y)$ Factoring the difference of squares

38. $8(x + y)(x - y)$

39. $4xy^4 - 4xz^4 = 4x(y^4 - z^4)$ Removing the common factor
    $= 4x[(y^2)^2 - (z^2)^2]$
    $= 4x(y^2 + z^2)(y^2 - z^2)$ Factoring the difference of squares
    $= 4x(y^2 + z^2)(y + z)(y - z)$ Factoring $y^2 - z^2$

40. $25a(b^2 + z^2)(b + z)(b - z)$

41. $4a^3 - 49a = a(4a^2 - 49)$
$$= a[(2a)^2 - 7^2]$$
$$= a(2a + 7)(2a - 7)$$

42. $x(3x + 5)(3x - 5)$

43. $3x^8 - 3y^8 = 3(x^8 - y^8)$
$$= 3[(x^4)^2 - (y^4)^2]$$
$$= 3(x^4 + y^4)(x^4 - y^4)$$
$$= 3(x^4 + y^4)[(x^2)^2 - (y^2)^2]$$
$$= 3(x^4 + y^4)(x^2 + y^2)(x^2 - y^2)$$
$$= 3(x^4 + y^4)(x^2 + y^2)(x + y)(x - y)$$

44. $a^2(3a + b)(3a - b)$

45. $9a^4 - 25a^2b^4 = a^2(9a^2 - 25b^4)$
$$= a^2[(3a)^2 - (5b^2)^2]$$
$$= a^2(3a + 5b^2)(3a - 5b^2)$$

46. $x^2(4x^2 + 11y^2)(4x^2 - 11y^2)$

47. $\dfrac{1}{25} - x^2 = \left(\dfrac{1}{5}\right)^2 - x^2$
$$= \left(\dfrac{1}{5} + x\right)\left(\dfrac{1}{5} - x\right)$$

48. $\left(\dfrac{1}{4} + y\right)\left(\dfrac{1}{4} - y\right)$

49. $0.04x^2 - 0.09y^2 = (0.2x)^2 - (0.3y)^2$
$$= (0.2x + 0.3y)(0.2x - 0.3y)$$

50. $(0.1x + 0.2y)(0.1x - 0.2y)$

51. $m^3 - 7m^2 - 4m + 28 = m^2(m - 7) - 4(m - 7)$
Factoring by grouping
$$= (m - 7)(m^2 - 4)$$
$$= (m - 7)(m + 2)(m - 2)$$
Factoring the difference of squares

52. $(x + 8)(x + 1)(x - 1)$

53. $a^3 - ab^2 - 2a^2 + 2b^2 = a(a^2 - b^2) - 2(a^2 - b^2)$
Factoring by grouping
$$= (a^2 - b^2)(a - 2)$$
$$= (a + b)(a - b)(a - 2)$$
Factoring the difference of squares

54. $(p + 5)(p - 5)(q + 3)$

55. $(a + b)^2 - 100 = (a + b)^2 - 10^2$
$$= (a + b + 10)(a + b - 10)$$

56. $(p + 5)(p - 19)$

57. $a^2 + 2ab + b^2 - 9 = (a^2 + 2ab + b^2) - 9$ Grouping as a difference of squares
$$= (a + b)^2 - 3^2$$
$$= (a + b + 3)(a + b - 3)$$

58. $(x - y + 5)(x - y - 5)$

59. $r^2 - 2r + 1 - 4s^2 = (r^2 - 2r + 1) - 4s^2$ Grouping as a difference of squares
$$= (r - 1)^2 - (2s)^2$$
$$= (r - 1 + 2s)(r - 1 - 2s)$$

60. $(c + 2d + 3p)(c + 2d - 3p)$

61. $2m^2 + 4mn + 2n^2 - 50b^2$
$= 2(m^2 + 2mn + n^2 - 25b^2)$ Removing the common factor
$= 2[(m^2 + 2mn + n^2) - 25b^2]$ Grouping as a difference of squares
$= 2[(m + n)^2 - (5b)^2]$
$= 2(m + n + 5b)(m + n - 5b)$

62. $3(2x + 1 + y)(2x + 1 - y)$

63. $9 - (a^2 + 2ab + b^2) = 9 - (a + b)^2$
$$= [3 + (a + b)][3 - (a + b)]$$
$$= (3 + a + b)(3 - a - b)$$

64. $(4 + x - y)(4 - x + y)$

65. $x - y + z = 6,$ (1)
$$2x + y - z = 0,$$ (2)
$$x + 2y + z = 3$$ (3)

Add (1) and (2).
$$x - y + z = 6 \quad (1)$$
$$\underline{2x + y - z = 0} \quad (2)$$
$$3x \qquad\quad = 6 \quad \text{Adding}$$
$$x = 2$$

Add (2) and (3).
$$2x + y - z = 0 \quad (2)$$
$$\underline{x + 2y + z = 3} \quad (3)$$
$$3x + 3y \qquad = 3 \quad (4)$$

Substitute 2 for x in (4).
$$3(2) + 3y = 3$$
$$6 + 3y = 3$$
$$3y = -3$$
$$y = -1$$

Substitute 2 for x and −1 for y in (1).
$$2 - (-1) + z = 6$$
$$3 + z = 6$$
$$z = 3$$

The solution is $(2, -1, 3)$.

66. $\left\{x \mid x \leqslant -\frac{4}{7} \text{ or } x \geqslant 2\right\}$, or $\left[-\infty, -\frac{4}{7}\right] \cup [2, \infty)$

67. $|5 - 7x| \leqslant 9$

$-9 \leqslant 5 - 7x \leqslant 9$

$-14 \leqslant -7x \leqslant 4$

$2 \geqslant x \geqslant -\frac{4}{7}$

The solution is $\left\{x \mid -\frac{4}{7} \leqslant x \leqslant 2\right\}$, or $\left[-\frac{4}{7}, 2\right]$.

68. $\left\{x \mid x < \frac{14}{19}\right\}$, or $\left[-\infty, \frac{14}{19}\right]$

69.

$a^2 + b^2 = (-2)^2 + 3^2 = 4 + 9 = 13$

$(a + b)^2 = (-2 + 3)^2 = 1^2 = 1$

$a^2 - b^2 = (-2)^2 - 3^2 = 4 - 9 = -5$

$a^2 + 2ab + b^2 = (-2)^2 + 2(-2)(3) + 3^2$

$= 4 - 12 + 9 = 1$

$a^2 - 2ab + b^2 = (-2)^2 - 2(-2)(3) + 3^2$

$= 4 + 12 + 9 = 25$

$(a - b)(a + b) = (-2 - 3)(-2 + 3) = -5 \cdot 1 = -5$

70. 17; 9; 15; 9; 25; 15

71.

$-225x + x^3 = x^3 - 225x$

$= x(x^2 - 225)$

$= x(x + 15)(x - 15)$

72. $4y(y - 12)^2$

73.

$3xy^2 - 150xy + 1875x = 3x(y^2 - 50y + 625)$

$= 3x(y - 25)^2$

74. $7(x + 25y)(x - 25y)$

75.

$12x^2 - 72xy + 108y^2 = 12(x^2 - 6xy + 9y^2)$

$= 12(x - 3y)^2$

76. $-3\left[\frac{1}{2}p - \frac{2}{5}\right]^2$

77.

$-\frac{8}{27}r^2 - \frac{10}{9}rs - \frac{1}{6}s^2 + \frac{2}{3}rs$

$= -\frac{8}{27}r^2 - \frac{4}{9}rs - \frac{1}{6}s^2$ Collecting like terms

$= -\frac{1}{54}(16r^2 + 24rs + 9s^2)$

$= -\frac{1}{54}(4r + 3s)^2$

78. $(x^2y^2 - 4)^2$

79.

$-24ab + 16a^2 + 9b^2 = 16a^2 - 24ab + 9b^2$

$= (4a - 3b)^2$

80. $\left[\frac{1}{6}x^4 + \frac{2}{3}\right]^2$

81. $0.09x^2 + 0.48x + 0.64 = (0.3x + 0.8)^2$

82. $(x^a + y)(x^a - y)$

83. $x^{4a} - y^{2b} = (x^{2a})^2 - (y^b)^2 = (x^{2a} + y^b)(x^{2a} - y^b)$

84. $4(y^{2a} + 5)$

85.

$25y^{2a} - (x^{2b} - 2x^b + 1)$

$= (5y^a)^2 - (x^b - 1)^2$

$= [5y^a + (x^b - 1)][5y^a - (x^b - 1)]$

$= (5y^a + x^b - 1)(5y^a - x^b + 1)$

86. $8(a - 7)^2$

87.

$3(x + 1)^2 + 12(x + 1) + 12$

$= 3[(x + 1)^2 + 4(x + 1) + 4]$

$= 3[(x + 1) + 2]^2$

$= 3(x + 3)^2$

88. $5(c^{50} + 4d^{50})(c^{25} + 2d^{25})(c^{25} - 2d^{25})$

89.

$(x + h)^2 - x^2 = [(x + h) + x][(x + h) - x]$

$= (2x + h)h$, or $h(2x + h)$

90. $c(c^w + 1)^2$

91.

$9x^{2n} - 6x^n + 1 = (3x^n)^2 - 6x^n + 1$

$= (3x^n - 1)^2$

92. a) $\pi(R + r)(R - r)$

b) 263.76 cm²

Exercise Set 5.6

1.

$x^3 + 8 = x^3 + 2^3$

$= (x + 2)(x^2 - 2x + 4)$

$A^3 + B^3 = (A + B)(A^2 - AB + B^2)$

2. $(c + 3)(c^2 - 3c + 9)$

3.

$y^3 - 64 = y^3 - 4^3$

$= (y - 4)(y^2 + 4y + 16)$

$A^3 - B^3 = (A - B)(A^2 + AB + B^2)$

4. $(z - 1)(z^2 + z + 1)$

5.

$w^3 + 1 = w^3 + 1^3$

$= (w + 1)(w^2 - w + 1)$

$A^3 + B^3 = (A + B)(A^2 - AB + B^2)$

6. $(x + 5)(x^2 - 5x + 25)$

7.

$8a^3 + 1 = (2a)^3 + 1^3$

$= (2a + 1)(4a^2 - 2a + 1)$

$A^3 + B^3 = (A + B)(A^2 - AB + B^2)$

8. $(3x + 1)(9x^2 - 3x + 1)$

9. $y^3 - 8 = y^3 - 2^3$
    $= (y - 2)(y^2 + 2y + 4)$
    $A^3 - B^3 = (A - B)(A^2 + AB + B^2)$

10. $(p - 3)(p^2 + 3p + 9)$

11. $8 - 27b^3 = 2^3 - (3b)^3$
    $= (2 - 3b)(4 + 6b + 9b^2)$

12. $(4 - 5x)(16 + 20x + 25x^2)$

13. $64y^3 + 1 = (4y)^3 + 1^3$
    $= (4y + 1)(16y^2 - 4y + 1)$

14. $(5x + 1)(25x^2 - 5x + 1)$

15. $8x^3 + 27 = (2x)^3 + 3^3$
    $= (2x + 3)(4x^2 - 6x + 9)$

16. $(3y + 4)(9y^2 - 12y + 16)$

17. $a^3 - b^3 = (a - b)(a^2 + ab + b^2)$

18. $(x - y)(x^2 + xy + y^2)$

19. $a^3 + \dfrac{1}{8} = a^3 + \left[\dfrac{1}{2}\right]^3$
    $= \left[a + \dfrac{1}{2}\right]\left[a^2 - \dfrac{1}{2}a + \dfrac{1}{4}\right]$

20. $\left[b + \dfrac{1}{3}\right]\left[b^2 - \dfrac{1}{3}b + \dfrac{1}{9}\right]$

21. $2y^3 - 128 = 2(y^3 - 64)$
    $= 2(y^3 - 4^3)$
    $= 2(y - 4)(y^2 + 4y + 16)$

22. $3(z - 1)(z^2 + z + 1)$

23. $24a^3 + 3 = 3(8a^3 + 1)$
    $= 3[(2a)^3 + 1^3]$
    $= 3(2a + 1)(4a^2 - 2a + 1)$

24. $2(3x + 1)(9x^2 - 3x + 1)$

25. $rs^3 + 64r = r(s^3 + 64)$
    $= r(s^3 + 4^3)$
    $= r(s + 4)(s^2 - 4s + 16)$

26. $a(b + 5)(b^2 - 5b + 25)$

27. $5x^3 - 40z^3 = 5(x^3 - 8z^3)$
    $= 5[x^3 - (2z)^3]$
    $= 5(x - 2z)(x^2 + 2xz + 4z^2)$

28. $2(y - 3z)(y^2 + 3yz + 9z^2)$

29. $x^3 + 0.001 = x^3 + (0.1)^3$
    $= (x + 0.1)(x^2 - 0.1x + 0.01)$

30. $(y + 0.5)(y^2 - 0.5y + 0.25)$

31. $64x^6 - 8t^6 = 8(8x^6 - t^6)$
    $= 8[(2x^2)^3 - (t^2)^3]$
    $= 8(2x^2 - t^2)(4x^4 + 2x^2t^2 + t^4)$

32. $(5c^2 - 2d^2)(25c^4 + 10c^2d^2 + 4d^4)$

33. $2y^4 - 128y = 2y(y^3 - 64)$
    $= 2y(y^3 - 4^3)$
    $= 2y(y - 4)(y^2 + 4y + 16)$

34. $3z^2(z - 1)(z^2 + z + 1)$

35. $z^6 - 1 = (z^3)^2 - 1^2$ Writing as a difference of
                                  squares
    $= (z^3 + 1)(z^3 - 1)$ Factoring a difference
                              of squares
    $= (z + 1)(z^2 - z + 1)(z - 1)(z^2 + z + 1)$
                              Factoring a sum and a
                              difference of cubes

36. $(t^2 + 1)(t^4 - t^2 + 1)$

37. $t^6 + 64y^6 = (t^2)^3 + (4y^2)^3$
    $= (t^2 + 4y^2)(t^4 - 4t^2y^2 + 16y^4)$

38. $(p + q)(p^2 - pq + q^2)(p - q)(p^2 + pq + q^2)$

39. Familiarize. Let w represent the width and ℓ
    represent the length of the rectangle. If the
    width is increased by 2 ft, the new width is
    w + 2. Also, recall the formulas for the
    perimeter and area of a rectangle:

    $P = 2\ell + 2w$

    $A = \ell w$

    Translate.
      Width is length less 7 ft.
        w $=$ ℓ $- 7$

    When the length is ℓ and the width is w + 2, the
    perimeter is 66 ft, so we have

    $66 = 2\ell + 2(w + 2)$.

    We have a system of equations:

    $w = \ell - 7$,
    $66 = 2\ell + 2(w + 2)$

    Carry out. Solving the system we get (19,12).

    Check. The width, 12 ft, is 7 ft less than the
    length, 19 ft. When the width is increased by
    2 ft it becomes 14 ft, and the perimeter of the
    rectangle is $2 \cdot 19 + 2 \cdot 14 = 38 + 28 = 66$ ft. The
    numbers check.

    State. The area of the original rectangle is
    19 ft·12 ft, or 228 ft².

40. {27,-27}

41. $|5x - 6| \leqslant 39$

 $-39 \leqslant 5x - 6 \leqslant 39$

 $-33 \leqslant 5x \leqslant 45$

 $-\frac{33}{5} \leqslant x \leqslant 9$

 The solution set is $\left\{x \middle| -\frac{33}{5} \leqslant x \leqslant 9\right\}$, or $\left[-\frac{33}{5}, 9\right]$.

42. $\left\{x \middle| x < -\frac{33}{5} \text{ or } x > 9\right\}$, or $\left[-\infty, -\frac{33}{5}\right] \cup (9, \infty)$

43. $(a + b)^3 = (-2 + 3)^3 = 1^3 = 1$

 $a^3 + b^3 = (-2)^3 + (3)^3 = -8 + 27 = 19$

 $(a + b)(a^2 - ab + b^2)$

 $= (-2 + 3)[(-2)^2 - (-2)(3) + (3)^2]$

 $= 1(4 + 6 + 9)$

 $= 19$

 $(a + b)(a^2 + ab + b^2)$

 $= (-2 + 3)[(-2)^2 + (-2)(3) + (3)^2]$

 $= 1(4 - 6 + 9)$

 $= 7$

 $(a + b)(a - b)(a - b) = (-2 + 3)(-2 - 3)(-2 - 3)$

 $= 1(-5)(-5)$

 $= 25$

44. 27; 63; 63; 39; 75

45. $x^{6a} + y^{3b} = (x^{2a})^3 + (y^b)^3$

 $= (x^{2a} + y^b)(x^{4a} - x^{2a}y^b + y^{2b})$

46. $(ax - by)(a^2x^2 + axby + b^2y^2)$

47. $3x^{3a} + 24y^{3b} = 3(x^{3a} + 8y^{3b})$

 $= 3[(x^a)^3 + (2y^b)^3]$

 $= 3(x^a + 2y^b)(x^{2a} - 2x^ay^b + 4y^{2b})$

48. $\left[\frac{2}{3}x + \frac{1}{4}y\right]\left[\frac{4}{9}x^2 - \frac{1}{6}xy + \frac{1}{16}y^2\right]$

49. $\frac{1}{24}x^3y^3 + \frac{1}{3}z^3 = \frac{1}{3}\left[\frac{1}{8}x^3y^3 + z^3\right]$

 $= \frac{1}{3}\left[\left[\frac{1}{2}xy\right]^3 + z^3\right]$

 $= \frac{1}{3}\left[\frac{1}{2}xy + z\right]\left[\frac{1}{4}x^2y^2 - \frac{1}{2}xyz + z^2\right]$

50. $\frac{1}{2}\left[\frac{1}{2}x^a + y^{2a}z^{3b}\right]\left[\frac{1}{4}x^{2a} - \frac{1}{2}x^ay^{2a}z^{3b} + y^{4a}z^{6b}\right]$

51. $7x^3 + \frac{7}{8} = 7\left[x^3 + \frac{1}{8}\right]$

 $= 7\left[x^3 + \left[\frac{1}{2}\right]^3\right]$

 $= 7\left[x + \frac{1}{2}\right]\left[x^2 - \frac{1}{2}x + \frac{1}{4}\right]$

52. $(c - 2d)^2(c^2 - cd + d^2)^2$

53. $(x + y)^3 - x^3$

 $= [(x + y) - x][(x + y)^2 + x(x + y) + x^2]$

 $= (x + y - x)(x^2 + 2xy + y^2 + x^2 + xy + x^2)$

 $= y(3x^2 + 3xy + y^2)$

54. $(x - 1)^3(x - 2)(x^2 - x + 1)$

55. $(a + 2)^3 - (a - 2)^3$

 $= [(a + 2) - (a - 2)][(a + 2)^2 + (a + 2)(a - 2) + (a - 2)^2]$

 $= (a + 2 - a + 2)(a^2 + 4a + 4 + a^2 - 4 + a^2 - 4a + 4)$

 $= 4(3a^2 + 4)$

56. $(y - 8)(y - 1)(y^2 + y + 1)$

57. $P(x) + Q(x) = x^3 + 8$

 $= (x + 2)(x^2 - 2x + 4)$

58. $h(3x^2 + 3xh + h^2)$

59.

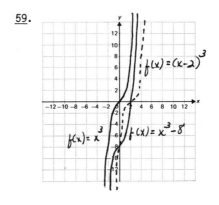

Exercise Set 5.7

1. $x^2 - 144$

 $= x^2 - 12^2$ Difference of squares

 $= (x + 12)(x - 12)$

2. $(y + 9)(y - 9)$

3. $2x^2 + 11x + 12$

 $= (2x + 3)(x + 4)$ FOIL or grouping method

4. $(4a - 1)(2a + 5)$

5. $3x^4 - 12$

 $= 3(x^4 - 4)$ Difference of squares

 $= 3(x^2 + 2)(x^2 - 2)$

6. $2x(y + 5)(y - 5)$

165

<u>7.</u> $a^2 + 25 + 10a$
      $= a^2 + 10a + 25$ Trinomial square
      $= (a + 5)^2$

<u>8.</u> $(p + 8)^2$

<u>9.</u> $2x^2 - 10x - 132$
      $= 2(x^2 - 5x - 66)$
      $= 2(x - 11)(x + 6)$ Trial and error

<u>10.</u> $3(y - 12)(y + 7)$

<u>11.</u> $9x^2 - 25y^2$
      $= (3x)^2 - (5y)^2$ Difference of squares
      $= (3x + 5y)(3x - 5y)$

<u>12.</u> $(4a + 9b)(4a - 9b)$

<u>13.</u> $m^6 - 1$
      $= (m^3)^2 - 1^2$ Difference of squares
      $= (m^3 + 1)(m^3 - 1)$ Sum and difference of cubes
      $= (m + 1)(m^2 - m + 1)(m - 1)(m^2 + m + 1)$

<u>14.</u> $(2t + 1)(4t^2 - 2t + 1)(2t - 1)(4t^2 + 2t + 1)$

<u>15.</u> $x^2 + 6x - y^2 + 9$
      $= x^2 + 6x + 9 - y^2$
      $= (x + 3)^2 - y^2$ Difference of squares
      $= [(x + 3) + y][(x + 3) - y]$
      $= (x + y + 3)(x - y + 3)$

<u>16.</u> $(t + p + 5)(t - p + 5)$

<u>17.</u> $250x^3 - 128y^3$
      $= 2(125x^3 - 64y^3)$
      $= 2[(5x)^3 - (4y)^3]$ Difference of cubes
      $= 2(5x - 4y)(25x^2 + 20xy + 16y^2)$

<u>18.</u> $(3a - 7b)(9a^2 + 21ab + 49b^2)$

<u>19.</u> $8m^3 + m^6 - 20$
      $= (m^3)^2 + 8m^3 - 20$
      $= (m^3 - 2)(m^3 + 10)$ Trial and error

<u>20.</u> $(x + 6)(x - 6)(x + 1)(x - 1)$

<u>21.</u> $ac + cd - ab - bd$
      $= c(a + d) - b(a + d)$ Factoring by grouping
      $= (c - b)(a + d)$

<u>22.</u> $(w + z)(x - y)$

<u>23.</u> $4c^2 - 4cd + d^2$ Trinomial square
      $= (2c - d)^2$

<u>24.</u> $(10b + a)(7b - a)$

<u>25.</u> $-7x^2 + 2x^3 + 4x - 14$
      $= 2x^3 - 7x^2 + 4x - 14$
      $= x^2(2x - 7) + 2(2x - 7)$ Factoring by grouping
      $= (x^2 + 2)(2x - 7)$

<u>26.</u> $(3m^2 + 8)(m + 3)$

<u>27.</u> $2x^3 + 6x^2 - 8x - 24$
      $= 2(x^3 + 3x^2 - 4x - 12)$
      $= 2[x^2(x + 3) - 4(x + 3)]$ Factoring by grouping
      $= 2(x^2 - 4)(x + 3)$ Difference of squares
      $= 2(x + 2)(x - 2)(x + 3)$

<u>28.</u> $3(x + 3)(x - 3)(x + 2)$

<u>29.</u> $16x^3 + 54y^3$
      $= 2(8x^3 + 27y^3)$
      $= 2[(2x)^3 + (3y)^3]$ Sum of cubes
      $= 2(2x + 3y)(4x^2 - 6xy + 9y^2)$

<u>30.</u> $2(5a + 3b)(25a^2 - 15ab + 9b^2)$

<u>31.</u> $36y^2 - 35 + 12y$
      $= 36y^2 + 12y - 35$
      $= (6y - 5)(6y + 7)$ FOIL or grouping method

<u>32.</u> $-2b(7a + 1)(2a - 1)$

<u>33.</u> $a^8 - b^8$ Difference of squares
      $= (a^4 + b^4)(a^4 - b^4)$ Difference of squares
      $= (a^4 + b^4)(a^2 + b^2)(a^2 - b^2)$ Difference of
                                        squares
      $= (a^4 + b^4)(a^2 + b^2)(a + b)(a - b)$

<u>34.</u> $2(x^2 + 4)(x + 2)(x - 2)$

<u>35.</u> $a^3b - 16ab^3$
      $= ab(a^2 - 16b^2)$ Difference of squares
      $= ab(a + 4b)(a - 4b)$

<u>36.</u> $xy(x + 5y)(x - 5y)$

<u>37.</u> $a(b - 2) + c(b - 2)$
      $= (b - 2)(a + c)$ Removing a common factor

<u>38.</u> $(x - 2)(2x + 13)$

<u>39.</u> $7a^4 - 14a^3 + 21a^2 - 7a$
      $= 7a(a^3 - 2a^2 + 3a - 1)$ Removing a common factor

<u>40.</u> $(a + b)^2(a - b)$

<u>41.</u> $42ab + 27a^2b^2 + 8$
      $= 27a^2b^2 + 42ab + 8$
      $= (9ab + 2)(3ab + 4)$ FOIL or grouping method

42. $(4xy - 3)(5xy - 2)$

43. $8y^4 - 125y$

    $= y(8y^3 - 125)$

    $= y[(2y)^3 - 5^3]$ Difference of cubes

    $= y(2y - 5)(4y^2 + 10y + 25)$

44. $p(4p - 1)(16p^2 + 4p + 1)$

45. Familiarize. It is helpful to organize the information in a table. We let x represent the number of correct answers and y the number of wrong answers.

	Number of questions	Point value
Correct answers	x	$2x$
Wrong answers	y	$-\frac{1}{2}y$
Total	75	100

    Translate. The total number of questions is 75. This gives us one statement.

    $$x + y = 75$$

    The total point value (score) of 100 gives us a second statement.

    $$2x - \frac{1}{2}y = 100$$

    We now have a system of equations.

    $$x + y = 75 \quad \text{or} \quad x + y = 75$$
    $$2x - \frac{1}{2}y = 100 \quad \text{or} \quad 4x - y = 200$$

    Carry out. Solving the system we get $(55, 20)$.

    Check. The total number of questions is $55 + 20$, or 75. The total score is $2(55) - \frac{1}{2}(20)$, or $110 - 10$, or 100. The numbers check.

    State. There were 55 correct answers and 20 wrong answers.

46. $\dfrac{80}{7}$

47. $20x^2 - 4x - 39$

    $= (10x + 13)(2x - 3)$ FOIL or grouping method

48. $(5y^2 - 12x)(6y^2 - 5x)$

49. $72a^2 + 284a - 160$

    $= 4(18a^2 + 71a - 40)$

    $= 4(9a + 40)(2a - 1)$ FOIL or grouping method

50. $z(3xy + 4z)(xy + 7z)$

51. $7x^3 - \dfrac{7}{8}$

    $= 7\left(x^3 - \dfrac{1}{8}\right)$ Difference of cubes

    $= 7\left(x - \dfrac{1}{2}\right)\left(x^2 + \dfrac{1}{2}x + \dfrac{1}{4}\right)$

52. $(11 + y^2)(2 + y)(2 - y)$

53. $[a^3 - (a + b)^3]$ Difference of cubes

    $= [a - (a + b)][a^2 + a(a + b) + (a + b)^2]$

    $= (a - a - b)(a^2 + a^2 + ab + a^2 + 2ab + b^2)$

    Simplifying

    $= -b(3a^2 + 3ab + b^2)$

54. $x(x - 2p)$

55. $x^4 - 50x^2 + 49$

    $= (x^2)^2 - 50x^2 + 49$

    $= (x^2 - 1)(x^2 - 49)$ Trial and error

    $= (x + 1)(x - 1)(x + 7)(x - 7)$ Difference of squares

56. $y(y - 1)^2(y - 2)$

57. $s^6 - 729t^6$

    $= (s^3)^2 - (27t^3)^2$ Difference of squares

    $= (s^3 + 27t^3)(s^3 - 27t^3)$ Sum and difference of cubes

    $= (s + 3t)(s^2 - 3st + 9t^2)(s - 3t)(s^2 + 3st + 9t^2)$

58. $(x - 1)^3(x^2 + 1)(x + 1)$

59. $27x^{6s} + 64y^{3t}$

    $= (3x^{2s})^3 + (4y^t)^3$ Sum of cubes

    $= (3x^{2s} + 4y^t)(9x^{4s} - 12x^{2s}y^t + 16y^{2t})$

60. $4(4a^{10} + b^{10})(2a^5 + b^5)(2a^5 - b^5)$

61. $4x^2 + 4xy + y^2 - r^2 + 6rs - 9s^2$

    $= (4x^2 + 4xy + y^2) - (r^2 - 6rs + 9s^2)$ Grouping

    $= (2x + y)^2 - (r - 3s)^2$ Difference of squares

    $= [(2x + y) + (r - 3s)][(2x + y) - (r - 3s)]$

    $= (2x + y + r - 3s)(2x + y - r + 3s)$

62. $(y + 2x)(y^2 + xy + x^2)$

63. $c^4d^4 - a^{16}$

    $= (c^2d^2)^2 - (a^8)^2$ Difference of squares

    $= (c^2d^2 + a^8)(c^2d^2 - a^8)$ Difference of squares

    $= (c^2d^2 + a^8)(cd + a^4)(cd - a^4)$

64. $x(1 - x)^3(x^2 - 3x + 3)$

65. $c^{2w+1} - 2c^{w+1} + c$

= $c^{2w} \cdot c - 2c^w \cdot c + c$

= $c(c^{2w} - 2c^w + 1)$

= $c[(c^w)^2 - 2(c^w) + 1]$ Trinomial square

= $c(c^w - 1)^2$

66. $6(2x^a + 1)(2x^a - 1)$

67. $y^9 - y$

= $y(y^8 - 1)$ Difference of squares

= $y(y^4 + 1)(y^4 - 1)$ Difference of squares

= $y(y^4 + 1)(y^2 + 1)(y^2 - 1)$ Difference of squares

= $y(y^4 + 1)(y^2 + 1)(y + 1)(y - 1)$

68. $\left(1 - \frac{x^9}{10}\right)\left(1 + \frac{x^9}{10} + \frac{x^{18}}{100}\right)$

69. $3a^2 + 3b^2 - 3c^2 - 3d^2 + 6ab - 6cd$

= $3(a^2 + b^2 - c^2 - d^2 + 2ab - 2cd)$

= $3[(a^2 + 2ab + b^2) - (c^2 + 2cd + d^2)]$ Grouping

= $3[(a + b)^2 - (c + d)^2]$ Difference of squares

= $3(a + b + c + d)(a + b - c - d)$

70. $3x(x + 5)$

71. $(m - 1)^3 + (m + 1)^3$ Sum of cubes

= $[(m - 1) + (m + 1)][(m - 1)^2 - (m - 1)(m + 1) + (m + 1)^2]$

= $(m - 1 + m + 1)(m^2 - 2m + 1 - m^2 + 1 + m^2 + 2m + 1)$

= $2m(m^2 + 3)$

72. $3(a - 7)^2$

73. $\left(x + \frac{2}{x}\right)^2 = 6$

$x^2 + 4 + \frac{4}{x^2} = 6$

$x^2 - 2 + \frac{4}{x^2} = 0$

Now consider $x^3 + \frac{8}{x^3}$.

$x^3 + \frac{8}{x^3}$ Sum of cubes

= $\left(x + \frac{2}{x}\right)\left(x^2 - 2 + \frac{4}{x^2}\right)$

= $\left(x + \frac{2}{x}\right)(0)$ Substituting 0 for $x^2 - 2 + \frac{4}{x^2}$

= 0

Exercise Set 5.8

1. $x^2 + 3x = 28$

$x^2 + 3x - 28 = 0$ Getting 0 on one side

$(x + 7)(x - 4) = 0$ Factoring

$x + 7 = 0$ or $x - 4 = 0$ Principle of zero products.

$x = -7$ or $x = 4$

The solutions are -7 and 4. The solution set is {-7,4}.

2. {9,-5}

3. $y^2 + 16 = 8y$

$y^2 - 8y + 16 = 0$ Getting 0 on one side

$(y - 4)(y - 4) = 0$ Factoring

$y - 4 = 0$ or $y - 4 = 0$ Principle of zero products

$y = 4$ or $y = 4$ We have a repeated root.

The solution is 4. The solution set is {4}.

4. {1}

5. $x^2 - 12x + 36 = 0$

$(x - 6)(x - 6) = 0$ Factoring

$x - 6 = 0$ or $x - 6 = 0$ Principle of zero products

$x = 6$ or $x = 6$ We have a repeated root.

The solution is 6. The solution set is {6}.

6. {-8}

7. $9x + x^2 + 20 = 0$

$x^2 + 9x + 20 = 0$ Changing order

$(x + 5)(x + 4) = 0$ Factoring

$x + 5 = 0$ or $x + 4 = 0$ Principle of zero products

$x = -5$ or $x = -4$

The solutions are -5 and -4. The solution set is {-5,-4}.

8. {-5,-3}

9. $x^2 + 8x = 0$

$x(x + 8) = 0$ Factoring

$x = 0$ or $x + 8 = 0$ Principle of zero products

$x = 0$ or $x = -8$

The solutions are 0 and -8. The solution set is {0,-8}.

10. {0,-9}

11. $x^2 - 9 = 0$
   $(x + 3)(x - 3) = 0$
   $x + 3 = 0$ or $x - 3 = 0$
      $x = -3$ or $x = 3$
   The solutions are -3 and 3. The solution set is {-3,3}.

12. {-4,4}

13. $z^2 = 36$
         $z^2 - 36 = 0$
   $(z + 6)(z - 6) = 0$
   $z + 6 = 0$ or $z - 6 = 0$
      $z = -6$ or $z = 6$
   The solutions are -6 and 6. The solution set is {-6,6}.

14. {-9,9}

15. $x^2 + 14x + 45 = 0$
   $(x + 9)(x + 5) = 0$
   $x + 9 = 0$ or $x + 5 = 0$
      $x = -9$ or $x = -5$
   The solutions are -9 and -5. The solution set is {-9,-5}.

16. {-8,-4}

17. $y^2 + 2y = 63$
      $y^2 + 2y - 63 = 0$
   $(y + 9)(y - 7) = 0$
   $y + 9 = 0$ or $y - 7 = 0$
      $y = -9$ or $y = 7$
   The solutions are -9 and 7. The solution set is {-9,7}.

18. {-8,5}

19. $p^2 - 11p = -28$
     $p^2 - 11p + 28 = 0$
   $(p - 7)(p - 4) = 0$
   $p - 7 = 0$ or $p - 4 = 0$
      $p = 7$ or $p = 4$
   The solutions are 7 and 4. The solution set is {7,4}.

20. {9,5}

21. $32 + 4x - x^2 = 0$
            $0 = x^2 - 4x - 32$
            $0 = (x - 8)(x + 4)$
   $x - 8 = 0$ or $x + 4 = 0$
      $x = 8$ or $x = -4$
   The solutions are 8 and -4. The solution set is {8,-4}.

22. {-9,-3}

23. $3b^2 + 8b + 4 = 0$
   $(3b + 2)(b + 2) = 0$
   $3b + 2 = 0$ or $b + 2 = 0$
      $3b = -2$ or $b = -2$
      $b = -\frac{2}{3}$ or $b = -2$
   The solutions are $-\frac{2}{3}$ and -2. The solution set is $\left\{-\frac{2}{3},-2\right\}$.

24. $\left\{-\frac{4}{3},-\frac{1}{3}\right\}$

25. $8y^2 - 10y + 3 = 0$
   $(4y - 3)(2y - 1) = 0$
   $4y - 3 = 0$ or $2y - 1 = 0$
      $4y = 3$ or $2y = 1$
      $y = \frac{3}{4}$ or $y = \frac{1}{2}$
   The solutions are $\frac{3}{4}$ and $\frac{1}{2}$. The solution set is $\left\{\frac{3}{4},\frac{1}{2}\right\}$.

26. $\left\{-\frac{3}{4},-2\right\}$

27. $6z - z^2 = 0$
            $0 = z^2 - 6z$
            $0 = z(z - 6)$
   $z = 0$ or $z - 6 = 0$
   $z = 0$ or $z = 6$
   The solutions are 0 and 6. The solution set is {0,6}.

28. {0,8}

29. $12z^2 + z = 6$
      $12z^2 + z - 6 = 0$
   $(4z + 3)(3z - 2) = 0$
   $4z + 3 = 0$ or $3z - 2 = 0$
      $4z = -3$ or $3z = 2$
      $z = -\frac{3}{4}$ or $z = \frac{2}{3}$
   The solutions are $-\frac{3}{4}$ and $\frac{2}{3}$. The solution set is $\left\{-\frac{3}{4},\frac{2}{3}\right\}$.

30. $\left\{-\frac{5}{6},2\right\}$

31.
$$5x^2 - 20 = 0$$
$$5(x^2 - 4) = 0$$
$$5(x + 2)(x - 2) = 0$$
$$x + 2 = 0 \quad \text{or} \quad x - 2 = 0$$
$$x = -2 \quad \text{or} \quad x = 2$$
The solutions are -2 and 2. The solution set is {-2,2}.

32. {-3,3}

33.
$$2x^2 - 15x = -7$$
$$2x^2 - 15x + 7 = 0$$
$$(2x - 1)(x - 7) = 0$$
$$2x - 1 = 0 \quad \text{or} \quad x - 7 = 0$$
$$2x = 1 \quad \text{or} \quad x = 7$$
$$x = \frac{1}{2} \quad \text{or} \quad x = 7$$

The solutions are $\frac{1}{2}$ and 7. The solution set is $\left\{\frac{1}{2}, 7\right\}$.

34. {8,1}

35.
$$21r^2 + r - 10 = 0$$
$$(3r - 2)(7r + 5) = 0$$
$$3r - 2 = 0 \quad \text{or} \quad 7r + 5 = 0$$
$$3r = 2 \quad \text{or} \quad 7r = -5$$
$$r = \frac{2}{3} \quad \text{or} \quad r = -\frac{5}{7}$$

The solutions are $\frac{2}{3}$ and $-\frac{5}{7}$. The solution set is $\left\{\frac{2}{3}, -\frac{5}{7}\right\}$.

36. $\left\{\frac{7}{4}, -\frac{4}{3}\right\}$

37.
$$15y^2 = 3y$$
$$15y^2 - 3y = 0$$
$$3y(5y - 1) = 0$$
$$3y = 0 \quad \text{or} \quad 5y - 1 = 0$$
$$y = 0 \quad \text{or} \quad 5y = 1$$
$$y = 0 \quad \text{or} \quad y = \frac{1}{5}$$

The solutions are 0 and $\frac{1}{5}$. The solution set is $\left\{0, \frac{1}{5}\right\}$.

38. $\left\{0, \frac{1}{2}\right\}$

39.
$$x^2 - \frac{1}{25} = 0$$
$$\left(x + \frac{1}{5}\right)\left(x - \frac{1}{5}\right) = 0$$
$$x + \frac{1}{5} = 0 \quad \text{or} \quad x - \frac{1}{5} = 0$$
$$x = -\frac{1}{5} \quad \text{or} \quad x = \frac{1}{5}$$

The solutions are $-\frac{1}{5}$ and $\frac{1}{5}$. The solution set is $\left\{-\frac{1}{5}, \frac{1}{5}\right\}$.

40. $\left\{-\frac{1}{8}, \frac{1}{8}\right\}$

41.
$$2x^3 - 2x^2 = 12x$$
$$2x^3 - 2x^2 - 12x = 0$$
$$2x(x^2 - x - 6) = 0$$
$$2x(x - 3)(x + 2) = 0$$
$$2x = 0 \quad \text{or} \quad x - 3 = 0 \quad \text{or} \quad x + 2 = 0$$
$$x = 0 \quad \text{or} \quad x = 3 \quad \text{or} \quad x = -2$$
The solutions are 0, 3, and -2. The solution set is {0,3,-2}.

42. {0,5,2}

43.
$$2x^3 = 128x$$
$$2x^3 - 128x = 0$$
$$2x(x^2 - 64) = 0$$
$$2x(x + 8)(x - 8) = 0$$
$$x = 0 \quad \text{or} \quad x + 8 = 0 \quad \text{or} \quad x - 8 = 0$$
$$x = 0 \quad \text{or} \quad x = -8 \quad \text{or} \quad x = 8$$
The solutions are 0, -8, and 8. The solution set is {0,-8,8}.

44. {0,-7,7}

45. **Familiarize**. Let r represent the speed of the faster car and d represent its distance. Then r - 15 and 651 - d represent the speed and distance of the slower car, respectively. We organize the information in a table.

	Speed	Time	Distance
Faster car	r	7	d
Slower car	r - 15	7	651 - d

Translate. We use the formula rt = d. Each row of the table gives us an equation.
$$7r = d$$
$$7(r - 15) = 651 - d$$

Carry out. We use the substitution method substituting 7r for d in the second equation and solving for r.
$$7(r - 15) = 651 - 7r \quad \text{Substituting}$$
$$7r - 105 = 651 - 7r$$
$$14r = 756$$
$$r = 54$$

45. (continued)

Check. If r = 54, then the speed of the faster car is 54 mph and the speed of the slower car is 54 - 15, or 39 mph. The distance the faster car travels is 54·7, or 378 miles. The distance the slower car travels is 39·7, or 273 miles. The total of the two distances is 378 + 273, or 651 miles. The values check.

State. The speed of the faster car is 54 mph. The speed of the slower car is 39 mph.

46. 36 color, 42 black and white

47. $2x - 14 + 9x > -8x + 16 + 10x$

$\quad\quad 11x - 14 > 2x + 16$ Collecting like terms

$\quad\quad\quad 9x - 14 > 16$ Adding $-2x$

$\quad\quad\quad\quad 9x > 30$ Adding 14

$\quad\quad\quad\quad x > \dfrac{10}{3}$ Multiplying by $\dfrac{1}{9}$

The solution set is $\left\{x \middle| x > \dfrac{10}{3}\right\}$, or $\left[\dfrac{10}{3}, \infty\right)$.

48. $(2, -2, -4)$

49. $\quad\quad x(x + 8) = 16(x - 1)$

$\quad\quad\quad x^2 + 8x = 16x - 16$

$\quad\quad x^2 - 8x + 16 = 0$

$\quad (x - 4)(x - 4) = 0$

$\quad x - 4 = 0 \quad\text{or}\quad x - 4 = 0$

$\quad\quad x = 4 \quad\text{or}\quad\quad x = 4$

The solution is 4. The solution set is {4}.

50. $\{-5, 4\}$

51. $\quad\quad (a - 5)^2 = 36$

$\quad a^2 - 10a + 25 = 36$

$\quad a^2 - 10a - 11 = 0$

$\quad (a - 11)(a + 1) = 0$

$\quad a - 11 = 0 \quad\text{or}\quad a + 1 = 0$

$\quad\quad a = 11 \quad\text{or}\quad\quad a = -1$

The solutions are 11 and -1. The solution set is {11, -1}.

52. $\{15, -3\}$

53. $(3x^2 - 7x - 20)(x - 5) = 0$

$\quad (3x + 5)(x - 4)(x - 5) = 0$

$3x + 5 = 0 \quad\text{or}\quad x - 4 = 0 \quad\text{or}\quad x - 5 = 0$

$\quad 3x = -5 \quad\text{or}\quad\quad x = 4 \quad\text{or}\quad\quad x = 5$

$\quad x = -\dfrac{5}{3} \quad\text{or}\quad\quad x = 4 \quad\text{or}\quad\quad x = 5$

The solutions are $-\dfrac{5}{3}$, 4, and 5. The solution set is $\left\{-\dfrac{5}{3}, 4, 5\right\}$.

54. $\left\{-\dfrac{11}{8}, \dfrac{2}{3}, -\dfrac{1}{4}\right\}$

55. $\quad\quad (x + 1)^3 = (x - 1)^3 + 26$

$x^3 + 3x^2 + 3x + 1 = (x^3 - 3x^2 + 3x - 1) + 26$

$x^3 + 3x^2 + 3x + 1 = x^3 - 3x^2 + 3x + 25$

$\quad\quad\quad 6x^2 - 24 = 0$

$\quad\quad\quad 6(x^2 - 4) = 0$

$\quad\quad 6(x + 2)(x - 2) = 0$

$\quad x + 2 = 0 \quad\text{or}\quad x - 2 = 0$

$\quad\quad x = -2 \quad\text{or}\quad\quad x = 2$

The solutions are -2 and 2. The solution set is {-2, 2}.

56. {1}

57. $\quad 3x^3 + 6x^2 - 27x - 54 = 0$

$\quad 3(x^3 + 2x^2 - 9x - 18) = 0$

$3[x^2(x + 2) - 9(x + 2)] = 0$

$\quad\quad 3(x + 2)(x^2 - 9) = 0$

$\quad 3(x + 2)(x + 3)(x - 3) = 0$

$x + 2 = 0 \quad\text{or}\quad x + 3 = 0 \quad\text{or}\quad x - 3 = 0$

$\quad x = -2 \quad\text{or}\quad\quad x = -3 \quad\text{or}\quad\quad x = 3$

The solutions are -2, -3, and 3. The solution set is {-2, -3, 3}.

58. $\{-3, -2, 2\}$

59. $(x + 6)(x - 3) + b = (x - 5)(x - 2) - 3b$

$\quad x^2 + 3x - 18 + b = x^2 - 7x + 10 - 3b$

$\quad\quad 3x - 18 + b = -7x + 10 - 3b$

$\quad\quad\quad\quad 10x = 28 - 4b$

$\quad\quad\quad\quad x = \dfrac{28 - 4b}{10}$

$\quad\quad\quad\quad x = \dfrac{14 - 2b}{5}$

60. $\dfrac{b^2 - 3b - c}{2b - 3}$

61. Use a compute software package or a graphing calculator to graph the function.

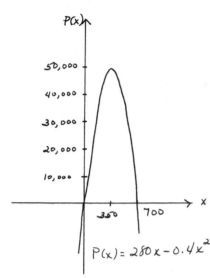

$$P(x) = 280x - 0.4x^2$$

From the graph we see that $P(x) = 0$ for $x = 0$ and $x = 700$.

62. $P(x) = 0$ for $x \approx -3.10, -0.65, 0.65, 3.10$

63. Use a compute software package or a graphing calculator to graph the function.

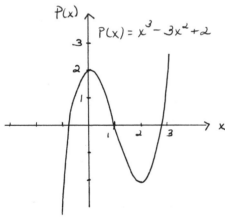

$$P(x) = x^3 - 3x^2 + 2$$

$P(x) = 0$ for $x \approx -0.73$ and 2.73.

64. $P(x) = 0$ for $x = -2$, $x = 1$, and $x \approx -1.41$ and 1.41.

Exercise Set 5.9

1. Familiarize. Let x represent the number.
Translate.

4 times the square of a number is 21 more than 8 times the number.

$$4 \cdot x^2 = 21 + 8 \cdot x$$

Carry out. We solve the equation:
$$4x^2 = 21 + 8x$$
$$4x^2 - 8x - 21 = 0$$
$$(2x + 3)(2x - 7) = 0$$
$$2x + 3 = 0 \quad \text{or} \quad 2x - 7 = 0$$
$$2x = -3 \quad \text{or} \quad 2x = 7$$
$$x = -\frac{3}{2} \quad \text{or} \quad x = \frac{7}{2}$$

Check. For $-\frac{3}{2}$: Four times the square of $-\frac{3}{2}$, or $4 \cdot \frac{9}{4}$, or 9, is 21 more than 8 times $-\frac{3}{2}$, or $8\left(-\frac{3}{2}\right)$, or -12.

For $\frac{7}{2}$: Four times the square of $\frac{7}{2}$, or $4 \cdot \frac{49}{4}$, or 49, is 21 more than 8 times $\frac{7}{2}$, or $8 \cdot \frac{7}{2}$, or 28. Both numbers check.

State. The number is $-\frac{3}{2}$ or $\frac{7}{2}$.

2. $\frac{9}{2}, -\frac{5}{2}$

3. Familiarize. Let x represent the number.
Translate.

Square of number plus number is 132.

$$x^2 + x = 132$$

Carry out. We solve the equation:
$$x^2 + x = 132$$
$$x^2 + x - 132 = 0$$
$$(x - 11)(x + 12) = 0$$
$$x - 11 = 0 \quad \text{or} \quad x + 12 = 0$$
$$x = 11 \quad \text{or} \quad x = -12$$

Check. The square of 11, which is 121, plus 11 is 132. The square of -12, which is 144, plus -12 is 132. Both numbers check.
State. The number is 11 or -12.

4. -13, 12

5. <u>Familiarize</u>. We let w represent the width and w + 5 represent the length. We make a drawing and label it.

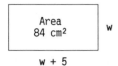

Recall that the formula for the area of a rectangle is A = length × width.

<u>Translate</u>.

    Area   is 84 cm².

  w(w + 5)   = 84

<u>Carry out</u>. We solve the equation:

$$w(w + 5) = 84$$
$$w^2 + 5w = 84$$
$$w^2 + 5w - 84 = 0$$
$$(w + 12)(w - 7) = 0$$
$$w + 12 = 0 \quad \text{or} \quad w - 7 = 0$$
$$w = -12 \quad \text{or} \quad w = 7$$

<u>Check</u>. The number -12 is not a solution, because width cannot be negative. If the width is 7 cm and the length is 5 cm more, or 12 cm, then the area is 12·7, or 84 cm². This is a solution.

<u>State</u>. The length is 12 cm, and the width is 7 cm.

6. ℓ = 12 cm, w = 8 cm

7. <u>Familiarize</u>. Let w represent the width and w + 25 represent the length. Make a drawing.

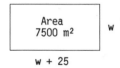

Recall that the formula for the area of a rectangle is A = length × width.

<u>Translate</u>.

    Area   is 7500 m².

  w(w + 25)   = 7500

<u>Carry out</u>. We solve the equation:

$$w(w + 25) = 7500$$
$$w^2 + 25w = 7500$$
$$w^2 + 25w - 7500 = 0$$
$$(w + 100)(w - 75) = 0$$
$$w + 100 = 0 \quad \text{or} \quad w - 75 = 0$$
$$w = -100 \quad \text{or} \quad w = 75$$

<u>Check</u>. The number -100 is not a solution because width cannot be negative. If the width is 75 m and the length is 25 m more, or 100 m, then the area will be 100·75, or 7500 m². This is a solution.

<u>State</u>. The dimensions will be 100 m by 75 m.

8. 9 m by 12 m

9. <u>Familiarize</u>. Let x represent the first positive odd integer. Then x + 2 represents the next positive odd integer.

<u>Translate</u>.

Square of first integer	plus	Square of second integer	is	202.
x^2	+	$(x + 2)^2$	=	202

<u>Carry out</u>. We solve the equation:

$$x^2 + (x + 2)^2 = 202$$
$$x^2 + x^2 + 4x + 4 = 202$$
$$2x^2 + 4x - 198 = 0$$
$$2(x^2 + 2x - 99) = 0$$
$$2(x + 11)(x - 9) = 0$$
$$x + 11 = 0 \quad \text{or} \quad x - 9 = 0$$
$$x = -11 \quad \text{or} \quad x = 9$$

<u>Check</u>. We only check 9 since the problem asks for consecutive <u>positive</u> odd integers. If x = 9, then x + 2 = 11, and 9 and 11 are consecutive positive odd integers. The sum of the squares of 9 and 11 is 81 + 121, or 202. The numbers check.

<u>State</u>. The integers are 9 and 11.

10. 13 and 15

11. <u>Familiarize</u>. We make a drawing and label it. We let x represent the length of a side of the original square.

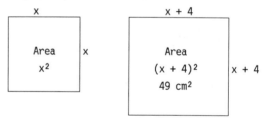

<u>Translate</u>.

    Area of new square is 49 cm².

    (x + 4)²   =   49

<u>Carry out</u>. We solve the equation:

$$(x + 4)^2 = 49$$
$$x^2 + 8x + 16 = 49$$
$$x^2 + 8x - 33 = 0$$
$$(x + 11)(x - 3) = 0$$
$$x + 11 = 0 \quad \text{or} \quad x - 3 = 0$$
$$x = -11 \quad \text{or} \quad x = 3$$

<u>Check</u>. We only check 3 since the length of a side cannot be negative. If we increase the length by 4, the new length is 3 + 4, or 7 cm. The new area is 7·7, or 49 cm². We have a solution.

<u>State</u>. The length of a side of the original square is 3 cm.

12. 6 m

13. Familiarize. Let h represent the height of the triangle. Then h + 9 represents the base. Recall that the formula for the area of a triangle is $A = \frac{1}{2} \times$ base $\times$ height.

    Translate.
    Area is 56 cm².

    $$\frac{1}{2}(h + 9)h = 56$$

    Carry out. We solve the equation:

    $$\frac{1}{2}(h + 9)h = 56$$
    $$(h + 9)h = 112 \quad \text{Multiplying by 2}$$
    $$h^2 + 9h = 112$$
    $$h^2 + 9h - 112 = 0$$
    $$(h + 16)(h - 7) = 0$$
    $$h + 16 = 0 \quad \text{or} \quad h - 7 = 0$$
    $$h = -16 \quad \text{or} \quad h = 7$$

    Check. We only check 7, since height cannot be negative. If the height is 7 cm, the base is 7 + 9, or 16 cm and the area is $\frac{1}{2} \cdot 16 \cdot 7$, or 56 cm². We have a solution.

    State. The height is 7 cm, and the base is 16 cm.

14. Height: 9 cm, base: 4 cm

15. Familiarize. Let x represent the length of a side of the square. Then the perimeter is x + x + x + x, or 4x, and the area is x·x, or x².
    Translate.
    Perimeter is 4 more than the area.
    $$4x = 4 + x^2$$

    Carry out. We solve the equation:
    $$4x = 4 + x^2$$
    $$0 = x^2 - 4x + 4$$
    $$0 = (x - 2)(x - 2)$$
    $$x - 2 = 0 \quad \text{or} \quad x - 2 = 0$$
    $$x = 2 \quad \text{or} \quad x = 2$$

    Check. If the length of a side is 2, the perimeter is 4·2 or 8, the area is 2·2 or 4, and 8 is four more than 4. The value checks.
    State. The length of a side is 2.

16. 6

17. Familiarize. We let x represent the first integer, x + 2 the second, and x + 4 the third.
    Translate.

Square of the first	+	Square of the third	= 136
x²	+	(x + 4)²	= 136

    Carry out. We solve the equation:
    $$x^2 + (x + 4)^2 = 136$$
    $$x^2 + x^2 + 8x + 16 = 136$$
    $$2x^2 + 8x - 120 = 0$$
    $$2(x^2 + 4x - 60) = 0$$
    $$2(x + 10)(x - 6) = 0$$
    $$x + 10 = 0 \quad \text{or} \quad x - 6 = 0$$
    $$x = -10 \quad \text{or} \quad x = 6$$

    Check. If x = -10, the consecutive even integers are -10, -8, and -6. The square of -10 is 100; the square of -6 is 36. The sum of 100 and 36 is 136. If x = 6, the consecutive even integers are 6, 8, and 10. The square of 6 is 36; the square of 10 is 100. The sum of 36 and 100 is 136. Both sets check.

    State. The three consecutive even integers can be -10, -8, and -6 or 6, 8, and 10.

18. 16, 18, and 20

19. Familiarize. Let x represent the first integer, x + 1 the second, and x + 2 the third.
    Translate.

First	·	Third	−	Second	=	1	+	10·Third
x	·	(x + 2)	−	(x + 1)	=	1	+	10(x + 2)

    Carry out. We solve the equation:
    $$x(x + 2) - (x + 1) = 1 + 10(x + 2)$$
    $$x^2 + 2x - x - 1 = 1 + 10x + 20$$
    $$x^2 - 9x - 22 = 0$$
    $$(x - 11)(x + 2) = 0$$
    $$x - 11 = 0 \quad \text{or} \quad x + 2 = 0$$
    $$x = 11 \quad \text{or} \quad x = -2$$

    Check. If x = 11, the consecutive integers are 11, 12, and 13.

First · Third − Second	1 + 10·Third
11·13 − 12	1 + 10·13
143 − 12	1 + 130
131	131

    If x = -2, the consecutive integers are -2, -1, and 0.

First · Third − Second	1 + 10·Third
-2·0 − (-1)	1 + 10·0
0 + 1	1 + 0
1	1

    Both sets of integers check.

    State. The three consecutive integers can be 11, 12, and 13 or -2, -1, and 0.

20. 3, 4, and 5 or 9, 10, and 11

21. <u>Familiarize</u>. Recall that in order for a firm to break even, its revenue must equal its production costs.

<u>Translate</u>.

Revenue equals cost.

$\frac{4}{9}x^2 + x = \frac{1}{3}x^2 + x + 1$

<u>Carry out</u>. We solve the equation:

$\frac{4}{9}x^2 + x = \frac{1}{3}x^2 + x + 1$

$4x^2 + 9x = 3x^2 + 9x + 9$ Multiplying by 9

$x^2 - 9 = 0$

$(x + 3)(x - 3) = 0$

$x + 3 = 0$ or $x - 3 = 0$

$x = -3$ or $x = 3$

<u>Check</u>. The number -3 cannot be a solution, since the number of clocks cannot be negative. If 3 clocks are produced and sold the revenue is $R(3) = \frac{4}{9} \cdot 3^2 + 3 = 7$ hundred dollars and the cost is $C(3) = \frac{1}{3} \cdot 3^2 + 3 + 1 = 7$ hundred dollars. We have a solution.

<u>State</u>. The firm breaks even when 3 clocks are produced and sold.

22. 6

23. <u>Familiarize</u>. Let x represent the other leg of the triangle. Then x + 1 represents the hypotenuse.

<u>Translate</u>. We use the Pythagorean theorem.

$a^2 + b^2 = c^2$

$9^2 + x^2 = (x + 1)^2$

<u>Carry out</u>. We solve the equation:

$9^2 + x^2 = (x + 1)^2$

$81 + x^2 = x^2 + 2x + 1$

$80 = 2x$

$40 = x$

<u>Check</u>. When x = 40, then x + 1 = 41, and $9^2 + 40^2 = 1681 = 41^2$. The numbers check.

<u>State</u>. The other sides have lengths of 40 m and 41 m.

24. 24 cm, 26 cm

25. <u>Familiarize</u>. We will use the function $h(t) = -16t^2 + 80t + 224$.

<u>Translate</u>.

Height   is 0 ft.

$-16t^2 + 80t + 224 = 0$

<u>Carry out</u>. We solve the equation:

$-16(t^2 - 5t - 14) = 0$

$-16(t - 7)(t + 2) = 0$

$t - 7 = 0$ or $t + 2 = 0$

$t = 7$ or $t = -2$

25. (continued)

<u>Check</u>. The number -2 is not a solution, since time cannot be negative. When t = 7, $h(7) = -16\cdot 7^2 + 80\cdot 7 + 224 = 0$. We have a solution.

<u>State</u>. The object reaches the ground after 7 sec.

26. 11 sec

27. <u>Familiarize</u>. We will use the formula $A = \$1000(1 + i)^2$.

<u>Translate</u>.

Amount   is $1144.90.

$\$1000(1 + i)^2 = \$1144.90$

<u>Carry out</u>. We solve the equation.

$10,000(1 + i)^2 = 11,449$ Multiplying by 10

$10,000(1 + 2i + i^2) = 11,449$

$10,000 + 20,000i + 10,000i^2 = 11,449$

$10,000i^2 + 20,000i - 1449 = 0$

$(100i + 207)(100i - 7) = 0$

$100i + 207 = 0$ or $100i - 7 = 0$

$100i = -207$ or $100i = 7$

$i = -2.07$ or $i = 0.07$

<u>Check</u>. The number -2.07 is not a solution, since the interest rate cannot be negative. If i = 0.07, then the amount in the account after 2 years is $\$1000(1.07)^2 = \$1000(1.1449) = \$1144.90$. We have a solution.

<u>State</u>. The interest rate is 0.07, or 7%.

28. 8%

29. <u>Familiarize</u>. Let x represent the length of a side of the square and y represent the length of a side of the triangle.

<u>Translate</u>.

Area of the square is 9 cm².

$x^2 = 9$

Perimeter of the square equals perimeter of the triangle.

$4x = 3y$

We have a system of equations:

$x^2 = 9,$ (1)

$4x = 3y$ (2)

<u>Carry out</u>. Solve (1).

$x^2 = 9$

$x^2 - 9 = 0$

$(x + 3)(x - 3) = 0$

$x + 3 = 0$ or $x - 3 = 0$

$x = -3$ or $x = 3$

29. (continued)

Length cannot be negative, so we only consider x = 3. Substitute 3 for x in (2).

$$4x = 3y$$
$$4 \cdot 3 = 3y$$
$$4 = y$$

Check. The perimeter of a square with a side of 3 cm is $4 \cdot 3$, or 12 cm, and the area is 3^2, or 9 cm². The perimeter of an equilateral triangle with a side of length 4 cm is $3 \cdot 4$, or 12 cm. The numbers check.

State. The length of a side of the triangle is 4 cm.

30. 3 and 14

31. Familiarize. Let x represent the width. Then 2x represents the length. We make a drawing and label it.

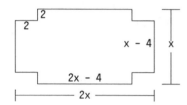

The dimensions of the box will be 2x - 4 by x - 4 by x. Recall the formula for the volume of a rectangular solid, V = length × width × height.

Translate.

Volume   is   480 cm³.

$$(2x - 4)(x - 4)(2) = 480$$

Carry out. We solve the equation.

$$(2x - 4)(x - 4)(2) = 480$$
$$(2x - 4)(x - 4) = 240 \quad \text{Multiplying by } \frac{1}{2}$$
$$2x^2 - 12x + 16 = 240$$
$$2x^2 - 12x - 224 = 0$$
$$2(x^2 - 6x - 112) = 0$$
$$2(x - 14)(x + 8) = 0$$
$$x - 14 = 0 \quad \text{or} \quad x + 8 = 0$$
$$x = 14 \quad \text{or} \quad x = -8$$

Check. The number -8 is not a solution, since the width cannot be negative. If the width of the piece of tin is 14 cm, then the length is 28 cm, and the dimensions of the box are 24 cm by 10 cm by 2 cm. The volume of the box will be 480 cm³. We have a solution.

State. The piece of tin is 28 cm by 14 cm.

32. 40 in. by 30 in. by 20 in.

33. Familiarize. We first make a drawing and let x represent the uniform width around the pool.

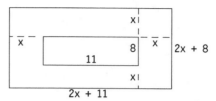

The dimensions of the pool are 11 and 8. The dimensions of the yard are 2x + 11 and 2x + 8.

Translate.

The area of the backyard is 1120 m².

$$(2x + 11)(2x + 8) = 1120$$

Carry out. We solve the equation.

$$4x^2 + 38x + 88 = 1120$$
$$2x^2 + 19x + 44 = 560 \quad \text{Multiplying by } \frac{1}{2}$$
$$2x^2 + 19x - 516 = 0$$
$$(2x + 43)(x - 12) = 0$$
$$2x + 43 = 0 \quad \text{or} \quad x - 12 = 0$$
$$x = -\frac{43}{2} \quad \text{or} \quad x = 12$$

Check. We only check 12 since the width cannot be negative. If x = 12, then the length is $2 \cdot 12 + 11$, or 35 m, and the width is $2 \cdot 12 + 8$, or 32 m. The area is $35 \cdot 32$, or 1120 m². The values check.

State. The strip is 12 m wide.

34. 54 cm²

Exercise Set 6.1

1. $v(t) = \dfrac{4t^2 - 5t + 2}{t + 3}$

$v(0) = \dfrac{4 \cdot 0^2 - 5 \cdot 0 + 2}{0 + 3} = \dfrac{0 - 0 + 2}{0 + 3} = \dfrac{2}{3}$

$v(3) = \dfrac{4 \cdot 3^2 - 5 \cdot 3 + 2}{3 + 3} = \dfrac{36 - 15 + 2}{3 + 3} = \dfrac{23}{6}$

$v(7) = \dfrac{4 \cdot 7^2 - 5 \cdot 7 + 2}{7 + 3} = \dfrac{196 - 35 + 2}{7 + 3} = \dfrac{163}{10}$

2. $42; \ -\dfrac{11}{7}; \ 15$

3. $r(y) = \dfrac{3y^3 - 2y}{y - 5}$

$r(0) = \dfrac{3 \cdot 0^3 - 2 \cdot 0}{0 - 5} = \dfrac{0 - 0}{0 - 5} = \dfrac{0}{-5} = 0$

$r(4) = \dfrac{3 \cdot 4^3 - 2 \cdot 4}{4 - 5} = \dfrac{192 - 8}{-1} = \dfrac{184}{-1} = -184$

$r(5) = \dfrac{3 \cdot 5^3 - 2 \cdot 5}{5 - 5} = \dfrac{375 - 10}{0}$

Since division by zero is not defined, $r(5)$ does not exist.

4. $25.12; \ 27.475;$ does not exist

5. $g(x) = \dfrac{2x^3 - 9}{x^2 - 4x + 4}$

$g(0) = \dfrac{2 \cdot 0^3 - 9}{0^2 - 4 \cdot 0 + 4} = \dfrac{0 - 9}{0 - 0 + 4} = -\dfrac{9}{4}$

$g(2) = \dfrac{2 \cdot 2^3 - 9}{2^2 - 4 \cdot 2 + 4} = \dfrac{16 - 9}{4 - 8 + 4} = \dfrac{7}{0}$

Since division by zero is not defined, $g(2)$ does not exist.

$g(-1) = \dfrac{2(-1)^3 - 9}{(-1)^2 - 4(-1) + 4} = \dfrac{-2 - 9}{1 + 4 + 4} = -\dfrac{11}{9}$

6. $0; \ \dfrac{2}{5}; \ \dfrac{40}{7}$

7. $f(t) = \dfrac{9 - t^2}{5 - 6t + t^2}$

$f(-3) = \dfrac{9 - (-3)^2}{5 - 6(-3) + (-3)^2} = \dfrac{9 - 9}{5 + 18 + 9} = \dfrac{0}{32} = 0$

$f(0) = \dfrac{9 - 0^2}{5 - 6 \cdot 0 + 0^2} = \dfrac{9 - 0}{5 - 0 + 0} = \dfrac{9}{5}$

$f(1) = \dfrac{9 - 1^2}{5 - 6 \cdot 1 + 1^2} = \dfrac{9 - 1}{5 - 6 + 1} = \dfrac{8}{0}$

Since division by zero is not defined, $f(1)$ does not exist.

8. $0;$ does not exist; $-\dfrac{4}{9}$

9. $f(x) = \dfrac{7}{3 - x}$

To avoid division by zero, we must determine which x-value causes $3 - x$ to be zero. We set

$3 - x = 0$

$3 = x.$ Adding x on both sides.

The domain of $f = \{x | x$ is a real number and $x \neq 3\}$.

10. $\{t | t$ is a real number and $t \neq 5\}$

11. $v(t) = \dfrac{t - 7}{t^2 - 4t}$

We set

$t^2 - 4t = 0$

$t(t - 4) = 0$

$t = 0 \ $ or $\ t - 4 = 0$

$t = 0 \ $ or $\quad t = 4.$

The domain of $v = \{t | t$ is a real number and $t \neq 0$ and $t \neq 4\}$.

12. $\{x | x$ is a real number and $x \neq 0$ and $x \neq 3\}$

13. $s(x) = \dfrac{5}{x^2 - 4}$

We set

$x^2 - 4 = 0$

$(x + 2)(x - 2) = 0$

$x + 2 = 0 \ $ or $\ x - 2 = 0$

$x = -2 \ $ or $\quad x = 2.$

The domain of $s = \{x | x$ is a real number and $x \neq -2$ and $x \neq 2\}$.

14. $\{y | y$ is a real number and $y \neq -5$ and $y \neq 5\}$

15. $F(x) = \dfrac{x^2 - 4}{x^2 - 8x + 12}$

We set

$x^2 - 8x + 12 = 0$

$(x - 6)(x - 2) = 0$

$x - 6 = 0 \ $ or $\ x - 2 = 0$

$x = 6 \ $ or $\quad x = 2.$

The domain of $F = \{x | x$ is a real number and $x \neq 6$ and $x \neq 2\}$.

16. $\{x | x$ is a real number and $x \neq 4$ and $x \neq 2\}$

17. $\dfrac{3x}{3x} \cdot \dfrac{x + 1}{x + 3} = \dfrac{3x(x + 1)}{3x(x + 3)}$

18. $\dfrac{(4 - y^2)(-1)}{(6 - y)(-1)}$

19. $\dfrac{t - 3}{t + 2} \cdot \dfrac{t + 3}{t + 3} = \dfrac{(t - 3)(t + 3)}{(t + 2)(t + 3)}$

20. $\dfrac{(p - 4)(p + 5)}{(p - 5)(p + 5)}$

21. $\dfrac{x^2 - 3}{x - 6} \cdot \dfrac{x + 6}{x + 6} = \dfrac{(x^2 - 3)(x + 6)}{(x - 6)(x + 6)}$

22. $\dfrac{(t^2 - 9)(t + 2)}{(3 - t)(t + 2)}$

23. $\dfrac{t^2 - 3}{t^2 - 3} \cdot \dfrac{t^2 + 3}{t^2 - 4} = \dfrac{(t^2 - 3)(t^2 + 3)}{(t^2 - 3)(t^2 - 4)}$

24. $\dfrac{(x^2 - 5)(x^2 - 3)}{(x^2 - 5)(x^2 + 5)}$

25. $\dfrac{9y^2}{15y} = \dfrac{3y \cdot 3y}{3y \cdot 5} = \dfrac{3y}{3y} \cdot \dfrac{3y}{5} = \dfrac{3y}{5}$

26. $\dfrac{x}{3}$

27. $\dfrac{8t^3}{4t^7} = \dfrac{4t^3 \cdot 2}{4t^3 \cdot t^4} = \dfrac{4t^3}{4t^3} \cdot \dfrac{2}{t^4} = \dfrac{2}{t^4}$

28. $\dfrac{3}{2y^2}$

29. $\dfrac{2a - 6}{2} = \dfrac{2(a - 3)}{2 \cdot 1} = \dfrac{2}{2} \cdot \dfrac{a - 3}{1} = a - 3$

30. $a - 2$

31. $\dfrac{6x - 9}{12} = \dfrac{3(2x - 3)}{3 \cdot 4} = \dfrac{3}{3} \cdot \dfrac{2x - 3}{4} = \dfrac{2x - 3}{4}$

32. $\dfrac{5a - 6}{3}$

33. $\dfrac{4y - 12}{4y + 12} = \dfrac{4(y - 3)}{4(y + 3)} = \dfrac{4}{4} \cdot \dfrac{y - 3}{y + 3} = \dfrac{y - 3}{y + 3}$

34. $\dfrac{x + 2}{x - 2}$

35. $\dfrac{6x - 12}{5x - 10} = \dfrac{6(x - 2)}{5(x - 2)} = \dfrac{6}{5} \cdot \dfrac{x - 2}{x - 2} = \dfrac{6}{5}$

36. $\dfrac{7}{3}$

37. $\dfrac{12 - 6x}{5x - 10} = \dfrac{-6(-2 + x)}{5(x - 2)} = \dfrac{-6(x - 2)}{5(x - 2)} = \dfrac{-6}{5} \cdot \dfrac{x - 2}{x - 2}$

$\qquad = -\dfrac{6}{5}$

38. $-\dfrac{7}{3}$

39. $\dfrac{t^2 - 16}{t^2 - 8t + 16} = \dfrac{(t + 4)(t - 4)}{(t - 4)(t - 4)} = \dfrac{t + 4}{t - 4} \cdot \dfrac{t - 4}{t - 4}$

$\qquad = \dfrac{t + 4}{t - 4}$

40. $\dfrac{p - 5}{p + 5}$

41. $\dfrac{x^2 + 9x + 8}{x^2 - 3x - 4} = \dfrac{(x + 1)(x + 8)}{(x + 1)(x - 4)} = \dfrac{x + 1}{x + 1} \cdot \dfrac{x + 8}{x - 4}$

$\qquad = \dfrac{x + 8}{x - 4}$

42. $\dfrac{t - 9}{t + 4}$

43. $\dfrac{16 - t^2}{t^2 - 8t + 16} = \dfrac{16 - t^2}{16 - 8t + t^2} = \dfrac{(4 + t)(4 - t)}{(4 - t)(4 - t)}$

$\qquad = \dfrac{4 + t}{4 - t} \cdot \dfrac{4 - t}{4 - t} = \dfrac{4 + t}{4 - t}$

44. $\dfrac{5 - p}{5 + p}$

45. $\dfrac{5x^2}{3t^5} \cdot \dfrac{9t^8}{25x} = \dfrac{5x^2 \cdot 9t^8}{3t^5 \cdot 25x}$

$\qquad = \dfrac{5x \cdot x \cdot 3t^5 \cdot 3t^3}{3t^5 \cdot 5x \cdot 5}$

$\qquad = \dfrac{5x \cdot 3t^5}{5x \cdot 3t^5} \cdot \dfrac{x \cdot 3t^3}{5}$

$\qquad = \dfrac{3t^3 x}{5}$

46. $\dfrac{7a^2}{6b^4}$

47. $\dfrac{3x - 6}{5x} \cdot \dfrac{x^3}{5x - 10} = \dfrac{(3x - 6)(x^3)}{5x(5x - 10)}$

$\qquad = \dfrac{3(x - 2)(x)(x^2)}{5 \cdot x \cdot 5(x - 2)}$

$\qquad = \dfrac{(x - 2)(x)}{(x - 2)(x)} \cdot \dfrac{3 \cdot x^2}{5 \cdot 5}$

$\qquad = \dfrac{3x^2}{25}$

48. $\dfrac{3t^2}{4}$

49. $\dfrac{y^2 - 16}{2y + 6} \cdot \dfrac{y + 3}{y - 4} = \dfrac{(y^2 - 16)(y + 3)}{(2y + 6)(y - 4)}$

$\qquad = \dfrac{(y + 4)(y - 4)(y + 3)}{2(y + 3)(y - 4)}$

$\qquad = \dfrac{(y - 4)(y + 3)}{(y - 4)(y + 3)} \cdot \dfrac{y + 4}{2}$

$\qquad = \dfrac{y + 4}{2}$

50. $\dfrac{m + n}{4}$

51. $\dfrac{x^2 - 16}{x^2} \cdot \dfrac{x^2 - 4x}{x^2 - x - 12} = \dfrac{(x^2 - 16)(x^2 - 4x)}{x^2(x^2 - x - 12)}$

$\qquad = \dfrac{(x + 4)(x - 4)(x)(x - 4)}{x \cdot x(x - 4)(x + 3)}$

$\qquad = \dfrac{x(x - 4)}{x(x - 4)} \cdot \dfrac{(x + 4)(x - 4)}{x(x + 3)}$

$\qquad = \dfrac{(x + 4)(x - 4)}{x(x + 3)}$

52. $\dfrac{y(y + 5)}{y - 3}$

53. $\dfrac{6 - 2t}{t^2 + 4t + 4} \cdot \dfrac{t^3 + 2t^2}{t^2 - 9} = \dfrac{(6 - 2t)(t^3 + 2t^2)}{(t^2 + 4t + 4)(t^2 - 9)}$

$\qquad = \dfrac{-2(-3 + t)(t^2)(t + 2)}{(t + 2)(t + 2)(t + 3)(t - 3)}$

$\qquad = \dfrac{(t - 3)(t + 2)}{(t - 3)(t + 2)} \cdot \dfrac{-2 \cdot t^2}{(t + 2)(t + 3)}$

$\qquad = -\dfrac{2t^2}{(t + 2)(t + 3)}$

54. $-\dfrac{(x + 3)(x - 3)}{4x^2}$

55. $\dfrac{x^2 - 2x - 35}{2x^3 - 3x^2} \cdot \dfrac{4x^3 - 9x}{7x - 49}$

$\qquad = \dfrac{(x^2 - 2x - 35)(4x^3 - 9x)}{(2x^3 - 3x^2)(7x - 49)}$

$\qquad = \dfrac{(x - 7)(x + 5)(x)(2x + 3)(2x - 3)}{x \cdot x(2x - 3)(7)(x - 7)}$

$\qquad = \dfrac{(x - 7)(x)(2x - 3)}{(x - 7)(x)(2x - 3)} \cdot \dfrac{(x + 5)(2x + 3)}{x \cdot 7}$

$\qquad = \dfrac{(x + 5)(2x + 3)}{7x}$

56. $\dfrac{1}{y + 1}$

57. $\dfrac{c^3 + 8}{c^2 - 4} \cdot \dfrac{c^2 - 4c + 4}{c^2 - 2c + 4}$

$= \dfrac{(c^3 + 8)(c^2 - 4c + 4)}{(c^2 - 4)(c^2 - 2c + 4)}$

$= \dfrac{(c + 2)(c^2 - 2c + 4)(c - 2)(c - 2)}{(c + 2)(c - 2)(c^2 - 2c + 4)\cdot 1}$

$= \dfrac{(c + 2)(c^2 - 2c + 4)(c - 2)}{(c + 2)(c^2 - 2c + 4)(c - 2)} \cdot \dfrac{c - 2}{1}$

$= c - 2$

58. $\dfrac{(x - 3)(x - 3)}{x + 3}$

59. $\dfrac{a^3 - b^3}{3a^2 + 9ab + 6b^2} \cdot \dfrac{a^2 + 2ab + b^2}{a^2 - b^2}$

$= \dfrac{(a^3 - b^3)(a^2 + 2ab + b^2)}{(3a^2 + 9ab + 6b^2)(a^2 - b^2)}$

$= \dfrac{(a - b)(a^2 + ab + b^2)(a + b)(a + b)}{3(a + b)(a + 2b)(a + b)(a - b)}$

$= \dfrac{(a - b)(a + b)(a + b)}{(a - b)(a + b)(a + b)} \cdot \dfrac{a^2 + ab + b^2}{3(a + 2b)}$

$= \dfrac{a^2 + ab + b^2}{3(a + 2b)}$

60. $\dfrac{x^2 - xy + y^2}{3(x + 3y)}$

61. $\dfrac{4x^2 - 9y^2}{8x^3 - 27y^3} \cdot \dfrac{4x^2 + 6xy + 9y^2}{4x^2 + 12xy + 9y^2}$

$= \dfrac{(4x^2 - 9y^2)(4x^2 + 6xy + 9y^2)}{(8x^3 - 27y^3)(4x^2 + 12xy + 9y^2)}$

$= \dfrac{(2x + 3y)(2x - 3y)(4x^2 + 6xy + 9y^2)\cdot 1}{(2x - 3y)(4x^2 + 6xy + 9y^2)(2x + 3y)(2x + 3y)}$

$= \dfrac{(2x + 3y)(2x - 3y)(4x^2 + 6xy + 9y^2)}{(2x + 3y)(2x - 3y)(4x^2 + 6xy + 9y^2)} \cdot \dfrac{1}{2x + 3y}$

$= \dfrac{1}{2x + 3y}$

62. $\dfrac{(x - y)(2x + 3y)}{2(x + y)(9x^2 + 6xy + 4y^2)}$

63. $\dfrac{16a^7}{3b^5} \div \dfrac{8a^3}{6b} = \dfrac{16a^7}{3b^5} \cdot \dfrac{6b}{8a^3}$ Multiplying by the reciprocal of the divisor

$= \dfrac{16a^7(6b)}{3b^5(8a^3)}$

$= \dfrac{8a^3 \cdot 2a^4 \cdot 3b \cdot 2}{3b \cdot b^4 \cdot 8a^3}$

$= \dfrac{8a^3 \cdot 3b}{8a^3 \cdot 3b} = \dfrac{2a^4 \cdot 2}{b^4}$

$= \dfrac{4a^4}{b^4}$

64. $6x^4y^7$

65. $\dfrac{3y + 15}{y} \div \dfrac{y + 5}{y} = \dfrac{3y + 15}{y} \cdot \dfrac{y}{y + 5}$

$= \dfrac{(3y + 15)(y)}{y(y + 5)}$

$= \dfrac{3(y + 5)(y)}{y(y + 5)\cdot 1}$

$= \dfrac{(y + 5)(y)}{(y + 5)(y)} \cdot \dfrac{3}{1}$

$= 3$

66. $6x^2$

67. $\dfrac{y^2 - 9}{y} \div \dfrac{y + 3}{y + 2} = \dfrac{y^2 - 9}{y} \cdot \dfrac{y + 2}{y + 3}$

$= \dfrac{(y^2 - 9)(y + 2)}{y(y + 3)}$

$= \dfrac{(y + 3)(y - 3)(y + 2)}{y(y + 3)}$

$= \dfrac{y + 3}{y + 3} \cdot \dfrac{(y - 3)(y + 2)}{y}$

$= \dfrac{(y - 3)(y + 2)}{y}$

68. $\dfrac{(x + 2)(x + 4)}{x}$

69. $\dfrac{4a^2 - 1}{a^2 - 4} \div \dfrac{2a - 1}{a - 2} = \dfrac{4a^2 - 1}{a^2 - 4} \cdot \dfrac{a - 2}{2a - 1}$

$= \dfrac{(4a^2 - 1)(a - 2)}{(a^2 - 4)(2a - 1)}$

$= \dfrac{(2a + 1)(2a - 1)(a - 2)}{(a + 2)(a - 2)(2a - 1)}$

$= \dfrac{(2a - 1)(a - 2)}{(2a - 1)(a - 2)} \cdot \dfrac{2a + 1}{a + 2}$

$= \dfrac{2a + 1}{a + 2}$

70. $\dfrac{5x + 2}{x - 3}$

71. $\dfrac{x^2 - y^2}{4x + 4y} \div \dfrac{3y - 3x}{12x^2} = \dfrac{x^2 - y^2}{4x + 4y} \cdot \dfrac{12x^2}{3y - 3x}$

$= \dfrac{(x^2 - y^2)(12x^2)}{(4x + 4y)(3y - 3x)}$

$= \dfrac{(x + y)(x - y)(3)(4)(x^2)}{4(x + y)(-1)(3)(-y + x)}$

$= \dfrac{(x + y)(x - y)(3)(4)}{(x + y)(x - y)(3)(4)} \cdot \dfrac{x^2}{-1}$

$= -x^2$

72. $-\dfrac{1}{y^3}$

73. $\dfrac{x^2 - 16}{x^2 - 10x + 25} \div \dfrac{3x - 12}{x^2 - 3x - 10}$

$= \dfrac{x^2 - 16}{x^2 - 10x + 25} \cdot \dfrac{x^2 - 3x - 10}{3x - 12}$

$= \dfrac{(x^2 - 16)(x^2 - 3x - 10)}{(x^2 - 10x + 25)(3x - 12)}$

$= \dfrac{(x + 4)(x - 4)(x - 5)(x + 2)}{(x - 5)(x - 5)(3)(x - 4)}$

$= \dfrac{(x - 4)(x - 5)}{(x - 4)(x - 5)} \cdot \dfrac{(x + 4)(x + 2)}{(x - 5)(3)}$

$= \dfrac{(x + 4)(x + 2)}{3(x - 5)}$

74. $\dfrac{(y + 6)(y + 3)}{3(y - 4)}$

75. $\dfrac{y^3 + 3y}{y^2 - 9} \div \dfrac{y^2 + 5y - 14}{y^2 + 4y - 21}$

$= \dfrac{y^3 + 3y}{y^2 - 9} \cdot \dfrac{y^2 + 4y - 21}{y^2 + 5y - 14}$

$= \dfrac{(y^3 + 3y)(y^2 + 4y - 21)}{(y^2 - 9)(y^2 + 5y - 14)}$

$= \dfrac{y(y^2 + 3)(y + 7)(y - 3)}{(y + 3)(y - 3)(y + 7)(y - 2)}$

$= \dfrac{(y + 7)(y - 3)}{(y + 7)(y - 3)} \cdot \dfrac{y(y^2 + 3)}{(y + 3)(y - 2)}$

$= \dfrac{y(y^2 + 3)}{(y + 3)(y - 2)}$

76. $\dfrac{a(a^2 + 4)}{(a + 4)(a + 3)}$

77. $\dfrac{x^3 - 64}{x^3 + 64} \div \dfrac{x^2 - 16}{x^2 - 4x + 16}$

$= \dfrac{x^3 - 64}{x^3 + 64} \cdot \dfrac{x^2 - 4x + 16}{x^2 - 16}$

$= \dfrac{(x^3 - 64)(x^2 - 4x + 16)}{(x^3 + 64)(x^2 - 16)}$

$= \dfrac{(x - 4)(x^2 + 4x + 16)(x^2 - 4x + 16)}{(x + 4)(x^2 - 4x + 16)(x + 4)(x - 4)}$

$= \dfrac{(x - 4)(x^2 - 4x + 16)}{(x - 4)(x^2 - 4x + 16)} \cdot \dfrac{x^2 + 4x + 16}{(x + 4)(x + 4)}$

$= \dfrac{x^2 + 4x + 16}{(x + 4)(x + 4)}$, or $\dfrac{x^2 + 4x + 16}{(x + 4)^2}$

78. $\dfrac{4y^2 + 6y + 9}{(4y - 1)(2y + 3)}$

79. $\dfrac{8a^3 + b^3}{2a^2 + 3ab + b^2} \div \dfrac{8a^2 - 4ab + 2b^2}{4a^2 + 4ab + b^2}$

$= \dfrac{8a^3 + b^3}{2a^2 + 3ab + b^2} \cdot \dfrac{4a^2 + 4ab + b^2}{8a^2 - 4ab + 2b^2}$

$= \dfrac{(8a^3 + b^3)(4a^2 + 4ab + b^2)}{(2a^2 + 3ab + b^2)(8a^2 - 4ab + 2b^2)}$

$= \dfrac{(2a + b)(4a^2 - 2ab + b^2)(2a + b)(2a + b)}{(2a + b)(a + b)(2)(4a^2 - 2ab + b^2)}$

$= \dfrac{(2a + b)(4a^2 - 2ab + b^2)}{(2a + b)(4a^2 - 2ab + b^2)} \cdot \dfrac{(2a + b)(2a + b)}{(a + b)(2)}$

$= \dfrac{(2a + b)(2a + b)}{2(a + b)}$, or $\dfrac{(2a + b)^2}{2(a + b)}$

80. $\dfrac{2(2x - y)}{x}$

81. $3x + y = 13,$ (1)

        $x = y + 1$ (2)

$3(y + 1) + y = 13$ Substituting $y + 1$ for x in (1)

   $3y + 3 + y = 13$

       $4y + 3 = 13$

          $4y = 10$

            $y = \dfrac{10}{4}$, or $\dfrac{5}{2}$

$x = \dfrac{5}{2} + 1$ Substituting $\dfrac{5}{2}$ for y in (2)

$x = \dfrac{7}{2}$

The solution is $\left(\dfrac{7}{2}, \dfrac{5}{2}\right)$.

82.   29

83. $\dfrac{2}{3}(3x - 4) = 8$

$\dfrac{3}{2} \cdot \dfrac{2}{3}(3x - 4) = \dfrac{3}{2} \cdot 8$ Multiplying by $\dfrac{3}{2}$

        $3x - 4 = 12$

           $3x = 16$

            $x = \dfrac{16}{3}$

84. $8

85.   To find the domain of f we set

    $x - 3 = 0$

       $x = 3.$

The domain of f = {x|x is a real number and $x \neq 3$}.

Simplify: $f(x) = \dfrac{x^2 - 9}{x - 3} = \dfrac{(x + 3)(x - 3)}{(x - 3) \cdot 1}$

                $= \dfrac{x - 3}{x - 3} \cdot \dfrac{x + 3}{1} = x + 3$

Graph $f(x) = x + 3$ using the domain found above.

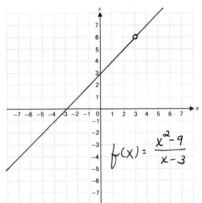

$f(x) = \dfrac{x^2 - 9}{x - 3}$

86.   a) $\dfrac{2x + 2h + 3}{4x + 4h - 1}$

     b) $\dfrac{2x + 3}{8x - 9}$

     c) $\dfrac{x + 5}{4x - 1}$

87. $\left[\dfrac{r^2 - 4s^2}{r + 2s} \div (r + 2s)\right] \cdot \dfrac{2s}{r - 2s}$

$= \left[\dfrac{r^2 - 4s^2}{r + 2s} \cdot \dfrac{1}{r + 2s}\right] \cdot \dfrac{2s}{r - 2s}$

$= \dfrac{(r + 2s)(r - 2s)(2s)}{(r + 2s)(r + 2s)(r - 2s)}$

$= \dfrac{(r + 2s)(r - 2s)}{(r + 2s)(r - 2s)} \cdot \dfrac{2s}{r + 2s}$

$= \dfrac{2s}{r + 2s}$

88. $\dfrac{(d - 1)(d - 5)}{5d(d + 5)}$

89. $\dfrac{834x}{y - 427.2} \cdot \dfrac{26.3x}{y + 427.2} = \dfrac{(834x)(26.3x)}{(y - 427.2)(y + 427.2)}$

$\qquad\qquad = \dfrac{21{,}934.2x^2}{y^2 - 182{,}499.84}$

90. $\dfrac{246{,}636}{x^2 - 8811.5769}$

91. $\dfrac{0.0049t}{t + 0.007} \cdot \dfrac{27{,}000t}{t - 0.007} = \dfrac{(0.0049t)(27{,}000t)}{(t + 0.007)(t - 0.007)}$

$\qquad\qquad = \dfrac{132.3t^2}{t^2 - 0.000049}$

92. $\dfrac{y^2 - 854{,}885.16}{0.001263}$

93. $\dfrac{x(x + 1) - 2(x + 3)}{(x + 1)(x + 2)(x + 3)} = \dfrac{x^2 + x - 2x - 6}{(x + 1)(x + 2)(x + 3)}$

$\qquad\qquad = \dfrac{x^2 - x - 6}{(x + 1)(x + 2)(x + 3)}$

$\qquad\qquad = \dfrac{(x - 3)(x + 2)}{(x + 1)(x + 2)(x + 3)}$

$\qquad\qquad = \dfrac{x + 2}{x + 2} \cdot \dfrac{x - 3}{(x + 1)(x + 3)}$

$\qquad\qquad = \dfrac{x - 3}{(x + 1)(x + 3)}$

94. $-\dfrac{4}{x - 2}$, or $\dfrac{4}{2 - x}$

95. $\dfrac{m^2 - t^2}{m^2 + t^2 + m + t + 2mt} = \dfrac{m^2 - t^2}{(m^2 + 2mt + t^2) + (m + t)}$

$\qquad\qquad = \dfrac{(m + t)(m - t)}{(m + t)^2 + (m + t)}$

$\qquad\qquad = \dfrac{(m + t)(m - t)}{(m + t)[(m + t) + 1]}$

$\qquad\qquad = \dfrac{m + t}{m + t} \cdot \dfrac{m - t}{m + t + 1}$

$\qquad\qquad = \dfrac{m - t}{m + t + 1}$

96. $\dfrac{a^2 + 2}{a^2 - 3}$

97. $\dfrac{x^3 + x^2 - y^3 - y^2}{x^2 - 2xy + y^2}$

$= \dfrac{(x^3 - y^3) + (x^2 - y^2)}{x^2 - 2xy + y^2}$

$= \dfrac{(x - y)(x^2 + xy + y^2) + (x + y)(x - y)}{(x - y)^2}$

$= \dfrac{(x - y)[(x^2 + xy + y^2) + (x + y)]}{(x - y)(x - y)}$

$= \dfrac{x - y}{x - y} \cdot \dfrac{x^2 + xy + y^2 + x + y}{x - y}$

$= \dfrac{x^2 + xy + y^2 + x + y}{x - y}$

98. $\dfrac{(u^2 - uv + v^2)^2}{u - v}$

99. $\dfrac{x^5 - x^3 + x^2 - 1 - (x^3 - 1)(x + 1)^2}{(x^2 - 1)^2}$

$= \dfrac{x^5 - x^3 + (x^2 - 1) - [(x^3 - 1)(x + 1)^2]}{(x^2 - 1)^2}$

$= \dfrac{x^3(x^2-1) + (x^2-1) - [(x-1)(x^2+x+1)(x+1)(x+1)]}{(x^2 - 1)^2}$

$= \dfrac{x^3(x^2-1) + (x^2-1) - [(x^2-1)(x+1)(x^2+x+1)]}{(x^2 - 1)^2}$

$= \dfrac{(x^2 - 1)[x^3 + 1 - (x + 1)(x^2 + x + 1)]}{(x^2 - 1)^2}$

$= \dfrac{-2x^2 - 2x}{x^2 - 1}$

$= \dfrac{-2x(x + 1)}{(x + 1)(x - 1)}$

$= \dfrac{-2x}{x - 1}$

100. Answers may vary.

101. a) $(f \cdot g)(x) = \dfrac{4}{x^2 - 1} \cdot \dfrac{4x^2 + 8x + 4}{x^3 - 1}$

$= \dfrac{4 \cdot 4(x + 1)(x + 1)}{(x + 1)(x - 1)(x - 1)(x^2 + x + 1)}$

$= \dfrac{x + 1}{x + 1} \cdot \dfrac{4 \cdot 4(x + 1)}{(x - 1)(x - 1)(x^2 + x + 1)}$

$= \dfrac{16(x + 1)}{(x - 1)^2(x^2 + x + 1)}$

(Note that $x \neq -1$ is an additional restriction, since -1 is not in the domain of f.)

b) $(f/g)(x) = \dfrac{4}{x^2 - 1} \div \dfrac{4x^2 + 8x + 4}{x^3 - 1}$

$= \dfrac{4}{x^2 - 1} \cdot \dfrac{x^3 - 1}{4x^2 + 8x + 4}$

$= \dfrac{4(x - 1)(x^2 + x + 1)}{(x + 1)(x - 1)(4)(x + 1)(x + 1)}$

$= \dfrac{4(x - 1)}{4(x - 1)} \cdot \dfrac{x^2 + x + 1}{(x + 1)(x + 1)(x + 1)}$

$= \dfrac{x^2 + x + 1}{(x + 1)^3}$

(Note that $x \neq 1$ is an additional restriction, since 1 is not in the domain of g.)

Exercise Set 6.2

1. $\dfrac{3}{2a} + \dfrac{5}{2a} = \dfrac{8}{2a} = \dfrac{2 \cdot 4}{2a} = \dfrac{2}{2} \cdot \dfrac{4}{a} = \dfrac{4}{a}$

2. $\dfrac{4}{y}$

3. $\dfrac{3}{4a^2b} - \dfrac{7}{4a^2b} = \dfrac{-4}{4a^2b} = \dfrac{-1 \cdot 4}{4a^2b} = \dfrac{4}{4} \cdot \dfrac{-1}{a^2b} = -\dfrac{1}{a^2b}$

4. $\dfrac{1}{3m^2n^2}$

5. $\dfrac{a - 3b}{a + 5} + \dfrac{a + 5b}{a + b} = \dfrac{a - 3b + a + 5b}{a + b}$

$= \dfrac{2a + 2b}{a + b} = \dfrac{2(a + b)}{a + b}$

$= \dfrac{2}{1} \cdot \dfrac{a + b}{a + b} = 2$

6. 2

7. $\dfrac{4y + 2}{y - 2} - \dfrac{y - 3}{y - 2} = \dfrac{4y + 2 - (y - 3)}{y - 2}$

$= \dfrac{4y + 2 - y + 3}{y - 2}$

$= \dfrac{3y + 5}{y - 2}$

8. $\dfrac{2t + 4}{t - 4}$

9. $\dfrac{3x - 4}{x^2 - 5x + 4} + \dfrac{3 - 2x}{x^2 - 5x + 4} = \dfrac{3x - 4 + 3 - 2x}{x^2 - 5x + 4}$

$= \dfrac{x - 1}{(x - 4)(x - 1)}$

$= \dfrac{x - 1}{x - 1} \cdot \dfrac{1}{x - 4} = \dfrac{1}{x - 4}$

10. $\dfrac{1}{x - 7}$

11. $\dfrac{3a - 8}{a^2 - 9} - \dfrac{2a - 5}{a^2 - 9} = \dfrac{3a - 8 - (2a - 5)}{a^2 - 9}$

$= \dfrac{3a - 8 - 2a + 5}{a^2 - 9}$

$= \dfrac{a - 3}{(a + 3)(a - 3)} = \dfrac{1}{a + 3}$

12. $\dfrac{1}{a + 5}$

13. $\dfrac{a^2}{a - b} + \dfrac{b^2}{b - a} = \dfrac{a^2}{a - b} + \dfrac{-1}{-1} \cdot \dfrac{b^2}{b - a}$

$= \dfrac{a^2}{a - b} + \dfrac{-b^2}{a - b}$

$= \dfrac{a^2 - b^2}{a - b} = \dfrac{(a + b)(a - b)}{a - b}$

$= \dfrac{a + b}{1} \cdot \dfrac{a - b}{a - b} = a + b$

14. $-(s + r)$

15. $\dfrac{3}{x} - \dfrac{8}{-x} = \dfrac{3}{x} - \dfrac{-1}{-1} \cdot \dfrac{8}{-x} = \dfrac{3}{x} + \dfrac{8}{x} = \dfrac{11}{x}$

16. $\dfrac{7}{a}$

17. $\dfrac{2x - 9}{x^2 - 25} - \dfrac{4 - x}{25 - x^2} = \dfrac{2x - 9}{x^2 - 25} - \dfrac{-1}{-1} \cdot \dfrac{4 - x}{25 - x^2}$

$= \dfrac{2x - 9}{x^2 - 25} + \dfrac{4 - x}{x^2 - 25}$

$= \dfrac{x - 5}{x^2 - 25} = \dfrac{x - 5}{(x + 5)(x - 5)}$

$= \dfrac{1}{x + 5}$

18. $-\dfrac{2}{y^2 - 16}$

19. $\dfrac{t^2 + 3}{t^4 - 16} + \dfrac{7}{16 - t^4} = \dfrac{t^2 + 3}{t^4 - 16} + \dfrac{-1}{-1} \cdot \dfrac{7}{16 - t^4}$

$= \dfrac{t^2 + 3}{t^4 - 16} + \dfrac{-7}{t^4 - 16}$

$= \dfrac{t^2 - 4}{t^4 - 16}$

$= \dfrac{(t + 2)(t - 2)}{(t^2 + 4)(t + 2)(t - 2)}$

$= \dfrac{(t + 2)(t - 2)}{(t + 2)(t - 2)} \cdot \dfrac{1}{t^2 + 4}$

$= \dfrac{1}{t^2 + 4}$

20. $\dfrac{1}{y^2 + 9}$

21. $\dfrac{m - 3n}{m^3 - n^3} - \dfrac{2n}{n^3 - m^3} = \dfrac{m - 3n}{m^3 - n^3} - \dfrac{-1}{-1} \cdot \dfrac{2n}{n^3 - m^3}$

$= \dfrac{m - 3n}{m^3 - n^3} + \dfrac{2n}{m^3 - n^3}$

$= \dfrac{m - n}{m^3 - n^3}$

$= \dfrac{m - n}{(m - n)(m^2 + mn + n^2)}$

$= \dfrac{m - n}{m - n} \cdot \dfrac{1}{m^2 + mn + n^2}$

$= \dfrac{1}{m^2 + mn + n^2}$

22. $\dfrac{1}{r^2 + rs + s^2}$

23. $\dfrac{y - 2}{y + 4} + \dfrac{y + 3}{y - 5}$

[LCM is $(y + 4)(y - 5)$.]

$= \dfrac{y - 2}{y + 4} \cdot \dfrac{y - 5}{y - 5} + \dfrac{y + 3}{y - 5} \cdot \dfrac{y + 4}{y + 4}$

$= \dfrac{(y^2 - 7y + 10) + (y^2 + 7y + 12)}{(y + 4)(y - 5)}$

$= \dfrac{2y^2 + 22}{(y + 4)(y - 5)}$

24. $\dfrac{2x^2 - x + 14}{(x + 3)(x - 4)}$

25. $2 + \dfrac{x - 3}{x + 1} = \dfrac{2}{1} + \dfrac{x - 3}{x + 1}$

[LCM is $x + 1$.]

$= \dfrac{2}{1} \cdot \dfrac{x + 1}{x + 1} + \dfrac{x - 3}{x + 1}$

$= \dfrac{(2x + 2) + (x - 3)}{x + 1}$

$= \dfrac{3x - 1}{x + 1}$

26. $\dfrac{4y - 13}{y - 5}$

27. $\dfrac{4xy}{x^2 - y^2} + \dfrac{x - y}{x + y}$

$= \dfrac{4xy}{(x + y)(x - y)} + \dfrac{x - y}{x + y}$

[LCM is $(x + y)(x - y)$.]

$= \dfrac{4xy}{(x + y)(x - y)} + \dfrac{x - y}{x + y} \cdot \dfrac{x - y}{x - y}$

$= \dfrac{4xy + x^2 - 2xy + y^2}{(x + y)(x - y)}$

$= \dfrac{x^2 + 2xy + y^2}{(x + y)(x - y)} = \dfrac{(x + y)(x + y)}{(x + y)(x - y)}$

$= \dfrac{x + y}{x + y} \cdot \dfrac{x + y}{x - y} = \dfrac{x + y}{x - y}$

28. $\dfrac{a^2 + 7ab + b^2}{(a + b)(a - b)}$

29. $\dfrac{9x + 2}{3x^2 - 2x - 8} + \dfrac{7}{3x^2 + x - 4}$

$= \dfrac{9x + 2}{(3x + 4)(x - 2)} + \dfrac{7}{(3x + 4)(x - 1)}$

[LCM is $(3x + 4)(x - 2)(x - 1)$.]

$= \dfrac{9x + 2}{(3x + 4)(x - 2)} \cdot \dfrac{x - 1}{x - 1} + \dfrac{7}{(3x + 4)(x - 1)} \cdot \dfrac{x - 2}{x - 2}$

$= \dfrac{9x^2 - 7x - 2 + 7x - 14}{(3x + 4)(x - 2)(x - 1)}$

$= \dfrac{9x^2 - 16}{(3x + 4)(x - 2)(x - 1)} = \dfrac{(3x + 4)(3x - 4)}{(3x + 4)(x - 2)(x - 1)}$

$= \dfrac{3x + 4}{3x + 4} \cdot \dfrac{3x - 4}{(x - 2)(x - 1)} = \dfrac{3x - 4}{(x - 2)(x - 1)}$

30. $\dfrac{3y^2 + 7y + 14}{(2y - 5)(y + 2)(y - 1)}$

31. $\dfrac{4}{x + 1} + \dfrac{x + 2}{x^2 - 1} + \dfrac{3}{x - 1}$

$= \dfrac{4}{x + 1} + \dfrac{x + 2}{(x + 1)(x - 1)} + \dfrac{3}{x - 1}$

[LCM is $(x + 1)(x - 1)$.]

$= \dfrac{4}{x + 1} \cdot \dfrac{x - 1}{x - 1} + \dfrac{x + 2}{(x + 1)(x - 1)} + \dfrac{3}{x - 1} \cdot \dfrac{x + 1}{x + 1}$

$= \dfrac{4x - 4 + x + 2 + 3x + 3}{(x + 1)(x - 1)}$

$= \dfrac{8x + 1}{(x + 1)(x - 1)}$

32. $\dfrac{4y + 17}{(y + 2)(y - 2)}$

33. $\dfrac{x - 1}{3x + 15} - \dfrac{x + 3}{5x + 25}$

$= \dfrac{x - 1}{3(x + 5)} - \dfrac{x + 3}{5(x + 5)}$

[LCM is $3 \cdot 5(x + 5)$, or $15(x + 5)$.]

$= \dfrac{x - 1}{3(x + 5)} \cdot \dfrac{5}{5} - \dfrac{x + 3}{5(x + 5)} \cdot \dfrac{3}{3}$

$= \dfrac{5x - 5 - (3x + 9)}{15(x + 5)}$

$= \dfrac{5x - 5 - 3x - 9}{15(x + 5)}$

$= \dfrac{2x - 14}{15(x + 5)}$

34. $\dfrac{y - 34}{20(y + 2)}$

35. $\dfrac{5ab}{a^2 - b^2} - \dfrac{a - b}{a + b}$

$= \dfrac{5ab}{(a + b)(a - b)} - \dfrac{a - b}{a + b}$

[LCM is $(a + b)(a - b)$.]

$= \dfrac{5ab}{(a + b)(a - b)} - \dfrac{a - b}{a + b} \cdot \dfrac{a - b}{a - b}$

$= \dfrac{5ab - (a^2 - 2ab + b^2)}{(a + b)(a - b)}$

$= \dfrac{5ab - a^2 + 2ab - b^2}{(a + b)(a - b)}$

$= \dfrac{-a^2 + 7ab - b^2}{(a + b)(a - b)}$

36. $\dfrac{-x^2 + 4xy - y^2}{(x + y)(x - y)}$

37. $\dfrac{x}{x^2 + 9x + 20} - \dfrac{4}{x^2 + 7x + 12}$

$= \dfrac{x}{(x + 5)(x + 4)} - \dfrac{4}{(x + 3)(x + 4)}$

[LCM is $(x + 5)(x + 4)(x + 3)$.]

$= \dfrac{x}{(x + 5)(x + 4)} \cdot \dfrac{x + 3}{x + 3} - \dfrac{4}{(x + 3)(x + 4)} \cdot \dfrac{x + 5}{x + 5}$

$= \dfrac{x^2 + 3x - (4x + 20)}{(x + 5)(x + 4)(x + 3)}$

$= \dfrac{x^2 + 3x - 4x - 20}{(x + 5)(x + 4)(x + 3)}$

$= \dfrac{x^2 - x - 20}{(x + 5)(x + 4)(x + 3)}$

$= \dfrac{(x - 5)(x + 4)}{(x + 5)(x + 4)(x + 3)}$

$= \dfrac{x + 4}{x + 4} \cdot \dfrac{x - 5}{(x + 5)(x + 3)}$

$= \dfrac{x - 5}{(x + 5)(x + 3)}$

38. $\dfrac{x - 6}{(x + 6)(x + 4)}$

39. $\dfrac{3y}{y^2 - 7y + 10} - \dfrac{2y}{y^2 - 8y + 15}$

$= \dfrac{3y}{(y - 5)(y - 2)} - \dfrac{2y}{(y - 5)(y - 3)}$

[LCM is $(y - 5)(y - 2)(y - 3)$.]

$= \dfrac{3y}{(y - 5)(y - 2)} \cdot \dfrac{y - 3}{y - 3} - \dfrac{2y}{(y - 5)(y - 3)} \cdot \dfrac{y - 2}{y - 2}$

$= \dfrac{3y^2 - 9y - (2y^2 - 4y)}{(y - 5)(y - 2)(y - 3)}$

$= \dfrac{3y^2 - 9y - 2y^2 + 4y}{(y - 5)(y - 2)(y - 3)}$

$= \dfrac{y^2 - 5y}{(y - 5)(y - 2)(y - 3)} = \dfrac{y(y - 5)}{(y - 5)(y - 2)(y - 3)}$

$= \dfrac{y - 5}{y - 5} \cdot \dfrac{y}{(y - 2)(y - 3)} = \dfrac{y}{(y - 2)(y - 3)}$

40. $\dfrac{2x^2 + 21x}{(x - 4)(x - 2)(x + 3)}$

41. $\dfrac{y}{y^2 - y - 20} + \dfrac{2}{y + 4}$

$= \dfrac{y}{(y - 5)(y + 4)} + \dfrac{2}{y + 4}$

[LCM is $(y - 5)(y + 4)$.]

$= \dfrac{y}{(y - 5)(y + 4)} + \dfrac{2}{y + 4} \cdot \dfrac{y - 5}{y - 5}$

$= \dfrac{y + 2y - 10}{(y - 5)(y + 4)}$

$= \dfrac{3y - 10}{(y - 5)(y + 4)}$

42. $\dfrac{3}{t - 3}$

43. $\dfrac{3y + 2}{y^2 + 5y - 24} + \dfrac{7}{y^2 + 4y - 32}$

$= \dfrac{3y + 2}{(y + 8)(y - 3)} + \dfrac{7}{(y + 8)(y - 4)}$

[LCM is $(y + 8)(y - 3)(y - 4)$.]

$= \dfrac{3y + 2}{(y + 8)(y - 3)} \cdot \dfrac{y - 4}{y - 4} + \dfrac{7}{(y + 8)(y - 4)} \cdot \dfrac{y - 3}{y - 3}$

$= \dfrac{3y^2 - 10y - 8 + 7y - 21}{(y + 8)(y - 3)(y - 4)}$

$= \dfrac{3y^2 - 3y - 29}{(y + 8)(y - 3)(y - 4)}$

44. $\dfrac{5x^2 - 11x - 6}{(x - 5)(x - 2)(x - 3)}$

45. $\dfrac{3x - 1}{x^2 + 2x - 3} - \dfrac{x + 4}{x^2 - 9}$

$= \dfrac{3x - 1}{(x + 3)(x - 1)} - \dfrac{x + 4}{(x + 3)(x - 3)}$

[LCM is $(x + 3)(x - 1)(x - 3)$.]

$= \dfrac{3x - 1}{(x + 3)(x - 1)} \cdot \dfrac{x - 3}{x - 3} - \dfrac{x + 4}{(x + 3)(x - 3)} \cdot \dfrac{x - 1}{x - 1}$

$= \dfrac{3x^2 - 10x + 3 - (x^2 + 3x - 4)}{(x + 3)(x - 1)(x - 3)}$

$= \dfrac{3x^2 - 10x + 3 - x^2 - 3x + 4}{(x + 3)(x - 1)(x - 3)}$

$= \dfrac{2x^2 - 13x + 7}{(x + 3)(x - 1)(x - 3)}$

46. $\dfrac{2p^2 + 7p + 10}{(p + 6)(p - 4)(p + 4)}$

47. $\dfrac{2}{a^2 - 5a + 4} + \dfrac{-2}{a^2 - 4}$

$= \dfrac{2}{(a - 4)(a - 1)} + \dfrac{-2}{(a + 2)(a - 2)}$

[LCM is $(a - 4)(a - 1)(a + 2)(a - 2)$.]

$= \dfrac{2}{(a - 4)(a - 1)} \cdot \dfrac{(a + 2)(a - 2)}{(a + 2)(a - 2)} +$

$\dfrac{-2}{(a + 2)(a - 2)} \cdot \dfrac{(a - 4)(a - 1)}{(a - 4)(a - 1)}$

$= \dfrac{2(a^2 - 4) - 2(a^2 - 5a + 4)}{(a - 4)(a - 1)(a + 2)(a - 2)}$

$= \dfrac{2a^2 - 8 - 2a^2 + 10a - 8}{(a - 4)(a - 1)(a + 2)(a - 2)}$

$= \dfrac{10a - 16}{(a - 4)(a - 1)(a + 2)(a - 2)}$, or

$\dfrac{10a - 16}{(a^2 - 5a + 4)(a^2 - 4)}$

48. $\dfrac{21a - 45}{(a - 6)(a - 1)(a + 3)(a - 3)}$

49. $3 + \dfrac{t}{t + 2} - \dfrac{2}{t^2 - 4} = \dfrac{3}{1} + \dfrac{t}{t + 2} - \dfrac{2}{(t + 2)(t - 2)}$

[LCM is $(t + 2)(t - 2)$.]

$= \dfrac{3}{1} \cdot \dfrac{(t+2)(t-2)}{(t+2)(t-2)} + \dfrac{t}{t+2} \cdot \dfrac{t-2}{t-2} - \dfrac{2}{(t+2)(t-2)}$

$= \dfrac{3t^2 - 12 + t^2 - 2t - 2}{(t + 2)(t - 2)}$

$= \dfrac{4t^2 - 2t - 14}{(t + 2)(t - 2)}$

50. $\dfrac{3t^2 + 3t - 21}{(t + 3)(t - 3)}$

51. $\dfrac{1}{x + 1} - \dfrac{x}{x - 2} + \dfrac{x^2 + 2}{x^2 - x - 2}$

$= \dfrac{1}{x + 1} - \dfrac{x}{x - 2} + \dfrac{x^2 + 2}{(x - 2)(x + 1)}$

[LCM is $(x + 1)(x - 2)$.]

$= \dfrac{1}{x + 1} \cdot \dfrac{x - 2}{x - 2} - \dfrac{x}{x - 2} \cdot \dfrac{x + 1}{x + 1} + \dfrac{x^2 + 2}{(x - 2)(x + 1)}$

$= \dfrac{x - 2 - (x^2 + x) + x^2 + 2}{(x + 1)(x - 2)}$

$= \dfrac{x - 2 - x^2 - x + x^2 + 2}{(x + 1)(x - 2)}$

$= \dfrac{0}{(x + 1)(x - 2)} = 0$

52. $-\dfrac{y}{(y + 3)(y - 1)}$

53. $\dfrac{4x}{x^2 - 1} + \dfrac{3x}{1 - x} - \dfrac{4}{x - 1}$

$= \dfrac{4x}{x^2 - 1} + \dfrac{-1}{-1} \cdot \dfrac{3x}{1 - x} - \dfrac{4}{x - 1}$

$= \dfrac{4x}{(x + 1)(x - 1)} + \dfrac{-3x}{x - 1} - \dfrac{4}{x - 1}$

[LCM is $(x + 1)(x - 1)$.]

$= \dfrac{4x}{(x + 1)(x - 1)} + \dfrac{-3x}{x - 1} \cdot \dfrac{x + 1}{x + 1} - \dfrac{4}{x - 1} \cdot \dfrac{x + 1}{x + 1}$

$= \dfrac{4x - 3x^2 - 3x - 4x - 4}{(x + 1)(x - 1)}$

$= \dfrac{-3x^2 - 3x - 4}{(x + 1)(x - 1)}$

54. $\dfrac{-14y^2 - 3y + 3}{(2y + 1)(2y - 1)}$

55. $\dfrac{1}{t^2 + 5t + 6} - \dfrac{2}{t^2 + 3t + 2} + \dfrac{1}{t^2 - 3t - 4}$

$= \dfrac{1}{(t + 3)(t + 2)} - \dfrac{2}{(t + 2)(t + 1)} + \dfrac{1}{(t - 4)(t + 1)}$

[LCM is $(t + 3)(t + 2)(t + 1)(t - 4)$.]

$= \dfrac{1}{(t + 3)(t + 2)} \cdot \dfrac{(t + 1)(t - 4)}{(t + 1)(t - 4)} -$

$\dfrac{2}{(t + 2)(t + 1)} \cdot \dfrac{(t + 3)(t - 4)}{(t + 3)(t - 4)} +$

$\dfrac{1}{(t - 4)(t + 1)} \cdot \dfrac{(t + 3)(t + 2)}{(t + 3)(t + 2)}$

$= \dfrac{(t^2 - 3t - 4) - 2(t^2 - t - 12) + (t^2 + 5t + 6)}{(t + 3)(t + 2)(t + 1)(t - 4)}$

$= \dfrac{t^2 - 3t - 4 - 2t^2 + 2t + 24 + t^2 + 5t + 6}{(t + 3)(t + 2)(t + 1)(t - 4)}$

$= \dfrac{4t + 26}{(t + 3)(t + 2)(t + 1)(t - 4)}$

56. $\dfrac{-6x + 42}{(x - 3)(x - 2)(x + 1)(x + 3)}$

57. $\dfrac{15x^{-7}y^{12}z^4}{35x^{-2}y^6z^{-3}} = \dfrac{15}{35}x^{-7-(-2)}y^{12-6}z^{4-(-3)}$

$= \dfrac{3}{7}x^{-5}y^6z^7 = \dfrac{3}{7} \cdot \dfrac{1}{x^5} \cdot y^6z^7$

$= \dfrac{3y^6z^7}{7x^5}$

58. $y = \dfrac{5}{4}x + \dfrac{11}{2}$

59. Familiarize. We let x, y, and z represent the number of rolls of dimes, nickels, and quarters, respectively.

Coins	Number of rolls	Value per roll	Total Value
Dimes	x	50 × 0.10, or 5.00	5x
Nickels	y	40 × 0.05, or 2.00	2y
Quarters	z	40 × 0.25, or 10.00	10z
Total	12		$70.00

Translate. The number of rolls of nickels is three more than the number of rolls of dimes. This gives us one equation.

$y = x + 3$

From the table we get two more equations.

$x + y + z = 12$

$5x + 2y + 10z = 70$

Carry out. Solving the system we get the ordered triple (2,5,5).

59. (continued)

Check. 2 rolls of dimes = 2 × $5, or $10

5 rolls of nickels = 5 × $2, or $10

5 rolls of quarters = 5 × $10, or $50

The total value is $10 + $10 + $50 = $70.

The total number of rolls of coins is 2 + 5 + 5, or 12, and the number of rolls of nickels, 5, is three more than the number of rolls of dimes, 2. The numbers check.

State. Robert has 2 rolls of dimes, 5 rolls of nickels, and 5 rolls of quarters.

60. 4 30-min tapes, 8 60-min tapes

61. $2x^{-2} + 3x^{-2}y^{-2} - 7xy^{-1}$

$= \dfrac{2}{x^2} + \dfrac{3}{x^2y^2} - \dfrac{7x}{y}$

[LCM is x^2y^2.]

$= \dfrac{2}{x^2} \cdot \dfrac{y^2}{y^2} + \dfrac{3}{x^2y^2} - \dfrac{7x}{y} \cdot \dfrac{x^2y}{x^2y}$

$= \dfrac{2y^2 + 3 - 7x^3y}{x^2y^2}$

62. $\dfrac{9x^2 + 28x + 15}{(x - 3)(x + 3)^2}$

63. $4(y-1)(2y-5)^{-1} + 5(2y+3)(5-2y)^{-1} + (y-4)(2y-5)^{-1}$

$= \dfrac{4(y - 1)}{2y - 5} + \dfrac{-1}{-1} \cdot \dfrac{5(2y + 3)}{5 - 2y} + \dfrac{y - 4}{2y - 5}$

$= \dfrac{4y - 4 - 10y - 15 + y - 4}{2y - 5}$

$= \dfrac{-5y - 23}{2y - 5}$, or $\dfrac{5y + 23}{5 - 2y}$

64. $\dfrac{1}{2x(x - 5)}$

65. $\dfrac{x^2 - 7x + 12}{x^2 - x - 29/3}\left[\dfrac{3x + 2}{x^2 + 5x - 24} + \dfrac{7}{x^2 + 4x - 32}\right]$

$= \dfrac{x^2 - 7x + 12}{x^2 - x - 29/3}\left[\dfrac{3x + 2}{(x + 8)(x - 3)} + \dfrac{7}{(x + 8)(x - 4)}\right]$

$= \dfrac{x^2-7x+12}{x^2-x-29/3}\left[\dfrac{3x+2}{(x+8)(x-3)} \cdot \dfrac{x-4}{x-4} + \dfrac{7}{(x+8)(x-4)} \cdot \dfrac{x-3}{x-3}\right]$

$= \dfrac{x^2 - 7x + 12}{x^2 - x - 29/3}\left[\dfrac{3x^2 - 10x - 8 + 7x - 21}{(x + 8)(x - 3)(x - 4)}\right]$

$= \dfrac{x^2 - 7x + 12}{x^2 - x - 29/3}\left[\dfrac{3x^2 - 3x - 29}{(x + 8)(x - 3)(x - 4)}\right]$

$= \dfrac{(x - 4)(x - 3)(3)(x^2 - x - 29/3)}{(x^2 - x - 29/3)(x + 8)(x - 3)(x - 4)}$

$= \dfrac{(x - 4)(x - 3)(x^2 - x - 29/3)}{(x - 4)(x - 3)(x^2 - x - 29/3)} \cdot \dfrac{3}{x + 8}$

$= \dfrac{3}{x + 8}$

66. $-4t^4$

67. $\dfrac{9t^3}{3t^3 - 12t^2 + 9t} \div \left[\dfrac{t + 4}{t^2 - 9} - \dfrac{3t - 1}{t^2 + 2t - 3}\right]$

$= \dfrac{9t^3}{3t^3 - 12t^2 + 9t} \div \left[\dfrac{t + 4}{(t + 3)(t - 3)} - \dfrac{3t - 1}{(t + 3)(t - 1)}\right]$

$= \dfrac{9t^3}{3t^3 - 12t^2 + 9t} \div \left[\dfrac{t+4}{(t+3)(t-3)} \cdot \dfrac{t-1}{t-1} - \dfrac{3t-1}{(t+3)(t-1)} \cdot \dfrac{t-3}{t-3}\right]$

$= \dfrac{9t^3}{3t^3 - 12t^2 + 9t} \div \left[\dfrac{t^2 + 3t - 4 - (3t^2 - 10t + 3)}{(t + 3)(t - 3)(t - 1)}\right]$

$= \dfrac{9t^3}{3t^3 - 12t^2 + 9t} \div \left[\dfrac{t^2 + 3t - 4 - 3t^2 + 10t - 3}{(t + 3)(t - 3)(t - 1)}\right]$

$= \dfrac{9t^3}{3t(t^2 - 4t + 3)} \div \dfrac{-2t^2 + 13t - 7}{(t + 3)(t - 3)(t - 1)}$

$= \dfrac{3t \cdot 3t^2}{3t(t - 3)(t - 1)} \cdot \dfrac{(t + 3)(t - 3)(t - 1)}{-2t^2 + 13t - 7}$

$= \dfrac{3t(t - 3)(t - 1)}{3t(t - 3)(t - 1)} \cdot \dfrac{3t^2(t + 3)}{-2t^2 + 13t - 7}$

$= \dfrac{3t^2(t + 3)}{-2t^2 + 13t - 7}$

68. $\dfrac{x - 6}{(x + 6)(x + 4)}$

69. $\dfrac{A + B}{A - B} - \dfrac{A - B}{A + B}$

$= \dfrac{(x + y) + (x - y)}{(x + y) - (x - y)} - \dfrac{(x + y) - (x - y)}{(x + y) + (x - y)}$

$= \dfrac{2x}{x + y - x + y} - \dfrac{x + y - x + y}{2x}$

$= \dfrac{2x}{2y} - \dfrac{2y}{2x} = \dfrac{2}{2} \cdot \dfrac{x}{y} - \dfrac{2}{2} \cdot \dfrac{y}{x}$

$= \dfrac{x}{y} - \dfrac{y}{x}$

$= \dfrac{x}{y} \cdot \dfrac{x}{x} - \dfrac{y}{x} \cdot \dfrac{y}{y}$

$= \dfrac{x^2 - y^2}{xy}$

70. $\dfrac{x - y + x^2 + xy}{x + y - x^2 + xy}$

71. a) $(f + g)(x) = \dfrac{x^3}{x^2 - 4} + \dfrac{x^2}{x^2 + 3x - 10}$

$= \dfrac{x^3}{(x + 2)(x - 2)} + \dfrac{x^2}{(x + 5)(x - 2)}$

$= \dfrac{x^3(x + 5) + x^2(x + 2)}{(x + 2)(x - 2)(x + 5)}$

$= \dfrac{x^4 + 5x^3 + x^3 + 2x^2}{(x + 2)(x - 2)(x + 5)}$

$= \dfrac{x^4 + 6x^3 + 2x^3}{(x + 2)(x - 2)(x + 5)}$

b) $(f - g)(x) = \dfrac{x^3}{x^2 - 4} - \dfrac{x^2}{x^2 + 3x - 10}$

$= \dfrac{x^3}{(x + 2)(x - 2)} - \dfrac{x^2}{(x + 5)(x - 2)}$

$= \dfrac{x^3(x + 5) - x^2(x + 2)}{(x + 2)(x - 2)(x + 5)}$

$= \dfrac{x^4 + 5x^3 - x^3 - 2x^2}{(x + 2)(x - 2)(x + 5)}$

$= \dfrac{x^4 + 4x^3 - 2x^2}{(x + 2)(x - 2)(x + 5)}$

71. (continued)

c) $(f \cdot g)(x) = \dfrac{x^3}{x^2 - 4} \cdot \dfrac{x^2}{x^2 + 3x - 10}$

$= \dfrac{x^5}{(x^2 - 4)(x^2 + 3x - 10)}$

d) $(f/g)(x) = \dfrac{x^3}{x^2 - 4} \div \dfrac{x^2}{x^2 + 3x - 10}$

$= \dfrac{x^2 \cdot x}{(x + 2)(x - 2)} \cdot \dfrac{(x + 5)(x - 2)}{x^2}$

$= \dfrac{x^2(x - 2)}{x^2(x - 2)} \cdot \dfrac{x(x + 5)}{x + 2}$

$= \dfrac{x(x + 5)}{x + 2}$

(Note that $x \neq 0$, $x \neq -5$, and $x \neq 2$ are additional restrictions, since $g(0) = 0$, -5 is not in the domain of g, and 2 is not in the domain of either f or g.)

e) We find the values of x that are in the domain of f _and_ the domain of g. We set

$x^2 - 4 = 0$	and	$x^2 + 3x - 10 = 0$
$(x + 2)(x - 2) = 0$	and	$(x + 5)(x - 2) = 0$
$(x + 2 = 0$ or $x - 2 = 0)$	and	$(x + 5 = 0$ or $x - 2 = 0)$
$(x = -2$ or $x = 2)$	and	$(x = -5$ or $x = 2)$.

The domain of $f + g = \{x \mid x$ is a real number and $x \neq -2$ and $x \neq 2$ and $x \neq -5\}$.

72. $x^4(x^2 + 1)(x + 1)(x - 1)(x^2 + x + 1)(x^2 - x + 1)$.

73. $2a^3 + 2a^2b + 2ab^2 = 2a(a^2 + ab + b^2)$

$a^6 - b^6 = (a^3 + b^3)(a^3 - b^3)$

$\qquad = (a + b)(a^2 - ab + b^2)(a - b)(a^2 + ab + b^2)$

$2b^2 + ab - 3a^2 =$
$(b - a)(2b + 3a)$, or $-(a - b)(2b + 3a)$

$2a^2b + 4ab^2 + 2b^3 = 2b(a^2 + 2ab + b^2)$

$\qquad\qquad = 2b(a + b)^2$

The LCM is
$2ab(a^2 + ab + b^2)(a + b)^2(a^2 - ab + b^2)(a - b)(2b + 3a)$.

74. $8a^4$, $8a^4b$, $8a^4b^2$, $8a^4b^3$, $8a^4b^4$, $8a^4b^5$, $8a^4b^6$, $8a^4b^7$

75. We find the LCM of the number of years it takes the planets to revolve around the sun.

Earth: 1

Jupiter: $12 = 2 \cdot 2 \cdot 3$

Saturn: $30 = 2 \cdot 3 \cdot 5$

The LCM is $2 \cdot 2 \cdot 3 \cdot 5$, or 60.

The earth, Jupiter, and Saturn line up with each other every 60 years.

76. 420 years

Exercise Set 6.3

1. $\dfrac{\frac{1}{x} + 4}{\frac{1}{x} - 3} = \dfrac{\frac{1}{x} + 4}{\frac{1}{x} - 3} \cdot \dfrac{x}{x}$ Using the LCM

$\quad = \dfrac{\frac{1}{x} \cdot x + 4 \cdot x}{\frac{1}{x} \cdot x - 3 \cdot x}$

$\quad = \dfrac{1 + 4x}{1 - 3x}$

2. $\dfrac{1 + 7y}{1 - 5y}$

3. $\dfrac{x - \frac{1}{x}}{x + \frac{1}{x}} = \dfrac{x - \frac{1}{x}}{x + \frac{1}{x}} \cdot \dfrac{x}{x}$ Using the LCM

$\quad = \dfrac{x \cdot x - \frac{1}{x} \cdot x}{x \cdot x + \frac{1}{x} \cdot x}$

$\quad = \dfrac{x^2 - 1}{x^2 + 1}$

4. $\dfrac{y^2 + 1}{y^2 - 1}$

5. $\dfrac{\frac{3}{x} + \frac{4}{y}}{\frac{4}{x} - \frac{3}{y}} = \dfrac{\frac{3}{x} + \frac{4}{y}}{\frac{4}{x} - \frac{3}{y}} \cdot \dfrac{xy}{xy}$ Using the LCM

$\quad = \dfrac{\frac{3}{x} \cdot xy + \frac{4}{y} \cdot xy}{\frac{4}{x} \cdot xy - \frac{3}{y} \cdot xy}$

$\quad = \dfrac{3y + 4x}{4y - 3x}$

6. $\dfrac{2z + 5y}{z - 4y}$

7. $\dfrac{\frac{x^2 - y^2}{xy}}{\frac{x - y}{y}} = \dfrac{x^2 - y^2}{xy} \cdot \dfrac{y}{x - y}$ Multiplying by the reciprocal of the divisor

$\quad = \dfrac{(x + y)(x - y)}{xy} \cdot \dfrac{y}{x - y}$

$\quad = \dfrac{y(x - y)}{y(x - y)} \cdot \dfrac{x + y}{x}$

$\quad = \dfrac{x + y}{x}$

8. $\dfrac{a + b}{a}$

9. $\dfrac{a - \frac{3a}{b}}{b - \frac{b}{a}} = \dfrac{a - \frac{3a}{b}}{b - \frac{b}{a}} \cdot \dfrac{ab}{ab}$ Using the LCM

$\quad = \dfrac{a(ab) - \frac{3a}{b} \cdot ab}{b(ab) - \frac{b}{a} \cdot ab}$

$\quad = \dfrac{a^2b - 3a^2}{ab^2 - b^2}$

$\quad = \dfrac{a^2(b - 3)}{b^2(a - 1)}$

10. $\dfrac{3}{3x + 2}$

11. $\dfrac{\frac{1}{a} + \frac{1}{b}}{\frac{a^2 - b^2}{ab}} = \dfrac{\frac{1}{a} + \frac{1}{b}}{\frac{a^2 - b^2}{ab}} \cdot \dfrac{ab}{ab}$ Using the LCM

$\quad = \dfrac{\frac{1}{a} \cdot ab + \frac{1}{b} \cdot ab}{\frac{a^2 - b^2}{ab} \cdot ab}$

$\quad = \dfrac{b + a}{a^2 - b^2} = \dfrac{b + a}{(a + b)(a - b)}$

$\quad = \dfrac{a + b}{a + b} \cdot \dfrac{1}{a - b}$ $(b + a = a + b)$

$\quad = \dfrac{1}{a - b}$

12. $\dfrac{1}{x - y}$

13. $\dfrac{\frac{1}{x + h} - \frac{1}{x}}{h} = \dfrac{\frac{1}{x + h} \cdot \frac{x}{x} - \frac{1}{x} \cdot \frac{x + h}{x + h}}{h}$ Adding in the numerator

$\quad = \dfrac{\frac{x - x - h}{x(x + h)}}{h} = \dfrac{\frac{-h}{x(x + h)}}{h}$

$\quad = \dfrac{-h}{x(x + h)} \cdot \dfrac{1}{h}$ Multiplying by the reciprocal of the divisor

$\quad = \dfrac{h}{h} \cdot \dfrac{-1}{x(x + h)}$

$\quad = -\dfrac{1}{x(x + h)}$

14. $\dfrac{1}{a(a - h)}$

15. $\dfrac{\frac{y^2 - y - 6}{y^2 - 5y - 14}}{\frac{y^2 + 6y + 5}{y^2 - 6y - 7}} = \dfrac{y^2 - y - 6}{y^2 - 5y - 14} \cdot \dfrac{y^2 - 6y - 7}{y^2 + 6y + 5}$

$\quad = \dfrac{(y - 3)(y + 2)(y - 7)(y + 1)}{(y - 7)(y + 2)(y + 5)(y + 1)}$

$\quad = \dfrac{(y + 2)(y - 7)(y + 1)}{(y + 2)(y - 7)(y + 1)} \cdot \dfrac{y - 3}{y + 5}$

$\quad = \dfrac{y - 3}{y + 5}$

16. $\dfrac{(x - 4)(x - 7)}{(x - 5)(x + 6)}$

17. $\dfrac{\dfrac{1}{x-2}+\dfrac{3}{x-1}}{\dfrac{2}{x-1}+\dfrac{5}{x-2}}$

$=\dfrac{\dfrac{1}{x-2}+\dfrac{3}{x-1}}{\dfrac{2}{x-1}+\dfrac{5}{x-2}}\cdot\dfrac{(x-2)(x-1)}{(x-2)(x-1)}$ Using the LCM

$=\dfrac{\dfrac{1}{x-2}\cdot(x-2)(x-1)+\dfrac{3}{x-1}\cdot(x-2)(x-1)}{\dfrac{2}{x-1}\cdot(x-2)(x-1)+\dfrac{5}{x-2}\cdot(x-2)(x-1)}$

$=\dfrac{x-1+3(x-2)}{2(x-2)+5(x-1)}$

$=\dfrac{x-1+3x-6}{2x-4+5x-5}=\dfrac{4x-7}{7x-9}$

18. $\dfrac{3y-1}{7y-5}$

19. $\dfrac{\dfrac{a}{a+3}-\dfrac{2}{a-1}}{\dfrac{a}{a+3}-\dfrac{1}{a-1}}$

$=\dfrac{\dfrac{a}{a+3}-\dfrac{2}{a-1}}{\dfrac{a}{a+3}-\dfrac{1}{a-1}}\cdot\dfrac{(a+3)(a-1)}{(a+3)(a-1)}$ Using the LCM

$=\dfrac{\dfrac{a}{a+3}\cdot(a+3)(a-1)-\dfrac{2}{a-1}\cdot(a+3)(a-1)}{\dfrac{a}{a+3}\cdot(a+3)(a-1)-\dfrac{1}{a-1}\cdot(a+3)(a-1)}$

$=\dfrac{a(a-1)-2(a+3)}{a(a-1)-(a+3)}$

$=\dfrac{a^2-a-2a-6}{a^2-a-a-3}=\dfrac{a^2-3a-6}{a^2-2a-3}$

20. $\dfrac{a^2-6a-6}{a^2-4a-2}$

21. $\dfrac{\dfrac{x}{x^2+3x-4}-\dfrac{1}{x^2+3x-4}}{\dfrac{x}{x^2+6x+8}+\dfrac{3}{x^2+6x+8}}=\dfrac{\dfrac{x-1}{x^2+3x-4}}{\dfrac{x+3}{x^2+6x+8}}$

Adding in the numerator and the denominator

$=\dfrac{x-1}{x^2+3x-4}\cdot\dfrac{x^2+6x+8}{x+3}$

$=\dfrac{x-1}{(x+4)(x-1)}\cdot\dfrac{(x+4)(x+2)}{x+3}$

$=\dfrac{(x-1)(x+4)}{(x-1)(x+4)}\cdot\dfrac{x+2}{x+3}=\dfrac{x+2}{x+3}$

22. $\dfrac{x-4}{x-2}$

23. $\dfrac{\dfrac{y}{y^2-1}+\dfrac{3}{1-y^2}}{\dfrac{y}{y^2-1}+\dfrac{9}{1-y^2}}=\dfrac{\dfrac{y}{y^2-1}+\dfrac{-1}{-1}\cdot\dfrac{3}{1-y^2}}{\dfrac{y}{y^2-1}+\dfrac{-1}{-1}\cdot\dfrac{9}{1-y^2}}$

$=\dfrac{\dfrac{y}{y^2-1}-\dfrac{3}{y^2-1}}{\dfrac{y}{y^2-1}-\dfrac{9}{y^2-1}}$

$=\dfrac{\dfrac{y-3}{y^2-1}}{\dfrac{y^2-9}{y^2-1}}$ Adding in the numerator and the denominator

$=\dfrac{y-3}{y^2-1}\cdot\dfrac{y^2-1}{y^2-9}$

$=\dfrac{(y-3)(y^2-1)}{(y^2-1)(y+3)(y-3)}$

$=\dfrac{(y-3)(y^2-1)}{(y-3)(y^2-1)}\cdot\dfrac{1}{y+3}$

$=\dfrac{1}{y+3}$

24. $\dfrac{1}{y+5}$

25. $\dfrac{\dfrac{2}{a^2-1}+\dfrac{1}{a+1}}{\dfrac{3}{a^2-1}+\dfrac{2}{a-1}}=\dfrac{\dfrac{2}{(a+1)(a-1)}+\dfrac{1}{a+1}}{\dfrac{3}{(a+1)(a-1)}+\dfrac{2}{a-1}}$

$=\dfrac{\dfrac{2}{(a+1)(a-1)}+\dfrac{1}{a+1}}{\dfrac{3}{(a+1)(a-1)}+\dfrac{2}{a-1}}\cdot\dfrac{(a+1)(a-1)}{(a+1)(a-1)}$ Using the LCM

$=\dfrac{\dfrac{2}{(a+1)(a-1)}\cdot(a+1)(a-1)+\dfrac{1}{a+1}\cdot(a+1)(a-1)}{\dfrac{3}{(a+1)(a-1)}\cdot(a+1)(a-1)+\dfrac{2}{a-1}\cdot(a+1)(a-1)}$

$=\dfrac{2+a-1}{3+2(a+1)}=\dfrac{a+1}{3+2a+2}=\dfrac{a+1}{2a+5}$

26. $\dfrac{2a-3}{a+1}$

27. $\dfrac{\dfrac{5}{x^2-4}-\dfrac{3}{x-2}}{\dfrac{4}{x^2-4}-\dfrac{2}{x+2}}=\dfrac{\dfrac{5}{(x+2)(x-2)}-\dfrac{3}{x-2}}{\dfrac{4}{(x+2)(x-2)}-\dfrac{2}{x+2}}$

$=\dfrac{\dfrac{5}{(x+2)(x-2)}-\dfrac{3}{x-2}}{\dfrac{4}{(x+2)(x-2)}-\dfrac{2}{x+2}}\cdot\dfrac{(x+2)(x-2)}{(x+2)(x-2)}$ Using the LCM

$=\dfrac{\dfrac{5}{(x+2)(x-2)}\cdot(x+2)(x-2)-\dfrac{3}{x-2}\cdot(x+2)(x-2)}{\dfrac{4}{(x+2)(x-2)}\cdot(x+2)(x-2)-\dfrac{2}{x+2}\cdot(x+2)(x-2)}$

$=\dfrac{5-3(x+2)}{4-2(x-2)}=\dfrac{5-3x-6}{4-2x+4}=\dfrac{-1-3x}{8-2x}$

28. $\dfrac{7-3x}{3-2x}$

29. $\dfrac{\dfrac{y^2}{y^2-9}-\dfrac{y}{y+3}}{\dfrac{y}{y^2-9}-\dfrac{1}{y-3}}=\dfrac{\dfrac{y^2}{(y+3)(y-3)}-\dfrac{y}{y+3}}{\dfrac{y}{(y+3)(y-3)}-\dfrac{1}{y-3}}$

$=\dfrac{\dfrac{y^2}{(y+3)(y-3)}-\dfrac{y}{y+3}}{\dfrac{y}{(y+3)(y-3)}-\dfrac{1}{y-3}}\cdot\dfrac{(y+3)(y-3)}{(y+3)(y-3)}\quad\begin{array}{l}\text{Using}\\\text{the LCM}\end{array}$

$=\dfrac{\dfrac{y^2}{(y+3)(y-3)}\cdot(y+3)(y-3)-\dfrac{y}{y+3}\cdot(y+3)(y-3)}{\dfrac{y}{(y+3)(y-3)}\cdot(y+3)(y-3)-\dfrac{1}{y-3}\cdot(y+3)(y-3)}$

$=\dfrac{y^2-y(y-3)}{y-(y+3)}=\dfrac{y^2-y^2+3y}{y-y-3}=\dfrac{3y}{-3}=\dfrac{3}{3}\cdot\dfrac{y}{-1}=-y$

30. $-y$

31. $\dfrac{\dfrac{a}{a+3}+\dfrac{4}{5a}}{\dfrac{a}{2a+6}+\dfrac{3}{a}}=\dfrac{\dfrac{a}{a+3}+\dfrac{4}{5a}}{\dfrac{a}{2(a+3)}+\dfrac{3}{a}}$

$=\dfrac{\dfrac{a}{a+3}+\dfrac{4}{5a}}{\dfrac{a}{2(a+3)}+\dfrac{3}{a}}\cdot\dfrac{10a(a+3)}{10a(a+3)}\quad\text{Using the LCM}$

$=\dfrac{\dfrac{a}{a+3}\cdot10a(a+3)+\dfrac{4}{5a}\cdot10a(a+3)}{\dfrac{a}{2(a+3)}\cdot10a(a+3)+\dfrac{3}{a}\cdot10a(a+3)}$

$=\dfrac{10a^2+8(a+3)}{5a^2+30(a+3)}=\dfrac{10a^2+8a+24}{5a^2+30a+90}$

$=\dfrac{2(5a^2+4a+12)}{5(a^2+6a+18)}$

32. $\dfrac{6(a^2+5a+10)}{3a^2+2a+4}$

33. $\dfrac{\dfrac{x}{x+y}+\dfrac{x}{y}}{\dfrac{x}{3x+3y}+\dfrac{y}{x}}=\dfrac{\dfrac{x}{x+y}+\dfrac{x}{y}}{\dfrac{x}{3(x+y)}+\dfrac{y}{x}}$

$=\dfrac{\dfrac{x}{x+y}+\dfrac{x}{y}}{\dfrac{x}{3(x+y)}+\dfrac{y}{x}}\cdot\dfrac{3xy(x+y)}{3xy(x+y)}\quad\text{Using the LCM}$

$=\dfrac{\dfrac{x}{x+y}\cdot3xy(x+y)+\dfrac{x}{y}\cdot3xy(x+y)}{\dfrac{x}{3(x+y)}\cdot3xy(x+y)+\dfrac{y}{x}\cdot3xy(x+y)}$

$=\dfrac{3x^2y+3x^2(x+y)}{x^2y+3y^2(x+y)}$

$=\dfrac{3x^2y+3x^3+3x^2y}{x^2y+3xy^2+3y^3}=\dfrac{3x^3+6x^2y}{x^2y+3xy^2+3y^3}$

$=\dfrac{3x^2(x+2y)}{y(x^2+3xy+3y^2)}$

34. $\dfrac{5x^2y+5xy^2+5y^3}{6x^2y+5x^3}$

35. $\dfrac{\dfrac{1}{x^2-1}+\dfrac{1}{x^2+4x+3}}{\dfrac{1}{x^2-1}+\dfrac{1}{x^2-3x+2}}=\dfrac{\dfrac{1}{(x+1)(x-1)}+\dfrac{1}{(x+3)(x+1)}}{\dfrac{1}{(x+1)(x-1)}+\dfrac{1}{(x-2)(x-1)}}$

$=\dfrac{\dfrac{1}{(x+1)(x-1)}+\dfrac{1}{(x+3)(x+1)}}{\dfrac{1}{(x+1)(x-1)}+\dfrac{1}{(x-2)(x-1)}}\cdot\dfrac{(x+1)(x-1)(x+3)(x-2)}{(x+1)(x-1)(x+3)(x-2)}$

$\qquad\qquad\qquad\qquad\qquad\qquad\qquad\text{Using the LCM}$

$=\dfrac{(x+3)(x-2)+(x-1)(x-2)}{(x+3)(x-2)+(x+1)(x+3)}$

$=\dfrac{x^2+x-6+x^2-3x+2}{x^2+x-6+x^2+4x+3}$

$=\dfrac{2x^2-2x-4}{2x^2+5x-3}$

36. $\dfrac{2x^2-5x+2}{2x^2+2x}$

37. $\dfrac{\dfrac{y}{y^2-4}-\dfrac{2y}{y^2+y-6}}{\dfrac{2y}{y^2-4}-\dfrac{y}{y^2+5y+6}}=\dfrac{\dfrac{y}{(y+2)(y-2)}-\dfrac{2y}{(y+3)(y-2)}}{\dfrac{2y}{(y+2)(y-2)}-\dfrac{y}{(y+3)(y+2)}}$

$=\dfrac{\dfrac{y}{(y+2)(y-2)}-\dfrac{2y}{(y+3)(y-2)}}{\dfrac{2y}{(y+2)(y-2)}-\dfrac{y}{(y+3)(y+2)}}\cdot\dfrac{(y+2)(y-2)(y+3)}{(y+2)(y-2)(y+3)}\quad\begin{array}{l}\text{Using}\\\text{the LCM}\end{array}$

$=\dfrac{y(y+3)-2y(y+2)}{2y(y+3)-y(y-2)}$

$=\dfrac{y^2+3y-2y^2-4y}{2y^2+6y-y^2+2y}=\dfrac{-y^2-y}{y^2+8y}$

$=\dfrac{y(-y-1)}{y(y+8)}=\dfrac{y}{y}\cdot\dfrac{-y-1}{y+8}=\dfrac{-y-1}{y+8}$

38. $\dfrac{-2y^2+13y-21}{2(y^2-y-20)}$

39. $\dfrac{\dfrac{1}{a^2+7a+12}+\dfrac{1}{a^2+a-6}}{\dfrac{1}{a^2+2a-8}+\dfrac{1}{a^2+5a+4}}=\dfrac{\dfrac{1}{(a+3)(a+4)}+\dfrac{1}{(a+3)(a-2)}}{\dfrac{1}{(a+4)(a-2)}+\dfrac{1}{(a+4)(a+1)}}$

$=\dfrac{\dfrac{1}{(a+3)(a+4)}+\dfrac{1}{(a+3)(a-2)}}{\dfrac{1}{(a+4)(a-2)}+\dfrac{1}{(a+4)(a+1)}}\cdot\dfrac{(a+3)(a+4)(a-2)(a+1)}{(a+3)(a+4)(a-2)(a+1)}$

$\qquad\qquad\qquad\qquad\qquad\qquad\qquad\text{Using the LCM}$

$=\dfrac{(a-2)(a+1)+(a+4)(a+1)}{(a+3)(a+1)+(a+3)(a-2)}$

$=\dfrac{a^2-a-2+a^2+5a+4}{a^2+4a+3+a^2+a-6}$

$=\dfrac{2a^2+4a+2}{2a^2+5a-3}$

40. $\dfrac{2a^2+7a-15}{2a^2-2a-12}$

41. $\dfrac{\dfrac{2}{x^2-7x+12} - \dfrac{1}{x^2+7x+10}}{\dfrac{2}{x^2-x-6} - \dfrac{1}{x^2+x-20}} = \dfrac{\dfrac{2}{(x-3)(x-4)} - \dfrac{1}{(x+5)(x+2)}}{\dfrac{2}{(x-3)(x+2)} - \dfrac{1}{(x+5)(x-4)}}$

$= \dfrac{\dfrac{2}{(x-3)(x-4)} - \dfrac{1}{(x+5)(x+2)}}{\dfrac{2}{(x-3)(x+2)} - \dfrac{1}{(x+5)(x-4)}} \cdot \dfrac{(x-3)(x-4)(x+5)(x+2)}{(x-3)(x-4)(x+5)(x+2)}$

Using the LCM

$= \dfrac{2(x+5)(x+2) - (x-3)(x-4)}{2(x-4)(x+5) - (x-3)(x+2)}$

$= \dfrac{2(x^2+7x+10) - (x^2-7x+12)}{2(x^2+x-20) - (x^2-x-6)}$

$= \dfrac{2x^2+14x+20 - x^2+7x-12}{2x^2+2x-40 - x^2+x+6}$

$= \dfrac{x^2+21x+8}{x^2+3x-34}$

42. $\dfrac{2x^2-11x-27}{2x^2+21x+13}$

43. $\dfrac{a}{x+y} = b$

$(x+y)\dfrac{a}{x+y} = b(x+y)$ Multiplying by $x+y$

$a = bx + by$ Simplifying

$a - by = bx$ Adding $-by$

$\dfrac{1}{b}(a-by) = \dfrac{1}{b} \cdot bx$ Multiplying by $\dfrac{1}{b}$

$\dfrac{1}{b} \cdot a - \dfrac{1}{b} \cdot by = x$ Using a distributive law

$\dfrac{a}{b} - y = x$ Simplifying

44. 11

45. Familiarize. We let ℓ and w represent the length and width of the second rectangle. Then $\ell - 3$ and $w - 4$ represent the length and width of the first rectangle. The perimeter of the second rectangle is $2\ell + 2w$; the perimeter of the first rectangle is $2(\ell - 3) + 2(w - 4)$, or $2\ell + 2w - 14$.

Translate.

Perimeter of second	=	2	·	Perimeter of first	−	1
$2\ell + 2w$	=	2	·	$(2\ell + 2w - 14)$	−	1

Carry out. We first solve for $2\ell + 2w$.

$2\ell + 2w = 2(2\ell + 2w - 14) - 1$

$2\ell + 2w = 4\ell + 4w - 28 - 1$

$2\ell + 2w = 4\ell + 4w - 29$

$29 = 2\ell + 2w$

If $2\ell + 2w = 29$, then $2\ell + 2w - 14 = 29 - 14$, or 15.

Check. The perimeter of the second rectangle is 1 less than twice the perimeter of the first rectangle:

$29 = 2 \cdot 15 - 1$

The values check.

State. The perimeter of the first rectangle is 15; the perimeter of the second rectangle is 29.

46. $17

47. $\dfrac{5x^{-1} - 5y^{-1} + 10x^{-1}y^{-1}}{6x^{-1} - 6y^{-1} + 12x^{-1}y^{-1}} = \dfrac{\dfrac{5}{x} - \dfrac{5}{y} + \dfrac{10}{xy}}{\dfrac{6}{x} - \dfrac{6}{y} + \dfrac{12}{xy}}$

$= \dfrac{\dfrac{5}{x} - \dfrac{5}{y} + \dfrac{10}{xy}}{\dfrac{6}{x} - \dfrac{6}{y} + \dfrac{12}{xy}} \cdot \dfrac{xy}{xy}$

$= \dfrac{5y - 5x + 10}{6y - 6x + 12} = \dfrac{5(y - x + 2)}{6(y - x + 2)} = \dfrac{5}{6}$

48. $\dfrac{-4a - 4}{8a - 5}$

49. $2 + \dfrac{2}{2 + \dfrac{2}{2 + \dfrac{2}{2 + \dfrac{2}{x}}}} = 2 + \dfrac{2}{2 + \dfrac{2}{2 + \dfrac{2}{\dfrac{2x+2}{x}}}}$

$= 2 + \dfrac{2}{2 + \dfrac{2}{2 + \dfrac{2x}{2x+2}}} = 2 + \dfrac{2}{2 + \dfrac{2}{\dfrac{6x+4}{2x+2}}}$

$= 2 + \dfrac{2}{2 + \dfrac{4x+4}{6x+4}} = 2 + \dfrac{2}{\dfrac{16x+12}{6x+4}}$

$= 2 + \dfrac{12x+8}{16x+12} = \dfrac{44x+32}{16x+12}$

$= \dfrac{4(11x+8)}{4(4x+3)} = \dfrac{11x+8}{4x+3}$

50. $\dfrac{x^4}{81}$

51. $\dfrac{(a^2b^{-1} + b^2a^{-1})(a^{-2} - b^{-2})}{(a^2 - ab + b^2)(a^{-2} + 2a^{-1}b^{-1} + b^{-2})}$

$= \dfrac{\left[\dfrac{a^2}{b} + \dfrac{b^2}{a}\right]\left[\dfrac{1}{a^2} - \dfrac{1}{b^2}\right]}{(a^2 - ab + b^2)\left[\dfrac{1}{a^2} + \dfrac{2}{ab} + \dfrac{1}{b^2}\right]}$

$= \dfrac{\left[\dfrac{a^3 + b^3}{ab}\right]\left[\dfrac{b^2 - a^2}{a^2b^2}\right]}{(a^2 - ab + b^2)\left[\dfrac{b^2 + 2ab + a^2}{a^2b^2}\right]}$

$= \dfrac{(a+b)(a^2-ab+b^2)(b+a)(b-a)}{a^3b^3} \cdot \dfrac{a^2b^2}{(a^2-ab+b^2)(b+a)^2}$

$= \dfrac{b - a}{ab}$

52. $\dfrac{x}{x^3 - 1}$

53. $\dfrac{a - 1}{1 - \dfrac{1}{a}} = \dfrac{a - 1}{\dfrac{a - 1}{a}} = \dfrac{a - 1}{1} \cdot \dfrac{a}{a - 1} = a$

54. $\dfrac{1}{(a^3 - b^3)(a^2 - ab + b^2)}$

55. First simplify the given expression.

$$1 + \cfrac{1}{1 + \cfrac{1}{1 + \cfrac{1}{1 + \frac{1}{x}}}} = 1 + \cfrac{1}{1 + \cfrac{1}{1 + \frac{x+1}{x}}}$$

$$= 1 + \cfrac{1}{1 + \cfrac{1}{1 + \frac{x}{x+1}}} = 1 + \cfrac{1}{1 + \cfrac{1}{\frac{2x+1}{x+1}}}$$

$$= 1 + \cfrac{1}{1 + \frac{x+1}{2x+1}} = 1 + \cfrac{1}{\frac{3x+2}{2x+1}}$$

$$= 1 + \frac{2x+1}{3x+2} = \frac{5x+3}{3x+2}$$

The reciprocal is $\frac{3x+2}{5x+3}$.

56. $\frac{x-1}{x}$; x

57. $f(x) = \frac{3}{x^2}$, $f(x+h) = \frac{3}{(x+h)^2}$

$$\frac{f(x+h) - f(x)}{h} = \frac{\frac{3}{(x+h)^2} - \frac{3}{x^2}}{h}$$

$$= \frac{3x^2 - 3(x+h)^2}{x^2(x+h)^2} \cdot \frac{1}{h}$$

$$= \frac{3x^2 - 3x^2 - 6xh - 3h^2}{x^2(x+h)^2} \cdot \frac{1}{h}$$

$$= \frac{-6xh - 3h^2}{x^2(x+h)^2 h}$$

$$= \frac{(-6x - 3h)h}{x^2(x+h)^2 h}$$

$$= \frac{-6x - 3h}{x^2(x+h)^2}$$

58. $\frac{-5}{x(x+h)}$

59. $f(x) = \frac{1}{1-x}$, $f(x+h) = \frac{1}{1-x-h}$

$$\frac{f(x+h) - f(x)}{h} = \frac{\frac{1}{1-x-h} - \frac{1}{1-x}}{h}$$

$$= \frac{1 - x - (1-x-h)}{(1-x-h)(1-x)} \cdot \frac{1}{h}$$

$$= \frac{1 - x - 1 + x + h}{(1-x-h)(1-x)} \cdot \frac{1}{h}$$

$$= \frac{h}{(1-x-h)(1-x)h}$$

$$= \frac{1}{(1-x-h)(1-x)}$$

60. $\frac{1}{(1+x+h)(1+x)}$

61. To avoid division by zero in $\frac{1}{x}$ and $\frac{8}{x^2}$ we must exclude 0 from the domain of $F(x \neq 0)$. To avoid division by zero in the complex fraction we set

$$2 - \frac{8}{x^2} = 0$$

$$2x^2 - 8 = 0$$

$$2(x^2 - 4) = 0$$

$$2(x+2)(x-2) = 0$$

$$x + 2 = 0 \quad \text{or} \quad x - 2 = 0$$

$$x = -2 \quad \text{or} \quad x = 2.$$

The domain of $F = \{x | x$ is a real number and $x \neq 0$ and $x \neq -2$ and $x \neq 2\}$.

62. $\{x | x$ is a real number and $x \neq 1$ and $x \neq -1$ and $x \neq 4$ and $x \neq -4$ and $x \neq 5$ and $x \neq -5\}$.

Exercise Set 6.4

1. $\frac{2}{5} + \frac{7}{8} = \frac{y}{20}$, LCM is 40   Check:

$$40\left(\frac{2}{5} + \frac{7}{8}\right) = 40 \cdot \frac{y}{20}$$

$$40 \cdot \frac{2}{5} + 40 \cdot \frac{7}{8} = 40 \cdot \frac{y}{20}$$

$$16 + 35 = 2y$$

$$51 = 2y$$

$$\frac{51}{2} = y$$

Check:
$$\frac{2}{5} + \frac{7}{8} = \frac{y}{20}$$

$$\frac{16}{40} + \frac{35}{40} \mid \frac{\frac{51}{2}}{20}$$

$$\frac{51}{40} \mid \frac{51}{2} \cdot \frac{1}{20}$$

$$\mid \frac{51}{40}$$

The solution is $\frac{51}{2}$.

2. $\frac{51}{5}$

3. $\frac{x}{3} - \frac{x}{4} = 12$, LCM is 12   Check:

$$12\left(\frac{x}{3} - \frac{x}{4}\right) = 12 \cdot 12$$

$$12 \cdot \frac{x}{3} - 12 \cdot \frac{x}{4} = 12 \cdot 12$$

$$4x - 3x = 144$$

$$x = 144$$

Check:
$$\frac{x}{3} - \frac{x}{4} = 12$$

$$\frac{144}{3} - \frac{144}{4} \mid 12$$

$$48 - 36 \mid$$

$$12 \mid$$

The solution is 144.

4. $-\frac{225}{2}$

5. $\frac{1}{3} - \frac{5}{6} = \frac{1}{x}$, LCM is 6x   Check:

$$6x\left(\frac{1}{3} - \frac{5}{6}\right) = 6x \cdot \frac{1}{x}$$

$$6x \cdot \frac{1}{3} - 6x \cdot \frac{5}{6} = 6x \cdot \frac{1}{x}$$

$$2x - 5x = 6$$

$$-3x = 6$$

$$x = -2$$

Check:
$$\frac{1}{3} - \frac{5}{6} = \frac{1}{x}$$

$$\frac{2}{6} - \frac{5}{6} \mid \frac{1}{-2}$$

$$-\frac{3}{6} \mid -\frac{1}{2}$$

$$-\frac{1}{2} \mid$$

The solution is -2.

6. $\frac{40}{9}$

7.
$$\frac{2}{3} - \frac{1}{5} = \frac{7}{3x}, \text{ LCM is } 3 \cdot 5 \cdot x$$

$$3 \cdot 5 \cdot x \left(\frac{2}{3} - \frac{1}{5} \right) = 3 \cdot 5 \cdot x \cdot \frac{7}{3x}$$

$$3 \cdot 5 \cdot x \cdot \frac{2}{3} - 3 \cdot 5 \cdot x \cdot \frac{1}{5} = 3 \cdot 5 \cdot x \cdot \frac{7}{3x}$$

$$10x - 3x = 35$$

$$7x = 35$$

$$x = 5$$

Check:

$$\frac{\frac{2}{3} - \frac{1}{5} = \frac{7}{3x}}{\begin{array}{c|c} \frac{2}{3} - \frac{1}{5} & \frac{7}{3 \cdot 5} \\ \frac{10}{15} - \frac{3}{15} & \frac{7}{15} \\ \frac{7}{15} & \end{array}}$$

The solution is 5.

8. 7

9.
$$\frac{2}{6} + \frac{1}{2x} = \frac{1}{3}, \text{ LCM is } 2 \cdot 3 \cdot x$$

$$2 \cdot 3 \cdot x \left(\frac{2}{6} + \frac{1}{2x} \right) = 2 \cdot 3 \cdot x \cdot \frac{1}{3}$$

$$2 \cdot 3 \cdot x \cdot \frac{2}{6} + 2 \cdot 3 \cdot x \cdot \frac{1}{2x} = 2 \cdot 3 \cdot x \cdot \frac{1}{3}$$

$$2x + 3 = 2x$$

$$3 = 0$$

We get a false equation. The given equation has no solution.

10. No solution

11.
$$\frac{4}{z} + \frac{2}{z} = 3, \text{ LCM is } z \qquad \text{Check:}$$

$$z \left(\frac{4}{z} + \frac{2}{z} \right) = z \cdot 3$$

$$z \cdot \frac{4}{z} + z \cdot \frac{2}{z} = 3z \qquad \frac{\frac{4}{z} + \frac{2}{z} = 3}{\begin{array}{c|c} \frac{4}{2} + \frac{2}{2} & 3 \\ 2 + 1 & \\ 3 & \end{array}}$$

$$4 + 2 = 3z$$

$$6 = 3z$$

$$2 = z$$

The solution is 2.

12. $-\frac{1}{2}$

13.
$$y + \frac{5}{y} = -6, \text{ LCM is } y$$

$$y \left(y + \frac{5}{y} \right) = y(-6)$$

$$y \cdot y + y \cdot \frac{5}{y} = -6y$$

$$y^2 + 5 = -6y$$

$$y^2 + 6y + 5 = 0$$

$$(y + 5)(y + 1) = 0$$

$$y + 5 = 0 \quad \text{or} \quad y + 1 = 0$$

$$y = -5 \quad \text{or} \qquad y = -1$$

Check:

For -5:

$$\frac{y + \frac{5}{y} = -6}{\begin{array}{c|c} -5 + \frac{5}{-5} & -6 \\ -5 - 1 & \\ -6 & \end{array}}$$

For -1:

$$\frac{y + \frac{5}{y} = -6}{\begin{array}{c|c} -1 + \frac{5}{-1} & -6 \\ -1 - 5 & \\ -6 & \end{array}}$$

The solutions are -5 and -1.

14. $-4, -1$

15.
$$2x - \frac{6}{x} = 1, \text{ LCM is } x$$

$$x \left(2x - \frac{6}{x} \right) = x \cdot 1$$

$$x \cdot 2x - x \cdot \frac{6}{x} = x$$

$$2x^2 - 6 = x$$

$$2x^2 - x - 6 = 0$$

$$(x - 2)(2x + 3) = 0$$

$$x - 2 = 0 \quad \text{or} \quad 2x + 3 = 0$$

$$x = 2 \quad \text{or} \qquad 2x = -3$$

$$x = 2 \quad \text{or} \qquad x = -\frac{3}{2}$$

Check:

For 2:

$$\frac{2x - \frac{6}{x} = 1}{\begin{array}{c|c} 2 \cdot 2 - \frac{6}{2} & 1 \\ 4 - 3 & \\ 1 & \end{array}}$$

For $-\frac{3}{2}$:

$$\frac{2x - \frac{6}{x} = 1}{\begin{array}{c|c} 2 \left(-\frac{3}{2} \right) - \frac{6}{-\frac{3}{2}} & 1 \\ -3 + 4 & \\ 1 & \end{array}}$$

The solutions are 2 and $-\frac{3}{2}$.

16. $3, -\frac{5}{2}$

17.
$$\frac{y - 1}{y - 3} = \frac{2}{y - 3}, \text{ LCM is } y - 3$$

$$(y - 3) \cdot \frac{y - 1}{y - 3} = (y - 3) \cdot \frac{2}{y - 3} \qquad \text{Check:}$$

$$y - 1 = 2 \qquad \frac{\frac{y - 1}{y - 3} = \frac{2}{y - 3}}{\begin{array}{c|c} \frac{3 - 1}{3 - 3} & \frac{2}{3 - 3} \\ \frac{2}{0} & \frac{2}{0} \end{array}}$$

$$y = 3$$

We know that 3 is not a solution of the original equation because it results in division by 0. The equation has no solution.

18. No solution

19. $\frac{x+1}{x} = \frac{3}{2}$, LCM is $2x$ Check:

$2x \cdot \frac{x+1}{x} = 2x \cdot \frac{3}{2}$

$2(x+1) = x \cdot 3$

$2x + 2 = 3x$

$2 = x$

The solution is 2.

$$\frac{\frac{x+1}{x} = \frac{3}{2}}{\frac{2+1}{2} \bigg| \frac{3}{2}}$$

$$\frac{3}{2}$$

20. 3

21. $\frac{x-3}{x+2} = \frac{1}{5}$, LCM is $5(x+2)$ Check:

$5(x+2) \cdot \frac{x-3}{x+2} = 5(x+2) \cdot \frac{1}{5}$

$5(x-3) = x+2$

$5x - 15 = x + 2$

$4x = 17$

$x = \frac{17}{4}$

The solution is $\frac{17}{4}$.

$$\frac{\frac{x-3}{x+2} = \frac{1}{5}}{\frac{17}{4} - \frac{12}{4} \bigg| \frac{1}{5}}$$

$$\frac{17}{4} + \frac{8}{4}$$

$$\frac{5}{4} \cdot \frac{4}{25}$$

$$\frac{1}{5}$$

22. 14

23. $\frac{3}{y+1} = \frac{2}{y-3}$, LCM is $(y+1)(y-3)$

$(y+1)(y-3) \cdot \frac{3}{y+1} = (y+1)(y-3) \cdot \frac{2}{y-3}$

$3(y-3) = 2(y+1)$

$3y - 9 = 2y + 2$

$y = 11$

Check:

$$\frac{\frac{3}{y+1} = \frac{2}{y-3}}{\frac{3}{11+1} \bigg| \frac{2}{11-3}}$$

$$\frac{3}{12} \bigg| \frac{2}{8}$$

$$\frac{1}{4} \bigg| \frac{1}{4}$$

The solution is 11.

24. -11

25. $\frac{7}{5x-2} = \frac{5}{4x}$, LCM is $4x(5x-2)$

$4x(5x-2) \cdot \frac{7}{5x-2} = 4x(5x-2) \cdot \frac{5}{4x}$

$4x \cdot 7 = 5(5x-2)$

$28x = 25x - 10$

$3x = -10$

$x = -\frac{10}{3}$

Since $-\frac{10}{3}$ checks, it is the solution.

26. 11

27. $\frac{2}{x} - \frac{3}{x} + \frac{4}{x} = 5$, LCM is x Check:

$x\left(\frac{2}{x} - \frac{3}{x} + \frac{4}{x}\right) = x \cdot 5$

$2 - 3 + 4 = 5x$

$3 = 5x$

$\frac{3}{5} = x$

The solution is $\frac{3}{5}$.

$$\frac{\frac{2}{x} - \frac{3}{x} + \frac{4}{x} = 5}{\frac{2}{\frac{3}{5}} - \frac{3}{\frac{3}{5}} + \frac{4}{\frac{3}{5}} \bigg| \frac{15}{3}}$$

$$\frac{10}{3} - \frac{15}{3} + \frac{20}{3}$$

$$\frac{15}{3}$$

28. $\frac{3}{4}$

29. $\frac{1}{2} - \frac{4}{9x} = \frac{4}{9} - \frac{1}{6x}$, LCM is $18x$

$18x\left(\frac{1}{2} - \frac{4}{9x}\right) = 18x\left(\frac{4}{9} - \frac{1}{6x}\right)$

$9x - 8 = 8x - 3$

$x = 5$

Since 5 checks, it is the solution.

30. -1

31. $\frac{z}{z-1} = \frac{6}{z+1}$, LCM is $(z-1)(z+1)$

$(z-1)(z+1) \cdot \frac{z}{z-1} = (z-1)(z+1) \cdot \frac{6}{z+1}$

$(z+1) \cdot z = (z-1) \cdot 6$

$z^2 + z = 6z - 6$

$z^2 - 5z + 6 = 0$

$(z-3)(z-2) = 0$

$z - 3 = 0$ or $z - 2 = 0$

$z = 3$ or $z = 2$

Both 3 and 2 check. The solutions are 3 and 2.

32. 2, 3

33. $\frac{60}{x} - \frac{60}{x-5} = \frac{2}{x}$, LCM is $x(x-5)$

$x(x-5)\left(\frac{60}{x} - \frac{60}{x-5}\right) = x(x-5) \cdot \frac{2}{x}$

$60(x-5) - 60x = 2(x-5)$

$60x - 300 - 60x = 2x - 10$

$-300 = 2x - 10$

$-290 = 2x$

$-145 = x$

Since -145 checks, it is the solution.

34. -23

35.
$$\frac{x}{x-2} + \frac{x}{x^2-4} = \frac{x+3}{x+2}$$

$$\frac{x}{x-2} + \frac{x}{(x+2)(x-2)} = \frac{x+3}{x+2}, \text{ LCM is } (x+2)(x-2)$$

$$(x+2)(x-2)\left[\frac{x}{x-2} + \frac{x}{(x+2)(x-2)}\right] = (x+2)(x-2) \cdot \frac{x+3}{x+2}$$

$$x(x+2) + x = (x-2)(x+3)$$

$$x^2 + 2x + x = x^2 + x - 6$$

$$3x = x - 6$$

$$2x = -6$$

$$x = -3$$

Since −3 checks, it is the solution.

36. 4

37.
$$\frac{a}{2a-6} - \frac{3}{a^2 - 6a + 9} = \frac{a-2}{3a-9}$$

$$\frac{a}{2(a-3)} - \frac{3}{(a-3)(a-3)} = \frac{a-2}{3(a-3)}$$

$$\text{LCM is } 2 \cdot 3(a-3)(a-3)$$

$$6(a-3)(a-3)\left[\frac{a}{2(a-3)} - \frac{3}{(a-3)(a-3)}\right] = 6(a-3)(a-3) \cdot \frac{a-2}{3(a-3)}$$

$$3a(a-3) - 6 \cdot 3 = 2(a-3)(a-2)$$

$$3a^2 - 9a - 18 = 2(a^2 - 5a + 6)$$

$$3a^2 - 9a - 18 = 2a^2 - 10a + 12$$

$$a^2 + a - 30 = 0$$

$$(a+6)(a-5) = 0$$

$$a + 6 = 0 \quad \text{or} \quad a - 5 = 0$$

$$a = -6 \quad \text{or} \quad a = 5$$

Both −6 and 5 check. The solutions are −6 and 5.

38. 3

39.
$$\frac{2x+3}{x-1} = \frac{10}{x^2-1} + \frac{2x-3}{x+1}$$

$$\frac{2x+3}{x-1} = \frac{10}{(x-1)(x+1)} + \frac{2x-3}{x+1}$$

$$\text{LCM is } (x-1)(x+1)$$

$$(x-1)(x+1) \cdot \frac{2x+3}{x-1} = (x-1)(x+1)\left[\frac{10}{(x-1)(x+1)} + \frac{2x-3}{x+1}\right]$$

$$(x+1)(2x+3) = 10 + (x-1)(2x-3)$$

$$2x^2 + 5x + 3 = 10 + 2x^2 - 5x + 3$$

$$5x + 3 = 13 - 5x$$

$$10x = 10$$

$$x = 1$$

We know that 1 is not a solution of the original equation because it results in division by 0. The equation has no solution.

40. No solution

41. $81x^4 - y^4 = (9x^2 + y^2)(9x^2 - y^2)$
$$= (9x^2 + y^2)(3x + y)(3x - y)$$

42. a) inconsistent

b) consistent

43. <u>Familiarize.</u> Let x, y, and z represent the number of multiple-choice, true-false and fill-in questions, respectively.

<u>Translate.</u> The total number of questions is 70.

x + y + z = 70

Number of true-false is twice number of fill-ins.

y = 2z

Number of multiple-choice is 5 less than number of true-false.

x = y − 5

<u>Carry out.</u> Solving the system of three equations we get (25,30,15).

<u>Check.</u> The sum of 25, 30, and 15 is 70. The number of true-false, 30, is twice the number of fill-ins, 15. The number of multiple-choice, 25, is 5 less than the number of true-false, 30.

<u>State.</u> On the test there are 25 multiple-choice, 30 true-false, and 15 fill-in questions.

44. 16 and 18 or −18 and −16

45.
$$\left[\frac{1}{1+x} + \frac{x}{1-x}\right] \div \left[\frac{x}{1+x} - \frac{1}{1-x}\right] = -1$$

$$\frac{1 \cdot (1-x) + x(1+x)}{(1+x)(1-x)} \div \frac{x(1-x) - 1 \cdot (1+x)}{(1+x)(1-x)} = -1$$

$$\frac{x^2 + 1}{(1+x)(1-x)} \cdot \frac{(1+x)(1-x)}{-x^2 - 1} = -1$$

$$\frac{x^2 + 1}{-x^2 - 1} = -1$$

$$-\frac{x^2 + 1}{x^2 + 1} = -1$$

$$-1 = -1$$

We know that 1 and −1 cannot be solutions of the original equation, because they result in division by 0. Since −1 = −1 is true for all values of x, all real numbers except 1 and −1 are solutions.

46. $-\frac{7}{2}$

47.
$$\frac{2.315}{y} - \frac{12.6}{17.4} = \frac{6.71}{7} + 0.763$$

$$\text{LCM is } 7(17.4)y$$

$$7(17.4)y\left[\frac{2.315}{y} - \frac{12.6}{17.4}\right] = 7(17.4)y\left[\frac{6.71}{7} + 0.763\right]$$

$$7(17.4)(2.315) - 7(12.6)y = (17.4)(6.71)y + 7(17.4)(0.763)y$$

$$281.967 - 88.2y = 116.754y + 92.9334y$$

$$281.967 = 88.2y + 116.754y + 92.9334y$$

$$281.967 = 297.8874y$$

$$0.9465556 \approx y$$

Since 0.9465556 checks, it is the solution.

48. 0.0854697

49. $\frac{x^3 + 8}{x + 2} = x^2 - 2x + 4$, LCM is $x + 2$

$x^3 + 8 = (x + 2)(x^2 - 2x + 4)$

$x^3 + 8 = x^3 + 8$

$8 = 8$

We know that -2 is not a solution of the original equation, because it results in division by 0. Since 8 = 8 is true for all values of x, all real numbers except -2 are solutions.

50. All real numbers except 3

51. $\frac{x^2 + 6x - 16}{x - 2} = x + 8$

$\frac{(x + 8)(x - 2)}{x - 2} = x + 8$

$x + 8 = x + 8$

$8 = 8$

Since 8 = 8 is true for all values of x, the original equation is true for all meaningful replacements of the variable. It is an identity.

52. Identity

53. $\frac{x - \frac{3}{2}}{x + \frac{2}{3}} = \frac{x + \frac{1}{2}}{x - \frac{2}{3}}$

$\frac{\frac{2x - 3}{2}}{\frac{3x + 2}{3}} = \frac{\frac{2x + 1}{2}}{\frac{3x - 2}{3}}$

$\frac{2x - 3}{2} \cdot \frac{3}{3x + 2} = \frac{2x + 1}{2} \cdot \frac{3}{3x - 2}$,

LCM is $2(3x + 2)(3x - 2)$

$2(3x + 2)(3x - 2) \cdot \frac{2x - 3}{2} \cdot \frac{3}{3x + 2} =$

$2(3x + 2)(3x - 2) \cdot \frac{2x + 1}{2} \cdot \frac{3}{3x - 2}$

$3(3x - 2)(2x - 3) = 3(3x + 2)(2x + 1)$

$(3x - 2)(2x - 3) = (3x + 2)(2x + 1)$

$6x^2 - 13x + 6 = 6x^2 + 7x + 2$

$-13x + 6 = 7x + 2$

$-20x = -4$

$x = \frac{1}{5}$

Since $\frac{1}{5}$ checks, it is the solution.

54. -8

55. Graph $f(x) = \frac{1}{x - 1} + 2$ and visually check for an x-value that is paired with the value 1 for $f(x)$.

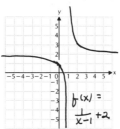

It appears from the graph that $f(x) = 1$ when $x = 0$.

Exercise Set 6.5

1. We let x represent the number and translate directly to an equation.

   Translate.

   The reciprocal of 5 plus the reciprocal of 7 is the reciprocal of the number.

   $\frac{1}{5}$ + $\frac{1}{7}$ = $\frac{1}{x}$

   Carry out. We solve the equation.

   $\frac{1}{5} + \frac{1}{7} = \frac{1}{x}$, LCM = 35x

   $35x\left(\frac{1}{5} + \frac{1}{7}\right) = 35x \cdot \frac{1}{x}$

   $7x + 5x = 35$

   $12x = 35$

   $x = \frac{35}{12}$

   Check. The reciprocal of $\frac{35}{12}$ is $\frac{12}{35}$. The sum of $\frac{1}{5} + \frac{1}{7}$ is $\frac{7}{35} + \frac{5}{35}$, or $\frac{12}{35}$, so the value checks.

   State. The number is $\frac{35}{12}$.

2. 2

3. We let x represent the number and translate directly to an equation.

   Translate.

   A number plus 6 times its reciprocal is -5.

   x +   6   · $\frac{1}{x}$ =   -5

   Carry out. We solve the equation.

   $x + \frac{6}{x} = -5$, LCM is x

   $x\left(x + \frac{6}{x}\right) = x(-5)$

   $x^2 + 6 = -5x$

   $x^2 + 5x + 6 = 0$

   $(x + 3)(x + 2) = 0$

   $x + 3 = 0$ or $x + 2 = 0$

   $x = -3$ or $x = -2$

3. (continued)

Check. The possible solutions are -3 and -2. We check -3 in the conditions of the problem.

Number: -3

6 times the reciprocal of the number: $6\left[-\frac{1}{3}\right] = -2$

Sum of the number and 6 times its reciprocal: $-3 + (-2) = -5$

The number -3 checks. So does -2, but that check is left to the student.

State. The numbers -3 and -2 both satisfy the conditions of the problem.

4. -3, -7

5. Familiarize. We let x represent the first integer. Then x + 1 represents the second, and their product is x(x + 1).

Translate.

Reciprocal of the product   is $\frac{1}{72}$.

$$\frac{1}{x(x + 1)} = \frac{1}{72}$$

Carry out. We solve the equation.

$$72x(x + 1) \cdot \frac{1}{x(x + 1)} = 72x(x + 1) \cdot \frac{1}{72}$$
$$72 = x(x + 1)$$
$$72 = x^2 + x$$
$$0 = x^2 + x - 72$$
$$0 = (x + 9)(x - 8)$$

$$x + 9 = 0 \quad \text{or} \quad x - 8 = 0$$
$$x = -9 \quad \text{or} \quad x = 8$$

Check. When x = -9, x + 1 = -8, so -9 and -8 is a possible solution. When x = 8, x + 1 = 9, so 8 and 9 is also a possible solution. We check -9 and -8. Their product is -9(-8), or 72, and the reciprocal of the product is $\frac{1}{72}$. These numbers check. Now we check 8 and 9. Their product is 8·9, or 72, and the reciprocal of the product is $\frac{1}{72}$. Both possible solutions check.

State. The numbers are -9 and -8 or 8 and 9.

6. 6 and 7 or -7 and -6

7. Familiarize. The job takes Sam 5 hours working alone and Willy 9 hours working alone. Then in 1 hour, Sam does $\frac{1}{5}$ of the job and Willy does $\frac{1}{9}$ of the job. Working together, they can do $\frac{1}{5} + \frac{1}{9}$ of the job in 1 hour. Let t represent the time required for Sam and Willy, working together, to do the job.

Translate. We want to find t such that

$$t\left[\frac{1}{5}\right] + t\left[\frac{1}{9}\right] = 1, \text{ or } \frac{t}{5} + \frac{t}{9} = 1,$$

where 1 represents one entire job.

Carry out. We solve the equation.

$$45\left[\frac{t}{5} + \frac{t}{9}\right] = 45 \cdot 1$$
$$9t + 5t = 45$$
$$14t = 45$$
$$t = \frac{45}{14}$$

Check. The possible solution is $\frac{45}{14}$ hours. If Sam works $\frac{45}{14}$ hours, he will do $\frac{1}{5} \cdot \frac{45}{14}$, or $\frac{9}{14}$ of the job. If Willy works $\frac{45}{14}$ hours, he will do $\frac{1}{9} \cdot \frac{45}{14}$, or $\frac{5}{14}$ of the job. Together, they will do $\frac{9}{14} + \frac{5}{14}$ of the job, or one complete job.

State. Working together it will take them $\frac{45}{14}$, or $3\frac{3}{14}$ hours.

8. $1\frac{5}{7}$ hr

9. Familiarize. The pool can be filled in 12 hours with only the pipe and in 30 hours with only the hose. Then in 1 hour, the pipe fills $\frac{1}{12}$ of the pool, and the hose fills $\frac{1}{30}$ of the pool. Using both the pipe and the hose, $\frac{1}{12} + \frac{1}{30}$ of the pool can be filled in 1 hour. Suppose that it takes t hours to fill the pool using both the pipe and hose.

Translate. We want to find t such that

$$t\left[\frac{1}{12}\right] + t\left[\frac{1}{30}\right] = 1, \text{ or } \frac{t}{12} + \frac{t}{30} = 1,$$

where 1 represents one entire job.

Carry out. We solve the equation.

$$60\left[\frac{t}{12} + \frac{t}{30}\right] = 60 \cdot 1$$
$$5t + 2t = 60$$
$$7t = 60$$
$$t = \frac{60}{7}$$

9. (continued)

Check. The possible solution is $\frac{60}{7}$ hours. If the pipe is used $\frac{60}{7}$ hours, it fills $\frac{1}{12} \cdot \frac{60}{7}$, or $\frac{5}{7}$ of the pool. If the hose is used $\frac{60}{7}$ hours, it fills $\frac{1}{30} \cdot \frac{60}{7}$, or $\frac{2}{7}$ of the pool. Using both, $\frac{5}{7} + \frac{2}{7}$ of the pool, or all of it, will be filled in $\frac{60}{7}$ hours.

State. Using both the pipe and the hose, it will take $\frac{60}{7}$, or $8\frac{4}{7}$ hours, to fill the pool.

10. 9.9 hr

11. Familiarize. The job takes Bill 5.5 hours working alone and his partner 7.5 hours working alone. Then in 1 hour, Bill does $\frac{1}{5.5}$ of the job and his partner does $\frac{1}{7.5}$ of the job. Working together, they can do $\frac{1}{5.5} + \frac{1}{7.5}$ of the job in 1 hour. Suppose it takes them t hours working together.

Translate. We want to find t such that
$$t\left(\frac{1}{5.5}\right) + t\left(\frac{1}{7.5}\right) = 1, \text{ or } \frac{t}{5.5} + \frac{t}{7.5} = 1.$$

Carry out. We solve the equation.
$$5.5(7.5)\left[\frac{t}{5.5} + \frac{t}{7.5}\right] = 5.5(7.5)(1)$$
$$7.5t + 5.5t = 41.25$$
$$13t = 41.25$$
$$t = \frac{41.25}{13}, \text{ or } \frac{41.25}{13} \cdot \frac{4}{4}$$
$$t = \frac{165}{52}, \text{ or } 3\frac{9}{52}$$

Check. The possible solution is $3\frac{9}{52}$ hours. If Bill works $3\frac{9}{52}$ hours, he will do $\frac{1}{5.5}\left(\frac{165}{52}\right) = \frac{2}{11} \cdot \frac{165}{52}$, or $\frac{15}{26}$ of the job. If his partner works $3\frac{9}{52}$ hours, he will do $\frac{1}{7.5}\left(\frac{165}{52}\right) = \frac{2}{15} \cdot \frac{165}{52}$, or $\frac{11}{26}$ of the job. Together they will do $\frac{15}{26} + \frac{11}{26}$ of the job, or all of it.

State. Working together, it will take them $3\frac{9}{52}$ hours.

12. 2.475 hr

13. Familiarize. Let t represent the time it takes A to paint the house alone. B takes 4 times as long as A, or 4t. Thus in 1 day A does $\frac{1}{t}$ of the job and B does $\frac{1}{4t}$ of the job.

Translate. In 8 days A and B will complete one entire job, so we have
$$8\left(\frac{1}{t}\right) + 8\left(\frac{1}{4t}\right) = 1, \text{ or } \frac{8}{t} + \frac{2}{t} = 1.$$

Carry out. We solve the equation.
$$t\left[\frac{8}{t} + \frac{2}{t}\right] = t \cdot 1$$
$$8 + 2 = t$$
$$10 = t$$

Check. In 1 day, A will do $\frac{1}{10}$ of the job and B will do $\frac{1}{40}$ of the job. Together they will do $\frac{1}{10} + \frac{1}{40} = \frac{4}{40} + \frac{1}{40} = \frac{5}{40} = \frac{1}{8}$ of the job. Thus, in 8 days they will do $8 \cdot \frac{1}{8} = 1$ job. The answer checks.

State. It would take A 10 days and B 40 days working alone.

14. A: $1\frac{1}{3}$ hr, B: 4 hr

15. Familiarize. Working alone, Rosita does $\frac{1}{2}$ of the job in 1 hr. Let t represent the time it takes Helga to wax the car, working alone. Then in 1 hr she does $\frac{1}{t}$ of the job. Represent 45 min as $\frac{3}{4}$ hr.

Translate. In $\frac{3}{4}$ hr they do 1 entire job, working together, so we have
$$\frac{3}{4}\left(\frac{1}{2}\right) + \frac{3}{4}\left(\frac{1}{t}\right) = 1, \text{ or } \frac{3}{8} + \frac{3}{4t} = 1.$$

Carry out. We solve the equation.
$$8t\left[\frac{3}{8} + \frac{3}{4t}\right] = 8t \cdot 1$$
$$3t + 6 = 8t$$
$$6 = 5t$$
$$\frac{6}{5} = t$$

Check. In $\frac{3}{4}$ hr, Rosita will do $\frac{3}{4} \cdot \frac{1}{2}$, or $\frac{3}{8}$, of the job, and Helga will do $\frac{3}{4}\left(\frac{1}{\frac{6}{5}}\right)$, or $\frac{3}{4} \cdot \frac{5}{6} = \frac{5}{8}$, of the job. Together they do $\frac{3}{8} + \frac{5}{8} = 1$ job. The answer checks.

State. It would take Helga $\frac{6}{5}$, or $1\frac{1}{5}$ hr, working alone.

16. 6 hr

17. <u>Familiarize</u>. Let t represent the time it takes Jake, working alone. Since it takes Jake half the time it takes Skyler, 2t represents Skyler's time, working alone. In 1 hr, Jake does $\frac{1}{t}$ and Skyler does $\frac{1}{2t}$ of the job.

<u>Translate</u>. Working together, they can do the entire job in 4 hr, so we want to find t such that

$$4\left(\frac{1}{t}\right) + 4\left(\frac{1}{2t}\right) = 1, \text{ or } \frac{4}{t} + \frac{2}{t} = 1.$$

<u>Carry out</u>. We solve the equation.

$$t\left(\frac{4}{t} + \frac{2}{t}\right) = t \cdot 1$$
$$4 + 2 = t$$
$$6 = t$$

<u>Check</u>. If Jake does the job in 6 hr, then in 4 hr he does $4 \cdot \frac{1}{6}$, or $\frac{2}{3}$, of the job. It takes Skyler $2 \cdot 6$, or 12 hr, so in 4 hr he does $4 \cdot \frac{1}{12}$, or $\frac{1}{3}$, of the job. Working together, they do $\frac{2}{3} + \frac{1}{3} = 1$ job in 4 hr. The answer checks.

<u>State</u>. It takes Jake 6 hr and Skyler 12 hr working alone.

18. Sarah: 22.5 min, Lucia: 45 min

19. <u>Familiarize</u>. We first make a drawing. We let r represent the speed of the boat in still water. Then r − 3 is the speed upstream and r + 3 is the speed downstream.

Upstream   4 miles   r − 3 mph

10 miles   r + 3 mph   Downstream

We organize the information in a table. The time is the same both upstream and downstream so we use t for each time.

	Distance	Speed	Time
Upstream	4	r − 3	t
Downstream	10	r + 3	t

<u>Translate</u>. Using $t = \frac{d}{r}$ we get two different equations from the rows of the table.

$$t = \frac{4}{r-3} \text{ and } t = \frac{10}{r+3}$$

<u>Carry out</u>. Since both rational expressions represent the same time, t, we can set them equal to each other and solve.

$$\frac{4}{r-3} = \frac{10}{r+3}, \text{ LCM is } (r-3)(r+3)$$
$$(r-3)(r+3) \cdot \frac{4}{r-3} = (r-3)(r+3) \cdot \frac{10}{r+3}$$
$$4(r+3) = 10(r-3)$$
$$4r + 12 = 10r - 30$$
$$42 = 6r$$
$$7 = r$$

19. (continued)

<u>Check</u>. If r = 7 mph, then r − 3 is 4 mph and r + 3 is 10 mph. The time upstream is $\frac{4}{4}$, or 1 hour. The time downstream is $\frac{10}{10}$, or 1 hour. The times are the same. The values check.

<u>State</u>. The speed of the boat in still water is 7 mph.

20. 12 mph

21. <u>Familiarize</u>. We first make a drawing. We let r represent the person's rate on a nonmoving sidewalk. Then r + 7 represents the rate walking forward on the moving sidewalk, and r − 7 represents the rate walking in the opposite direction.

Forward   80 ft   r + 7 ft/sec

r − 7 ft/sec   15 ft   Opposite Direction

We organize the information in a table. Both distances are covered in the same amount of time. We let t represent the time.

	Distance	Rate	Time
Forward	80	r + 7	t
Opposite Direction	15	r − 7	t

<u>Translate</u>. Using $t = \frac{d}{r}$ we get two equations.

$$t = \frac{80}{r+7} \text{ and } t = \frac{15}{r-7}$$

<u>Carry out</u>. Set the rational expressions equal to each other and solve.

$$\frac{80}{r+7} = \frac{15}{r-7}, \text{ LCM is } (r+7)(r-7)$$
$$(r+7)(r-7) \cdot \frac{80}{r+7} = (r+7)(r-7) \cdot \frac{15}{r-7}$$
$$80(r-7) = 15(r+7)$$
$$80r - 560 = 15r + 105$$
$$65r = 665$$
$$r = \frac{665}{65} = \frac{133}{13}$$

<u>Check</u>. If the answer checks, the rate on a nonmoving sidewalk is $\frac{133}{13}$ ft/sec. We calculate:

Walking forward: Rate is $\frac{133}{13} + 7 = \frac{224}{13}$ ft/sec

In opposite direction: Rate is $\frac{133}{13} - 7 = \frac{42}{13}$ ft/sec

Now we calculate time, using $t = \frac{d}{r}$.

At $\frac{224}{13}$ ft/sec: $t = \frac{80}{\frac{224}{13}} = \frac{80}{1} \cdot \frac{13}{224} = \frac{65}{14}$ sec

21. (continued)

At $\frac{42}{13}$ ft/sec: $t = \frac{15}{\frac{42}{13}} = \frac{15}{1} \cdot \frac{13}{42} = \frac{65}{14}$ sec

The times are the same, so the answer checks.

State. The person would walk $\frac{133}{13}$, or $10\frac{3}{13}$ ft/sec, on a nonmoving sidewalk.

22. 10.8 ft/sec

23. Familiarize. We first make a drawing. Let r represent Rosanna's speed. Then r + 2 represents Simone's speed.

Rosanna   5 mi   r mph

Simone   8 mi   r + 2 mph

We organize the information in a table. Both distances are covered in the same amount of time. We let t represent the time.

	Distance	Rate	Time
Rosanna	5	r	t
Simone	8	r + 2	t

Translate. Using $t = \frac{d}{r}$ we get two equations.

$t = \frac{5}{r}$ and $t = \frac{8}{r+2}$

Carry out. Set the rational expressions equal to each other and solve.

$\frac{5}{r} = \frac{8}{r+2}$, LCM is r(r + 2)

$r(r + 2) \cdot \frac{5}{r} = r(r + 2) \cdot \frac{8}{r+2}$

$5(r + 2) = 8r$

$5r + 10 = 8r$

$10 = 3r$

$\frac{10}{3} = r$

Check. If the answer checks, Rosanna's speed is $\frac{10}{3}$ mph, and Simone's speed is $\frac{10}{3} + 2$, or $\frac{16}{3}$ mph. We calculate time, using $t = \frac{d}{r}$.

Rosanna: $t = \frac{5}{\frac{10}{3}} = \frac{5}{1} \cdot \frac{3}{10} = \frac{3}{2}$ hr

Simone: $t = \frac{8}{\frac{16}{3}} = \frac{8}{1} \cdot \frac{3}{16} = \frac{3}{2}$ hr

The times are the same, so the answer checks.

State. Rosanna's speed is $\frac{10}{3}$, or $3\frac{1}{3}$ mph, and Simone's speed is $\frac{16}{3}$, or $5\frac{1}{3}$ mph.

24. Local: 35 mph, express: 42 mph

25. Familiarize. We let r represent the speed of Train B. Then r - 12 represents the speed of Train A. The times are the same. We use t for each. We organize the information in a table.

	Distance	Rate	Time
Train A	230	r - 12	t
Train B	290	r	t

Translate. Using $t = \frac{d}{r}$, we get two equations.

$t = \frac{230}{r-12}$ and $t = \frac{290}{r}$

Carry out. Set the rational expressions equal to each other and solve.

$\frac{230}{r-12} = \frac{290}{r}$, LCM is r(r - 12)

$r(r - 12) \cdot \frac{230}{r-12} = r(r - 12) \cdot \frac{290}{r}$

$230r = 290(r - 12)$

$230r = 290r - 3480$

$-60r = -3480$

$r = 58$

Check. If the speed of Train B is 58 mph, then the speed of Train A is 58 - 12, or 46 mph. The time for Train A is $\frac{230}{46}$, or 5 hours. The time for Train B is $\frac{290}{58}$, or 5 hours. The times are the same. The values check.

State. The speed of Train A is 46 mph; the speed of Train B is 58 mph.

26. Passenger train: 80 mph, freight train: 66 mph

27. Familiarize. Let r represent the speed of motorboat B. Then r + 10 represents the speed of motorboat A. Both distances are covered in the same amount of time. We let t represent the time. We organize the information in a table.

	Distance	Rate	Time
A	75	r + 10	t
B	50	r	t

Translate. Using $t = \frac{d}{r}$, we get two equations.

$t = \frac{75}{r+10}$ and $t = \frac{50}{r}$

Carry out. We set the rational expressions equal to each other and solve.

$\frac{75}{r+10} = \frac{50}{r}$, LCM is r(r + 10)

$r(r + 10) \cdot \frac{75}{r+10} = r(r + 10) \cdot \frac{50}{r}$

$75r = 50(r + 10)$

$75r = 50r + 500$

$25r = 500$

$r = 20$

27. (continued)

   Check. If the speed of motorboat B is 20 km/h, then the speed of motorboat A is 20 + 10, or 30 km/h. The time for A is $\frac{75}{30}$, or $\frac{5}{2}$ hr. The time for B is $\frac{50}{20}$, or $\frac{5}{2}$ hr. The times are the same, so the answer checks.

   State. The speed of motorboat A is 30 km/h, and the speed of motorboat B is 20 km/h.

28. Jaime: 23 km/h, Mara: 15 km/h

29. Familiarize. We first make a drawing. We let r represent the speed of the river. Then 15 + r is her speed downstream and 15 − r is her speed upstream.

   140 km   15 + r   t hours   Downstream
   ←————————————————————————————•

   35 km   15 − r   t hours   Upstream
   •————————————————————————→

   The times are the same. Let t represent the time. We organize the information in a table.

	Distance	Speed	Time
Downstream	140	15 + r	t
Upstream	35	15 − r	t

   Translate. Using $t = \frac{d}{r}$, we get two equations from the table.

   $$t = \frac{140}{15 + r} \text{ and } t = \frac{35}{15 - r}$$

   Carry out. Set the rational expressions equal to each other and solve.

   $$\frac{140}{15 + r} = \frac{35}{15 - r}, \text{ LCM is } (15 + r)(15 - r)$$

   $$(15 + r)(15 - r) \cdot \frac{140}{15 + r} = (15 + r)(15 - r) \cdot \frac{35}{15 - r}$$

   $$140(15 - r) = 35(15 + r)$$
   $$2100 - 140r = 525 + 35r$$
   $$1575 = 175r$$
   $$9 = r$$

   Check. If r = 9, then the speed downstream is 15 + 9, or 24 km/h and the speed upstream is 15 − 9, or 6 km/h. The time for the trip downstream is $\frac{140}{24}$, or $5\frac{5}{6}$ hours. The time for the trip upstream is $\frac{35}{6}$, or $5\frac{5}{6}$ hours. The times are same. The values check.

   State. The speed of the river is 9 km/h.

30. $1\frac{1}{5}$ km/h

31. Familiarize. Let n represent the number.
   Translate.
       8% of what number is 480.
       0.08 ·   n   = 480

   Carry out. We solve the equation.
       0.08n = 480
           n = 6000

   Check. 8% of 6000 is 0.08(6000), or 480. The answer checks.

   State. 8% of 6000 is 480. The number is 6000.

32. {x|x is a real number and x ≠ 5 and x ≠ −1}

33. |x − 2| = 9
    x − 2 = 9 or x − 2 = −9
       x = 11 or   x = −7
    The solutions are 11 and −7.

34. $2y − 8xy^2 + 6xy$

35. Familiarize. It helps to first make a drawing.

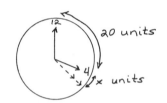

   The minute hand moves 60 units per hour while the hour hand moves 5 units per hour, where one unit represents one minute on the face of the clock. When the hands are in the same position the first time, the hour hand will have moved x units and the minute hand will have moved x + 20 units. The times are the same. We use t for time.

	Distance	Speed	Time
Minute	x + 20	60	t
Hour	x	5	t

   Translate. Using $t = \frac{d}{r}$, we get two equations from the table.

   $$t = \frac{x + 20}{60} \text{ and } t = \frac{x}{5}$$

35. (continued)

Carry out. Set the rational expressions equal to each other and solve.

$$\frac{x + 20}{60} = \frac{x}{5}, \text{ LCM is } 60$$

$$60 \cdot \frac{x + 20}{60} = 60 \cdot \frac{x}{5}$$

$$x + 20 = 12x$$

$$20 = 11x$$

$$\frac{20}{11} = x$$

$$\text{or } x = 1\frac{9}{11}$$

Check. If the hour hand moves $1\frac{9}{11}$ units, then the minute hand moves $1\frac{9}{11} + 20$, or $21\frac{9}{11}$ units $\left[21\frac{9}{11} \text{ minutes after } 4\right]$. The time for the hour hand is $\frac{20}{11} \div 5$, or $\frac{4}{11}$ hour. The time for the minute hand is $\frac{240}{11} \div 60$, or $\frac{4}{11}$ hour. The times are the same. The values check.

State. At $21\frac{9}{11}$ minutes after 4:00, the hands will be in the same position.

36. $8\frac{2}{11}$ min after 10:30

37. Familiarize. We first make a drawing. We let x represent the speed of the boat in still water and r represent the speed of the stream. Then x + r represents the speed downstream and x - r represents the speed upstream.

```
              96 km      4 hr     x + r km/h
           •─────────────────────────────────→
Downstream
           ←─────────────────────────────────•
              28 km      7 hr     x - r km/h
                                       Upstream
```

We organize the information in a table.

	Distance	Speed	Time
Downstream	96	x + r	4
Upstream	28	x - r	7

Translate. Using d = rt, we get a system of equations from the table.

$$96 = (x + r)4 \text{ or } x + r = 24$$
$$28 = (x - r)7 \text{ or } x - r = 4$$

Carry out. Solving the system we get (10,14).

Check. The speed downstream is 14 + 10, or 24 km/h. The distance downstream is 24·4, or 96 km. The speed upstream is 14 - 10, or 4 km/h. The distance upstream is 4·7, or 28 km. The values check.

State. The speed of the boat is 14 km/h; the speed of the stream is 10 km/h.

38. 700 mi away from the airport

39. Familiarize. We let x represent the speed of the current and 3x represent the speed of the boat. Then the speed up the river is 3x - x, or 2x, and the speed down the river is 3x + x, or 4x. The total distance is 100 km; thus the distance each way is 50 km. Using $t = \frac{d}{r}$, we can use $\frac{50}{2x}$ for the time up the river and $\frac{50}{4x}$ for the time down the river.

Translate. Since the total of the times is 10 hours, we have the following equation.

$$\frac{50}{2x} + \frac{50}{4x} = 10$$

Carry out. We solve the equation. The LCM is 4x.

$$4x\left(\frac{50}{2x} + \frac{50}{4x}\right) = 4x \cdot 10$$

$$100 + 50 = 40x$$

$$150 = 40x$$

$$\frac{15}{5} = x$$

$$\text{or } x = 3\frac{3}{4}$$

Check. If the speed of the current is $\frac{15}{4}$ km/h, then the speed of the boat is $3 \cdot \frac{15}{4}$, or $\frac{45}{4}$. The speed up the river is $\frac{45}{4} - \frac{15}{4}$, or $\frac{15}{2}$ km/h, and the time traveling up the river is $50 \div \frac{15}{2}$, or $6\frac{2}{3}$ hr. The speed down the river is $\frac{45}{4} + \frac{15}{4}$, or 15 km/h, and the time traveling down the river is $50 \div 15$, or $3\frac{1}{3}$ hr. The total time for the trip is $6\frac{2}{3} + 3\frac{1}{3}$, or 10 hr. The value checks.

State. The speed of the current is $3\frac{3}{4}$ km/h.

40. 30 mi

41. Familiarize. If the drain is closed, $\frac{1}{9}$ of the tank is filled in 1 hr. If the tank is not being filled, $\frac{1}{11}$ of the tank is drained in 1 hr. If the tank is being filled with the drain left open, $\frac{1}{9} - \frac{1}{11}$ of the tank is filled in 1 hr. Let t represent the time it takes to fill the tank with the drain left open.

Translate. We want to find t such that

$$t\left(\frac{1}{9} - \frac{1}{11}\right) = 1, \text{ or } \frac{t}{9} - \frac{t}{11} = 1.$$

Carry out. We solve the equation.

$$9 \cdot 11\left(\frac{t}{9} - \frac{t}{11}\right) = 9 \cdot 11 \cdot 1$$

$$11t - 9t = 99$$

$$2t = 99$$

$$t = \frac{99}{2}$$

41. (continued)

Check. In $\frac{99}{2}$ hr, the tank is $\frac{99}{2} \cdot \frac{1}{9}$, or $\frac{11}{2}$ full. In $\frac{99}{2}$ hr, $\frac{99}{2} \cdot \frac{1}{11}$, or $\frac{9}{2}$ of the tank, is drained. This leaves $\frac{11}{2} - \frac{9}{2} = \frac{2}{2} = 1$ full tank.

State. It will take $\frac{99}{2}$, or $49\frac{1}{2}$ hr, to fill the tank.

42. 40 min

43. Familiarize. We make a drawing and organize the information in a table. Remember: $t = \frac{d}{r}$.

```
|←————————————— 200 km —————————————→|
|←——— 100 km ———→|←——— 100 km ———→|
•————————————————|————————————————|
     40 km/h            60 km/h
```

	Distance	Speed	Time
1st part	100	40	$\frac{100}{40}$, or $\frac{5}{2}$
2nd part	100	60	$\frac{100}{60}$, or $\frac{5}{3}$

The total distance is 200 km.

The total time is $\frac{5}{2} + \frac{5}{3}$, or $\frac{25}{6}$ hr.

Translate. Average speed $= \dfrac{\text{Total distance}}{\text{Total time}}$

$$= \frac{200 \text{ km}}{\frac{25}{6} \text{ hr}}$$

Carry out. We simplify.

$$\text{Average speed} = \frac{200 \text{ km}}{1} \cdot \frac{6}{25 \text{ hr}} = 48 \text{ km/h}$$

Check. Calculate the total time at 48 km/h. $\frac{200}{48} = \frac{25}{6}$ hr. The answer checks.

State. The average speed was 48 km/h.

44. $51\frac{3}{7}$ mph

45. Familiarize. Trucks A, B, and C, working together, move a load of sand in t hours. Thus, together, they do $\frac{1}{t}$ of the job in 1 hour. Truck A, alone, can move a load of sand in $t + 1$ hours. Thus, Truck A does $\frac{1}{t+1}$ of the job in 1 hour. Truck B, alone, can move a load of sand in $t + 6$ hours. Thus, Truck B does $\frac{1}{t+6}$ of the job in 1 hour. Truck C, alone, can move a load of sand in $t + t$, or $2t$ hours. Thus, Truck C does $\frac{1}{2t}$ of the job in 1 hour. Working together, they can do $\frac{1}{t+1} + \frac{1}{t+6} + \frac{1}{2t}$ of the job in 1 hour.

45. (continued)

Translate. We want to find t such that

$$t\left[\frac{1}{t+1} + \frac{1}{t+6} + \frac{1}{2t}\right] = 1, \text{ or}$$

$$\frac{t}{t+1} + \frac{t}{t+6} + \frac{1}{2} = 1$$

Carry out. We solve the equation.

$$\frac{t}{t+1} + \frac{t}{t+6} = \frac{1}{2} \quad \text{Adding } -\frac{1}{2}$$

$$2(t+1)(t+6)\left[\frac{t}{t+1} + \frac{t}{t+6}\right] = 2(t+1)(t+6) \cdot \frac{1}{2}$$

$$2t(t+6) + 2t(t+1) = (t+1)(t+6)$$

$$2t^2 + 12t + 2t^2 + 2t = t^2 + 7t + 6$$

$$4t^2 + 14t = t^2 + 7t + 6$$

$$3t^2 + 7t - 6 = 0$$

$$(3t - 2)(t + 3) = 0$$

$$3t - 2 = 0 \quad \text{or} \quad t + 3 = 0$$

$$3t = 2 \quad \text{or} \quad t = -3$$

$$t = \frac{2}{3} \quad \text{or} \quad t = -3$$

Check. Since time cannot be negative, we need only check $\frac{2}{3}$. If $t = \frac{2}{3}$, then $\frac{1}{t} = \frac{3}{2}$. If $t = \frac{2}{3}$, then working alone it takes A $\frac{2}{3} + 1$, or $\frac{5}{3}$ hr, B $\frac{2}{3} + 6$, or $\frac{20}{3}$ hr, and C $2 \cdot \frac{2}{3}$, or $\frac{4}{3}$ hr. We calculate.

$$\frac{1}{\frac{5}{3}} + \frac{1}{\frac{20}{3}} + \frac{1}{\frac{4}{3}}$$

$$= \frac{3}{5} + \frac{3}{20} + \frac{3}{4} = \frac{12}{20} + \frac{3}{20} + \frac{15}{20} = \frac{30}{20} = \frac{3}{2}$$

State. Working together, it takes $\frac{2}{3}$ hour.

46. 12 mi

Exercise Set 6.6

1. $\dfrac{30x^8 - 15x^6 + 40x^4}{5x^4}$

$= \dfrac{30x^8}{5x^4} - \dfrac{15x^6}{5x^4} + \dfrac{40x^4}{5x^4}$

$= 6x^4 - 3x^2 + 8$

2. $4y^4 + 3y^3 - 6$

3. $\dfrac{-14a^3 + 28a^2 - 21a}{7a}$

$= \dfrac{-14a^3}{7a} + \dfrac{28a^2}{7a} - \dfrac{21a}{7a}$

$= -2a^2 + 4a - 3$

4. $-8x^3 - 6x^2 - 3x$

5. $(9y^4 - 18y^3 + 27y^2) \div 9y$

$$= \frac{9y^4}{9y} - \frac{18y^3}{9y} + \frac{27y^2}{9y}$$

$$= y^3 - 2y^2 + 3y$$

6. $12a^2 + 14a - 10$

7. $(36x^6 - 18x^4 - 12x^2) \div -6x$

$$= \frac{36x^6}{-6x} - \frac{18x^4}{-6x} - \frac{12x^2}{-6x}$$

$$= -6x^5 - (-3x^3) - (-2x)$$

$$= -6x^5 + 3x^3 + 2x$$

8. $-6y^5 + 9y^2 + 1$

9. $(a^2b - a^3b^3 - a^5b^5) \div a^2b$

$$= \frac{a^2b}{a^2b} - \frac{a^3b^3}{a^2b} - \frac{a^5b^5}{a^2b}$$

$$= 1 - ab^2 - a^3b^4$$

10. $x - xy - x^2$

11. $(6p^2q^2 - 9p^2q + 12pq^2) \div -3pq$

$$= \frac{6p^2q^2}{-3pq} - \frac{9p^2q}{-3pq} + \frac{12pq^2}{-3pq}$$

$$= -2pq - (-3p) + (-4q)$$

$$= -2pq + 3p - 4q$$

12. $4z - 2y^2z^3 + 3y^4z^2$

13.
```
           x  +  7
  x + 3 ) x² + 10x + 21
          x² +  3x
                7x + 21    (x² + 10x) − (x² + 3x) = 7x
                7x + 21
                     0
```
The answer is $x + 7$.

14. $y - 4$

15.
```
           a  -  12
  a + 4 ) a² −  8a − 16
          a² +  4a
              −12a − 16    (a² − 8a) − (a² + 4a) = −12a
              −12a − 48
                      32   (−12a − 16) − (−12a − 48) = 32
```
The answer is $a - 12$, R 32, or $a - 12 + \dfrac{32}{a + 4}$.

16. $y - 5 + \dfrac{-50}{y - 5}$

17.
```
             x  −  6
  x − 5 ) x² − 11x + 23
          x² −  5x
               −6x + 23
               −6x + 30
                     −7
```
The answer is $x - 6$, R -7, or $x - 6 + \dfrac{-7}{x - 5}$.

18. $x - 4 + \dfrac{-5}{x - 7}$

19.
```
           y  −  5
  y + 5 ) y² + 0y − 25       Writing in the missing term
          y² + 5y
              −5y − 25
              −5y − 25
                    0
```
The answer is $y - 5$.

20. $a + 9$

21.
```
           y²  −  2y  −  1
  y − 2 ) y³ − 4y² + 3y − 6
          y³ − 2y²
             −2y² + 3y
             −2y² + 4y
                   −y − 6
                   −y + 2
                       −8
```
The answer is $y^2 - 2y - 1$, R -8, or
$y^2 - 2y - 1 + \dfrac{-8}{y - 2}$.

22. $x^2 - 2x - 2 + \dfrac{-13}{x - 3}$

23.
```
            2x²  −  x  +  1
  x + 2 ) 2x³ + 3x² − x − 3
          2x³ + 4x²
               −x² −  x
               −x² − 2x
                     x − 3
                     x + 2
                        −5
```
The answer is $2x^2 - x + 1$, R -5, or
$2x^2 - x + 1 + \dfrac{-5}{x + 2}$.

24. $3x^2 + x - 1 + \dfrac{-4}{x - 2}$

25.

$$
\begin{array}{r}
a^2 + 4a \;+\; 15 \\
a - 4 \,\overline{\smash{\big)}\, a^3 + 0a^2 - \;\; a + 12} \\
\underline{a^3 - 4a^2} \\
4a^2 - \;\; a \\
\underline{4a^2 - 16a} \\
15a + 12 \\
\underline{15a - 60} \\
72
\end{array}
$$

The answer is $a^2 + 4a + 15$, R 72, or $a^2 + 4a + 15 + \dfrac{72}{a - 4}$.

26. $x^2 - 2x + 3$

27.

$$
\begin{array}{r}
4x^2 - \;\; 6x \;+\; 9 \\
2x + 3 \,\overline{\smash{\big)}\, 8x^3 + \;\; 0x^2 + \;\; 0x + 27} \\
\underline{8x^3 + 12x^2} \\
-12x^2 + \;\; 0x \\
\underline{-12x^2 - 18x} \\
18x + 27 \\
\underline{18x + 27} \\
0
\end{array}
$$

The answer is $4x^2 - 6x + 9$.

28. $16y^2 + 8y + 4$

29.

$$
\begin{array}{r}
x^2 \;+\; 6 \\
x^2 - 7 \,\overline{\smash{\big)}\, x^4 - \;\; x^2 - 42} \\
\underline{x^4 - 7x^2} \\
6x^2 - 42 \\
\underline{6x^2 - 42} \\
0
\end{array}
$$

The answer is $x^2 + 6$.

30. $y^2 + 2 + \dfrac{-48}{y^2 - 3}$

31.

$$
\begin{array}{r}
x^3 + \;\; x^2 - 1 \\
x - 1 \,\overline{\smash{\big)}\, x^4 + 0x^3 - x^2 - x + 2} \\
\underline{x^4 - \;\; x^3} \\
x^3 - \;\; x^2 \\
\underline{x^3 - \;\; x^2} \\
0 - x + 2 \\
\underline{-x + 1} \\
1
\end{array}
$$

The answer is $x^3 + x^2 - 1$, R 1, or $x^3 + x^2 - 1 + \dfrac{1}{x - 1}$.

32. $y^3 - y^2 - 1 + \dfrac{4}{y + 1}$

33.

$$
\begin{array}{r}
2y^2 + \;\; 2y \;-\; 1 \\
5y - 2 \,\overline{\smash{\big)}\, 10y^3 + \;\; 6y^2 - 9y + 10} \\
\underline{10y^3 - \;\; 4y^2} \\
10y^2 - 9y \\
\underline{10y^2 - 4y} \\
-5y + 10 \\
\underline{-5y + \;\; 2} \\
8
\end{array}
$$

The answer is $2y^2 + 2y - 1$, R 8, or $2y^2 + 2y - 1 + \dfrac{8}{5y - 2}$.

34. $3x^2 - x + 4 + \dfrac{10}{2x - 3}$

35.

$$
\begin{array}{r}
2x^2 - \;\; x \;-\; 9 \\
x^2 + 2 \,\overline{\smash{\big)}\, 2x^4 - x^3 - 5x^2 + \;\; x - \;\; 6} \\
\underline{2x^4 + 4x^2} \\
-x^3 - 9x^2 + \;\; x \\
\underline{-x^3 - 2x} \\
-9x^2 + 3x - \;\; 6 \\
\underline{-9x^2 - 18} \\
3x + 12
\end{array}
$$

The answer is $2x^2 - x - 9$, R $3x + 12$, or $2x^2 - x - 9 + \dfrac{3x + 12}{x^2 + 2}$.

36. $3x^2 + 2x - 5 + \dfrac{2x - 5}{x^2 - 2}$

37.

$$
\begin{array}{r}
2x^3 + x^2 - 1 \\
x^2 + 1 \,\overline{\smash{\big)}\, 2x^5 + x^4 + 2x^3 + 0x^2 + x} \\
\underline{2x^5 + 2x^3} \\
x^4 + 0x^2 \\
\underline{x^4 + \;\; x^2} \\
-x^2 + x \\
\underline{-x^2 - 1} \\
x + 1
\end{array}
$$

The answer is $2x^3 + x^2 - 1$, R $x + 1$, or $2x^3 + x^2 - 1 + \dfrac{x + 1}{x^2 + 1}$.

38. $2x^3 - x + 1 + \dfrac{-x + 5}{x^2 - 1}$

39.
$$x^2 - 5x = 0$$
$$x(x - 5) = 0$$

$x = 0 \;$ or $\; x - 5 = 0 \quad$ Principle of zero products

$x = 0 \;$ or $\qquad\quad x = 5$

The solution set is $\{0, 5\}$.

40. $-\dfrac{8}{5}, \dfrac{8}{5}$

41. **Familiarize**. Let x, $x + 1$, and $x + 2$ represent the three consecutive positive integers.

Translate. Rewording, we write an equation.

Product of first and second is product of second and third less 26.

$x(x + 1)$ $=$ $(x + 1)(x + 2)$ $-$ 26

Carry out. We solve the equation.

$$x^2 + x = x^2 + 3x + 2 - 26$$
$$x = 3x - 24$$
$$24 = 2x$$
$$12 = x$$

If the first integer is 12, the next two are 13 and 14.

Check. The product of 12 and 13 is 156. The product of 13 and 14 is 182, and $182 - 26 = 156$. The numbers check.

State. The three consecutive positive integers are 12, 13, and 14.

42. $-54a^3$

43.
$$\begin{array}{r} x^2 + 2y \\ x^2 - xy + y^2 \overline{)\, x^4 - x^3y + x^2y^2 + 2x^2y - 2xy^2 + 2y^3} \\ \underline{x^4 - x^3y + x^2y^2} \\ 0 + 2x^2y - 2xy^2 + 2y^3 \\ \underline{2x^2y - 2xy^2 + 2y^3} \\ 0 \end{array}$$

The answer is $x^2 + 2y$.

44. $a^2 + ab$

45.
$$\begin{array}{r} x^3 + x^2y + xy^2 + y^3 \\ x - y \overline{)\, x^4 \qquad\qquad\qquad - y^4} \\ \underline{x^4 - x^3y} \\ x^3y \\ \underline{x^3y - x^2y^2} \\ x^2y^2 \\ \underline{x^2y^2 - xy^3} \\ xy^3 - y^4 \\ \underline{xy^3 - y^4} \\ 0 \end{array}$$

The answer is $x^3 + x^2y + xy^2 + y^3$.

46. $a^6 - a^5b + a^4b^2 - a^3b^3 + a^2b^4 - ab^5 + b^6$

47.
$$\begin{array}{r} x^2 + (-k - 2)x + (2k + 7) \\ x + 2 \overline{)\, x^3 - \quad kx^2 + \qquad 3x + \qquad 7k} \\ \underline{x^3 + \qquad 2x^2} \\ (-k - 2)x^2 + \qquad 3x \\ \underline{(-k - 2)x^2 + (-2k - 4)x} \\ (2k + 7)x + \qquad 7k \\ \underline{(2k + 7)x + (4k + 14)} \end{array}$$
The remainder must be 0. ⟶

47. (continued)

Thus, we solve the following equation for k.
$$7k - (4k + 14) = 0$$
$$7k - 4k - 14 = 0$$
$$3k = 14$$
$$k = \frac{14}{3}$$

48. $-\dfrac{3}{2}$

49. a), b)
$$\begin{array}{r} 3 \\ x + 2 \overline{)\, 3x + 7} \\ \underline{3x + 6} \\ 1 \end{array}$$

$$f(x) = 3 + \frac{1}{x + 2}$$

c) The graph of f looks like the graph of g, shifted up 3 units. The graph of g looks like the graph of h, shifted to the left 2 units.

Exercise Set 6.7

1. $(x^3 - 2x^2 + 2x - 5) \div (x - 1)$

$$\begin{array}{r} 1| \quad 1 \quad -2 \quad 2 \quad -5 \\ \quad 1 \quad -1 \quad 1 \\ \hline 1 \quad -1 \quad 1 \mid -4 \end{array}$$

The quotient is $x^2 - x + 1$. The remainder is -4.

2. $x^2 - 3x + 5$, R -10

3. $(a^2 + 11a - 19) \div (a + 4) =$
$(a^2 + 11a - 19) \div [a - (-4)]$

$$\begin{array}{r} -4| \quad 1 \quad 11 \quad -19 \\ \quad -4 \quad -28 \\ \hline 1 \quad 7 \mid -47 \end{array}$$

The quotient is $a + 7$. The remainder is -47.

4. $a + 15$, R 41

5. $(x^3 - 7x^2 - 13x + 3) \div (x - 2)$

    ```
    2| 1   -7   -13    3
           2   -10   -46
       1   -5   -23 | -43
    ```

    The quotient is $x^2 - 5x - 23$. The remainder is -43.

6. $x^2 - 9x + 5$, R -7

7. $(3x^3 + 7x^2 - 4x + 3) \div (x + 3) =$
    $(3x^3 + 7x^2 - 4x + 3) \div [x - (-3)]$

    ```
    -3| 3    7   -4    3
            -9    6   -6
        3   -2    2 | -3
    ```

    The quotient is $3x^2 - 2x + 2$. The remainder is -3.

8. $3x^2 + 16x + 44$, R 135

9. $(y^3 - 3y + 10) \div (y - 2) =$
    $(y^3 + 0y^2 - 3y + 10) \div (y - 2)$

    ```
    2| 1    0   -3   10
            2    4    2
       1    2    1 | 12
    ```

    The quotient is $y^2 + 2y + 1$. The remainder is 12.

10. $x^2 - 4x + 8$, R -8

11. $(3x^4 - 25x^2 - 18) \div (x - 3) =$
     $(3x^4 + 0x^3 - 25x^2 + 0x - 18) \div (x - 3)$

    ```
    3| 3    0   -25    0   -18
            9    27    6    18
       3    9     2    6 |   0
    ```

     The quotient is $3x^3 + 9x^2 + 2x + 6$. The remainder is 0.

12. $6y^3 - 3y^2 + 9y + 1$, R 3

13. $(x^3 - 27) \div (x - 3) =$
     $(x^3 + 0x^2 + 0x - 27) \div (x - 3)$

    ```
    3| 1    0    0   -27
            3    9    27
       1    3    9 |   0
    ```

     The quotient is $x^2 + 3x + 9$. The remainder is 0.

14. $y^2 - 3y + 9$

15. $(y^5 - 1) \div (y - 1) =$
     $(y^5 + 0y^4 + 0y^3 + 0y^2 + 0y - 1) \div (y - 1)$

    ```
    1| 1    0    0    0    0   -1
            1    1    1    1    1
       1    1    1    1    1 |  0
    ```

     The quotient is $y^4 + y^3 + y^2 + y + 1$. The remainder is 0.

16. $x^4 + 2x^3 + 4x^2 + 8x + 16$

17. $(3x^4 + 8x^3 + 2x^2 - 7x - 4) \div (x + 2) =$
     $(3x^4 + 8x^3 + 2x^2 - 7x - 4) \div [(x - (-2)]$

    ```
    -2| 3    8    2   -7   -4
            -6   -4    4    6
        3    2   -2   -3 |  2
    ```

     The quotient is $3x^3 + 2x^2 - 2x - 3$. The remainder is 2.

18. $2x^3 - 3x^2 - 2x + 3$, R 4

19. $(3x^3 + 7x^2 - x + 1) \div \left[x + \frac{1}{3}\right] =$
     $(3x^3 + 7x^2 - x + 1) \div \left[x - \left[-\frac{1}{3}\right]\right]$

    ```
    -1/3| 3    7   -1    1
              -1   -2    1
          3    6   -3 |  2
    ```

     The quotient is $3x^2 + 6x - 3$. The remainder is 2.

20. $8x^2 - 2x + 6$, R 2

21. Graph: $2x - 3y > 6$

     First graph the line $2x - 3y = 6$. The intercepts are (0,-2) and (3,0). We draw the line dashed since the inequality is >. Since the ordered pair (0,0) is <u>not</u> a solution of the inequality ($2 \cdot 0 - 3 \cdot 0 > 6$ is false), we shade the lower half-plane.

     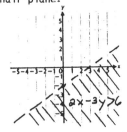

22.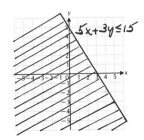

23. Graph: y > 4

    First graph the line y = 4. The line is parallel to the x-axis with y-intercept (0,4). We draw the line dashed since the inequality is >. Since the ordered pair (0,0) is <u>not</u> a solution of the inequality (0 > 4 is false), we shade the upper half-plane.

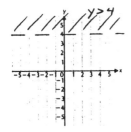

24.

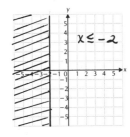

25. $(3.41x^4 - 24.25x^2 - 13.47) \div (x - 2.41) =$
     $(3.41x^4 + 0x^3 - 24.25x^2 + 0x - 13.47) \div (x - 2.41)$

    2.41| 3.41 0   -24.25   0   -13.47
              8.2181 19.805621 -10.7109533 -25.813397
         3.41 8.2181 -4.444379 -10.7109533 |-39.283397

    The answer is $3.41x^3 + 8.2181x^2 - 4.444379x - 10.7109533$, R -39.283397

26. $5.032x^4 - 85.89624x^3 + 1477.6628x^2 - 25,223.704x + 430,568.62$, R $-7,349,598$

27. a) $(8x^5 - 3x^4 + 7x - 4) \div (x - 2)$

    2| 8   -3   0   0   7   -4
          16   26   52   104   222
       8   13   26   52   111 | 218

    The remainder is 218.

    b) $P(x) = 8x^5 - 3x^4 + 0x^3 + 0x^2 + 7x - 4$
        $= x(x(x(x(8x - 3) + 0) + 0) + 7) - 4$

    $P(2) = 2(2(2(2(8 \cdot 2 - 3) + 0) + 0) + 7) - 4$
    $= 2(2(2(2(13) + 0) + 0) + 7) - 4$
    $= 2(2(2(26 + 0) + 0) + 7) - 4$
    $= 2(2(52 + 0) + 7) - 4$
    $= 2(104 + 7) - 4$
    $= 222 - 4$
    $= 218$

27. (continued)

    c) The answers and the computations are the same.

28. Convert ax + b to $a\left[x - \left(-\dfrac{b}{a}\right)\right]$. Perform synthetic division with $\left[x - \left(-\dfrac{b}{a}\right)\right]$ as the divisor. Divide all the coefficients in the answer by a.

29. a) 4| 1   -5   5   -4
                 4   -4   4
             1   -1   1 | 0

    The remainder is 0. Therefore, we know that $x^3 - 5x^2 + 5x - 4 = (x - 4)(x^2 - x + 1)$.

    b) $f(x) = (x - 4)(x^2 - x + 1)$, so
        $f(4) = (4 - 4)(4^2 - 4 + 1) = 0$

    c) $f(4) = 4^3 - 5 \cdot 4^2 + 5 \cdot 4 - 4$
        $= 64 - 80 + 20 - 4$
        $= 0$

Exercise Set 6.8

1. $\dfrac{W_1}{W_2} = \dfrac{d_1}{d_2}$

    $\dfrac{d_2 W_1}{W_2} = d_1$ Multiplying by d_2

2. $W_1 = \dfrac{d_1 W_2}{d_2}$

3. $s = \dfrac{(v_1 + v_2)t}{2}$

    $2s = (v_1 + v_2)t$ Multiplying by 2

    $\dfrac{2s}{v_1 + v_2} = t$ Multiplying by $\dfrac{1}{v_1 + v_2}$

4. $v_1 = \dfrac{2s}{t} - v_2$

5. $\dfrac{1}{R} = \dfrac{1}{r_1} + \dfrac{1}{r_2}$

    $Rr_1 r_2 \cdot \dfrac{1}{R} = Rr_1 r_2 \left[\dfrac{1}{r_1} + \dfrac{1}{r_2}\right]$ Multiplying by $Rr_1 r_2$

    $r_1 r_2 = Rr_2 + Rr_1$

    $r_1 r_2 - Rr_1 = Rr_2$ Adding $-Rr_1$

    $r_1(r_2 - R) = Rr_2$ Factoring

    $r_1 = \dfrac{Rr_2}{r_2 - R}$ Multiplying by $\dfrac{1}{r_2 - R}$

6. $R = \dfrac{r_1 r_2}{r_2 + r_1}$

7. $R = \dfrac{gs}{g + s}$

    $R(g + s) = gs$ Multiplying by $g + s$

    $Rg + Rs = gs$ Removing parentheses

    $Rg = gs - Rs$ Adding $-Rs$

    $Rg = s(g - R)$ Factoring

    $\dfrac{Rg}{g - R} = s$ Multiplying by $\dfrac{1}{g - R}$

8. $g = \dfrac{Rs}{s - R}$

9. $I = \dfrac{2V}{R + 2r}$

$I(R + 2r) = 2V$	Multiplying by $R + 2r$
$IR + 2Ir = 2V$	Removing parentheses
$2Ir = 2V - IR$	Adding $-IR$
$r = \dfrac{2V - IR}{2I}$	Multiplying by $\dfrac{1}{2I}$

10. $R = \dfrac{2V}{I} - 2r$

11. $\dfrac{1}{p} + \dfrac{1}{q} = \dfrac{1}{f}$

$pqf\left(\dfrac{1}{p} + \dfrac{1}{q}\right) = pqf \cdot \dfrac{1}{f}$	Multiplying by pqf
$qf + pf = pq$	
$pf = pq - qf$	Adding $-qf$
$pf = q(p - f)$	Factoring
$\dfrac{pf}{p - f} = q$	Multiplying by $\dfrac{1}{p - f}$

12. $p = \dfrac{qf}{q - f}$

13. $I = \dfrac{nE}{R + nr}$

$I(R + nr) = nE$	Multiplying by $R + nr$
$IR + Inr = nE$	Removing parentheses
$Inr = nE - IR$	Adding $-IR$
$r = \dfrac{nE - IR}{In}$	Multiplying by $\dfrac{1}{In}$

14. $n = \dfrac{IR}{E - Ir}$

15. $S = \dfrac{H}{m(t_1 - t_2)}$

$m(t_1 - t_2)S = H$	Multiplying by $m(t_1 - t_2)$

16. $t_1 = \dfrac{H}{Sm} + t_2$

17. $\dfrac{E}{e} = \dfrac{R + r}{r}$

$er \cdot \dfrac{E}{e} = er \cdot \dfrac{R + r}{r}$	Multiplying by er
$rE = e(R + r)$	
$\dfrac{rE}{R + r} = e$	Multiplying by $\dfrac{1}{R + r}$

18. $r = \dfrac{er}{E - e}$

19. $S = \dfrac{a - ar^n}{1 - r}$

$S(1 - r) = a - ar^n$	Multiplying by $1 - r$
$S(1 - r) = a(1 - r^n)$	Factoring
$\dfrac{S - Sr}{1 - r^n} = a$	Multiplying by $\dfrac{1}{1 - r^n}$

20. $r = 1 - \dfrac{a}{S}$

21. Familiarize and Translate. We use the formula given.

    Carry out. First solve the formula for R.

    $$A = \dfrac{9R}{I}$$
    $$AI = 9R$$
    $$\dfrac{AI}{9} = R$$

    Then substitute 2.4 for A and 45 for I.

    $$\dfrac{2.4(45)}{9} = R$$
    $$12 = R$$

    Check. We substitute in the original formula.

    $$\begin{array}{c|c} & A = 9R/I \\ \hline 2.4 & \dfrac{9 \cdot 12}{45} \\ & \dfrac{108}{45} \\ & 2.4 \end{array}$$

    The answer checks.

    State. 12 earned runs were given up.

22. $5\dfrac{5}{23}$ ohms

23. Familiarize. We want to find r_2 using the formula $\dfrac{1}{R} = \dfrac{1}{r_1} + \dfrac{1}{r_2}$. We know $R = 5$ ohms and $r_1 = 50$ ohms.

    Translate. Using a result from Example 4, we have

    $$r_2 = \dfrac{Rr_1}{r_1 - R}.$$

    Carry out. We substitute and compute.

    $$r_2 = \dfrac{5 \cdot 50}{50 - 5}$$
    $$= \dfrac{250}{45}$$
    $$= \dfrac{50}{9}, \text{ or } 5\dfrac{5}{9}$$

    Check. We substitute into the original formula.

    $$\begin{array}{c|c} \dfrac{1}{R} = & \dfrac{1}{r_1} + \dfrac{1}{r_2} \\ \hline \dfrac{1}{5} & \dfrac{1}{50} + \dfrac{1}{\frac{50}{9}} \\ & \dfrac{1}{50} + \dfrac{9}{50} \\ & \dfrac{10}{50} \\ & \dfrac{1}{5} \end{array}$$

    The answer checks.

    State. A resistor with a resistance of $5\dfrac{5}{9}$ ohms should be used.

24. 6 cm

25.
$$\frac{1}{t} = \frac{1}{u} + \frac{1}{v}$$

$tuv \cdot \frac{1}{t} = tuv\left(\frac{1}{u} + \frac{1}{v}\right)$ Multiplying by tuv

$uv = tv + tu$

$uv = t(v + u)$ Factoring

$\frac{uv}{v + u} = t$ Multiplying by $\frac{1}{v + u}$

26. $Q = \frac{2Tt - 2AT}{A - q}$

27.
$$v = \frac{d_2 - d_1}{t_2 - t_1}$$

$(t_2 - t_1)v = (t_2 - t_1) \cdot \frac{d_2 - d_1}{t_2 - t_1}$ Multiplying by $(t_2 - t_1)$

$(t_2 - t_1)v = d_2 - d_1$

$t_2 - t_1 = \frac{d_2 - d_1}{v}$ Multiplying by $\frac{1}{v}$

$t_2 = \frac{d_2 - d_1}{v} + t_1$ Adding t_1

28. $r = \frac{A}{P} - 1$

29. Familiarize. We want to find r using the formula $P = \frac{A}{1 + r}$. We know $P = \$1600$ and $A = \$1712$.

Translate. Using the result of Exercise 28, we have $r = \frac{A}{P} - 1$.

Carry out. Substitute and compute.

$r = \frac{1712}{1600} - 1$

$r = 1.07 - 1$

$r = 0.07$

Check. We substitute in the original formula.

$$P = \frac{A}{1 + r}$$

$$1600 \;\Big|\; \frac{1712}{1 + 0.07}$$

$$\frac{1712}{1.07}$$

$$1600$$

The answer checks.

State. The interest rate is 0.07, or 7%.

30. 3:45 A.M.

31.
$$I_t = \frac{I_f}{1 - T}$$

$(1 - T)I_t = I_f$ Multiplying by $(1 - T)$

$1 - T = \frac{I_f}{I_t}$ Multiplying by $\frac{1}{I_t}$

$-T = \frac{I_f}{I_t} - 1$ Adding -1

$-1 \cdot (-T) = -1 \cdot \left[\frac{I_f}{I_t} - 1\right]$ Multiplying by -1

$T = -\frac{I_f}{I_t} + 1$

32. $d = \frac{LD}{R + L}$

33.
$$\frac{V^2}{R^2} = \frac{2g}{R + h}$$

$(R + h) \cdot \frac{V^2}{R^2} = (R + h) \cdot \frac{2g}{R + h}$ Multiplying by $(R + h)$

$\frac{(R + h)V^2}{R^2} = 2g$

$R + h = \frac{2gR^2}{V^2}$ Multiplying by $\frac{R^2}{V^2}$

$h = \frac{2gR^2}{V^2} - R$ Adding $-R$

34. $t_1 = t_2 - \frac{v_2 - v_1}{a}$

35. Familiarize and Translate. We want to find T using the formula $A = \frac{2Tt + Qq}{2T + Q}$. We know $t = 79$, $q = 90$, $A = 84$, and $Q = 5$.

Carry out. We first solve the formula for T.

$(2T + Q)A = 2Tt + Qq$

$2AT + AQ = 2Tt + Qq$

$2AT - 2Tt = Qq - AQ$

$T(2A - 2t) = Qq - AQ$

$T = \frac{Qq - AQ}{2A - 2t}$

Now we substitute and compute.

$T = \frac{(5 \cdot 90) - (84 \cdot 5)}{(2 \cdot 84) - (2 \cdot 79)}$

$T = \frac{30}{10}$

$T = 3$

Check. We substitute in the original formula.

$$A = \frac{2Tt + Qq}{2T + Q}$$

$$84 \;\Big|\; \frac{(2 \cdot 3 \cdot 79) + (5 \cdot 90)}{(2 \cdot 3) + 5}$$

$$\frac{924}{11}$$

$$84$$

The answer checks.

State. The student took 3 tests.

36. $(-15,2)$ and $(-30,8)$

37. Graph: $6x - y < 6$

First graph the line $6x - y = 6$. The intercepts are $(0,-6)$ and $(1,0)$. We draw the line dashed since the inequality is $<$. Since the ordered pair $(0,0)$ is a solution of the inequality ($6 \cdot 0 - 0 < 6$ is true), we shade the upper half plane.

38. $8a^3 - 2a$

39. $t^3 + 8b^3 = t^3 + (2b)^3 = (t + 2b)(t^2 - 2tb + 4b^2)$

40. $-\dfrac{5}{3}, \dfrac{7}{2}$

41. <u>Familiarize</u>. We can use the formula in Exercise 33 to find h, the satellite's height above the earth. Then to find the satellite's distance from the center of the earth, we will compute R + h, the sum of the earth's radius and the satellite's height above earth. We know V = 6.5 mi/sec, R = 3960 mi, and g = 32.2 ft/sec².

<u>Translate</u>. Using the result of Exercise 33, we have $h = \dfrac{2gR^2}{V^2} - R$. Then $R + h = R + \dfrac{2gR^2}{V^2} - R = \dfrac{2gR^2}{V^2}$.

<u>Carry out</u>. We must convert 32.2 ft/sec² to mi/sec² so all units of length are the same.

$$32.2 \ \frac{ft}{sec^2} \cdot \frac{1 \ mi}{5280 \ ft} \approx 0.0060984 \ \frac{mi}{sec^2}$$

Now we substitute and compute.

$$R + h = \frac{2(0.0060984)(3960)^2}{(6.5)^2}$$

$$R + h \approx 4526.99$$

<u>Check</u>. We go over our calculations. The answer checks.

<u>State</u>. The satellite is 4526.99 mi from the center of the earth.

42. $M = \dfrac{2ab}{b + a}$

43.
$$x^2\left[1 - \frac{2pq}{x}\right] = \frac{2p^2q^3 - pq^2x}{-q}$$

$\quad\quad x^2 - 2pqx = -2p^2q^2 + pqx \quad$ Multiplying on the left; dividing on the right

$x^2 - 3pqx + 2p^2q^2 = 0$

$(x - pq)(x - 2pq) = 0$

$x - pq = 0 \quad$ or $\quad x - 2pq = 0$

$\quad\quad x = pq \quad$ or $\quad\quad x = 2pq$

The solution set is {pq,2pq}.

44. $t_1 = t_2 + \dfrac{(d_2 - d_1)(t_4 - t_3)}{a(t_4 - t_2)(t_4 - t_3) + d_3 - d_4}$

Exercise Set 7.1

1. The square roots of 16 are 4 and -4, because
$4^2 = 16$ and $(-4)^2 = 16$.

2. 15, -15

3. The square roots of 144 are 12 and -12, because
$12^2 = 144$ and $(-12)^2 = 144$.

4. 3, -3

5. The square roots of 400 are 20 and -20, because
$20^2 = 400$ and $(-20)^2 = 400$.

6. 9, -9

7. The square roots of 49 are 7 and -7, because
$7^2 = 49$ and $(-7)^2 = 49$.

8. 30, -30

9. $-\sqrt{\dfrac{49}{36}} = -\dfrac{7}{6}$ Since $\sqrt{\dfrac{49}{36}} = \dfrac{7}{6}$, $-\sqrt{\dfrac{49}{36}} = -\dfrac{7}{6}$.

10. $-\dfrac{19}{3}$

11. $\sqrt{196} = 14$ Remember, $\sqrt{}$ indicates the principal square root.

12. 21

13. $-\sqrt{\dfrac{16}{81}} = -\dfrac{4}{9}$ Since $\sqrt{\dfrac{16}{81}} = \dfrac{4}{9}$, $-\sqrt{\dfrac{16}{81}} = -\dfrac{4}{9}$.

14. $-\dfrac{9}{12}$, or $-\dfrac{3}{4}$

15. $\sqrt{0.09} = 0.3$

16. 0.6

17. $-\sqrt{0.0049} = -0.07$

18. 0.12

19. $5\sqrt{p^2 + 4}$
The radical is the expression written under the radical sign, $p^2 + 4$.

20. $y^2 - 8$

21. $x^2y^2\sqrt{\dfrac{x}{y + 4}}$
The radicand is the expression written under the radical sign, $\dfrac{x}{y + 4}$.

22. $\dfrac{a}{a^2 - b}$

23. $f(y) = \sqrt{5y - 10}$

$f(6) = \sqrt{5 \cdot 6 - 10} = \sqrt{20}$

$f(2) = \sqrt{5 \cdot 2 - 10} = \sqrt{0} = 0$

$f(0) = \sqrt{5 \cdot 0 - 10} = \sqrt{-10}$
Since negative numbers do not have real-number square roots, $f(0)$ does not exist.

$f(-3) = \sqrt{5(-3) - 10} = \sqrt{-25}$
Since negative numbers do not have real-number square roots, $f(-3)$ does not exist.

24. 0; 0; does not exist; $\sqrt{11}$; $\sqrt{11}$

25. $t(x) = -\sqrt{2x + 1}$

$t(4) = -\sqrt{2 \cdot 4 + 1} = -\sqrt{9} = -3$

$t(-4) = -\sqrt{2(-4) + 1} = -\sqrt{-7}$; $t(-4)$ does not exist

$t(0) = -\sqrt{2 \cdot 0 + 1} = -\sqrt{1} = -1$

$t(12) = -\sqrt{2 \cdot 12 + 1} = -\sqrt{25} = -5$

26. Does not exist; $\sqrt{30}$; $\sqrt{30}$; $\sqrt{180}$; $\sqrt{180}$

27. $f(t) = \sqrt{t^2 + 1}$

$f(5) = \sqrt{5^2 + 1} = \sqrt{26}$

$f(-5) = \sqrt{(-5)^2 + 1} = \sqrt{26}$

$f(0) = \sqrt{0^2 + 1} = \sqrt{1} = 1$

$f(10) = \sqrt{10^2 + 1} = \sqrt{101}$

$f(-10) = \sqrt{(-10)^2 + 1} = \sqrt{101}$

28. -2; 0; -4; -2; -6; -4

29. $g(x) = \sqrt{x^3 + 9}$

$g(2) = \sqrt{2^3 + 9} = \sqrt{17}$

$g(-2) = \sqrt{(-2)^3 + 9} = \sqrt{1} = 1$

$g(3) = \sqrt{3^3 + 9} = \sqrt{36} = 6$

$g(-3) = \sqrt{(-3)^3 + 9} = \sqrt{-18}$; $g(-3)$ does not exist

30. Does not exist; does not exist; $\sqrt{17}$; does not exist

31. $f(x) = \sqrt{x + 7}$
We need to find all x-values such that $x + 7$ is nonnegative. We solve an inequality.
$x + 7 \geqslant 0$
$x \geqslant -7$ Adding -7 on both sides
Domain of $f = \{x | x \geqslant -7\}$

32. $\{x | x \geqslant 10\}$

33. $g(t) = \sqrt{2t - 9}$

    We solve an inequality.

    $2t - 9 \geqslant 0$

    $2t \geqslant 9$ Adding 9 on both sides

    $t \geqslant \frac{9}{2}$ Multiplying by $\frac{1}{2}$ on both sides

    Domain of $g = \left\{ t \middle| t \geqslant \frac{9}{2} \right\}$

34. $\left\{ t \middle| t \geqslant -\frac{5}{2} \right\}$

35. $h(z) = -\sqrt{5z + 3}$

    We solve an inequality.

    $5z + 3 \geqslant 0$

    $5z \geqslant -3$

    $z \geqslant -\frac{3}{5}$

    Domain of $h = \left\{ z \middle| z \geqslant -\frac{3}{5} \right\}$

36. $\left\{ x \middle| x \geqslant \frac{5}{7} \right\}$

37. $f(t) = 7 + 2\sqrt{3t - 5}$

    We solve an inequality.

    $3t - 5 \geqslant 0$

    $3t \geqslant 5$

    $t \geqslant \frac{5}{3}$

    Domain of $f = \left\{ t \middle| t \geqslant \frac{5}{3} \right\}$

38. $\left\{ t \middle| t \geqslant \frac{4}{5} \right\}$

39. $\sqrt{16x^2} = \sqrt{(4x)^2} = |4x| = 4|x|$

    Since x might be negative, absolute-value notation is necessary.

40. $5|t|$

41. $\sqrt{(-7c)^2} = |-7c| = |-7| \cdot |c| = 7|c|$

    Since c might be negative, absolute-value notation is necessary.

42. $6|b|$

43. $\sqrt{(a + 1)^2} = |a + 1|$

    Since a + 1 might be negative, absolute-value notation is necessary.

44. $|5 - b|$

45. $\sqrt{x^2 - 4x + 4} = \sqrt{(x - 2)^2} = |x - 2|$

    Since x - 2 might be negative, absolute-value notation is necessary.

46. $|y + 8|$

47. $\sqrt{4x^2 + 28x + 49} = \sqrt{(2x + 7)^2} = |2x + 7|$

    Since 2x + 7 might be negative, absolute-value notation is necessary.

48. $|3x - 5|$

49. $\sqrt{16x^2} = \sqrt{(4x)^2} = 4x$ Assuming x is nonnegative

50. $5t$

51. $\sqrt{(-6b)^2} = \sqrt{36b^2} = 6b$ Assuming b is nonnegative

52. $7c$

53. $\sqrt{(a + 1)^2} = a + 1$ Assuming a + 1 is nonnegative

54. $5 + b$

55. $\sqrt{4x^2 + 8x + 4} = \sqrt{4(x^2 + 2x + 1)} = \sqrt{[2(x + 1)]^2} = 2(x + 1)$, or $2x + 2$

56. $3x + 6$

57. $\sqrt{9t^2 - 12t + 4} = \sqrt{(3t - 2)^2} = 3t - 2$

58. $5t - 2$

59. $\sqrt[3]{27} = 3$ $[3^3 = 27]$

60. -4

61. $\sqrt[3]{-64x^3} = -4x$ $[(-4x)^3 = -64x^3]$

62. $-5y$

63. $\sqrt[3]{-216} = -6$ $[(-6)^3 = -216]$

64. 10

65. $-\sqrt[3]{-125y^3} = -(-5y)$ $[(-5y)^3 = -125y^3]$

    $= 5y$

66. $4x$

67. $\sqrt[3]{0.343(x + 1)^3} = 0.7(x + 1)$

    $[(0.7(x + 1))^3 = 0.343(x + 1)^3]$

68. $0.02(y - 2)$

69. $f(x) = \sqrt[3]{x + 1}$

    $f(7) = \sqrt[3]{7 + 1} = \sqrt[3]{8} = 2$

    $f(26) = \sqrt[3]{26 + 1} = \sqrt[3]{27} = 3$

    $f(-9) = \sqrt[3]{-9 + 1} = \sqrt[3]{-8} = -2$

    $f(-65) = \sqrt[3]{-65 + 1} = \sqrt[3]{-64} = -4$

70. $1; 5; 3; -5$

71. $g(t) = \sqrt[4]{t - 3}$

    $g(19) = \sqrt[4]{19 - 3} = \sqrt[4]{16} = 2$

    $g(-13) = \sqrt[4]{-13 - 3} = \sqrt[4]{-16}$; $g(-13)$ does not
                                       exist

    $g(1) = \sqrt[4]{1 - 3} = \sqrt[4]{-2}$; $g(1)$ does not exist

    $g(84) = \sqrt[4]{84 - 3} = \sqrt[4]{81} = 3$

72. 1; 2; does not exist; 3

73. $\sqrt[4]{625} = 5$ Since $5^4 = 625$

74. -4

75. $\sqrt[5]{-1} = -1$ Since $(-1)^5 = -1$

76. 2

77. $\sqrt[5]{-\dfrac{32}{243}} = -\dfrac{2}{3}$ Since $\left(-\dfrac{2}{3}\right)^5 = -\dfrac{32}{243}$

78. $-\dfrac{1}{2}$

79. $\sqrt[6]{x^6} = |x|$

    The index is even. Use absolute-value notation
    since x could have a negative value.

80. $|y|$

81. $\sqrt[4]{(5a)^4} = |5a| = 5|a|$

    The index is even. Use absolute-value notation
    since a could have a negative value.

82. $7|b|$

83. $\sqrt[10]{(-6)^{10}} = |-6| = 6$

84. 10

85. $\sqrt[414]{(a + b)^{414}} = |a + b|$

    The index is even. Use absolute-value notation
    since a + b could have a negative value.

86. $|2a + b|$

87. $\sqrt[7]{y^7} = y$

    We do not use absolute-value notation when the
    index is odd.

88. -6

89. $\sqrt[5]{(x - 2)^5} = x - 2$

    We do not use absolute-value notation when the
    index is odd.

90. $2xy$

91. $(a^3b^2c^5)^3 = a^{3\cdot3}b^{2\cdot3}c^{5\cdot3} = a^9b^6c^{15}$

92. $10a^{10}b^9$

93. $(x - 3)(x + 3) = x^2 - 3^2 = x^2 - 9$

94. $a^2 - b^2x^2$

95. $N = 2.5\sqrt{A}$

    a) $N = 2.5\sqrt{25} = 2.5(5) = 12.5 \approx 13$

    b) $N = 2.5\sqrt{36} = 2.5(6) = 15$

    c) $N = 2.5\sqrt{49} = 2.5(7) = 17.5 \approx 18$

    d) $N = 2.5\sqrt{64} = 2.5(8) = 20$

96.

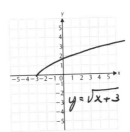

97. $y = \sqrt{x} + 3$

    Make a table of values. Note that x must be
    nonnegative.

x	y
0	3
1	4
2	4.4
3	4.7
4	5
5	5.2

Plot these points and draw the graph.

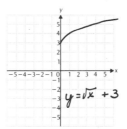

98.

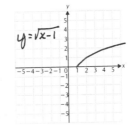

99. $y = \sqrt{x} - 2$

Make a table of values. Note that x must be nonnegative.

x	y
0	-2
1	-1
3	-0.3
4	0
5	0.3

Plot these points and draw the graph.

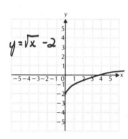

100. $\{x \mid x \geqslant -1\}$

101. $g(x) = \sqrt[4]{5 - x}$

The radicand, 5 - x, must be nonnegative. We solve an inequality.

$5 - x \geqslant 0$

$5 \geqslant x$

Domain of g = $\{x \mid x \leqslant 5\}$

102. $\left\{t \mid t \geqslant \dfrac{5}{2}\right\}$

103. $f(t) = \sqrt[8]{3t + 7}$

The radicand, 3t + 7, must be nonnegative. We solve an inequality.

$3t + 7 \geqslant 0$

$3t \geqslant -7$

$t \geqslant -\dfrac{7}{3}$

Domain of f = $\left\{t \mid t \geqslant -\dfrac{7}{3}\right\}$

104. $\{x \mid x \geqslant 3\}$

105. $q(x) = \sqrt[6]{x^3 - 8}$

We must solve the inequality $x^3 - 8 \geqslant 0$. Sketch a graph of $y = x^3 - 8$.

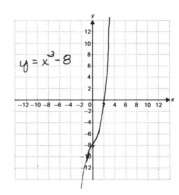

The graph shows that $x^3 - 8 \geqslant 0$ when $x \geqslant 2$.

Domain of q = $\{x \mid x \geqslant 2\}$

Exercise Set 7.2

1. $\sqrt{3} \sqrt{2} = \sqrt{3 \cdot 2} = \sqrt{6}$

2. $\sqrt{35}$

3. $\sqrt[3]{2} \ \sqrt[3]{5} = \sqrt[3]{2 \cdot 5} = \sqrt[3]{10}$

4. $\sqrt[3]{14}$

5. $\sqrt[4]{8} \ \sqrt[4]{9} = \sqrt[4]{8 \cdot 9} = \sqrt[4]{72}$

6. $\sqrt[4]{18}$

7. $\sqrt{3a} \ \sqrt{10b} = \sqrt{3a \cdot 10b} = \sqrt{30ab}$

8. $\sqrt{26xy}$

9. $\sqrt[5]{9t^2} \ \sqrt[5]{2t} = \sqrt[5]{9t^2 \cdot 2t} = \sqrt[5]{18t^3}$

10. $\sqrt[5]{80y^4}$

11. $\sqrt{x - a} \ \sqrt{x + a} = \sqrt{(x - a)(x + a)} = \sqrt{x^2 - a^2}$

12. $\sqrt{y^2 - b^2}$

13. $\sqrt[3]{0.3x} \ \sqrt[3]{0.2x} = \sqrt[3]{0.3x \cdot 0.2x} = \sqrt[3]{0.06x^2}$

14. $\sqrt[3]{0.21y^2}$

15. $\sqrt[4]{x - 1} \ \sqrt[4]{x^2 + x + 1} = \sqrt[4]{(x - 1)(x^2 + x + 1)}$

   $= \sqrt[4]{x^3 - 1}$

16. $\sqrt[5]{(x - 2)^3}$

17. $\sqrt{\dfrac{6}{x}} \ \sqrt{\dfrac{y}{5}} = \sqrt{\dfrac{6}{x} \cdot \dfrac{y}{5}} = \sqrt{\dfrac{6y}{5x}}$

18. $\sqrt{\dfrac{7s}{11t}}$

19. $\sqrt[7]{\dfrac{x-3}{4}} \sqrt[7]{\dfrac{5}{x+2}} = \sqrt[7]{\dfrac{x-3}{4} \cdot \dfrac{5}{x+2}} = \sqrt[7]{\dfrac{5x-15}{4x+8}}$

20. $\sqrt[6]{\dfrac{3a}{b^2-4}}$

21. $\sqrt{27} = \sqrt{9\cdot3}$ 9 is the largest perfect square factor
    $= \sqrt{9} \cdot \sqrt{3}$
    $= 3\sqrt{3}$

22. $2\sqrt{7}$

23. $\sqrt{45} = \sqrt{9\cdot5}$ 9 is the largest perfect square factor
    $= \sqrt{9} \cdot \sqrt{5}$
    $= 3\sqrt{5}$

24. $2\sqrt{3}$

25. $\sqrt{8} = \sqrt{4\cdot2} = \sqrt{4} \cdot \sqrt{2} = 2\sqrt{2}$

26. $3\sqrt{2}$

27. $\sqrt{24} = \sqrt{4\cdot6} = \sqrt{4} \cdot \sqrt{6} = 2\sqrt{6}$

28. $2\sqrt{5}$

29. $\sqrt{180x^4} = \sqrt{36\cdot5\cdot x^4}$ Factoring the radicand
    $= \sqrt{36\cdot x^4\cdot5}$
    $= \sqrt{36} \sqrt{x^4} \sqrt{5}$ Factoring into several radicals
    $= 6x^2 \sqrt{5}$ Taking square roots.

30. $5y^3\sqrt{7}$

31. $\sqrt[3]{800} = \sqrt[3]{8\cdot100}$ 8 is the largest perfect cube factor
    $= \sqrt[3]{8} \cdot \sqrt[3]{100}$
    $= 2\sqrt[3]{100}$

32. $3\sqrt[3]{10}$

33. $\sqrt[3]{-16x^6} = \sqrt[3]{-8\cdot2\cdot x^6}$ Factoring the radicand
    $= \sqrt[3]{-8\cdot x^6\cdot2}$
    $= \sqrt[3]{-8} \sqrt[3]{x^6} \sqrt[3]{2}$ Factoring into several radicals
    $= -2x^2 \sqrt[3]{2}$ Taking cube roots

34. $-2a^2 \sqrt[3]{4}$

35. $\sqrt[3]{54x^8} = \sqrt[3]{27\cdot2\cdot x^6\cdot x^2} = \sqrt[3]{27\cdot x^6\cdot2\cdot x^2} =$
    $\sqrt[3]{27} \sqrt[3]{x^6} \sqrt[3]{2x^2} = 3x^2 \sqrt[3]{2x^2}$

36. $2y\sqrt[3]{5}$

37. $\sqrt[3]{80x^8} = \sqrt[3]{8\cdot10\cdot x^6\cdot x^2} = \sqrt[3]{8\cdot x^6\cdot10\cdot x^2} =$
    $\sqrt[3]{8} \sqrt[3]{x^6} \sqrt[3]{10x^2} = 2x^2 \sqrt[3]{10x^2}$

38. $3m\sqrt[3]{4m^2}$

39. $\sqrt[4]{32} = \sqrt[4]{16\cdot2} = \sqrt[4]{16} \cdot \sqrt[4]{2} = 2\sqrt[4]{2}$

40. $2\sqrt[4]{5}$

41. $\sqrt[4]{810} = \sqrt[4]{81\cdot10} = \sqrt[4]{81} \cdot \sqrt[4]{10} = 3\sqrt[4]{10}$

42. $2\sqrt[4]{10}$

43. $\sqrt[4]{96a^8} = \sqrt[4]{16\cdot6\cdot a^8} = \sqrt[4]{16\cdot a^8\cdot6} =$
    $\sqrt[4]{16} \sqrt[4]{a^8} \sqrt[4]{6} = 2a^2 \sqrt[4]{6}$

44. $2x^2 \sqrt[4]{15}$

45. $\sqrt[4]{162c^4d^6} = \sqrt[4]{81\cdot2\cdot c^4\cdot d^4\cdot d^2} =$
    $\sqrt[4]{81\cdot c^4\cdot d^4\cdot2\cdot d^2} = \sqrt[4]{81} \sqrt[4]{c^4} \sqrt[4]{d^4} \sqrt[4]{2d^2} =$
    $3cd\sqrt[4]{2d^2}$

46. $3x^2y^2 \sqrt[4]{3y^2}$

47. $\sqrt[3]{(x+y)^4} = \sqrt[3]{(x+y)^3(x+y)} =$
    $\sqrt[3]{(x+y)^3} \sqrt[3]{x+y} = (x+y)\sqrt[3]{x+y}$

48. $(a-b)\sqrt[3]{(a-b)^2}$

49. $\sqrt[3]{8000(m+n)^8} = \sqrt[3]{8000(m+n)^6(m+n)^2} =$
    $\sqrt[3]{8000} \sqrt[3]{(m+n)^6} \sqrt[3]{(m+n)^2} = 20(m+n)^2 \sqrt[3]{(m+n)^2}$

50. $-10(x+y)^3 \sqrt[3]{x+y}$

51. $\sqrt[5]{-a^6b^{11}c^{17}} = \sqrt[5]{-1\cdot a^5\cdot a\cdot b^{10}\cdot b\cdot c^{15}\cdot c^2} =$
    $\sqrt[5]{-1\cdot a^5\cdot b^{10}\cdot c^{15}\cdot a\cdot b\cdot c^2} =$
    $\sqrt[5]{-1} \sqrt[5]{a^5} \sqrt[5]{b^{10}} \sqrt[5]{c^{15}} \sqrt[5]{abc^2} = -ab^2c^3 \sqrt[5]{abc^2}$

52. $x^2yz^4 \sqrt[5]{x^3y^3z^2}$

53. $\sqrt{3} \sqrt{6} = \sqrt{3\cdot6} = \sqrt{18} = \sqrt{9\cdot2} = 3\sqrt{2}$

54. $5\sqrt{2}$

55. $\sqrt{15} \sqrt{12} = \sqrt{15\cdot12} = \sqrt{180} = \sqrt{36\cdot5} = 6\sqrt{5}$

56. 8

57. $\sqrt{6} \sqrt{8} = \sqrt{6\cdot8} = \sqrt{48} = \sqrt{16\cdot3} = 4\sqrt{3}$

58. $6\sqrt{7}$

59. $\sqrt[3]{3} \sqrt[3]{18} = \sqrt[3]{3\cdot18} = \sqrt[3]{54} = \sqrt[3]{27\cdot2} = 3\sqrt[3]{2}$

60. $30\sqrt{3}$

61. $\sqrt{5b^3}\,\sqrt{10c^4} = \sqrt{5b^3\cdot10c^4}$ ⎫ Multiplying
    $\qquad = \sqrt{50b^3c^4}$ ⎭ radicands
    $\qquad = \sqrt{25b^2c^4\cdot2b}$ Factoring the radicand
    $\qquad = \sqrt{25b^2c^4}\,\sqrt{2b}$ Factoring into two radicals
    $\qquad = 5bc^2\sqrt{2b}$ Taking the square root of $25b^2c^4$

62. $-2a\sqrt[3]{15a^2}$

63. $\sqrt[3]{10x^5}\,\sqrt[3]{-75x^2} = \sqrt[3]{10x^5\cdot(-75x^2)}$ ⎫ Multiplying
    $\qquad = \sqrt[3]{-750x^7}$ ⎭ radicands
    $\qquad = \sqrt[3]{-125x^6\cdot6x}$ Factoring the radicand
    $\qquad = \sqrt[3]{-125x^6}\,\sqrt[3]{6x}$ Factoring into two radicals
    $\qquad = -5x^2\,\sqrt[3]{6x}$ Taking the cube root of $-125x^6$

64. $2x^2y\sqrt{6}$

65. $\sqrt[3]{y^4}\,\sqrt[3]{16y^5} = \sqrt[3]{y^4\cdot16y^5} = \sqrt[3]{16y^9} = \sqrt[3]{8y^9\cdot2} = 2y^3\,\sqrt[3]{2}$

66. $25t^3\,\sqrt[3]{t}$

67. $\sqrt[3]{(b+3)^4}\,\sqrt[3]{(b+3)^2} = \sqrt[3]{(b+3)^4(b+3)^2} = \sqrt[3]{(b+3)^6} = (b+3)^2$

68. $(x+y)^2\,\sqrt[3]{(x+y)^2}$

69. $\sqrt{12a^3b}\,\sqrt{8a^4b^2} = \sqrt{12a^3b\cdot8a^4b^2} = \sqrt{96a^7b^3} = \sqrt{16a^6b^2\cdot6ab} = \sqrt{16a^6b^2}\,\sqrt{6ab} = 4a^3b\sqrt{6ab}$

70. $6a^2b^4\sqrt{15ab}$

71. $\sqrt[5]{a^2(b+c)^4}\,\sqrt[5]{a^4(b+c)^7} =$
    $\sqrt[5]{a^2(b+c)^4\cdot a^4(b+c)^7} = \sqrt[5]{a^6(b+c)^{11}} =$
    $\sqrt[5]{a^5(b+c)^{10}\cdot a(b+c)} =$
    $\sqrt[5]{a^5(b+c)^{10}}\,\sqrt[5]{a(b+c)} = a(b+c)^2\,\sqrt[5]{a(b+c)}$

72. $x(y-z)^3\,\sqrt[5]{x^4(y-z)}$

73. Unless a number is a perfect square, its square root is irrational. Thus, $\sqrt{18}$ is irrational.

74. Rational

75. $8.23\overline{23}$ is rational, because decimal notation repeats.

76. Irrational

77. $\sqrt{36}$ is rational, because 36 is a perfect square.

78. Irrational

79. $2.101001...$ is irrational, because decimal notation neither terminates nor repeats.

80. Irrational

81. $\sqrt{252} = \sqrt{36\cdot7}$ Factoring the radicand
    $\qquad = \sqrt{36}\cdot\sqrt{7}$ Factoring the expression
    $\qquad = 6\sqrt{7}$
    $\qquad \approx 6\times2.646$ Using Table 1
    $\qquad \approx 15.9$ Multiplying and rounding

82. 11.8

83. $\sqrt{189} = \sqrt{9\cdot21}$ Factoring the radicand
    $\qquad = \sqrt{9}\cdot\sqrt{21}$ Factoring the expression
    $\qquad = 3\sqrt{21}$
    $\qquad \approx 3\times4.583$ Using Table 1
    $\qquad \approx 13.7$ Multiplying and rounding

84. 22.1

85. $\sqrt{350} = \sqrt{25\cdot14} = \sqrt{25}\,\sqrt{14} = 5\sqrt{14} \approx 5\times3.742 \approx 18.7$

86. 17.9

87. $\sqrt{891} = \sqrt{81\cdot11} = \sqrt{81}\,\sqrt{11} = 9\sqrt{11} \approx 9\times3.317 \approx 29.8$

88. 19.4

89. $\sqrt{240} = \sqrt{16\cdot15} = \sqrt{16}\,\sqrt{15} = 4\sqrt{15} \approx 4\times3.873 \approx 15.5$

90. 19.2

91. $\sqrt{24,500,000,000} = \sqrt{245\cdot10^8}$ Factoring the radicand
    $\qquad = \sqrt{245}\cdot\sqrt{10^8}$ Factoring the expression
    $\qquad \approx 15.65247584\cdot10^4$ Approximating $\sqrt{245}$ with a calculator and finding $\sqrt{10^8}$
    $\qquad \approx 156,524.7584$ Multiplying

92. $128,452.3258$

93. $\sqrt{468,200,000,000} = \sqrt{4682\cdot10^8}$
    $\qquad = \sqrt{4682}\cdot\sqrt{10^8}$
    $\qquad \approx 68.42514158\cdot10^4$
    $\qquad \approx 684,251.4158$

94. 315,277.6554

95. $\sqrt{175,420,000,000} = \sqrt{175,420 \cdot 10^6}$

$\qquad = \sqrt{175,420} \cdot \sqrt{10^6}$

$\qquad \approx 418.8317084 \cdot 10^3$

$\qquad \approx 418,831.7084$

96. 0.002151975836

97. $\sqrt{0.0000000395} = \sqrt{3.95 \cdot 10^{-8}}$

$\qquad = \sqrt{3.95} \cdot \sqrt{10^{-8}}$

$\qquad \approx 1.987460691 \cdot 10^{-4}$

$\qquad \approx 0.0001987460691$

98. 0.0003928103868

99. $\sqrt{0.0000005001} = \sqrt{50.01 \cdot 10^{-8}}$

$\qquad = \sqrt{50.01} \cdot \sqrt{10^{-8}}$

$\qquad \approx 7.071774883 \cdot 10^{-4}$

$\qquad \approx 0.0007071774883$

100. 0.003178207042

101. <u>Familiarize</u>. Let x and y represent the number of 30-sec and 60-sec commercials, respectively. Then the total number of minutes of commercial time during the show is $\frac{30x + 60y}{60}$, or $\frac{x}{2} + y$. (We divide by 60 to convert seconds to minutes.)

<u>Translate</u>. Rewording where necessary, we write two equations.

Total number of commercials　is　12.

$\qquad x + y \qquad = \qquad 12$

Number of 30-sec commercials　is　total minutes of commercial time　less 6.

$\qquad x \qquad = \qquad \frac{x}{2} + y \qquad - 6$

<u>Carry out</u>. Solving the system of equations, we get (4,8).

<u>Check</u>. If there are 4 30-sec and 8 60-sec commercials, the total number of commercials is 12. The total amount of commercial time is $4 \cdot 30$ sec $+ 8 \cdot 60$ sec $= 600$ sec, or 10 min. Then the number of 30-sec commercials is 6 less than the total number of minutes of commercial time. The values check.

<u>State</u>. 8 60-sec commercials were used.

102. $4x^2 - 9$

103. $4x^2 - 49 = (2x)^2 - 7^2 = (2x + 7)(2x - 7)$

104. $2(x - 9)(x - 4)$

105. $\qquad r = 2\sqrt{5L}$

a) $r = 2\sqrt{5 \cdot 20} = 2\sqrt{100} = 2 \cdot 10 = 20$ mph

b) $r = 2\sqrt{5 \cdot 70} = 2\sqrt{350}$

$\qquad \approx 2 \times 18.708$　Using Table 1 or a calculator

$\qquad \approx 37.4$ mph　Multiplying and rounding

c) $r = 2\sqrt{5 \cdot 90} = 2\sqrt{450}$

$\qquad \approx 2 \times 21.213$　Using Table 1 or a calculator

$\qquad \approx 42.4$ mph　Multiplying and rounding

106. a) $-3.3°$ C

b) $-16.6°$ C

c) $-25.5°$ C

d) $-54.0°$ C

107. $\sqrt[3]{5x^{k+1}} \ \sqrt[3]{25x^k} = 5x^7$

$\sqrt[3]{5x^{k+1} \cdot 25x^k} = 5x^7$

$\sqrt[3]{125x^{2k+1}} = 5x^7$

$\sqrt[3]{125} \ \sqrt[3]{x^{2k+1}} = 5x^7$

$5\sqrt[3]{x^{2k+1}} = 5x^7$

$\sqrt[3]{x^{2k+1}} = x^7$

$(\sqrt[3]{x^{2k+1}})^3 = (x^7)^3$

$x^{2k+1} = x^{21}$

Since the base is the same, the exponents must be equal. We have:

$2k + 1 = 21$

$2k = 20$

$k = 10$

108. 6

109. We must assume $2x + 3$ is nonnegative.

$2x + 3 \geqslant 0$

$2x \geqslant -3$

$x \geqslant -\frac{3}{2}$

Thus we must assume $x \geqslant -\frac{3}{2}$.

110.

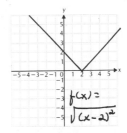

$f(x) = \sqrt{(x-2)^2}$

{x|x is a real number}

Exercise Set 7.3

1. $\sqrt{\dfrac{16}{25}} = \dfrac{\sqrt{16}}{\sqrt{25}} = \dfrac{4}{5}$

2. $\dfrac{10}{9}$

3. $\sqrt[3]{\dfrac{64}{27}} = \dfrac{\sqrt[3]{64}}{\sqrt[3]{27}} = \dfrac{4}{3}$

4. $\dfrac{7}{8}$

5. $\sqrt{\dfrac{49}{y^2}} = \dfrac{\sqrt{49}}{\sqrt{y^2}} = \dfrac{7}{y}$

6. $\dfrac{11}{x}$

7. $\sqrt{\dfrac{25y^3}{x^4}} = \dfrac{\sqrt{25y^3}}{\sqrt{x^4}} = \dfrac{\sqrt{25y^2 \cdot y}}{\sqrt{x^4}} = \dfrac{\sqrt{25y^2}\sqrt{y}}{\sqrt{x^4}} = \dfrac{5y\sqrt{y}}{x^2}$

8. $\dfrac{6a^2\sqrt{a}}{b^3}$

9. $\sqrt[3]{\dfrac{8x^5}{27y^3}} = \dfrac{\sqrt[3]{8x^5}}{\sqrt[3]{27y^3}} = \dfrac{\sqrt[3]{8x^3 \cdot x^2}}{\sqrt[3]{27y^3}} = \dfrac{\sqrt[3]{8x^3}\sqrt[3]{x^2}}{\sqrt[3]{27y^3}} = \dfrac{2x\sqrt[3]{x^2}}{3y}$

10. $\dfrac{2x^2\sqrt[3]{x}}{3y^2}$

11. $\sqrt[4]{\dfrac{16a^4}{81}} = \dfrac{\sqrt[4]{16a^4}}{\sqrt[4]{81}} = \dfrac{2a}{3}$

12. $\dfrac{3x}{y^2}$

13. $\sqrt[4]{\dfrac{a^5b^8}{c^{10}}} = \dfrac{\sqrt[4]{a^5b^8}}{\sqrt[4]{c^{10}}} = \dfrac{\sqrt[4]{a^4b^8 \cdot a}}{\sqrt[4]{c^8 \cdot c^2}} = \dfrac{\sqrt[4]{a^4b^8}\sqrt[4]{a}}{\sqrt[4]{c^8}\sqrt[4]{c^2}} =$

$\dfrac{ab^2\sqrt[4]{a}}{c^2\sqrt[4]{c^2}}$, or $\dfrac{ab^2}{c^2}\sqrt[4]{\dfrac{a}{c^2}}$

14. $\dfrac{x^2y^3}{z}\sqrt[4]{\dfrac{x}{z^2}}$

15. $\sqrt[5]{\dfrac{32x^6}{y^{11}}} = \dfrac{\sqrt[5]{32x^6}}{\sqrt[5]{y^{11}}} = \dfrac{\sqrt[5]{32x^5 \cdot x}}{\sqrt[5]{y^{10} \cdot y}} = \dfrac{\sqrt[5]{32x^5} \cdot \sqrt[5]{x}}{\sqrt[5]{y^{10}}\sqrt[5]{y}} =$

$\dfrac{2x\sqrt[5]{x}}{y^2\sqrt[5]{y}}$, or $\dfrac{2x}{y^2}\sqrt[5]{\dfrac{x}{y}}$

16. $\dfrac{3a}{b^2}\sqrt[5]{\dfrac{a^4}{b^3}}$

17. $\sqrt[6]{\dfrac{x^6y^8}{z^{15}}} = \dfrac{\sqrt[6]{x^6y^8}}{\sqrt[6]{z^{15}}} = \dfrac{\sqrt[6]{x^6y^6 \cdot y^2}}{\sqrt[6]{z^{12} \cdot z^3}} = \dfrac{\sqrt[6]{x^6y^6}\sqrt[6]{y^2}}{\sqrt[6]{z^{12}}\sqrt[6]{z^3}} =$

$\dfrac{xy\sqrt[6]{y^2}}{z^2\sqrt[6]{z^3}}$, or $\dfrac{xy}{z^2}\sqrt[6]{\dfrac{y^2}{z^3}}$

18. $\dfrac{ab^2}{c^2}\sqrt[6]{\dfrac{a^3}{c}}$

19. $\dfrac{\sqrt{21a}}{\sqrt{3a}} = \sqrt{\dfrac{21a}{3a}} = \sqrt{7}$

20. $\sqrt{7}$

21. $\dfrac{\sqrt[3]{54}}{\sqrt[3]{2}} = \sqrt[3]{\dfrac{54}{2}} = \sqrt[3]{27} = 3$

22. 2

23. $\dfrac{\sqrt{40xy^3}}{\sqrt{8x}} = \sqrt{\dfrac{40xy^3}{8x}} = \sqrt{5y^3} = \sqrt{y^2 \cdot 5y} = \sqrt{y^2}\sqrt{5y} =$

$y\sqrt{5y}$

24. $2b\sqrt{2b}$

25. $\dfrac{\sqrt[3]{96a^4b^2}}{\sqrt[3]{12a^2b}} = \sqrt[3]{\dfrac{96a^4b^2}{12a^2b}} = \sqrt[3]{8a^2b} = \sqrt[3]{8}\ \sqrt[3]{a^2b} =$

$2\sqrt[3]{a^2b}$

26. $3xy\sqrt[3]{y^2}$

27. $\dfrac{\sqrt{72xy}}{2\sqrt{2}} = \dfrac{1}{2} \cdot \sqrt{\dfrac{72xy}{2}} = \dfrac{1}{2}\sqrt{36xy} = \dfrac{1}{2}\sqrt{36}\sqrt{xy} =$

$\dfrac{1}{2} \cdot 6\sqrt{xy} = 3\sqrt{xy}$

28. $\dfrac{5}{3}\sqrt{ab}$

29. $\dfrac{\sqrt[4]{48x^3y^{11}}}{\sqrt[4]{3x^{-1}y^2}} = \sqrt[4]{\dfrac{48x^3y^{11}}{3x^{-1}y^2}} = \sqrt[4]{16x^4y^9} = \sqrt[4]{16x^4y^8 \cdot y} =$

$\sqrt[4]{16x^4y^8}\sqrt[4]{y} = 2xy^2\sqrt[4]{y}$

30. $\dfrac{a^2b}{3}$

31. a) $\sqrt{(6a)^3} = \sqrt{6^3a^2} = \sqrt{6^2a^2}\sqrt{6a} = 6a\sqrt{6a}$

    b) $(\sqrt{6a})^3 = \sqrt{6a}\sqrt{6a}\sqrt{6a} = 6a\sqrt{6a}$

32. $7y\sqrt{7y}$

33. a) $(\sqrt{16b^2})^3 = \sqrt{16b^2}\sqrt{16b^2}\sqrt{16b^2} =$

    $16b^2\sqrt{16b^2} = 16b^2 \cdot 4b = 64b^3$

    b) $\sqrt{(16b^2)^3} = \sqrt{(2^4b^2)^3} = \sqrt{2^{12}b^6} = 2^6b^3$, or $64b^3$

34. $125r^3$

35. a) $\sqrt{(18a^2b)^3} = \sqrt{5832a^6b^3} = \sqrt{2916a^6b^2 \cdot 2b} =$

54a^3b$\sqrt{2b}$

 b) $(\sqrt{18a^2b})^3 = \sqrt{18a^2b}\,\sqrt{18a^2b}\,\sqrt{18a^2b} =$

$18a^2b\sqrt{18a^2b} = 18a^2b\sqrt{9a^2 \cdot 2b} = 18a^2b \cdot 3a\sqrt{2b} =$

$54a^3b\sqrt{2b}$

36. $24x^3y\sqrt{3y}$

37. a) $(\sqrt[3]{3c^2d})^4 = \sqrt[3]{3c^2d}\,\sqrt[3]{3c^2d}\,\sqrt[3]{3c^2d}\,\sqrt[3]{3c^2d} =$

$3c^2d\,\sqrt[3]{3c^2d}$

 b) $\sqrt[3]{(3c^2d)^4} = \sqrt[3]{81c^8d^4} = \sqrt[3]{27c^6d^3 \cdot 3c^2d} =$

$3c^2d\,\sqrt[3]{3c^2d}$

38. $2x^2y\,\sqrt[3]{2x^2y}$

39. a) $\sqrt[3]{(5x^2y)^2} = \sqrt[3]{25x^4y} = \sqrt[3]{x^3 \cdot 25xy} = x\sqrt[3]{25xy}$

 b) $(\sqrt[3]{5x^2y})^2 = \sqrt[3]{5x^2y}\,\sqrt[3]{5x^2y} = \sqrt[3]{25x^4y} =$

$\sqrt[3]{x^3 \cdot 25xy} = x\sqrt[3]{25xy}$

40. $b\sqrt[3]{36a^2b}$

41. a) $\sqrt[4]{(x^2y)^3} = \sqrt[4]{x^6y^3} = \sqrt[4]{x^4 \cdot x^2y^3} = x\sqrt[4]{x^2y^3}$

 b) $(\sqrt[4]{x^2y})^3 = \sqrt[4]{x^2y}\,\sqrt[4]{x^2y}\,\sqrt[4]{x^2y} = \sqrt[4]{x^6y^3} =$

$\sqrt[4]{x^4 \cdot x^2y^3} = x\sqrt[4]{x^2y^3}$

42. $a^2\,\sqrt[4]{8a}$

43. a) $(\sqrt[3]{8a^4b})^2 = (\sqrt[3]{8a^3 \cdot ab})^2 = (2a\sqrt[3]{ab})^2 =$

$2a\sqrt[3]{ab} \cdot 2a\sqrt[3]{ab} = 4a^2\,\sqrt[3]{a^2b^2}$

 b) $\sqrt[3]{(8a^4b)^2} = \sqrt[3]{64a^8b^2} = \sqrt[3]{64a^6 \cdot a^2b^2} = 4a^2\,\sqrt[3]{a^2b^2}$

44. $9y^3\,\sqrt[3]{x^2y}$

45. a) $(\sqrt[4]{16x^2y^3})^2 = (2\sqrt[4]{x^2y^3})^2 =$

$2\sqrt[4]{x^2y^3} \cdot 2\sqrt[4]{x^2y^3} = 4\sqrt[4]{x^4y^6} = 4\sqrt[4]{x^4y^4 \cdot y^2} =$

$4xy\sqrt[4]{y^2}$

 b) $\sqrt[4]{(16x^2y^3)^2} = \sqrt[4]{256x^4y^6} = \sqrt[4]{256x^4y^4 \cdot y^2} =$

$4xy\sqrt[4]{y^2}$

46. $8y^3\,\sqrt[4]{x^3y^3}$

47.

$$\frac{12x}{x-4} - \frac{3x^2}{x+4} = \frac{384}{x^2-16}$$

$$\frac{12x}{x-4} - \frac{3x^2}{x+4} = \frac{384}{(x+4)(x-4)}$$

The LCM is $(x+4)(x-4)$.

$$(x+4)(x-4)\left[\frac{12x}{x-4} - \frac{3x^2}{x+4}\right] = (x+4)(x-4) \cdot \frac{384}{(x+4)(x-4)}$$

$12x(x+4) - 3x^2(x-4) = 384$

$12x^2 + 48x - 3x^3 + 12x^2 = 384$

$-3x^3 + 24x^2 + 48x - 384 = 0$

$-3x^2(x-8) + 48(x-8) = 0$

$(x-8)(-3x^2+48) = 0$

$x - 8 = 0$ or $\quad -3x^2 + 48 = 0$

$x = 8$ or $\quad -3(x^2 - 16) = 0$

$x = 8$ or $-3(x+4)(x-4) = 0$

$x = 8$ or $x + 4 = 0$ or $x - 4 = 0$

$x = 8$ or $\quad x = -4$ or $\quad x = 4$

Check: For $x = 8$:

$$\frac{12x}{x-4} - \frac{3x^2}{x+4} = \frac{384}{x^2-16}$$

$$\begin{array}{c|c} \dfrac{12\cdot 8}{8-4} - \dfrac{3\cdot 8^2}{8+4} & \dfrac{384}{8^2-16} \\[2mm] \dfrac{96}{4} - \dfrac{192}{12} & \dfrac{384}{48} \\[2mm] 24 - 16 & 8 \\[1mm] 8 & \end{array}$$

8 is a solution.

For $x = -4$:

$$\frac{12x}{x-4} - \frac{3x^2}{x+4} = \frac{384}{x^2-16}$$

$$\begin{array}{c|c} \dfrac{12(-4)}{-4-4} - \dfrac{3(-4)^2}{-4+4} & \dfrac{384}{(-4)^2-16} \\[2mm] \dfrac{-48}{-8} - \dfrac{48}{0} & \dfrac{384}{16-16} \end{array}$$

-4 is not a solution.

For $x = 4$:

$$\frac{12x}{x-4} - \frac{3x^2}{x+4} = \frac{384}{x^2-16}$$

$$\begin{array}{c|c} \dfrac{12\cdot 4}{4-4} - \dfrac{3\cdot 4^2}{4+4} & \dfrac{384}{4^2-16} \\[2mm] \dfrac{48}{0} - \dfrac{48}{8} & \dfrac{384}{16-16} \end{array}$$

4 is not a solution.

The solution is 8.

48. $\dfrac{15}{2}$

49. _Familiarize_. Let x and y represent the width and length of the rectangle, respectively.

Translate. We write two equations.

The width is one-fourth the length.

$$x \;=\; \frac{1}{4} \cdot y$$

The area is twice the perimeter.

$$xy \;=\; 2 \cdot (2x + 2y)$$

Carry out. Solving the system of equations we get (5,20).

Check. The width, 5, is one-fourth the length, 20. The area is $5 \cdot 20$, or 100. The perimeter is $2 \cdot 5 + 2 \cdot 20$, or 50. Since $100 = 2 \cdot 50$, the area is twice the perimeter. The values check.

State. The width is 5, and the length is 20.

50. a) 1.62 sec

  b) 1.99 sec

  c) 2.20 sec

51. $\dfrac{7\sqrt{a^2 b}\ \sqrt{25xy}}{5\sqrt{a^{-4}b^{-1}}\ \sqrt{49x^{-1}y^{-3}}} = \dfrac{7\sqrt{25a^2 bxy}}{5\sqrt{49a^{-4}b^{-1}x^{-1}y^{-3}}} =$

$\dfrac{7}{5}\sqrt{\dfrac{25a^2 bxy}{49a^{-4}b^{-1}x^{-1}y^{-3}}} = \dfrac{7}{5}\sqrt{\dfrac{25}{49} \cdot a^6 b^2 x^2 y^4} =$

$\dfrac{7}{5} \cdot \dfrac{5}{7} \cdot a^3 bxy^2 = a^3 bxy^2$

52. $9\sqrt[3]{9n^2}$

53. $\dfrac{\sqrt{44x^2 y^9 z}\ \sqrt{22y^9 z^6}}{(\sqrt{11xy^8 z^2})^2} = \dfrac{\sqrt{44 \cdot 22x^2 y^{18} z^7}}{\sqrt{11 \cdot 11x^2 y^{16} z^4}} =$

$\sqrt{\dfrac{44 \cdot 22x^2 y^{18} z^7}{11 \cdot 11x^2 y^{16} z^4}} = \sqrt{4 \cdot 2y^2 z^3} = \sqrt{4y^2 z^2 \cdot 2z} = 2yz\sqrt{2z}$

54. $\sqrt{x^2 + xy + y^2}$

55. Odd roots exist for all real numbers.

56. x^3 is negative for $x < 0$, but x^2 is never negative.

Exercise Set 7.4

1. $6\sqrt{3} + 2\sqrt{3} = (6 + 2)\sqrt{3} = 8\sqrt{3}$

2. $17\sqrt{5}$

3. $9\sqrt[3]{5} - 6\sqrt[3]{5} = (9 - 6)\sqrt[3]{5} = 3\sqrt[3]{5}$

4. $8\sqrt[5]{2}$

5. $4\sqrt[3]{y} + 9\sqrt[3]{y} = (4 + 9)\sqrt[3]{y} = 13\sqrt[3]{y}$

6. $3\sqrt[4]{t}$

7. $8\sqrt{2} - 6\sqrt{2} + 5\sqrt{2} = (8 - 6 + 5)\sqrt{2} = 7\sqrt{2}$

8. $7\sqrt{6}$

9. $4\sqrt[3]{3} - \sqrt{5} + 2\sqrt[3]{3} + \sqrt{5} =$

$(4 + 2)\sqrt[3]{3} + (-1 + 1)\sqrt{5} = 6\sqrt[3]{3}$

10. $6\sqrt{7} + \sqrt[4]{11}$

11. $8\sqrt{27} - 3\sqrt{3} = 8\sqrt{9 \cdot 3} - 3\sqrt{3}$ ⎱ Factoring the first radical

$= 8\sqrt{9} \cdot \sqrt{3} - 3\sqrt{3}$

$= 8 \cdot 3\sqrt{3} - 3\sqrt{3}$ Taking the square root

$= 24\sqrt{3} - 3\sqrt{3}$

$= (24 - 3)\sqrt{3}$ Factoring out $\sqrt{3}$

$= 21\sqrt{3}$

12. $41\sqrt{2}$

13. $8\sqrt{45} + 7\sqrt{20} = 8\sqrt{9 \cdot 5} + 7\sqrt{4 \cdot 5}$ ⎱ Factoring the radicals

$= 8\sqrt{9} \cdot \sqrt{5} + 7\sqrt{4} \cdot \sqrt{5}$

$= 8 \cdot 3\sqrt{5} + 7 \cdot 2\sqrt{5}$ Taking the square roots

$= 24\sqrt{5} + 14\sqrt{5}$

$= (24 + 14)\sqrt{5}$ Factoring out $\sqrt{5}$

$= 38\sqrt{5}$

14. $66\sqrt{3}$

15. $18\sqrt{72} + 2\sqrt{98} = 18\sqrt{36 \cdot 2} + 2\sqrt{49 \cdot 2} =$

$18\sqrt{36} \cdot \sqrt{2} + 2\sqrt{49} \cdot \sqrt{2} = 18 \cdot 6\sqrt{2} + 2 \cdot 7\sqrt{2} =$

$108\sqrt{2} + 14\sqrt{2} = (108 + 14)\sqrt{2} = 122\sqrt{2}$

16. $4\sqrt{5}$

17. $3\sqrt[3]{16} + \sqrt[3]{54} = 3\sqrt[3]{8 \cdot 2} + \sqrt[3]{27 \cdot 2} =$

$3\sqrt[3]{8} \cdot \sqrt[3]{2} + \sqrt[3]{27} \cdot \sqrt[3]{2} = 3 \cdot 2\sqrt[3]{2} + 3\sqrt[3]{2} =$

$6\sqrt[3]{2} + 3\sqrt[3]{2} = (6 + 3)\sqrt[3]{2} = 9\sqrt[3]{2}$

18. -7

19. $2\sqrt{128} - \sqrt{18} + 4\sqrt{32} =$

$2\sqrt{64 \cdot 2} - \sqrt{9 \cdot 2} + 4\sqrt{16 \cdot 2} =$

$2\sqrt{64} \cdot \sqrt{2} - \sqrt{9} \cdot \sqrt{2} + 4\sqrt{16} \cdot \sqrt{2} =$

$2 \cdot 8\sqrt{2} - 3\sqrt{2} + 4 \cdot 4\sqrt{2} = 16\sqrt{2} - 3\sqrt{2} + 16\sqrt{2} =$

$(16 - 3 + 16)\sqrt{2} = 29\sqrt{2}$

20. $55\sqrt{2}$

21. $\sqrt{5a} + 2\sqrt{45a^3} = \sqrt{5a} + 2\sqrt{9a^2 \cdot 5a} =$

$\sqrt{5a} + 2\sqrt{9a^2} \cdot \sqrt{5a} = \sqrt{5a} + 2\cdot 3a\sqrt{5a} =$

$\sqrt{5a} + 6a\sqrt{5a} = (1 + 6a)\sqrt{5a}$

22. $(4x - 2)\sqrt{3x}$

23. $\sqrt[3]{24x} - \sqrt[3]{3x^4} = \sqrt[3]{8\cdot 3x} - \sqrt[3]{x^3\cdot 3x} =$

$\sqrt[3]{8} \cdot \sqrt[3]{3x} - \sqrt[3]{x^3} \cdot \sqrt[3]{3x} = 2\sqrt[3]{3x} - x\sqrt[3]{3x} =$

$(2 - x)\sqrt[3]{3x}$

24. $(3 - x)\sqrt[3]{2x}$

25. $\sqrt{8y - 8} + \sqrt{2y - 2} = \sqrt{4(2y - 2)} + \sqrt{2y - 2} =$

$\sqrt{4} \cdot \sqrt{2y - 2} + \sqrt{2y - 2} = 2\sqrt{2y - 2} + \sqrt{2y - 2} =$

$(2 + 1)\sqrt{2y - 2} = 3\sqrt{2y - 2}$

26. $3\sqrt{3t + 3}$

27. $\sqrt{x^3 - x^2} + \sqrt{9x - 9} = \sqrt{x^2(x - 1)} + \sqrt{9(x - 1)} =$

$\sqrt{x^2} \cdot \sqrt{x - 1} + \sqrt{9} \cdot \sqrt{x - 1} = x\sqrt{x - 1} + 3\sqrt{x - 1} =$

$(x + 3)\sqrt{x - 1}$

28. $(2 - x)\sqrt{x - 1}$

29. $6\sqrt{8} + 3\sqrt{18} - 4\sqrt{32} = 6\sqrt{4\cdot 2} + 3\sqrt{9\cdot 2} - 4\sqrt{16\cdot 2} =$

$6\sqrt{4} \cdot \sqrt{2} + 3\sqrt{9} \cdot \sqrt{2} - 4\sqrt{16} \cdot \sqrt{2} =$

$6\cdot 2\sqrt{2} + 3\cdot 3\sqrt{2} - 4\cdot 4\sqrt{2} = 12\sqrt{2} + 9\sqrt{2} - 16\sqrt{2} =$

$(12 + 9 - 16)\sqrt{2} = 5\sqrt{2}$

30. $6\sqrt{3}$

31. $5\sqrt[3]{32} - \sqrt[3]{108} + 2\sqrt[3]{256} =$

$5\sqrt[3]{8\cdot 4} - \sqrt[3]{27\cdot 4} + 2\sqrt[3]{64\cdot 4} =$

$5\sqrt[3]{8} \cdot \sqrt[3]{4} - \sqrt[3]{27} \cdot \sqrt[3]{4} + 2\sqrt[3]{64} \cdot \sqrt[3]{4} =$

$5\cdot 2\sqrt[3]{4} - 3\sqrt[3]{4} + 2\cdot 4\sqrt[3]{4} = 10\sqrt[3]{4} - 3\sqrt[3]{4} + 8\sqrt[3]{4} =$

$(10 - 3 + 8)\sqrt[3]{4} = 15\sqrt[3]{4}$

32. $2\sqrt[3]{x}$

33. $9\sqrt{45} - 10\sqrt{20} - 4\sqrt{80} =$

$9\sqrt{9\cdot 5} - 10\sqrt{4\cdot 5} - 4\sqrt{16\cdot 5} =$

$9\sqrt{9} \cdot \sqrt{5} - 10\sqrt{4} \cdot \sqrt{5} - 4\sqrt{16} \cdot \sqrt{5} =$

$9\cdot 3\sqrt{5} - 10\cdot 2\sqrt{5} - 4\cdot 4\sqrt{5} =$

$27\sqrt{5} - 20\sqrt{5} - 16\sqrt{5} = (27 - 20 - 16)\sqrt{5} = -9\sqrt{5}$

34. $8\sqrt{7}$

35. $\sqrt{x^3 + x^2} + \sqrt{4x^3 + 4x^2} - \sqrt{9x^3 + 9x^2} =$

$\sqrt{x^2(x + 1)} + \sqrt{4x^2(x + 1)} - \sqrt{9x^2(x + 1)} =$

$\sqrt{x^2} \cdot \sqrt{x + 1} + \sqrt{4x^2} \cdot \sqrt{x + 1} - \sqrt{9x^2} \cdot \sqrt{x + 1} =$

$x\sqrt{x + 1} + 2x\sqrt{x + 1} - 3x\sqrt{x + 1} =$

$(x + 2x - 3x)\sqrt{x + 1} = 0$

36. $-8\sqrt{5x^2 + 4}$

37. $\sqrt[3]{32x^4} - \sqrt[3]{5x^2} - \sqrt[3]{108x^7} + \sqrt[3]{5x^5} =$

$\sqrt[3]{8x^3\cdot 4x} - \sqrt[3]{5x^2} - \sqrt[3]{27x^6\cdot 4x} + \sqrt[3]{x^3\cdot 5x^2} =$

$\sqrt[3]{8x^3}\cdot\sqrt[3]{4x} - \sqrt[3]{5x^2} - \sqrt[3]{27x^6}\cdot\sqrt[3]{4x} + \sqrt[3]{x^3}\cdot\sqrt[3]{5x^2} =$

$2x\sqrt[3]{4x} - \sqrt[3]{5x^2} - 3x^2 \sqrt[3]{4x} + x\sqrt[3]{5x^2} =$

$(2x - 3x^2)\sqrt[3]{4x} + (-1 + x)\sqrt[3]{5x^2}$, or

$(-3x^2 + 2x)\sqrt[3]{4x} + (x - 1)\sqrt[3]{5x^2}$

38. $(2a - 5a^2)\sqrt[3]{3a^2} + (2a - a^2)\sqrt[3]{2a}$

39. $\sqrt[4]{x^5 - x^4} + 3\sqrt[4]{x^9 - x^8} = \sqrt[4]{x^4(x - 1)} + 3\sqrt[4]{x^8(x - 1)} =$

$\sqrt[4]{x^4} \cdot \sqrt[4]{x - 1} + 3\sqrt[4]{x^8} \sqrt[4]{x - 1} = x\sqrt[4]{x - 1} + 3x^2 \sqrt[4]{x - 1} =$

$(x + 3x^2)\sqrt[4]{x - 1}$, or $(3x^2 + x)\sqrt[4]{x - 1}$

40. $(2a - 2a^2)\sqrt[4]{1 + a}$

41. $(4x + 2)(x - 5) = 4x^2 - 18x - 10$ Using FOIL

42. 12 ft by 15 ft

43. $\dfrac{3}{x - 2} + \dfrac{2}{x + 2} = \dfrac{x + 2}{x + 2} \cdot \dfrac{3}{x - 2} + \dfrac{x - 2}{x - 2} \cdot \dfrac{2}{x + 2}$

LCM is $(x - 2)(x + 2)$

$= \dfrac{3x + 6}{(x + 2)(x - 2)} + \dfrac{2x - 4}{(x - 2)(x + 2)}$

$= \dfrac{3x + 6 + 2x - 4}{(x - 2)(x + 2)}$

$= \dfrac{5x + 2}{(x - 2)(x + 2)}$

44. $\dfrac{10x + 9}{(x - 4)(x - 1)(x + 3)}$

45. $\sqrt{432} - \sqrt{6125} + \sqrt{845} - \sqrt{4800} =$

$\sqrt{144\cdot 3} - \sqrt{1225\cdot 5} + \sqrt{169\cdot 5} - \sqrt{1600\cdot 3} =$

$12\sqrt{3} - 35\sqrt{5} + 13\sqrt{5} - 40\sqrt{3} = -28\sqrt{3} - 22\sqrt{5}$

46. $(25x - 30y - 9xy)\sqrt{2xy}$

47. $\frac{1}{2}\sqrt{36a^5bc^4} - \frac{1}{2}\sqrt[3]{64a^4bc^6} + \frac{1}{6}\sqrt{144a^3bc^2} =$

$\frac{1}{2}\sqrt{36a^4c^4 \cdot ab} - \frac{1}{2}\sqrt[3]{64a^3c^6 \cdot ab} + \frac{1}{6}\sqrt{144a^2c^2 \cdot ab} =$

$\frac{1}{2}(6a^2c^2)\sqrt{ab} - \frac{1}{2}(4ac^2)\sqrt[3]{ab} + \frac{1}{6}(12ac)\sqrt{ab} =$

$3a^2c^2\sqrt{ab} - 2ac^2\sqrt[3]{ab} + 2ac\sqrt{ab} =$

$(3a^2c^2 + 2ac)\sqrt{ab} - 2ac^2\sqrt[3]{ab}$, or

$ac[(3ac + 2)\sqrt{ab} - 2c\sqrt[3]{ab}]$

48. $(7x^2 - 2y^2)\sqrt{x + y}$

49. $\sqrt{a} + \sqrt{b} = \sqrt{a + b}$ is true if $a = 0$ and $b \geqslant 0$, or if $a \geqslant 0$ and $b = 0$. Three pairs for which the equation is false are $a = 1$, $b = 1$; $a = 2$, $b = 7$; $a = 9$, $b = 4$. There are many others.

50. 24

Exercise Set 7.5

1. $\sqrt{6}(2 - 3\sqrt{6}) = \sqrt{6} \cdot 2 - \sqrt{6} \cdot 3\sqrt{6} = 2\sqrt{6} - 3 \cdot 6 =$

$2\sqrt{6} - 18$

2. $4\sqrt{3} + 3$

3. $\sqrt{2}(\sqrt{3} - \sqrt{5}) = \sqrt{2} \cdot \sqrt{3} - \sqrt{2} \cdot \sqrt{5} = \sqrt{6} - \sqrt{10}$

4. $5 - \sqrt{10}$

5. $\sqrt{3}(2\sqrt{5} - 3\sqrt{4}) = \sqrt{3}(2\sqrt{5} - 3 \cdot 2) =$

$\sqrt{3} \cdot 2\sqrt{5} - \sqrt{3} \cdot 6 = 2\sqrt{15} - 6\sqrt{3}$

6. $6\sqrt{5} - 4$

7. $\sqrt[3]{2}(\sqrt[3]{4} - 2\sqrt[3]{32}) = \sqrt[3]{2} \cdot \sqrt[3]{4} - \sqrt[3]{2} \cdot 2\sqrt[3]{32} =$

$\sqrt[3]{8} - 2\sqrt[3]{64} = 2 - 2 \cdot 4 = 2 - 8 = -6$

8. $3 - 4\sqrt[3]{63}$

9. $\sqrt[3]{a}(\sqrt[3]{2a^2} + \sqrt[3]{16a^2}) = \sqrt[3]{a} \cdot \sqrt[3]{2a^2} + \sqrt[3]{a} \cdot \sqrt[3]{16a^2} =$

$\sqrt[3]{2a^3} + \sqrt[3]{16a^3} = \sqrt[3]{a^3 \cdot 2} + \sqrt[3]{8a^3 \cdot 2} = a\sqrt[3]{2} + 2a\sqrt[3]{2} =$

$3a\sqrt[3]{2}$

10. $-2x\sqrt[3]{3}$

11. $\sqrt[4]{x}(\sqrt[4]{x^7} + \sqrt[4]{3x^2}) = \sqrt[4]{x} \cdot \sqrt[4]{x^7} + \sqrt[4]{x} \cdot \sqrt[4]{3x^2} =$

$\sqrt[4]{x^8} + \sqrt[4]{3x^3} = x^2 + \sqrt[4]{3x^3}$

12. $\sqrt[4]{2a^2} - a^3$

13. $\sqrt[5]{2a^3}(\sqrt[5]{16a^3} - \sqrt[5]{3a}) = \sqrt[5]{2a^3} \cdot \sqrt[5]{16a^3} - \sqrt[5]{2a^3} \cdot \sqrt[5]{3a} =$

$\sqrt[5]{32a^6} - \sqrt[5]{6a^4} = \sqrt[5]{32a^5 \cdot a} - \sqrt[5]{6a^4} = 2a\sqrt[5]{a} - \sqrt[5]{6a^4}$

14. $3x + \sqrt[5]{6x^3}$

15. $(5 - \sqrt{7})(5 + \sqrt{7}) = 5^2 - (\sqrt{7})^2 = 25 - 7 = 18$

16. 4

17. $(2 + \sqrt{3})(2 - \sqrt{3}) = 2^2 - (\sqrt{3})^2 = 4 - 3 = 1$

18. 47

19. $(\sqrt{3} - \sqrt{2})(\sqrt{3} + \sqrt{2}) = (\sqrt{3})^2 - (\sqrt{2})^2 =$

$3 - 2 = 1$

20. -1

21. $(\sqrt{5} + \sqrt{8})(\sqrt{5} - \sqrt{8}) = (\sqrt{5})^2 - (\sqrt{8})^2 =$

$5 - 8 = -3$

22. -2

23. $(3 - 2\sqrt{7})(3 + 2\sqrt{7}) = 3^2 - (2\sqrt{7})^2 = 9 - 4 \cdot 7 =$

$9 - 28 = -19$

24. -2

25. $(\sqrt{8} + 2\sqrt{5})(\sqrt{8} - 2\sqrt{5}) = (\sqrt{8})^2 - (2\sqrt{5})^2 =$

$8 - 4 \cdot 5 = 8 - 20 = -12$

26. -45

27. $(\sqrt{a} + \sqrt{b})(\sqrt{a} - \sqrt{b}) = (\sqrt{a})^2 - (\sqrt{b})^2 = a - b$

28. $x - y$

29. $(3 - \sqrt{5})(2 + \sqrt{5})$

$= 3 \cdot 2 + 3\sqrt{5} - 2\sqrt{5} - (\sqrt{5})^2$ Using FOIL

$= 6 + 3\sqrt{5} - 2\sqrt{5} - 5$

$= 1 + \sqrt{5}$ Simplifying

30. $2 + 2\sqrt{6}$

31. $(\sqrt{3} + 1)(2\sqrt{3} + 1)$

$= \sqrt{3} \cdot 2\sqrt{3} + \sqrt{3} \cdot 1 + 1 \cdot 2\sqrt{3} + 1^2$ Using FOIL

$= 2 \cdot 3 + \sqrt{3} + 2\sqrt{3} + 1$

$= 7 + 3\sqrt{3}$ Simplifying

32. $2 - 3\sqrt{3}$

33. $(2\sqrt{7} - 4\sqrt{2})(3\sqrt{7} + 6\sqrt{2}) =$

$2\sqrt{7} \cdot 3\sqrt{7} + 2\sqrt{7} \cdot 6\sqrt{2} - 4\sqrt{2} \cdot 3\sqrt{7} - 4\sqrt{2} \cdot 6\sqrt{2} =$

$6 \cdot 7 + 12\sqrt{14} - 12\sqrt{14} - 24 \cdot 2 =$

$42 + 12\sqrt{14} - 12\sqrt{14} - 48 = -6$

34. $24 - 7\sqrt{15}$

35. $(\sqrt{a} + \sqrt{2})(\sqrt{a} + \sqrt{3}) =$

$(\sqrt{a})^2 + \sqrt{a}\cdot\sqrt{3} + \sqrt{2}\cdot\sqrt{a} + \sqrt{2}\cdot\sqrt{3} =$

$a + \sqrt{3a} + \sqrt{2a} + \sqrt{6}$

36. $2 - 3\sqrt{x} + x$

37. $(2\sqrt[3]{3} + \sqrt[3]{2})(\sqrt[3]{3} - 2\sqrt[3]{2}) =$

$2\sqrt[3]{3}\cdot\sqrt[3]{3} - 2\sqrt[3]{3}\cdot2\sqrt[3]{2} + \sqrt[3]{2}\cdot\sqrt[3]{3} - \sqrt[3]{2}\cdot2\sqrt[3]{2} =$

$2\sqrt[3]{9} - 4\sqrt[3]{6} + \sqrt[3]{6} - 2\sqrt[3]{4} = 2\sqrt[3]{9} - 3\sqrt[3]{6} - 2\sqrt[3]{4}$

38. $6\sqrt[4]{63} - 9\sqrt[4]{42} + 2\sqrt[4]{54} - 3\sqrt[4]{36}$

39. $(2 + \sqrt{3})^2 = 2^2 + 4\sqrt{3} + (\sqrt{3})^2$ Squaring a binomial

$= 4 + 4\sqrt{3} + 3$

$= 7 + 4\sqrt{3}$

40. $6 + 2\sqrt{5}$

41. $(a + \sqrt{b})^2 = a^2 + 2a\sqrt{b} + (\sqrt{b})^2$ Squaring a binomial

$= a^2 + 2a\sqrt{b} + b$

42. $x^2 - 2x\sqrt{y} + y$

43. $(2x - \sqrt{y})^2 = (2x)^2 - 2\cdot2x\sqrt{y} + (\sqrt{y})^2$

$= 4x^2 - 4x\sqrt{y} + y$

44. $9a^2 + 6a\sqrt{b} + b$

45. $(\sqrt{m} + \sqrt{n})^2 = (\sqrt{m})^2 + 2\sqrt{m}\cdot\sqrt{n} + (\sqrt{n})^2$

$= m + 2\sqrt{mn} + n$

46. $r - 2\sqrt{rs} + s$

47. $\dfrac{5}{x - 1} + \dfrac{9}{x^2 + x + 1} = \dfrac{15}{x^3 - 1}$

$\dfrac{5}{x-1} + \dfrac{9}{x^2+x+1} = \dfrac{15}{(x-1)(x^2+x+1)}$

The LCM is $(x - 1)(x^2 + x + 1)$.

$(x-1)(x^2+x+1)\left[\dfrac{5}{x-1} + \dfrac{9}{x^2+x+1}\right] =$

$\qquad (x-1)(x^2+x+1) \cdot \dfrac{15}{(x-1)(x^2+x+1)}$

$5(x^2 + x + 1) + 9(x - 1) = 15$

$5x^2 + 5x + 5 + 9x - 9 = 15$

$5x^2 + 14x - 4 = 15$

$5x^2 + 14x - 19 = 0$

$(5x + 19)(x - 1) = 0$

$5x + 19 = 0 \quad$ or $\quad x - 1 = 0$

$5x = -19 \quad$ or $\qquad x = 1$

$x = -\dfrac{19}{5}$ or $\qquad x = 1$

Only $-\dfrac{19}{5}$ checks. The solution is $-\dfrac{19}{5}$.

48. $\dfrac{x - 2}{x + 3}$

49. a) $(3 + \sqrt{2})^2 \approx (3 + 1.414213562)^2 \approx$
$(4.414213562)^2 \approx 19.48528137$

    b) $(3 + \sqrt{2})^2 = 9 + 6\sqrt{2} + 2 \approx$
$9 + 6(1.414213562) + 2 \approx 9 + 8.48528137 + 2 = 19.48528137$

50. 6

51. $(\sqrt{x + 2} - \sqrt{x - 2})^2 =$

$(\sqrt{x + 2})^2 - 2\sqrt{x + 2}\,\sqrt{x - 2} + (\sqrt{x - 2})^2 =$

$x + 2 - 2\sqrt{x^2 - 4} + x - 2 = 2x - 2\sqrt{x^2 - 4}$

52. $14 + 2\sqrt{15} - 6\sqrt{2} - 2\sqrt{30}$

53. $\sqrt[3]{y}(1 - \sqrt[3]{y})(1 + \sqrt[3]{y}) = \sqrt[3]{y}[1 - (\sqrt[3]{y})^2] =$

$\sqrt[3]{y}(1 - \sqrt[3]{y^2}) = \sqrt[3]{y} - \sqrt[3]{y^3} = \sqrt[3]{y} - y$

54. $\sqrt[3]{91}$

55. $\left[\sqrt{3 + \sqrt{2 + \sqrt{1}}}\,\right]^4 = \left[\sqrt{3 + \sqrt{2 + 1}}\,\right]^4 =$

$\left[\sqrt{3 + \sqrt{3}}\,\right]^4 = \sqrt{(3 + \sqrt{3})^4} = (3 + \sqrt{3})^2 =$

$9 + 6\sqrt{3} + 3 = 12 + 6\sqrt{3}$

56. y

57. $(8\sqrt{a + 3} - a\sqrt{3})(2\sqrt{a + 3} + 5a\sqrt{3}) =$

$16(a + 3) + 40a\sqrt{3a + 9} - 2a\sqrt{3a + 9} - 15a^2 =$

$16(a + 3) + 38a\sqrt{3a + 9} - 15a^2$

58. $48\sqrt{21} - 16\sqrt{15} + 30\sqrt{14} - 10\sqrt{10}$

59. $(\sqrt{17} + \sqrt{2} - \sqrt{19})(\sqrt{17} + \sqrt{2} + \sqrt{19}) =$
$[(\sqrt{17} + \sqrt{2}) - \sqrt{19}][(\sqrt{17} + \sqrt{2}) + \sqrt{19}] =$
$(\sqrt{17} + \sqrt{2})^2 - (\sqrt{19})^2 = 17 + 2\sqrt{34} + 2 - 19 =$
$2\sqrt{34}$

60. $7 + 4\sqrt{3} = (2 + \sqrt{3})^2$ and $7 - 4\sqrt{3} = (2 - \sqrt{3})^2$ so
$\sqrt{7 + 4\sqrt{3}} - \sqrt{7 - 4\sqrt{3}} =$
$\sqrt{(2 + \sqrt{3})^2} - \sqrt{(2 - \sqrt{3})^2} = 2 + \sqrt{3} - (2 - \sqrt{3}) =$
$2 + \sqrt{3} - 2 + \sqrt{3} = 2\sqrt{3}.$

Exercise Set 7.6

1. $\sqrt{\dfrac{6}{5}} = \sqrt{\dfrac{6}{5} \cdot \dfrac{5}{5}} = \sqrt{\dfrac{30}{25}} = \dfrac{\sqrt{30}}{\sqrt{25}} = \dfrac{\sqrt{30}}{5}$

2. $\dfrac{\sqrt{66}}{6}$

3. $\sqrt{\dfrac{10}{7}} = \sqrt{\dfrac{10}{7} \cdot \dfrac{7}{7}} = \sqrt{\dfrac{70}{49}} = \dfrac{\sqrt{70}}{\sqrt{49}} = \dfrac{\sqrt{70}}{7}$

4. $\dfrac{\sqrt{66}}{3}$

5. $\dfrac{6\sqrt{5}}{5\sqrt{3}} = \dfrac{6\sqrt{5}}{5\sqrt{3}} \cdot \dfrac{\sqrt{3}}{\sqrt{3}} = \dfrac{6\sqrt{15}}{5 \cdot 3} = \dfrac{2\sqrt{15}}{5}$

6. $\dfrac{\sqrt{6}}{5}$

7. $\sqrt[3]{\dfrac{16}{9}} = \sqrt[3]{\dfrac{16}{9} \cdot \dfrac{3}{3}} = \sqrt[3]{\dfrac{48}{27}} = \dfrac{\sqrt[3]{8 \cdot 6}}{\sqrt[3]{27}} = \dfrac{2\sqrt[3]{6}}{3}$

8. $\dfrac{\sqrt[3]{6}}{3}$

9. $\dfrac{\sqrt[3]{3a}}{\sqrt[3]{5c}} = \dfrac{\sqrt[3]{3a}}{\sqrt[3]{5c}} \cdot \dfrac{\sqrt[3]{(5c)^2}}{\sqrt[3]{(5c)^2}} = \dfrac{\sqrt[3]{3a \cdot 25c^2}}{\sqrt[3]{(5c)^3}} = \dfrac{\sqrt[3]{75ac^2}}{5c}$

10. $\dfrac{\sqrt[3]{63xy^2}}{3y}$

11. $\dfrac{\sqrt[3]{5y^4}}{\sqrt[3]{6x^4}} = \dfrac{\sqrt[3]{5y^4}}{\sqrt[3]{6x^4}} \cdot \dfrac{\sqrt[3]{36x^2}}{\sqrt[3]{36x^2}} = \dfrac{\sqrt[3]{y^3 \cdot 180x^2y}}{\sqrt[3]{216x^6}} = \dfrac{y\sqrt[3]{180x^2y}}{6x^2}$

12. $\dfrac{a\sqrt[3]{147ab}}{7b}$

13. $\dfrac{1}{\sqrt[3]{xy}} = \dfrac{1}{\sqrt[3]{xy}} \cdot \dfrac{\sqrt[3]{(xy)^2}}{\sqrt[3]{(xy)^2}} = \dfrac{\sqrt[3]{(xy)^2}}{\sqrt[3]{(xy)^3}} = \dfrac{\sqrt[3]{x^2y^2}}{xy}$

14. $\dfrac{\sqrt[3]{a^2b^2}}{ab}$

15. $\sqrt{\dfrac{7a}{18}} = \sqrt{\dfrac{7a}{18} \cdot \dfrac{2}{2}} = \sqrt{\dfrac{14a}{36}} = \dfrac{\sqrt{14a}}{\sqrt{36}} = \dfrac{\sqrt{14a}}{6}$

16. $\dfrac{\sqrt{30x}}{10}$

17. $\sqrt{\dfrac{9}{20x^2y}} = \sqrt{\dfrac{9}{20x^2y} \cdot \dfrac{5y}{5y}} = \sqrt{\dfrac{9 \cdot 5y}{100x^2y^2}} = \dfrac{\sqrt{9 \cdot 5y}}{\sqrt{100x^2y^2}} =$
$\dfrac{3\sqrt{5y}}{10xy}$

18. $\dfrac{\sqrt{10a}}{8ab}$

19. $\sqrt[3]{\dfrac{9}{100x^2y^5}} = \sqrt[3]{\dfrac{9}{100x^2y^5} \cdot \dfrac{10xy}{10xy}} = \sqrt[3]{\dfrac{90xy}{1000x^3y^6}} =$
$\dfrac{\sqrt[3]{90xy}}{\sqrt[3]{1000x^3y^6}} = \dfrac{\sqrt[3]{90xy}}{10xy^2}$

20. $\dfrac{\sqrt[3]{42a^2b^2}}{6a^2b}$

21. $\dfrac{\sqrt{7}}{\sqrt{3x}} = \dfrac{\sqrt{7}}{\sqrt{3x}} \cdot \dfrac{\sqrt{7}}{\sqrt{7}} = \dfrac{7}{\sqrt{21x}}$

22. $\dfrac{6}{\sqrt{30x}}$

23. $\sqrt{\dfrac{14}{21}} = \sqrt{\dfrac{2}{3}} = \sqrt{\dfrac{2}{3} \cdot \dfrac{2}{2}} = \sqrt{\dfrac{4}{6}} = \dfrac{\sqrt{4}}{\sqrt{6}} = \dfrac{2}{\sqrt{6}}$

24. $\dfrac{2}{\sqrt{5}}$

25. $\dfrac{4\sqrt{13}}{3\sqrt{7}} = \dfrac{4\sqrt{13}}{3\sqrt{7}} \cdot \dfrac{\sqrt{13}}{\sqrt{13}} = \dfrac{4 \cdot 13}{3\sqrt{91}} = \dfrac{52}{3\sqrt{91}}$

26. $\dfrac{105}{2\sqrt{105}}$

27. $\dfrac{\sqrt[3]{7}}{\sqrt[3]{2}} = \dfrac{\sqrt[3]{7}}{\sqrt[3]{2}} \cdot \dfrac{\sqrt[3]{49}}{\sqrt[3]{49}} = \dfrac{7}{\sqrt[3]{98}}$

28. $\dfrac{5}{\sqrt[3]{100}}$

29. $\sqrt{\dfrac{7x}{3y}} = \sqrt{\dfrac{7x}{3y} \cdot \dfrac{7x}{7x}} = \dfrac{\sqrt{49x^2}}{\sqrt{21xy}} = \dfrac{7x}{\sqrt{21xy}}$

30. $\dfrac{6a}{\sqrt{30ab}}$

31. $\dfrac{\sqrt[3]{5y^4}}{\sqrt[3]{6x^5}} = \dfrac{\sqrt[3]{5y^4}}{\sqrt[3]{6x^5}} \cdot \dfrac{\sqrt[3]{25y^2}}{\sqrt[3]{25y^2}} = \dfrac{\sqrt[3]{125y^6}}{\sqrt[3]{150x^5y^2}} =$
$\dfrac{5y^2}{\sqrt{x^3 \cdot 150x^2y^2}} = \dfrac{5y^2}{x\sqrt[3]{150x^2y^2}}$

32. $\dfrac{3a^2}{\sqrt[3]{63ab^2}}$

33. $\dfrac{\sqrt{ab}}{3} = \dfrac{\sqrt{ab}}{3} \cdot \dfrac{\sqrt{ab}}{\sqrt{ab}} = \dfrac{ab}{3\sqrt{ab}}$

34. $\dfrac{xy}{5\sqrt{xy}}$

35. $\sqrt{\dfrac{x^3y}{2}} = \sqrt{\dfrac{x^3y}{2} \cdot \dfrac{xy}{xy}} = \sqrt{\dfrac{x^4y^2}{2xy}} = \dfrac{\sqrt{x^4y^2}}{\sqrt{2xy}} = \dfrac{x^2y}{\sqrt{2xy}}$

36. $\dfrac{ab^3}{\sqrt{3ab}}$

37. $\dfrac{\sqrt[3]{a^2b}}{\sqrt[3]{5}} = \dfrac{\sqrt[3]{a^2b}}{\sqrt[3]{5}} \cdot \dfrac{\sqrt[3]{ab^2}}{\sqrt[3]{ab^2}} = \dfrac{\sqrt[3]{a^3b^3}}{\sqrt[3]{5ab^2}} = \dfrac{ab}{\sqrt[3]{5ab^2}}$

38. $\dfrac{xy}{\sqrt[3]{7x^2y}}$

39. $\sqrt[3]{\dfrac{x^4y^2}{3}} = \sqrt[3]{\dfrac{x^4y^2}{3} \cdot \dfrac{x^2y}{x^2y}} = \sqrt[3]{\dfrac{x^6y^3}{3x^2y}} = \dfrac{\sqrt[3]{x^6y^3}}{\sqrt[3]{3x^2y}} = \dfrac{x^2y}{\sqrt[3]{3x^2y}}$

40. $\dfrac{a^2b}{\sqrt[3]{2ab^2}}$

41. $\dfrac{5}{8 - \sqrt{6}} = \dfrac{5}{8 - \sqrt{6}} \cdot \dfrac{8 + \sqrt{6}}{8 + \sqrt{6}} = \dfrac{5(8 + \sqrt{6})}{8^2 - (\sqrt{6})^2} =$

$\dfrac{5(8 + \sqrt{6})}{64 - 6} = \dfrac{5(8 + \sqrt{6})}{58}$

42. $\dfrac{7(9 - \sqrt{10})}{71}$

43. $\dfrac{-4\sqrt{7}}{\sqrt{5} - \sqrt{3}} = \dfrac{-4\sqrt{7}}{\sqrt{5} - \sqrt{3}} \cdot \dfrac{\sqrt{5} + \sqrt{3}}{\sqrt{5} + \sqrt{3}} =$

$\dfrac{-4\sqrt{7}(\sqrt{5} + \sqrt{3})}{5 - 3} = \dfrac{-4\sqrt{7}(\sqrt{5} + \sqrt{3})}{2} =$

$-2\sqrt{7}(\sqrt{5} + \sqrt{3})$

44. $\dfrac{3\sqrt{2}(\sqrt{3} + \sqrt{5})}{2}$

45. $\dfrac{\sqrt{5} - 2\sqrt{6}}{\sqrt{3} - 4\sqrt{5}} = \dfrac{\sqrt{5} - 2\sqrt{6}}{\sqrt{3} - 4\sqrt{5}} \cdot \dfrac{\sqrt{3} + 4\sqrt{5}}{\sqrt{3} + 4\sqrt{5}} =$

$\dfrac{\sqrt{15} + 4 \cdot 5 - 2\sqrt{18} - 8\sqrt{30}}{(\sqrt{3})^2 - (4\sqrt{5})^2} =$

$\dfrac{\sqrt{15} + 20 - 2\sqrt{9 \cdot 2} - 8\sqrt{30}}{3 - 16 \cdot 5} =$

$\dfrac{\sqrt{15} + 20 - 6\sqrt{2} - 8\sqrt{30}}{-77}$

46. $\dfrac{3\sqrt{2} + 2\sqrt{42} - 3\sqrt{15} - 6\sqrt{35}}{-25}$

47. $\dfrac{\sqrt{x} - \sqrt{y}}{\sqrt{x} + \sqrt{y}} = \dfrac{\sqrt{x} - \sqrt{y}}{\sqrt{x} + \sqrt{y}} \cdot \dfrac{\sqrt{x} - \sqrt{y}}{\sqrt{x} - \sqrt{y}} =$

$\dfrac{x - \sqrt{xy} - \sqrt{xy} + y}{x - y} = \dfrac{x - 2\sqrt{xy} + y}{x - y}$

48. $\dfrac{a + 2\sqrt{ab} + b}{a - b}$

49. $\dfrac{5\sqrt{3} - 3\sqrt{2}}{3\sqrt{2} - 2\sqrt{3}} = \dfrac{5\sqrt{3} - 3\sqrt{2}}{3\sqrt{2} - 2\sqrt{3}} \cdot \dfrac{3\sqrt{2} + 2\sqrt{3}}{3\sqrt{2} + 2\sqrt{3}} =$

$\dfrac{15\sqrt{6} + 10 \cdot 3 - 9 \cdot 2 - 6\sqrt{6}}{9 \cdot 2 - 4 \cdot 3} = \dfrac{12 + 9\sqrt{6}}{6} =$

$\dfrac{3(4 + 3\sqrt{6})}{3 \cdot 2} = \dfrac{3}{3} \cdot \dfrac{4 + 3\sqrt{6}}{2} = \dfrac{4 + 3\sqrt{6}}{2}$

50. $\dfrac{4\sqrt{6} + 9}{3}$

51. $\dfrac{3\sqrt{x} + \sqrt{y}}{2\sqrt{x} + 3\sqrt{y}} = \dfrac{3\sqrt{x} + \sqrt{y}}{2\sqrt{x} + 3\sqrt{y}} \cdot \dfrac{2\sqrt{x} - 3\sqrt{y}}{2\sqrt{x} - 3\sqrt{y}} =$

$\dfrac{6x - 9\sqrt{xy} + 2\sqrt{xy} - 3y}{4x - 9y} = \dfrac{6x - 7\sqrt{xy} - 3y}{4x - 9y}$

52. $\dfrac{6a - 7\sqrt{ab} + 2b}{9a - 4b}$

53. $\dfrac{\sqrt{3} + 5}{8} = \dfrac{\sqrt{3} + 5}{8} \cdot \dfrac{\sqrt{3} - 5}{\sqrt{3} - 5} = \dfrac{3 - 25}{8(\sqrt{3} - 5)} =$

$\dfrac{-22}{8(\sqrt{3} - 5)} = \dfrac{2(-11)}{2 \cdot 4(\sqrt{3} - 5)} = \dfrac{2}{2} \cdot \dfrac{-11}{4(\sqrt{3} - 5)} =$

$\dfrac{-11}{4(\sqrt{3} - 5)}$

54. $\dfrac{7}{5(3 + \sqrt{2})}$

55. $\dfrac{\sqrt{3} - 5}{\sqrt{2} + 5} = \dfrac{\sqrt{3} - 5}{\sqrt{2} + 5} \cdot \dfrac{\sqrt{3} + 5}{\sqrt{3} + 5} =$

$\dfrac{3 - 25}{\sqrt{6} + 5\sqrt{2} + 5\sqrt{3} + 25} = \dfrac{-22}{\sqrt{6} + 5\sqrt{2} + 5\sqrt{3} + 25}$

56. $\dfrac{-3}{3\sqrt{2} + 3\sqrt{3} + 7\sqrt{6} + 21}$

57. $\dfrac{\sqrt{x} - \sqrt{y}}{\sqrt{x} + \sqrt{y}} = \dfrac{\sqrt{x} - \sqrt{y}}{\sqrt{x} + \sqrt{y}} \cdot \dfrac{\sqrt{x} + \sqrt{y}}{\sqrt{x} + \sqrt{y}} =$

$\dfrac{x - y}{x + \sqrt{xy} + \sqrt{xy} + y} = \dfrac{x - y}{x + 2\sqrt{xy} + y}$

58. $\dfrac{x - y}{x - 2\sqrt{xy} + y}$

59. $\dfrac{4\sqrt{6} - 5\sqrt{3}}{2\sqrt{3} + 7\sqrt{6}} = \dfrac{4\sqrt{6} - 5\sqrt{3}}{2\sqrt{3} + 7\sqrt{6}} \cdot \dfrac{4\sqrt{6} + 5\sqrt{3}}{4\sqrt{6} + 5\sqrt{3}} =$

$\dfrac{16\cdot6 - 25\cdot3}{8\sqrt{18} + 10\cdot3 + 28\cdot6 + 35\sqrt{18}} = \dfrac{96 - 75}{43\sqrt{18} + 30 + 168} =$

$\dfrac{21}{43\sqrt{9\cdot2} + 198} = \dfrac{21}{43\cdot3\sqrt{2} + 198} = \dfrac{3\cdot7}{3(43\sqrt{2} + 66)} =$

$\dfrac{7}{43\sqrt{2} + 66}$

60. $\dfrac{53}{75\sqrt{6} - 187}$

61. $\dfrac{\sqrt{3} + 2\sqrt{x}}{\sqrt{3} - \sqrt{x}} = \dfrac{\sqrt{3} + 2\sqrt{x}}{\sqrt{3} - \sqrt{x}} \cdot \dfrac{\sqrt{3} - 2\sqrt{x}}{\sqrt{3} - 2\sqrt{x}} =$

$\dfrac{3 - 4x}{3 - 2\sqrt{3x} - \sqrt{3x} + 2x} = \dfrac{3 - 4x}{3 - 3\sqrt{3x} + 2x}$

62. $\dfrac{5 - 9x}{5 + 4\sqrt{5x} + 3x}$

63. $\dfrac{a + b\sqrt{c}}{a + \sqrt{c}} = \dfrac{a + b\sqrt{c}}{a + \sqrt{c}} \cdot \dfrac{a - b\sqrt{c}}{a - b\sqrt{c}} =$

$\dfrac{a^2 - b^2c}{a^2 - ab\sqrt{c} + a\sqrt{c} - bc} = \dfrac{a^2 - b^2c}{a^2 + (-ab + a)\sqrt{c} - bc}$,

or $\dfrac{a^2 - b^2c}{a^2 + (a - ab)\sqrt{c} - bc}$

64. $\dfrac{a^2b - c^2}{ab + (1 - a)c\sqrt{b} - c^2}$

65. $\dfrac{1}{2} - \dfrac{1}{3} = \dfrac{1}{t}$ LCM is 6t

$6t\left(\dfrac{1}{2} - \dfrac{1}{3}\right) = 6t\left(\dfrac{1}{t}\right)$

$3t - 2t = 6$

$t = 6$

Check:

$\dfrac{1}{2} - \dfrac{1}{3} = \dfrac{1}{t}$

$\begin{array}{c|c} \dfrac{1}{2} - \dfrac{1}{3} & \dfrac{1}{6} \\[2mm] \dfrac{3}{6} - \dfrac{2}{6} & \\[2mm] \dfrac{1}{6} & \end{array}$

The solution is 6.

66. 1

67. $\dfrac{a - \sqrt{a + b}}{\sqrt{a + b} - b} = \dfrac{a - \sqrt{a + b}}{\sqrt{a + b} - b} \cdot \dfrac{\sqrt{a + b} + b}{\sqrt{a + b} + b} =$

$\dfrac{a\sqrt{a + b} + ab - (a + b) - b\sqrt{a + b}}{a + b - b^2} =$

$\dfrac{ab + (a - b)\sqrt{a + b} - a - b}{a + b - b^2}$

68. $\dfrac{15y + 6\sqrt{y(z + y)} + 20y\sqrt{z} + 8\sqrt{yz(z + y)}}{21y - 4z}$

69. $\dfrac{b + \sqrt{b}}{1 + b + \sqrt{b}} = \dfrac{b + \sqrt{b}}{(1 + b) + \sqrt{b}} \cdot \dfrac{(1 + b) - \sqrt{b}}{(1 + b) - \sqrt{b}} =$

$\dfrac{b(1 + b) - b\sqrt{b} + \sqrt{b}(1 + b) - b}{(1 + b)^2 - b} =$

$\dfrac{b + b^2 - b\sqrt{b} + \sqrt{b} + b\sqrt{b} - b}{1 + 2b + b^2 - b} = \dfrac{b^2 + \sqrt{b}}{1 + b + b^2}$

70. $\dfrac{12(3 + \sqrt{3} - y)}{6 + y}$

71. $\dfrac{a - b}{a^2 + a\sqrt{a - b}} = \dfrac{a - b}{a^2 + a\sqrt{a - b}} \cdot \dfrac{a^2 - a\sqrt{a - b}}{a^2 - a\sqrt{a - b}} =$

$\dfrac{a^3 - a^2\sqrt{a - b} - a^2b + ab\sqrt{a - b}}{a^4 - a^2(a - b)} =$

$\dfrac{a^3 - a^2\sqrt{a - b} - a^2b + ab\sqrt{a - b}}{a^4 - a^3 + a^2b} =$

$\dfrac{a(a^2 - a\sqrt{a - b} - ab + b\sqrt{a - b})}{a(a^3 - a^2 + ab)} =$

$\dfrac{a^2 - a\sqrt{a - b} - ab + b\sqrt{a - b}}{a^3 - a^2 + ab}$, or

$\dfrac{a^2 - ab - a\sqrt{a - b} + b\sqrt{a - b}}{a^3 - a^2 + ab}$

72. $\dfrac{1}{\sqrt{y + 18} + \sqrt{y}}$

73. $\dfrac{\sqrt{x + 6} - 5}{\sqrt{x + 6} + 5} = \dfrac{\sqrt{x + 6} - 5}{\sqrt{x + 6} + 5} \cdot \dfrac{\sqrt{x + 6} + 5}{\sqrt{x + 6} + 5} =$

$\dfrac{x + 6 - 25}{x + 6 + 5\sqrt{x + 6} + 5\sqrt{x + 6} + 25} =$

$\dfrac{x - 19}{x + 10\sqrt{x + 6} + 31}$

74. $\dfrac{-3\sqrt{a^2 - 3}}{a^2 - 3}$

75. $5\sqrt{\dfrac{x}{y}} + 4\sqrt{\dfrac{y}{x}} - \dfrac{3}{\sqrt{xy}} = \dfrac{5\sqrt{x}}{\sqrt{y}} + \dfrac{4\sqrt{y}}{\sqrt{x}} - \dfrac{3}{\sqrt{xy}} =$

$\dfrac{5\sqrt{x}}{\sqrt{y}} \cdot \dfrac{\sqrt{x}}{\sqrt{x}} + \dfrac{4\sqrt{y}}{\sqrt{x}} \cdot \dfrac{\sqrt{y}}{\sqrt{y}} - \dfrac{3}{\sqrt{xy}} = \dfrac{5x}{\sqrt{xy}} + \dfrac{4y}{\sqrt{xy}} - \dfrac{3}{\sqrt{xy}} =$

$\dfrac{5x + 4y - 3}{\sqrt{xy}} = \dfrac{5x + 4y - 3}{\sqrt{xy}} \cdot \dfrac{\sqrt{xy}}{\sqrt{xy}} = \dfrac{(5x + 4y - 3)\sqrt{xy}}{xy}$

76. $1 - \sqrt{w}$

77. $\dfrac{1}{4 + \sqrt{3}} + \dfrac{1}{\sqrt{3}} + \dfrac{1}{\sqrt{3} - 4} =$

$\dfrac{1}{4+\sqrt{3}} \cdot \dfrac{\sqrt{3}(\sqrt{3}-4)}{\sqrt{3}(\sqrt{3}-4)} + \dfrac{1}{\sqrt{3}} \cdot \dfrac{(4+\sqrt{3})(\sqrt{3}-4)}{(4+\sqrt{3})(\sqrt{3}-4)} + \dfrac{1}{\sqrt{3}-4} \cdot \dfrac{\sqrt{3}(4+\sqrt{3})}{\sqrt{3}(4+\sqrt{3})} =$

$\dfrac{3 - 4\sqrt{3} - 16 + 3 + 4\sqrt{3} + 3}{\sqrt{3}(4 + \sqrt{3})(\sqrt{3} - 4)} = \dfrac{-7}{\sqrt{3}(-16 + 3)} =$

$\dfrac{-7}{-13\sqrt{3}} \cdot \dfrac{\sqrt{3}}{\sqrt{3}} = \dfrac{7\sqrt{3}}{39}$

78. $\dfrac{x^2 + 4x + (x + 2)\sqrt{2x} - (3x + 6)\sqrt{3x} + (3x + 6)\sqrt{6}}{3(x + 2)(x - 2)}$

Exercise Set 7.7

---

1. $x^{1/4} = \sqrt[4]{x}$

2. $\sqrt[5]{y}$

3. $(8)^{1/3} = \sqrt[3]{8} = 2$

4. 4

5. $81^{1/4} = \sqrt[4]{81} = 3$

6. 4

7. $9^{1/2} = \sqrt{9} = 3$

8. 5

9. $(xyz)^{1/3} = \sqrt[3]{xyz}$

10. $\sqrt[4]{ab}$

11. $(a^2b^2)^{1/5} = \sqrt[5]{a^2b^2}$

12. $\sqrt[4]{x^3y^3}$

13. $a^{2/3} = \sqrt[3]{a^2}$

14. $b\sqrt{b}$

15. $16^{3/4} = \sqrt[4]{16^3} = (\sqrt[4]{16})^3 = 2^3 = 8$

16. 128

17. $25^{3/2} = \sqrt{25^3} = (\sqrt{25})^3 = 5^3 = 125$

18. 81

19. $9^{5/2} = \sqrt{9^5} = (\sqrt{9})^5 = 3^5 = 243$

20. 729

21. $(81x)^{3/4} = \sqrt[4]{(81x)^3} = \sqrt[4]{81^3 \cdot x^3} = \sqrt[4]{81^3} \cdot \sqrt[4]{x^3} =$

$(\sqrt[4]{81})^3 \cdot \sqrt[4]{x^3} = 3^3 \sqrt[4]{x^3} = 27\sqrt[4]{x^3}$

22. $25\sqrt[3]{a^2}$

23. $(25x^4)^{3/2} = \sqrt{(25x^4)^3} = \sqrt{25^3 \cdot x^{12}} =$

$\sqrt{25^3} \cdot \sqrt{x^{12}} = (\sqrt{25})^3 x^6 = 5^3 x^6 = 125 x^6$

24. $27y^9$

25. $(8a^2b^4)^{2/3} = \sqrt[3]{(8a^2b^4)^2} = \sqrt[3]{64a^4b^8} =$

$\sqrt[3]{64a^3b^6 \cdot ab^2} = 4ab^2 \sqrt[3]{ab^2}$

26. $9a^3 \sqrt[3]{ab^2}$

27. $\sqrt[3]{20} = 20^{1/3}$

28. $19^{1/3}$

29. $\sqrt{17} = 17^{1/2}$

30. $6^{1/2}$

31. $\sqrt{x^3} = x^{3/2}$

32. $a^{5/2}$

33. $\sqrt[5]{m^2} = m^{2/5}$

34. $n^{4/5}$

35. $\sqrt[4]{cd} = (cd)^{1/4}$ Parentheses are required.

36. $(xy)^{1/5}$

37. $\sqrt[5]{xy^2z} = (xy^2z)^{1/5}$

38. $(x^3y^2z^2)^{1/7}$

39. $(\sqrt{3mn})^3 = (3mn)^{3/2}$

40. $(7xy)^{4/3}$

41. $(\sqrt[7]{8x^2y})^5 = (8x^2y)^{5/7}$

42. $(2a^5b)^{7/6}$

43. $x^{-1/3} = \dfrac{1}{x^{1/3}}$

44. $\dfrac{1}{y^{1/4}}$

45. $(2rs)^{-3/4} = \dfrac{1}{(2rs)^{3/4}}$

46. $\dfrac{1}{(5xy)^{5/6}}$

47. $\left(\dfrac{1}{10}\right)^{-2/3} = \dfrac{1}{\left(\frac{1}{10}\right)^{2/3}} = 1 \cdot \left(\dfrac{10}{1}\right)^{2/3} = 10^{2/3}$

48. $8^{3/4}$

49. $\dfrac{1}{x^{-2/3}} = x^{2/3}$

50. $x^{5/6}$

51. $5x^{-1/4} = 5 \cdot x^{-1/4} = 5 \cdot \dfrac{1}{x^{1/4}} = \dfrac{5}{x^{1/4}}$

52. $\dfrac{8}{a^{1/3}}$

53. $\dfrac{3}{5m^{-1/2}} = \dfrac{3}{5} \cdot \dfrac{1}{m^{-1/2}} = \dfrac{3}{5} \cdot m^{1/2} = \dfrac{3m^{1/2}}{5}$

54. $\dfrac{2x^{1/3}}{7}$

55. $5^{3/4} \cdot 5^{1/8} = 5^{3/4+1/8} = 5^{6/8+1/8} = 5^{7/8}$

We added exponents after finding a common denominator.

56. $11^{7/6}$

57. $\dfrac{7^{5/8}}{7^{3/8}} = 7^{5/8-3/8} = 7^{2/8} = 7^{1/4}$

We subtracted exponents and simplified using arithmetic.

58. $9^{2/11}$

59. $\dfrac{8.3^{3/4}}{8.3^{2/5}} = 8.3^{3/4-2/5} = 8.3^{15/20-8/20} = 8.3^{7/20}$

We subtracted exponents after finding a common denominator.

60. $3.9^{7/20}$

61. $(10^{3/5})^{2/5} = 10^{3/5 \cdot 2/5} = 10^{6/25}$

We multiplied exponents.

62. $5^{15/28}$

63. $a^{2/3} \cdot a^{5/4} = a^{2/3+5/4} = a^{8/12+15/12} = a^{23/12}$

We add exponents after finding a common denominator.

64. $x^{17/12}$

65. $(x^{2/3})^{3/7} = x^{2/3 \cdot 3/7} = x^{6/21} = x^{2/7}$

We multiplied exponents and simplified using arithmetic.

66. $a^{3/5}$

67. $(m^{2/3}n^{1/2})^{1/4} = m^{2/3 \cdot 1/4} \cdot n^{1/2 \cdot 1/4} = m^{2/12}n^{1/8} = m^{1/6}n^{1/8}$

We raised each factor to the power 1/4, multiplied exponents, and simplified using arithmetic.

68. $x^{1/12}y^{1/10}$

69. $(a^{-2/3}b^{-1/4})^{-6} = a^{-2/3 \cdot (-6)} \cdot b^{-1/4 \cdot (-6)} = a^{12/3}b^{6/4} = a^4 b^{3/2}$

We raised each factor to the power -6, multiplied exponents, and simplified using arithmetic.

70. $m^2 n^{25/3}$

71. $\sqrt[6]{a^4} = a^{4/6}$ Converting to an exponential expression

$= a^{2/3}$ Using arithmetic to simplify the exponent

$= \sqrt[3]{a^2}$ Converting back to radical notation

72. $\sqrt[3]{y}$

73. $\sqrt[3]{8y^6} = \sqrt[3]{2^3 y^6}$ Recognizing that 8 is 2^3

$= (2^3 y^6)^{1/3}$ Converting to an exponential expression

$= 2^{3/3}y^{6/3}$ Using the third and fourth laws of exponents

$= 2y^2$ Using arithmetic to simplify the exponents

74. $x^2 y^3$

75. $\sqrt[4]{32} = \sqrt[4]{2^5} = 2^{5/4} = 2^{4/4+1/4} = 2^{4/4} \cdot 2^{1/4} = 2\sqrt[4]{2}$

76. $\sqrt{3}$

77. $\sqrt[6]{4x^2} = \sqrt[6]{2^2 x^2} = \sqrt[6]{(2x)^2} = [(2x)^2]^{1/6} = (2x)^{2/6} = (2x)^{1/3} = \sqrt[3]{2x}$

78. $2x\sqrt{y}$

79. $\sqrt[5]{32c^{10}d^{15}} = \sqrt[5]{2^5 c^{10}d^{15}} = (2^5 c^{10}d^{15})^{1/5} = 2^{5/5}c^{10/5}d^{15/5} = 2c^2 d^3$

80. $2x^3 y^4$

81. $\sqrt[6]{\dfrac{m^{12}n^{24}}{64}} = \sqrt[6]{\dfrac{m^{12}n^{24}}{2^6}} = \left(\dfrac{m^{12}n^{24}}{2^6}\right)^{1/6} = \dfrac{m^{12/6}n^{24/6}}{2^{6/6}} = \dfrac{m^2 n^4}{2}$

82. $\dfrac{x^3 y^4}{2}$

83. $\sqrt[8]{r^4 s^2} = (r^4 s^2)^{1/8} = r^{4/8} \cdot s^{2/8} = r^{2/4} \cdot s^{1/4} = (r^2 s)^{1/4} = \sqrt[4]{r^2 s}$

84. $\sqrt{2ts}$

85. $\sqrt[3]{27a^3 b^9} = \sqrt[3]{3^3 a^3 b^9} = (3^3 a^3 b^9)^{1/3} = 3^{3/3}a^{3/3}b^{9/3} = 3ab^3$

86. $3x^2 y^2$

87. $\sqrt[3]{7} \cdot \sqrt{2} = 7^{1/3} \cdot 2^{1/2}$ Converting to exponential notation

   $= 7^{2/6} \cdot 2^{3/6}$ Rewriting so that exponents have a common denominator

   $= (7^2 \cdot 2^3)^{1/6}$ Using the third and fourth laws of exponents

   $= \sqrt[6]{7^2 \cdot 2^3}$ Converting back to radical notation

   $= \sqrt[6]{392}$ Multiplying under the radical

88. $\sqrt[12]{300,125}$

89. $\sqrt{x}\,\sqrt[3]{2x} = x^{1/2} \cdot (2x)^{1/3} = x^{1/2} \cdot 2^{1/3} \cdot x^{1/3} =$

   $x^{3/6} \cdot 2^{1/3} \cdot x^{2/6} = 2^{1/3} \cdot x^{5/6} = 2^{2/6} x^{5/6} =$

   $(2^2 \cdot x^5)^{1/6} = \sqrt[6]{2^2 \cdot x^5} = \sqrt[6]{4x^5}$

90. $\sqrt[15]{27y^8}$

91. $\sqrt{x}\,\sqrt[3]{x-2} = x^{1/2} \cdot (x-2)^{1/3} = x^{3/6} \cdot (x-2)^{2/6} =$

   $[x^3(x-2)^2]^{1/6} = \sqrt[6]{x^3(x-2)^2} =$

   $\sqrt[6]{x^3(x^2-4x+4)} = \sqrt[6]{x^5 - 4x^4 + 4x^3}$

92. $\sqrt[4]{3xy^2 + 24xy + 48x}$

93. $\dfrac{\sqrt[3]{(a+b)^2}}{\sqrt{a+b}} = \dfrac{(a+b)^{2/3}}{(a+b)^{1/2}} = (a+b)^{2/3 - 1/2} =$

   $(a+b)^{4/6 - 3/6} = (a+b)^{1/6} = \sqrt[6]{a+b}$

94. $\sqrt[12]{(x+y)^{-1}}$

95. $a^{2/3} \cdot b^{3/4} = a^{8/12} \cdot b^{9/12} = (a^8 b^9)^{1/12} =$

   $\sqrt[12]{a^8 b^9}$

96. $\sqrt[12]{x^4 y^3 z^2}$

97. $\dfrac{s^{7/12} \cdot t^{5/6}}{s^{1/3} \cdot t^{-1/6}} = s^{7/12 - 1/3} \cdot t^{5/6 - (-1/6)} =$

   $s^{7/12 - 4/12} \cdot t^{5/6 + 1/6} = s^{3/12} t = s^{1/4} t = t\sqrt[4]{s}$

98. $y \sqrt[5]{x}$

99. $\sqrt[5]{yx^2}\,\sqrt{xy} = (yx^2)^{1/5} (xy)^{1/2} = y^{1/5} x^{2/5} x^{1/2} y^{1/2} =$

   $x^{2/5 + 1/2} y^{1/5 + 1/2} = x^{4/10 + 5/10} y^{2/10 + 5/10} =$

   $x^{9/10} y^{7/10} = (x^9 y^7)^{1/10} = \sqrt[10]{x^9 y^7}$

100. $a^{10}\sqrt{ab^7}$

101. $\sqrt{a^3bc^4}\,\sqrt[3]{ab^2c} = (a^3bc^4)^{1/2}(ab^2c)^{1/3} =$

   $a^{3/2} b^{1/2} c^2 a^{1/3} b^{2/3} c^{1/3} = a^{3/2 + 1/3} b^{1/2 + 2/3} c^{2 + 1/3} =$

   $a^{11/6} b^{7/6} c^{14/6} = \sqrt[6]{a^{11} b^7 c^{14}} = \sqrt[6]{a^6 b^6 c^{12} \cdot a^5 b c^2} =$

   $abc^2 \sqrt[6]{a^5 b c^2}$

102. $xyz \sqrt[6]{xy^5 z^2}$

103. $\quad x^2 - 1 = 8$

   $\quad x^2 - 9 = 0$

   $(x+3)(x-3) = 0$

   $x+3 = 0 \quad \text{or} \quad x-3 = 0$

   $\quad x = -3 \quad \text{or} \qquad x = 3$

   Both values check. The solutions are -3 and 3.

104. $-\dfrac{11}{2}$

105. $\dfrac{1}{x} + 2 = 5$

   $\qquad \dfrac{1}{x} = 3$

   $x \cdot \dfrac{1}{x} = x \cdot 3$

   $\qquad 1 = 3x$

   $\qquad \dfrac{1}{3} = x$

   This value checks. The solution is $\dfrac{1}{3}$.

106. $\$93,500$

107. $\sqrt[5]{yx^2 \sqrt{xy}} = \sqrt[5]{yx^2(xy)^{1/2}} = \sqrt[5]{yx^2 x^{1/2} y^{1/2}} =$

   $\sqrt[5]{x^{5/2} y^{3/2}} = (x^{5/2} y^{3/2})^{1/5} = x^{5/10} y^{3/10} =$

   $(x^5 y^3)^{1/10} = \sqrt[10]{x^5 y^3}$

108. $x^3 \sqrt[6]{x}$

109. $\dfrac{\sqrt{(a+b)^3}\,\sqrt[3]{(a+b)^2}}{\sqrt[4]{a+b}} = \dfrac{(a+b)^{3/2}(a+b)^{2/3}}{(a+b)^{1/4}} =$

   $(a+b)^{3/2 + 2/3 - 1/4} = (a+b)^{18/12 + 8/12 - 3/12} =$

   $(a+b)^{23/12} = \sqrt[12]{(a+b)^{23}} =$

   $\sqrt[12]{(a+b)^{12}(a+b)^{11}} = (a+b)\sqrt[12]{(a+b)^{11}}$

110. $\sqrt[4]{2xy^2}$

111. $(-\sqrt[4]{7}\,\sqrt[3]{w})^{12} = (-1 \cdot 7^{1/4} \cdot w^{1/3})^{12} = (-1)^{12} 7^3 w^4 =$
   $343w^4$

112. $\sqrt[6]{p+q}$

113. $\dfrac{1}{\sqrt[3]{3} - \sqrt[3]{2}}$

   First observe that $3 - 2 = (\sqrt[3]{3})^3 - (\sqrt[3]{2})^3 =$

   $(\sqrt[3]{3} - \sqrt[3]{2})[(\sqrt[3]{3})^2 + \sqrt[3]{3}\,\sqrt[3]{2} + (\sqrt[3]{2})^2]$, or

   $(\sqrt[3]{3} - \sqrt[3]{2})(\sqrt[3]{9} + \sqrt[3]{6} + \sqrt[3]{4})$. That is, we can factor $3 - 2$ as the difference of two cubes. We can use this to rationalize the denominator of the given expression.

   $\dfrac{1}{\sqrt[3]{3} - \sqrt[3]{2}} = \dfrac{1}{\sqrt[3]{3} - \sqrt[3]{2}} \cdot \dfrac{\sqrt[3]{9} + \sqrt[3]{6} + \sqrt[3]{4}}{\sqrt[3]{9} + \sqrt[3]{6} + \sqrt[3]{4}} =$

   $\dfrac{\sqrt[3]{9} + \sqrt[3]{6} + \sqrt[3]{4}}{3 - 2} = \sqrt[3]{9} + \sqrt[3]{6} + \sqrt[3]{4}$

114. x^3

115. $\sqrt[p]{x^5 p y^7 p^{+1} z^{p+3}} = \sqrt[p]{x^5 p y^7 p y z p z^3} = \sqrt[p]{x^5 p y^7 p z^p \cdot yz^3} =$
      $(x^5 p y^7 p z p y z^3)^{1/p} = x^5 y^7 z (yz^3)^{1/p} = x^5 y^7 z \sqrt[p]{yz^3}$

116. a) 27.4 m

      b) 45.9 m

      c) 21.9 m

      d) 79.4 m

117. Make a table of values for $f(x) = x^{3/2}$, $x \geqslant 0$.

x	f(x)
0	0
1	1
4	8
5	11.2

We plot these points and draw the graph. We also graph $y = x$ and $y = x^2$ on the same axes.

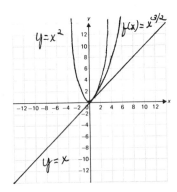

118. 7.937×10^{-13} to 1

Exercise Set 7.8

1. $\sqrt{2x - 3} = 1$

       $(\sqrt{2x - 3})^2 = 1^2$ Principle of powers (squaring)
           $2x - 3 = 1$
               $2x = 4$
                $x = 2$

Check: $\dfrac{\sqrt{2x - 3} = 1}{\begin{array}{c|c} \sqrt{2 \cdot 2 - 3} & 1 \\ \sqrt{1} & \\ 1 & \end{array}}$

The solution is 2.

2. 33

3. $\sqrt{3x} + 1 = 7$

       $\sqrt{3x} = 6$ Adding to isolate the radical

    $(\sqrt{3x})^2 = 6^2$ Principle of powers (squaring)
          $3x = 36$
           $x = 12$

Check: $\dfrac{\sqrt{3x} + 1 = 7}{\begin{array}{c|c} \sqrt{3 \cdot 12} + 1 & 7 \\ 6 + 1 & \\ 7 & \end{array}}$

The solution is 12.

4. 32

5. $\sqrt{y + 1} - 5 = 8$

       $\sqrt{y + 1} = 13$ Adding to isolate the radical

    $(\sqrt{y + 1})^2 = 13^2$ Principle of powers (squaring)
          $y + 1 = 169$
              $y = 168$

Check: $\dfrac{\sqrt{y + 1} - 5 = 8}{\begin{array}{c|c} \sqrt{168 + 1} - 5 & 8 \\ 13 - 5 & \\ 8 & \end{array}}$

The solution is 168.

6. 11

7. $\sqrt{y - 3} + 4 = 2$

       $\sqrt{y - 3} = -2$

At this point we might observe that this equation has no real-number solution, because the principle square root of a number is never negative. However, we will continue with the solution.

    $(\sqrt{y - 3})^2 = (-2)^2$
          $y - 3 = 4$
              $y = 7$

Check: $\dfrac{\sqrt{y - 3} + 4 = 2}{\begin{array}{c|c} \sqrt{7 - 3} + 4 & 2 \\ \sqrt{4} + 4 & \\ 6 & \end{array}}$

The number 7 does not check. There is no solution.

8. -3

9. $\sqrt[3]{x + 5} = 2$ Check: $\dfrac{\sqrt[3]{x + 5} = 2}{\begin{array}{c|c} \sqrt[3]{3 + 5} & 2 \\ \sqrt[3]{8} & \\ 2 & \end{array}}$
    $(\sqrt[3]{x + 5})^3 = 2^3$
           $x + 5 = 8$
               $x = 3$

The solution is 3.

10. 29

11. $\sqrt[4]{y-3} = 2$ Check: $\dfrac{\sqrt[4]{y-3} = 2}{\begin{array}{c|c} \sqrt[4]{19-3} & 2 \\ & \\ \sqrt[4]{16} & \\ & 2 \end{array}}$

     $(\sqrt[4]{y-3})^4 = 2^4$

           $y - 3 = 16$

               $y = 19$

     The solution is 19.

12.   78

13. $\sqrt{3y+1} = 9$ Check: $\dfrac{\sqrt{3y+1} = 9}{\begin{array}{c|c} \sqrt{3 \cdot \frac{80}{3} + 1} & 9 \\ & \\ \sqrt{81} & \\ & 9 \end{array}}$

     $(\sqrt{3y+1})^2 = 9^2$

           $3y + 1 = 81$

             $3y = 80$

              $y = \dfrac{80}{3}$

     The solution is $\dfrac{80}{3}$.

14.   84

15. $3\sqrt{x} = 6$

       $\sqrt{x} = 2$ Multiplying by $\dfrac{1}{3}$

    $(\sqrt{x})^2 = 2^2$

          $x = 4$

     Check: $\dfrac{3\sqrt{x} = 6}{\begin{array}{c|c} 3\sqrt{4} & 6 \\ 3 \cdot 2 & \\ 6 & \end{array}}$

     The solution is 4.

16. $\dfrac{1}{16}$

17. $2\sqrt{y} - 7 = 9$ Check: $\dfrac{2\sqrt{y} - 7 = 9}{\begin{array}{c|c} 2\sqrt{64} - 7 & 9 \\ 2 \cdot 8 - 7 & \\ & 9 \end{array}}$

         $2\sqrt{y} = 16$

          $\sqrt{y} = 8$

    $(\sqrt{y})^2 = 8^2$

           $y = 64$

     The solution is 64.

18.   No solution

19. $\sqrt[3]{x} = -3$ Check: $\dfrac{\sqrt[3]{x} = -3}{\begin{array}{c|c} \sqrt[3]{-27} & -3 \\ & -3 \end{array}}$

    $(\sqrt[3]{x})^3 = (-3)^3$

          $x = -27$

     The solution is -27.

20.   -64

21. $\sqrt{y+3} - 20 = 0$ Check: $\dfrac{\sqrt{y+3} - 20 = 0}{\begin{array}{c|c} \sqrt{397+3} - 20 & 0 \\ & \\ \sqrt{400} - 20 & \\ & 0 \end{array}}$

         $\sqrt{y+3} = 20$

   $(\sqrt{y+3})^2 = 20^2$

          $y + 3 = 400$

             $y = 397$

     The solution is 397.

22.   117

23. $\sqrt{x+2} = -4$

     We might observe that this equation has no real-number solution, since the principal square root of a number is never negative. However, we will go through the solution process.

    $(\sqrt{x+2})^2 = (-4)^2$ Check: $\dfrac{\sqrt{x+2} = -4}{\begin{array}{c|c} \sqrt{14+2} & -4 \\ & 4 \end{array}}$

        $x + 2 = 16$

             $x = 14$

     The number 14 does not check. The equation has no solution.

24.   No solution

25. $\sqrt{2x+3} - 5 = -2$ Check: $\dfrac{\sqrt{2x+3} - 5 = -2}{\begin{array}{c|c} \sqrt{2 \cdot 3 + 3} - 5 & -2 \\ & \\ \sqrt{9} - 5 & \\ & -2 \end{array}}$

        $\sqrt{2x+3} = 3$

   $(\sqrt{2x+3})^2 = 3^2$

        $2x + 3 = 9$

           $2x = 6$

             $x = 3$

     The solution is 3.

26. $\dfrac{8}{3}$

27. $8 = \dfrac{1}{\sqrt{x}}$ Check: $8 = \dfrac{1}{\sqrt{x}}$

    $8 \cdot \sqrt{x} = \dfrac{1}{\sqrt{x}} \cdot \sqrt{x}$ $\begin{array}{c|c} & \dfrac{1}{\sqrt{\frac{1}{64}}} \\ 8 & \\ & \dfrac{1}{\frac{1}{8}} \\ & 8 \end{array}$

       $8\sqrt{x} = 1$

   $(8\sqrt{x})^2 = 1^2$

       $64x = 1$

         $x = \dfrac{1}{64}$

     The solution is $\dfrac{1}{64}$.

28. $\dfrac{1}{9}$

29. $\sqrt[3]{6x+9} + 8 = 5$ Check: $\dfrac{\sqrt[3]{6x+9} + 8 = 5}{\begin{array}{c|c} \sqrt[3]{6(-6)+9} + 8 & 5 \\ & \\ \sqrt[3]{-27} + 8 & \\ & 5 \end{array}}$

       $\sqrt[3]{6x+9} = -3$

   $(\sqrt[3]{6x+9})^3 = (-3)^3$

        $6x + 9 = -27$

           $6x = -36$

            $x = -6$

     The solution is -6.

30. $-\dfrac{5}{3}$

31. $\sqrt{3y + 1} = \sqrt{2y + 6}$ One radical is already isolated.

$(\sqrt{3y + 1})^2 = (\sqrt{2y + 6})^2$ Squaring both sides

$3y + 1 = 2y + 6$

$y = 5$

The number 5 checks and is the solution.

32. 2

33. $2\sqrt{1 - x} = \sqrt{5}$ One radical is already isolated.

$(2\sqrt{1 - x})^2 = (\sqrt{5})^2$ Squaring both sides

$4(1 - x) = 5$

$4 - 4x = 5$

$-4x = 1$

$x = -\dfrac{1}{4}$

The number $-\dfrac{1}{4}$ checks and is the solution.

34. 3

35. $2\sqrt{t - 1} = \sqrt{3t - 1}$

$(2\sqrt{t - 1})^2 = (\sqrt{3t - 1})^2$

$4(t - 1) = 3t - 1$

$4t - 4 = 3t - 1$

$t = 3$

The number 3 checks and is the solution.

36. -1

37. $\sqrt{y - 5} + \sqrt{y} = 5$

$\sqrt{y - 5} = 5 - \sqrt{y}$ Adding $-\sqrt{y}$; this isolates one of the radical terms

$(\sqrt{y - 5})^2 = (5 - \sqrt{y})^2$ Squaring both sides

$y - 5 = 25 - 10\sqrt{y} + y$

$-30 = -10\sqrt{y}$ Isolating the remaining radical term

$3 = \sqrt{y}$ Multiplying by $-\dfrac{1}{10}$

$3^2 = (\sqrt{y})^2$ Squaring both sides

$9 = y$

The number 9 checks and is the solution.

38. No solution

39. $3 + \sqrt{z - 6} = \sqrt{z + 9}$ One radical is already isolated.

$(3 + \sqrt{z - 6})^2 = (\sqrt{z + 9})^2$ Squaring both sides

$9 + 6\sqrt{z - 6} + z - 6 = z + 9$

$6\sqrt{z - 6} = 6$

$\sqrt{z - 6} = 1$ Multiplying by $\dfrac{1}{6}$

$(\sqrt{z - 6})^2 = 1^2$ Squaring both sides

$z - 6 = 1$

$z = 7$

The number 7 checks and is the solution.

40. 7, 3

41. $\sqrt{20 - x} + 8 = \sqrt{9 - x} + 11$

$\sqrt{20 - x} = \sqrt{9 - x} + 3$ Adding -8

$(\sqrt{20 - x})^2 = (\sqrt{9 - x} + 3)^2$ Squaring both sides

$20 - x = 9 - x + 6\sqrt{9 - x} + 9$

$2 = 6\sqrt{9 - x}$ Isolating the remaining radical

$1 = 3\sqrt{9 - x}$ Multiplying by $\dfrac{1}{2}$

$1^2 = (3\sqrt{9 - x})^2$ Squaring both sides

$1 = 9(9 - x)$

$1 = 81 - 9x$

$-80 = -9x$

$\dfrac{80}{9} = x$

The number $\dfrac{80}{9}$ checks and is the solution.

42. $\dfrac{15}{4}$

43. $\sqrt{x + 2} + \sqrt{3x + 4} = 2$

$\sqrt{x + 2} = 2 - \sqrt{3x + 4}$ Isolating one radical

$(\sqrt{x + 2})^2 = (2 - \sqrt{3x + 4})^2$

$x + 2 = 4 - 4\sqrt{3x + 4} + 3x + 4$

$-2x - 6 = -4\sqrt{3x + 4}$ Isolating the remaining radical

$x + 3 = 2\sqrt{3x + 4}$ Multiplying by $-\dfrac{1}{2}$

$(x + 3)^2 = (2\sqrt{3x + 4})^2$

$x^2 + 6x + 9 = 4(3x + 4)$

$x^2 + 6x + 9 = 12x + 16$

$x^2 - 6x - 7 = 0$

$(x - 7)(x + 1) = 0$

$x - 7 = 0$ or $x + 1 = 0$

$x = 7$ or $x = -1$

43. (continued)

Check: For 7: $\dfrac{\sqrt{x + 2} + \sqrt{3x + 4} = 2}{\sqrt{7 + 2} + \sqrt{3 \cdot 7 + 4}\ \big|\ 2}$

$$\sqrt{9} + \sqrt{25}$$
$$8 \,\big|$$

For -1: $\dfrac{\sqrt{x + 2} + \sqrt{3x + 4} = 2}{\sqrt{-1 + 2} + \sqrt{3(-1) + 4}\ \big|\ 2}$

$$\sqrt{1} + \sqrt{1}$$
$$2 \,\big|$$

Since -1 checks but 7 does not, the solution is -1.

44. $\frac{1}{3}$, -1

45. $\sqrt{4y + 1} - \sqrt{y - 2} = 3$

$$\sqrt{4y + 1} = 3 + \sqrt{y - 2}$$
$$(\sqrt{4y + 1})^2 = (3 + \sqrt{y - 2})^2$$
$$4y + 1 = 9 + 6\sqrt{y - 2} + y - 2$$
$$3y - 6 = 6\sqrt{y - 2}$$
$$y - 2 = 2\sqrt{y - 2}$$
$$(y - 2)^2 = (2\sqrt{y - 2})^2$$
$$y^2 - 4y + 4 = 4(y - 2)$$
$$y^2 - 4y + 4 = 4y - 8$$
$$y^2 - 8y + 12 = 0$$
$$(y - 6)(y - 2) = 0$$
$$y - 6 = 0 \text{ or } y - 2 = 0$$
$$y = 6 \text{ or } \qquad y = 2$$

The numbers 6 and 2 check and are the solutions.

46. 1

47. $\sqrt{3x - 5} + \sqrt{2x + 3} + 1 = 0$

$$\sqrt{3x - 5} + 1 = -\sqrt{2x + 3}$$
$$(\sqrt{3x - 5} + 1)^2 = (-\sqrt{2x + 3})^2$$
$$3x - 5 + 2\sqrt{3x - 5} + 1 = 2x + 3$$
$$2\sqrt{3x - 5} = -x + 7$$
$$(2\sqrt{3x - 5})^2 = (-x + 7)^2$$
$$4(3x - 5) = x^2 - 14x + 49$$
$$12x - 20 = x^2 - 14x + 49$$
$$0 = x^2 - 26x + 69$$
$$0 = (x - 23)(x - 3)$$
$$x - 23 = 0 \text{ or } x - 3 = 0$$
$$x = 23 \text{ or } \qquad x = 3$$

Neither number checks. There is no solution.

48. 2

49. $2\sqrt{3x + 6} - \sqrt{4x + 9} = 5$

$$2\sqrt{3x + 6} = 5 + \sqrt{4x + 9}$$
$$(2\sqrt{3x + 6})^2 = (5 + \sqrt{4x + 9})^2$$
$$4(3x + 6) = 25 + 10\sqrt{4x + 9} + 4x + 9$$
$$12x + 24 = 34 + 10\sqrt{4x + 9} + 4x$$
$$8x - 10 = 10\sqrt{4x + 9}$$
$$4x - 5 = 5\sqrt{4x + 9}$$
$$(4x - 5)^2 = (5\sqrt{4x + 9})^2$$
$$16x^2 - 40x + 25 = 25(4x + 9)$$
$$16x^2 - 40x + 25 = 100x + 225$$
$$16x^2 - 140x - 200 = 0$$
$$4(4x + 5)(x - 10) = 0$$
$$4x + 5 = 0 \text{ or } x - 10 = 0$$
$$x = -\frac{5}{4} \text{ or } \qquad x = 10$$

Since 10 checks and $-\frac{5}{4}$ does not, 10 is the solution.

50. 7

51. $3\sqrt{t + 1} - \sqrt{2t - 5} = 7$

$$3\sqrt{t + 1} = 7 + \sqrt{2t - 5}$$
$$(3\sqrt{t + 1})^2 = (7 + \sqrt{2t - 5})^2$$
$$9(t + 1) = 49 + 14\sqrt{2t - 5} + 2t - 5$$
$$9t + 9 = 44 + 14\sqrt{2t - 5} + 2t$$
$$7t - 35 = 14\sqrt{2t - 5}$$
$$t - 5 = 2\sqrt{2t - 5}$$
$$(t - 5)^2 = (2\sqrt{2t - 5})^2$$
$$t^2 - 10t + 25 = 4(2t - 5)$$
$$t^2 - 10t + 25 = 8t - 20$$
$$t^2 - 18t + 45 = 0$$
$$(t - 15)(t - 3) = 0$$
$$t - 15 = 0 \text{ or } t - 3 = 0$$
$$t = 15 \qquad t = 3$$

Since 15 checks and 3 does not, the solution is 15.

52. 5

53. $\qquad \dfrac{3}{2x} + \dfrac{1}{x} = \dfrac{2x + 3.5}{3x} \qquad$ LCM is $6x$

$$6x\left[\dfrac{3}{2x} + \dfrac{1}{x}\right] = 6x\left[\dfrac{2x + 3.5}{3x}\right]$$
$$9 + 6 = 4x + 7$$
$$8 = 4x$$
$$2 = x$$

The number 2 checks and is the solution.

54. Height: 7 in., base: 9 in.

55. $V = 1.2\sqrt{h}$

    $V = 1.2\sqrt{30,000}$

    $V \approx 208$ mi   Using a calculator to compute

56. 72.25 ft

57. $\dfrac{x + \sqrt{x + 1}}{x - \sqrt{x + 1}} = \dfrac{5}{11}$

    $11(x + \sqrt{x + 1}) = 5(x - \sqrt{x + 1})$

    $11x + 11\sqrt{x + 1} = 5x - 5\sqrt{x + 1}$

    $16\sqrt{x + 1} = -6x$

    $8\sqrt{x + 1} = -3x$

    $(8\sqrt{x + 1})^2 = (-3x)^2$

    $64(x + 1) = 9x^2$

    $64x + 64 = 9x^2$

    $0 = 9x^2 - 64x - 64$

    $0 = (9x + 8)(x - 8)$

    $9x + 8 = 0$ or $x - 8 = 0$

    $9x = -8$ or $x = 8$

    $x = -\dfrac{8}{9}$ or $x = 8$

    Since $-\dfrac{8}{9}$ checks but 8 does not, the solution is $-\dfrac{8}{9}$.

58. 6912

59. $\sqrt[4]{z^2 + 17} = 3$

    $(\sqrt[4]{z^2 + 17})^4 = 3^4$

    $z^2 + 17 = 81$

    $z^2 - 64 = 0$

    $(z + 8)(z - 8) = 0$

    $z + 8 = 0$ or $z - 8 = 0$

    $z = -8$ or $z = 8$

    The numbers -8 and 8 check and are the solutions.

60. 0

61. $\sqrt[3]{x^2 + x + 15} - 3 = 0$

    $(\sqrt[3]{x^2 + x + 15})^3 = 3^3$

    $x^2 + x + 15 = 27$

    $x^2 + x - 12 = 0$

    $(x + 4)(x - 3) = 0$

    $x + 4 = 0$ or $x - 3 = 0$

    $x = -4$ or $x = 3$

    The numbers -4 and 3 check and are the solutions.

62. -1, 6

63. $\sqrt{8 - b} = b\sqrt{8 - b}$

    $(\sqrt{8 - b})^2 = (b\sqrt{8 - b})^2$

    $(8 - b) = b^2(8 - b)$

    $0 = b^2(8 - b) - (8 - b)$

    $0 = (b^2 - 1)(8 - b)$

    $0 = (b + 1)(b - 1)(8 - b)$

    $b + 1 = 0$ or $b - 1 = 0$ or $8 - b = 0$

    $b = -1$ or $b = 1$ or $8 = b$

    Since the numbers 1 and 8 check but -1 does not, 1 and 8 are the solutions.

64. 2

65. $6\sqrt{y} + 6y^{-1/2} = 37$

    $6\sqrt{y} + \dfrac{6}{\sqrt{y}} = 37$

    $\sqrt{y}\left[6\sqrt{y} + \dfrac{6}{\sqrt{y}}\right] = \sqrt{y} \cdot 37$

    $6y + 6 = 37\sqrt{y}$

    $(6y + 6)^2 = (37\sqrt{y})^2$

    $36y^2 + 72y + 36 = 1369y$

    $36y^2 - 1297y + 36 = 0$

    $(36y - 1)(y - 36) = 0$

    $36y - 1 = 0$ or $y - 36 = 0$

    $36y = 1$ or $y = 36$

    $y = \dfrac{1}{36}$ or $y = 36$

    The numbers $\dfrac{1}{36}$ and 36 check and are the solutions.

66. $0, \dfrac{125}{4}$

67. $\sqrt{\sqrt{x} + 4} = \sqrt{x} - 2$

    $\left[\sqrt{\sqrt{x} + 4}\right]^2 = (\sqrt{x} - 2)^2$

    $\sqrt{x} + 4 = x - 4\sqrt{x} + 4$

    $5\sqrt{x} = x$

    $(5\sqrt{x})^2 = x^2$

    $25x = x^2$

    $0 = x^2 - 25x$

    $0 = x(x - 25)$

    $x = 0$ or $x - 25 = 0$

    $x = 0$ or $x = 25$

    Since 25 checks but 0 does not, the solution is 25.

68. 2

69. $\sqrt{x+1} - \dfrac{2}{\sqrt{x+1}} = 1$

$\sqrt{x+1}\left(\sqrt{x+1} - \dfrac{2}{\sqrt{x+1}}\right) = \sqrt{x+1} \cdot 1$

$x + 1 - 2 = \sqrt{x+1}$

$(x - 1)^2 = (\sqrt{x+1})^2$

$x^2 - 2x + 1 = x + 1$

$x^2 - 3x = 0$

$x(x - 3) = 0$

$x = 0$ or $x - 3 = 0$

$x = 0$ or $\quad\quad x = 3$

Since 3 checks but 0 does not, the solution is 3.

Exercise Set 7.9

1. $\sqrt{-15} = \sqrt{-1 \cdot 15} = \sqrt{-1} \cdot \sqrt{15} = i\sqrt{15}$, or $\sqrt{15}i$

2. $\sqrt{17}i$

3. $\sqrt{-16} = \sqrt{-1 \cdot 16} = \sqrt{-1} \cdot \sqrt{16} = i \cdot 4 = 4i$

4. $5i$

5. $-\sqrt{-36} = -\sqrt{-1 \cdot 36} = -\sqrt{-1} \cdot \sqrt{36} = -i \cdot 6 = -6i$

6. $-7i$

7. $-\sqrt{-12} = -\sqrt{-1 \cdot 4 \cdot 3} = -\sqrt{-1} \cdot \sqrt{4} \cdot \sqrt{3} = -i \cdot 2 \cdot \sqrt{3} =$
   $-2\sqrt{3}i$

8. $-2\sqrt{5}i$

9. $\sqrt{-250} = \sqrt{-1 \cdot 25 \cdot 10} = \sqrt{-1} \cdot \sqrt{25} \cdot \sqrt{10} = i \cdot 5 \cdot \sqrt{10} =$
   $5\sqrt{10}i$

10. $6\sqrt{5}i$

11. $(3 + 2i) + (5 - i) = (3 + 5) + (2 - 1)i$
    Collecting the real and
    the imaginary parts
    $= 8 + i$

12. $5 + 11i$

13. $(4 - 3i) + (5 - 2i) = (4 + 5) + (-3 - 2)i$
    Collecting the real and
    the imaginary parts
    $= 9 - 5i$

14. $-1 - 8i$

15. $(9 - i) + (-2 + 5i) = (9 - 2) + (-1 + 5)i$
    $= 7 + 4i$

16. $8 + i$

17. $(3 - i) - (5 + 2i) = (3 - 5) + (-1 - 2)i$
    $= -2 - 3i$

18. $-9 + 5i$

19. $(4 - 2i) - (5 - 3i) = (4 - 5) + [-2 - (-3)]i$
    $= -1 + i$

20. $-3 + 2i$

21. $(9 + 5i) - (-2 - i) = [9 - (-2)] + [5 - (-1)]i$
    $= 11 + 6i$

22. $4 - 7i$

23. $(-5 - 2i) + (-7 - 4i) = (-5 - 7) + (-2 - 4)i$
    $= -12 - 6i$

24. $-7 - 12i$

25. $(2 + 3i) + (1 + 2i) + (4 + i) =$
    $(2 + 1 + 4) + (3 + 2 + 1)i = 7 + 6i$

26. $12 + 6i$

27. $(5 - 2i) + (3 + 4i) - (2 + 7i) =$
    $(5 + 3 - 2) + (-2 + 4 - 7)i = 6 - 5i$

28. $-7i$

29. $(5 - 9i) - (3 - 4i) - (-2 + i) =$
    $[5 - 3 - (-2)] + [-9 - (-4) - 1]i = 4 - 6i$

30. $7 - 5i$

31. $\sqrt{-25}\,\sqrt{-36} = \sqrt{-1} \cdot \sqrt{25} \cdot \sqrt{-1} \cdot \sqrt{36}$
    $= i \cdot 5 \cdot i \cdot 6$
    $= i^2 \cdot 30$
    $= -1 \cdot 30 \quad\quad i^2 = -1$
    $= -30$

32. -63

33. $\sqrt{-6}\,\sqrt{-5} = \sqrt{-1} \cdot \sqrt{6} \cdot \sqrt{-1} \cdot \sqrt{5} = i \cdot \sqrt{6} \cdot i \cdot \sqrt{5}$
    $= i^2 \cdot \sqrt{30} = -1 \cdot \sqrt{30} = -\sqrt{30}$

34. $-\sqrt{70}$

35. $\sqrt{-50}\,\sqrt{-3} = \sqrt{-1 \cdot 25 \cdot 2} \cdot \sqrt{-1 \cdot 3} =$
    $\sqrt{-1} \cdot \sqrt{25} \cdot \sqrt{2} \cdot \sqrt{-1} \cdot \sqrt{3} = i \cdot 5 \cdot \sqrt{2} \cdot i \cdot \sqrt{3} =$
    $i^2 \cdot 5 \cdot \sqrt{6} = -1 \cdot 5 \cdot \sqrt{6} = -5\sqrt{6}$

36. $-6\sqrt{6}$

37. $\sqrt{-48}\,\sqrt{-6} = \sqrt{-1\cdot16\cdot3}\cdot\sqrt{-1\cdot3\cdot2}$

$\qquad = \sqrt{-1}\cdot\sqrt{16}\cdot\sqrt{3}\cdot\sqrt{-1}\cdot\sqrt{3}\cdot\sqrt{2}$

$\qquad = i\cdot4\cdot3\cdot i\cdot\sqrt{2} \qquad\qquad \sqrt{3}\cdot\sqrt{3} = 3$

$\qquad = i^2\cdot12\cdot\sqrt{2}$

$\qquad = -1\cdot12\cdot\sqrt{2}$

$\qquad = -12\sqrt{2}$

38. $-15\sqrt{5}$

39. $5i\cdot8i = 40\cdot i^2$

$\qquad = 40\cdot(-1) \qquad i^2 = -1$

$\qquad = -40$

40. -54

41. $5i\cdot(-7i) = -35\cdot i^2$

$\qquad = -35\cdot(-1) \qquad i^2 = 1$

$\qquad = 35$

42. 28

43. $5i(3 - 2i) = 5i\cdot3 + 5i(-2i)$ Using the distributive law

$\qquad = 15i - 10i^2$

$\qquad = 15i + 10 \qquad i^2 = -1$

$\qquad = 10 + 15i$

44. $24 + 20i$

45. $-3i(7 - 4i) = -3i\cdot7 + (-3i)(-4i) = -21i + 12i^2 =$
$-21i - 12 = -12 - 21i$

46. $-21 - 63i$

47. $(3 + 2i)(1 + i) = 3 + 3i + 2i + 2i^2$ Using FOIL

$\qquad = 3 + 3i + 2i - 2 \qquad i^2 = -1$

$\qquad = 1 + 5i$

48. $-7 + 26i$

49. $(2 + 3i)(6 - 2i) = 12 - 4i + 18i - 6i^2$ Using FOIL

$\qquad = 12 - 4i + 18i + 6 \qquad i^2 = -1$

$\qquad = 18 + 14i$

50. $16 + 7i$

51. $(6 - 5i)(3 + 4i) = 18 + 24i - 15i - 20i^2 =$
$18 + 24i - 15i + 20 = 38 + 9i$

52. $40 + 13i$

53. $(7 - 2i)(2 - 6i) = 14 - 42i - 4i + 12i^2 =$
$14 - 42i - 4i - 12 = 2 - 46i$

54. $8 + 31i$

55. $(5 - 3i)(4 - 5i) = 20 - 25i - 12i + 15i^2 =$
$20 - 25i - 12i - 15 = 5 - 37i$

56. $7 - 61i$

57. $(-2 + 3i)(-2 + 5i) = 4 - 10i - 6i + 15i^2 =$
$4 - 10i - 6i - 15 = -11 - 16i$

58. $-15 - 30i$

59. $(-5 - 4i)(3 + 7i) = -15 - 35i - 12i - 28i^2 =$
$-15 - 35i - 12i + 28 = 13 - 47i$

60. $39 - 37i$

61. $(3 - 2i)^2 = 3^2 - 2\cdot3\cdot2i + (2i)^2$ Squaring a binomial

$\qquad = 9 - 12i + 4i^2$

$\qquad = 9 - 12i - 4 \qquad i^2 = -1$

$\qquad = 5 - 12i$

62. $21 - 20i$

63. $(2 + 3i)^2 = 2^2 + 2\cdot2\cdot3i + (3i)^2$ Squaring a binomial

$\qquad = 4 + 12i + 9i^2$

$\qquad = 4 + 12i - 9$

$\qquad = -5 + 12i$

64. $12 + 16i$

65. $(-2 + 3i)^2 = 4 - 12i + 9i^2 = 4 - 12i - 9 =$
$-5 - 12i$

66. $21 + 20i$

67. $i^7 = i^6\cdot i = (i^2)^3\cdot i = (-1)^3\cdot i = -i$

68. i

69. $i^{40} = (i^2)^{20} = (-1)^{20} = 1$

70. -1

71. $i^{53} = i^{52}\cdot i = (i^2)^{26}\cdot i = (-1)^{26}\cdot i = i$

72. 1

73. $i^{62} = (i^2)^{31} = (-1)^{31} = -1$

74. $-i$

75. $5 - i^{22} = 5 - (i^2)^{11} = 5 - (-1)^{11} = 5 - (-1) =$
$5 + 1 = 6$

76. 8

<u>77.</u> $9i^2 + 23i^{32} = 9(-1) + 23(i^2)^{16} = -9 + 23(-1)^{16} =$
$-9 + 23 = 14$

<u>78.</u> 18

<u>79.</u> $i^{35} - i^{48} = i^{34} \cdot i - (i^2)^{24} = (i^2)^{17} \cdot i - (-1)^{24} =$
$(-1)^{17} \cdot i - 1 = -i - 1 = -1 - i$

<u>80.</u> $-1 - i$

<u>81.</u> $-5i^{17} + i^{25} = -5 \cdot i^{16} \cdot i + i^{24} \cdot i =$
$-5(i^2)^8 \cdot i + (i^2)^{12} \cdot i = -5(-1)^8 \cdot i + (-1)^{12} \cdot i =$
$-5i + i = -4i$

<u>82.</u> $-6i$

<u>83.</u> $(3 + 7i)(3 - 7i) = 3^2 - (7i)^2$ Multiplying
                                              conjugates
$= 9 - 49i^2$
$= 9 + 49$ $i^2 = -1$
$= 58$

<u>84.</u> 29

<u>85.</u> $(8 - 5i)(8 + 5i) = 8^2 - (5i)^2 = 64 - 25i^2 =$
$64 + 25 = 89$

<u>86.</u> 29

<u>87.</u> $(-3 + 4i)(-3 - 4i) = (-3)^2 - (4i)^2 = 9 - 16i^2 =$
$9 + 16 = 25$

<u>88.</u> 34

<u>89.</u> $\dfrac{5}{3 - i} = \dfrac{5}{3 - i} \cdot \dfrac{3 + i}{3 + i}$ Multiplying by 1, using
                                                                          the conjugate

$= \dfrac{15 + 5i}{10}$ Performing the multiplication

$= \dfrac{15}{10} + \dfrac{5}{10}i$, or $\dfrac{3}{2} + \dfrac{1}{2}i$ Writing in the
                                                                           form a + bi

<u>90.</u> $\dfrac{15}{26} - \dfrac{3}{26}i$

<u>91.</u> $\dfrac{2i}{7 + 3i} = \dfrac{2i}{7 + 3i} \cdot \dfrac{7 - 3i}{7 - 3i}$ Multiplying by 1,
                                                                                using the conjugate

$= \dfrac{14i + 6}{58}$ Performing the multiplication

$= \dfrac{6}{58} + \dfrac{14}{58}i$, or $\dfrac{3}{29} + \dfrac{7}{29}i$ Writing in the
                                                                            form a + bi

<u>92.</u> $-\dfrac{20}{29} + \dfrac{8}{29}i$

<u>93.</u> $\dfrac{7}{6i} = \dfrac{7}{6i} \cdot \dfrac{-6i}{-6i} = \dfrac{-42i}{36} = -\dfrac{7}{6}i$

<u>94.</u> $-\dfrac{3}{10}i$

<u>95.</u> $\dfrac{8 - 3i}{7i} = \dfrac{8 - 3i}{7i} \cdot \dfrac{-7i}{-7i} = \dfrac{-56i - 21}{49} = -\dfrac{21}{49} - \dfrac{56}{49}i =$
$-\dfrac{3}{7} - \dfrac{8}{7}i$

<u>96.</u> $\dfrac{8}{5} - \dfrac{3}{5}i$

<u>97.</u> $\dfrac{3 + 2i}{2 + i} = \dfrac{3 + 2i}{2 + i} \cdot \dfrac{2 - i}{2 - i} = \dfrac{8 + i}{5} = \dfrac{8}{5} + \dfrac{1}{5}i$

<u>98.</u> $\dfrac{15}{26} + \dfrac{29}{26}i$

<u>99.</u> $\dfrac{5 - 2i}{2 + 5i} = \dfrac{5 - 2i}{2 + 5i} \cdot \dfrac{2 - 5i}{2 - 5i} = \dfrac{-29i}{29} = -i$

<u>100.</u> $\dfrac{6}{25} - \dfrac{17}{25}i$

<u>101.</u> $\dfrac{3 - 5i}{3 - 2i} = \dfrac{3 - 5i}{3 - 2i} \cdot \dfrac{3 + 2i}{3 + 2i} = \dfrac{19 - 9i}{13} = \dfrac{19}{13} - \dfrac{9}{13}i$

<u>102.</u> $\dfrac{38}{41} - \dfrac{27}{41}i$

<u>103.</u> $\dfrac{196}{x^2 - 7x + 49} - \dfrac{2x}{x + 7} = \dfrac{2058}{x^3 + 343}$

$\dfrac{196}{x^2 - 7x + 49} - \dfrac{2x}{x + 7} = \dfrac{2058}{(x + 7)(x^2 - 7x + 49)}$

The LCM is $(x + 7)(x^2 - 7x + 49)$. Using the
LCM to clear of fractions, we have

$196(x + 7) - 2x(x^2 - 7x + 49) = 2058$
$196x + 1372 - 2x^3 + 14x^2 - 98x = 2058$
$-2x^3 + 14x^2 + 98x - 686 = 0$
$-2(x^3 - 7x^2 - 49x + 343) = 0$
$-2[x^2(x - 7) - 49(x - 7)] = 0$
$-2(x - 7)(x^2 - 49) = 0$
$-2(x - 7)(x + 7)(x - 7) = 0$

$x - 7 = 0$ or $x + 7 = 0$ or $x - 7 = 0$
$x = 7$ or $x = -7$ or $x = 7$

Since 7 checks and -7 does not, the solution is
7.

<u>104.</u> $\dfrac{70}{29}$

<u>105.</u> $28 = 3x^2 - 17x$
$0 = 3x^2 - 17x - 28$
$0 = (3x + 4)(x - 7)$

$3x + 4 = 0$ or $x - 7 = 0$
$3x = -4$ or $x = 7$
$x = -\dfrac{4}{3}$ or $x = 7$

Both values check. The solutions are $-\dfrac{4}{3}$
and 7.

<u>106.</u> $-4 - 8i$; $-2 + 4i$; $8 - 6i$

107. $\dfrac{1}{\dfrac{1-i}{10} - \left(\dfrac{1-i}{10}\right)^2} = \dfrac{1}{\dfrac{1-i}{10} - \left(\dfrac{-2i}{100}\right)} = \dfrac{1}{\dfrac{1-i}{10} + \dfrac{i}{50}} =$

$\dfrac{1}{\dfrac{1-i}{10} + \dfrac{i}{50}} \cdot \dfrac{50}{50} = \dfrac{50}{5 - 5i + i} = \dfrac{50}{5 - 4i} =$

$\dfrac{50}{5 - 4i} \cdot \dfrac{5 + 4i}{5 + 4i} = \dfrac{250 + 200i}{41} = \dfrac{250}{41} + \dfrac{200}{41}i$

108. $-3 - 4i$

109. $12\sqrt{-\dfrac{1}{32}} = 12\sqrt{-1 \cdot \dfrac{1}{16} \cdot \dfrac{1}{2}} = 12 \cdot i \cdot \dfrac{1}{4}\sqrt{\dfrac{1}{2}} =$

$3\sqrt{\dfrac{1}{2}}i = 3\sqrt{\dfrac{1}{2} \cdot \dfrac{2}{2}}i = \dfrac{3\sqrt{2}}{2}i$

110. $-88i$

111. $\dfrac{i^5 + i^6 + i^7 + i^8}{(1 - i)^4} =$

$\dfrac{(i^2)^2 \cdot i + (i^2)^3 + (i^2)^3 \cdot i + (i^2)^4}{(1 - i)^2(1 - i)^2} =$

$\dfrac{(-1)^2 \cdot i + (-1)^3 + (-1)^3 \cdot i + (-1)^4}{-2i(-2i)} =$

$\dfrac{i - 1 - i + 1}{-4} = 0$

112. 8

113. $\dfrac{5 - \sqrt{5}i}{\sqrt{5}i} = \dfrac{5 - \sqrt{5}i}{\sqrt{5}i} \cdot \dfrac{-\sqrt{5}i}{-\sqrt{5}i} = \dfrac{-5\sqrt{5}i - 5}{5} =$

$-\dfrac{5}{5} - \dfrac{5\sqrt{5}}{5}i = -1 - \sqrt{5}i$

114. $\dfrac{3}{5} + \dfrac{9}{5}i$

115. $\left(\dfrac{1}{2} - \dfrac{1}{3}i\right)^2 - \left(\dfrac{1}{2} + \dfrac{1}{3}i\right)^2 =$

$\dfrac{1}{4} - \dfrac{1}{3}i - \dfrac{1}{9} - \left(\dfrac{1}{4} + \dfrac{1}{3}i - \dfrac{1}{9}\right) =$

$\dfrac{1}{4} - \dfrac{1}{3}i - \dfrac{1}{9} - \dfrac{1}{4} - \dfrac{1}{3}i + \dfrac{1}{9} = -\dfrac{2}{3}i$

116. 1

Exercise Set 8.1

$\underline{1}$. $4x^2 = 20$

   $x^2 = 5$ Multiplying by $\frac{1}{4}$

   $x = \sqrt{5}$ or $x = -\sqrt{5}$ Taking square roots

   Check: $\dfrac{4x^2 = 20}{\begin{array}{c|c} 4(\pm\sqrt{5})^2 & 20 \\ 4\cdot 5 & \\ 20 & \end{array}}$

   The solutions are $\sqrt{5}$ and $-\sqrt{5}$, or $\pm\sqrt{5}$.

$\underline{2}$. $\pm\sqrt{7}$

$\underline{3}$. $10x^2 = 0$

   $x^2 = 0$ Multiplying by $\frac{1}{10}$

   $x = 0$ Taking the square root

   Check: $\dfrac{10x^2 = 0}{\begin{array}{c|c} 10\cdot 0^2 & 0 \\ 10\cdot 0 & \\ 0 & \end{array}}$

   The solution is 0.

$\underline{4}$. 0

$\underline{5}$. $16x^2 = 1$ Check: $\dfrac{16x^2 = 1}{\begin{array}{c|c} 16\left[\pm\frac{1}{4}\right]^2 & 1 \\ 16\cdot\frac{1}{16} & \\ 1 & \end{array}}$

   $x^2 = \frac{1}{16}$

   $x = \frac{1}{4}$ or $x = -\frac{1}{4}$

   The solutions are $\frac{1}{4}$ and $-\frac{1}{4}$, or $\pm\frac{1}{4}$.

$\underline{6}$. $\pm\frac{3}{5}$

$\underline{7}$. $2x^2 - 18 = 0$ Check: $\dfrac{2x^2 - 18 = 0}{\begin{array}{c|c} 2(\pm 3)^2 - 18 & 0 \\ 2\cdot 9 - 18 & \\ 18 - 18 & \\ 0 & \end{array}}$

   $2x^2 = 18$

   $x^2 = 9$

   $x = 3$ or $x = -3$

   The solutions are 3 and -3, or ± 3.

$\underline{8}$. ± 5

$\underline{9}$. $2x^2 - 3 = 0$ Check:

   $2x^2 = 3$ $\dfrac{2x^2 - 3 = 0}{\begin{array}{c|c} 2\left[\pm\frac{\sqrt{6}}{2}\right]^2 - 3 & 0 \\ 2\cdot\frac{6}{4} - 3 & \\ 3 - 3 & \\ 0 & \end{array}}$

   $x^2 = \frac{3}{2}$

   $x = \sqrt{\frac{3}{2}}$ or $x = -\sqrt{\frac{3}{2}}$

   $x = \sqrt{\frac{3}{2}\cdot\frac{2}{2}}$ or $x = -\sqrt{\frac{3}{2}\cdot\frac{2}{2}}$

   $x = \frac{\sqrt{6}}{2}$ or $x = -\frac{\sqrt{6}}{2}$

   The solutions are $\frac{\sqrt{6}}{2}$ and $-\frac{\sqrt{6}}{2}$, or $\pm\frac{\sqrt{6}}{2}$.

$\underline{10}$. $\pm\frac{\sqrt{21}}{3}$

$\underline{11}$. $-3x^2 + 5 = 0$ Check:

   $-3x^2 = -5$ $\dfrac{-3x^2 + 5 = 0}{\begin{array}{c|c} -3\left[\pm\frac{\sqrt{15}}{3}\right]^2 + 5 & 0 \\ -3\cdot\frac{15}{9} + 5 & \\ -5 + 5 & \\ 0 & \end{array}}$

   $x^2 = \frac{5}{3}$

   $x = \sqrt{\frac{5}{3}}$ or $x = -\sqrt{\frac{5}{3}}$

   $x = \sqrt{\frac{5}{3}\cdot\frac{3}{3}}$ or $x = -\sqrt{\frac{5}{3}\cdot\frac{3}{3}}$

   $x = \frac{\sqrt{15}}{3}$ or $x = -\frac{\sqrt{15}}{3}$

   The solutions are $\frac{\sqrt{15}}{3}$ and $-\frac{\sqrt{15}}{3}$, or $\pm\frac{\sqrt{15}}{3}$.

$\underline{12}$. $\pm\frac{\sqrt{2}}{2}$

$\underline{13}$. $x^2 + 100 = 0$

   $x^2 = -100$ Adding -100

   $x = \sqrt{-100}$ or $x = -\sqrt{-100}$ Taking square roots

   $x = 10i$ or $x = -10i$ Simplifying

   Check: $\dfrac{x^2 + 100 = 0}{\begin{array}{c|c} (\pm 10i)^2 + 100 & 0 \\ -100 + 100 & \\ 0 & \end{array}}$

   The solutions are $10i$ and $-10i$, or $\pm 10i$.

$\underline{14}$. $\pm 9i$

$\underline{15}$. $x^2 + 5 = 0$

   $x^2 = -5$ Adding -5

   $x = \sqrt{-5}$ or $x = -\sqrt{-5}$ Taking square roots

   $x = i\sqrt{5}$ or $x = -i\sqrt{5}$ Simplifying

   Check: $\dfrac{x^2 + 5 = 0}{\begin{array}{c|c} (\pm i\sqrt{5})^2 + 5 & \\ -5 + 5 & \\ 0 & \end{array}}$

   The solutions are $i\sqrt{5}$ and $-i\sqrt{5}$, or $\pm i\sqrt{5}$.

16. $\pm i\sqrt{6}$

17.
$$0 = 4 + 25x^2$$
$$-25x^2 = 4$$
$$x^2 = -\frac{4}{25}$$
$$x = \sqrt{-\frac{4}{25}} \quad \text{or} \quad x = -\sqrt{-\frac{4}{25}}$$
$$x = \frac{2}{5}i \quad \text{or} \quad x = -\frac{2}{5}i$$

Check:

$$
\begin{array}{c|c}
25x^2 + 4 = 0 & \\
\hline
25\left[\pm\frac{2}{5}i\right] + 4 & 0 \\
25\left[-\frac{4}{25}\right] + 4 & \\
-4 + 4 & \\
& 0
\end{array}
$$

The solutions are $\frac{2}{5}i$ and $-\frac{2}{5}i$, or $\pm\frac{2}{5}i$.

18. $\pm\frac{4}{3}i$

19.
$$2x^2 + 14 = 0$$
$$2x^2 = -14$$
$$x^2 = -7$$
$$x = \sqrt{-7} \quad \text{or} \quad x = -\sqrt{-7}$$
$$x = i\sqrt{7} \quad \text{or} \quad x = -i\sqrt{7}$$

These numbers check, so the solutions are $i\sqrt{7}$ and $-i\sqrt{7}$, or $\pm i\sqrt{7}$.

20. $\pm i\sqrt{5}$

21.
$$\frac{4}{9}x^2 = 1$$
$$x^2 = \frac{9}{4}$$
$$x = \sqrt{\frac{9}{4}} \quad \text{or} \quad x = -\sqrt{\frac{9}{4}}$$
$$x = \frac{3}{2} \quad \text{or} \quad x = -\frac{3}{2}$$

Both numbers check, so the solutions are $\frac{3}{2}$ and $-\frac{3}{2}$, or $\pm\frac{3}{2}$.

22. $\pm\frac{5}{4}$

23.
$$x^2 - 5x = 0$$
$$x(x - 5) = 0 \quad \text{Factoring}$$
$$x = 0 \quad \text{or} \quad x - 5 = 0 \quad \text{Principle of zero products}$$
$$x = 0 \quad \text{or} \quad x = 5$$

Check: For 0:

$$
\begin{array}{c|c}
x^2 - 5x = 0 & \\
\hline
0^2 - 5\cdot0 & 0 \\
0 - 0 & \\
0 &
\end{array}
$$

For 5:

$$
\begin{array}{c|c}
x^2 - 5x = 0 & \\
\hline
5^2 - 5\cdot5 & 0 \\
25 - 25 & \\
0 &
\end{array}
$$

The solutions are 0 and 5.

24. 0, 6

25.
$$5x^2 + 10x = 0$$
$$x^2 + 2x = 0 \quad \text{Multiplying by } \frac{1}{5}$$
$$x(x + 2) = 0 \quad \text{Factoring}$$
$$x = 0 \quad \text{or} \quad x + 2 = 0 \quad \text{Principle of zero products}$$
$$x = 0 \quad \text{or} \quad x = -2$$

The solutions are 0 and -2.

26. 0, -4

27.
$$3x^2 - 2x = 0$$
$$x(3x - 2) = 0$$
$$x = 0 \quad \text{or} \quad 3x - 2 = 0$$
$$x = 0 \quad \text{or} \quad x = \frac{2}{3}$$

The solutions are 0 and $\frac{2}{3}$.

28. 0, $\frac{3}{7}$

29.
$$14x^2 + 9x = 0$$
$$x(14x + 9) = 0$$
$$x = 0 \quad \text{or} \quad 14x + 9 = 0$$
$$x = 0 \quad \text{or} \quad x = -\frac{9}{14}$$

The solutions are 0 and $-\frac{9}{14}$.

30. 0, $-\frac{8}{19}$

31.
$$9x^2 - 11x = 0$$
$$x(9x - 11) = 0$$
$$x = 0 \quad \text{or} \quad 9x - 11 = 0$$
$$x = 0 \quad \text{or} \quad x = \frac{11}{9}$$

The solutions are 0 and $\frac{11}{9}$.

32. 0, $\frac{13}{7}$

33.
$$x^2 - 4x + 3 = 0$$
$$(x - 1)(x - 3) = 0 \quad \text{Factoring}$$
$$x - 1 = 0 \quad \text{or} \quad x - 3 = 0 \quad \text{Principle of zero products}$$
$$x = 1 \quad \text{or} \quad x = 3$$

Check: For 1:

$$
\begin{array}{c|c}
x^2 - 4x + 3 = 0 & \\
\hline
1^2 - 4\cdot1 + 3 & 0 \\
1 - 4 + 3 & \\
0 &
\end{array}
$$

For 3:

$$
\begin{array}{c|c}
x^2 - 4x + 3 = 0 & \\
\hline
3^2 - 4\cdot3 + 3 & 0 \\
9 - 12 + 3 & \\
0 &
\end{array}
$$

The solutions are 1 and 3.

34. 5, 1

35.
$$x^2 + 6 = 7x$$
$$x^2 - 7x + 6 = 0 \quad \text{Finding standard form}$$
$$(x - 6)(x - 1) = 0$$
$$x - 6 = 0 \quad \text{or} \quad x - 1 = 0$$
$$x = 6 \quad \text{or} \quad x = 1$$

The solutions are 6 and 1.

36. 5, -1

37.
$$x^2 - 6x - 7 = 0$$
$$(x - 7)(x + 1) = 0$$
$$x - 7 = 0 \quad \text{or} \quad x + 1 = 0$$
$$x = 7 \quad \text{or} \quad x = -1$$

The solutions are 7 and -1.

38. -5, -3

39.
$$x^2 + 9x + 14 = 0$$
$$(x + 7)(x + 2) = 0$$
$$x + 7 = 0 \quad \text{or} \quad x + 2 = 0$$
$$x = -7 \quad \text{or} \quad x = -2$$

The solutions are -7 and -2.

40. -6, -5

41.
$$5 = 14x + 3x^2$$
$$0 = 3x^2 + 14x - 5$$
$$0 = (3x - 1)(x + 5)$$
$$3x - 1 = 0 \quad \text{or} \quad x + 5 = 0$$
$$x = \frac{1}{3} \quad \text{or} \quad x = -5$$

The solutions are $\frac{1}{3}$ and -5.

42. $\frac{5}{2}$, -3

43.
$$2x^2 + 13x + 15 = 0$$
$$(2x + 3)(x + 5) = 0$$
$$2x + 3 = 0 \quad \text{or} \quad x + 5 = 0$$
$$x = -\frac{3}{2} \quad \text{or} \quad x = -5$$

The solutions are $-\frac{3}{2}$ and -5.

44. $\frac{2}{3}$, $-\frac{1}{2}$

45.
$$3y^2 = 8 - 10y$$
$$3y^2 + 10y - 8 = 0$$
$$(3y - 2)(y + 4) = 0$$
$$3y - 2 = 0 \quad \text{or} \quad y + 4 = 0$$
$$y = \frac{2}{3} \quad \text{or} \quad y = -4$$

The solutions are $\frac{2}{3}$ and -4.

46. $-\frac{4}{3}$, $-\frac{1}{3}$

47.
$$10r^2 + r = 3$$
$$10r^2 + r - 3 = 0$$
$$(5r + 3)(2r - 1) = 0$$
$$5r + 3 = 0 \quad \text{or} \quad 2r - 1 = 0$$
$$r = -\frac{3}{5} \quad \text{or} \quad r = \frac{1}{2}$$

The solutions are $-\frac{3}{5}$ and $\frac{1}{2}$.

48. $-\frac{1}{3}$, $\frac{5}{2}$

49.
$$3x^2 + 7x = 20$$
$$3x^2 + 7x - 20 = 0$$
$$(3x - 5)(x + 4) = 0$$
$$3x - 5 = 0 \quad \text{or} \quad x + 4 = 0$$
$$x = \frac{5}{3} \quad \text{or} \quad x = -4$$

The solutions are $\frac{5}{3}$ and -4.

50. $-\frac{5}{3}$, -1

51.
$$t(2t + 9) = -7$$
$$2t^2 + 9t = -7 \quad \text{Multiplying}$$
$$2t^2 + 9t + 7 = 0 \quad \text{Finding standard form}$$
$$(2t + 7)(t + 1) = 0 \quad \text{Factoring}$$
$$2t + 7 = 0 \quad \text{or} \quad t + 1 = 0 \quad \text{Principle of zero products}$$
$$t = -\frac{7}{2} \quad \text{or} \quad t = -1$$

The solutions are $-\frac{7}{2}$ and -1.

52. $-\frac{1}{4}$, $\frac{3}{2}$

53.
$$16(t - 1) = t(t + 8)$$
$$16t - 16 = t^2 + 8t$$
$$0 = t^2 - 8t + 16$$
$$0 = (t - 4)(t - 4)$$
$$t - 4 = 0 \quad \text{or} \quad t - 4 = 0$$
$$t = 4 \quad \text{or} \quad t = 4$$

The solution is 4.

54. 10, -3

55. $(7x + 5)(x - 2) = (x - 2)(3x - 3)$

$7x^2 - 9x - 10 = 3x^2 - 9x + 6$ Multiplying

$4x^2 - 16 = 0$

$x^2 - 4 = 0$ Multiplying by $\frac{1}{4}$

$x^2 = 4$

$x = 2$ or $x = -2$

The solutions are 2 and -2, or ±2.\

56. $-\frac{5}{3}$, -5

57. $14(x - 4) - (x + 2) = (x + 2)(x - 4)$

$14x - 56 - x - 2 = x^2 - 2x - 8$

$13x - 58 = x^2 - 2x - 8$

$0 = x^2 - 15x + 50$

$0 = (x - 10)(x - 5)$

$x - 10 = 0$ or $x - 5 = 0$

$x = 10$ or $x = 5$

The solutions are 10 and 5.

58. -2, -1

59. Familiarize. We make a drawing and label it with both known and unknown information. We let x represent the width of the frame.

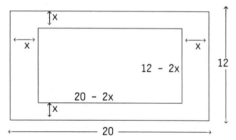

The length and width of the picture that shows are represented by 20 - 2x and 12 - 2x. The area of the picture that shows is 84 cm².

Translate. Using the formula for the area of a rectangle, A = ℓ·w, we have

$84 = (20 - 2x)(12 - 2x).$

Carry out. We solve the equation.

$84 = 240 - 64x + 4x^2$

$84 = 4(60 - 16x + x^2).$

$21 = 60 - 16x + x^2$

$0 = x^2 - 16x + 39$

$0 = (x - 3)(x - 13)$

$x - 3 = 0$ or $x - 13 = 0$

$x = 3$ or $x = 13$

59. (continued)

Check. We see that 13 is not a solution because when x = 13, 20 - 2x = -6 and 12 - 2x = -14, and the length and width of the frame cannot be negative. We check 3. When x = 3, 20 - 2x = 14 and 12 - 2x = 6 and 14·6 = 84. The area is 84. The value checks.

State. The width of the frame is 3 cm.

60. 2 cm

61. Familiarize. We make a drawing and label it. We let x represent the length of the rectangle. Then x - 5 represents the width.

Translate. We use the formula for the area of a rectangle, A = ℓ·w.

$24 = x(x - 5)$

Carry out. We solve the equation.

$24 = x^2 - 5x$

$0 = x^2 - 5x - 24$

$0 = (x - 8)(x + 3)$

$x - 8 = 0$ or $x + 3 = 0$

$x = 8$ or $x = -3$

Check. We only check 8 since the length of a rectangle cannot be negative. If x = 8, then x - 5 = 3 and the area is 8·3, or 24. The value checks.

State. The length is 8 m, and the width is 3 m.

62. Length: 6 m, width: 2 m

63. Familiarize. We make a drawing and label it. We let x represent the width. Then 2x represents the length.

Translate. We use the formula for the area of a rectangle, A = ℓ·w, we have

$338 = 2x \cdot x.$

Carry out. We solve the equation.

$338 = 2x^2$

$169 = x^2$

$x = 13$ or $x = -13$

Check. We only check 13 since the width of a rectangle cannot be negative. If x = 13, then 2x = 26 and the area is 26·13, or 338. The value checks.

State. The length is 26 km, and the width is 13 km.

64. Length: 24 m, width: 12 m

65. Familiarize. We first make a drawing. We let x represent the length of the shorter leg. Then x + 7 represents the longer leg.

Since we have a right triangle, we can use the Pythagorean property:

$$a^2 + b^2 = c^2$$

Translate. Substituting, we have

$$x^2 + (x + 7)^2 = 13^2.$$

Carry out.

$$x^2 + x^2 + 14x + 49 = 169$$
$$2x^2 + 14x - 120 = 0$$
$$x^2 + 7x - 60 = 0$$
$$(x + 12)(x - 5) = 0$$

$$x + 12 = 0 \quad \text{or} \quad x - 5 = 0$$
$$x = -12 \quad \text{or} \quad x = 5$$

Check. We only check 5 since the length of a leg cannot be negative. If x = 5, then x + 7 = 12. $5^2 + 12^2 = 25 + 144 = 169$, and $169 = 13^2$, so the answer checks.

State. The lengths of the legs are 5 m and 12 m.

66. 4 m and 3 m

67. Familiarize. We first make a drawing. We let x represent the length of the longer leg. Then x - 17 represents the length of the shorter leg.

We use the Pythagorean property:

$$a^2 + b^2 = c^2$$

Translate. Substituting, we have

$$x^2 + (x - 17)^2 = 25^2.$$

Carry out. We solve the equation.

$$x^2 + x^2 - 34x + 289 = 625$$
$$2x^2 - 34x - 336 = 0$$
$$x^2 - 17x - 168 = 0$$
$$(x + 7)(x - 24) = 0$$

$$x + 7 = 0 \quad \text{or} \quad x - 24 = 0$$
$$x = -7 \quad \text{or} \quad x = 24$$

Check. We only check 24 since the length of a leg cannot be negative. If x = 24, then x - 17 = 7. $24^2 + 7^2 = 576 + 49 = 625$, and $625 = 25^2$, so the answer checks.

67. (continued)

State. The lengths of the legs are 24 km and 7 km.

68. 10 m and 24 m

69. Familiarize. We make a drawing and label it with both known and unknown information. We let x represent the width of the sidewalk.

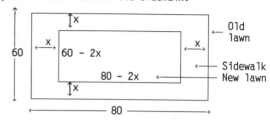

The area of the old lawn is 80·60, or 4800 ft². Then the area of the new lawn is $\frac{1}{6}$ · 4800, or 800 ft².

Translate. Rewording, we have

Area of new lawn is 800 ft².

$$(80 - 2x)(60 - 2x) = 800$$

Carry out. We solve the equation.

$$4800 - 280x + 4x^2 = 800$$
$$4x^2 - 280x + 4000 = 0$$
$$x^2 - 70x + 1000 = 0$$
$$(x - 20)(x - 50) = 0$$

$$x = 20 \quad \text{or} \quad x = 50$$

Check. If the sidewalk is 20 ft wide, the new lawn will have length 80 - 2·20, or 40 ft. The width will be 60 - 2·20, or 20 ft. The area of the new lawn will be 40·20, or 800 ft². This is $\frac{1}{6}$ of 4800 ft², the area of the old lawn, so this answer checks.

If the sidewalk is 50 ft wide, the new lawn will have length 80 - 2·50, or -20. Since length cannot be negative, this is not a solution.

State. The sidewalk is 20 ft wide.

70. 5 ft

71. Familiarize. We first make a drawing. Let x represent the distance from the bottom of the ladder to the base of the wall. The x + 7 represents the distance from the top of the ladder to the base of the wall.

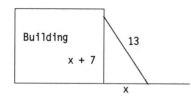

We will use the Pythagorean property:
$$a^2 + b^2 = c^2$$

Translate. Substituting, we have
$$x^2 + (x + 7)^2 = 13^2.$$

Carry out. We solve the equation.
$$x^2 + x^2 + 14x + 49 = 169$$
$$2x^2 + 14x - 120 = 0$$
$$x^2 + 7x - 60 = 0$$
$$(x + 12)(x - 5) = 0$$

$x + 12 = 0$ or $x - 5 = 0$
$x = -12$ or $x = 5$

Check. We only need to check 5, since distance cannot be negative. If the bottom of the ladder is 5 ft from the base of the building, then the top of the ladder is 5 + 7, or 12 ft, from the base of the building.

$5^2 + 12^2 = 25 + 144 = 169$, and $169 = 13^2$, so the answer checks.

State. The bottom of the ladder is 5 ft from the wall.

72. Length: 200 ft, width: 150 ft

73. $s = 16t^2$
$s = 16(2)^2$ Substituting 2 for t
$s = 16 \cdot 4$
$s = 64$

The object will fall 64 ft in 2 sec.

74. 400 ft

75. $s = 16t^2$
$1377 = 16t^2$ Substituting 1377 for s
$\dfrac{1377}{16} = t^2$
$\sqrt{\dfrac{1377}{16}} = t$ Taking the positive square root, since time cannot be negative
$9.277 \approx t$ Using a calculator

It would take about 9.277 sec for an object to fall from the top.

76. 8.318 sec

77. Graph: $y = 2x + 1$

x	y
-3	-5
0	1
2	5

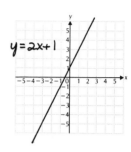

78. $2\sqrt{22}$

79. $14 - \sqrt{88} \approx 14 - 9.3803$ Using Table 1 or a calculator
≈ 4.6

80. $\dfrac{\sqrt{10}}{5}$

81. $25.55x^2 - 1635.2 = 0$
$25.55x^2 = 1635.2$
$x^2 = \dfrac{1635.2}{25.55}$
$x^2 = 64$
$x = 8$ or $x = -8$

The solutions are 8 and -8, or ±8.

82. $-\dfrac{5}{3}$, 4, $\dfrac{5}{2}$

83. $x(2x^2 + 9x - 56)(3x + 10) = 0$
$x(2x - 7)(x + 8)(3x + 10) = 0$

$x = 0$ or $2x - 7 = 0$ or $x + 8 = 0$ or $3x + 10 = 0$
$x = 0$ or $x = \dfrac{7}{2}$ or $x = -8$ or $x = -\dfrac{10}{3}$

The solutions are -8, $-\dfrac{10}{3}$, 0, and $\dfrac{7}{2}$.

84. $\dfrac{1}{3}$, $\dfrac{1}{18}$

85. <u>Familiarize</u>. It is helpful to list the information in a table and to make a drawing. We let r represent the speed, in km/h, of boat A. Then r − 7 represents the speed of boat B.

Boat	Speed	Time	Distance
A	r	4	4r
B	r − 7	4	4(r − 7)

Recall that Distance = Rate × Time.

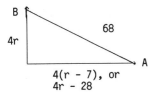

We will use the Pythagorean property:
$$a^2 + b^2 = c^2$$

<u>Translate</u>. Substituting, we have
$$(4r - 28)^2 + (4r)^2 = 68^2.$$

<u>Carry out</u>. We solve the equation.
$$16r^2 - 224r + 784 + 16r^2 = 4624$$
$$32r^2 - 224r - 3840 = 0$$
$$r^2 - 7r - 120 = 0$$
$$(r + 8)(r - 15) = 0$$

$$r + 8 = 0 \quad \text{or} \quad r - 15 = 0$$
$$r = -8 \quad \text{or} \quad r = 15$$

<u>Check</u>. We only check 15 since the speeds of the boats cannot be negative. If the speed of Boat A is 15 km/h, then the speed of Boat B is 15 − 7, or 8 km/h, and the distances they travel are 4·15, or 60 km, and 4·8, or 32 km, respectively.
$60^2 + 32^2 = 3600 + 1024 = 4624$, and $4624 = 68^2$, so the answer checks.

<u>State</u>. The speed of boat A is 15 km/h, and the speed of boat B is 8 km/h.

86. 5, 6, and 7

87. $ax^2 - bx = 0$
$x(ax - b) = 0$

$x = 0 \quad \text{or} \quad ax - b = 0$
$x = 0 \quad \text{or} \quad ax = b$
$x = 0 \quad \text{or} \quad x = \frac{b}{a}$

The solutions are 0 and $\frac{b}{a}$.

88. $\pm\sqrt{\frac{b}{a}}$, or $\pm\frac{\sqrt{ab}}{a}$

89. We first make a drawing. Let s represent the length of a side of the triangle, and let h represent its height, or altitude. Recall that the altitude of an equilateral triangle bisects its base. Also, recall that the formula for the circumference of a circle is $C = 2\pi r$. Thus, for this circle:

89. (continued)
$$6\pi = 2\pi r$$
$$3 = r$$

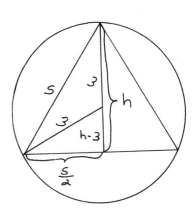

Use the Pythagorean property and the right triangle with legs $\frac{s}{2}$ and h and hypotenuse s to find h in terms of s:
$$\left(\frac{s}{2}\right)^2 + h^2 = s^2$$
$$h^2 = s^2 - \frac{s^2}{4}$$
$$h^2 = \frac{3s^2}{4}$$
$$h = \frac{s\sqrt{3}}{2}$$

Next consider the smaller right triangle with legs $\frac{s}{2}$ and h − 3 and hypotenuse 3. Since $h = \frac{s\sqrt{3}}{2}$, then $h - 3 = \frac{s\sqrt{3}}{2} - 3 = \frac{s\sqrt{3} - 6}{2}$.

Use the Pythagorean property again to find s:
$$\left(\frac{s}{2}\right)^2 + \left(\frac{s\sqrt{3} - 6}{2}\right)^2 = 3^2$$
$$\frac{s^2}{4} + \frac{3s^2 - 12s\sqrt{3} + 36}{4} = 9$$
$$s^2 + 3s^2 - 12s\sqrt{3} + 36 = 36 \quad \text{Multiplying by 4}$$
$$4s^2 - 12s\sqrt{3} = 0$$
$$s^2 - 3s\sqrt{3} = 0$$
$$s(s - 3\sqrt{3}) = 0$$

$$s = 0 \quad \text{or} \quad s = 3\sqrt{3}$$

The length of a side cannot be 0, so we only consider $3\sqrt{3}$. If $s = 3\sqrt{3}$, then
$$h = \frac{s\sqrt{3}}{2} = \frac{3\sqrt{3} \cdot \sqrt{3}}{2} = \frac{9}{2}.$$

Now we can find the area of the triangle. The base is s, or $3\sqrt{3}$, and the height is $\frac{9}{2}$.
$$A = \frac{1}{2}bh$$
$$A = \frac{1}{2} \cdot 3\sqrt{3} \cdot \frac{9}{2}$$
$$A = \frac{27\sqrt{3}}{4}$$

The area of the triangle is $\frac{27\sqrt{3}}{4}$ in².

Exercise Set 8.2

1. $x^2 + 10x$

    We take half the coefficient of x and square it:
    Half of 10 is 5, and $5^2 = 25$. We add 25.
    $x^2 + 10x + 25$, $(x + 5)^2$

2. $x^2 + 16x + 64$, $(x + 8)^2$

3. $x^2 - 8x$

    We take half the coefficient of x and square it:
    Half of -8 is -4, and $(-4)^2 = 16$. We add 16.
    $x^2 - 8x + 16$, $(x - 4)^2$

4. $x^2 - 6x + 9$, $(x - 3)^2$

5. $x^2 - 24x$

    We take half the coefficient of x and square it:
    $\frac{1}{2}(-24) = -12$ and $(-12)^2 = 144$. We add 144.
    $x^2 - 24x + 144$, $(x - 12)^2$

6. $x^2 - 18x + 81$, $(x - 9)^2$

7. $x^2 + 9x$

    $\frac{1}{2} \cdot 9 = \frac{9}{2}$, and $\left(\frac{9}{2}\right)^2 = \frac{81}{4}$. We add $\frac{81}{4}$.
    $x^2 + 9x + \frac{81}{4}$, $\left(x + \frac{9}{2}\right)^2$

8. $x^2 + 3x + \frac{9}{4}$, $\left(x + \frac{3}{2}\right)^2$

9. $x^2 - 7x$

    $\frac{1}{2}(-7) = -\frac{7}{2}$, and $\left(-\frac{7}{2}\right)^2 = \frac{49}{4}$. We add $\frac{49}{4}$.
    $x^2 - 7x + \frac{49}{4}$, $\left(x - \frac{7}{2}\right)^2$

10. $x^2 - 11x + \frac{121}{4}$, $\left(x - \frac{11}{2}\right)^2$

11. $x^2 + \frac{2}{3}x$

    $\frac{1}{2} \cdot \frac{2}{3} = \frac{1}{3}$, and $\left(\frac{1}{3}\right)^2 = \frac{1}{9}$. We add $\frac{1}{9}$.
    $x^2 + \frac{2}{3}x + \frac{1}{9}$, $\left(x + \frac{1}{3}\right)^2$

12. $x^2 + \frac{2}{5}x + \frac{1}{25}$, $\left(x + \frac{1}{5}\right)^2$

13. $x^2 - \frac{5}{6}x$

    $\frac{1}{2}\left(-\frac{5}{6}\right) = -\frac{5}{12}$, and $\left(-\frac{5}{12}\right)^2 = \frac{25}{144}$. We add $\frac{25}{144}$.
    $x^2 - \frac{5}{6}x + \frac{25}{144}$, $\left(x - \frac{5}{12}\right)^2$

14. $x^2 - \frac{5}{3}x + \frac{25}{36}$, $\left(x - \frac{5}{6}\right)^2$

15. $x^2 + \frac{9}{5}x$

    $\frac{1}{2} \cdot \frac{9}{5} = \frac{9}{10}$, and $\left(\frac{9}{10}\right)^2 = \frac{81}{100}$. We add $\frac{81}{100}$.
    $x^2 + \frac{9}{5}x + \frac{81}{100}$, $\left(x + \frac{9}{10}\right)^2$

16. $x^2 + \frac{9}{4}x + \frac{81}{64}$, $\left(x + \frac{9}{8}\right)^2$

17. $\quad x^2 + 8x = -7$

    $x^2 + 8x + 16 = -7 + 16$ Adding 16 on both sides to complete the square

    $\quad (x + 4)^2 = 9$ Factoring

    $\quad\quad x + 4 = \pm 3$ Principle of positive and negative roots

    $\quad\quad\quad x = -4 \pm 3$

    $x = -4 - 3$ or $x = -4 + 3$

    $x = -7$ or $x = -1$

    The solutions are -7 and -1.

18. -7, 1

19. $\quad\quad x^2 - 10x = 22$

    $x^2 - 10x + 25 = 22 + 25$ Adding 25 on both sides to complete the square

    $\quad (x - 5)^2 = 47$

    $\quad\quad x - 5 = \pm\sqrt{47}$ Principle of positive and negative roots

    $\quad\quad\quad x = 5 \pm \sqrt{47}$

    The solutions are $5 \pm \sqrt{47}$.

20. $4 \pm \sqrt{7}$

21. $x^2 + 6x + 5 = 0$

    $x^2 + 6x \quad = -5$ Adding -5 on both sides

    $x^2 + 6x + 9 = -5 + 9$ Completing the square

    $\quad (x + 3)^2 = 4$

    $\quad\quad x + 3 = \pm 2$

    $\quad\quad\quad x = -3 \pm 2$

    $x = -3 - 2$ or $x = -3 + 2$

    $x = -5$ or $x = -1$

    The solutions are -5 and -1.

22. -9, -1

23. $x^2 - 10x + 21 = 0$

    $x^2 - 10x \quad = -21$

    $x^2 - 10x + 25 = -21 + 25$

    $\quad (x - 5)^2 = 4$

    $\quad\quad x - 5 = \pm 2$

    $\quad\quad\quad x = 5 \pm 2$

    $x = 5 - 2$ or $x = 5 + 2$

    $x = 3$ or $x = 7$

    The solutions are 3 and 7.

24. 4, 6

25. $x^2 + 5x + 6 = 0$

    $x^2 + 5x = -6$

    $x^2 + 5x + \dfrac{25}{4} = -6 + \dfrac{25}{4}$

    $\left(x + \dfrac{5}{2}\right)^2 = \dfrac{1}{4}$

    $x + \dfrac{5}{2} = \pm \dfrac{1}{2}$

    $x = -\dfrac{5}{2} \pm \dfrac{1}{2}$

    $x = -\dfrac{5}{2} - \dfrac{1}{2}$ or $x = -\dfrac{5}{2} + \dfrac{1}{2}$

    $x = -3$ or $x = -2$

    The solutions are -3 and -2.

26. -4, -3

27. $x^2 + 4x + 1 = 0$

    $x^2 + 4x = -1$

    $x^2 + 4x + 4 = -1 + 4$

    $(x + 2)^2 = 3$

    $x + 2 = \pm \sqrt{3}$

    $x = -2 \pm \sqrt{3}$

    The solutions are $-2 \pm \sqrt{3}$.

28. $-3 \pm \sqrt{2}$

29. $x^2 - 10x + 23 = 0$

    $x^2 - 10x = -23$

    $x^2 - 10x + 25 = -23 + 25$

    $(x - 5)^2 = 2$

    $x - 5 = \pm \sqrt{2}$

    $x = 5 \pm \sqrt{2}$

    The solutions are $5 \pm \sqrt{2}$.

30. $3 \pm \sqrt{5}$

31. $x^2 + 6x + 13 = 0$

    $x^2 + 6x = -13$

    $x^2 + 6x + 9 = -13 + 9$

    $(x + 3)^2 = -4$

    $x + 3 = \pm 2i$

    $x = -3 \pm 2i$

    The solutions are $-3 \pm 2i$.

32. $-4 \pm 3i$

33. $2x^2 - 5x - 3 = 0$

    $2x^2 - 5x = 3$

    $x^2 - \dfrac{5}{2}x = \dfrac{3}{2}$ Dividing by 2 on both sides

    $x^2 - \dfrac{5}{2}x + \dfrac{25}{16} = \dfrac{3}{2} + \dfrac{25}{16}$

    $\left(x - \dfrac{5}{4}\right)^2 = \dfrac{49}{16}$

    $x - \dfrac{5}{4} = \pm \dfrac{7}{4}$

    $x = \dfrac{5}{4} \pm \dfrac{7}{4}$

    $x = \dfrac{5}{4} - \dfrac{7}{4}$ or $x = \dfrac{5}{4} + \dfrac{7}{4}$

    $x = -\dfrac{1}{2}$ or $x = 3$

    The solutions are $-\dfrac{1}{2}$ and 3.

34. $-2, \dfrac{1}{3}$

35. $4x^2 + 8x + 3 = 0$

    $4x^2 + 8x = -3$

    $x^2 + 2x = -\dfrac{3}{4}$

    $x^2 + 2x + 1 = -\dfrac{3}{4} + 1$

    $(x + 1)^2 = \dfrac{1}{4}$

    $x + 1 = \pm \dfrac{1}{2}$

    $x = -1 \pm \dfrac{1}{2}$

    $x = -1 - \dfrac{1}{2}$ or $x = -1 + \dfrac{1}{2}$

    $x = -\dfrac{3}{2}$ or $x = -\dfrac{1}{2}$

    The solutions are $-\dfrac{3}{2}$ and $-\dfrac{1}{2}$.

36. $-\dfrac{4}{3}, -\dfrac{2}{3}$

37. $6x^2 - x - 2 = 0$

    $6x^2 - x = 2$

    $x^2 - \dfrac{1}{6}x = \dfrac{1}{3}$

    $x^2 - \dfrac{1}{6}x + \dfrac{1}{144} = \dfrac{1}{3} + \dfrac{1}{144}$

    $\left(x - \dfrac{1}{12}\right)^2 = \dfrac{49}{144}$

    $x - \dfrac{1}{12} = \pm \dfrac{7}{12}$

    $x = \dfrac{1}{12} \pm \dfrac{7}{12}$

    $x = \dfrac{1}{12} - \dfrac{7}{12}$ or $x = \dfrac{1}{12} + \dfrac{7}{12}$

    $x = -\dfrac{1}{2}$ or $x = \dfrac{2}{3}$

    The solutions are $-\dfrac{1}{2}$ and $\dfrac{2}{3}$.

<u>38.</u> $-\frac{3}{2}, \frac{5}{3}$

<u>39.</u> $2x^2 + 4x + 1 = 0$

$2x^2 + 4x \qquad = -1$

$x^2 + 2x \qquad = -\frac{1}{2}$

$x^2 + 2x + 1 = -\frac{1}{2} + 1$

$(x + 1)^2 = \frac{1}{2}$

$x + 1 = \pm\sqrt{\frac{1}{2}}$

$x + 1 = \pm\frac{\sqrt{2}}{2}$ Rationalizing the denominator

$x = -1 \pm \frac{\sqrt{2}}{2}$

The solutions are $-1 \pm \frac{\sqrt{2}}{2}$.

<u>40.</u> $-2, -\frac{1}{2}$

<u>41.</u> $3x^2 - 5x - 3 = 0$

$3x^2 - 5x \qquad = 3$

$x^2 - \frac{5}{3}x \qquad = 1$

$x^2 - \frac{5}{3}x + \frac{25}{36} = 1 + \frac{25}{36}$

$\left(x - \frac{5}{6}\right)^2 = \frac{61}{36}$

$x - \frac{5}{6} = \pm\frac{\sqrt{61}}{6}$

$x = \frac{5 \pm \sqrt{61}}{6}$

The solutions are $\frac{5 \pm \sqrt{61}}{6}$.

<u>42.</u> $\frac{3 \pm \sqrt{13}}{4}$

<u>43.</u> $x^2 + 6x + 4 = 0$

$a = 1, \quad b = 6, \quad c = 4$

$x = \frac{-b \pm \sqrt{b^2 - 4ac}}{2a}$

$x = \frac{-6 \pm \sqrt{6^2 - 4 \cdot 1 \cdot 4}}{2 \cdot 1} = \frac{-6 \pm \sqrt{36 - 16}}{2}$

$x = \frac{-6 \pm \sqrt{20}}{2} = \frac{-6 \pm 2\sqrt{5}}{2}$

$x = \frac{2(-3 \pm \sqrt{5})}{2} = -3 \pm \sqrt{5}$

The solutions are $-3 + \sqrt{5}$ and $-3 - \sqrt{5}$.

<u>44.</u> $3 \pm \sqrt{13}$

<u>45.</u> $x^2 + 4x - 5 = 0$

$(x + 5)(x - 1) = 0$

$x + 5 = 0 \quad \text{or} \quad x - 1 = 0$

$x = -5 \quad \text{or} \qquad x = 1$

The solutions are -5 and 1.

<u>46.</u> $5, -3$

<u>47.</u> $y^2 + 7y = 30$

$y^2 + 7y - 30 = 0$

$(y + 10)(y - 3) = 0$

$y + 10 = 0 \quad \text{or} \quad y - 3 = 0$

$y = -10 \quad \text{or} \qquad y = 3$

The solutions are -10 and 3.

<u>48.</u> $10, -3$

<u>49.</u> $2t^2 - 3t - 2 = 0$

$(2t + 1)(t - 2) = 0$

$2t + 1 = 0 \quad \text{or} \quad t - 2 = 0$

$t = -\frac{1}{2} \quad \text{or} \qquad t = 2$

The solutions are $-\frac{1}{2}$ and 2.

<u>50.</u> $\frac{2}{5}, -1$

<u>51.</u> $3p^2 = -8p - 5$

$3p^2 + 8p + 5 = 0$

$(3p + 5)(p + 1) = 0$

$3p + 5 = 0 \quad \text{or} \quad p + 1 = 0$

$p = -\frac{5}{3} \quad \text{or} \qquad p = -1$

The solutions are $-\frac{5}{3}$ and -1.

<u>52.</u> $3 \pm \sqrt{7}$

<u>53.</u> $x^2 - x + 1 = 0$

$a = 1, \quad b = -1, \quad c = 1$

$x = \frac{-(-1) \pm \sqrt{(-1)^2 - 4 \cdot 1 \cdot 1}}{2 \cdot 1} = \frac{1 \pm \sqrt{1 - 4}}{2}$

$x = \frac{1 \pm \sqrt{-3}}{2} = \frac{1 \pm i\sqrt{3}}{2}$

The solutions are $\frac{1 + i\sqrt{3}}{2}$ and $\frac{1 - i\sqrt{3}}{2}$.

<u>54.</u> $\frac{-1 \pm i\sqrt{7}}{2}$

<u>55.</u> $x^2 + 13 = 4x$

$x^2 - 4x + 13 = 0$ Finding standard form

$a = 1, \quad b = -4, \quad c = 13$

$x = \frac{-(-4) \pm \sqrt{(-4)^2 - 4 \cdot 1 \cdot 13}}{2 \cdot 1} = \frac{4 \pm \sqrt{16 - 52}}{2}$

$x = \frac{4 \pm \sqrt{-36}}{2} = \frac{4 \pm 6i}{2} = 2 \pm 3i$

The solutions are $2 + 3i$ and $2 - 3i$.

<u>56.</u> $3 \pm 2i$

57. $z^2 + 5 = 0$

$\quad\quad z^2 = -5$

$\quad\quad z = \pm \sqrt{-5}$

$\quad\quad z = \pm i\sqrt{5}$

The solutions are $i\sqrt{5}$ and $-i\sqrt{5}$, or $\pm i\sqrt{5}$.

58. $\pm i\sqrt{3}$

59. $\quad r^2 + 3r = 8$

$r^2 + 3r - 8 = 0$ Finding standard form

$a = 1, \quad b = 3, \quad c = -8$

$r = \dfrac{-3 \pm \sqrt{3^2 - 4 \cdot 1 \cdot (-8)}}{2 \cdot 1} = \dfrac{-3 \pm \sqrt{9 + 32}}{2}$

$r = \dfrac{-3 \pm \sqrt{41}}{2}$

The solutions are $\dfrac{-3 + \sqrt{41}}{2}$ and $\dfrac{-3 - \sqrt{41}}{2}$.

60. $3 \pm \sqrt{5}$

61. $1 + \dfrac{2}{x} + \dfrac{5}{x^2} = 0$

$x^2 + 2x + 5 = 0$ Multiplying by x^2, the LCM of the denominators

$a = 1, \quad b = 2, \quad c = 5$

$x = \dfrac{-2 \pm \sqrt{2^2 - 4 \cdot 1 \cdot 5}}{2 \cdot 1} = \dfrac{-2 \pm \sqrt{4 - 20}}{2}$

$x = \dfrac{-2 \pm \sqrt{-16}}{2} = \dfrac{-2 \pm 4i}{2} = -1 \pm 2i$

The solutions are $-1 + 2i$ and $-1 - 2i$.

62. $1 \pm 2i$

63. $x^2 - 2x + 5 = 0$

$a = 1, \quad b = -2, \quad c = 5$

$x = \dfrac{-(-2) \pm \sqrt{(-2)^2 - 4 \cdot 1 \cdot 5}}{2 \cdot 1} = \dfrac{2 \pm \sqrt{4 - 20}}{2}$

$x = \dfrac{2 \pm \sqrt{-16}}{2} = \dfrac{2 \pm 4i}{2} = 1 \pm 2i$

The solutions are $1 + 2i$ and $1 - 2i$.

64. $2 \pm i$

65. $5 = 2x^2$

$\dfrac{5}{2} = x^2$

$x = \sqrt{\dfrac{5}{2}} \quad$ or $\quad x = -\sqrt{\dfrac{5}{2}}$

$x = \sqrt{\dfrac{5}{2} \cdot \dfrac{2}{2}} \quad$ or $\quad x = -\sqrt{\dfrac{5}{2} \cdot \dfrac{2}{2}}$

$x = \dfrac{\sqrt{10}}{2} \quad$ or $\quad x = -\dfrac{\sqrt{10}}{2}$

The solutions are $\dfrac{\sqrt{10}}{2}$ and $-\dfrac{\sqrt{10}}{2}$, or $\pm \dfrac{\sqrt{10}}{2}$.

66. $\pm \dfrac{\sqrt{6}}{3}$

67. $x(x - 2) = -3x$

$\quad x^2 - 2x = -3x$

$\quad x^2 + x = 0$

$x(x + 1) = 0$

$x = 0 \quad$ or $\quad x + 1 = 0$

$x = 0 \quad$ or $\quad\quad x = -1$

The solutions are 0 and -1.

68. $0, -1$

69. $\quad\quad 2x + 1 = -5x^2$

$5x^2 + 2x + 1 = 0$

$a = 5, \quad b = 2, \quad c = 1$

$x = \dfrac{-2 \pm \sqrt{2^2 - 4 \cdot 5 \cdot 1}}{2 \cdot 5} = \dfrac{-2 \pm \sqrt{4 - 20}}{10}$

$x = \dfrac{-2 \pm \sqrt{-16}}{10} = \dfrac{-2 \pm 4i}{10}$

$x = \dfrac{2(-1 \pm 2i)}{2 \cdot 5} = \dfrac{-1 \pm 2i}{5}$

The solutions are $\dfrac{-1 + 2i}{5}$ and $\dfrac{-1 - 2i}{5}$.

70. $\dfrac{-1 \pm i\sqrt{23}}{6}$

71. $\quad\quad (2t - 3)^2 + 17t = 15$

$4t^2 - 12t + 9 + 17t = 15$

$\quad\quad 4t^2 + 5t - 6 = 0$

$\quad\quad (4t - 3)(t + 2) = 0$

$4t - 3 = 0 \quad$ or $\quad t + 2 = 0$

$t = \dfrac{3}{4} \quad$ or $\quad\quad t = -2$

The solutions are $\dfrac{3}{4}$ and -2.

72. $-3, 2$

73. $\quad\quad (x - 2)^2 + (x + 1)^2 = 0$

$x^2 - 4x + 4 + x^2 + 2x + 1 = 0$

$\quad\quad\quad 2x^2 - 2x + 5 = 0$

$a = 2, \quad b = -2, \quad c = 5$

$x = \dfrac{-(-2) \pm \sqrt{(-2)^2 - 4 \cdot 2 \cdot 5}}{2 \cdot 2} = \dfrac{2 \pm \sqrt{4 - 40}}{4}$

$x = \dfrac{2 \pm \sqrt{-36}}{4} = \dfrac{2 \pm 6i}{4}$

$x = \dfrac{2(1 \pm 3i)}{2 \cdot 2} = \dfrac{1 \pm 3i}{2}$

The solutions are $\dfrac{1 + 3i}{2}$ and $\dfrac{1 - 3i}{2}$.

74. $-1 \pm 2i$

75.
$$x + \frac{1}{x} = \frac{13}{6}$$
$$6x^2 + 6 = 13x \quad \text{Multiplying by } 6x$$
$$6x^2 - 13x + 6 = 0$$
$$(3x - 2)(2x - 3) = 0$$
$$3x - 2 = 0 \quad \text{or} \quad 2x - 3 = 0$$
$$x = \frac{2}{3} \quad \text{or} \quad x = \frac{3}{2}$$

The solutions are $\frac{2}{3}$ and $\frac{3}{2}$.

76. $\frac{3}{2}$, 6

77. $x^2 + 4x - 7 = 0$
$a = 1, \quad b = 4, \quad c = -7$
$$x = \frac{-4 \pm \sqrt{4^2 - 4 \cdot 1 \cdot (-7)}}{2 \cdot 1} = \frac{-4 \pm \sqrt{16 + 28}}{2}$$
$$x = \frac{-4 \pm \sqrt{44}}{2} = \frac{-4 \pm 2\sqrt{11}}{2} = -2 \pm \sqrt{11}$$

Using a calculator we find that $\sqrt{11} \approx 3.317$.
$-2 + \sqrt{11} \approx -2 + 3.317 \approx 1.317 \approx 1.3$
$-2 - \sqrt{11} \approx -2 - 3.317 \approx -5.317 \approx -5.3$

The solutions are approximately 1.3 and -5.3.

78. -0.8, -5.2

79. $x^2 - 6x + 4 = 0$
$a = 1, \quad b = -6, \quad c = 4$
$$x = \frac{-(-6) \pm \sqrt{(-6)^2 - 4 \cdot 1 \cdot 4}}{2 \cdot 1} = \frac{6 \pm \sqrt{36 - 16}}{2}$$
$$x = \frac{6 \pm \sqrt{20}}{2} = \frac{6 \pm 2\sqrt{5}}{2} = 3 \pm \sqrt{5}$$

Using a calculator we find that $\sqrt{5} \approx 2.236$.
$3 + \sqrt{5} \approx 3 + 2.236 \approx 5.236 \approx 5.2$
$3 - \sqrt{5} \approx 3 - 2.236 \approx 0.764 \approx 0.8$

The solutions are approximately 5.2 and 0.8.

80. 3.7, 0.3

81. $2x^2 - 3x - 7 = 0$
$a = 2, \quad b = -3, \quad c = -7$
$$x = \frac{-(-3) \pm \sqrt{(-3)^2 - 4 \cdot 2 \cdot (-7)}}{2 \cdot 2} = \frac{3 \pm \sqrt{9 + 56}}{4}$$
$$x = \frac{3 \pm \sqrt{65}}{4}$$

Using a calculator we find that $\sqrt{65} \approx 8.062$.
$\frac{3 + \sqrt{65}}{4} \approx \frac{3 + 8.062}{4} \approx \frac{11.062}{4} \approx 2.7655 \approx 2.8$
$\frac{3 - \sqrt{65}}{4} \approx \frac{3 - 8.062}{4} \approx \frac{-5.062}{4} \approx -1.2655 \approx -1.3$

The solutions are approximately 2.8 and -1.3.

82. -0.5, 1.5

83. **Familiarize.** Let x represent the number of pounds of coffee A to be used, and let y represent the number of pounds of coffee B. We organize the information in a table.

Coffee	Price per pound	Number of pounds	Total cost
A	$1.50	x	$1.50x
B	$2.50	y	$2.50y
Blend	$1.90	50	1.90 × 50, or $95

Translate. From the last two columns of the table we get a system of equations.
$$x + y = 50,$$
$$1.50x + 2.50y = 95$$

Carry out. Solving the system of equations, we get (30,20).

Check. The total number of pounds in the blend is 30 + 20, or 50. The total cost of the blend is 1.50(30) + 2.50(20), or 45 + 50, or $95. The values check.

State. The blend should consist of 30 pounds of coffee A and 20 pounds of coffee B.

84. $\frac{1}{3}$

85. $2.2x^2 + 0.5x - 1 = 0$
$22x^2 + 5x - 10 = 0 \quad \text{Multiplying by } 10$
$a = 22, \quad b = 5, \quad c = -10$
$$x = \frac{-5 \pm \sqrt{5^2 - 4 \cdot 22 \cdot (-10)}}{2 \cdot 22} = \frac{-5 \pm \sqrt{25 + 880}}{44}$$
$$x = \frac{-5 \pm \sqrt{905}}{44} \approx \frac{-5 \pm 30.08321791}{44}$$
$$\frac{-5 + 30.08321791}{44} \approx 0.5700731$$
$$\frac{-5 - 30.08321791}{44} \approx -0.7973459$$

The solutions are approximately 0.5700731 and -0.7973459.

86. 1.8692840, -0.3251940

87. $t^2 + 0.2t - 0.3 = 0$
$10t^2 + 2t - 3 = 0 \quad \text{Multiplying by } 10$
$a = 10, \quad b = 2, \quad c = -3$
$$t = \frac{-2 \pm \sqrt{2^2 - 4 \cdot 10 \cdot (-3)}}{2 \cdot 10} = \frac{-2 \pm \sqrt{4 + 120}}{20}$$
$$t = \frac{-2 \pm \sqrt{124}}{20} = \frac{-2 \pm 2\sqrt{31}}{20}$$
$$t = \frac{-1 \pm \sqrt{31}}{10} \approx \frac{-1 \pm 5.567764}{10}$$
$$\frac{-1 + 5.567764}{10} \approx 0.4567764$$
$$\frac{-1 - 5.567764}{10} \approx -0.6567764$$

The solutions are approximately 0.4567764 and -0.6567764.

88. 0.3216991, -0.6216991

89. $x^2 - 0.75x - 0.5 = 0$

   $100x^2 - 75x - 50 = 0$ Multiplying by 100

   $4x^2 - 3x - 2 = 0$ Multiplying by $\frac{1}{25}$

   $a = 4,\ \ b = -3,\ \ c = -2$

   $x = \dfrac{-(-3) \pm \sqrt{(-3)^2 - 4 \cdot 4 \cdot (-2)}}{2 \cdot 4} = \dfrac{3 \pm \sqrt{9 + 32}}{8}$

   $x = \dfrac{3 \pm \sqrt{41}}{8} \approx \dfrac{3 \pm 6.403124237}{8}$

   $\dfrac{3 + 6.403124237}{8} \approx 1.1753905$

   $\dfrac{3 - 6.403124237}{8} \approx -0.4253905$

   The solutions are approximately 1.1753905 and -0.4253905.

90. 0.3392101, -1.1792101

91. $x^2 + x - \sqrt{2} = 0$

   $a = 1,\ \ b = 1,\ \ c = -\sqrt{2}$

   $x = \dfrac{-1 \pm \sqrt{1^2 - 4 \cdot 1 \cdot (-\sqrt{2})}}{2 \cdot 1} = \dfrac{-1 \pm \sqrt{1 + 4\sqrt{2}}}{2}$

92. $\dfrac{1 \pm \sqrt{1 + 4\sqrt{3}}}{2}$

93. $x^2 + \sqrt{5}x - \sqrt{3} = 0$

   $a = 1,\ \ b = \sqrt{5},\ \ c = -\sqrt{3}$

   $x = \dfrac{-\sqrt{5} \pm \sqrt{(\sqrt{5})^2 - 4 \cdot 1 \cdot (-\sqrt{3})}}{2 \cdot 1} = \dfrac{-\sqrt{5} \pm \sqrt{5 + 4\sqrt{3}}}{2}$

94. $\dfrac{-5\sqrt{2} \pm \sqrt{34}}{4}$

95. $x^2 + 3x + i = 0$

   $a = 1,\ \ b = 3,\ \ c = i$

   $x = \dfrac{-3 \pm \sqrt{3^2 - 4 \cdot 1 \cdot i}}{2 \cdot 1} = \dfrac{-3 \pm \sqrt{9 - 4i}}{2}$

96. $\dfrac{1 \pm \sqrt{1 - i}}{i}$, or $-i \pm i\sqrt{1 - i}$

97. $3x^2 + xy + 4y^2 = 9$

   $3x^2 + yx + (4y^2 - 9) = 0$

   $a = 3,\ \ b = y,\ \ c = 4y^2 - 9$

   $x = \dfrac{-y \pm \sqrt{y^2 - 4 \cdot 3 \cdot (4y^2 - 9)}}{2 \cdot 3}$

   $x = \dfrac{-y \pm \sqrt{y^2 - 48y^2 + 108}}{6} = \dfrac{-y \pm \sqrt{-47y^2 + 108}}{6}$

98. $\dfrac{22 \pm 4i\sqrt{31}}{7}$

99. $2x^2 - x - \sqrt{5} = 0$

   $a = 2,\ \ b = -1,\ \ c = -\sqrt{5}$

   $x = \dfrac{-(-1) \pm \sqrt{(-1)^2 - 4 \cdot 2 \cdot (-\sqrt{5})}}{2 \cdot 2} = \dfrac{1 \pm \sqrt{1 + 8\sqrt{5}}}{4}$

100. $-\sqrt{3} \pm \sqrt{2}$

101. $ix^2 - x - 1 = 0$

   $a = i,\ \ b = -1,\ \ c = -1$

   $x = \dfrac{-(-1) \pm \sqrt{(-1)^2 - 4 \cdot i \cdot (-1)}}{2 \cdot i} = \dfrac{1 \pm \sqrt{1 + 4i}}{2i}$

   $= \dfrac{i \pm i\sqrt{1 + 4i}}{2i^2} = \dfrac{-i \pm i\sqrt{1 + 4i}}{2}$

102. $\sqrt{3},\ \dfrac{3 - \sqrt{3}}{2}$

103. $kx^2 + 3x - k = 0$

   $k(-2)^2 + 3(-2) - k = 0$ Substituting -2 for x

   $4k - 6 - k = 0$

   $3k = 6$

   $k = 2$

   $2x^2 + 3x - 2 = 0$ Substituting 2 for k

   $(2x - 1)(x + 2) = 0$

   $2x - 1 = 0$ or $x + 2 = 0$

   $x = \dfrac{1}{2}$ or $x = -2$

   The other solution is $\dfrac{1}{2}$.

104. a) $\left[\dfrac{-1 + \sqrt{33}}{2}, 0\right]$, $\left[\dfrac{-1 - \sqrt{33}}{2}, 0\right]$

    b) $\dfrac{-1 + \sqrt{69}}{2}$, $\dfrac{-1 - \sqrt{69}}{2}$

Exercise Set 8.3

1. $\dfrac{1}{x} = \dfrac{x - 2}{24}$, LCM is $24x$

   $24x \cdot \dfrac{1}{x} = 24x \cdot \dfrac{x - 2}{24}$ Multiplying by the LCM

   $24 = x(x - 2)$

   $24 = x^2 - 2x$

   $0 = x^2 - 2x - 24$ Standard form

   $0 = (x - 6)(x + 4)$ Factoring

   $x = 6$ or $x = -4$ Principle of zero products

   Check: For 6:   For -4:

$\dfrac{1}{x} = \dfrac{x - 2}{24}$		$\dfrac{1}{x} = \dfrac{x - 2}{24}$	
$\dfrac{1}{6}$	$\dfrac{6 - 2}{24}$	$\dfrac{1}{-4}$	$\dfrac{-4 - 2}{24}$
	$\dfrac{4}{24}$	$-\dfrac{1}{4}$	$\dfrac{-6}{24}$
	$\dfrac{1}{6}$		$-\dfrac{1}{4}$

   The solutions are 6 and -4.

2. $-7, 4$

3.
$$\frac{1}{2x - 1} - \frac{1}{2x + 1} = \frac{1}{4}$$
$$\text{LCM is } 4(2x - 1)(2x + 1)$$
$$4(2x-1)(2x+1)\left[\frac{1}{2x-1} - \frac{1}{2x+1}\right] = 4(2x-1)(2x+1) \cdot \frac{1}{4}$$
$$4(2x + 1) - 4(2x - 1) = (2x - 1)(2x + 1)$$
$$8x + 4 - 8x + 4 = 4x^2 - 1$$
$$8 = 4x^2 - 1$$
$$9 = 4x^2$$
$$\frac{9}{4} = x^2$$
$$x = \frac{3}{2} \quad \text{or} \quad x = -\frac{3}{2}$$

Both numbers check. The solutions are $\frac{3}{2}$ and $-\frac{3}{2}$, or $\pm\frac{3}{2}$.

4. $-8, 2$

5.
$$\frac{50}{x} - \frac{50}{x - 5} = -\frac{1}{2}, \text{ LCM is } 2x(x - 5)$$
$$2x(x - 5)\left[\frac{50}{x} - \frac{50}{x - 5}\right] = 2x(x - 5)\left[-\frac{1}{2}\right]$$
$$100(x - 5) - 100x = -x(x - 5)$$
$$100x - 500 - 100x = -x^2 + 5x$$
$$x^2 - 5x - 500 = 0$$
$$(x - 25)(x + 20) = 0$$
$$x - 25 = 0 \quad \text{or} \quad x + 20 = 0$$
$$x = 25 \quad \text{or} \quad x = -20$$

Both numbers check. The solutions are 25 and -20.

6. $2, -1$

7.
$$\frac{x + 2}{x} = \frac{x - 1}{2}, \text{ LCM is } 2x$$
$$2x \cdot \frac{x + 2}{x} = 2x \cdot \frac{x - 1}{2}$$
$$2(x + 2) = x(x - 1)$$
$$2x + 4 = x^2 - x$$
$$0 = x^2 - 3x - 4$$
$$0 = (x - 4)(x + 1)$$
$$x - 4 = 0 \quad \text{or} \quad x + 1 = 0$$
$$x = 4 \quad \text{or} \quad x = -1$$

Both numbers check. The solutions are 4 and -1.

8. $6, -3$

9.
$$x - 6 = \frac{1}{x + 6}, \text{ LCM is } x + 6$$
$$(x + 6)(x - 6) = (x + 6) \cdot \frac{1}{x + 6}$$
$$x^2 - 36 = 1$$
$$x^2 = 37$$
$$x = \sqrt{37} \quad \text{or} \quad x = -\sqrt{37}$$

Both numbers check. The solutions are $\pm\sqrt{37}$.

10. $\pm 5\sqrt{2}$

11.
$$\frac{2}{x} = \frac{x + 3}{5}, \text{ LCM is } 5x$$
$$5x \cdot \frac{2}{x} = 5x \cdot \frac{x + 3}{5}$$
$$5 \cdot 2 = x(x + 3)$$
$$10 = x^2 + 3x$$
$$0 = x^2 + 3x - 10$$
$$0 = (x + 5)(x - 2)$$
$$x + 5 = 0 \quad \text{or} \quad x - 2 = 0$$
$$x = -5 \quad \text{or} \quad x = 2$$

Both numbers check. The solutions are -5 and 2.

12. $\dfrac{7 \pm \sqrt{85}}{2}$

13.
$$x + 5 = \frac{3}{x - 5}, \text{ LCM is } x - 5$$
$$(x - 5)(x + 5) = (x - 5) \cdot \frac{3}{x - 5}$$
$$x^2 - 25 = 3$$
$$x^2 = 28$$
$$x = \pm\sqrt{28}, \quad \text{or} \quad \pm 2\sqrt{7}$$

Both numbers check. The solutions are $\pm 2\sqrt{7}$.

14. $\pm\sqrt{65}$

15.
$$\frac{40}{x} - \frac{20}{x - 3} = \frac{8}{7}, \text{ LCM} = 7x(x - 3)$$
$$7x(x - 3)\left[\frac{40}{x} - \frac{20}{x - 3}\right] = 7x(x - 3) \cdot \frac{8}{7}$$
$$280(x - 3) - 140x = 8x(x - 3)$$
$$280x - 840 - 140x = 8x^2 - 24x$$
$$0 = 8x^2 - 164x + 840$$
$$0 = 2x^2 - 41x + 210$$
$$0 = (2x - 21)(x - 10)$$
$$2x - 21 = 0 \quad \text{or} \quad x - 10 = 0$$
$$x = \frac{21}{2} \quad \text{or} \quad x = 10$$

Both numbers check. The solutions are $\frac{21}{2}$ and 10.

16. $-\dfrac{11}{9}, 2$

17.
$$\frac{5}{x+2} + \frac{3x}{x+6} = 2, \text{ LCM is} \atop (x+2)(x+6)$$

$$(x+2)(x+6)\left[\frac{5}{x+2} + \frac{3x}{x+6}\right] = (x+2)(x+6)\cdot 2$$

$$5(x+6) + 3x(x+2) = 2(x^2 + 8x + 12)$$

$$5x + 30 + 3x^2 + 6x = 2x^2 + 16x + 24$$

$$x^2 - 5x + 6 = 0$$

$$(x-3)(x-2) = 0$$

$x = 3$ or $x = 2$

Both numbers check. The solutions are 3 and 2.

18. $-\frac{21}{4}, -1$

19.
$$\frac{3}{3x+1} + \frac{6x}{11x-1} = 1,$$
$$\text{LCM is } (3x+1)(11x-1)$$

$$(3x+1)(11x-1)\left[\frac{3}{3x+1} + \frac{6x}{11x-1}\right] = (3x+1)(11x-1)\cdot 1$$

$$3(11x-1) + 6x(3x+1) = (3x+1)(11x-1)$$

$$33x - 3 + 18x^2 + 6x = 33x^2 + 8x - 1$$

$$0 = 15x^2 - 31x + 2$$

$$0 = (15x-1)(x-2)$$

$15x - 1 = 0$ or $x - 2 = 0$

$x = \frac{1}{15}$ or $x = 2$

Both numbers check. The solutions are $\frac{1}{15}$ and 2.

20. $\frac{3}{2}, 1$

21. $\frac{16}{5(x-2)(x+2)} + \frac{9}{5(x+2)(x+3)} = \frac{-x}{(x-2)(x+3)}$,
$$\text{LCM is } 5(x-2)(x+2)(x+3)$$

$$5(x-2)(x+2)(x+3)\left[\frac{16}{5(x-2)(x+2)} + \frac{9}{5(x+2)(x+3)}\right] =$$

$$5(x-2)(x+2)(x+3)\left[\frac{-x}{(x-2)(x+3)}\right]$$

$$16(x+3) + 9(x-2) = -5x(x+2)$$

$$16x + 48 + 9x - 18 = -5x^2 - 10x$$

$$5x^2 + 35x + 30 = 0$$

$$x^2 + 7x + 6 = 0 \quad \text{Multiplying by } \frac{1}{5}$$

$$(x+6)(x+1) = 0$$

$x = -6$ or $x = -1$

Both numbers check. The solutions are -6 and -1.

22. 4

23.
$$\frac{6}{x^2-4x-5} - \frac{6}{x^2-2x-3} = \frac{x}{x^2-8x+15}$$

$$\frac{6}{(x-5)(x+1)} - \frac{6}{(x-3)(x+1)} = \frac{x}{(x-5)(x-3)},$$
$$\text{LCM is } (x-5)(x+1)(x-3)$$

$$(x-5)(x+1)(x-3)\left[\frac{6}{(x-5)(x+1)} - \frac{6}{(x-3)(x+1)}\right] =$$

$$(x-5)(x+1)(x-3)\left[\frac{x}{(x-5)(x-3)}\right]$$

$$6(x-3) - 6(x-5) = x(x+1)$$

$$6x - 18 - 6x + 30 = x^2 + x$$

$$0 = x^2 + x - 12$$

$$0 = (x+4)(x-3)$$

$x = -4$ or $x = 3$

Only -4 checks. The solution is -4.

24. 5, -4

25. Familiarize. We first make a drawing, labeling it with the known and unknown information. We can also organize the information in a table. We let r represent the speed and t the time for the first part of the trip.

r mph   t hr   r - 5 mph   3 - t hr
───────────────●───────────────
    80 mi   35 mi

Canoe trip	Distance	Speed	Time
1st part	80	r	t
2nd part	35	r - 5	3 - t

Translate. Using $r = \frac{d}{t}$, we get two equations from the table, $r = \frac{80}{t}$ and $r - 5 = \frac{35}{3-t}$.

Carry out. We substitute $\frac{80}{t}$ for r in the second equation and solve for t.

$$\frac{80}{t} - 5 = \frac{35}{3-t}, \text{ LCM is } t(3-t)$$

$$t(3-t)\left[\frac{80}{t} - 5\right] = t(3-t)\cdot\frac{35}{3-t}$$

$$80(3-t) - 5t(3-t) = 35t$$

$$5t^2 - 130t + 240 = 0 \quad \text{Standard form}$$

$$t^2 - 26t + 48 = 0 \quad \text{Multiplying by } \frac{1}{5}$$

$$(t-24)(t-2) = 0$$

$t = 24$ or $t = 2$

Check. Since the time cannot be negative (If $t = 24$, $3 - t = -21$.), we only check 2 hr. If $t = 2$, then $3 - t = 1$. The speed of the first part is $\frac{80}{2}$, or 40 mph. The speed of the second part is $\frac{35}{1}$, or 35 mph. The speed of the second part is 5 mph slower than the first part. The value checks.

State. The speed of the first part was 40 mph, and the speed of the second part was 35 mph.

26. First part: 60 mph, second part: 50 mph

27. Familiarize. We first make a drawing. We also organize the information in a table. We let r represent the speed and t the time of the slower trip.

280 mi   r mph   t hr

280 mi   r + 5 mph   t - 1 hr

Trip	Distance	Speed	Time
Slower	280	r	t
Faster	280	r + 5	t - 1

Translate.

Using $t = \frac{d}{r}$, we get two equations from the table, $t = \frac{280}{r}$, and $t - 1 = \frac{280}{r + 5}$.

Carry out. We substitute $\frac{280}{r}$ for t in the second equation and solve for r.

$$\frac{280}{r} - 1 = \frac{280}{r + 5}, \text{ LCM is } r(r + 5)$$

$$r(r + 5)\left[\frac{280}{r} - 1\right] = r(r + 5) \cdot \frac{280}{r + 5}$$

$$280(r + 5) - r(r + 5) = 280r$$

$$0 = r^2 + 5r - 1400$$

$$0 = (r - 35)(r + 40)$$

r = 35 or r = -40

Check. Since negative speed has no meaning in this problem, we only check 35. If r = 35, then the time for the slow trip is $\frac{280}{35}$, or 8 hours. If r = 35, then r + 5 = 40 and the time for the fast trip is $\frac{280}{40}$, or 7 hours. This is 1 hour less time than the slow trip took, so we have an answer to the problem.

State. The speed is 35 mph.

28. 40 mph

29. Familiarize. We make a drawing and then organize the information in a table. We let r represent the speed and t the time of plane A.

2800 km   r km/h   t hr

2000 km   r + 50 km/h   t - 3 hr

Plane	Distance	Rate	Time
A	2800	r	t
B	2000	r + 50	t - 3

29. (continued)

Translate. Using $r = \frac{d}{t}$, we get two equations from the table,

$$r = \frac{2800}{t} \quad \text{and} \quad r + 50 = \frac{2000}{t - 3}.$$

Carry out. We substitute $\frac{2800}{t}$ for r in the second equation and solve for t.

$$\frac{2800}{t} + 50 = \frac{2000}{t - 3}, \text{ LCM is } t(t - 3)$$

$$t(t - 3)\left[\frac{2800}{t} + 50\right] = t(t - 3) \cdot \frac{2000}{t - 3}$$

$$2800(t - 3) + 50t(t - 3) = 2000t$$

$$50t^2 + 650t - 8400 = 0$$

$$t^2 + 13t - 168 = 0$$

$$(t + 21)(t - 8) = 0$$

t = -21 or t = 8

Check. Since negative time has no meaning in this problem, we only check 8 hours. If t = 8, then t - 3 = 5. The speed of plane A is $\frac{2800}{8}$, or 350 km/h. The speed of plane B is $\frac{2000}{5}$, or 400 km/h. Since the speed of plane B is 50 km/h faster than the speed of plane A, the value checks.

State. The speed of plane A is 350 km/h; the speed of plane B is 400 km/h.

30. A: 200 mph, B: 250 mph; or A: 150 mph, B: 200 mph

31. We make a drawing and then organize the information in a table. Let r represent the speed and t the time of the trip out of Homeville to Fartown.

Homeville   Fartown
600 mi   r mph   t hr

600 mi   r - 10 mph   22 - t hr

Trip	Distance	Speed	Time
Out	600	r	t
Back	600	r - 10	22 - t

Translate. Using $t = \frac{d}{r}$, we get two equations from the table,

$$t = \frac{600}{r} \quad \text{and} \quad 22 - t = \frac{600}{r - 10}.$$

31. (continued)

Carry out. We substitute $\frac{600}{r}$ for t in the second equation and solve for r.

$$22 - \frac{600}{r} = \frac{600}{r - 10}, \text{ LCM is } r(r - 10)$$

$$r(r - 10)\left[22 - \frac{600}{r}\right] = r(r - 10) \cdot \frac{600}{r - 10}$$

$$22r(r - 10) - 600(r - 10) = 600r$$

$$22r^2 - 1420r + 6000 = 0$$

$$11r^2 - 710r + 3000 = 0$$

$$(11r - 50)(r - 60) = 0$$

$$r = \frac{50}{11} \text{ or } r = 60$$

Check. Since negative speed has no meaning in this problem $\left[\text{If } r = \frac{50}{11}, \text{ then } r - 10 = -\frac{60}{11}.\right]$, we only check 60 mph. If r = 60, then the time of the trip out is $\frac{600}{60}$, or 10 hr. The speed of the trip back is 60 - 10, or 50 mph, and the time is $\frac{600}{50}$, or 12 hr. The total time for the round trip is 10 + 12, or 22 hr. The value checks.

State. The speed out of Homeville was 60 mph, and the speed back was 50 mph.

32. 10 mph to Valleytown, 4 mph coming back

33. We make a drawing and organize the information in a table. Let r represent the speed of the boat in still water, and let t represent the time of the trip upstream.

```
  24 mi        r - 2 mph        t hr
  ──────────────────────────────────→ Upstream

            24 mi      r + 2 mph      5 - t hr
  Downstream ←──────────────────────────────────●
```

Trip	Distance	Rate	Time
Upstream	24	r - 2	t
Downstream	24	r + 2	5 - t

Translate. Using $t = \frac{d}{r}$, we get two equations from the table,

$$t = \frac{24}{r - 2} \text{ and } 5 - t = \frac{24}{r + 2}.$$

Carry out. We substitute $\frac{24}{r - 2}$ for t in the second equation and solve for r.

$$5 - \frac{24}{r - 2} = \frac{24}{r + 2}, \text{ LCM is } (r - 2)(r + 2)$$

$$(r-2)(r+2)\left[5 - \frac{24}{r-2}\right] = (r-2)(r+2) \cdot \frac{24}{r+2}$$

$$5(r-2)(r+2) - 24(r+2) = 24(r-2)$$

$$5r^2 - 48r - 20 = 0$$

$$(5r + 2)(r - 10) = 0$$

$$r = -\frac{2}{5} \text{ or } r = 10$$

33. (continued)

Check. Since negative speed has no meaning in this problem, we only check 10 mph. If r = 10, then the speed upstream is 10 - 2, or 8 mph, and the time is $\frac{24}{8}$, or 3 hr. The speed downstream is 10 + 2, or 12 mph, and the time is $\frac{24}{12}$, or 2 hr. The total time of the round trip is 3 + 2, or 5 hr. The value checks.

State. The speed of the boat in still water is 10 mph.

34. 15 mph

35. Familiarize. Let x represent the time it takes the smaller pipe to fill the tank. Then x - 3 represents the time it takes the larger pipe to fill the tank. It takes them 2 hr to fill the tank when both pipes are working together, so they can fill $\frac{1}{2}$ of the tank in 1 hr. The smaller pipe will fill $\frac{1}{x}$ of the tank in 1 hr, and the larger pipe will fill $\frac{1}{x - 3}$ of the tank in 1 hr.

Translate. We have an equation.

$$\frac{1}{x} + \frac{1}{x - 3} = \frac{1}{2}$$

Carry out. We solve the equation. We multiply by the LCM which is 2x(x - 3).

$$2x(x - 3)\left[\frac{1}{x} + \frac{1}{x - 3}\right] = 2x(x - 3) \cdot \frac{1}{2}$$

$$2(x - 3) + 2x = x(x - 3)$$

$$0 = x^2 - 7x + 6$$

$$0 = (x - 6)(x - 1)$$

$$x = 6 \text{ or } x = 1$$

Check. Since negative time has no meaning in this problem, 1 is not a solution (1 - 3 = -2). We only check 6 hr. This is the time it would take the smaller pipe working alone. Then the larger pipe would take 6 - 3, or 3 hr working alone. The larger pipe would fill $2\left(\frac{1}{3}\right)$, or $\frac{2}{3}$, of the tank in 2 hr, and the smaller pipe would fill $2\left(\frac{1}{6}\right)$, or $\frac{1}{3}$, or the tank in 2 hr. Thus in 2 hr they would fill $\frac{2}{3} + \frac{1}{3}$ of the tank. This is all of it, so the numbers check.

State. It takes the smaller pipe, working alone, 6 hr to fill the tank.

36. 12 hr

37. **Familiarize.** We make a drawing and then organize the information in a table. We let r represent the speed of the boat in still water. Then r − 2 is the speed upstream and r + 2 is the speed downstream. Using $t = \frac{d}{r}$, we let $\frac{1}{r-2}$ represent the time upstream and $\frac{1}{r+2}$ represent the time downstream.

1 km   r − 2 km/h
●————————————————→ Upstream

     1 km   r + 2 km/h
Downstream ←————————————————●

Trip	Distance	Speed	Time
Upstream	1	r − 2	$\frac{1}{r-2}$
Downstream	1	r + 2	$\frac{1}{r+2}$

Translate. The time for the round trip is 1 hour. We now have an equation.

$$\frac{1}{r-2} + \frac{1}{r+2} = 1$$

Carry out. We solve the equation. We multiply by the LCM, (r − 2)(r + 2).

$$(r-2)(r+2)\left[\frac{1}{r-2} + \frac{1}{r+2}\right] = (r-2)(r+2)\cdot 1$$
$$(r+2) + (r-2) = (r-2)(r+2)$$
$$2r = r^2 - 4$$
$$0 = r^2 - 2r - 4$$
$$a = 1,\ b = -2,\ c = -4$$

$$r = \frac{-(-2) \pm \sqrt{(-2)^2 - 4\cdot 1 (-4)}}{2\cdot 1}$$

$$r = \frac{2 \pm \sqrt{4 + 16}}{2} = \frac{2 \pm \sqrt{20}}{2}$$

$$r = \frac{2 \pm 2\sqrt{5}}{2} = 1 \pm \sqrt{5}$$

$$1 + \sqrt{5} \approx 1 + 2.236 \approx 3.24$$
$$1 - \sqrt{5} \approx 1 - 2.236 \approx -1.24$$

Check. Since negative speed has no meaning in this problem, we only check 3.24 km/h. If r ≈ 3.24, then r − 2 ≈ 1.24 and r + 2 ≈ 5.24. The time it takes to travel upstream is approximately $\frac{1}{1.24}$, or 0.806 hr, and the time it takes to travel downstream is approximately $\frac{1}{5.24}$, or 0.191 hr. The total time is 0.997 which is approximately 1 hour. The value checks.

State. The speed of the boat in still water is approximately 3.24 km/h.

38. 9.34 mph

39.
$$\sqrt{3x + 1} = \sqrt{2x - 1} + 1$$
$$3x + 1 = 2x - 1 + 2\sqrt{2x - 1} + 1$$

          Squaring both sides

$$x + 1 = 2\sqrt{2x - 1}$$
$$x^2 + 2x + 1 = 4(2x - 1)\quad \text{Squaring both sides again}$$
$$x^2 + 2x + 1 = 8x - 4$$
$$x^2 - 6x + 5 = 0$$
$$(x - 1)(x - 5) = 0$$
$$x = 1 \ \text{ or } \ x = 5$$

Both numbers check. The solutions are 1 and 5.

40. $\frac{1}{x-2}$

41. $\sqrt[3]{18y^3}\ \sqrt[3]{4x^2} = \sqrt[3]{72x^2y^3} = \sqrt[3]{8y^3 \cdot 9x^2} = 2y\sqrt[3]{9x^2}$

42. A: 23.95 hr, B: 51.01 hr

43. **Familiarize.** Let t represent the number of hours it takes Chester to make the crusts. Then it takes Ron t − 1.2 hr. In one hour Chester will do $\frac{1}{t}$ of the job and Ron will do $\frac{1}{t-1.2}$ of the job. Working together it takes them 1.8 hr, so they will do $\frac{1}{1.8}$ of the job in 1 hr.

Translate. We have an equation.
$$\frac{1}{t} + \frac{1}{t - 1.2} = \frac{1}{1.8}$$

Carry out. We solve the equation. The LCM is 1.8t(t − 1.2).

$$1.8t(t-1.2)\left[\frac{1}{t} + \frac{1}{t-1.2}\right] = 1.8t(t-1.2)\cdot\frac{1}{1.8}$$
$$1.8(t - 1.2) + 1.8t = t(t - 1.2)$$
$$0 = t^2 - 4.8t + 2.16$$

Use the quadratic formula.
$$t = \frac{-(-4.8) \pm \sqrt{(-4.8)^2 - 4\cdot 1\cdot(2.16)}}{2\cdot 1}$$

$$t \approx 4.3 \ \text{ or } \ t \approx 0.5$$

Check. Since negative time has no meaning in this problem (If t ≈ 0.5, then t − 1.2 ≈ −0.7), we only check 4.3. This is the time for Chester to do the job alone. Then it would take Ron about 4.3 − 1.2, or 3.1 hr. In 1.8 hr, Chester would do about $\frac{1.8}{4.3}$ of the job, and Ron would do about $\frac{1.8}{3.1}$ of it. Together in 1.8 hr, they would do $\frac{1.8}{4.3} + \frac{1.8}{3.1}$, or 0.419 + 0.581, of the job. This is all of it so the numbers check.

State. It takes Chester about 4.3 hr and Ron about 3.1 hr to do the job alone.

44. $\frac{-3 \pm \sqrt{57}}{2}$

45.
$$\frac{x^2}{x-2} - \frac{x+4}{2} + \frac{2-4x}{x-2} + 1 = 0$$

LCM is $2(x-2)$

$$2(x-2)\left[\frac{x^2}{x-2} - \frac{x+4}{2} + \frac{2-4x}{x-2} + 1\right] = 2(x-2)\cdot 0$$

$$2x^2 - (x-2)(x+4) + 2(2-4x) + 2(x-2) = 0$$

$$2x^2 - x^2 - 2x + 8 + 4 - 8x + 2x - 4 = 0$$

$$x^2 - 8x + 8 = 0$$

Use the quadratic formula.

$$x = \frac{-(-8) \pm \sqrt{(-8)^2 - 4\cdot 1\cdot 8}}{2\cdot 1}$$

$$x = \frac{8 \pm \sqrt{32}}{2} = \frac{8 \pm 4\sqrt{2}}{2} = 4 \pm 2\sqrt{2}$$

Both numbers check. The solutions are $4 \pm 2\sqrt{2}$.

46. $\dfrac{-i \pm \sqrt{23}}{4}$

47.
$$\frac{1}{a-1} = a + 1, \text{ LCM is } a - 1$$

$$(a-1)\cdot\frac{1}{a-1} = (a-1)(a+1)$$

$$1 = a^2 - 1$$

$$2 = a^2$$

$$\pm\sqrt{2} = a$$

Both values check. The solution is $\pm\sqrt{2}$.

48. $2.50

Exercise Set 8.4

1. $x^2 - 6x + 9 = 0$

$a = 1,\ b = -6,\ c = 9$

We compute the discriminant.

$b^2 - 4ac = (-6)^2 - 4\cdot 1\cdot 9$

$\qquad = 36 - 36 = 0$

Since $b^2 - 4ac = 0$, there is just one solution, and it is a real number.

2. One real

3. $x^2 + 7 = 0$

$a = 1,\ b = 0,\ c = 7$

We compute the discriminant.

$b^2 - 4ac = 0^2 - 4\cdot 1\cdot 7$

$\qquad = -28$

Since $b^2 - 4ac < 0$, there are two nonreal solutions.

4. Two nonreal

5. $x^2 - 2 = 0$

$a = 1,\ b = 0,\ c = -2$

We compute the discriminant.

$b^2 - 4ac = 0^2 - 4\cdot 1\cdot(-2)$

$\qquad = 8$

Since $b^2 - 4ac > 0$, there are two real solutions.

6. Two real

7. $4x^2 - 12x + 9 = 0$

$a = 4,\ b = -12,\ c = 9$

We compute the discriminant.

$b^2 - 4ac = (-12)^2 - 4\cdot 4\cdot 9$

$\qquad = 144 - 144 = 0$

Since $b^2 - 4ac = 0$, there is just one solution, and it is a real number.

8. Two real

9. $x^2 - 2x + 4 = 0$

$a = 1,\ b = -2,\ c = 4$

We compute the discriminant.

$b^2 - 4ac = (-2)^2 - 4\cdot 1\cdot 4$

$\qquad = 4 - 16 = -12$

Since $b^2 - 4ac < 0$, there are two nonreal solutions.

10. Two nonreal

11. $a^2 - 10a + 21 = 0$

$a = 1,\ b = -10,\ c = 21$

We compute the discriminant.

$b^2 - 4ac = (-10)^2 - 4\cdot 1\cdot 21$

$\qquad = 100 - 84 = 16$

Since $b^2 - 4ac > 0$, there are two real solutions.

12. One real

13. $6x^2 + 5x - 4 = 0$

$a = 6,\ b = 5,\ c = -4$

We compute the discriminant.

$b^2 - 4ac = 5^2 - 4\cdot 6\cdot(-4)$

$\qquad = 25 + 96 = 121$

Since $b^2 - 4ac > 0$, there are two real solutions.

14. Two real

15. $9t^2 - 3t = 0$

$a = 9,\ b = -3,\ c = 0$

We compute the discriminant.

$b^2 - 4ac = (-3)^2 - 4\cdot 9\cdot 0$

$\qquad = 9 - 0 = 9$

Since $b^2 - 4ac > 0$, there are two real solutions.

16. Two real

17. $x^2 + 5x = 7$

    $x^2 + 5x - 7 = 0$ Standard form

    $a = 1$, $b = 5$, $c = -7$

    We compute the discriminant.

    $b^2 - 4ac = 5^2 - 4\cdot1\cdot(-7)$

    $\qquad\qquad = 25 + 28 = 53$

    Since $b^2 - 4ac > 0$, there are two real solutions.

18. Two nonreal

19.
    $$y^2 = \tfrac{1}{2}y - \tfrac{3}{5}$$

    $y^2 - \tfrac{1}{2}y + \tfrac{3}{5} = 0$ Standard form

    $a = 1$, $b = -\tfrac{1}{2}$, $c = \tfrac{3}{5}$

    We compute the discriminant.

    $b^2 - 4ac = \left(-\tfrac{1}{2}\right)^2 - 4\cdot1\cdot\tfrac{3}{5}$

    $\qquad\qquad = \tfrac{1}{4} - \tfrac{12}{5} = -\tfrac{43}{20}$

    Since $b^2 - 4ac < 0$, there are two nonreal solutions.

20. Two real

21. $4x^2 - 4\sqrt{3}x + 3 = 0$

    $a = 4$, $b = -4\sqrt{3}$, $c = 3$

    We compute the discriminant.

    $b^2 - 4ac = (-4\sqrt{3})^2 - 4\cdot4\cdot3$

    $\qquad\qquad = 48 - 48$

    $\qquad\qquad = 0$

    Since $b^2 - 4ac = 0$, there is just one solution, and it is a real number.

22. Two real

23. $3t^2 - 5\sqrt{2}t + 7 = 0$

    $a = 3$, $b = -5\sqrt{2}$, $c = 7$

    We compute the discriminant.

    $b^2 - 4ac = (-5\sqrt{2})^2 - 4\cdot3\cdot7$

    $\qquad\qquad = 50 - 84 = -34$

    Since $b^2 - 4ac < 0$, there are two nonreal solutions.

24. Two nonreal

25. The solutions are -11 and 9.

    $x = -11$ or $x = 9$

    $x + 11 = 0$ or $x - 9 = 0$

    $(x + 11)(x - 9) = 0$ Principle of zero products

    $x^2 + 2x - 99 = 0$ FOIL

26. $x^2 - 16 = 0$

27. The only solution is 7. It must be a double solution.

    $x = 7$ or $x = 7$

    $x - 7 = 0$ or $x - 7 = 0$

    $(x - 7)(x - 7) = 0$ Principle of zero products

    $x^2 - 14x + 49 = 0$ FOIL

28. $x^2 + 10x + 25 = 0$

29. The solutions are -3 and -5.

    $x = -3$ or $x = -5$

    $x + 3 = 0$ or $x + 5 = 0$

    $(x + 3)(x + 5) = 0$

    $x^2 + 8x + 15 = 0$

30. $x^2 + 9x + 14 = 0$

31. The solutions are 4 and $\tfrac{2}{3}$.

    $x = 4$ or $x = \tfrac{2}{3}$

    $x - 4 = 0$ or $x - \tfrac{2}{3} = 0$

    $(x - 4)\left(x - \tfrac{2}{3}\right) = 0$

    $x^2 - \tfrac{2}{3}x - 4x + \tfrac{8}{3} = 0$

    $x^2 - \tfrac{14}{3}x + \tfrac{8}{3} = 0$

    $3x^2 - 14x + 8 = 0$ Multiplying by 3

32. $4x^2 - 23x + 15 = 0$

33. The solutions are $\tfrac{1}{2}$ and $\tfrac{1}{3}$.

    $x = \tfrac{1}{2}$ or $x = \tfrac{1}{3}$

    $x - \tfrac{1}{2} = 0$ or $x - \tfrac{1}{3} = 0$

    $\left(x - \tfrac{1}{2}\right)\left(x - \tfrac{1}{3}\right) = 0$

    $x^2 - \tfrac{1}{3}x - \tfrac{1}{2}x + \tfrac{1}{6} = 0$

    $x^2 - \tfrac{5}{6}x + \tfrac{1}{6} = 0$

    $6x^2 - 5x + 1 = 0$ Multiplying by 6

34. $8x^2 + 6x + 1 = 0$

35. The solutions are $-\frac{2}{5}$ and $\frac{6}{5}$.

$$x = -\frac{2}{5} \quad \text{or} \quad x = \frac{6}{5}$$

$$x + \frac{2}{5} = 0 \quad \text{or} \quad x - \frac{6}{5} = 0$$

$$\left(x + \frac{2}{5}\right)\left(x - \frac{6}{5}\right) = 0$$

$$x^2 - \frac{4}{5}x - \frac{12}{25} = 0$$

$$25x^2 - 20x - 12 = 0 \quad \text{Multiplying by 25}$$

36. $49x^2 + 7x - 6 = 0$

37. The solutions are $\sqrt{2}$ and $3\sqrt{2}$.

$$x = \sqrt{2} \quad \text{or} \quad x = 3\sqrt{2}$$

$$x - \sqrt{2} = 0 \quad \text{or} \quad x - 3\sqrt{2} = 0$$

$$(x - \sqrt{2})(x - 3\sqrt{2}) = 0$$

$$x^2 - 4\sqrt{2}x + 6 = 0$$

38. $x^2 - \sqrt{3}x - 6 = 0$

39. The solutions are $-\sqrt{5}$ and $-2\sqrt{5}$.

$$x = -\sqrt{5} \quad \text{or} \quad x = -2\sqrt{5}$$

$$x + \sqrt{5} = 0 \quad \text{or} \quad x + 2\sqrt{5} = 0$$

$$(x + \sqrt{5})(x + 2\sqrt{5}) = 0$$

$$x^2 + 3\sqrt{5}x + 10 = 0$$

40. $x^2 + 4\sqrt{6}x + 18 = 0$

41. The solutions are $3i$ and $-3i$.

$$x = 3i \quad \text{or} \quad x = -3i$$

$$x - 3i = 0 \quad \text{or} \quad x + 3i = 0$$

$$(x - 3i)(x + 3i) = 0$$

$$x^2 - (3i)^2 = 0$$

$$x^2 + 9 = 0$$

42. $x^2 + 16 = 0$

43. The solutions are $-1 + i$ and $-1 - i$.

$$x = -1 + i \quad \text{or} \quad x = -1 - i$$

$$x + 1 - i = 0 \quad \text{or} \quad x + 1 + i = 0$$

$$[x + (1 - i)][x + (1 + i)] = 0$$

$$x^2 + x(1 + i) + x(1 - i) + (1 - i)(1 + i) = 0$$

$$x^2 + x + xi + x - xi + 1 - i^2 = 0$$

$$x^2 + 2x + 2 = 0$$

$$(i^2 = -1)$$

44. $x^2 + 6x + 10 = 0$

45. The solutions are $5 - 2i$ and $5 + 2i$.

$$x = 5 - 2i \quad \text{or} \quad x = 5 + 2i$$

$$x - 5 + 2i = 0 \quad \text{or} \quad x - 5 - 2i = 0$$

$$[x + (-5 + 2i)][x + (-5 - 2i)] = 0$$

$$x^2 + x(-5-2i) + x(-5+2i) + (-5+2i)(-5-2i) = 0$$

$$x^2 - 5x - 2xi - 5x + 2xi + 25 - 4i^2 = 0$$

$$x^2 - 10x + 29 = 0$$

$$(i^2 = -1)$$

46. $x^2 - 4x + 53 = 0$

47. The solutions are $\frac{1 + 3i}{2}$ and $\frac{1 - 3i}{2}$.

$$x = \frac{1 + 3i}{2} \quad \text{or} \quad x = \frac{1 - 3i}{2}$$

$$x - \frac{1 + 3i}{2} = 0 \quad \text{or} \quad x - \frac{1 - 3i}{2} = 0$$

$$\left(x - \frac{1 + 3i}{2}\right)\left(x - \frac{1 - 3i}{2}\right) = 0$$

$$x^2 - \frac{x - 3xi}{2} - \frac{x + 3xi}{2} + \frac{1 - 9i^2}{4} = 0$$

$$x^2 - \frac{x}{2} + \frac{3xi}{2} - \frac{x}{2} - \frac{3xi}{2} + \frac{10}{4} = 0$$

$$x^2 - x + \frac{5}{2} = 0$$

$$2x^2 - 2x + 5 = 0$$

48. $9x^2 - 12x + 5 = 0$

49. <u>Familiarize</u>. Let x and y represent the number of 30-sec and 60-sec commercials, respectively.

Then the amount of time for the 30-sec commercials was $30x$ sec, or $\frac{30x}{60} = \frac{x}{2}$ min. The amount of time for the 60-sec commercials was $60x$ sec, or $\frac{60x}{60} = x$ min.

<u>Translate</u>. Rewording, we write two equations. We will express time in minutes.

Total number of commercials is 12.
$$x + y = 12$$

Time for 30-sec commercials	is	total commercial time	less 6 min.
$\frac{x}{2}$	$=$	$\frac{x}{2} + x$	$- 6$

<u>Carry out</u>. Solving the system of equations we get (6,6).

<u>Check</u>. If there are six 30-sec and six 60-sec commercials, the total number of commercials is 12. The amount of time for six 30-sec commercials is 180 sec, or 3 min, and for six 60-sec commercials is 360 sec, or 6 min. The total commercial time is 9 min, and the amount of time for 30-sec commercials is 6 min less than this. The numbers check.

<u>State</u>. There were six 30-sec and six 60-sec commercials.

50. a) 2; b) $\frac{11}{2}$

51. a) $kx^2 - 2x + k = 0$; one solution is -3

We first find k by substituting -3 for x.

$k(-3)^2 - 2(-3) + k = 0$

$9k + 6 + k = 0$

$10k = -6$

$k = -\dfrac{6}{10}$

$k = -\dfrac{3}{5}$

b) Now substitute $-\dfrac{3}{5}$ for k in the original equation.

$-\dfrac{3}{5}x^2 - 2x + \left(-\dfrac{3}{5}\right) = 0$

$3x^2 + 10x + 3 = 0$ Multiplying by -5

$(3x + 1)(x + 3) = 0$

$x = -\dfrac{1}{3}$ or $x = -3$

The other solution is $-\dfrac{1}{3}$.

52. a) 2; b) $1 - i$

53. a) $x^2 - (6 + 3i)x + k = 0$; one solution is 3.

We first find k by substituting 3 for x.

$3^2 - (6 + 3i)3 + k = 0$

$9 - 18 - 9i + k = 0$

$-9 - 9i + k = 0$

$k = 9 + 9i$

b) Now we substitute $9 + 9i$ for k in the original equation.

$x^2 - (6 + 3i)x + (9 + 9i) = 0$

$x^2 - (6 + 3i)x + 3(3 + 3i) = 0$

$[x - (3 + 3i)][x - 3] = 0$

$x = 3 + 3i$ or $x = 3$

The other solution is $3 + 3i$.

54. -1

55. Consider a quadratic equation in standard form, $ax^2 + bx + c = 0$. The solutions are

$x = \dfrac{-b \pm \sqrt{b^2 - 4ac}}{2a}$.

The sum of the solutions is

$\dfrac{-b + \sqrt{b^2 - 4ac}}{2a} + \dfrac{-b - \sqrt{b^2 - 4ac}}{2a} = \dfrac{-2b}{2a} = -\dfrac{b}{a}$.

The product of the solutions is

$\left[\dfrac{-b + \sqrt{b^2 - 4ac}}{2a}\right]\left[\dfrac{-b - \sqrt{b^2 - 4ac}}{2a}\right] =$

$\dfrac{(-b)^2 - (\sqrt{b^2 - 4ac})^2}{(2a^2)} = \dfrac{b^2 - (b^2 - 4ac)}{4a^2} = \dfrac{4ac}{4a^2} = \dfrac{c}{a}$.

Thus $-\dfrac{b}{a} = \sqrt{3}$ and $\dfrac{c}{a} = 8$.

55. (continued)

Multiplying $ax^2 + bx + c = 0$ by $\dfrac{1}{a}$ we have:

$x^2 + \dfrac{b}{a}x + \dfrac{c}{a} = 0$

$x^2 - \left(-\dfrac{b}{a}\right)x + \dfrac{c}{a} = 0$

$x^2 - \sqrt{3}x + 8 = 0$ Substituting

56. $a = 1$, $b = 2$, $c = -3$.

57. a), b) See Exercise 55.

58. a) $4x^2 - 8x + 1 = 0$;

b) $ghx^2 + (h^2 - g^2)x - gh = 0$;

c) $x^2 - 4x + 29 = 0$

59. $3x^2 - hx + 4k = 0$

$a = 3$, $b = -h$, $c = 4k$

The sum of the solutions is -12.

$-\dfrac{b}{a} = -\dfrac{-h}{3} = \dfrac{h}{3}$ (See Exercise 55.)

$\dfrac{h}{3} = -12$, so $h = -36$.

The product of the solutions is 20.

$\dfrac{c}{a} = \dfrac{4k}{3}$ (See Exercise 55.)

$\dfrac{4k}{3} = 20$, so $4k = 60$ and $k = 15$.

Thus, $h = -36$ and $k = 15$.

60. 1

61. The solutions of $x^2 + 2kx - 5 = 0$ are

$x = \dfrac{-2k \pm \sqrt{(2k)^2 - 4 \cdot 1 \cdot (-5)}}{2 \cdot 1}$

$= \dfrac{-2k \pm \sqrt{4k^2 + 20}}{2}$

$= \dfrac{-2k \pm 2\sqrt{k^2 + 5}}{2}$

$= -k \pm \sqrt{k^2 + 5}$

The sum of the squares of the solutions is 26.

$(-k + \sqrt{k^2 + 5})^2 + (-k - \sqrt{k^2 + 5})^2 = 26$

$k^2 - 2k\sqrt{k^2+5} + (k^2+5) + k^2 + 2k\sqrt{k^2+5} + (k^2+5) = 26$

$4k^2 + 10 = 26$

$4k^2 = 16$

$k^2 = 4$

$k = \pm 2$

$|k| = 2$

62. -1

63. $x^2 + kx + 8 = 0$ $x^2 - kx + 8 = 0$

$$x = \frac{-k \pm \sqrt{k^2 - 4 \cdot 1 \cdot 8}}{2 \cdot 1} \qquad x = \frac{-(-k) \pm \sqrt{(-k)^2 - 4 \cdot 1 \cdot 8}}{2 \cdot 1}$$

$$x = \frac{-k \pm \sqrt{k^2 - 32}}{2} \qquad x = \frac{k \pm \sqrt{k^2 - 32}}{2}$$

$$\frac{-k + \sqrt{k^2 - 32}}{2} + 6 = \frac{k + \sqrt{k^2 - 32}}{2}$$

$$-k + \sqrt{k^2 - 32} + 12 = k + \sqrt{k^2 - 32}$$

$$-k + 12 = k$$
$$12 = 2k$$
$$6 = k$$

64. a) 4, -2; b) $1 \pm \sqrt{26}$

Exercise Set 8.5

---

1. $x^4 - 10x^2 + 25 = 0$

   Let $u = x^2$ and think of x^4 as $(x^2)^2$.

   $u^2 - 10u + 25 = 0$ Substituting u for x^2
   $(u - 5)(u - 5) = 0$

   $u - 5 = 0$ or $u - 5 = 0$
   $\quad u = 5$ or $\quad\quad u = 5$

   Now we substitute x^2 for u and solve the equation.
   $$x^2 = 5$$
   $$x = \pm \sqrt{5}$$

   Both $\sqrt{5}$ and $-\sqrt{5}$ check. They are the solutions.

2. $\pm \sqrt{2}, \pm 1$

3. $x^4 - 12x^2 + 27 = 0$

   Let $u = x^2$.

   $u^2 - 12u + 27 = 0$ Substituting u for x^2
   $(u - 9)(u - 3) = 0$
   $u = 9$ or $u = 3$

   Now substitute x^2 for u and solve these equations:
   $$x^2 = 9 \text{ or } x^2 = 3$$
   $$x = \pm 3 \text{ or } x = \pm \sqrt{3}$$

   The numbers 3, -3, $\sqrt{3}$, and $-\sqrt{3}$ check. They are the solutions.

4. $\pm 2, \pm \sqrt{5}$

5. $9x^4 - 14x^2 + 5 = 0$

   Let $u = x^2$.

   $9u^2 - 14u + 5 = 0$ Substituting u for x^2
   $(9u - 5)(u - 1) = 0$

   $u = \dfrac{5}{9}$ or $u = 1$

   $x^2 = \dfrac{5}{9}$ or $x^2 = 1$ Substituting x^2 for u

   $x = \pm \dfrac{\sqrt{5}}{3}$ or $x = \pm 1$

   The numbers $\dfrac{\sqrt{5}}{3}$, $-\dfrac{\sqrt{5}}{3}$, 1, and -1 check. They are the solutions.

6. $\pm \dfrac{\sqrt{3}}{2}, \pm 2$

7. $x - 10\sqrt{x} + 9 = 0$

   Let $u = \sqrt{x}$ and think of x as $(\sqrt{x})^2$.

   $u^2 - 10u + 9 = 0$ Substituting u for $\sqrt{x}$
   $(u - 9)(u - 1) = 0$

   $u - 9 = 0$ or $u - 1 = 0$
   $\quad u = 9$ or $\quad\quad u = 1$

   Now we substitute $\sqrt{x}$ for u and solve these equations:

   $\sqrt{x} = 9$ or $\sqrt{x} = 1$
   $\quad x = 81$ $\quad\quad x = 1$

   The numbers 81 and 1 both check. They are the solutions.

8. $\dfrac{1}{4}$, 16

9. $3x + 10\sqrt{x} - 8 = 0$

   Let $u = \sqrt{x}$.

   $3u^2 + 10u - 8 = 0$ Substituting u for $\sqrt{x}$
   $(3u - 2)(u + 4) = 0$

   $u = \dfrac{2}{3}$ or $u = -4$

   $\sqrt{x} = \dfrac{2}{3}$ or $\sqrt{x} = -4$ Substituting $\sqrt{x}$ for u

   Squaring the first equation, we get $x = \dfrac{4}{9}$. Note that $\sqrt{x} = -4$ has no real solution. The number $\dfrac{4}{9}$ checks and is the solution.

10. $\dfrac{4}{25}$

11. $(x^2 - 9)^2 + 3(x^2 - 9) + 2 = 0$

    Let $u = x^2 - 9$.

    $u^2 + 3u + 2 = 0$ Substituting u for $x^2 - 9$

    $(u + 2)(u + 1) = 0$

    $u = -2$ or $u = -1$

    $x^2 - 9 = -2$ or $x^2 - 9 = -1$ Substituting $x^2 - 9$ for u

    $x^2 = 7$ or $x^2 = 8$

    $x = \pm\sqrt{7}$ or $x = \pm\sqrt{8}$

    $x = \pm\sqrt{7}$ or $x = \pm 2\sqrt{2}$

    The numbers $\sqrt{7}$, $-\sqrt{7}$, $2\sqrt{2}$, and $-2\sqrt{2}$ check. They are the solutions.

12. ± 1, $\pm\sqrt{2}$

13. $(x^2 - 6x) - 2(x^2 - 6x) - 35 = 0$

    Let $u = x^2 - 6x$.

    $u^2 - 2u - 35 = 0$ Substituting u for $x^2 - 6x$

    $(u - 7)(u + 5) = 0$

    $u - 7 = 0$ or $u + 5 = 0$

    $u = 7$ or $u = -5$

    $x^2 - 6x = 7$ or $x^2 - 6x = -5$

    Substituting $x^2 - 6x$ for u

    $x^2 - 6x - 7 = 0$ or $x^2 - 6x + 5 = 0$

    $(x - 7)(x + 1) = 0$ or $(x - 5)(x - 1) = 0$

    $x = 7$ or $x = -1$ or $x = 5$ or $x = 1$

    The numbers -1, 1, 5, and 7 check. They are the solutions.

14. $\dfrac{3 \pm \sqrt{33}}{2}$, 4, -1

15. $(3 + \sqrt{x})^2 - 3(3 + \sqrt{x}) - 10 = 0$

    Let $u = 3 + \sqrt{x}$

    $u^2 - 3u - 10 = 0$ Substituting u for $3 + \sqrt{x}$

    $(u - 5)(u + 2) = 0$

    $u = 5$ or $u = -2$

    $3 + \sqrt{x} = 5$ or $3 + \sqrt{x} = -2$ Substituting $3 + \sqrt{x}$ for u

    $\sqrt{x} = 2$ or $\sqrt{x} = -5$

    $x = 4$ No real solution

    The number 4 checks and is the solution.

16. 1

17. $(y^2 - 5y)^2 - 2(y^2 - 5y) - 24 = 0$

    Let $u = y^2 - 5y$.

    $u^2 - 2u - 24 = 0$ Substituting u for $y^2 - 5y$

    $(u - 6)(u + 4) = 0$

    $u = 6$ or $u = -4$

    $y^2 - 5y = 6$ or $y^2 - 5y = -4$

    Substituting $y^2 - 5y$ for u

    $y^2 - 5y - 6 = 0$ or $y^2 - 5y + 4 = 0$

    $(y - 6)(y + 1) = 0$ or $(y - 4)(y - 1) = 0$

    $y = 6$ or $y = -1$ or $y = 4$ or $y = 1$

    The numbers -1, 1, 4, and 6 check. They are the solutions.

18. $-\dfrac{3}{2}$, 1, $\dfrac{1}{2}$, -1

19. $x^{-2} - x^{-1} - 6 = 0$

    Let $u = x^{-1}$ and think of x^{-2} as $(x^{-1})^2$.

    $u^2 - u - 6 = 0$ Substituting u for x^{-1}

    $(u - 3)(u + 2) = 0$

    $u = 3$ or $u = -2$

    Now we substitute x^{-1} for u and solve these equations:

    $x^{-1} = 3$ or $x^{-1} = -2$

    $\dfrac{1}{x} = 3$ or $\dfrac{1}{x} = -2$

    $\dfrac{1}{3} = x$ or $-\dfrac{1}{2} = x$

    Both $\dfrac{1}{3}$ and $-\dfrac{1}{2}$ check. They are the solutions.

20. $\dfrac{4}{5}$, -1

21. $2x^{-2} + x^{-1} - 1 = 0$

    Let $u = x^{-1}$.

    $2u^2 + u - 1 = 0$ Substituting u for x^{-1}

    $(2u - 1)(u + 1) = 0$

    $2u = 1$ or $u = -1$

    $u = \dfrac{1}{2}$ or $u = -1$

    $x^{-1} = \dfrac{1}{2}$ or $x^{-1} = -1$ Substituting x^{-1} for u

    $\dfrac{1}{x} = \dfrac{1}{2}$ or $\dfrac{1}{x} = -1$

    $x = 2$ or $x = -1$

    Both 2 and -1 check. They are the solutions.

22. $-\dfrac{1}{10}$, 1

23. $t^{2/3} + t^{1/3} - 6 = 0$

    Let $u = t^{1/3}$ and think of $t^{2/3}$ as $(t^{1/3})^2$.

    $u^2 + u - 6 = 0$ Substituting u for $t^{1/3}$

    $(u + 3)(u - 2) = 0$

    $u = -3$ or $u = 2$

    Now we substitute $t^{1/3}$ for u and solve these equations:

    $t^{1/3} = -3$ or $t^{1/3} = 2$

    $t = (-3)^3$ or $t = 2^3$ Raising to the third power

    $t = -27$ or $t = 8$

    Both -27 and 8 check. They are the solutions.

24. 64, -8

25. $z^{1/2} - z^{1/4} - 2 = 0$

    Let $u = z^{1/4}$.

    $u^2 - u - 2 = 0$ Substituting u for $z^{1/4}$

    $(u - 2)(u + 1) = 0$

    $u = 2$ or $u = -1$

    $z^{1/4} = 2$ or $z^{1/4} = -1$ Substituting $z^{1/4}$ for u

    $\sqrt[4]{z} = 2$ or $\sqrt[4]{z} = -1$

    $z = 16$ This equation has no real solution since principal fourth roots are never negative.

    The number 16 checks, so it is the solution.

26. 729

27. $w^4 - 4w^2 - 2 = 0$

    Let $u = w^2$.

    $u^2 - 4u - 2 = 0$ Substituting u for w^2

    $u = \dfrac{-(-4) \pm \sqrt{(-4)^2 - 4 \cdot 1 \cdot (-2)}}{2 \cdot 1}$

    $u = \dfrac{4 \pm \sqrt{24}}{2}$

    $u = \dfrac{4 \pm 2\sqrt{6}}{2}$

    $u = 2 \pm \sqrt{6}$

    Now we substitute w^2 for u and solve these equations:

    $w^2 = 2 + \sqrt{6}$ or $w^2 = 2 - \sqrt{6}$

    $w = \pm\sqrt{2 + \sqrt{6}}$ or $w = \pm\sqrt{2 - \sqrt{6}}$

    All four numbers check. They are the solutions.

28. $\pm\sqrt{\dfrac{5 + \sqrt{5}}{2}}, \pm\sqrt{\dfrac{5 - \sqrt{5}}{2}}$

29. $x^{2/5} + x^{1/5} - 6 = 0$

    Let $u = x^{1/5}$.

    $u^2 + u - 6 = 0$ Substituting u for $x^{1/5}$

    $(u + 3)(u - 2) = 0$

    $u = -3$ or $u = 2$

    $x^{1/5} = -3$ or $x^{1/5} = 2$ Substituting $x^{1/5}$ for u

    $x = -243$ or $x = 32$ Raising to the fifth power

    Both -243 and 32 check. They are the solutions.

30. 81

31. $t^{1/3} + 2t^{1/6} = 3$

    $t^{1/3} + 2t^{1/6} - 3 = 0$

    Let $u = t^{1/6}$.

    $u^2 + 2u - 3 = 0$ Substituting u for $t^{1/6}$

    $(u + 3)(u - 1) = 0$

    $u = -3$ or $u = 1$

    $t^{1/6} = -3$ or $t^{1/6} = 1$ Substituting $t^{1/6}$ for u

    No real $t = 1$
    solution

    The number 1 checks and is the solution.

32. 81, 16

33. $\sqrt{3x^2}\,\sqrt[3]{3x^3} = \sqrt{3x^2 \cdot 3x^3} = \sqrt{9x^5} = \sqrt{9x^4 \cdot x} = 3x^2\sqrt{x}$

34. 4 L of A, 8 L of B

35. $\dfrac{x + 1}{x - 1} - \dfrac{x + 1}{x^2 + x + 1}$, LCM is $(x - 1)(x^2 + x + 1)$

    $= \dfrac{x + 1}{x - 1} \cdot \dfrac{x^2 + x + 1}{x^2 + x + 1} - \dfrac{x + 1}{x^2 + x + 1} \cdot \dfrac{x - 1}{x - 1}$

    $= \dfrac{(x^3 + 2x^2 + 2x + 1) - (x^2 - 1)}{(x - 1)(x^2 + x + 1)}$

    $= \dfrac{x^3 + x^2 + 2x + 2}{x^3 - 1}$

36. $(\sqrt{7},0), (-\sqrt{7},0), (1,0), (-1,0)$

263

37. $6.75x - 35\sqrt{x} - 5.36 = 0$

Let $u = \sqrt{x}$.

$6.75u^2 - 35u - 5.36 = 0$

$u = \dfrac{-(-35) \pm \sqrt{(-35)^2 - 4(6.75)(-5.36)}}{2(6.75)}$

$u = \dfrac{35 \pm \sqrt{1225 + 144.72}}{13.5}$

$u \approx \dfrac{35 \pm 37.01}{13.5}$

$u \approx \dfrac{72.01}{13.5}$ or $u \approx \dfrac{-2.01}{13.5}$

$u \approx 5.334$ or $u \approx -0.149$

$\sqrt{x} \approx 5.334$ or $\sqrt{x} \approx -0.149$

$x \approx 28.5$ No real solution

The number 28.5 checks, so it is the solution.

38. ± 2.0

39. $\left(\dfrac{y^2 - 1}{y}\right)^2 - 4\left(\dfrac{y^2 - 1}{y}\right) - 12 = 0$

Let $u = \dfrac{y^2 - 1}{y}$.

$u^2 - 4u - 12 = 0$

$(u - 6)(u + 2) = 0$

$u = 6$ or $u = -2$

$\dfrac{y^2 - 1}{y} = 6$ or $\dfrac{y^2 - 1}{y} = -2$

$y^2 - 1 = 6y$ or $y^2 - 1 = -2y$

$y^2 - 6y - 1 = 0$ or $y^2 + 2y - 1 = 0$

$y = \dfrac{6 \pm \sqrt{40}}{2}$ $y = \dfrac{-2 \pm \sqrt{8}}{2}$

$y = 3 \pm \sqrt{10}$ $y = -1 \pm \sqrt{2}$

The numbers $3 \pm \sqrt{10}$ and $-1 \pm \sqrt{2}$ check. They are the solutions.

40. $\dfrac{432}{143}$

41. $\left(\dfrac{x^2 - 1}{x}\right)^2 - \left(\dfrac{x^2 - 1}{x}\right) - 2 = 0$

Let $u = \dfrac{x^2 - 1}{x}$.

$u^2 - u - 2 = 0$

$(u - 2)(u + 1) = 0$

$u = 2$ or $u = -1$

$\dfrac{x^2 - 1}{x} = 2$ or $\dfrac{x^2 - 1}{x} = -1$

$x^2 - 1 = 2x$ $x^2 - 1 = -x$

$x^2 - 2x - 1 = 0$ $x^2 + x - 1 = 0$

$x = \dfrac{2 \pm \sqrt{8}}{2}$ $x = \dfrac{-1 \pm \sqrt{5}}{2}$

$x = 1 \pm \sqrt{2}$

The numbers $1 \pm \sqrt{2}$ and $\dfrac{-1 \pm \sqrt{5}}{2}$ check. They are the solutions.

42. $\dfrac{9 \pm \sqrt{89}}{2}$, $-1 \pm \sqrt{3}$

43. $\left(\dfrac{x^2 + 1}{x}\right)^2 - 8\left(\dfrac{x^2 + 1}{x}\right) + 15 = 0$

Let $u = \dfrac{x^2 + 1}{x}$.

$u^2 - 8u + 15 = 0$

$(u - 5)(u - 3) = 0$

$u = 5$ or $u = 3$

$\dfrac{x^2 + 1}{x} = 5$ or $\dfrac{x^2 + 1}{x} = 3$

$x^2 + 1 = 5x$ $x^2 + 1 = 3x$

$x^2 - 5x + 1 = 0$ $x^2 - 3x + 1 = 0$

$x = \dfrac{5 \pm \sqrt{21}}{2}$ $x = \dfrac{3 \pm \sqrt{5}}{2}$

The numbers $\dfrac{5 \pm \sqrt{21}}{2}$ and $\dfrac{3 \pm \sqrt{5}}{2}$ check. They are the solutions.

44. $\dfrac{100}{99}$

45. $\left(\dfrac{x + 1}{x - 1}\right)^2 + \left(\dfrac{x + 1}{x - 1}\right) - 2 = 0$

Let $u = \dfrac{x + 1}{x - 1}$.

$u^2 + u - 2 = 0$

$(u + 2)(u - 1) = 0$

$u = -2$ or $u = 1$

$\dfrac{x + 1}{x - 1} = -2$ or $\dfrac{x + 1}{x - 1} = 1$

$x + 1 = -2(x - 1)$ $x + 1 = x - 1$

$x + 1 = -2x + 2$ $1 = -1$

$3x = 1$ Since $1 \neq -1$, there is no solution of $\dfrac{x + 1}{x - 1} = 1$.

$x = \dfrac{1}{3}$

The number $\dfrac{1}{3}$ checks, so it is the solution.

46. $-\dfrac{6}{7}$

47. $9x^{3/2} - 8 = x^3$

$0 = x^3 - 9x^{3/2} + 8$

Let $u = x^{3/2}$.

$0 = u^2 - 9u + 8$

$0 = (u - 8)(u - 1)$

$u = 8$ or $u = 1$

$x^{3/2} = 8$ or $x^{3/2} = 1$

$x^3 = 64$ or $x^3 = 1$

$x = 4$ or $x = 1$

Both 4 and 1 check. They are the solutions.

48. $-\dfrac{3}{2}$, -1

49. $\sqrt{x-3} - \sqrt[4]{x-3} = 2$

Let $u = \sqrt[4]{x-3}$.

$u^2 - u - 2 = 0$

$(u-2)(u+1) = 0$

$u = 2$ or $u = -1$

$\sqrt[4]{x-3} = 2$ or $\sqrt[4]{x-3} = -1$

$x - 3 = 16$ No real solution

$x = 19$

The number 19 checks, so it is the solution.

50. 9

51.
$$\frac{2x+1}{x} = 3 + 7\sqrt{\frac{2x+1}{x}}$$

$$\frac{2x+1}{x} - 7\sqrt{\frac{2x+1}{x}} - 3 = 0$$

Let $u = \sqrt{\frac{2x+1}{x}}$.

$u^2 - 7u - 3 = 0$

$$u = \frac{7 \pm \sqrt{61}}{2}$$

$$\sqrt{\frac{2x+1}{x}} = \frac{7 + \sqrt{61}}{2}$$

$$\frac{2x+1}{x} = \frac{110 + 14\sqrt{61}}{4}$$

$$8x + 4 = 110x + 14\sqrt{61}x$$

$$4 = 102x + 14\sqrt{61}x$$

$$\frac{4}{102 + 14\sqrt{61}} = x$$

$$\frac{2}{51 + 7\sqrt{61}} = x$$

or

$$\sqrt{\frac{2x+1}{x}} = \frac{7 - \sqrt{61}}{2}$$

$$\frac{2x+1}{x} = \frac{110 - 14\sqrt{61}}{4}$$

$$8x + 4 = 110x - 14\sqrt{61}x$$

$$4 = 102x - 14\sqrt{61}x$$

$$\frac{4}{102 - 14\sqrt{61}} = x$$

$$\frac{2}{51 - 7\sqrt{61}} = x$$

Both numbers check. The solutions are $\dfrac{2}{51 \pm 7\sqrt{61}}$.
The solutions can also be expressed as
$\dfrac{-51 \pm 7\sqrt{61}}{194}$.

1. $A = 6s^2$

$\dfrac{A}{6} = s^2$

$\sqrt{\dfrac{A}{6}} = s$

2. $r = \dfrac{1}{2}\sqrt{\dfrac{A}{\pi}}$

3. $F = \dfrac{Gm_1 m_2}{r^2}$

$Fr^2 = Gm_1 m_2$

$r^2 = \dfrac{Gm_1 m_2}{F}$

$r = \sqrt{\dfrac{Gm_1 m_2}{F}}$

4. $s = \sqrt{\dfrac{kQ_1Q_2}{N}}$

5. $E = mc^2$

$\dfrac{E}{m} = c^2$

$\sqrt{\dfrac{E}{m}} = c$

6. $r = \sqrt{\dfrac{A}{\pi}}$

7. $a^2 + b^2 = c^2$

$b^2 = c^2 - a^2$

$b = \sqrt{c^2 - a^2}$

8. $c = \sqrt{d^2 - a^2 - b^2}$

9. $N = \dfrac{k^2 - 3k}{2}$

$2N = k^2 - 3k$

$0 = k^2 - 3k - 2N$

$a = 1, \quad b = -3, \quad c = -2N$

$$k = \frac{-(-3) \pm \sqrt{(-3)^2 - 4\cdot 1 \cdot (-2N)}}{2\cdot 1}$$

$$= \frac{3 \pm \sqrt{9 + 8N}}{2}$$

Since taking the negative square root would result in a negative answer, we take the positive one.

$$k = \frac{3 + \sqrt{9 + 8N}}{2}$$

10. $t = \dfrac{-v_0 + \sqrt{v_0^2 + 2gs}}{g}$

11. $A = 2\pi r^2 + 2\pi rh$

$0 = 2\pi r^2 + 2\pi rh - A$

$a = 2\pi, \quad b = 2\pi h, \quad c = -A$

$r = \dfrac{-2\pi h \pm \sqrt{(2\pi h)^2 - 4 \cdot 2\pi \cdot (-A)}}{2 \cdot 2\pi}$

$\quad = \dfrac{-2\pi h \pm \sqrt{4\pi^2 h^2 + 8\pi A}}{4\pi}$

$\quad = \dfrac{-2\pi h \pm 2\sqrt{\pi^2 h^2 + 2\pi A}}{4\pi}$

$\quad = \dfrac{-\pi h \pm \sqrt{\pi^2 h^2 + 2\pi A}}{2\pi}$

Since taking the negative square root would result in a negative answer, we take the positive one.

$r = \dfrac{-\pi h + \sqrt{\pi^2 h^2 + 2\pi A}}{2\pi}$

12. $r = \dfrac{-\pi s + \sqrt{\pi^2 s^2 + 4\pi A}}{2\pi}$

13. $N = \frac{1}{2}(n^2 - n)$

$N = \frac{1}{2}n^2 - \frac{1}{2}n$

$0 = \frac{1}{2}n^2 - \frac{1}{2}n - N$

$a = \frac{1}{2}, \quad b = -\frac{1}{2}, \quad c = -N$

$n = \dfrac{-\left(-\frac{1}{2}\right) \pm \sqrt{\left(-\frac{1}{2}\right)^2 - 4 \cdot \frac{1}{2} \cdot (-N)}}{2\left(\frac{1}{2}\right)}$

$\quad = \frac{1}{2} \pm \sqrt{\frac{1}{4} + 2N}$

$\quad = \frac{1}{2} \pm \sqrt{\dfrac{1 + 8N}{4}}$

$\quad = \frac{1}{2} \pm \frac{1}{2}\sqrt{1 + 8N}$

Since taking the negative square root would result in a negative answer, we take the positive one.

$n = \frac{1}{2} + \frac{1}{2}\sqrt{1 + 8N}, \quad \text{or} \quad \dfrac{1 + \sqrt{1 + 8N}}{2}$

14. $r = 1 \pm \sqrt{\dfrac{A}{A_0}}$

15. $A = 2w^2 + 4\ell w$

$0 = 2w^2 + 4\ell w - A$

$a = 2, \quad b = 4\ell, \quad c = -A$

$w = \dfrac{-4\ell \pm \sqrt{(4\ell)^2 - 4 \cdot 2 \cdot (-A)}}{2 \cdot 2}$

$\quad = \dfrac{-4\ell \pm \sqrt{16\ell^2 + 8A}}{4}$

$\quad = \dfrac{-4\ell \pm 2\sqrt{4\ell^2 + 2A}}{4}$

$\quad = \dfrac{-2\ell \pm \sqrt{4\ell^2 + 2A}}{2}$

Since taking the negative square root would result in a negative answer, we take the positive one.

$w = \dfrac{-2\ell + \sqrt{4\ell^2 + 2A}}{2}$

16. $r = \dfrac{-\pi h + \sqrt{\pi^2 h^2 + 4\pi A}}{4\pi}$

17. $T = 2\pi\sqrt{\dfrac{\ell}{g}}$

$\dfrac{T}{2\pi} = \sqrt{\dfrac{\ell}{g}} \qquad \text{Multiplying by } \frac{1}{2\pi}$

$\dfrac{T^2}{4\pi^2} = \dfrac{\ell}{g} \qquad \text{Squaring}$

$gT^2 = 4\pi^2\ell \qquad \text{Multiplying by } 4\pi^2 g$

$g = \dfrac{4\pi^2\ell}{T^2} \qquad \text{Multiplying by } \frac{1}{T^2}$

18. $L = \dfrac{1}{W^2 C}$

19. $P_1 - P_2 = \dfrac{32LV}{gD^2}$

$gD^2(P_1 - P_2) = 32LV$

$D^2 = \dfrac{32LV}{g(P_1 - P_2)}$

$D = \sqrt{\dfrac{32LV}{g(P_1 - P_2)}}$

20. $R = A\sqrt{\dfrac{6 \cdot 2}{p(N + p)}}$

21. $m = \dfrac{m_0}{\sqrt{1 - \dfrac{v^2}{c^2}}}$

$m^2 = \dfrac{m_0^2}{1 - \dfrac{v^2}{c^2}}$

$m^2\left(1 - \dfrac{v^2}{c^2}\right) = m_0^2$

$m^2 - \dfrac{m^2 v^2}{c^2} = m_0^2$

$m^2 - m_0^2 = \dfrac{m^2 v^2}{c^2}$

$c^2(m^2 - m_0^2) = m^2 v^2$

$\dfrac{c^2(m^2 - m_0^2)}{m^2} = v^2$

$\sqrt{\dfrac{c^2(m^2 - m_0^2)}{m^2}} = v$

$\dfrac{c}{m}\sqrt{m^2 - m_0^2} = v$

22. $c = \dfrac{mv}{\sqrt{m^2 - m_0^2}}$

23. a) <u>Familiarize and Translate</u>. From Example 5, we know

$$t = \sqrt{\frac{s}{4.9}}.$$

<u>Carry out</u>. Substituting 75 for s, we have

$$t = \sqrt{\frac{75}{4.9}}$$

$$t \approx 3.9$$

<u>Check</u>. Substitute 3.9 in the original formula. (See Example 5.)

$$s = 4.9t^2 = 4.9(3.9)^2 \approx 75$$

The answer checks.

<u>State</u>. It takes about 3.9 sec to reach the ground.

b) <u>Familiarize and Translate</u>. We will use the formula $s(t) = 4.9t^2 + v_0 t$.

<u>Carry out</u>. Solve the formula for t.

$$0 = 4.9t^2 + v_0 t - s$$

$$t = \frac{-v_0 \pm \sqrt{v_0^2 - 4(4.9)(-s)}}{2(4.9)}$$

Taking the positive square root, we have

$$t = \frac{-v_0 + \sqrt{v_0^2 + 19.6s}}{9.8}.$$

Now substitute 75 for s and 30 for v_0.

$$t = \frac{-30 + \sqrt{30^2 + 19.6(75)}}{9.8}$$

$$t \approx 1.9$$

<u>Check</u>. Substitute 30 for v_0 and 1.9 for t in the original formula.

$$s = 4.9t^2 + v_0 t = 4.9(1.9)^2 + (30)(1.9)$$

$$\approx 75$$

The answer checks.

<u>State</u>. It takes about 1.9 sec to reach the ground.

c) <u>Familiarize and Translate</u>. We will use the formula $s(t) = 4.9t^2 + v_0 t$.

<u>Carry out</u>. Substitute 2 for t and 30 for v_0.

$$s(2) = 4.9(2)^2 + 30(2) = 79.6$$

<u>Check</u>. We can substitute 30 for v_0 and 79.6 for s in the form of the formula we used in part b).

$$t = \frac{-v_0 + \sqrt{v_0^2 + 19.6s}}{9.8} =$$

$$\frac{-30 + \sqrt{(30)^2 + 19.6(79.6)}}{9.8} = 2$$

The answer checks.

<u>State</u>. The object will fall 79.6 m.

24. a) 10.1 sec; b) 7.49 sec; c) 272.5 m

25. <u>Familiarize and Translate</u>. We will use the formula in Exercise 11, $A = 2\pi r^2 + 2\pi rh$.

<u>Carry out</u>. From Exercise 11, we know that

$$r = \frac{-\pi h + \sqrt{\pi^2 h^2 + 2\pi A}}{2\pi}.$$

Substitute 3 for h and 8π for A.

$$r = \frac{-\pi(3) + \sqrt{\pi^2(3)^2 + 2\pi(8\pi)}}{2\pi}$$

$$= \frac{-3\pi + \sqrt{25\pi^2}}{2\pi}$$

$$= \frac{-3\pi + 5\pi}{2\pi}$$

$$= \frac{2\pi}{2\pi}$$

$$= 1$$

<u>Check</u>. Substitute 1 for r and 3 for h in the original formula.

$$A = 2\pi(1)^2 + 2\pi(1)(3) = 2\pi + 6\pi = 8\pi$$

The solution checks.

<u>State</u>. The radius is 1 m.

26. 1 ft

27. <u>Familiarize</u>. We first make a drawing. Let d represent the distance the lower end of the ladder would have to be pulled away, and let h represent the height of the top of the ladder above the ground (on the wall).

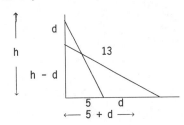

<u>Translate</u>. We will first use the Pythagorean formula to find h: $h^2 + 5^2 = 13^2$

Then we will use the Pythagorean formula again to find d: $(h - d)^2 + (5 + d)^2 = 13^2$

<u>Carry out</u>. Solve the first formula for h.

$$h^2 + 5^2 = 13^2$$

$$h^2 = 13^2 - 5^2$$

$$h = \sqrt{13^2 - 5^2} = \sqrt{169 - 25} = \sqrt{144} = 12$$

Substitute 12 for h in the second formula and solve for d.

$$(12 - d)^2 + (5 + d)^2 = 13^2$$

$$144 - 24d + d^2 + 25 + 10d + d^2 = 169$$

$$2d^2 - 14d = 0$$

$$2d(d - 7) = 0$$

$$d = 0 \quad \text{or} \quad d = 7$$

27. (continued)

    <u>Check</u>. Since d = 0 has no meaning in this problem (d = 0 means the ladder is not moved), we check d = 7. If d = 7, then the top of the ladder is (12 - 7, or 5 ft, above the ground and the bottom of the ladder is 5 + 7, or 12 ft, from the wall. Using the Pythagorean formula we see that $5^2 + 12^2 = 13^2$, so the solution checks.

    <u>State</u>. The lower end would have to be pulled away 7 ft.

28. 2 ft

29. <u>Familiarize</u>. We first make a drawing. Let r represent the radius of the semicircle.

    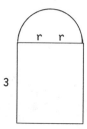

    <u>Translate</u>. The area of the window is the area of the rectangle plus the area of the semicircle. The rectangle has length 3 and width 2r, and the semicircle has radius r. Thus, $A = 3(2r) + \frac{\pi r^2}{2}$.

    <u>Carry out</u>. Substitute 18.28 for A and 3.14 for π and solve for r.

    $$18.28 = 3(2r) + \frac{3.14r^2}{2}$$

    $$0 = 1.57r^2 + 6r - 18.28$$

    Using the quadratic formula with a = 1.57, b = 6, and c = -18.28 we get r = 2 (taking the positive square root). Then 2r, the width of the window, is 2·2, or 4.

    <u>Check</u>. Substitute 2 for r and 3.14 for π in the original formula. The solution checks.

    <u>State</u>. The window is 4 ft wide.

30. 4.685 cm

31. <u>Familiarize</u>. We make a drawing. Let d represent the distance from second base to home.

    <u>Translate</u>. Use the Pythagorean formula.
    $$d^2 = 90^2 + 90^2$$
    <u>Carry out</u>. Solve for d.
    $$d^2 = 8100 + 8100$$
    $$d = \sqrt{16,200} \approx 127.3$$

31. (continued)

    <u>Check</u>. Recheck the calculations. The solution checks.

    <u>State</u>. It is approximately 127.3 ft from second base to home.

32. Train A: 15 mph, Train B: 20 mph

33. <u>Familiarize and Translate</u>. We will use the formula in Exercise 13. (If we did not know that this is the correct formula, we could derive it ourselves.)

    $$N = \frac{1}{2}(n^2 - n)$$

    <u>Carry out</u>. Substitute 91 for N and solve for n.

    $$91 = \frac{1}{2}(n^2 - n)$$
    $$182 = n^2 - n$$
    $$0 = n^2 - n - 182$$
    $$0 = (n - 14)(n + 13)$$

    $$n = 14 \quad \text{or} \quad n = -13$$

    <u>Check</u>. Since a negative solution has no meaning in this problem, we check 14. Substitute 14 for n in the original formula.

    $$N = \frac{1}{2}(14^2 - 14) = \frac{1}{2}(182) = 91$$

    The solution checks.

    <u>State</u>. There are 14 teams in the league.

34. 12

35. <u>Familiarize</u>. The profit is the difference between the revenue and the cost. Let P(x) be a function representing the profit, in thousands of dollars.

    <u>Translate</u>.
    $$P(x) = R(x) - C(x)$$
    $$P(x) = (3x^2 + 5x - 1) - (2x^2 + 15x + 3)$$
    $$P(x) = x^2 - 10x - 4$$

    <u>Carry out</u>. Substitute 20 for P(x) and solve for x.
    $$20 = x^2 - 10x - 4$$
    $$0 = x^2 - 10x - 24$$
    $$0 = (x - 12)(x + 2)$$

    $$x = 12 \quad \text{or} \quad x = -2$$

    <u>Check</u>. Since a negative solution has no meaning in this problem, we check 12. Substitute in P(x).
    $$P(12) = 12^2 - 10·12 - 4 = 144 - 120 - 4 = 20$$
    The solution checks.

    <u>State</u>. The dealer must sell 12 cars.

36. 4

37. <u>Familiarize and Translate.</u> We will use the formula $s(t) = 4.9t^2 + v_0t$.

<u>Carry out.</u> Solve the formula for v_0.

$$s - 4.9t^2 = v_0t$$

$$\frac{s - 4.9t^2}{t} = v_0$$

Now substitute 51.6 for s and 3 for t.

$$\frac{51.6 - 4.9(3)^2}{3} = v_0$$

$$2.5 = v_0$$

<u>Check.</u> Substitute 3 for t and 2.5 for v_0 in the original formula.

$$s = 4.9(3)^2 + 2.5(3) = 51.6$$

The solution checks.

<u>State.</u> The initial velocity is 2.5 m/sec.

38. 3.2 m/sec

39. <u>Familiarize and Translate.</u> From Example 6 we know that

$$i = -1 + \frac{-P_2 + \sqrt{P_2^2 + 4P_1A}}{2P_1},$$

where A is the total amount in the account after two years, P_1 is the amount of the original deposit, P_2 is deposited at the beginning of the second year, and i is the annual interest rate.

<u>Carry out.</u> Substitute 3000 for P_1, 1700 for P_2, and 5253.70 for A.

$$i = -1 + \frac{-1700 + \sqrt{(1700)^2 + 4(3000)(5253.70)}}{2(3000)}$$

Using a calculator, we have i = 0.07.

<u>Check.</u> Substitute in the original formula in Example 6.

$$P_1(1 + i)^2 + P_2(1 + i) = A$$
$$3000(1.07)^2 + 1700(1.07) = A$$
$$5253.70 = A$$

The answer checks.

<u>State.</u> The annual interest rate is 0.07, or 7%.

40. 8.5%

41. $\sqrt{-20} = \sqrt{-4 \cdot 5} = 2i\sqrt{5}$

42. $\dfrac{x - 9}{(x + 9)(x + 7)}$

43. $\sqrt{x^2} = -20$

Since the principal square root cannot be negative, there is no solution.

44. 8.9%

45. <u>Familiarize.</u> Let i represent the semiannual interest rate. If P_1 represents the original deposit, then after 24 months (or 4 compounding periods) it will be worth $P_1(1 + i)^4$. If P_2 represents the amount deposited after 12 months, then at the end of 24 months it will be worth $P_2(1 + i)^2$.

<u>Translate.</u> Let A represent the total amount in the account after 24 months.

$$A = P_1(1 + i)^4 + P_2(1 + i)^2$$

Let $u = (1 + i)^2$, and substitute.

$$A = P_1u^2 + P_2u$$
$$0 = P_1u^2 + P_2u - A$$

$$u = \frac{-P_2 \pm \sqrt{P_2^2 - 4P_1(-A)}}{2P_1}$$

$$u = \frac{-P_2 + \sqrt{P_2^2 + 4P_1A}}{2P_1} \quad \text{Taking the positive square root}$$

$$(1 + i)^2 = \frac{-P_2 + \sqrt{P_2^2 + 4P_1A}}{2P_1}$$

$$1 + i = \sqrt{\frac{-P_2 + \sqrt{P_2^2 + 4P_1A}}{2P_1}} \quad \text{Taking the positive square root}$$

$$i = -1 + \sqrt{\frac{-P_2 + \sqrt{P_2^2 + 4P_1A}}{2P_1}}$$

Substitute 6000 for P_1, 8000 for P_2, and 16,113.04 for A.

$$i = -1 + \sqrt{\frac{-8000 + \sqrt{(8000)^2 + 4(6000)(16,113.04)}}{2(6000)}}$$

Using a calculator, we have i ≈ 0.05.

<u>Check.</u> Substitute in the original formula.

$$A = 6000(1.05)^4 + 8000(1.05)^2 \approx 16,113.04$$

The answer checks.

<u>State.</u> The semiannual interest rate is 0.05, or 5%.

46. $n = \pm\sqrt{\dfrac{r^2 \pm \sqrt{r^4 + 4m^4r^2p - 4mp}}{2m}}$

47. $rt^2 - rt - st^2 + s^2r - st = 0$

$(r - s)t^2 + (-r - s)t + s^2r = 0$ Collecting like terms

$$t = \frac{-(-r - s) \pm \sqrt{(-r - s)^2 - 4(r - s)(s^2r)}}{2(r - s)}$$

$$t = \frac{r + s \pm \sqrt{r^2 + 2rs + s^2 - 4r^2s^2 + 4s^3r}}{2r - 2s}$$

48. $d = \dfrac{-\pi h + \sqrt{\pi^2h^2 + 2\pi A}}{\pi}$

49. Let s represent the length of a side of the cube, let V represent the volume of the cube, and let L represent the length of the cube's three-dimensional diagonal. From the Pythagorean formula in three dimensions we know that $L^2 = s^2 + s^2 + s^2$, or $L^2 = 3s^2$. From the formula for the volume of a cube we know that $V = s^3$, or $V^{1/3} = s$. Then

$$L^2 = 3s^2 = 3(V^{1/3})^2 = 3V^{2/3}, \text{ and}$$

$$L = \sqrt{3V^{2/3}}, \text{ or } \sqrt{3}\sqrt[3]{V}.$$

50. $L = \sqrt{\dfrac{A}{2}}$, where A represents the surface area of the cube and L the length of a three dimensional diagonal.

Exercise Set 8.7

---

1. $y = kx$

$24 = k \cdot 3$ Substituting

$8 = k$

The variation constant is 8.
The equation of variation is $y = 8x$.

2. $k = \dfrac{5}{12}$, $y = \dfrac{5}{12}x$

3. $y = kx$

$3.6 = k \cdot 1$ Substituting

$3.6 = k$

The variation constant is 3.6.
The equation of variation is $y = 3.6x$.

4. $k = \dfrac{2}{5}$, $y = \dfrac{2}{5}x$

5. $y = kx$

$15 = k \cdot 3$ Substituting

$5 = k$

The variation constant is 5.
The equation of variation is $y = 5x$.

6. $k = \dfrac{1}{2}$, $y = \dfrac{1}{2}x$

7. $y = kx$

$30 = k \cdot 8$ Substituting

$\dfrac{30}{8} = k$

$\dfrac{15}{4} = k$

The variation constant is $\dfrac{15}{4}$.
The equation of variation is $y = \dfrac{15}{4}x$.

8. $k = 3$, $y = 3x$

9. $y = kx$

$0.8 = k(0.5)$ Substituting

$8 = k \cdot 5$ Clearing of decimals

$\dfrac{8}{5} = k$

$1.6 = k$

The variation constant is 1.6.
The equation of variation is $y = 1.6x$.

10. $k = 1.5$, $y = 1.5x$

11. Familiarize. Because of the phrase "I. . .varies directly as. . .V," we express the current as a function of the voltage. Thus we have $I(V) = kV$. We know that $I(12) = 4$.

Translate. We find the variation constant and then find the equation of variation.

$I(V) = kV$

$I(12) = k \cdot 12$ Replacing V with 12

$4 = k \cdot 12$ Substituting

$\dfrac{4}{12} = k$

$\dfrac{1}{3} = k$

The equation of variation is $I(V) = \dfrac{1}{3}V$.

Carry out. We compute $I(18)$.

$I(V) = \dfrac{1}{3}V$

$I(18) = \dfrac{1}{3} \cdot 18$ Replacing V with 18

$= 6$

Check. Reexamine the calculations. Note that the answer seems reasonable since 12/4 = 18/6.

State. The current is 6 amperes when 18 volts is applied.

12. $66\dfrac{2}{3}$ cm

13. Familiarize. Because N varies directly as t, we write N as a function of t: $N(t) = kt$. We know that $N(8) = 20{,}000$.

Translate.

$N(t) = kt$

$N(8) = k \cdot 8$ Replacing t with 8

$20{,}000 = k \cdot 8$ Substituting

$2500 = k$ Variation constant

$N(t) = 2500t$ Equation of variation

Carry out. Find $N(50)$.

$N(t) = 2500t$

$N(50) = 2500 \cdot 50$

$= 125{,}000$

Check. Reexamine the calculations.

State. The machine can produce 125,000 straws in 50 hr.

14. 204,000,000

15. **Familiarize.** Because A varies directly as t, we write A(t) = kt, where A is in tons. We know that A(60,000) = 42,600.

Translate.

$$A(t) = kt$$

$$A(60,000) = k \cdot 60,000 \quad \text{Replacing t with 60,000}$$

$$42,600 = k \cdot 60,000 \quad \text{Substituting}$$

$$\frac{42,600}{60,000} = k$$

$$0.71 = k \quad \text{Variation constant}$$

$$A(t) = 0.71t \quad \text{Equation of variation}$$

Carry out. Find A(750,000).

$$A(t) = 0.71t$$

$$A(750,000) = 0.71(750,000)$$

$$= 532,500$$

Check. Reexamine the calculations.

State. 532,500 tons of pollutants would enter the atmosphere in a city with a population of 750,000.

16. 16.8 lb

17. **Familiarize.** Because M varies directly as E, we write M(E) = kE. We know that M(95) = 38.

Translate.

$$M(E) = kE$$

$$M(95) = k \cdot 95 \quad \text{Replacing E with 95}$$

$$38 = k \cdot 95 \quad \text{Substituting}$$

$$\frac{38}{95} = k$$

$$0.4 = k \quad \text{Variation constant}$$

$$M(E) = 0.4E \quad \text{Equation of variation}$$

Carry out. Find M(100).

$$M(100) = 0.4(100)$$

$$= 40$$

Check. Reexamine the calculations.

State. A 100-lb person would weigh 40 lb on Mars.

18. 50 kg

19. **Familiarize.** Because V varies directly as I, we write V(I) = kI. We know that V(3) = 10.

Translate.

$$V(I) = kI$$

$$V(3) = k \cdot 3 \quad \text{Replacing I with 3}$$

$$10 = k \cdot 3 \quad \text{Substituting}$$

$$\frac{10}{3} = k \quad \text{Variation constant}$$

$$V(I) = \frac{10}{3}I \quad \text{Equation of variation}$$

Carry out. Find V(15).

$$V(15) = \frac{10}{3} \cdot 15$$

$$= 50$$

Check. Reexamine the calculations.

State. The voltage is 50 volts when the current is 15 amperes.

20. 3.36

21. $y = \frac{k}{x}$

$$6 = \frac{k}{10} \quad \text{Substituting}$$

$$60 = k$$

The variation constant is 60.

The equation of variation is $y = \frac{60}{x}$.

22. $k = 64, y = \frac{64}{x}$

23. $y = \frac{k}{x}$

$$4 = \frac{k}{3} \quad \text{Substituting}$$

$$12 = k$$

The variation constant is 12.

The equation of variation is $y = \frac{12}{x}$.

24. $k = 36, y = \frac{36}{x}$

25. $y = \frac{k}{x}$

$$12 = \frac{k}{3} \quad \text{Substituting}$$

$$36 = k$$

The variation constant is 36.

The equation of variation is $y = \frac{36}{x}$.

26. $k = 45, y = \frac{45}{x}$

27. $y = \frac{k}{x}$

$$27 = \frac{k}{\frac{1}{3}} \quad \text{Substituting}$$

$$9 = k$$

The variation constant is 9.

The equation of variation is $y = \frac{9}{x}$.

28. $k = 9, y = \frac{9}{x}$

29. <u>Familiarize</u>. Because I varies inversely as R, we express I as a function of R. Thus we write I(R) = k/R. We know that I(240) = 1/2.

<u>Translate</u>.

$$I(R) = \frac{k}{R}$$

$$I(240) = \frac{k}{240} \quad \text{Replacing R with 240}$$

$$\frac{1}{2} = \frac{k}{240} \quad \text{Substituting}$$

$$120 = k \quad \text{Variation constant}$$

$$I(R) = \frac{120}{R} \quad \text{Equation of variation}$$

<u>Carry out</u>. Find I(540).

$$I(540) = \frac{120}{540}$$

$$= \frac{2}{9}$$

<u>Check</u>. Reexamine the calculations. Note that, as expected, when the resistance increases the current decreases.

<u>State</u>. The current is $\frac{2}{9}$ ampere.

30. 27 min

31. <u>Familiarize</u>. Because V varies inversely as P, we write V(P) = k/P. We know that V(32) = 200.

<u>Translate</u>.

$$V(P) = \frac{k}{P}$$

$$V(32) = \frac{k}{32} \quad \text{Replacing P with 32}$$

$$200 = \frac{k}{32} \quad \text{Substituting}$$

$$6400 = k \quad \text{Variation constant}$$

$$V(P) = \frac{6400}{P} \quad \text{Equation of variation}$$

<u>Carry out</u>. Find V(40).

$$V(40) = \frac{6400}{40}$$

$$= 160$$

<u>Check</u>. Reexamine the calculations.

<u>State</u>. The volume will be 160 cm³.

32. 3.5 hr

33. <u>Familiarize</u>. Because t varies inversely as r, we write t(r) = k/r. We know that t(80) = 5.

<u>Translate</u>.

$$t(r) = \frac{k}{r}$$

$$t(80) = \frac{k}{80} \quad \text{Replacing r with 80}$$

$$5 = \frac{k}{80} \quad \text{Substituting}$$

$$400 = k \quad \text{Variation constant}$$

$$t(r) = \frac{400}{r} \quad \text{Equation of variation}$$

<u>Carry out</u>. Find t(60).

$$t(60) = \frac{400}{60}$$

$$= 6\frac{2}{3}$$

<u>Check</u>. Reexamine the calculations.

<u>State</u>. It will take $6\frac{2}{3}$ hr.

34. 450 m

35. $y = kx^2$

$$0.15 = k(0.1)^2 \quad \text{Substituting}$$
$$0.15 = 0.01k$$
$$\frac{0.15}{0.01} = k$$
$$15 = k \quad \text{Variation constant}$$
The equation of variation is $y = 15x^2$.

36. $y = \frac{2}{3}x^2$

37. $y = \frac{k}{x^2}$

$$0.15 = \frac{k}{(0.1)^2} \quad \text{Substituting}$$
$$0.15 = \frac{k}{0.01}$$
$$0.15(0.01) = k$$
$$0.0015 = k \quad \text{Variation constant}$$
The equation of variation is $y = \frac{0.0015}{x^2}$.

38. $y = \frac{54}{x^2}$

39. $y = kxz$

$$56 = k \cdot 7 \cdot 8 \quad \text{Substituting 56 for y, 7 for x, and 8 for z}$$
$$56 = 56k$$
$$1 = k \quad \text{Variation constant}$$
The equation of variation is $y = xz$.

40. $y = \frac{5x}{z}$

41. $y = kxz^2$

$105 = k \cdot 14 \cdot 5^2$ Substituting 105 for y,
                                   14 for x, and 5 for z

$105 = 350k$

$\dfrac{105}{350} = k$

$0.3 = k$

The equation of variation is $y = 0.3xz^2$.

42. $y = \dfrac{xz}{w}$

43. $y = k \cdot \dfrac{x}{z^2}$

$1.2 = k \cdot \dfrac{14}{5^2}$ Substituting

$\dfrac{30}{14} = k$

$\dfrac{15}{7} = k$ Variation constant

$y = \dfrac{15}{7} \cdot \dfrac{x}{z^2}$,   or $\dfrac{15x}{7z^2}$

44. $y = \dfrac{3x^2}{z}$

45. $y = k \cdot \dfrac{wx^2}{z}$

$49 = k \cdot \dfrac{3 \cdot 7^2}{12}$ Substituting

$4 = k$ Variation constant

$y = \dfrac{4wx^2}{z}$

46. $y = \dfrac{6x}{wz^2}$

47. $y = k \cdot \dfrac{xz}{wp}$

$\dfrac{3}{28} = k \cdot \dfrac{3 \cdot 10}{7 \cdot 8}$ Substituting

$\dfrac{3}{28} = k \cdot \dfrac{30}{56}$

$\dfrac{3}{28} \cdot \dfrac{56}{30} = k$

$\dfrac{1}{5} = k$ Variation constant

The equation of variation is $y = \dfrac{xz}{5wp}$.

48. $y = \dfrac{5xz}{4w^2}$

49. Familiarize. Because d varies directly as the square of r, we write $d = kr^2$. We know that d = 200 when r = 60.

Translate. We first find k.

$d = kr^2$

$200 = k(60)^2$

$\dfrac{1}{18} = k$

$d = \dfrac{1}{18}r^2$ Equation of variation

49. (continued)

Carry out. Substitute 72 for d and solve for r.

$72 = \dfrac{1}{18}r^2$

$1296 = r^2$

$36 = r$

Check. Recheck the calculations and perhaps make an estimate to see if the answer seems reasonable.

State. The car can go 36 mph.

50. 7 in.

51. Familiarize. I varies inversely as d^2, so we write $I = k/d^2$. We know that I = 25 when d = 2.

Translate. First we find k.

$I = \dfrac{k}{d^2}$

$25 = \dfrac{k}{2^2}$

$100 = k$

$I = \dfrac{100}{d^2}$ Equation of variation

Carry out. Substitute 2.56 for I and solve for d.

$2.56 = \dfrac{100}{d^2}$

$d^2 = \dfrac{100}{2.56}$

$d = 6.25$

Check. Recheck the calculations.

State. You are 6.25 km from the transmitter.

52. 5 sec

53. Familiarize. W varies inversely as d^2, so we write $W = k/d^2$. We know that W = 100 when d = 6400. Note that when the astronaut is 200 km above the surface of the earth, she is 6400 + 200, or 6600 km, from the center of the earth.

Translate. Find k.

$W = \dfrac{k}{d^2}$

$100 = \dfrac{k}{(6400)^2}$

$4{,}096{,}000{,}000 = k$

$W = \dfrac{4{,}096{,}000{,}000}{d^2}$ Equation of variation

Carry out. Substitute 6600 for d and solve for W.

$W = \dfrac{4{,}096{,}000{,}000}{(6600)^2}$

$W \approx 94.03$

Check. Recheck the calculations.

State. Her weight is about 94.03 lb.

54. 22.5 W/m²

55. Familiarize. A varies directly as R and inversely as I, so we write A = kR/I. We know that A = 2.92 when R = 85 and I = 262.

    Translate. Find k.

    $$A = \frac{kR}{I}$$

    $$2.92 = \frac{k \cdot 85}{262}$$

    $$9 \approx k$$

    $$A = \frac{9R}{I} \quad \text{Equation of variation}$$

    Carry out. Substitute 2.92 for A and 300 for I and solve for R.

    $$2.92 = \frac{9R}{300}$$

    $$97 \approx R \quad \text{Rounding}$$

    Check. Recheck the calculations.

    State. 97 earned runs would be given up.

56. 220 cm³

57. Familiarize. R varies directly as ℓ and inversely as d², so we write R = kℓ/d². We know that R = 0.1 when ℓ = 50 and d = 1.

    Translate. Find k.

    $$R = \frac{k\ell}{d^2}$$

    $$0.1 = \frac{k \cdot 50}{1^2}$$

    $$0.002 = k$$

    $$R = \frac{0.002\ell}{d^2} \quad \text{Equation of variation}$$

    Carry out. Substitute 1 for R and 2000 for ℓ and solve for d.

    $$1 = \frac{0.002(2000)}{d^2}$$

    $$d^2 = 4$$

    $$d = 2$$

    Check. Recheck the calculations.

    State. The diameter is 2 mm.

58. 4.8 cm

59. Familiarize. Let v represent the boat's velocity. F varies jointly as A and v², so we write F = kAv². We know that F = 86 when A = 41.2 and v = 6.5.

    Translate. Find k.

    $$F = kAv^2$$

    $$86 = k(41.2)(6.5)^2$$

    $$0.0494 \approx k$$

    $$F = 0.0494Av^2 \quad \text{Equation of variation}$$

    Carry out. Substitute 28.5 for A and 94 for F and solve for v.

    $$94 = 0.0494(28.5)v^2$$

    $$66.7661 \approx v^2$$

    $$8.17 \approx v$$

    Check. Recheck the calculations.

    State. The boat must go about 8.17 mph.

60. 57.43 mph

61. Use the slope-intercept form, y = mx + b, where m is the slope and b is the y-intercept.

    $$y = -\frac{2}{3}x - 5$$

62. $\dfrac{c - 2a}{3c + 4a}$

63. $f(x) = x^3 - 2x^2$

    $f(3) = 3^3 - 2 \cdot 3^2 = 27 - 18 = 9$

64. $9x^2 - 12xy + 4y^2$

65. Write y as a function of x, and then substitute 3x for x.

    $$y(x) = kx$$

    $$y(3x) = k \cdot 3x = 3 \cdot kx = 3 \cdot y(x)$$

    y is tripled.

66. y is multiplied by $\frac{1}{3}$

67. Write y as a function of x, and then substitute nx for x.

    $$y(x) = \frac{k}{x^2}$$

    $$y(nx) = \frac{k}{(nx)^2} = \frac{k}{n^2 x^2} = \frac{1}{n^2} \cdot \frac{k}{x^2} = \frac{1}{n^2} \cdot y(x)$$

    y is multiplied by $\frac{1}{n^2}$.

68. y is multiplied by n^2

69. We are told A = kd², and we know A = πr² so we have:

    $$kd^2 = \pi r^2$$

    $$kd^2 = \pi \left(\frac{d}{2}\right)^2 \qquad r = \frac{d}{2}$$

    $$kd^2 = \frac{\pi d^2}{4}$$

    $$k = \frac{\pi}{4} \quad \text{Variation constant}$$

70. $7.20

71. Write y as a function of x, and then substitute 0.5x for x.

$$y(x) = \frac{k}{x^3}$$

$$y(0.5x) = \frac{k}{(0.5x)^3} = \frac{k}{0.125x^3} = \frac{1}{0.125} \cdot \frac{k}{x^3}$$

$$= 8 \cdot y(x)$$

y is multiplied by 8.

72. a) $k = 0.001$, $N = \frac{0.001P_1P_2}{d^2}$; b) 1173 km

73. <u>Familiarize.</u> We write $T = km\ell^2f^2$. We know that $T = 100$ when $m = 5$, $\ell = 2$, and $f = 80$.

<u>Translate.</u> Find k.

$$T = km\ell^2f^2$$
$$100 = k(5)(2)^2(80)^2$$
$$0.00078125 = k$$
$$T = 0.00078125m\ell^2f^2$$

<u>Carry out.</u> Substitute 72 for T, 5 for m, and 80 for f and solve for ℓ.

$$72 = 0.00078125(5)\ell^2(80)^2$$
$$2.88 = \ell^2$$
$$1.697 \approx \ell$$

<u>Check.</u> Recheck the calculations.

<u>State.</u> The string should be about 1.697 m long.

74. $d = \frac{28}{s}$; 70 yd

75. Let C represent the number of complaints, and let E represent the number of employees. Write C as a function of E.

$$C(E) = \frac{k}{E}$$

Now $C(5) = \frac{k}{5}$ and $C(10) = \frac{k}{10} = \frac{1}{2} \cdot \frac{k}{5}$, so $C(10) = \frac{1}{2} \cdot C(5)$, or expanding from 5 to 10 employees will result in half as many complaints. Also, $C(20) = \frac{k}{20}$ and $C(25) = \frac{k}{25} = \frac{4}{5} \cdot \frac{k}{20}$, so $C(25) = \frac{4}{5} \cdot C(20)$, or expanding from 20 to 25 employees will result in four-fifths as many complaints. Thus, the firm will benefit more by expanding from 5 to 10 employees. A graph of this function is shaped like the following:

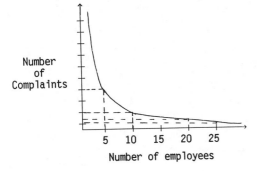

Number of Complaints

Number of employees

75. (continued)

Projecting up and across, we see that the number of complaints decreases more as the number of employees changes from 5 to 10 than from 20 to 25. Thus, the graph also shows that the firm will benefit more by expanding from 5 to 10 employees.

Exercise Set 8.8

1. $f(x) = x^2$

See Example 1 in the text.

$f(x) = x^2$

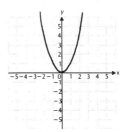

2. $f(x) = -x^2$

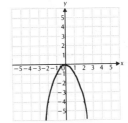

3. $f(x) = -4x^2$

We choose some numbers for x and compute $f(x)$ for each one. Then we plot the ordered pairs $(x, f(x))$ and connect them with a smooth curve.

x	$f(x) = -4x^2$
0	0
1	-4
2	-16
-1	-4
-2	-16

$f(x) = -4x^2$

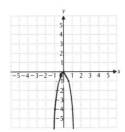

4.

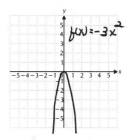

5. $g(x) = \frac{1}{4}x^2$

x	$g(x) = \frac{1}{4}x^2$
0	0
1	$\frac{1}{4}$
2	1
3	$\frac{9}{4}$
-1	$\frac{1}{4}$
-2	1
-3	$\frac{9}{4}$

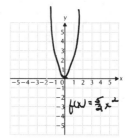

6.

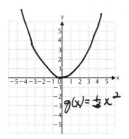

7. $h(x) = -\frac{1}{3}x^2$

x	$h(x) = -\frac{1}{3}x^2$
0	0
1	$-\frac{1}{3}$
2	$-\frac{4}{3}$
3	-3
-1	$-\frac{1}{3}$
-2	$-\frac{4}{3}$
-3	-3

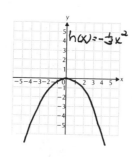

8.

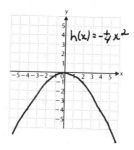

9. $f(x) = \frac{3}{2}x^2$

x	$f(x) = \frac{3}{2}x^2$
0	0
1	$\frac{3}{2}$
2	6
-1	$\frac{3}{2}$
-2	6

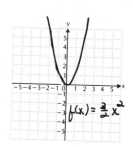

10.

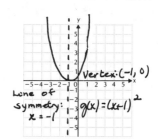

11. $g(x) = (x + 1)^2 = [x - (-1)]^2$

We know that the graph of $g(x) = (x + 1)^2$ looks like the graph of $f(x) = x^2$ (see Exercise 1) but moved to the left 1 unit.

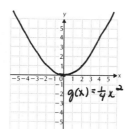

12.

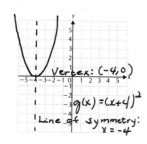

16.

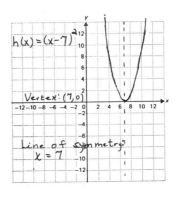

13. $f(x) = (x - 4)^2$

The graph of $f(x) = (x - 4)^2$ looks like the graph of $f(x) = x^2$ (see Exercise 1) but moved to the right 4 units.

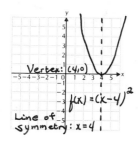

17. $f(x) = -(x + 4)^2 = -[x - (-4)]^2$

The graph of $f(x) = -(x + 4)^2$ looks like the graph of $f(x) = x^2$ (see Exercise 1) but moved to the left 4 units. It will also open downward because of the negative coefficient, -1.

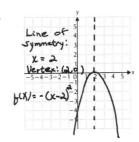

14.

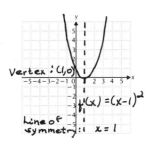

18.

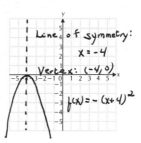

15. $h(x) = (x - 3)^2$

The graph of $h(x) = (x - 3)^2$ looks like the graph of $f(x) = x^2$ (see Exercise 1) but moved to the right 3 units.

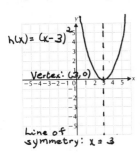

19. $g(x) = -(x - 1)^2$

The graph of $g(x) = -(x - 1)^2$ looks like the graph of $f(x) = x^2$ (see Exercise 1) but moved to the right 1 unit. It will also open downward because of the negative coefficient, -1.

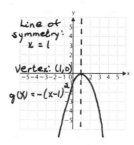

20.

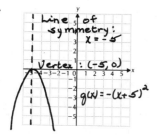

24.

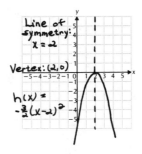

21. $f(x) = 2(x - 1)^2$

The graph of $f(x) = 2(x - 1)^2$ will look like the graph of $h(x) = 2x^2$ (see Example 1) but moved to the right 1 unit.

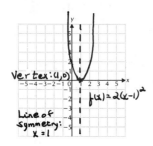

25. $f(x) = \frac{1}{2}(x + 1)^2 = \frac{1}{2}[x - (-1)]^2$

The graph of $f(x) = \frac{1}{2}(x + 1)^2$ looks like the graph of $g(x) = \frac{1}{2}x^2$ (see Example 1) but moved to the left 1 unit.

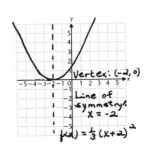

22.

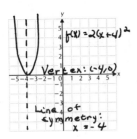

26.

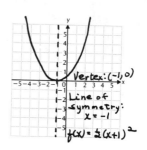

23. $h(x) = -\frac{1}{2}(x - 3)^2$

The graph of $h(x) = -\frac{1}{2}(x - 3)^2$ looks like the graph of $g(x) = \frac{1}{2}x^2$ (see Example 1) but moved to the right 3 units. It will also open downward because of the negative coefficient, $-\frac{1}{2}$.

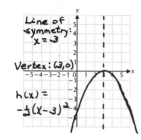

27. $g(x) = -3(x - 2)^2$

The graph of $g(x) = -3(x - 2)^2$ looks like the graph of $f(x) = -3x^2$ (see Exercise 4) but moved to the right 2 units.

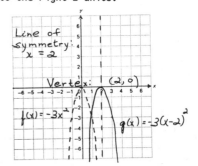

28.

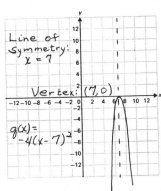

29. f(x) = -2(x + 9)² = -2[x - (-9)]²

The graph of f(x) = -2(x + 9)² looks like the
graph of h(x) = 2x² (see Example 1) but moved to
the left 9 units. It will also open downward
because of the negative coefficient, -2.

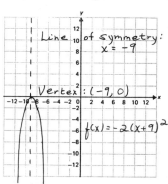

30.

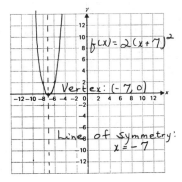

31. h(x) = $-3\left(x - \frac{1}{2}\right)^2$

The graph of h(x) = $-3\left(x - \frac{1}{2}\right)^2$ looks like the
graph of f(x) = -3x² (see Exercise 4) but moved to
the right $\frac{1}{2}$ unit.

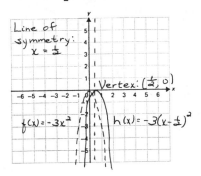

32.

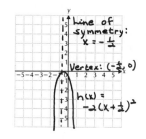

33. f(x) = (x - 3)² + 1

We know that the graph looks like the graph of
f(x) = x² (see Example 1) but moved to the right
3 units and up 1 unit. The vertex is (3,1), and
the line of symmetry is x = 3. Since the
coefficient of (x - 3)² is positive (1 > 0), there
is a minimum function value, 1.

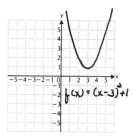

34. f(x) = (x + 2)² - 3

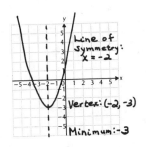

35. $f(x) = (x + 1)^2 - 2$

We know that the graph looks like the graph of $f(x) = x^2$ (see Example 1) but moved to the left 1 unit and down 2 units. The vertex is $(-1,2)$, and the line of symmetry is $x = -1$. Since the coefficient of $(x + 1)^2$ is positive $(1 > 0)$, there is a minimum function value, -2.

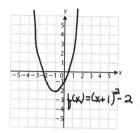

36. $f(x) = (x - 1)^2 + 2$

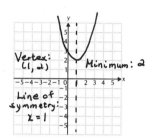

37. $g(x) = (x + 4)^2 + 1$

We know that the graph looks like the graph of $f(x) = x^2$ (see Example 1) but moved to the left 4 units and up 1 unit. The vertex is $(-4,1)$, and the line of symmetry is $x = -4$. Since the coefficient of $(x + 4)^2$ is positive $(1 > 0)$, there is a minimum function value, 1.

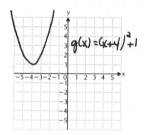

38. $g(x) = -(x - 2)^2 - 4$

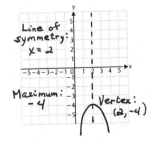

39. $f(x) = \frac{1}{2}(x - 5)^2 + 2$

We know that the graph looks like the graph of $g(x) = \frac{1}{2}x^2$ but moved to the right 5 units and up 2 units. The vertex is $(5,2)$, and the line of symmetry is $x = 5$. Since the coefficient of $(x - 5)^2$ is positive $\left[\frac{1}{2} > 0\right]$, there is a minimum function value, 2.

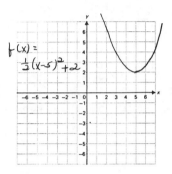

40. $f(x) = \frac{1}{2}(x + 1)^2 - 2$

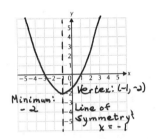

41. $h(x) = -2(x - 1)^2 - 3$

We know that the graph looks like the graph of $h(x) = 2x^2$ but moved to the right 1 unit and down 3 units and turned upside down. The vertex is $(1,-3)$, and the line of symmetry is $x = 1$. The maximum function value is -3.

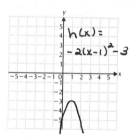

42. $h(x) = -2(x + 1)^2 + 4$

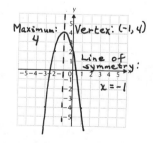

43. $f(x) = -3(x + 4)^2 + 1$

   We know that the graph looks like the graph of $f(x) = -3x^2$ (see Exercise 4) but moved to the left 4 units and up 1 unit. The vertex is (-4,1), the line of symmetry is $x = -4$, and the maximum function value is 1.

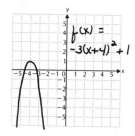

44. $f(x) = -2(x - 5)^2 - 3$

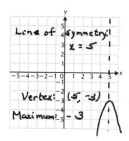

45. $g(x) = -\frac{3}{2}(x - 1)^2 + 2$

   We know that the graph looks like the graph of $f(x) = \frac{3}{2}x^2$ (see Exercise 9) but moved to the right 1 unit and up 2 units and turned upside down. The vertex is (1,2), the line of symmetry is $x = 1$, and the maximum function value is 2.

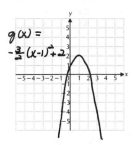

46. $g(x) = \frac{3}{2}(x + 2)^2 - 1$

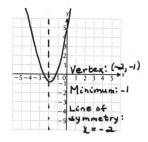

47. $f(x) = 8(x - 9)^2 + 5$

   This function is of the form $f(x) = a(x - h)^2 + k$ with $a = 8$, $h = 9$, and $k = 5$. The vertex is (h,k), or (9,5). The line of symmetry is $x = h$, or $x = 9$. Since $a > 0$, then k, or 5, is the minimum function value.

48. Vertex: (-5,-8)

   Line of symmetry: $x = -5$

   Minimum: -8

49. $h(x) = -\frac{2}{7}(x + 6)^2 + 11$

   This function is of the form $f(x) = a(x - h)^2 + k$ with $a = -\frac{2}{7}$, $h = -6$, and $k = 11$. The vertex is (h,k), or (-6,11). The line of symmetry is $x = h$, or $x = -6$. Since $a < 0$, then k, or 11, is the maximum function value.

50. Vertex: (7,-9)

   Line of symmetry: $x = 7$

   Maximum: -9

51. $f(x) = 5\left(x + \frac{1}{4}\right)^2 - 13$

   This function is of the form $f(x) = a(x - h)^2 + k$ with $a = 5$, $h = -\frac{1}{4}$, and $k = -13$. The vertex is (h,k), or $\left(-\frac{1}{4}, -13\right)$. The line of symmetry is $x = h$, or $x = -\frac{1}{4}$. Since $a > 0$, then k, or -13, is the minimum function value.

52. Vertex: $\left(\frac{1}{4}, 19\right)$

   Line of symmetry: $x = \frac{1}{4}$

   Minimum: 19

53. $f(x) = -7(x - 10)^2 - 20$

   This function is of the form $f(x) = a(x - h)^2 + k$ with $a = -7$, $h = 10$, and $k = -20$. The vertex is (h,k), or (10,-20). The line of symmetry is $x = h$, or $x = 10$. Since $a < 0$, then k, or -20, is the maximum function value.

54. Vertex: (-12,23)

   Line of symmetry: $x = -12$

   Maximum: 23

55. $f(x) = \sqrt{2}(x + 4.58)^2 + 65\pi$

   This function is of the form $f(x) = a(x - h)^2 + k$ with $a = \sqrt{2}$, $h = -4.58$, and $k = 65\pi$. The vertex is (h,k), or (-4.58,65π). The line of symmetry is $x = h$, or $x = -4.58$. Since $a > 0$, then k, or 65π, is the minimum function value.

56. Vertex: $(38.2, -\sqrt{34})$

   Line of symmetry: $x = 38.2$

   Minimum: $-\sqrt{34}$

57. 500 = 4a + 2b + c   (1)
     300 = a + b + c   (2)
       0 = c   (3)
     Substitute 0 for c in (1) and (2).
     500 = 4a + 2b   (4)
     300 = a + b   (5)
     Multiply (5) by –2 and add the resulting equation
     to (4).

     $$\begin{array}{rl} 500 = & 4a + 2b \\ \underline{-600 = } & \underline{-2a - 2b} \\ -100 = & 2a \qquad \text{Adding} \\ -50 = & a \end{array}$$

     Substitute –50 for a in (5).
         300 = –50 + b
         350 = b
     The solution is (–50,350,0).

58. $\frac{1}{6}$, 2

59. Since there is a maximum at (0,4), the parabola
     will have the same shape as f(x) = –2x². It will
     be of the form f(x) = –2(x – h)² + k with h = 0
     and k = 4: f(x) = –2x² + 4

60. f(x) = 2(x – 2)²

61. Since there is a minimum at (6,0), the parabola
     will have the same shape as f(x) = 2x². It will
     be of the form f(x) = 2(x – h)² + k with h = 6
     and k = 0: f(x) = 2(x – 6)²

62. f(x) = –2x² + 3

63. Since there is a maximum at (3,8), the parabola
     will have the same shape as f(x) = –2x². It will
     be of the form f(x) = –2(x – h)² + k with h = 3
     and k = 8: f(x) = –2(x – 3)² + 8

64. f(x) = 2(x + 2)² + 3

65. Since there is a minimum at (–3,6), the parabola
     will have the same shape as f(x) = 2x². It will
     be of the form f(x) = 2(x – h)² + k with h = –3
     and k = 6: f(x) = 2[x – (–3)]² + 6, or
     f(x) = 2(x + 3)² + 6

66. f(x) = –2(x + 4)² – 3

67. Since there is a minimum at (2,–3), the parabola
     will have the same shape as f(x) = 2x². It will
     be of the form f(x) = 2(x – h)² + k with h = 2
     and k = –3: f(x) = 2(x – 2)² – 3

68. f(x) = 6(x – 4)²

69. Since the parabola has the same shape as
     f(x) = $-\frac{1}{2}$(x – 2)² + 4, it is of the form
     g(x) = $-\frac{1}{2}$(x – h)² + k. Since it has a maximum
     value at the same point as f(x) = –2(x – 1)² – 6,
     its vertex is (1,–6). That is, h = 1 and k = –6.
     The equation is g(x) = $-\frac{1}{2}$(x – 1)² – 6.

70.
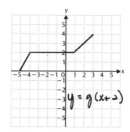

71. y = g(x – 3)
     The graph looks just like the graph of y = g(x)
     but moved 3 units to the right.
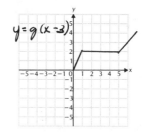

72.
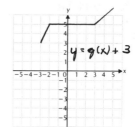

73. y = g(x) + 4
     The graph looks just like the graph of y = g(x)
     but moved up 4 units.
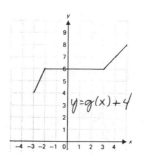

74.

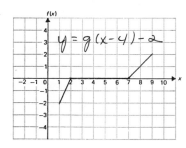

y = g(x-4) - 2

75. y = g(x - 2) + 3

The graph looks just like the graph of y = g(x) but moved to the right 2 units and up 3 units.

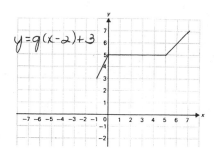

y = g(x-2)+3

Exercise Set 8.9

---

1. f(x) = x² - 2x - 3

    = (x² - 2x + 1) - 3 - 1   Adding and subtracting 1

    = (x - 1)² - 4

The vertex is (1,-4), the line of symmetry is x = 1, and the graph opens upward since the coefficient 1 is positive. We plot a few points as a check and draw the curve.

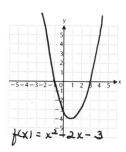

f(x) = x² + 2x - 3

2.

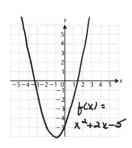

f(x) = x² + 2x - 5

3. g(x) = x² + 6x + 13

    = (x² + 6x + 9) + 13 - 9   Adding and subtracting 9

    = (x + 3)² + 4

The vertex is (-3,4), the line of symmetry is x = -3, and the graph opens upward since the coefficient 1 is positive. We plot a few points as a check and draw the curve.

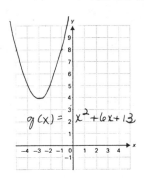

g(x) = x² + 6x + 13

4.

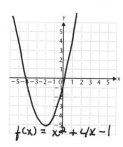

g(x) = x² - 4x + 5

5. f(x) = x² + 4x - 1

    = (x² + 4x + 4) - 1 - 4   Adding and subtracting 4

    = (x + 2)² - 5

The vertex is (-2,-5), the line of symmetry is x = -2, and the graph opens upward since the coefficient 1 is positive.

f(x) = x² + 4x - 1

6.

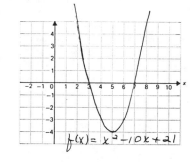

f(x) = x² - 10x + 21

7. $h(x) = 2x^2 + 16x + 25$

      $= 2(x^2 + 8x) + 25$ Factoring 2 from the
                                    first two terms

      $= 2(x^2 + 8x + 16) + 25 - 2\cdot16$ Adding and
                                          subtracting
                                          $2\cdot16$, or 32

      $= 2(x + 4)^2 - 7$

The vertex is $(-4,-7)$, the line of symmetry is
$x = -4$, and the graph opens upward since the
coefficient 2 is positive.

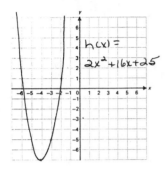

8.

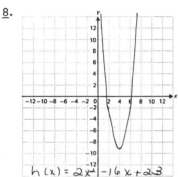

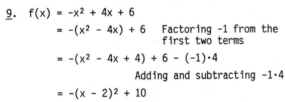

9. $f(x) = -x^2 + 4x + 6$

      $= -(x^2 - 4x) + 6$ Factoring -1 from the
                                  first two terms

      $= -(x^2 - 4x + 4) + 6 - (-1)\cdot4$

                          Adding and subtracting $-1\cdot4$

      $= -(x - 2)^2 + 10$

The vertex is $(2,10)$, the line of symmetry is
$x = 2$, and the graph opens downward since the
coefficient -1 is negative.

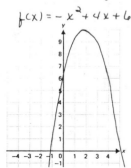

10.

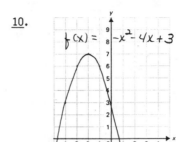

11. $g(x) = x^2 + 3x - 10$

      $= \left(x^2 + 3x + \dfrac{9}{4}\right) - 10 - \dfrac{9}{4}$

      $= \left(x + \dfrac{3}{2}\right)^2 - \dfrac{49}{4}$

The vertex is $\left(-\dfrac{3}{2}, -\dfrac{49}{4}\right)$, the line of symmetry
is $x = -\dfrac{3}{2}$, and the graph opens upward since
the coefficient 1 is positive.

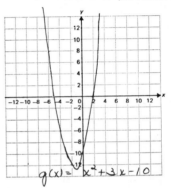

12.

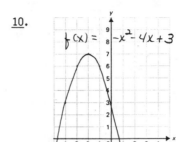

13. $f(x) = 3x^2 - 24x + 50$

      $= 3(x^2 - 8x) + 50$ Factoring

      $= 3(x^2 - 8x + 16) + 50 - 3\cdot16$

                        Adding and subtracting $3\cdot16$

      $= 3(x - 4)^2 + 2$

The vertex is $(4,2)$, the line of symmetry is
$x = 4$, and the graph opens upward since the
coefficient 3 is positive.

13. (continued)

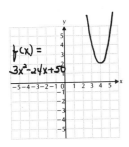

$$f(x) = 3x^2 - 24x + 50$$

14.

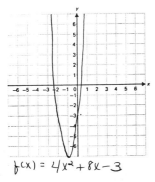

$$f(x) = 4x^2 + 8x - 3$$

15. $h(x) = x^2 - 9x$

$$= \left[x^2 - 9x + \frac{81}{4}\right] - \frac{81}{4}$$

$$= \left(x - \frac{9}{2}\right)^2 - \frac{81}{4}$$

The vertex is $\left(\frac{9}{2}, -\frac{81}{4}\right)$, the line of symmetry is $x = \frac{9}{2}$, and the graph opens upward since the coefficient 1 is positive.

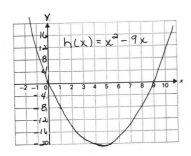

$$h(x) = x^2 - 9x$$

16.

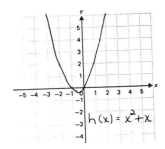

$$h(x) = x^2 + x$$

17. $f(x) = -2x^2 - 4x - 6$

$$= -2(x^2 + 2x) - 6 \quad \text{Factoring}$$

$$= -2(x^2 + 2x + 1) - 6 - (-2)\cdot 1$$

$$\qquad\qquad \text{Adding and subtracting } -2\cdot 1$$

$$= -2(x + 1)^2 - 4$$

The vertex is $(-1, -4)$, the line of symmetry is $x = -1$, and the graph opens downward since the coefficient -2 is negative.

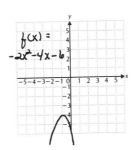

$$f(x) = -2x^2 - 4x - 6$$

18.

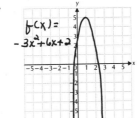

$$f(x) = -3x^2 + 6x + 2$$

19. $g(x) = 2x^2 - 10x + 14$

$$= 2(x^2 - 5x) + 14 \quad \text{Factoring}$$

$$= 2\left[x^2 - 5x + \frac{25}{4}\right] + 14 - 2 \cdot \frac{25}{4}$$

$$\qquad\qquad \text{Adding and subtracting } 2 \cdot \frac{25}{4}$$

$$= 2\left(x - \frac{5}{2}\right)^2 + \frac{3}{2}$$

The vertex is $\left(\frac{5}{2}, \frac{3}{2}\right)$, the line of symmetry is $x = \frac{5}{2}$, and the graph opens upward since the coefficient 2 is positive.

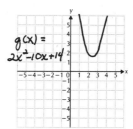

$$g(x) = 2x^2 - 10x + 14$$

20.

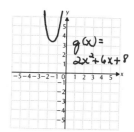

21. $f(x) = -3x^2 - 3x + 1$

$\quad = -3(x^2 + x) + 1$ Factoring

$\quad = -3\left[x^2 + x + \frac{1}{4}\right] + 1 - (-3) \cdot \frac{1}{4}$

$\qquad\qquad$ Adding and subtracting $-3 \cdot \frac{1}{4}$

$\quad = -3\left[x + \frac{1}{2}\right]^2 + \frac{7}{4}$

The vertex is $\left[-\frac{1}{2}, \frac{7}{4}\right]$, the line of symmetry is $x = -\frac{1}{2}$, and the graph opens downward since the coeffient -3 is negative.

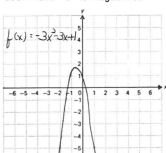

22.

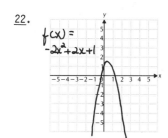

23. $h(x) = \frac{1}{2}x^2 + 4x + \frac{19}{3}$

$\quad = \frac{1}{2}(x^2 + 8x) + \frac{19}{3}$ Factoring

$\quad = \frac{1}{2}(x^2 + 8x + 16) + \frac{19}{3} - \frac{1}{2} \cdot 16$

$\qquad\qquad$ Adding and subtracting $\frac{1}{2} \cdot 16$

$\quad = \frac{1}{2}(x + 4)^2 - \frac{5}{3}$

The vertex is $\left[-4, -\frac{5}{3}\right]$, the line of symmetry is $x = -4$, and the graph opens upward since the coefficient $\frac{1}{2}$ is positive.

23. (continued)

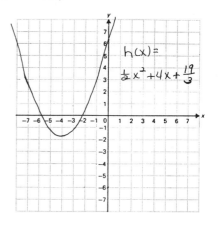

24.

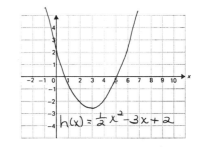

25. Solve $0 = x^2 - 4x + 1$. Use the quadratic formula.

$\quad x = \dfrac{-(-4) \pm \sqrt{(-4)^2 - 4 \cdot 1 \cdot 1}}{2 \cdot 1}$

$\quad x = \dfrac{4 \pm \sqrt{12}}{2} = \dfrac{4 \pm 2\sqrt{3}}{2} = 2 \pm \sqrt{3}$

The x-intercepts are $(2 + \sqrt{3}, 0)$ and $(2 - \sqrt{3}, 0)$.

26. None exist.

27. Solve: $0 = -x^2 + 2x + 3$

$\qquad\quad 0 = x^2 - 2x - 3$ Multiplying by -1

$\qquad\quad 0 = (x - 3)(x + 1)$

$\qquad\quad x = 3$ or $x = -1$

The x-intercepts are $(3,0)$ and $(-1,0)$.

28. $(1 + \sqrt{6}, 0)$, $(1 - \sqrt{6}, 0)$

29. Solve: $0 = x^2 - 3x - 4$

$\qquad\quad 0 = (x - 4)(x + 1)$

$\qquad\quad x = 4$ or $x = -1$

The x-intercepts are $(4,0)$ and $(-1,0)$.

30. $(4 + \sqrt{11}, 0)$, $(4 - \sqrt{11}, 0)$

31. Solve: $0 = -x^2 + 3x - 2$

$\qquad\quad 0 = x^2 - 3x + 2$

$\qquad\quad 0 = (x - 2)(x - 1)$

$\qquad\quad x = 2$ or $x = 1$

The x-intercepts are $(2,0)$ and $(1,0)$.

32. None exist.

33. Solve $0 = 2x^2 + 4x - 1$. Using the quadratic formula we get $x = \frac{-2 \pm \sqrt{6}}{2}$. The x-intercepts are $\left(\frac{-2 + \sqrt{6}}{2}, 0\right)$ and $\left(\frac{-2 - \sqrt{6}}{2}, 0\right)$.

34. None exist.

35. Solve: $0 = x^2 - x + 1$

    $x = \frac{-(-1) \pm \sqrt{(-1)^2 - 4 \cdot 1 \cdot 1}}{2 \cdot 1}$

    $x = \frac{1 \pm \sqrt{-3}}{2} = \frac{1 \pm i\sqrt{3}}{2}$

    Since the equation has no real solutions, no x-intercepts exist.

36. $\left(-\frac{3}{2}, 0\right)$

37. Solve: $0 = -x^2 - 3x - 3$

    $x = \frac{-(-3) \pm \sqrt{(-3)^2 - 4(-1)(-3)}}{2(-1)}$

    $x = \frac{3 \pm \sqrt{-3}}{-2} = \frac{3 \pm i\sqrt{3}}{-2}$

    Since the equation has no real solutions, no x-intercepts exist.

38. $\left(\frac{5 + \sqrt{73}}{6}, 0\right)$, $\left(\frac{5 - \sqrt{73}}{6}, 0\right)$

39. $\sqrt{4x - 4} = \sqrt{x + 4} + 1$

    $4x - 4 = x + 4 + 2\sqrt{x + 4} + 1$ Squaring both sides

    $3x - 9 = 2\sqrt{x + 4}$

    $9x^2 - 54x + 81 = 4(x + 4)$ Squaring both sides again

    $9x^2 - 54x + 81 = 4x + 16$

    $9x^2 - 58x + 65 = 0$

    $(9x - 13)(x - 5) = 0$

    $x = \frac{13}{9}$ or $x = 5$

    Check. For $x = \frac{13}{9}$:

$\sqrt{4x - 4} = \sqrt{x + 4} + 1$	
$\sqrt{4\left(\frac{13}{9}\right) - 4}$	$\sqrt{\frac{13}{9} + 4} + 1$
$\sqrt{\frac{16}{9}}$	$\sqrt{\frac{49}{9}} + 1$
$\frac{4}{3}$	$\frac{7}{3} + 1$
	$\frac{10}{3}$

39. (continued)

    For $x = 5$:

$\sqrt{4x - 4} = \sqrt{x + 4} + 1$	
$\sqrt{4 \cdot 5 - 4}$	$\sqrt{5 + 4} + 1$
$\sqrt{16}$	$\sqrt{9} + 1$
4	$3 + 1$
	4

   5 checks, but $\frac{13}{9}$ does not. The solution is 5.

40. 4

41. $f(x) = 2.31x^2 - 3.135x - 5.89$

    $= 2.31(x^2 - 1.357142857x) - 5.89$

    $= 2.31(x^2 - 1.357142857x + 0.460459183) -$
    $\qquad 5.89 - 1.063660714$

    $= 2.31(x - 0.678571428)^2 - 6.953660714$

    Since the coefficient 2.31 is positive, the function has a minimum value. It is -6.953660714.

42. Maximum: 7.01412766

43. Solve: $0 = 0.05x^2 - 4.735x + 100.23$

    $x = \frac{-(-4.735) \pm \sqrt{(-4.735)^2 - 4(0.05)(100.23)}}{2(0.05)}$

    $x \approx \frac{4.735 \pm 1.540852037}{0.1}$

    $x \approx \frac{4.735 + 1.540852037}{0.1}$ or

    $\qquad\qquad x = \frac{4.735 - 1.540852037}{0.1}$

    $x \approx 62.758520$ or $x \approx 31.941480$

    The x-intercepts are (62.758520, 0) and (31.941480, 0).

44. (1.557155757, 0), (-4.042996465, 0)

45. Solve: $0 = 2.12x^2 + 3.21x + 9.73$

    $x = \frac{-3.21 \pm \sqrt{(3.21)^2 - 4(2.12)(9.73)}}{2(2.12)}$

    $x = \frac{-3.21 \pm \sqrt{-72.2063}}{4.24}$

    Since the radicand is negative, there are no real solutions and hence no x-intercepts.

46. (1.271900195, 0), (-0.725746349, 0)

47. f(x) = x² - x - 6

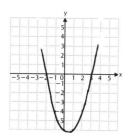

a) x² - x - 6 = 2

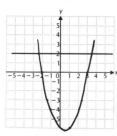

The solutions are approximately -2.4 and 3.4.

b) x² - x - 6 = -3

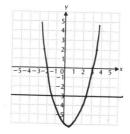

The solutions are approximately -1.3 and 2.3.

48. a) 3, 1; b) -4.5, 2.5; c) -5.5, 3.5

49. f(x) = ax² + bx + c

$$= a\left(x² + \frac{b}{a}x\right) + c$$

$$= a\left(x² + \frac{b}{a}x + \frac{b²}{4a²}\right) + c - a \cdot \frac{b²}{4a²}$$

$$= a\left(x + \frac{b}{2a}\right)² + c - \frac{b²}{4a}$$

$$= a\left[x - \left(-\frac{b}{2a}\right)\right]² + \frac{4ac - b²}{4a}$$

50. $f(x) = 3\left[x - \left(-\frac{m}{6}\right)\right]² + \frac{11m²}{12}$

51. f(x) = |x² - 1|

We plot some points and draw the curve. Note that it will lie entirely on or above the x-axis since absolute value is never negative. For positive values of x² - 1 it will look like the graph of f(x) = x² but moved down 1 unit.

x	f(x) = |x² - 1|
-2	3
-1	0
0	1
1	0
2	3

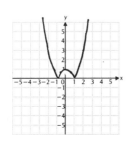

52. f(x) = |3 - 2x - x²|

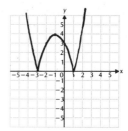

53. y < x² - 4x - 1

We first graph the parabola y = x² - 4x - 1. (See Exercise 25.) Since the inequality symbol is <, we draw it dashed. We determine whether to shade above or below the parabola by trying some point off the parabola. The point (0,0) is easy to check.

0 < 0² - 4·0 - 1   Substituting 0 for x and 0
                      for y

0 < -1

Since 0 < -1 is false, the point (0,0) is not in the graph. Thus, we shade below the parabola. The graph consists of the region below the parabola, but not the parabola.

y < x² - 4x - 1

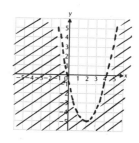

54. y ⩾ x² + 3x - 4

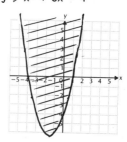

55. $y \leqslant x^2 + 5x + 6$

We first graph the parabola $y = x^2 + 5x + 6$. Since the inequality symbol is $\leqslant$, we draw it solid. We determine whether to shade above or below the parabola by trying some point off the parabola. The point $(0,0)$ is easy to check.

$0 \leqslant 0^2 + 5 \cdot 0 + 6$ Substituting 0 for x and 0 for y

$0 \leqslant 6$

Since $0 \leqslant 6$ is true, the point $(0,0)$ is in the graph. Thus, we shade below the parabola. The graph consists of the region below the parabola and the parabola.

$y \leqslant x^2 + 5x + 6$

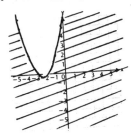

56. $y < -x^2 - 2x + 3$

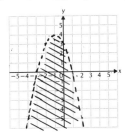

57. $y > 3x^2 + 6x + 2$

We first graph the parabola $y = 3x^2 + 6x + 2$. Since the inequality symbol is $<$, we draw it dashed. We determine whether to shade above or below the parabola by trying some point off the parabola. The point $(0,0)$ is easy to check.

$0 > 3 \cdot 0^2 + 6 \cdot 0 + 2$ Substituting 0 for x and 0 for y

$0 > 2$

Since $0 > 2$ is false, the point $(0,0)$ is not in the graph. Thus, we shade above the parabola. The graph consists of the region above the parabola, but not the parabola.

$y > 3x^2 + 6x + 2$

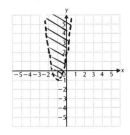

58. $y > 2x^2 + 4x - 2$

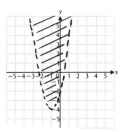

59. $y \geqslant 4x^2 + 8x + 3$

We first graph the parabola $y = 4x^2 + 8x + 3$. (See Exercise 14.) Since the inequality symbol is $\geqslant$, we draw it solid. We determine whether to shade above or below the parabola by trying some point off the parabola. The point $(0,0)$ is easy to check.

$0 \geqslant 4 \cdot 0^2 + 8 \cdot 0 + 3$ Substituting 0 for x and 0 for y

$0 \geqslant 3$

Since $0 \geqslant 3$ is false, the point $(0,0)$ is not in the graph. Thus, we shade above the parabola. The graph consists of the region above the parabola and the parabola.

$y \geqslant 4x^2 + 8x + 3$

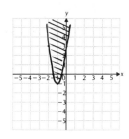

60. $y < 2x^2 - 4x + 2$

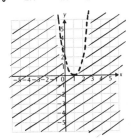

61. The equation can be written in the form $f(x) = a(x - h)^2 + k$ where $h = 1$ and $k = -8$. We have $f(x) = a(x - 1)^2 - 8$. Substitute the coordinates of one x-intercept to find a. We will use $(-3,0)$.

$0 = a(-3 - 1)^2 - 8$

$0 = 16a - 8$

$8 = 16a$

$\frac{1}{2} = a$

The equation is $f(x) = \frac{1}{2}(x - 1)^2 - 8$, or $f(x) = \frac{1}{2}x^2 - x - \frac{15}{2}$.

Exercise Set 8.10

1. Familiarize. We make a drawing and label it.

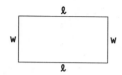

Perimeter: $2\ell + 2w = 76$ ft

Area: $A = \ell \cdot w$

Translate. We have a system of equations.

$2\ell + 2w = 76,$

$A = \ell w$

Carry out. Solving the first equation for ℓ, we get $\ell = 38 - w$. Substituting for ℓ in the second equation we get a quadratic function A:

$A = (38 - w)w$

$A = -w^2 + 38w$

Completing the square, we get

$A = -(w - 19)^2 + 361.$

The maximum function value is 361. It occurs when w is 19. When $w = 19$, $\ell = 38 - 19$, or 19.

Check. We check a function value for w less than 19 and for w greater than 19.

$A(18) = -(18)^2 + 38 \cdot 18 = 360$

$A(20) = -(20)^2 + 38 \cdot 20 = 360$

Since 361 is greater than these numbers, it looks as though we have a maximum.

State. The maximum area of 361 ft² occurs when the dimensions are 19 ft by 19 ft.

2. 17 ft by 17 ft; 289 ft²

3. Familiarize. We let x and y represent the numbers, and we let P represent their product.

Translate. We have two equations.

$x + y = 16,$

$P = xy$

Carry out. Solving the first equation for y, we get $y = 16 - x$. Substituting for y in the second equation we get a quadratic function P:

$P = x(16 - x)$

$P = -x^2 + 16x$

Completing the square, we get

$P = -(x - 8)^2 + 64.$

The maximum function value is 64. It occurs when $x = 8$. When $x = 8$, $y = 16 - 8$, or 8.

Check. We can check a function value for x less than 8 and for x greater than 8.

$P(7) = -(7)^2 + 16 \cdot 7 = 63$

$P(9) = -(9)^2 + 16 \cdot 9 = 63$

Since 64 is greater than these numbers, it looks as though we have a maximum.

State. The maximum product of 64 occurs for the numbers 8 and 8.

4. 196; 14 and 14

5. Familiarize. Let x and y represent the two numbers, and let P represent their product.

Translate. We have two equations.

$x + y = 22,$

$P = xy$

Carry out. Solve the first equation for y.

$y = 22 - x$

Substitute for y in the second equation.

$P = x(22 - x)$

$P = -x^2 + 22x$

Completing the square, we get

$P = -(x - 11)^2 + 121$

The maximum function value is 121. It occurs when $x = 11$. When $x = 11$, $y = 22 - 11$, or 11.

Check. Check a function value for x less than 11 and for x greater than 11.

$P(10) = -(10)^2 + 22 \cdot 10 = 120$

$P(12) = -(12)^2 + 22 \cdot 12 = 120$

Since 121 is greater than these numbers, it looks as though we have a maximum.

State. The maximum product of 121 occurs for the numbers 11 and 11.

6. $\frac{2025}{4}$; $\frac{45}{2}$ and $\frac{45}{2}$

7. Familiarize. Let x and y represent the two numbers, and let P represent their product.

Translate. We have two equations.

$x - y = 4,$

$P = xy$

Carry out. Solve the first equation for x.

$x = 4 + y$

Substitute for x in the second equation.

$P = (4 + y)y$

$P = y^2 + 4y$

Completing the square, we get

$P = (y + 2)^2 - 4.$

The minimum function value is -4. It occurs when $y = -2$. When $y = -2$, $x = 4 + (-2)$, or 2.

Check. Check a function value for y less than -2 and for y greater than -2.

$P(-3) = (-3)^2 + 4(-3) = -3$

$P(-1) = (-1)^2 + 4(-1) = -3$

Since -4 is less than these numbers, it looks as though we have a minimum.

State. The minimum product of -4 occurs for the numbers 2 and -2.

8. -25; 5 and -5

9. Familiarize. Let x and y represent the two numbers, and let P represent their product.

Translate. We have two equations.

$$x - y = 5,$$
$$P = xy$$

Carry out. Solve the first equation for x.

$$x = y + 5$$

Substitute for x in the second equation.

$$P = (y + 5)y$$
$$P = y^2 + 5y$$

Completing the square, we get

$$P = \left(y + \frac{5}{2}\right)^2 - \frac{25}{4}.$$

The minimum function value is $-\frac{25}{4}$. It occurs when $y = -\frac{5}{2}$. When $y = -\frac{5}{2}$, $x = -\frac{5}{2} + 5 = \frac{5}{2}$.

Check. Check a function value for y less than $-\frac{5}{2}$ and for y greater than $-\frac{5}{2}$.

$$P(-3) = (-3)^2 + 5(-3) = -6$$
$$P(-2) = (-2)^2 + 5(-2) = -6$$

Since $-\frac{25}{4}$ is less than these numbers, it looks as though we have a minimum.

State. The minimum product of $-\frac{25}{4}$ occurs for the numbers $\frac{5}{2}$ and $-\frac{5}{2}$.

10. $-\frac{49}{4}$; $\frac{7}{2}$ and $-\frac{7}{2}$

11. Familiarize. Let x and y represent the two numbers, and let P represent their product.

Translate. We have two equations.

$$x + y = -7,$$
$$P = xy$$

Carry out. Solve the first equation for y.

$$y = -x - 7$$

Substitute for y in the second equation.

$$P = x(-x - 7)$$
$$P = -x^2 - 7x$$

Completing the square, we get

$$P = -\left(x + \frac{7}{2}\right)^2 + \frac{49}{4}$$

The maximum function value is $\frac{49}{4}$. It occurs when $x = -\frac{7}{2}$. When $x = -\frac{7}{2}$, $y = -\left(-\frac{7}{2}\right) - 7 = -\frac{7}{2}$.

Check. Check a function value for x less than $-\frac{7}{2}$ and for x greater than $-\frac{7}{2}$.

$$P(-4) = -(-4)^2 - 7(-4) = 12$$
$$P(-3) = -(-3)^2 - 7(-3) = 12$$

Since $\frac{49}{4}$ is greater than these numbers, it looks as though we have a maximum.

11. (continued)

State. The maximum product of $\frac{49}{4}$ occurs for the numbers $-\frac{7}{2}$ and $-\frac{7}{2}$.

12. $\frac{81}{4}$; $-\frac{9}{2}$ and $-\frac{9}{2}$

13. Familiarize. We make a drawing and label it.

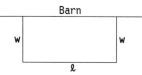

Translate. We have two equations.

$$\ell + 2w = 40,$$
$$A = \ell w$$

Carry out. Solve the first equation for ℓ.

$$\ell = 40 - 2w$$

Substitute for ℓ in the second equation.

$$A = (40 - 2w)w$$
$$A = -2w^2 + 40w$$

Completing the square, we get

$$A = -2(w - 10)^2 + 200.$$

The maximum function value of 200 occurs when $w = 10$. When $w = 10$, $\ell = 40 - 2 \cdot 10 = 20$.

Check. Check a function value for w less than 10 and for w greater than 10.

$$A(9) = -2 \cdot 9^2 + 40 \cdot 9 = 198$$
$$A(11) = -2 \cdot 11^2 + 40 \cdot 11 = 198$$

Since 200 is greater than these numbers, it looks as though we have a maximum.

State. The maximum area of 200 ft² will occur when the dimensions are 10 ft by 20 ft.

14. 450 ft²; 15 ft by 30 ft

15. We look for a function of the form $f(x) = ax^2 + bx + c$. Substituting the data points, we get

$$4 = a(1)^2 + b(1) + c,$$
$$-2 = a(-1)^2 + b(-1) + c,$$
$$13 = a(2)^2 + b(2) + c,$$

or

$$4 = a + b + c,$$
$$-2 = a - b + c,$$
$$13 = 4a + 2b + c.$$

Solving this system, we get

$$a = 2, \quad b = 3, \quad \text{and} \quad c = -1.$$

Therefore the function we are looking for is

$$f(x) = 2x^2 + 3x - 1.$$

16. $f(x) = 3x^2 - x + 2$

17. We look for a function of the form
$f(x) = ax^2 + bx + c$. Substituting the data points, we get

$5 = a(1)^2 + b(1) + c,$

$9 = a(2)^2 + b(2) + c,$

$7 = a(3)^2 + b(3) + c,$

or

$5 = a + b + c,$

$9 = 4a + 2b + c,$

$7 = 9a + 3b + c.$

Solving this system, we get

$a = -3$, $b = 13$, and $c = -5$

Therefore the function we are looking for is

$f(x) = -3x^2 + 13x - 5.$

18. $f(x) = x^2 - 5x$

19. We look for a function of the form
$f(x) = ax^2 + bx + c$. Substituting the data points, we get

$9 = a(-10)^2 + b(-10) + c,$

$-4 = a(-5)^2 + b(-5) + c,$

$-7 = a(0)^2 + b(0) + c,$

or

$9 = 100a - 10b + c,$

$-4 = 25a - 5b + c,$

$-7 = c.$

Solving this system, we get

$a = \frac{1}{5}$, $b = \frac{2}{5}$, $c = -7.$

Therefore the function we are looking for is

$f(x) = \frac{1}{5}x^2 + \frac{2}{5}x - 7.$

20. $f(x) = \frac{1}{3}x^2 + \frac{2}{3}x - 5$

21. We look for a function of the form
$f(x) = ax^2 + bx + c$. Substituting the data points, we get

$0 = a(2)^2 + b(2) + c,$

$3 = a(4)^2 + b(4) + c,$

$-5 = a(12)^2 + b(12) + c,$

or

$0 = 4a + 2b + c,$

$3 = 16a + 4b + c,$

$-5 = 144a + 12b + c.$

Solving this system, we get

$a = -\frac{1}{4}$, $b = 3$, $c = -5.$

Therefore the function we are looking for is

$f(x) = -\frac{1}{4}x^2 + 3x - 5.$

22. $f(x) = -\frac{1}{3}x^2 + 5x - 12$

23. a) Familiarize. We look for a function of the form $f(x) = ax^2 + bx + c$, where $f(x)$ represents the earnings for week x.

    Translate. We substitute the given values of x and $f(x)$.

    $38 = a(1)^2 + b(1) + c,$

    $66 = a(2)^2 + b(2) + c,$

    $86 = a(3)^2 + b(3) + c,$

    or

    $38 = a + b + c,$

    $66 = 4a + 2b + c,$

    $86 = 9a + 3b + c.$

    Carry out. Solving the system of equations, we get

    $a = -4$, $b = 40$, $c = 2.$

    Check. Recheck the calculations.

    State. The function $f(x) = -4x^2 + 40x + 2$ fits the data.

    b) Find $f(4)$.

    $f(4) = -4(4)^2 + 40(4) + 2 = 98$

    The predicted earnings for the fourth week are $98.

24. a) $f(x) = 2500x^2 - 6500x + 5000$; b) $19,000

25. a) Familiarize. We look for a function of the form $A(s) = as^2 + bs + c$, where $A(s)$ represents the number of daytime accidents (for every 200 million km) and s represents the travel speed (in km/h).

    Translate. We substitute the given values of s and $A(s)$.

    $100 = a(60)^2 + b(60) + c,$

    $130 = a(80)^2 + b(80) + c,$

    $200 = a(100)^2 + b(100) + c,$

    or

    $100 = 3600a + 60b + c,$

    $130 = 6400a + 80b + c,$

    $200 = 10,000a + 100b + c.$

    Carry out. Solving the system of equations, we get

    $a = 0.05$, $b = -5.5$, $c = 250.$

    Check. Recheck the calculations.

    State. The function $A(s) = 0.05s^2 - 5.5s + 250$ fits the data.

    b) Find $A(50)$.

    $A(50) = 0.05(50)^2 - 5.5(50) + 250 = 100$

    100 accidents occur at 50 km/h.

26. a) $A(s) = \frac{3}{16}s^2 - \frac{135}{4}s + 1750$; b) 531.25

27. a) Familiarize. We look for a function of the
form $f(x) = ax^2 + bx + c$, where $f(x)$
represents the price for a pizza with
diameter x.

Translate. Substitute the given values of x
and $f(x)$.

$$3 = a(8)^2 + b(8) + c,$$
$$4.25 = a(12)^2 + b(12) + c,$$
$$5.75 = a(16)^2 + b(16) + c,$$

or

$$3 = 64a + 8b + c,$$
$$4.25 = 144a + 12b + c,$$
$$5.75 = 256a + 16b + c.$$

Carry out. Solving the system of equations,
we get

$a = 0.0078125$, $b = 0.15625$, $c = 1.25$

Check. Recheck the calculations.

State. The function $f(x) = 0.0078125x^2 +$
$0.15625x + 1.25$ fits the data.

b) Find $f(14)$.

$f(14) = 0.0078125(14)^2 + 0.15625(14) + 1.25$
$= 4.96875$

The price of a 14-in. pizza is approximately
$4.97.

28. $P(x) = -(x - 490)^2 + 237,100$;
Maximum profit: $237,100 at $x = 490$

29. Find the total profit:

$$P(x) = R(x) - C(x)$$
$$P(x) = (200x - x^2) - (5000 + 8x)$$
$$P(x) = -x^2 + 192x - 5000$$

To find the maximum value of the total profit
and the value of x at which it occurs we
complete the square:

$$P(x) = -(x^2 - 192x) - 5000$$
$$= -(x^2 - 192x + 9216) - 5000 - (-1)(9216)$$
$$= -(x - 96)^2 + 4216$$

The maximum profit of $4216 occurs at $x = 96$.

30. $P(x) = -(x - 110)^2 + 12,050$;
Maximum profit: $12,050 at $x = 110$

31. $\sqrt[4]{5x^3y^5} \sqrt[4]{125x^2y^3} = \sqrt[4]{625x^5y^8} = \sqrt[4]{625x^4y^8 \cdot x} =$

$5xy^2 \sqrt[4]{x}$

32. $12a^2b^2$

33. $f(x) = 2.31x^2 - 3.105x - 5.98$

We complete the square.

$f(x) = 2.31(x^2 - 1.344155844x) - 5.98$

$= 2.31(x^2 - 1.344155844x + 0.451688733) -$

$5.98 - 1.043400974$

$= 2.31(x - 0.672077922)^2 - 7.023400974$

The minimum function value is -7.023400974.

34. Maximum: 6.638814972

35. We look for a function of the form
$f(x) = ax^2 + bx + c$. Substituting the data
points, we get

$$-5.86 = a(20.34)^2 + b(20.34) + c,$$
$$-6.02 = a(34.67)^2 + b(34.67) + c,$$
$$-8.46 = a(28.55)^2 + b(28.55) + c,$$

or

$$-5.86 = 413.7156a + 20.34b + c,$$
$$-6.02 = 1202.0089a + 34.67b + c,$$
$$-8.46 = 815.1025a + 28.55b + c.$$

Solving the system of equations, we get
$a = 0.0499218$, $b = -2.7573651$, $c = 29.571379$.
Therefore the function we are looking for is
$f(x) = 0.0499218x^2 - 2.7573651x + 29.571379$.

36. $f(x) = -691.95954x^2 + 1757.2503x - 710.71113$

37. Familiarize. Recall that the formula for the area
of a triangle is $A = \frac{1}{2}bh$.

Translate. We have two equations.

$$b + h = 38,$$
$$A = \frac{1}{2}bh$$

Carry out. Solve the first equation for h.

$h = 38 - b$

Substitute for h in the second equation.

$A = \frac{1}{2}b(38 - b)$

$A = -\frac{1}{2}b^2 + 19b$

Completing the square, we get

$A = -\frac{1}{2}(b - 19)^2 + \frac{361}{2}$

The maximum function value of $\frac{361}{2}$, or 180.5,
occurs when $b = 19$. When $b = 19$, $h = 38 - 19$, or
19.

Check. Check a function value for b less than 19
and for b greater than 19.

$A(18) = -\frac{1}{2}(18)^2 + 19(18) = 180$

$A(20) = -\frac{1}{2}(20)^2 + 19(20) = 180$

Since 180.5 is greater than these numbers, it
looks as though we have a maximum.

State. The maximum area of 180.5 cm² occurs when
the base and height are both 19 cm.

38. $11\sqrt{2}$ ft

<u>39.</u> <u>Familiarize</u>. Let x represent the number of 10¢
increases in the admission price. Then
2.00 + 0.1x represents the admission price and
100 - x represents the corresponding average
attendance.

<u>Translate</u>. Since total revenue is admission
price times the attendance, we have the
following function for the revenue.

$R(x) = (2.00 + 0.1x)(100 - x)$

$R(x) = -0.1x^2 + 8x + 200$

<u>Carry out</u>. Completing the square, we get

$R(x) = -0.1(x - 40)^2 + 360$

The maximum function value of 360 occurs when
x = 40. When x = 40 the admission price is
2.00 + 0.1(40), or $6.

<u>Check</u>. We check a function value for x less
than 40 and for x greater than 40.

$R(39) = -0.1(39)^2 + 8(39) + 200 = 359.9$

$R(41) = -0.1(41)^2 + 8(41) + 200 = 359.9$

Since 360 is greater than these numbers, it looks
as though we have a maximum.

<u>State</u>. In order to maximize revenue the theater
owner should charge $6.00 for admission.

<u>40.</u> 30

<u>41.</u> <u>Familiarize</u>. We want to find the maximum value
of a function of the form $h(t) = at^2 + bt + c$
that fits the following data.

Time (sec)	Height (ft)
0	64
3	64
3 + 2, or 5	0

<u>Translate</u>. Substitute the given values for t and
h(t).

$64 = a(0)^2 + b(0) + c,$

$64 = a(3)^2 + b(3) + c,$

$0 = a(5)^2 + b(5) + c,$

or

$64 = c,$

$64 = 9a + 3b + c,$

$0 = 25a + 5b + c.$

<u>Carry out</u>. Solving the system of equations, we
get a = -6.4, b = 19.2, c = 64. The function
$h(t) = -6.4t^2 + 19.2t + 64$ fits the data.

Completing the square, we get

$h(t) = -6.4(t - 1.5)^2 + 78.4.$

The maximum function value of 78.4 occurs at
t = 1.5.

<u>Check</u>. Recheck the calculations. Also check a
function value for t less than 1.5 and for t
greater than 1.5.

$h(1) = -6.4(1)^2 + 19.2(1) + 64 = 76.8$

$h(2) = -6.4(2)^2 + 19.2(2) + 64 = 76.8$

Since 78.4 is greater than these numbers, it
looks as though we have a maximum.

<u>41.</u> (continued)

<u>State</u>. The maximum height is 78.4 ft.

<u>42.</u> 158 ft

Exercise Set 9.1

<u>1</u>. Graph: $y = 2^x$

We compute some function values, thinking of y as f(x), and keep the results in a table.

$f(0) = 2^0 = 1$

$f(1) = 2^1 = 2$

$f(2) = 2^2 = 4$

$f(-1) = 2^{-1} = \frac{1}{2^1} = \frac{1}{2}$

$f(-2) = 2^{-2} = \frac{1}{2^2} = \frac{1}{4}$

x	y, or f(x)
0	1
1	2
2	4
-1	$\frac{1}{2}$
-2	$\frac{1}{4}$

Next we plot these points and connect them with a smooth curve.

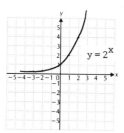

<u>2</u>.

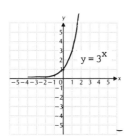

<u>3</u>. Graph: $y = 5^x$

We compute some function values, thinking of y as f(x), and keep the results in a table.

$f(0) = 5^0 = 1$

$f(1) = 5^1 = 5$

$f(2) = 5^2 = 25$

$f(-1) = 5^{-1} = \frac{1}{5^1} = \frac{1}{5}$

$f(-2) = 5^{-2} = \frac{1}{5^2} = \frac{1}{25}$

x	y, or f(x)
0	1
1	5
2	25
-1	$\frac{1}{5}$
-2	$\frac{1}{25}$

Next we plot these points and connect them with a smooth curve.

<u>3</u>. (continued)

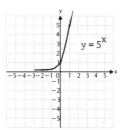

<u>4</u>.

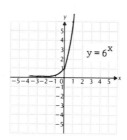

<u>5</u>. Graph: $y = 2^{x+1}$

We compute some function values, thinking of y as f(x), and keep the results in a table.

$f(0) = 2^{0+1} = 2^1 = 2$

$f(-1) = 2^{-1+1} = 2^0 = 1$

$f(-2) = 2^{-2+1} = 2^{-1} = \frac{1}{2^1} = \frac{1}{2}$

$f(-3) = 2^{-3+1} = 2^{-2} = \frac{1}{2^2} = \frac{1}{4}$

$f(1) = 2^{1+1} = 2^2 = 4$

$f(2) = 2^{2+1} = 2^3 = 8$

x	y, or f(x)
0	2
-1	1
-2	$\frac{1}{2}$
-3	$\frac{1}{4}$
1	4
2	8

Next we plot these points and connect them with a smooth curve.

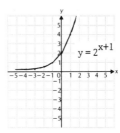

<u>6</u>.

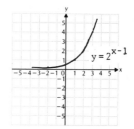

<u>7</u>. Graph: $y = 3^{x-2}$

We compute some function values, thinking of y as f(x), and keep the results in a table.

$f(0) = 3^{0-2} = 3^{-2} = \frac{1}{3^2} = \frac{1}{9}$

$f(1) = 3^{1-2} = 3^{-1} = \frac{1}{3^1} = \frac{1}{3}$

$f(2) = 3^{2-2} = 3^0 = 1$

$f(3) = 3^{3-2} = 3^1 = 3$

$f(4) = 3^{4-2} = 3^2 = 9$

$f(-1) = 3^{-1-2} = 3^{-3} = \frac{1}{3^3} = \frac{1}{27}$

$f(-2) = 3^{-2-2} = 3^{-4} = \frac{1}{3^4} = \frac{1}{81}$

x	y, or f(x)
0	$\frac{1}{9}$
1	$\frac{1}{3}$
2	1
3	3
4	9
-1	$\frac{1}{27}$
-2	$\frac{1}{81}$

Next we plot these points and connect them with a smooth curve.

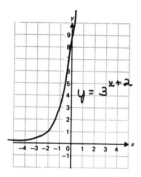

<u>8</u>.

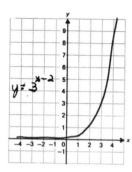

<u>9</u>. Graph: $y = 2^x - 3$

We construct a table of values, thinking of y as f(x). Then we plot the points and connect them with a smooth curve.

$f(0) = 2^0 - 3 = 1 - 3 = -2$

$f(1) = 2^1 - 3 = 2 - 3 = -1$

$f(2) = 2^2 - 3 = 4 - 3 = 1$

$f(3) = 2^3 - 3 = 8 - 3 = 5$

$f(-1) = 2^{-1} - 3 = \frac{1}{2} - 3 = -\frac{5}{2}$

$f(-2) = 2^{-2} - 3 = \frac{1}{4} - 3 = -\frac{11}{4}$

x	y, or f(x)
0	-2
1	-1
2	1
3	5
-1	$-\frac{5}{2}$
-2	$-\frac{11}{4}$

<u>9</u>. (continued)

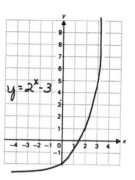

<u>10</u>.

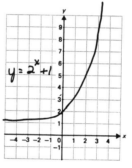

<u>11</u>. Graph: $y = 5^{x+3}$

We construct a table of values, thinking of y as f(x). Then we plot the points and connect them with a smooth curve.

$f(0) = 5^{0+3} = 5^3 = 125$

$f(-1) = 5^{-1+3} = 5^2 = 25$

$f(-2) = 5^{-2+3} = 5^1 = 5$

$f(-3) = 5^{-3+3} = 5^0 = 1$

$f(-4) = 5^{-4+3} = 5^{-1} = \frac{1}{5}$

$f(-5) = 5^{-5+3} = 5^{-2} = \frac{1}{25}$

x	y, or f(x)
0	125
-1	25
-2	5
-3	1
-4	$\frac{1}{5}$
-5	$\frac{1}{25}$

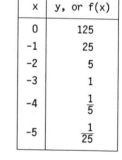

<u>12</u>.

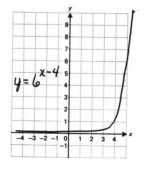

296

13. Graph: $y = \left(\frac{1}{2}\right)^x$

We construct a table of values, thinking of y as f(x). Then we plot the points and connect them with a smooth curve.

$f(0) = \left(\frac{1}{2}\right)^0 = 1$

$f(1) = \left(\frac{1}{2}\right)^1 = \frac{1}{2}$

$f(2) = \left(\frac{1}{2}\right)^2 = \frac{1}{4}$

$f(3) = \left(\frac{1}{2}\right)^3 = \frac{1}{8}$

$f(-1) = \left(\frac{1}{2}\right)^{-1} = \frac{1}{\left(\frac{1}{2}\right)^1} = \frac{1}{\frac{1}{2}} = 2$

$f(-2) = \left(\frac{1}{2}\right)^{-2} = \frac{1}{\left(\frac{1}{2}\right)^2} = \frac{1}{\frac{1}{4}} = 4$

$f(-3) = \left(\frac{1}{2}\right)^{-3} = \frac{1}{\left(\frac{1}{2}\right)^3} = \frac{1}{\frac{1}{8}} = 8$

x	y, or f(x)
0	1
1	$\frac{1}{2}$
2	$\frac{1}{4}$
3	$\frac{1}{8}$
-1	2
-2	4
-3	8

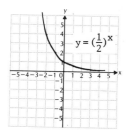

14.

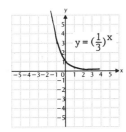

15. Graph: $y = \left(\frac{1}{5}\right)^x$

We construct a table of values, thinking of y as f(x). Then we plot the points and connect them with a smooth curve.

$f(0) = \left(\frac{1}{5}\right)^0 = 1$

$f(1) = \left(\frac{1}{5}\right)^1 = \frac{1}{5}$

$f(2) = \left(\frac{1}{5}\right)^2 = \frac{1}{25}$

$f(-1) = \left(\frac{1}{5}\right)^{-1} = \frac{1}{\frac{1}{5}} = 5$

$f(-2) = \left(\frac{1}{5}\right)^{-2} = \frac{1}{\frac{1}{25}} = 25$

x	y, or f(x)
0	1
1	$\frac{1}{5}$
2	$\frac{1}{25}$
-1	5
-2	25

15. (continued)

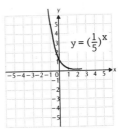

16.

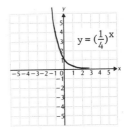

17. Graph: $y = 2^{2x-1}$

We construct a table of values, thinking of y as f(x). Then we plot the points and connect them with a smooth curve.

$f(0) = 2^{2 \cdot 0 - 1} = 2^{-1} = \frac{1}{2}$

$f(1) = 2^{2 \cdot 1 - 1} = 2^1 = 2$

$f(2) = 2^{2 \cdot 2 - 1} = 2^3 = 8$

$f(-1) = 2^{2(-1)-1} = 2^{-3} = \frac{1}{8}$

$f(-2) = 2^{2(-2)-1} = 2^{-5} = \frac{1}{32}$

x	y, or f(x)
0	$\frac{1}{2}$
1	2
2	8
-1	$\frac{1}{8}$
-2	$\frac{1}{32}$

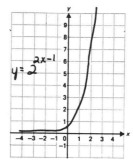

18.

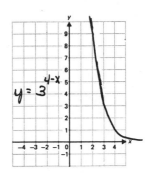

19. Graph: $y = 2^{x-1} - 3$

We construct a table of values, thinking of y as f(x). Then we plot the points and connect them with a smooth curve.

$f(0) = 2^{0-1} - 3 = 2^{-1} - 3 = \frac{1}{2} - 3 = -\frac{5}{2}$

$f(1) = 2^{1-1} - 3 = 2^{0} - 3 = 1 - 3 = -2$

$f(2) = 2^{2-1} - 3 = 2^{1} - 3 = 2 - 3 = -1$

$f(3) = 2^{3-1} - 3 = 2^{2} - 3 = 4 - 3 = 1$

$f(4) = 2^{4-1} - 3 = 2^{3} - 3 = 8 - 3 = 5$

$f(-1) = 2^{-1-1} - 3 = 2^{-2} - 3 = \frac{1}{4} - 3 = -\frac{11}{4}$

$f(-2) = 2^{-2-1} - 3 = 2^{-3} - 3 = \frac{1}{8} - 3 = -\frac{23}{8}$

x	y, or f(x)
0	$-\frac{5}{2}$
1	-2
2	-1
3	1
4	5
-1	$-\frac{11}{4}$
-2	$-\frac{23}{8}$

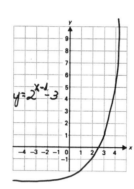

20.

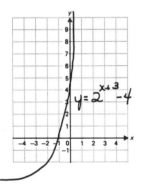

21. Graph: $x = 2^y$

We can find ordered pairs by choosing values for y and then computing values for x.

For $y = 0$, $x = 2^0 = 1$.

For $y = 1$, $x = 2^1 = 2$.

For $y = 2$, $x = 2^2 = 4$.

For $y = 3$, $x = 2^3 = 8$.

For $y = -1$, $x = 2^{-1} = \frac{1}{2^1} = \frac{1}{2}$.

For $y = -2$, $x = 2^{-2} = \frac{1}{2^2} = \frac{1}{4}$.

For $y = -3$, $x = 2^{-3} = \frac{1}{2^3} = \frac{1}{8}$.

21. (continued)

x	y
1	0
2	1
4	2
8	3
$\frac{1}{2}$	-1
$\frac{1}{4}$	-2
$\frac{1}{8}$	-3

(1) Choose values for y.

(2) Compute values for x.

We plot these points and connect them with a smooth curve.

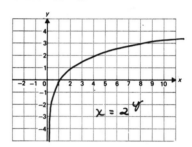

22.

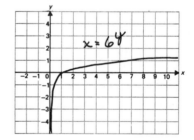

23. Graph: $x = \left(\frac{1}{2}\right)^y$

We can find ordered pairs by choosing values for y and then computing values for x. Then we plot these points and connect them with a smooth curve.

For $y = 0$, $x = \left(\frac{1}{2}\right)^0 = 1$.

For $y = 1$, $x = \left(\frac{1}{2}\right)^1 = \frac{1}{2}$.

For $y = 2$, $x = \left(\frac{1}{2}\right)^2 = \frac{1}{4}$.

For $y = 3$, $x = \left(\frac{1}{2}\right)^3 = \frac{1}{8}$.

For $y = -1$, $x = \left(\frac{1}{2}\right)^{-1} = \frac{1}{\frac{1}{2}} = 2$.

For $y = -2$, $x = \left(\frac{1}{2}\right)^{-2} = \frac{1}{\frac{1}{4}} = 4$.

For $y = -3$, $x = \left(\frac{1}{2}\right)^{-3} = \frac{1}{\frac{1}{8}} = 8$.

x	y
1	0
$\frac{1}{2}$	1
$\frac{1}{4}$	2
$\frac{1}{8}$	3
2	-1
4	-2
8	-3

23. (continued)

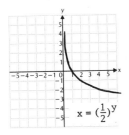

$x = (\frac{1}{2})^y$

24.

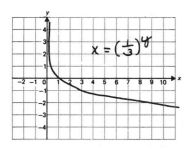

$x = (\frac{1}{3})^y$

25. Graph: $x = 5^y$

We can find ordered pairs by choosing values for y and then computing values for x. Then we plot these points and connect them with a smooth curve.

For $y = 0$, $x = 5^0 = 1.$
For $y = 1$, $x = 5^1 = 5.$
For $y = 2$, $x = 5^2 = 25.$
For $y = -1$, $x = 5^{-1} = \frac{1}{5}.$
For $y = -2$, $x = 5^{-2} = \frac{1}{25}.$

x	y
1	0
5	1
25	2
$\frac{1}{5}$	-1
$\frac{1}{25}$	-2

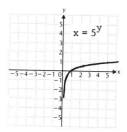

$x = 5^y$

26.

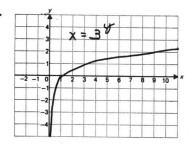

$x = 3^y$

27. Graph: $x = (\frac{2}{3})^y$

We can find ordered pairs by choosing values for y and then computing values for x. Then we plot these points and connect them with a smooth curve.

For $y = 0$, $x = (\frac{2}{3})^0 = 1.$
For $y = 1$, $x = (\frac{2}{3})^1 = \frac{2}{3}.$
For $y = 2$, $x = (\frac{2}{3})^2 = \frac{4}{9}.$
For $y = 3$, $x = (\frac{2}{3})^3 = \frac{8}{27}.$
For $y = -1$, $x = (\frac{2}{3})^{-1} = \frac{1}{\frac{2}{3}} = \frac{3}{2}.$
For $y = -2$, $x = (\frac{2}{3})^{-2} = \frac{1}{\frac{4}{9}} = \frac{9}{4}.$
For $y = -3$, $x = (\frac{2}{3})^{-3} = \frac{1}{\frac{8}{27}} = \frac{27}{8}.$

x	y
1	0
$\frac{2}{3}$	1
$\frac{4}{9}$	2
$\frac{8}{27}$	3
$\frac{3}{2}$	-1
$\frac{9}{4}$	-2
$\frac{27}{8}$	-3

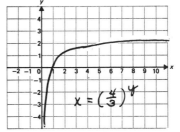

$x = (\frac{2}{3})^y$

28.

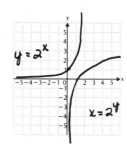

$x = (\frac{4}{3})^y$

29. Graph $y = 2^x$ (see Exercise 1) and $x = 2^y$ (see Exercise 21) using the same set of axes.

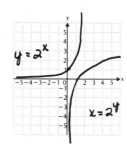

$y = 2^x$

$x = 2^y$

30.

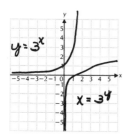

31. Graph $y = \left[\frac{1}{2}\right]^x$ (see Exercise 13) and $x = \left[\frac{1}{2}\right]^y$ (see Exercise 23) using the same set of axes.

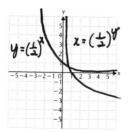

32.

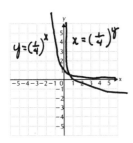

33. a) Substitute $50,000 for P and 9%, or 0.09, for i in the formula $A = P(1 + i)^t$:

$A(t) = \$50,000(1 + 0.09)^t = \$50,000(1.09)^t$

b) Substitute for t.

$A(0) = \$50,000(1.09)^0 = \$50,000(1) = \$50,000;$

$A(4) = \$50,000(1.09)^4 = \$50,000(1.41158161) \approx \$70,579.08;$

$A(8) = \$50,000(1.09)^8 \approx \$50,000(1.992562642) \approx \$99,628.13;$

$A(10) = \$50,000(1.09)^{10} \approx \$50,000(2.367363675) \approx \$118,368.18$

c) We use the function values computed in part (b) to draw the graph. Note that the axes are scaled differently because of the large numbers.

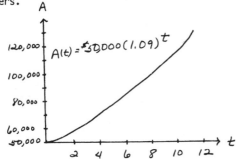

34. a) 2.6 lb, 3.7 lb, 10.9 lb, 14.6 lb; b) 22.6 lb

c)

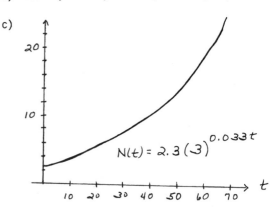

35. a) Substitute for t.

$N(0) = 250,000\left[\frac{1}{4}\right]^0 = 250,000(1) = 250,000;$

$N(1) = 250,000\left[\frac{1}{4}\right]^1 = 250,000\left[\frac{1}{4}\right] = 62,500;$

$N(4) = 250,000\left[\frac{1}{4}\right]^4 = 250,000\left[\frac{1}{256}\right] \approx 977;$

$N(10) = 250,000\left[\frac{1}{4}\right]^{10} = 250,000\left[\frac{1}{1,048,576}\right] \approx 0$

b) Use the function values computed in part (a) to draw the graph of the function.

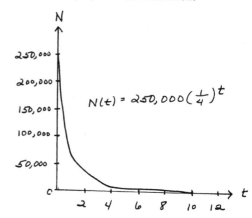

36. a) $5200, $4160, $3328, $1703.94, $558.35

b)

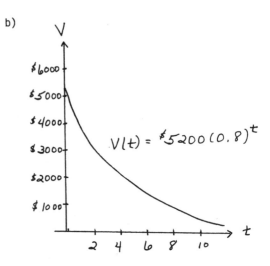

<u>37.</u> a) Keep in mind that t represents the number of years after 1985 and that N is given in millions.

For 1985, t = 0:

$N(0) = 7.5(6)^{0.5(0)} = 7.5(6)^0 = 7.5(1) =$ 7.5 million;

For 1986, t = 1:

$N(1) = 7.5(6)^{0.5(1)} = 7.5(6)^{0.5} \approx$ $7.5(2.449489743) \approx 18.4$ million;

For 1988, t = 1988 - 1985, or 3:

$N(3) = 7.5(6)^{0.5(3)} = 7.5(6)^{1.5} \approx$ $7.5(14.69693846) \approx 110.2$ million;

For 1990, t = 1990 - 1985, or 5:

$N(5) = 7.5(6)^{0.5(5)} = 7.5(6)^{2.5} \approx$ $7.5(88.18163074) \approx 661.4$ million;

For 1995, t = 1995 - 1985, or 10:

$N(10) = 7.5(6)^{0.5(10)} = 7.5(6)^5 =$ $7.5(7776) = 58,320$ million;

For 2000, t = 2000 - 1985, or 15:

$N(15) = 7.5(6)^{0.5(15)} = 7.5(6)^{7.5} \approx$ $7.5(685,700.3606) \approx$ $5,142,752.7$ million

b) Use the function values computed in part (a) to draw the graph of the function.

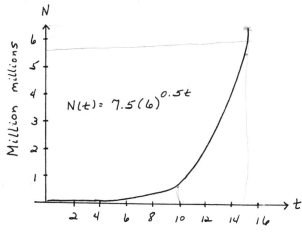

<u>38.</u> a) 4243, 6000, 8485, 12,000, 24,000

b)

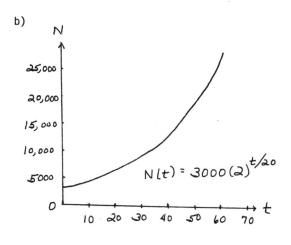

<u>39.</u> $x^{-5} \cdot x^3 = x^{-5+3} = x^{-2}$, or $\frac{1}{x^2}$

<u>40.</u> x^{-12}, or $\frac{1}{x^{12}}$

<u>41.</u> $\frac{x^{-3}}{x^4} = x^{-3-4} = x^{-7}$, or $\frac{1}{x^7}$

<u>42.</u> 1

<u>43.</u> Use a calculator.

a) $2^3 = 8$

b) $2^{3.1} = 8.574188$

c) $2^{3.14} \approx 8.815241$

d) $2^{3.141} \approx 8.821353$

e) $2^{3.1415} \approx 8.8244411$

f) $2^{3.14159} \approx 8.824962$

<u>44.</u> π^5

<u>45.</u> Since the bases are the same, the one with the larger exponent is the larger number. Thus $\pi^{2.4}$ is larger.

<u>46.</u> $8^{\sqrt{3}}$

<u>47.</u> Graph: $f(x) = (2.3)^x$

Use a calculator with a power key to construct a table of values. (We will round values of f(x) to the nearest hundredth.) Then plot these points and connect them with a smooth curve.

x	f(x)
0	1
1	2.3
2	5.29
3	12.17
-1	0.43
-2	0.19

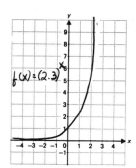

<u>48.</u>

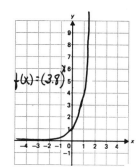

49. Graph: $g(x) = (0.125)^x$

Use the procedure described in Exercise 47, rounding values of $g(x)$ to the nearest thousandth.

x	g(x)
0	1
1	0.125
2	0.016
3	0.002
-1	8
-2	64

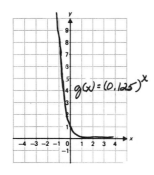

50.

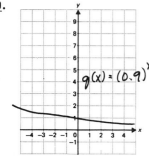

51. Graph: $y = 2^x + 2^{-x}$

Construct a table of values, thinking of y as $f(x)$. Then plot these points and connect them with a curve.

$f(0) = 2^0 + 2^{-0} = 1 + 1 = 2$

$f(1) = 2^1 + 2^{-1} = 2 + \frac{1}{2} = 2\frac{1}{2}$

$f(2) = 2^2 + 2^{-2} = 4 + \frac{1}{4} = 4\frac{1}{4}$

$f(3) = 2^3 + 2^{-3} = 8 + \frac{1}{8} = 8\frac{1}{8}$

$f(-1) = 2^{-1} + 2^{-(-1)} = \frac{1}{2} + 2 = 2\frac{1}{2}$

$f(-2) = 2^{-2} + 2^{-(-2)} = \frac{1}{4} + 4 = 4\frac{1}{4}$

$f(-3) = 2^{-3} + 2^{-(-3)} = \frac{1}{8} + 8 = 8\frac{1}{8}$

x	y, or f(x)
0	2
1	$2\frac{1}{2}$
2	$4\frac{1}{4}$
3	$8\frac{1}{8}$
-1	$2\frac{1}{2}$
-2	$4\frac{1}{4}$
-3	$8\frac{1}{8}$

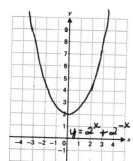

52.

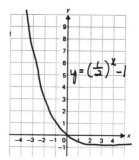

53. Graph: $y = 3^x + 3^{-x}$

We construct a table of values, thinking of y as $f(x)$. Then plot these points and connect them with a curve.

$f(0) = 3^0 + 3^{-0} = 1 + 1 = 2$

$f(1) = 3^1 + 3^{-1} = 3 + \frac{1}{3} = 3\frac{1}{3}$

$f(2) = 3^2 + 3^{-2} = 9 + \frac{1}{9} = 9\frac{1}{9}$

$f(-1) = 3^{-1} + 3^{-(-1)} = \frac{1}{3} + 3 = 3\frac{1}{3}$

$f(-2) = 3^{-2} + 3^{-(-2)} = \frac{1}{9} + 9 = 9\frac{1}{9}$

x	y, or f(x)
0	2
1	$3\frac{1}{3}$
2	$9\frac{1}{9}$
-1	$3\frac{1}{3}$
-2	$9\frac{1}{9}$

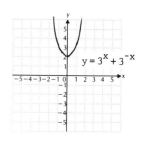

54.

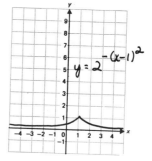

55. Graph: $y = |2x^2 - 1|$

We construct a table of values, thinking of y as f(x). Then we plot these points and connect them with a curve.

$f(0) = |2 \cdot 0^2 - 1| = |1 - 1| = 0$

$f(1) = |2 \cdot 1^2 - 1| = |2 - 1| = 1$

$f(2) = |2 \cdot 2^2 - 1| = |16 - 1| = 15$

$f(-1) = |2(-1)^2 - 1| = |2 - 1| = 1$

$f(-2) = |2(-2)^2 - 1| = |16 - 1| = 15$

x	y, or f(x)
0	0
1	1
2	15
-1	1
-2	15

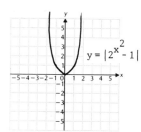

$y = |2x^2 - 1|$

56.

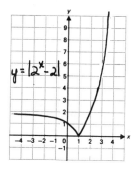

$y = |2^x - 2|$

57. $y = 3^{-(x-1)}$ $x = 3^{-(y-1)}$

x	y
0	3
1	1
2	$\frac{1}{3}$
3	$\frac{1}{9}$
-1	9

x	y
3	0
1	1
$\frac{1}{3}$	2
$\frac{1}{9}$	3
9	-1

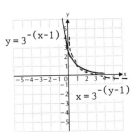

$y = 3^{-(x-1)}$

$x = 3^{-(y-1)}$

58.

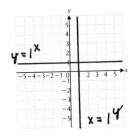

$y = 1^x$

$x = 1^y$

59. a) $S(10) = 200[1 - (0.99)^{10}] \approx 19$ words per minute;

$S(20) = 200[1 - (0.99)^{20}] \approx 36$ words per minute;

$S(40) = 200[1 - (0.99)^{40}] \approx 66$ words per minute;

$S(85) = 200[1 - (0.99)^{85}] \approx 115$ words per minute

b) Use the function values calculated in part (a) to draw the graph.

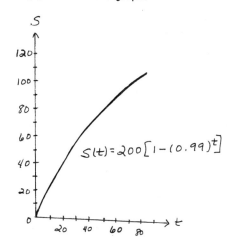

$S(t) = 200[1 - (0.99)^t]$

Exercise Set 9.2

1. $f \circ g(x) = f(g(x)) = f(2x - 1) = 3(2x - 1)^2 + 2 =$
   $3(4x^2 - 4x + 1) + 2 =$
   $12x^2 - 12x + 3 + 2 = 12x^2 - 12x + 5$

   $g \circ f(x) = g(f(x)) = g(3x^2 + 2) = 2(3x^2 + 2) - 1 =$
   $6x^2 + 4 - 1 = 6x^2 + 3$

2. $f \circ g(x) = 8x^2 - 17;$
   $g \circ f(x) = 32x^2 + 48x + 13$

3. $f \circ g(x) = f(g(x)) = f\left(\frac{2}{x}\right) = 4\left(\frac{2}{x}\right)^2 - 1 =$
   $4\left(\frac{4}{x^2}\right) - 1 = \frac{16}{x^2} - 1$

   $g \circ f(x) = g(f(x)) = g(4x^2 - 1) = \frac{2}{4x^2 - 1}$

4. $f \circ g(x) = \frac{3}{2x^2 + 3};$
   $g \circ f(x) = \frac{18}{x^2} + 3$

5. $f \circ g(x) = f(g(x)) = f(x^2 - 1) = (x^2 - 1)^2 + 1 =$
   $x^4 - 2x^2 + 1 + 1 = x^4 - 2x^2 + 2$

   $g \circ f(x) = g(f(x)) = g(x^2 + 1) = (x^2 + 1)^2 - 1 =$
   $x^4 + 2x^2 + 1 - 1 = x^4 + 2x^2$

6. $f \circ g(x) = \dfrac{1}{x^2 + 4x + 4}$;

    $g \circ f(x) = \dfrac{1}{x^2} + 2$

7. $h(x) = (5 - 3x)^2$

    This is 5 - 3x to the 2nd power, so the two most obvious functions are $f(x) = x^2$ and $g(x) = 5 - 3x$.

8. $f(x) = 4x^2 + 9$, $g(x) = 3x - 1$

9. $h(x) = (3x^2 - 7)^5$

    This is $3x^2 - 7$ to the 5th power, so the two most obvious functions are $f(x) = x^5$ and $g(x) = 3x^2 - 7$.

10. $f(x) = \sqrt{x}$, $g(x) = 5x + 2$

11. $h(x) = \dfrac{1}{x - 1}$

    This is the reciprocal of x - 1, so the two most obvious functions are $f(x) = \dfrac{1}{x}$ and $g(x) = x - 1$.

12. $f(x) = x + 4$, $g(x) = \dfrac{3}{x}$

13. $h(x) = \dfrac{1}{\sqrt{7x + 2}}$

    This is the reciprocal of the square root of 7x + 2. Two functions that can be used are $f(x) = \dfrac{1}{\sqrt{x}}$ and $g(x) = 7x + 2$.

14. $f(x) = \sqrt{x} - 3$, $g(x) = x - 7$

15. $h(x) = \dfrac{x^3 + 1}{x^3 - 1}$

    Two functions that can be used are $f(x) = \dfrac{x + 1}{x - 1}$ and $g(x) = x^3$.

16. $f(x) = x^4$, $g(x) = \sqrt{x} + 5$

17. The graph of $f(x) = 3x - 4$ is shown below.

    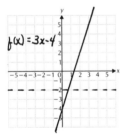

    Since there is no horizontal line that crosses the graph more than once, the function is one-to-one.

18. Yes

19. The graph of $f(x) = x^2 - 3$ is shown below.

    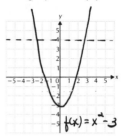

    There are many horizontal lines that cross the graph more than once. In particular, the line y = 4 crosses the graph more than once. The function is not one-to-one.

20. No

21. The graph of $g(x) = 3^x$ is shown below.

    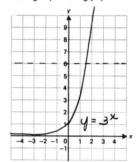

    Since no horizontal line crosses the graph more than once, the function is one-to-one.

22. Yes

23. The graph of $g(x) = |x|$ is shown below.

    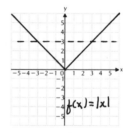

    There are many horizontal lines that cross the graph more than once. In particular, the line y = 3 crosses the graph more than once. The function is not one-to-one.

24. No

25. The graph of f(x) = |x + 3| is shown below.

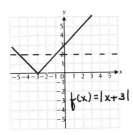

The line y = 2 is one of many lines that cross the graph more than once. The function is not one-to-one.

26. No

27. The graph of g(x) = $\frac{-2}{x}$ is shown below.

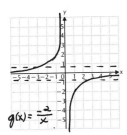

Since there is no horizontal line that crosses the graph more than once, the function is one-to-one.

28. Yes

29. a) The graph of f(x) = x + 2 is shown below. It passes the horizontal line test, so it is one-to-one.

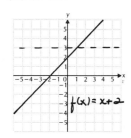

b) Replace f(x) by y: y = x + 2

Solve for x: y - 2 = x

Interchange x and y: x - 2 = y

Replace y by f⁻¹(x): f⁻¹(x) = x - 2

30. a) Yes; b) f⁻¹(x) = x - 7

31. a) The graph of f(x) = 5 - x is shown below. It passes the horizontal line test, so the function is one-to-one.

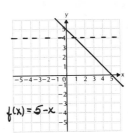

b) Replace f(x) by y: y = 5 - x

Solve for x: x = 5 - y

Interchange x and y: y = 5 - x

Replace y by f⁻¹(x): f⁻¹(x) = 5 - x

32. a) Yes; b) f⁻¹(x) = 9 - x

33. a) The graph of g(x) = x - 5 is shown below. It passes the horizontal line test, so the function is one-to-one.

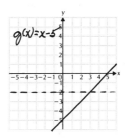

b) Replace g(x) by y: y = x - 5

Solve for x: y + 5 = x

Interchange x and y: x + 5 = y

Replace y by g⁻¹(x): g⁻¹(x) = x + 5

34. a) Yes; b) g⁻¹(x) = x + 8

35. a) The graph of f(x) = 3x is shown below. It passes the horizontal line test, so the function is one-to-one.

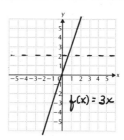

b) Replace f(x) by y: y = 3x

Solve for x: $\frac{y}{3}$ = x

Interchange x and y: $\frac{x}{3}$ = y

Replace y by f⁻¹(x): f⁻¹(x) = $\frac{x}{3}$

36. a) Yes; b) $f^{-1}(x) = \frac{x}{4}$

37. a) The graph of $g(x) = 3x + 2$ is shown below. It passes the horizontal line test, so the function is one-to-one.

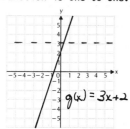

    b) Replace $g(x)$ by y: $y = 3x + 2$
    Solve for x: $y - 2 = 3x$
    $$\frac{y - 2}{3} = x$$
    Interchange variables: $\frac{x - 2}{3} = y$
    Replace y by $g^{-1}(x)$: $g^{-1}(x) = \frac{x - 2}{3}$

38. a) Yes; b) $g^{-1}(x) = \frac{x - 7}{4}$

39. a) The graph of $h(x) = \frac{4}{x + 3}$ is shown below. It passes the horizontal line test, so the function is one-to-one.

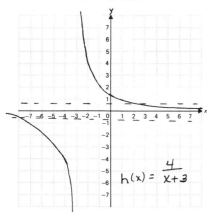

    b) Replace $h(x)$ by y: $y = \frac{4}{x + 3}$
    Solve for x: $y(x + 3) = 4$
    $$x + 3 = \frac{4}{y}$$
    $$x = \frac{4}{y} - 3, \text{ or } \frac{4 - 3y}{y}$$
    Interchange x and y: $y = \frac{4}{x} - 3, \text{ or } \frac{4 - 3x}{x}$
    Replace y by $h^{-1}(x)$: $h^{-1}(x) = \frac{4}{x} - 3,$
    $$\text{or } \frac{4 - 3x}{x}$$

40. a) Yes; b) $h^{-1}(x) = \frac{1}{x} + 8, \text{ or } \frac{1 + 8x}{x}$

41. a) The graph of $f(x) = \frac{1}{x}$ is shown below. It passes the horizontal line test, so the function is one-to-one.

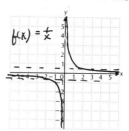

    b) Replace $f(x)$ by y: $y = \frac{1}{x}$
    Solve for x: $xy = 1$
    $$x = \frac{1}{y}$$
    Interchange x and y: $y = \frac{1}{x}$
    Replace y by $f^{-1}(x)$: $f^{-1}(x) = \frac{1}{x}$

42. a) Yes; b) $f^{-1}(x) = \frac{3}{x}$

43. a) The graph of $f(x) = \frac{2x + 1}{3}$ is shown below. It passes the horizontal line test, so the function is one-to-one.

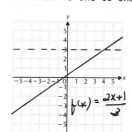

    b) Replace $f(x)$ by y: $y = \frac{2x + 1}{3}$
    Solve for x: $3y = 2x + 1$
    $$3y - 1 = 2x$$
    $$\frac{3y - 1}{2} = x$$
    Interchange x and y: $\frac{3x - 1}{2} = y$
    Replace y by $f^{-1}(x)$: $f^{-1}(x) = \frac{3x - 1}{2}$

44. a) Yes; b) $f^{-1}(x) = \frac{5x - 2}{3}$

45. a) The graph of $g(x) = \frac{x - 3}{x + 4}$ is shown below. It passes the horizontal line test, so the function is one-to-one.

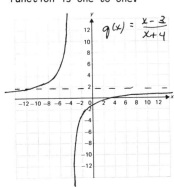

b) Replace $g(x)$ by y: $\quad y = \frac{x - 3}{x + 4}$

Solve for x: $(x + 4)y = x - 3$

$\qquad xy + 4y = x - 3$

$\qquad 3 + 4y = x - xy$

$\qquad 3 + 4y = x(1 - y)$

$\qquad \dfrac{3 + 4y}{1 - y} = x$

Interchange x and y: $\dfrac{3 + 4x}{1 - x} = y$

Replace y by $g^{-1}(x)$: $g^{-1}(x) = \dfrac{3 + 4x}{1 - x}$

46. a) Yes; b) $g^{-1}(x) = \dfrac{3x + 1}{2 - 5x}$

47. a) The graph of $f(x) = x^3 - 1$ is shown below. It passes the horizontal line test, so the function is one-to-one.

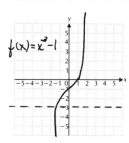

b) Replace $f(x)$ by y: $\quad y = x^3 - 1$

Solve for x: $\quad y = x^3 - 1$

$\qquad\qquad y + 1 = x^3$

$\qquad\qquad \sqrt[3]{y + 1} = x$

Interchange x and y: $\sqrt[3]{x + 1} = y$

Replace y by $f^{-1}(x)$: $f^{-1}(x) = \sqrt[3]{x + 1}$

48. a) Yes; b) $f^{-1}(x) = \sqrt[3]{x - 5}$

49. a) The graph of $g(x) = (x - 2)^3$ is shown below. It passes the horizontal line test, so the function is one-to-one.

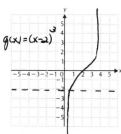

b) Replace $g(x)$ by y: $\quad y = (x - 2)^3$

Solve for x: $\quad \sqrt[3]{y} = x - 2$

$\qquad\qquad \sqrt[3]{y} + 2 = x$

Interchange x and y: $\sqrt[3]{x} + 2 = y$

Replace y by $g^{-1}(x)$: $g^{-1}(x) = \sqrt[3]{x} + 2$

50. a) Yes; b) $g^{-1}(x) = \sqrt[3]{x} - 7$

51. a) The graph of $f(x) = \sqrt[3]{x}$ is shown below. It passes the horizontal line test, so the function is one-to-one.

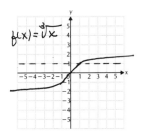

b) Replace $f(x)$ by y: $\quad y = \sqrt[3]{x}$

Solve for x: $\quad y^3 = x$

Interchange x and y: $\quad x^3 = y$

Replace y by $f^{-1}(x)$: $\quad f^{-1}(x) = x^3$

52. a) Yes; b) $f^{-1}(x) = x^3 + 4$

53. a) The graph of $f(x) = 2x^2 + 3$, $x \geqslant 0$, is shown below. It passes the horizontal line test, so the function is one-to-one.

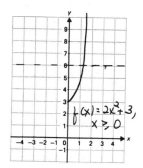

<u>53.</u> (continued)

  b) Replace f(x) by y: $y = 2x^2 + 3$
      Solve for x: $y - 3 = 2x^2$
      $$\frac{y - 3}{2} = x^2$$
      $$\sqrt{\frac{y - 3}{2}} = x$$

      (We take the principal square root since
      $x \geqslant 0.$)

      Interchange x and y: $\sqrt{\frac{x - 3}{2}} = y$

      Replace y by $f^{-1}(x)$: $f^{-1}(x) = \sqrt{\frac{x - 3}{2}}$

<u>54.</u> a) Yes; b) $f^{-1}(x) = \sqrt{\frac{x + 2}{3}}$

<u>55.</u> First graph $f(x) = \frac{1}{2}x - 3$. Then graph the inverse
      function by flipping the graph of $f(x) = \frac{1}{2}x - 3$
      over the line y = x. The graph of the inverse
      function can also be found by first finding a
      formula for the inverse and then substituting to
      find function values.

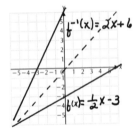

<u>56.</u>

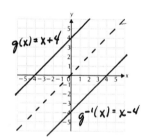

<u>57.</u> Follow the procedure described in Exercise 55
      to graph the function and its inverse.

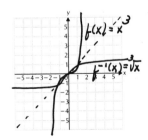

<u>58.</u>

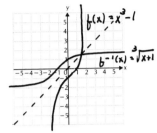

<u>59.</u> Use the procedure described in Exercise 55 to
      graph the function and its inverse.

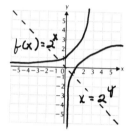

<u>60.</u>

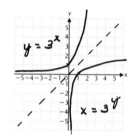

<u>61.</u> Use the procedure described in Exercise 55 to
      graph the function and its inverse.

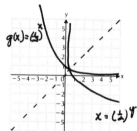

<u>62.</u>

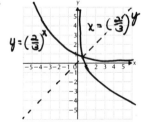

63. Use the procedure described in Exercise 55 to graph the function and its inverse.

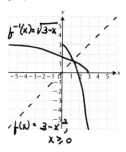

64.

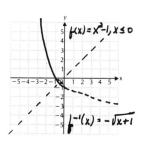

65. We check to see that $f^{-1} \circ f(x) = x$ and $f \circ f^{-1}(x) = x$.

   a) $f^{-1} \circ f(x) = f^{-1}(f(x)) = f^{-1}\left[\frac{4}{5}x\right] = \frac{5}{4} \cdot \frac{4}{5}x = x$

   b) $f \circ f^{-1}(x) = f(f^{-1}(x)) = f\left[\frac{5}{4}x\right] = \frac{4}{5} \cdot \frac{5}{4}x = x$

66. a) $f^{-1} \circ f(x) = 3\left[\frac{x + 7}{3}\right] - 7 = x + 7 - 7 = x$

   b) $f \circ f^{-1}(x) = \frac{(3x - 7) + 7}{3} = \frac{3x}{3} = x$

67. We check to see that $f^{-1} \circ f(x) = x$ and $f \circ f^{-1}(x) = x$.

   a) $f^{-1} \circ f(x) = f^{-1}(f(x)) = f^{-1}\left[\frac{1 - x}{x}\right] =$
   $\frac{1}{\frac{1 - x}{x} + 1} = \frac{1}{\frac{1 - x}{x} + 1} \cdot \frac{x}{x} = \frac{x}{1 - x + x} = \frac{x}{1} = x$

   b) $f \circ f^{-1}(x) = f(f^{-1}(x)) = f\left[\frac{1}{x + 1}\right] =$
   $\frac{1 - \frac{1}{x + 1}}{\frac{1}{x + 1}} = \frac{1 - \frac{1}{x + 1}}{\frac{1}{x + 1}} \cdot \frac{x + 1}{x + 1} = \frac{x + 1 - 1}{1} =$
   $\frac{x}{1} = x$

68. a) $f^{-1} \circ f(x) = \sqrt[3]{x^3 - 5 + 5} = \sqrt[3]{x^3} = x$

   b) $f \circ f^{-1}(x) = (\sqrt[3]{x + 5})^3 - 5 = x + 5 - 5 = x$

69. a) $f(8) = 8 + 32 = 40$

   Size 40 in France corresponds to size 8 in the U.S.

   $f(10) = 10 + 32 = 42$

   Size 42 in France corresponds to size 10 in the U.S.

   $f(14) = 14 + 32 = 46$

   Size 46 in France corresponds to size 14 in the U.S.

   $f(18) = 18 + 32 = 50$

   Size 50 in France corresponds to size 18 in the U.S.

   b) The graph of $f(x) = x + 32$ is shown below.

   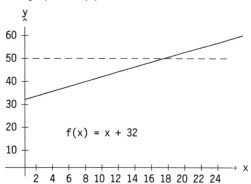

   It passes the horizontal line test, so the function is one-to-one and, hence, has an inverse that is a function. We now find a formula for the inverse.

   Replace $f(x)$ by y: $y = x + 32$

   Solve for x: $y - 32 = x$

   Interchange x and y: $x - 32 = y$

   Replace y by $f^{-1}(x)$: $f^{-1}(x) = x - 32$

   c) $f^{-1}(40) = 40 - 32 = 8$

   Size 8 in the U.S. corresponds to size 40 in France.

   $f^{-1}(42) = 42 - 32 = 10$

   Size 10 in the U.S. corresponds to size 42 in France.

   $f^{-1}(46) = 46 - 32 = 14$

   Size 14 in the U.S. corresponds to size 46 in France.

   $f^{-1}(50) = 50 - 32 = 18$

   Size 18 in the U.S. corresponds to size 50 in France.

70. a) 40, 44, 52, 60; b) $f^{-1}(x) = \frac{x - 24}{2}$;
   c) 8, 10, 14, 18

71. The graph of $f(x) = 4$ is shown below. Since the horizontal line $y = 4$ crosses the graph in more than one place, the function does not have an inverse that is a function.

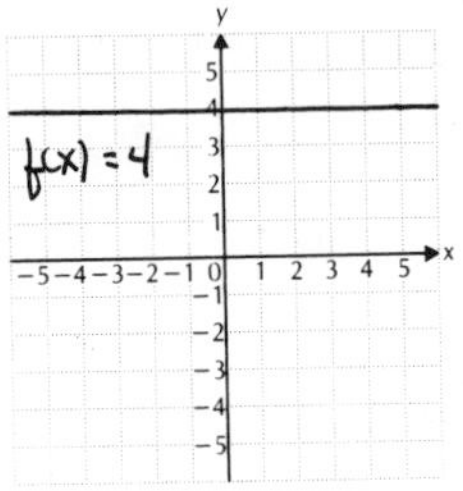

72. $C^{-1}(x) = \frac{100}{x - 5}$; $C^{-1}(x)$ gives the number of people in the group, where x is the cost per person, in dollars.

73. Check to see if $g \circ f(x) = x$ and $f \circ g(x) = x$.

a) $g \circ f(x) = g(f(x)) = g\left(\frac{1}{2}\right) = 2$

Since $g \circ f(x) \neq x$, the functions are not inverses of each other.

74. Yes

75. Check to see if $g \circ f(x) = x$ and $f \circ g(x) = x$.

a) $g \circ f(x) = g(f(x)) = g(\sqrt[4]{x}) = (\sqrt[4]{x})^4 = x$

b) $f \circ g(x) = f(g(x)) = f(x^4) = \sqrt[4]{x^4} = |x|$

For $x < 0$, $|x| = -x$ and $f \circ g(x) \neq x$.

The functions are not inverses of each other.

76. No

Exercise Set 9.3

1. Graph: $y = \log_2 x$

The equation $y = \log_2 x$ is equivalent to $2^y = x$. We can find ordered pairs by choosing values for y and computing the corresponding x-values.

For $y = 0$, $x = 2^0 = 1$.

For $y = 1$, $x = 2^1 = 2$.

For $y = 2$, $x = 2^2 = 4$.

For $y = 3$, $x = 2^3 = 8$.

For $y = -1$, $x = 2^{-1} = \frac{1}{2}$.

For $y = -2$, $x = 2^{-2} = \frac{1}{4}$.

x, or 2^y	y
1	0
2	1
4	2
8	3
$\frac{1}{2}$	-1
$\frac{1}{4}$	-2

(1) Select y.
(2) Compute x.

We plot the set of ordered pairs and connect the points with a smooth curve.

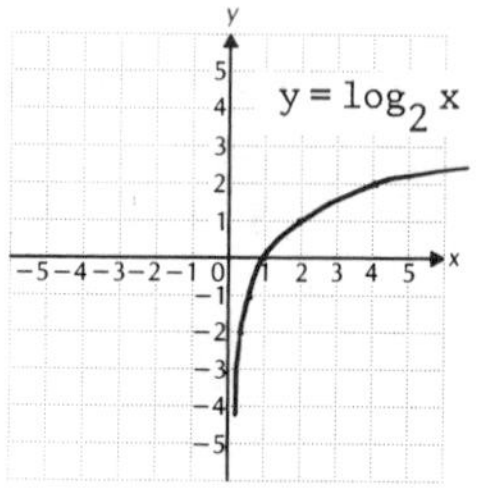

2.

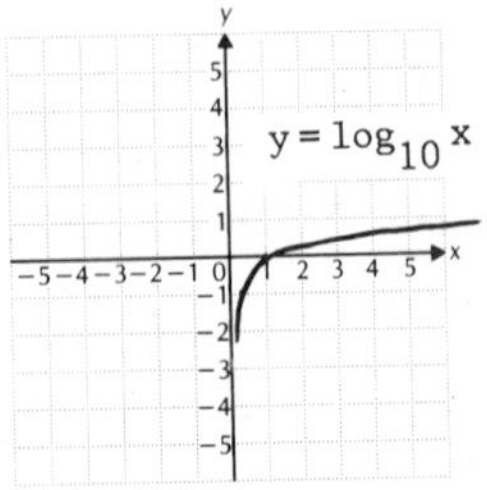

<u>3.</u> Graph: $y = \log_6 x$

The equation $y = \log_6 x$ is equivalent to $6^y = x$. We can find ordered pairs by choosing values for y and computing the corresponding x-values.

For $y = 0$, $x = 6^0 = 1$.

For $y = 1$, $x = 6^1 = 6$.

For $y = 2$, $x = 6^2 = 36$.

For $y = -1$, $x = 6^{-1} = \frac{1}{6}$.

For $y = -2$, $x = 6^{-2} = \frac{1}{36}$.

x, or 6^y	y
1	0
6	1
36	2
$\frac{1}{6}$	-1
$\frac{1}{36}$	-2

We plot the set of ordered pairs and connect the points with a smooth curve.

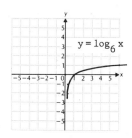

<u>4.</u>

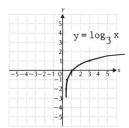

<u>5.</u> Graph: $f(x) = \log_4 x$

Think of $f(x)$ as y. Then $y = \log_4 x$ is equivalent to $4^y = x$. We find ordered pairs by choosing values for x and computing the corresponding x-values. Then we plot the points and connect them with a smooth curve.

For $y = 0$, $x = 4^0 = 1$.

For $y = 1$, $x = 4^1 = 4$.

For $y = 2$, $x = 4^2 = 16$.

For $y = -1$, $x = 4^{-1} = \frac{1}{4}$.

For $y = -2$, $x = 4^{-2} = \frac{1}{16}$.

x, or 4^y	y
1	0
4	1
16	2
$\frac{1}{4}$	-1
$\frac{1}{16}$	-2

<u>5.</u> (continued)

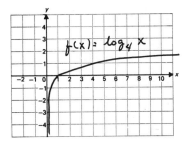

<u>6.</u>

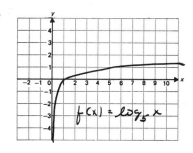

<u>7.</u> Graph: $f(x) = \log_{1/2} x$

Think of $f(x)$ as y. Then $y = \log_{1/2} x$ is equivalent to $\left(\frac{1}{2}\right)^y = x$. We construct a table of values, plot these points, and connect them with a smooth curve.

For $y = 0$, $x = \left(\frac{1}{2}\right)^0 = 1$.

For $y = 1$, $x = \left(\frac{1}{2}\right)^1 = \frac{1}{2}$.

For $y = 2$, $x = \left(\frac{1}{2}\right)^2 = \frac{1}{4}$.

For $y = -1$, $x = \left(\frac{1}{2}\right)^{-1} = 2$.

For $y = -2$, $x = \left(\frac{1}{2}\right)^{-2} = 4$.

For $y = -3$, $x = \left(\frac{1}{2}\right)^{-3} = 8$.

x, or $\left(\frac{1}{2}\right)^y$	y
1	0
$\frac{1}{2}$	1
$\frac{1}{4}$	2
2	-1
4	-2
8	-3

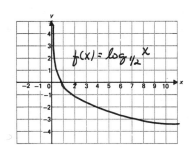

8.

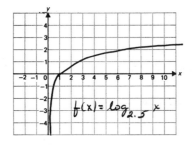

$$f(x) = \log_{2.5} x$$

9. Graph $f(x) = 3^x$ (see Exercise Set 9.1, Exercise 2) and $f^{-1}(x) = \log_3 x$ (see Exercise 4 above) on the same set of axes.

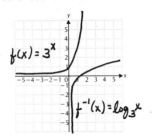

$$f(x) = 3^x$$
$$f^{-1}(x) = \log_3 x$$

10.

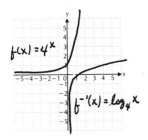

$$f(x) = 4^x$$
$$f^{-1}(x) = \log_4 x$$

11. $10^3 = 1000 \Longleftrightarrow 3 = \log_{10} 1000$ The exponent is the logarithm.
The base remains the same.

12. $2 = \log_{10} 100$

13. $5^{-3} = \dfrac{1}{125} \Longleftrightarrow -3 = \log_5 \dfrac{1}{125}$ The exponent is the logarithm.
The base remains the same.

14. $-5 = \log_4 \dfrac{1}{1024}$

15. $8^{1/3} = 2 \Longleftrightarrow \dfrac{1}{3} = \log_8 2$

16. $\dfrac{1}{4} = \log_{16} 2$

17. $10^{0.3010} = 2 \Longleftrightarrow 0.3010 = \log_{10} 2$

18. $0.4771 = \log_{10} 3$

19. $e^2 = t \Longleftrightarrow 2 = \log_e t$

20. $k = \log_p 3$

21. $Q^t = x \Longleftrightarrow t = \log_Q x$

22. $m = \log_p V$

23. $e^2 = 7.3891 \Longleftrightarrow 2 = \log_e 7.3891$

24. $3 = \log_e 20.0855$

25. $e^{-2} = 0.1353 \Longleftrightarrow -2 = \log_e 0.1353$

26. $-4 = \log_e 0.0183$

27. $t = \log_3 8 \Longleftrightarrow 3^t = 8$ The logarithm is the exponent.
The base remains the same.

28. $7^h = 10$

29. $\log_5 25 = 2 \Longleftrightarrow 5^2 = 25$ The logarithm is the exponent.
The base remains the same.

30. $6^1 = 6$

31. $\log_{10} 0.1 = -1 \Longleftrightarrow 10^{-1} = 0.1$

32. $10^{-2} = 0.01$

33. $\log_{10} 7 = 0.845 \Longleftrightarrow 10^{0.845} = 7$

34. $10^{0.4771} = 3$

35. $\log_e 20 = 2.9957 \Longleftrightarrow e^{2.9957} = 20$

36. $e^{2.3026} = 10$

37. $\log_t Q = k \Longleftrightarrow t^k = Q$

38. $m^a = P$

39. $\log_e 0.25 = -1.3863 \Longleftrightarrow e^{-1.3863} = 0.25$

40. $e^{-0.0111} = 0.989$

41. $\log_r T = -x \Longleftrightarrow r^{-x} = T$

42. $c^{-w} = M$

43. $\log_3 x = 2$
$\quad 3^2 = x$ Converting to an exponential equation
$\quad 9 = x$ Computing 3^2

44. 64

45. $\log_x 36 = 2$

$\quad\quad x^2 = 36$ Converting to an exponential equation

$x = 6$ or $x = -6$ Principle of positive and negative roots

$\log_6 36 = 2$ because $6^2 = 36$. Thus, 6 is a solution. Since all logarithm bases must be positive, $\log_{-6} 36$ is not defined and -6 is not a solution.

46. 4

47. $\log_2 x = -1$

$\quad 2^{-1} = x$ Converting to an exponential equation

$\quad \dfrac{1}{2} = x$ Simplifying

48. $\dfrac{1}{9}$

49. $\log_8 x = \dfrac{1}{3}$

$\quad 8^{1/3} = x$

$\quad\quad 2 = x$

50. 2

51. Let $\log_{10} 100 = x$.

Then $10^x = 100$

$\quad\quad 10^x = 10^2$

$\quad\quad\quad x = 2$.

Thus, $\log_{10} 100 = 2$.

52. 5

53. Let $\log_{10} 0.1 = x$.

Then $10^x = 0.1 = \dfrac{1}{10}$

$\quad\quad 10^x = 10^{-1}$

$\quad\quad\quad x = -1$.

Thus, $\log_{10} 0.1 = -1$.

54. -3

55. Let $\log_{10} 1 = x$.

Then $10^x = 1$

$\quad\quad 10^x = 10^0$ $(10^0 = 1)$

$\quad\quad\quad x = 0$.

Thus, $\log_{10} 1 = 0$.

56. 1

57. Let $\log_5 625 = x$.

Then $5^x = 625$

$\quad\quad 5^x = 5^4$

$\quad\quad\quad x = 4$.

Thus, $\log_5 625 = 4$.

58. 6

59. Let $\log_5 \dfrac{1}{25} = x$.

Then $5^x = \dfrac{1}{25}$

$\quad\quad 5^x = 5^{-2}$

$\quad\quad\quad x = -2$.

Thus, $\log_5 \dfrac{1}{25} = -2$.

60. -4

61. Let $\log_3 1 = x$.

Then $3^x = 1$

$\quad\quad 3^x = 3^0$ $(3^0 = 1)$

$\quad\quad\quad x = 0$.

Thus, $\log_3 1 = 0$.

62. 1

63. Let $\log_e e = x$.

Then $e^x = e$

$\quad\quad e^x = e^1$

$\quad\quad\quad x = 1$.

Thus, $\log_e e = 1$.

64. 0

65. Let $\log_{27} 9 = x$.

Then $27^x = 9$

$\quad (3^3)^x = 3^2$

$\quad\quad 3^{3x} = 3^2$

$\quad\quad\quad 3x = 2$

$\quad\quad\quad x = \dfrac{2}{3}$.

Thus, $\log_{27} 9 = \dfrac{2}{3}$.

66. $\dfrac{1}{3}$

67. Let $\log_e e^3 = x$.

Then $e^x = e^3$

$\quad\quad\quad x = 3$.

Thus, $\log_e e^3 = 3$.

68. -4

69. Let $\log_{10} 10^t = x$.

Then $10^x = 10^t$

$\quad\quad\quad x = t$.

Thus, $\log_{10} 10^t = t$.

70. p

71. We want to move the decimal point 3 places, between the 5 and 2, so we multiply by $10^{-3} \times 10^3$.

$\quad 5240 = 5240 \times 10^{-3} \times 10^3 = 5.24 \times 10^3$

72. 8.45×10^{-2}

73. $8^{-4} = \dfrac{1}{8^4}$, or $\dfrac{1}{4096}$

74. $\sqrt[5]{x^4}$

75. $t^{-2/3} = \dfrac{1}{t^{2/3}} = \dfrac{1}{\sqrt[3]{t^2}}$

76. 5

77. Graph: $y = \left(\dfrac{3}{2}\right)^x$ Graph: $y = \log_{3/2} x$, or $x = \left(\dfrac{3}{2}\right)^y$

x	y, or $\left(\dfrac{3}{2}\right)^x$
0	1
1	$\dfrac{3}{2}$
2	$\dfrac{9}{4}$
3	$\dfrac{27}{8}$
-1	$\dfrac{2}{3}$
-2	$\dfrac{4}{9}$

x, or $\left(\dfrac{3}{2}\right)^y$	y
1	0
$\dfrac{3}{2}$	1
$\dfrac{9}{4}$	2
$\dfrac{27}{8}$	3
$\dfrac{2}{3}$	-1
$\dfrac{4}{9}$	-2

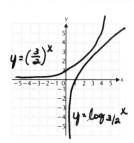

78.

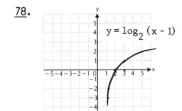

79. Graph: $y = \log_3 |x + 1|$

x	y
0	0
2	1
8	2
-2	0
-4	1
-9	2

80. $27, \dfrac{1}{27}$

81. $\log_{125} x = \dfrac{2}{3}$

$125^{2/3} = x$

$(5^3)^{2/3} = x$

$5^2 = x$

$25 = x$

82. 4

83. $\log_{\sqrt{5}} x = -3$

$(\sqrt{5})^{-3} = x$

$\dfrac{1}{(\sqrt{5})^3} = x$

$\dfrac{1}{5\sqrt{5}} = x$

$\dfrac{1}{5\sqrt{5}} \cdot \dfrac{\sqrt{5}}{\sqrt{5}} = x$

$\dfrac{\sqrt{5}}{25} = x$

84. $0, \dfrac{1}{2}$

85. $\log_4 (3x - 2) = 2$

$4^2 = 3x - 2$

$16 = 3x - 2$

$18 = 3x$

$6 = x$

86. $-\dfrac{7}{16}$

87. $\log_x \sqrt[5]{36} = \dfrac{1}{10}$

$x^{1/10} = \sqrt[5]{36}$

$x^{1/10} = 36^{1/5}$

$(x^{1/10})^{10} = (36^{1/5})^{10}$

$x = 36^2$

$x = 1296$

88. 4, -25

89. Let $\log_{1/4} \frac{1}{64} = x$.

Then $\left(\frac{1}{4}\right)^x = \frac{1}{64}$

$\left(\frac{1}{4}\right)^x = \left(\frac{1}{4}\right)^3$

$x = 3$.

Thus, $\log_{1/4} \frac{1}{64} = 3$.

90. 1

91. $\log_{10} (\log_4 (\log_3 81))$
= $\log_{10} (\log_4 4)$ ($\log_3 81 = 4$)
= $\log_{10} 1$ ($\log_4 4 = 1$)
= 0

92. 1

93. Let $\log_{\sqrt{3}} \frac{1}{81} = x$.

Then $(\sqrt{3})^x = \frac{1}{81}$

$(3^{1/2})^x = \frac{1}{3^4}$

$3^{x/2} = 3^{-4}$

$\frac{x}{2} = -4$

$x = -8$

Thus, $\log_{\sqrt{3}} \frac{1}{81} = -8$.

94. -2

Exercise Set 9.4

1. $\log_2 (32 \cdot 8) = \log_2 32 + \log_2 8$ Property 1

2. $\log_3 27 + \log_3 81$

3. $\log_4 (64 \cdot 16) = \log_4 64 + \log_4 16$ Property 1

4. $\log_5 25 + \log_5 125$

5. $\log_C Bx = \log_C B + \log_C x$ Property 1

6. $\log_t 5 + \log_t Y$

7. $\log_a 6 + \log_a 70 = \log_a (6 \cdot 70)$ Property 1

8. $\log_b (65 \cdot 2)$, or $\log_b 130$

9. $\log_C K + \log_C y = \log_C K \cdot y$ Property 1

10. $\log_t HM$

11. $\log_a x^3 = 3 \log_a x$ Property 2

12. $5 \log_b t$

13. $\log_C y^6 = 6 \log_C y$ Property 2

14. $7 \log_{10} y$

15. $\log_b C^{-3} = -3 \log_b C$ Property 2

16. $-5 \log_C M$

17. $\log_a \frac{67}{5} = \log_a 67 - \log_a 5$ Property 3

18. $\log_t T - \log_t 7$

19. $\log_b \frac{3}{4} = \log_b 3 - \log_b 4$ Property 3

20. $\log_a y - \log_a x$

21. $\log_a 15 - \log_a 7 = \log_a \frac{15}{7}$ Property 3

22. $\log_b \frac{42}{7}$, or $\log_b 6$

23. $\log_a x^2 y^3 z$
= $\log_a x^2 + \log_a y^3 + \log_a z$ Property 1
= $2 \log_a x + 3 \log_a y + \log_a z$ Property 2

24. $\log_a x + 4 \log_a y + 3 \log_a z$

25. $\log_b \frac{xy^2}{z^3}$
= $\log_b xy^2 - \log_b z^3$ Property 3
= $\log_b x + \log_b y^2 - \log_b z^3$ Property 1
= $\log_b x + 2 \log_b y - 3 \log_b z$ Property 2

26. $2 \log_b x + 5 \log_b y - 4 \log_b w - 7 \log_b z$

27. $\log_C \sqrt[3]{\frac{x^4}{y^3 z^2}}$

= $\log_C \left[\frac{x^4}{y^3 z^2}\right]^{1/3}$

= $\frac{1}{3} \log_C \frac{x^4}{y^3 z^2}$ Property 2

= $\frac{1}{3} (\log_C x^4 - \log_C y^3 z^2)$ Property 3

= $\frac{1}{3}[\log_C x^4 - (\log_C y^3 + \log_C z^2)]$ Property 1

= $\frac{1}{3}(\log_C x^4 - \log_C y^3 - \log_C z^2)$ Removing parentheses

= $\frac{1}{3}(4 \log_C x - 3 \log_C y - 2 \log_C z)$ Property 2

28. $\frac{1}{2}(6 \log_a x - 5 \log_a y - 8 \log_a z)$

29. $\log_a \sqrt[4]{\dfrac{x^8 y^{12}}{a^3 z^5}}$

    $= \dfrac{1}{4}\left[\log_a \dfrac{x^8 y^{12}}{a^3 z^5}\right]$ Property 2

    $= \dfrac{1}{4}(\log_a x^8 y^{12} - \log_a a^3 z^5)$ Property 3

    $= \dfrac{1}{4}[\log_a x^8 + \log_a y^{12} - (\log_a a^3 + \log_a z^5)]$

    Property 1

    $= \dfrac{1}{4}(\log_a x^8 + \log_a y^{12} - \log_a a^3 - \log_a z^5)$

    Removing parentheses

    $= \dfrac{1}{4}(\log_a x^8 + \log_a y^{12} - 3 - \log_a z^5)$ Property 4

    $= \dfrac{1}{4}(8 \log_a x + 12 \log_a y - 3 - 5 \log_a z)$

    Property 2

30. $\dfrac{1}{3}(6 \log_a x + 3 \log_a y - 2 - 7 \log_a z)$

31. $\dfrac{2}{3} \log_a x - \dfrac{1}{2} \log_a y$

    $= \log_a x^{2/3} - \log_a y^{1/2}$ Property 2

    $= \log_a \sqrt[3]{x^2} - \log_a \sqrt{y}$

    $= \log_a \dfrac{\sqrt[3]{x^2}}{\sqrt{y}}$ Property 3

    $= \log_a \dfrac{\sqrt[3]{x^2}\ \sqrt{y}}{y}$ Multiplying by $\dfrac{\sqrt{y}}{\sqrt{y}}$

32. $\log_a \dfrac{\sqrt{x}\ y^3}{x^2}$

33. $\log_a 2x + 3(\log_a x - \log_a y)$

    $= \log_a 2x + 3 \log_a x - 3 \log_a y$

    $= \log_a 2x + \log_a x^3 - \log_a y^3$ Property 2

    $= \log_a 2x^4 - \log_a y^3$ Property 1

    $= \log_a \dfrac{2x^4}{y^3}$ Property 3

34. $\log_a x$

35. $\log_a \dfrac{a}{\sqrt{x}} - \log_a \sqrt{ax}$

    $= \log_a ax^{-1/2} - \log_a a^{1/2}x^{1/2}$

    $= \log_a \dfrac{ax^{-1/2}}{a^{1/2}x^{1/2}}$ Property 3

    $= \log_a \dfrac{a^{1/2}}{x}$

    $= \log_a \dfrac{\sqrt{a}}{x}$

36. $\log_a (x + 2)$

37. $\log_b 15 = \log_b (3 \cdot 5)$

    $= \log_b 3 + \log_b 5$ Property 1

    $= 1.099 + 1.609$

    $= 2.708$

38. -0.51

39. $\log_b \dfrac{5}{3} = \log_b 5 - \log_b 3$ Property 3

    $= 1.609 - 1.099$

    $= 0.51$

40. -1.099

41. $\log_b \dfrac{1}{5} = \log_b 1 - \log_b 5$ Property 3

    $= 0 - 1.609$ $(\log_b 1 = 0)$

    $= -1.609$

42. $\dfrac{1}{2}$

43. $\log_b \sqrt{b^3} = \log_b b^{3/2} = \dfrac{3}{2}$ Property 4

44. 2.099

45. $\log_b 5b = \log_b 5 + \log_b b$ Property 1

    $= 1.609 + 1$ $(\log_b b = 1)$

    $= 2.609$

46. 2.198

47. $\log_b 25 = \log_b 5^2$

    $= 2 \log_b 5$ Property 2

    $= 2(1.609)$

    $= 3.218$

48. 4.317

49. $\log_t t^9 = 9$ Property 4

50. 4

51. $\log_e e^m = m$ Property 4

52. -2

53. $\log_3 3^4 = x$

    $4 = x$ Property 4

54. 7

55. $\log_e e^x = -7$

    $x = -7$ Property 4

56. 2.7

57. $i^{29} = i^{28} \cdot i = (i^4)^7 \cdot i = 1^7 \cdot i = 1 \cdot i = i$

58. 5

59. $\dfrac{2 + i}{2 - i} = \dfrac{2 + i}{2 - i} \cdot \dfrac{2 + i}{2 + i} = \dfrac{4 + 4i + i^2}{4 - i^2} = \dfrac{4 + 4i - 1}{4 - (-1)} =$

    $\dfrac{3 + 4i}{5} = \dfrac{3}{5} + \dfrac{4}{5}i$

60. $23 - 18i$

61. $\log_a (x^8 - y^8) - \log_a (x^2 + y^2)$

= $\log_a \dfrac{x^8 - y^8}{x^2 + y^2}$ Property 3

= $\log_a \dfrac{(x^4 + y^4)(x^2 + y^2)(x + y)(x - y)}{x^2 + y^2}$

                                                Factoring

= $\log_a [(x^4 + y^4)(x^2 - y^2)]$ Simplifying

= $\log_a (x^6 - x^4y^2 + x^2y^4 - y^6)$ Multiplying

62. $\log_a (x^3 + y^3)$

63. $\log_a \sqrt{1 - s^2}$

= $\log_a (1 - s^2)^{1/2}$

= $\dfrac{1}{2} \log_a (1 - s^2)$

= $\dfrac{1}{2} \log_a [(1 - s)(1 + s)]$

= $\dfrac{1}{2} \log_a (1 - s) + \dfrac{1}{2} \log_a (1 + s)$

64. $\dfrac{1}{2} \log_a (c - d) - \dfrac{1}{2} \log_a (c + d)$

65. $\log_a \dfrac{\sqrt[3]{x^2z}}{\sqrt[3]{y^2z^{-2}}}$

= $\log_a \left[\dfrac{x^2z^3}{y^2}\right]^{1/3}$

= $\dfrac{1}{3} (\log_a x^2z^3 - \log_a y^2)$

= $\dfrac{1}{3} (2 \log_a x + 3 \log_a z - 2 \log_a y)$

= $\dfrac{1}{3} [2 \cdot 2 + 3 \cdot 4 - 2 \cdot 3]$ Substituting

= $\dfrac{1}{3}(10)$

= $\dfrac{10}{3}$

66. $-2, 0$

67. $\log_a 5x = \log_a 5 + \log_a x$

$\log_a 5x = \log_a 5x$ Property 1

We get an equation that is true for all values of x for which the logarithm function is defined. Thus, the solution is $\{x | x > 0\}$, or $(0, \infty)$.

68. -2

69. $\log_a x = 2$ Given

       $a^2 = x$ Definition

Let $\log_{1/a} x = n$ and solve for n.

       $\log_{1/a} a^2 = n$ Substituting a^2 for x

       $\left[\dfrac{1}{a}\right]^n = a^2$

       $(a^{-1})^n = a^2$

       $a^{-n} = a^2$

       $-n = 2$

       $n = -2$

Thus, $\log_{1/a} x = -2$ when $\log_a x = 2$.

70.   False

71.   The statement is false. For example, let $a = 10$, $P = 100$, and $Q = 10$. Then

$\dfrac{\log_a P}{\log_a Q} = \dfrac{\log_{10} 100}{\log_{10} 10} = \dfrac{2}{1} = 2$, but

$\log_a P - \log_a Q = \log_{10} 100 - \log_{10} 10 = 2 - 1 = 1$.

72.   True

73.   The statement is false. For example, let $a = 3$ and $x = 9$. Then

$\log_a 3x = \log_3 3 \cdot 9 = \log_3 27 = 3$, but

$3 \log_a x = 3 \log_3 9 = 3 \cdot 2 = 6$.

74.   False

75.   The statement is true, by Property 2.

76. $\log_a \left[\dfrac{1}{x}\right] = \log_a x^{-1} = -1 \cdot \log_a x = -\log_a x$

77. $\log_a \left[\dfrac{x + \sqrt{x^2 - 3}}{3}\right] =$

$\log_a \left[\dfrac{x + \sqrt{x^2 - 3}}{3} \cdot \dfrac{x - \sqrt{x^2 - 3}}{x - \sqrt{x^2 - 3}}\right] =$

$\log_a \left[\dfrac{3}{3(x - \sqrt{x^2 - 3})}\right] = \log_a \left[\dfrac{1}{x - \sqrt{x^2 - 3}}\right] =$

$\log_a 1 - \log_a (x - \sqrt{x^2 - 3}) =$

$0 - \log_a (x - \sqrt{x^2 - 3}) = -\log_a (x - \sqrt{x^2 - 3})$

Exercise Set 9.5

1.   0.3010

2.   0.6990

3.   0.9031

4.   1.0414

5.   0.8021

6. 0.7007

7. 1.6532

8. 1.8692

9. 1.7952

10. 1.0569

11. 2.6405

12. 2.4698

13. 4.1271

14. 4.9689

15. -0.2441

16. -0.1612

17. -1.2840

18. -0.4123

19. -2.2069

20. -2.3161

21. 1000

22. 100,000

23. 501.1872

24. 6.3096×10^{14}

25. 3.0001

26. 1.1623

27. 0.2841

28. 0.4567

29. 0.0011

30. 79,104.2833

31. 0.6931

32. 1.0986

33. 2.0794

34. 2.4849

35. 4.1271

36. 3.4012

37. 8.3814

38. 6.8037

39. -5.0832

40. -7.2225

41. 36.7890

42. 138.5457

43. 0.0023

44. 0.1002

45. 1.0057

46. 1.0112

47. 5.8346×10^{14}

48. 2.0917×10^{24}

49. 8.1490

50. 9.1083

51. -3.3496

52. -5.3645

53. 1637.9488

54. 547.7396

55. 7.6331

56. 0.2520

57. We will use common logarithms for the conversion. Let a = 10, b = 6, and M = 100 and substitute into the change-of-base formula.

$$\log_b M = \frac{\log_a M}{\log_a b}$$

$$\log_6 100 = \frac{\log_{10} 100}{\log_{10} 6}$$

$$\approx \frac{2}{0.7782}$$

$$\approx 2.5702$$

58. 2.6309

59. We will use common logarithms for the conversion. Let a = 10, b = 2, and M = 10 and substitute in the change-of-base formula.

$$\log_2 10 = \frac{\log_{10} 10}{\log_{10} 2}$$

$$\approx \frac{1}{0.3010}$$

$$\approx 3.3219$$

60. 2.0104

61. We will use natural logarithms for the conversion. Let a = e, b = 200, and M = 30 and substitute in the change-of-base formula.

$$\log_{200} 30 = \frac{\ln 30}{\ln 200}$$

$$\approx \frac{3.4012}{5.2983}$$

$$\approx 0.6419$$

62. 0.7386

63. We will use natural logarithms for the conversion. Let a = e, b = 0.5, and M = 5 and substitute in the change-of-base formula.

$$\log_{0.5} 5 = \frac{\ln 5}{\ln 0.5}$$

$$\approx \frac{1.6094}{-0.6931}$$

$$\approx -2.3219$$

64. -0.4771

65. We will use common logarithms for the conversion. Let a = 10, b = 2, and M = 0.2 and substitute in the change-of-base formula.

$$\log_2 0.2 = \frac{\log_{10} 0.2}{\log_{10} 2}$$

$$\approx \frac{-0.6990}{0.3010}$$

$$\approx -2.3219$$

66. -3.6439

67. We will use natural logarithms for the conversion. Let a = e, b = π, and M = 58 and substitute in the change-of-base formula.

$$\log_\pi 58 = \frac{\ln 58}{\ln \pi}$$

$$\approx \frac{4.0604}{1.1447}$$

$$\approx 3.5471$$

68. 4.6284

69. $ax^2 - b = 0$

$$ax^2 = b$$

$$x^2 = \frac{b}{a}$$

$$x = \pm\sqrt{\frac{b}{a}}$$

The solution is $\pm\sqrt{\dfrac{b}{a}}$.

70. $0, \dfrac{b}{a}$

71. $x^{1/2} - 6x^{1/4} + 8 = 0$

Let $u = x^{1/4}$.

$$u^2 - 6u + 8 = 0 \qquad \text{Substituting}$$

$$(u - 4)(u - 2) = 0$$

| u = | 4 | or | u = 2 |

$x^{1/4} =$ 4   or $x^{1/4} = 2$

x = 256   or   x = 16   Raising both sides to the fourth power

Both numbers check. The solutions are 256 and 16.

72. $\dfrac{1}{4}$, 9

73. Use the change-of-base formula with a = 10 and b = e. We obtain

$$\ln M = \frac{\log M}{\log e}.$$

74. $\log M = \dfrac{\ln M}{\ln 10}$

75.

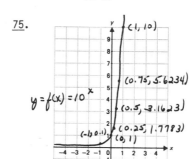

76.

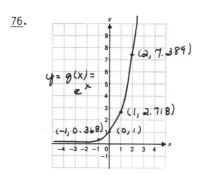

77.

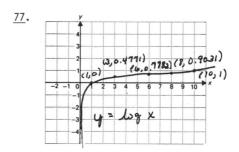

78.

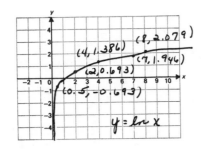

$y = \ln x$

79. $\dfrac{\log_3 8}{\log_3 5} = \log_5 8$ Change-of-base formula

80. $\log_{16} 47$

81.
$$\log 374x = 4.2931$$
$$\log 374 + \log x = 4.2931 \qquad \text{Property 1}$$
$$2.572871602 + \log x \approx 4.2931$$
$$\log x \approx 1.720228398$$
$$x \approx 52.5084 \qquad \text{Finding the antilogarithm}$$

82. 3.3112

83.
$$\log 692 + \log x = \log 3450$$
$$\log x = \log 3450 - \log 692$$
$$\log x = \log \frac{3450}{692} \qquad \text{Property 3}$$
$$x = \frac{3450}{692}$$
$$x \approx 4.9855$$

84. 1.5893

Exercise Set 9.6

1. $2^x = 8$
$2^x = 2^3$
$x = 3$ The exponents are the same.

2. 4

3. $4^x = 256$
$4^x = 4^4$
$x = 4$ The exponents are the same.

4. 3

5. $2^{2x} = 32$
$2^{2x} = 2^5$
$2x = 5$
$x = \dfrac{5}{2}$

6. 1

7. $3^{5x} = 27$
$3^{5x} = 3^3$
$5x = 3$
$x = \dfrac{3}{5}$

8. $\dfrac{4}{7}$

9. $2^x = 9$
$\log 2^x = \log 9$ Taking the common logarithm on both sides
$x \log 2 = \log 9$ Property 2
$x = \dfrac{\log 9}{\log 2}$ Solving for x
$x \approx 3.170$ Using a calculator

10. 4.907

11. $2^x = 10$
$\log 2^x = \log 10$ Taking the common logarithm on both sides
$x \log 2 = \log 10$ Property 2
$x = \dfrac{\log 10}{\log 2}$ Solving for x
$x \approx 3.322$ Using a calculator

12. 5.044

13. $5^{4x-7} = 125$
$5^{4x-7} = 5^3$
$4x - 7 = 3$ The exponents are the same.
$4x = 10$
$x = \dfrac{10}{4}, \text{ or } \dfrac{5}{2}$

14. -1

15. $3^{x^2} \cdot 3^{4x} = \dfrac{1}{27}$
$3^{x^2+4x} = 3^{-3}$
$x^2 + 4x = -3$
$x^2 + 4x + 3 = 0$
$(x + 3)(x + 1) = 0$
$x = -3 \text{ or } x = -1$

16. $\dfrac{1}{2}, -3$

17. $4^x = 7$
$\log 4^x = \log 7$
$x \log 4 = \log 7$
$x = \dfrac{\log 7}{\log 4}$
$x \approx \dfrac{0.8451}{0.6021}$
$x \approx 1.404$

18. 1.107

19. $e^t = 100$
$\ln e^t = \ln 100$ Taking ln on both sides
$t = \ln 100$ Property 4
$t \approx 4.605$ Using a calculator

20. 6.908

21. $e^{-t} = 0.1$
$\ln e^{-t} = \ln 0.1$ Taking ln on both sides
$-t = \ln 0.1$ Property 4
$-t \approx -2.303$
$t \approx 2.303$

22. 4.605

23. $e^{-0.02t} = 0.06$
$\ln e^{-0.02t} = \ln 0.06$ Taking ln on both sides
$-0.02t = \ln 0.06$ Property 4
$t = \dfrac{\ln 0.06}{-0.02}$
$t \approx \dfrac{-2.8134}{-0.02}$
$t \approx 140.671$

24. 9.902

25. $2^x = 3^{x-1}$
$\log 2^x = \log 3^{x-1}$
$x \log 2 = (x - 1) \log 3$
$x \log 2 = x \log 3 - \log 3$
$\log 3 = x \log 3 - x \log 2$
$\log 3 = x(\log 3 - \log 2)$
$\dfrac{\log 3}{\log 3 - \log 2} = x$
$\dfrac{0.4771}{0.4771 - 0.3010} \approx x$
$-2.710 \approx x$

26. 7.452

27. $(2.8)^x = 41$
$\log (2.8)^x = \log 41$
$x \log 2.8 = \log 41$
$x = \dfrac{\log 41}{\log 2.8}$
$x \approx \dfrac{1.6128}{0.4472}$
$x \approx 3.607$

28. 3.581

29. $20 - (1.7)^x = 0$
$20 = (1.7)^x$
$\log 20 = \log (1.7)^x$
$\log 20 = x \log 1.7$
$\dfrac{\log 20}{\log 1.7} = x$
$\dfrac{1.3010}{0.2304} \approx x$
$5.646 \approx x$

30. 3.210

31. $\log_3 x = 3$
$x = 3^3$ Writing an equivalent exponential expression
$x = 27$

32. 625

33. $\log_2 x = -3$
$x = 2^{-3}$ Writing an equivalent exponential expression
$x = \dfrac{1}{8}$

34. 2

35. $\log x = 1$ The base is 10.
$x = 10^1$
$x = 10$

36. 1000

37. $\log x = -2$ The base is 10.
$x = 10^{-2}$
$x = \dfrac{1}{100}$

38. $\dfrac{1}{1000}$

39. $\ln x = 2$
$x = e^2 \approx 7.389$

40. $e \approx 2.718$

41. $\ln x = -1$
$x = e^{-1}$
$x = \dfrac{1}{e} \approx 0.368$

42. $\dfrac{1}{e^3} \approx 0.050$

43. $\log_5 (2x - 7) = 3$
$\quad 2x - 7 = 5^3$
$\quad 2x - 7 = 125$
$\quad\quad 2x = 132$
$\quad\quad\; x = 66$
The answer checks. The solution is 66.

44. $-\dfrac{25}{6}$

45. $\log x + \log (x - 9) = 1$ The base is 10.
$\quad \log_{10} [x(x - 9)] = 1$ Property 1
$\quad\quad x(x - 9) = 10^1$
$\quad\quad x^2 - 9x = 10$
$\quad x^2 - 9x - 10 = 0$
$\quad (x - 10)(x + 1) = 0$

$x = 10 \quad \text{or} \quad x = -1$

Check: For 10:
$$\frac{\log x + \log (x - 9) = 1}{\log 10 + \log (10 - 9) \;\big|\; 1}$$
$$\log 10 + \log 1 \;\big|$$
$$1 + 0 \;\big|$$
$$1 \;\big|$$

For -1:
$$\frac{\log x + \log (x - 9) = 1}{\log (-1) + \log (-1 - 9) \;\big|\; 1}$$

The number -1 does not check, because negative numbers do not have logarithms. The solution is 10.

46. 1

47. $\log x - \log (x + 3) = -1$ The base is 10.
$\quad \log_{10} \dfrac{x}{x + 3} = -1$ Property 3
$\quad\quad \dfrac{x}{x + 3} = 10^{-1}$
$\quad\quad \dfrac{x}{x + 3} = \dfrac{1}{10}$
$\quad\quad 10x = x + 3$
$\quad\quad 9x = 3$
$\quad\quad x = \dfrac{1}{3}$

The answer checks. The solution is $\dfrac{1}{3}$.

48. 1

49. $\log_2 (x + 1) + \log_2 (x - 1) = 3$
$\quad \log_2 [(x + 1)(x - 1)] = 3$ Property 1
$\quad\quad (x + 1)(x - 1) = 2^3$
$\quad\quad\quad x^2 - 1 = 8$
$\quad\quad\quad x^2 = 9$
$\quad\quad\quad x = \pm 3$
The number 3 checks, but -3 does not. The solution is 3.

50. $\dfrac{83}{15}$

51. $\log_4 (x + 6) - \log_4 x = 2$
$\quad \log_4 \dfrac{x + 6}{x} = 2$ Property 3
$\quad\quad \dfrac{x + 6}{x} = 4^2$
$\quad\quad \dfrac{x + 6}{x} = 16$
$\quad\quad x + 6 = 16x$
$\quad\quad 6 = 15x$
$\quad\quad \dfrac{2}{5} = x$

The answer checks. The solution is $\dfrac{2}{5}$.

52. 4

53. $\log_4 (x + 3) + \log_4 (x - 3) = 2$
$\quad \log_4 [(x + 3)(x - 3)] = 2$ Property 1
$\quad\quad (x + 3)(x - 3) = 4^2$
$\quad\quad x^2 - 9 = 16$
$\quad\quad x^2 = 25$
$\quad\quad x = \pm 5$
The number 5 checks, but -5 does not. The solution is 5.

54. $\sqrt{41}$

55. $(125x^7 y^{-2} z^6)^{-2/3} =$
$(5^3)^{-2/3}(x^7)^{-2/3}(y^{-2})^{-2/3}(z^6)^{-2/3} =$
$5^{-2} x^{-14/3} y^{4/3} z^{-4} = \dfrac{1}{25} x^{-14/3} y^{4/3} z^{-4}$, or
$\dfrac{y^{4/3}}{25 x^{14/3} z^4}$

56. $-i$

57. $E = mc^2$
$\dfrac{E}{m} = c^2$
$\sqrt{\dfrac{E}{m}} = c$ Taking the principal square root

58. $\pm 10, \pm 2$

59.
$$8^X = 16^{3X+9}$$
$$(2^3)^X = (2^4)^{3X+9}$$
$$2^{3X} = 2^{12X+36}$$
$$3x = 12x + 36$$
$$-36 = 9x$$
$$-4 = x$$

60. $\dfrac{12}{5}$

61.
$$\log_6 (\log_2 x) = 0$$
$$\log_2 x = 6^0$$
$$\log_2 x = 1$$
$$x = 2^1$$
$$x = 2$$

62. $\sqrt[3]{3}$

63.
$$\log \sqrt[3]{x} = \sqrt{\log x}$$
$$\log x^{1/3} = \sqrt{\log x}$$
$$\tfrac{1}{3} \log x = \sqrt{\log x} \qquad \text{Property 3}$$
$$\tfrac{1}{9} (\log x)^2 = \log x \qquad \text{Squaring both sides}$$
$$\tfrac{1}{9} (\log x)^2 - \log x = 0$$

Let $u = \log x$.
$$\tfrac{1}{9}u^2 - u = 0$$
$$u\left[\tfrac{1}{9}u - 1\right] = 0$$
$$u = 0 \quad \text{or} \quad \tfrac{1}{9}u - 1 = 0$$
$$u = 0 \quad \text{or} \quad \tfrac{1}{9}u = 1$$
$$u = 0 \quad \text{or} \quad u = 9$$
$$\log x = 0 \quad \text{or} \quad \log x = 9 \qquad \begin{array}{l}\text{Substituting } \log x \\ \text{for } u\end{array}$$
$$x = 10^0 \quad \text{or} \quad x = 10^9$$
$$x = 1 \quad \text{or} \quad x = 10^9$$

Both answers check. The solutions are 1 and 10^9.

64. 1, 10^{16}

65.
$$\log_5 \sqrt{x^2 + 1} = 1$$
$$\sqrt{x^2 + 1} = 5^1$$
$$x^2 + 1 = 25 \qquad \text{Squaring both sides}$$
$$x^2 = 24$$
$$x = \pm\sqrt{24}, \text{ or } \pm 2\sqrt{6}$$

Both numbers check. The solutions are $\pm 2\sqrt{6}$.

66. $\pm\dfrac{\sqrt{2}}{4}$

67.
$$\log_5 \sqrt{x^2 - 9} = 1$$
$$\sqrt{x^2 - 9} = 5^1$$
$$x^2 - 9 = 25 \qquad \text{Squaring both sides}$$
$$x^2 = 34$$
$$x = \pm\sqrt{34}$$

Both numbers check. The solutions are $\pm\sqrt{34}$.

68. -1

69.
$$\log (\log x) = 5 \qquad \text{The base is 10.}$$
$$\log x = 10^5$$
$$\log x = 100{,}000$$
$$x = 10^{100{,}000}$$

The number checks. The solution is $10^{100{,}000}$.

70. $-3, -1$

71.
$$\log x^2 = (\log x)^2$$
$$2 \log x = (\log x)^2$$
$$0 = (\log x)^2 - 2 \log x$$

Let $u = \log x$.
$$0 = u^2 - 2u$$
$$0 = u(u - 2)$$
$$u = 0 \quad \text{or} \quad u = 2$$
$$\log x = 0 \quad \text{or} \quad \log x = 2$$
$$x = 10^0 \quad \text{or} \quad x = 10^2$$
$$x = 1 \quad \text{or} \quad x = 100$$

Both numbers check. The solutions are 1 and 100.

72. 625, -625

73.
$$\log x^{\log x} = 25$$
$$\log x (\log x) = 25 \qquad \text{Property 2}$$
$$(\log x)^2 = 25$$
$$\log x = \pm 5$$
$$x = 10^5 \quad \text{or} \quad x = 10^{-5}$$
$$x = 100{,}000 \quad \text{or} \quad x = \frac{1}{100{,}000}$$

Both numbers check. The solutions are 100,000 and $\dfrac{1}{100{,}000}$.

74. $\dfrac{1}{2}$, 5000

75.
$$\log_a a^{x^2+4x} = 21$$
$$x^2 + 4x = 21 \qquad \text{Property 4}$$
$$x^2 + 4x - 21 = 0$$
$$(x + 7)(x - 3) = 0$$
$$x = -7 \quad \text{or} \quad x = 3$$

Both numbers check. The solutions are -7 and 3.

76. a, $\frac{1}{a}$

77.
$$x^{\log_{10} x} = \frac{x^{-4}}{1000}$$

$$x^{\log_{10} x} \cdot x^4 = \frac{1}{1000} \qquad \text{Multiplying by } x^4$$

$$x^{\log_{10} x + 4} = \frac{1}{1000} \qquad \text{Adding exponents}$$

$$\log_{10} x^{\log_{10} x + 4} = \log_{10} \frac{1}{1000}$$

$$(\log_{10} x + 4)\log_{10} x = -3$$

$$(\log_{10} x)^2 + 4\log_{10} x + 3 = 0$$

Let $u = \log_{10} x$.

$$u^2 + 4u + 3 = 0$$

$$(u + 3)(u + 1) = 0$$

$$u = -3 \quad \text{or} \quad u = -1$$

$$\log_{10} x = -3 \quad \text{or} \quad \log_{10} x = -1$$

$$x = 10^{-3} \quad \text{or} \quad x = 10^{-1}$$

$$x = \frac{1}{1000} \quad \text{or} \quad x = \frac{1}{10}$$

Both numbers check. The solutions are $\frac{1}{1000}$ and $\frac{1}{10}$.

78. 1.465, 1

79.
$$(81^{x-2})(27^{x+1}) = 9^{2x-3}$$

$$\left[(3^4)^{x-2}\right]\left[(3^3)^{x+1}\right] = (3^2)^{2x-3}$$

$$(3^{4x-8})(3^{3x+3}) = 3^{4x-6}$$

$$3^{7x-5} = 3^{4x-6}$$

$$7x - 5 = 4x - 6$$

$$3x = -1$$

$$x = -\frac{1}{3}$$

80. $\frac{1}{2}$

81.
$$3^{2x} - 3^{2x-1} = 18$$

$$3^{2x}(1 - 3^{-1}) = 18 \qquad \text{Factoring}$$

$$3^{2x}\left[1 - \frac{1}{3}\right] = 18$$

$$3^{2x}\left[\frac{2}{3}\right] = 18$$

$$3^{2x} = 27 \qquad \text{Multiplying by } \frac{3}{2}$$

$$3^{2x} = 3^3$$

$$2x = 3$$

$$x = \frac{3}{2}$$

82. 38

83. $\log_5 125 = 3$ and $\log_{125} 5 = \frac{1}{3}$, so
$$x = \left(\log_{125} 5\right)^{\log_5 125} \text{ is equivalent to}$$
$$x = \left(\frac{1}{3}\right)^3 = \frac{1}{27}. \text{ Then } \log_3 x = \log_3 \frac{1}{27} = -3.$$

Exercise Set 9.7

1. a) We set A(t) = \$450,000 and solve for t:
$$450,000 = 50,000(1.06)^t$$

$$\frac{450,000}{50,000} = (1.06)^t$$

$$9 = (1.06)^t$$

$$\log 9 = \log (1.06)^t \qquad \begin{array}{l}\text{Taking the common} \\ \text{logarithm on both} \\ \text{sides}\end{array}$$

$$\log 9 = t \log 1.06 \qquad \text{Property 2}$$

$$t = \frac{\log 9}{\log 1.06} \approx \frac{0.95424}{0.02531} \approx 37.7$$

It will take about 37.7 years for the \$50,000 to grow to \$450,000.

b) We set A(t) = \$100,000 and solve for t:
$$100,000 = 50,000(1.06)^t$$

$$2 = (1.06)^t$$

$$\log 2 = \log (1.06)^t$$

$$\begin{array}{l}\text{Taking the common} \\ \text{logarithm on both sides}\end{array}$$

$$\log 2 = t \log 1.06 \qquad \text{Property 2}$$

$$t = \frac{\log 2}{\log 1.06} \approx \frac{0.30103}{0.02531} \approx 11.9$$

The doubling time is about 11.9 years.

2. a) 59.7 years;   b) 19.1 years

3. a) We set N(t) = 60,000 and solve for t:
$$60,000 = 250,000\left[\frac{1}{4}\right]^t$$

$$\frac{60,000}{250,000} = \left[\frac{1}{4}\right]^t$$

$$0.24 = (0.25)^t \qquad \left[\frac{1}{4} = 0.25\right]$$

$$\log 0.24 = \log (0.25)^t$$

$$\log 0.24 = t \log 0.25$$

$$t = \frac{\log 0.24}{\log 0.25} \approx \frac{-0.61979}{-0.60206} \approx 1.0$$

After about 1 year 60,000 cans will still be in use.

3. (continued)

b) We set N(t) = 10 and solve for t.

$$10 = 250,000 \left[\frac{1}{4}\right]^t$$

$$\frac{10}{250,000} = \left[\frac{1}{4}\right]^t$$

$$0.00004 = (0.25)^t$$

$$\log 0.00004 = \log (0.25)^t$$

$$\log 0.00004 = t \log 0.25$$

$$t = \frac{\log 0.00004}{\log 0.25} \approx \frac{-4.39794}{-0.60206} \approx 7.3$$

After about 7.3 years only 10 cans will still be in use.

4. a) 6.6 years; b) 3.1 years

5. a) One billion is 1000 millions, so we set N(t) = 1000 and solve for t:

$$1000 = 7.5(6)^{0.5t}$$

$$\frac{1000}{7.5} = (6)^{0.5t}$$

$$\log \frac{1000}{7.5} = \log(6)^{0.5t}$$

$$\log 1000 - \log 7.5 = 0.5t \log 6$$

$$t = \frac{\log 1000 - \log 7.5}{0.5 \log 6}$$

$$t \approx \frac{3 - 0.87506}{0.5(0.77815)} \approx 5.5$$

After about 5.5 years, one billion compact discs will be sold in a year.

b) When t = 0, N(t) = 7.5(6)$^{0.5(0)}$ = 7.5(6)^0 = 7.5(1) = 7.5. Twice this initial number is 15, so we set N(t) = 15 and solve for t:

$$15 = 7.5(6)^{0.5t}$$

$$2 = (6)^{0.5t}$$

$$\log 2 = \log(6)^{0.5t}$$

$$\log 2 = 0.5t \log 6$$

$$t = \frac{\log 2}{0.5 \log 6} \approx \frac{0.30103}{0.5(0.77815)} \approx 0.8$$

The doubling time is about 0.8 year.

6. a) 86.4 minutes;

b) 300.5 minutes;

c) 20 minutes

7. a) S(0) = 68 - 20 log (0 + 1) = 68 - 20 log 1 = 68 - 20(0) = 68%

b) S(4) = 68 - 20 log (4 + 1) = 68 - 20 log 5 ≈ 68 - 20(0.69897) ≈ 54%

S(24) = 68 - 20 log (24 + 1) = 68 - 20 log 25 ≈ 68 - 20 (1.39794) ≈ 40%

7. (continued)

c) Using the values we computed in parts (a) and (b) and any others we wish to calculate, we sketch the graph:

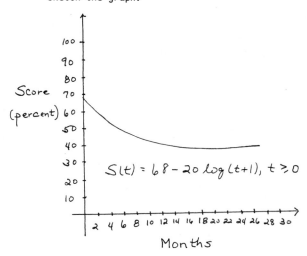

d) We set S(t) = 50 and solve for t:

$$50 = 68 - 20 \log (t + 1)$$

$$-18 = -20 \log (t + 1)$$

$$0.9 = \log (t + 1)$$

$$10^{0.9} = t + 1 \quad \text{Using the definintion of logarithms or taking the antilogarithm}$$

$$7.9 \approx t + 1$$

$$6.9 \approx t$$

After about 6.9 months, the average score was 50.

8. a) 78%; b) 68%, 57%

c)

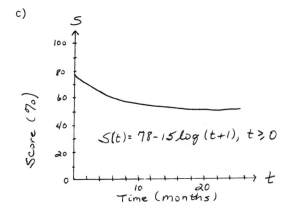

d) About 1584 months

9. a) N(1) = 1000 + 200 log 1 = 1000 + 200(0) = 1000 units

    b) N(5) = 1000 + 200 log 5 ≈ 1000 + 200(0.6990) ≈ 1140 units

    c) Using the values computed in parts (a) and (b) and any others we wish to calculate, we can sketch the graph:

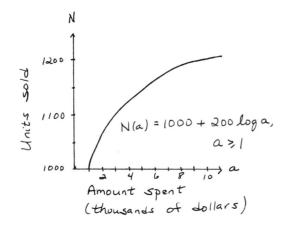

    d) Set N(a) = 1276 and solve for a.

        1276 = 1000 + 200 log a

        276 = 200 log a

        1.38 = log a

        $10^{1.38}$ = a   Using the definition of logarithms or taking the antilogarithm

        24 ≈ a

        About $24,000 would have to be spent.

10. a) 2000;

    b) 2452

    c)

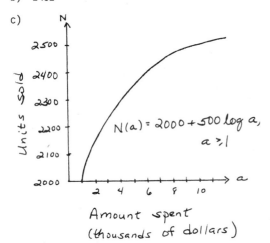

    d) $1,000,000,000

11. We substitute 6.3×10^{-7} for H^+ in the formula for pH.

    pH = -log [H^+] = -log [6.3×10^{-7}] =
    -[log 6.3 + log 10^{-7}] ≈ -[0.7993 - 7] =
    -[-6.2007] = 6.2007 ≈ 6.2

    The pH of a common brand of mouthwash is about 6.2.

12. 7.4

13. We substitute 1.6×10^{-8} for H^+ in the formula for pH.

    pH = -log [H^+] = -log [1.6×10^{-8}] =
    -[log 1.6 + log 10^{-8}] ≈ -[0.2041 - 8] =
    -[-7.7959] = 7.7959 ≈ 7.8

    The pH of eggs is about 7.8.

14. 4.2

15. We substitute 7 for pH in the formula and solve for H^+.

        7 = -log [H^+]

        -7 = log [H^+]

        10^{-7} = H^+ Using the definition of logarithm or taking the antilogarithm

    The hydrogen ion concentration of tap water is 10^{-7} moles/liter.

16. 4.0×10^{-6} moles/liter

17. We substitute 3.2 for pH in the formula and solve for H^+.

        3.2 = -log [H^+]

        -3.2 = log [H^+]

        $10^{-3.2}$ = H^+ Using the definition of logarithm or taking the antilogarithm

        0.00063 ≈ H^+

        6.3×10^{-4} ≈ H^+ Writing in scientific notation

    The hydrogen ion concentration of orange juice is about 6.3×10^{-4} moles/liter.

18. 1.6×10^{-5} moles/liter

19. We substitute into the formula.

    R = log $\dfrac{10^{8.25} I_0}{I_0}$ = log $10^{8.25}$ = 8.25

20. 5

21. We substitute into the formula.

    L = 10 log $\dfrac{2510 I_0}{I_0}$ = 10 log 2510 ≈ 10(3.399674) ≈ 34 decibels

22. 64 decibels

23. We substitute into the formula.

$$L = 10 \log \frac{10^6 \, I_0}{I_0} = 10 \log 10^6 = 10(6) =$$
60 decibels

24. 90 decibels

25. a) Substitute 100 for S(t) and solve for t.

$$100 = 200[1 - (0.99)^t]$$
$$0.5 = 1 - (0.99)^t$$
$$(0.99)^t = 0.5$$
$$\log(0.99)^t = \log 0.5$$
$$t \log 0.99 = \log 0.5$$
$$t = \frac{\log 0.5}{\log 0.99} \approx \frac{-0.3010}{-0.0044} \approx 69$$

Kristen's speed will be 100 words per minute after she has studied typing for about 69 hours.

b) Substitute 150 for S(t) and solve for t.

$$150 = 200[1 - (0.99)^t]$$
$$0.75 = 1 - (0.99)^t$$
$$(0.99)^t = 0.25$$
$$\log(0.99)^t = \log 0.25$$
$$t \log 0.99 = \log 0.25$$
$$t = \frac{\log 0.25}{\log 0.99} \approx \frac{-0.6021}{-0.0044} \approx 138$$

She studied about 138 hours.

26. a) Limits that ensure S(t) ≥ 0 (about 2511 months in Exercise 7 and 158,488 months in Exercise 8)

b) No. The limits exceed the human life-span.

Exercise Set 9.8

1. Graph: $f(x) = e^x$

We find some function values with a calculator. We use these values to plot points and draw the graph.

x	e^x
0	1
1	2.7
2	7.4
3	20.1
-1	0.4
-2	0.1

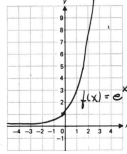

2.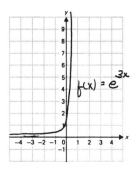

3. Graph: $f(x) = e^{-3x}$

We find some function values, plot points, and draw the graph.

x	e^{-3x}
0	1
1	0.05
2	0.002
-1	20.1
-2	403.4

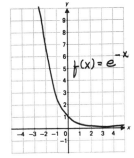

4.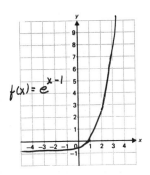

5. Graph: $f(x) = e^{x-1}$

We find some function values, plot points, and draw the graph.

x	e^{x-1}
0	0.4
1	1
2	2.7
3	7.4
4	20.1
-1	0.1
-2	0.05

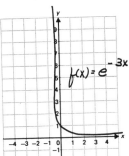

<u>6.</u>

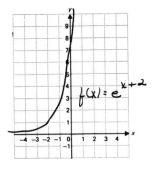

$f(x) = e^x + 2$

<u>7.</u> Graph: $f(x) = e^{-x} - 3$

We find some function values, plot points, and draw the graph.

x	$e^{-x} - 3$
0	-2
1	-2.6
2	-2.9
3	-3.0
-1	-0.3
-2	4.4
-3	17.1

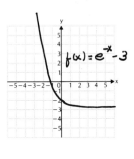

$f(x) = e^{-x} - 3$

<u>8.</u>

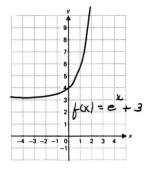

$f(x) = e^x + 3$

<u>9.</u> Graph: $f(x) = 5e^{0.2x}$

We find some function values, plot points, and draw the graph.

x	$5e^{0.2x}$
0	5
1	6.1
2	7.5
3	9.1
4	11.1
-1	4.1
-2	3.3
-3	2.7
-4	2.2

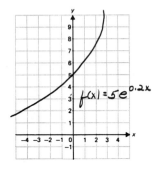

$f(x) = 5e^{0.2x}$

<u>10.</u>

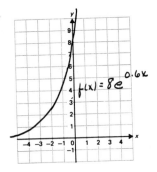

$f(x) = 8e^{0.6x}$

<u>11.</u> Graph: $f(x) = 20e^{-0.5x}$

We find some function values, plot points, and draw the graph.

x	$20e^{-0.5x}$
0	20
1	12.1
2	7.4
3	4.5
4	2.7
-1	33.0
-2	54.4
-3	89.6
-4	147.8

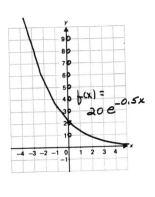

$f(x) = 20e^{-0.5x}$

<u>12.</u>

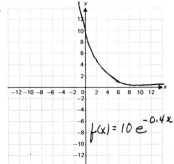

$f(x) = 10e^{-0.4x}$

<u>13.</u> Graph: $f(x) = \ln(x + 4)$

x	$\ln(x + 4)$
0	1.4
1	1.6
2	1.8
3	1.9
4	2.1
-1	1.1
-2	0.7
-3	0
-4	Undefined

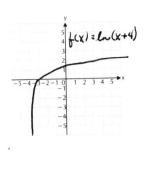

$f(x) = \ln(x + 4)$

14.

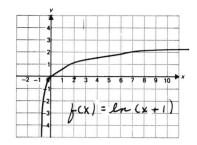

$f(x) = \ln (x + 1)$

18.

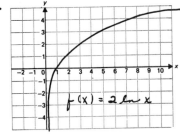

$f(x) = 2 \ln x$

15. Graph: $f(x) = 2 - \ln x$

x	2 - ln x
0.5	2.7
1	2
2	1.3
3	0.9
4	0.6
6	0.2
8	-0.1
10	-0.3

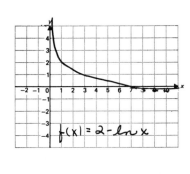

$f(x) = 2 - \ln x$

19. Graph: $f(x) = \ln (x - 2)$

x	ln (x - 2)
2.5	-0.7
3	0
4	0.7
5	1.1
7	1.6
9	1.9
10	2.1

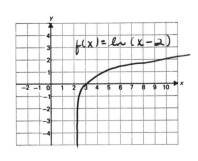

$f(x) = \ln (x - 2)$

16.

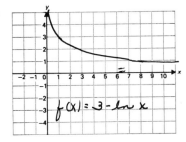

$f(x) = 3 - \ln x$

20.

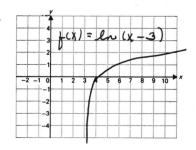

$f(x) = \ln (x - 3)$

17. Graph: $f(x) = 3 \ln x$

x	3 ln x
0.5	-2.1
1	0
2	2.1
3	3.3
4	4.2
5	4.8
6	5.4

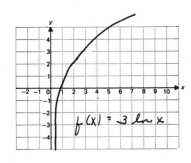

$f(x) = 3 \ln x$

21. Graph: $f(x) = \frac{1}{2} \ln x$

x	$\frac{1}{2}$ ln x
0.5	-0.3
1	0
2	0.3
5	0.8
7	0.97
8	1.04
10	1.15

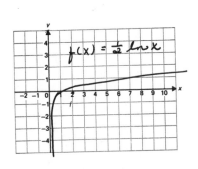

$f(x) = \frac{1}{2} \ln x$

22.

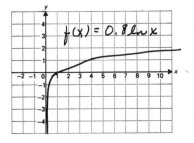

$f(x) = 0.8 \ln x$

23. Graph: $f(x) = \ln x - 3$

x	ln x - 3
0.5	-3.7
1	-2
2	-2.3
3	-1.9
4	-1.6
5	-1.4
6	-1.2

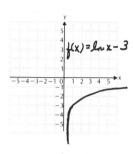

24.

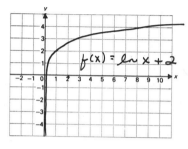

25. Graph: $f(x) = 1 - e^{-x}$

x	1 - e^{-x}
-2	-6.4
-1	-1.7
0	0
1	0.6
3	0.95
5	0.99
6	0.998

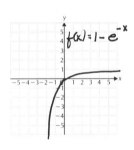

26.

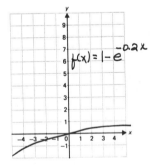

27. We substitute 175 for P in the function for walking speed, since P is in thousands.

$$R(P) = 0.37 \ln P + 0.05$$
$$R(175) = 0.37 \ln 175 + 0.05$$
$$\approx 0.37(5.1648) + 0.05 \quad \text{Finding } \ln 175 \text{ on a calculator}$$
$$\approx 1.96 \text{ ft/sec}$$

28. 0.91 ft/sec

29. We substitute 50.4 for P in the function, since P is in thousands.

$$R(50.4) = 0.37 \ln 50.4 + 0.05$$
$$\approx 0.37(3.9200) + 0.05$$
$$\approx 1.50 \text{ ft/sec}$$

30. 1.86 ft/sec

31. a) The equation $P(t) = P_0 e^{kt}$ can be used to model population growth. At t = 0 (1987), the population was 5.0 billion. We substitute 5 for P_0 and 2.8%, or 0.028, for k to obtain the exponential growth function:

$$P(t) = 5e^{0.028t}$$

b) In 1996, t = 1996 - 1987, or 9. To find the population in 1996, we substitute 9 for t:

$$P(9) = 5e^{0.028(9)}$$
$$= 5e^{0.252}$$
$$\approx 5(1.2866)$$
$$\approx 6.4$$

We can predict that the population will be about 6.4 billion in 1996.

In 2000, t = 2000 - 1987, or 13. To find the population in 2000, we substitute 13 for t:

$$P(13) = 5e^{0.028(13)}$$
$$= 5e^{0.364}$$
$$\approx 5(1.4391)$$
$$\approx 7.2$$

We can predict that the population will be about 7.2 billion in 2000.

c) We set $P(t) = 6$ and solve for t:

$$6 = 5e^{0.028t}$$
$$1.2 = e^{0.028t}$$
$$\ln 1.2 = \ln e^{0.028t}$$
$$\ln 1.2 = 0.028t \qquad \text{Property 4}$$
$$\frac{\ln 1.2}{0.028} = t$$
$$\frac{0.1823}{0.028} \approx t$$
$$7 \approx t$$

The population will be 6.0 billion about 7 years after 1987, or in 1994.

d) Using the ordered pairs found in parts (b) and (c) and any others we wish to compute, we graph the function:

31. (continued)

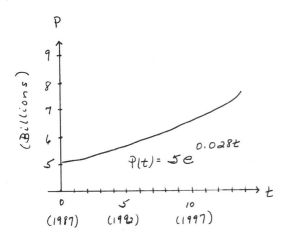

$$P(t) = 5e^{0.028t}$$

32. a) $C(t) = \$100\,e^{0.06t}$, where t is the number of years after 1967

    b) $536.56

    c) $724.27

    d)

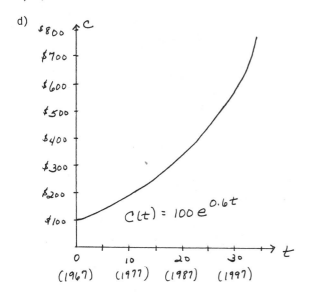

$$C(t) = 100\,e^{0.6t}$$

33. a) We can use the function $C(t) = C_0\,e^{kt}$ as a model. At $t = 0$(1962), the cost was 5¢, or $0.05. We substitute 0.05 for C_0 and 9.7%, or 0.097, for k:

    $$C(t) = 0.05\,e^{0.097t}$$

    b) In 1993, t = 1993 − 1962, or 3.1. We substitute 31 for t:

    $$C(31) = 0.05\,e^{0.097(31)}$$
    $$= 0.05\,e^{3.007}$$
    $$\approx 0.05(20.2266)$$
    $$\approx 1.01$$

    In 1993 a Hershey bar will cost about $1.01.

33. (continued)

    In 2000, t = 2000 − 1962, or 38. We substitute 38 for t:

    $$C(38) = 0.05\,e^{0.097(38)}$$
    $$= 0.05\,e^{3.686}$$
    $$\approx 0.05(39.8850)$$
    $$\approx 1.99$$

    In 2000 a Hershey bar will cost about $1.99.

    c) We set $C(t) = 5$ and solve for t:

    $$5 = 0.05\,e^{0.097t}$$
    $$100 = e^{0.097t}$$
    $$\ln 100 = \ln e^{0.097t}$$
    $$\ln 100 = 0.097t$$
    $$\frac{\ln 100}{0.097} = t$$
    $$\frac{4.6052}{0.097} \approx t$$
    $$47 \approx t$$

    A Hershey bar will cost $5 about 47 years after 1962, or in 2009.

    d) Using the ordered pairs found in parts (b) and (c) and any others we wish to compute, we draw the graph:

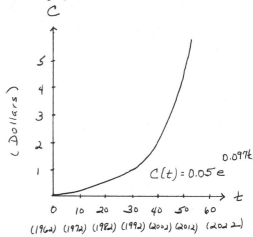

$$C(t) = 0.05\,e^{0.097t}$$

    e) To find the doubling time, we set $C(t) = 2(\$0.05)$, or $0.10 and solve for t:

    $$0.1 = 0.05\,e^{0.097t}$$
    $$2 = e^{0.097t}$$
    $$\ln 2 = \ln e^{0.097t}$$
    $$\ln 2 = 0.097t$$
    $$\frac{\ln 2}{0.097} = t$$

    (Note: We could also have used the expression relating the growth rate k and doubling time T: $T = \frac{\ln 2}{k}$.)

    $$\frac{0.6931}{0.097} \approx t$$
    $$7.1 \approx t$$

    The doubling time is about 7.1 years.

34. a) $P(t) = 100\ e^{0.117t}$

b) 227

c)

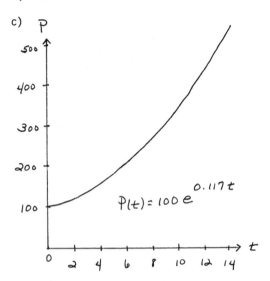

d) 5.9 days

35. a) We plot the ordered pairs given in the table and sketch the following graph. We draw a curve as close to the data points as possible. It is very close to the graph of an exponential function.

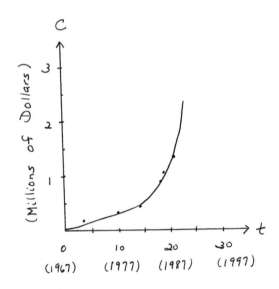

b) The exponential growth function is $C(t) = C_0\ e^{kt}$. Substituting 80 for C_0, we have

$$C(t) = 80\ e^{kt},$$

where t is the number of years after 1967.

35. (continued)

To find k using the data point C(21) = $1350 thousand, we substitute $1350 for C(t) and 21 for t and solve for k:

$$1350 = 80\ e^{k(21)}$$

$$16.875 = e^{21k}$$

$$\ln 16.875 = \ln e^{21k}$$

$$\ln 16.875 = 21k$$

$$\frac{\ln 16.875}{21} = k$$

$$\frac{2.8258}{21} \approx k$$

$$0.1346 \approx k$$

The exponential growth function for the cost of a Super Bowl commercial is

$$C(t) = 80\ e^{0.1346t}.$$

c) The year 1995 is 28 years from 1967. We let $t = 28$ and find C(28):

$$C(28) = 80\ e^{0.1346(28)} \approx 3466$$

The cost of a 60-second commercial will be about $3466 thousand, or $3,466,000 in 1995.

d) We set C(t) = 3000 ($3000 thousand is $3,000,000) and solve for t:

$$3000 = 80\ e^{0.1346t}$$

$$37.5 = e^{0.1346t}$$

$$\ln 37.5 = \ln e^{0.1346t}$$

$$\ln 37.5 = 0.1346t$$

$$\frac{\ln 37.5}{0.1346} = t$$

$$\frac{3.6243}{0.1346} \approx t$$

$$27 \approx t$$

The cost of a commercial will be $3,000,000 about 27 years after 1967, or in 1994.

e) We can set C(t) = 160 and solve for t, or we can use the expression relating growth rate k and doubling time T, $T = \frac{\ln 2}{k}$. We will use the latter.

$$T = \frac{\ln 2}{0.1346} \approx \frac{0.6931}{0.1346} \approx 5.1 \text{ years}$$

36. a) $k \approx 0.03$, $P(t) = 52\ e^{0.03t}$

b) $1.07

c) 23 years

d) 2028

37. a) Substitute 0.09 for k:
$$P(t) = P_0\, e^{0.09t}$$

b) To find the balance after one year, set $P_0 = 1000$ and $t = 1$. We find $P(1)$:

$P(1) = 1000\, e^{0.09(1)} = 1000\, e^{0.09} \approx$
$1000(1.094174284) \approx \$1094.17$

To find the balance after 2 years, set $P_0 = 1000$ and $t = 2$. We find $P(2)$:

$P(2) = 1000\, e^{0.09(2)} = 1000\, e^{0.18} \approx$
$1000(1.197217363) \approx \$1197.22$

c) We will use the expression relating growth rate k and doubling time T. (We could also set $P(t) = 2000$ and solve for t.)

$$T = \frac{\ln 2}{0.09} \approx \frac{0.6931}{0.09} \approx 7.7 \text{ years}$$

38. a) $P(t) = P_0\, e^{0.1t}$

b) $\$22,103.42$, $\$24,428.06$

c) 6.9 years

39. We will use the expression relating growth rate k and doubling time T:
$$T = \frac{\ln 2}{k}.$$

Substitute 1%, or 0.01, for k:

$$T = \frac{\ln 2}{0.01} \approx \frac{0.6931}{0.01} \approx 69.3 \text{ years}$$

40. 19.8 years

41. We will use the expression relating growth rate k and doubling time T:
$$k = \frac{\ln 2}{T}.$$

Substitute 8 for T:

$$k = \frac{\ln 2}{8} \approx \frac{0.6931}{8} \approx 0.087$$

The interest rate is about 8.7%.

42. 7.4%

43. a) The exponential growth function is $V(t) = V_0\, e^{kt}$. We will express $V(t)$ in thousands of dollars and t as the number of years after 1947. Since $V_0 = 84$ thousand, we have

$$V(t) = 84\, e^{kt}.$$

In 1987 ($t = 40$), we know that $V(t) = 53,900$ thousand. We substitute and solve for k.

$$53,900 = 84\, e^{k(40)}$$
$$\frac{1925}{3} = e^{40k}$$
$$\ln \frac{1925}{3} = \ln e^{40k}$$
$$\ln \frac{1925}{3} = 40k$$
$$\frac{\ln \dfrac{1925}{3}}{40} = k$$
$$\frac{6.4641}{40} \approx k$$
$$0.16 \approx k$$

The exponential growth rate is about 0.16, or 16%. The exponential growth function is $V(t) = 84\, e^{0.16t}$, where V is in thousands of dollars and t is the number of years after 1947.

b) In 1997, $t = 50$. We find $V(50)$.
$$V(50) = 84\, e^{0.16(50)} = 84\, e^8 \approx$$
$$84(2980.9580) \approx \$250,400$$

c) We will use the expression relating growth rate k and doubling time T. (We could also set $V(t) = 168$ and solve for t.)

$$T = \frac{\ln 2}{k} \approx \frac{0.6931}{0.16} \approx 4.3 \text{ years}$$

d) $1 billion = $1,000,000 thousand. We set $V(t) = 1,000,000$ and solve for t.
$$1,000,000 = 84\, e^{0.16t}$$
$$\frac{250,000}{21} = e^{0.16t}$$
$$\ln \frac{250,000}{21} = \ln e^{0.16t}$$
$$\ln \frac{250,000}{21} = 0.16t$$
$$\frac{\ln \dfrac{250,000}{21}}{0.16} = t$$
$$\frac{9.3847}{0.16} \approx t$$
$$58.7 \approx t$$

The value of the painting will be $1 billion about 58.7 years after 1947.

44. a) $k \approx 0.25$, $V(t) = 17.5\, e^{0.25t}$, where t is the number of years after 1983.

b) $\$351.50$, $\$1226.84$

c) 2.8 years

d) 2002

45. Using the expression relating growth rate k and doubling time T, we substitute 10%, or 0.1, for k:

$$T = \frac{\ln 2}{k} = \frac{0.6931}{0.1} \approx 7$$

The demand will be double that of 1990 about 7 years after 1990, or 1997.

46. 2007

47. a) The exponential growth function with $P_0 = 844{,}401$ is $P(t) = 844{,}401\ e^{kt}$, where t is the number of years after 1970. To find k we substitute 12 for t and 943,848 for P(t) and solve for k.

$$943{,}848 = 844{,}401\ e^{k(12)}$$

$$\frac{943{,}848}{844{,}401} = e^{12k}$$

$$\ln \frac{943{,}848}{844{,}401} = \ln e^{12k}$$

$$\ln \frac{943{,}848}{844{,}401} = 12k$$

$$\frac{\ln \frac{943{,}848}{844{,}401}}{12} = k$$

$$\frac{0.1113}{12} \approx k$$

$$0.0093 \approx k$$

The value of k is 0.0093, or 0.93%. The exponential growth function is

$$P(t) = 844{,}401\ e^{0.0093t},$$

where t is the number of years after 1970.

b) In 1996, t = 1996 - 1970, or 26. We find P(26).

$$P(26) = 844{,}401\ e^{0.0093(26)} = 844{,}401\ e^{0.2418} \approx 844{,}401(1.2735) \approx 1{,}075{,}378$$

In 1996, the population of Dallas will be about 1,075,378.

48. a) $k \approx 0.0035$, $P(t) = 623{,}988\ e^{0.0035t}$, where t is the number of years after 1970.

b) 693,070

49. We will use the function derived in Example 11:

$$P(t) = P_0\ e^{-0.00012t}$$

If the tusk has lost 20% of its carbon-14 from an initial amount P_0, then 80% (P_0) is the amount present. To find the age of the tusk t, we substitute 80% (P_0), or $0.8P_0$, for P(t) in the function above and solve for t.

$$0.8P_0 = P_0\ e^{-0.00012t}$$

$$0.8 = e^{-0.00012t}$$

$$\ln 0.8 = \ln e^{-0.00012t}$$

$$-0.2231 \approx -0.00012t$$

$$t \approx \frac{-0.2231}{-0.00012} \approx 1860$$

The tusk is about 1860 years old.

50. 878 years

51. The function $P(t) = P_0\ e^{-kt}$, k > 0, can be used to model decay. For iodine-131, k = 9.6%, or 0.096. to find the half-life we substitute 0.096 for k and $\frac{1}{2}\ P_0$ for P(t), and solve for t.

$$\frac{1}{2}\ P_0 = P_0\ e^{-0.096t},\ \text{or}\ \frac{1}{2} = e^{-0.096t}$$

$$\ln \frac{1}{2} = \ln e^{-0.096t} = -0.096t$$

$$t = \frac{\ln 0.5}{-0.096} \approx \frac{-0.6931}{-0.096} \approx 7.2\ \text{days}$$

52. 11 years

53. We use the function $P(t) = P_0\ e^{-kt}$, k > 0. When t = 3, $P(t) = \frac{1}{2}\ P_0$. We substitute and solve for k.

$$\frac{1}{2}\ P_0 = P_0\ e^{-k(3)},\ \text{or}\ \frac{1}{2} = e^{-3k}$$

$$\ln \frac{1}{2} = \ln e^{-3k} = -3k$$

$$k = \frac{\ln 0.5}{-3} \approx \frac{-0.6931}{-3} \approx 0.23$$

The decay rate is 0.23, or 23%, per minute.

54. 3.2% per year

55. a) The value of k in the function is 0.006, so the animal loses 0.6% of its weight each day.

b) Find W(30).

$$W(30) = W_0\ e^{-0.006(30)} = W_0\ e^{-0.18} \approx 0.835\ W_0$$

After 30 days, about 0.835, or 83.5%, of its initial weight W_0 remains.

56. a) 5.6 watts; b) 115.5 days; c) 268 days; d) 50 watts

57. a) Substitute 14.7 for P_0 and 2000 for a.
$$P = 14.7\ e^{-0.00005(2000)} = 14.7\ e^{-0.1} \approx 14.7(0.9048) \approx 13.3\ \text{lb/in}^2$$

b) Substitute 14.7 for P_0 and 30,000 for a.
$$P = 14.7\ e^{-0.00005(30{,}000)} = 14.7\ e^{-1.5} \approx 14.7(0.2231) \approx 3.3\ \text{lb/in}^2$$

c) Substitute 1.47 for P and 14.7 for P_0, and solve for a.
$$1.47 = 14.7\ e^{-0.00005a}$$
$$0.1 = e^{-0.00005a}$$
$$\ln 0.1 = \ln e^{-0.00005a}$$
$$\ln 0.1 = -0.00005a$$
$$a = \frac{\ln 0.1}{-0.00005} \approx \frac{-2.3026}{-0.00005} \approx 46{,}052\ \text{ft}$$

58. a) $28,000;

b) $3789.39

<u>59</u>.   Set $S(x) = D(x)$, and solve for x.

$$e^x = 162,755 \ e^{-x}$$

$$e^{2x} = 162,755 \quad \text{Multiplying by } e^x \text{ on both sides}$$

$$\ln e^{2x} = \ln 162,755$$

$$2x = \ln 162,755$$

$$x = \frac{\ln 162,755}{2}$$

$$x \approx \frac{12.0000}{2} \approx 6$$

To find the second coordinate of the equilibrium point, find $S(6)$ or $D(6)$. We will find $S(6)$.

$$S(6) = e^6 \approx 403$$

The equilibrium point is (6,$403).

<u>60</u>.   Measure the atmospheric pressure P at the top of building. Substitute that value in the equation, and solve for the height, or altitude, a. (Note: We assume that the base of the Empire State Building is essentially at sea level.)

Exercise Set 10.1

1. $d = \sqrt{(x_2 - x_1)^2 + (y_2 - y_1)^2}$ Distance formula

    $d = \sqrt{[1 - (-3)]^2 + [1 - (-2)]^2}$ Substituting

    $d = \sqrt{4^2 + 3^2}$

    $d = \sqrt{25}$

    $d = 5$

2. $3\sqrt{5} \approx 6.708$

3. $d = \sqrt{(x_2 - x_1)^2 + (y_2 - y_1)^2}$ Distance formula

    $d = \sqrt{(3 - 0)^2 + [-4 - (-7)]^2}$ Substituting

    $d = \sqrt{3^2 + 3^2}$

    $d = \sqrt{18}$

    $d = 3\sqrt{2} \approx 4.243$ Simplifying and approximating

4. $4\sqrt{2} \approx 5.657$

5. $d = \sqrt{(x_2 - x_1)^2 + (y_2 - y_1)^2}$ Distance formula

    $d = \sqrt{(6 - 9)^2 + (1 - 5)^2}$ Substituting

    $d = \sqrt{(-3)^2 + (-4)^2}$

    $d = \sqrt{25}$

    $d = 5$

6. 10

7. $d = \sqrt{(x_2 - x_1)^2 + (y_2 - y_1)^2}$

    $d = \sqrt{(5 - 5)^2 + (-2 - 6)^2}$

    $d = \sqrt{0^2 + (-8)^2}$

    $d = \sqrt{64}$

    $d = 8$

    (Since these points are on a vertical line, we could have found the distance between them by subtracting their second coordinates and taking the absolute value: $d = |-2 - 6| = |-8| = 8$)

8. 5

9. $d = \sqrt{(x_2 - x_1)^2 + (y_2 - y_1)^2}$

    $d = \sqrt{(-9.2 - 8.6)^2 + [-3.4 - (-3.4)]^2}$

    $d = \sqrt{(-17.8)^2 + 0^2}$

    $d = \sqrt{316.84}$

    $d = 17.8$

    (Since these points are on a horizontal line, we could have found the distance between them by subtracting their first coordinates and taking the absolute value: $d = |-9.2 - 8.6| = |-17.8| = 17.8$)

10. $\sqrt{37.33} \approx 6.110$

11. $d = \sqrt{(x_2 - x_1)^2 + (y_2 - y_1)^2}$

     $d = \sqrt{[6 - (-1)]^2 + (2k - 3k)^2}$

     $d = \sqrt{7^2 + (-k)^2}$

     $d = \sqrt{49 + k^2}$

12. $\sqrt{a^2 + 64}$

13. $d = \sqrt{(x_2 - x_1)^2 + (y_2 - y_1)^2}$

     $d = \sqrt{(\sqrt{6} - 0)^2 + (0 - \sqrt{7})^2}$

     $d = \sqrt{(\sqrt{6})^2 + (-\sqrt{7})^2}$

     $d = \sqrt{6 + 7}$

     $d = \sqrt{13} \approx 3.606$

14. $2\sqrt{3c}$

15. $d = \sqrt{(x_2 - x_1)^2 + (y_2 - y_1)^2}$

     $d = \sqrt{(-2m - 6m)^2 + [n - (-7n)]^2}$

     $d = \sqrt{(-8m)^2 + (8n)^2}$

     $d = \sqrt{64m^2 + 64n^2}$

     $d = \sqrt{64(m^2 + n^2)}$

     $d = 8\sqrt{m^2 + n^2}$

16. $\dfrac{\sqrt{41}}{7} \approx 0.915$

17. $d = \sqrt{(x_2 - x_1)^2 + (y_2 - y_1)^2}$

     $d = \sqrt{[\sqrt{3} - (-3\sqrt{3})]^2 + [1 + \sqrt{6} - (1 - \sqrt{6})]^2}$

     $d = \sqrt{(4\sqrt{3})^2 + (2\sqrt{6})^2}$

     $d = \sqrt{48 + 24}$ $[(4\sqrt{3})^2 = 16 \cdot 3 = 48, (2\sqrt{6})^2 = 4 \cdot 6 = 24]$

     $d = \sqrt{72}$

     $d = 6\sqrt{2} \approx 8.485$

18. $\sqrt{100.930954} \approx 10.046$

19. $d = \sqrt{(x_2 - x_1)^2 + (y_2 - y_1)^2}$

     $d = \sqrt{(a - 0)^2 + (b - 0)^2}$

     $d = \sqrt{a^2 + b^2}$

20. $\sqrt{5} \approx 2.236$

21. $d = \sqrt{(x_2 - x_1)^2 + (y_2 - y_1)^2}$

     $d = \sqrt{(-\sqrt{a} - \sqrt{a})^2 + (\sqrt{b} - \sqrt{b})^2}$

     $d = \sqrt{(-2\sqrt{a})^2 + 0^2}$

     $d = \sqrt{4a}$

     $d = 2\sqrt{a}$

22. $2\sqrt{d^2 + c^2}$

23. We use the midpoint formula:
$$\left(\frac{x_1 + x_2}{2}, \frac{y_1 + y_2}{2}\right) = \left(\frac{-3 + 2}{2}, \frac{6 + (-8)}{2}\right), \text{ or}$$
$$\left(\frac{-1}{2}, \frac{-2}{2}\right), \text{ or } \left(-\frac{1}{2}, -1\right)$$

24. $\left(\frac{13}{2}, -1\right)$

25. We use the midpoint formula:
$$\left(\frac{x_1 + x_2}{2}, \frac{y_1 + y_2}{2}\right) = \left(\frac{8 + (-1)}{2}, \frac{5 + 2}{2}\right), \text{ or } \left(\frac{7}{2}, \frac{7}{2}\right)$$

26. $\left(0, -\frac{1}{2}\right)$

27. We use the midpoint formula:
$$\left(\frac{x_1 + x_2}{2}, \frac{y_1 + y_2}{2}\right) = \left(\frac{-8 + 6}{2}, \frac{-5 + (-1)}{2}\right), \text{ or}$$
$$\left(\frac{-2}{2}, \frac{-6}{2}\right), \text{ or } (-1, -3)$$

28. $\left(\frac{5}{2}, 1\right)$

29. $\left(\frac{x_1 + x_2}{2}, \frac{y_1 + y_2}{2}\right) = \left(\frac{-3.4 + 2.9}{2}, \frac{8.1 - 8.7}{2}\right), \text{ or}$
$$\left(\frac{-0.5}{2}, \frac{-0.6}{2}\right), \text{ or } (-0.25, -0.3)$$

30. $(4.65, 0)$

31. $\left(\frac{x_1 + x_2}{2}, \frac{y_1 + y_2}{2}\right) = \left(\frac{\frac{1}{6} + \left(-\frac{1}{3}\right)}{2}, \frac{-\frac{3}{4} + \frac{5}{6}}{2}\right), \text{ or}$
$$\left(\frac{-\frac{1}{6}}{2}, \frac{\frac{1}{12}}{2}\right), \text{ or } \left(-\frac{1}{12}, \frac{1}{24}\right)$$

32. $\left(-\frac{27}{80}, \frac{1}{24}\right)$

33. $\left(\frac{x_1 + x_2}{2}, \frac{y_1 + y_2}{2}\right) = \left(\frac{\sqrt{2} + \sqrt{3}}{2}, \frac{-1 + 4}{2}\right), \text{ or}$
$$\left(\frac{\sqrt{2} + \sqrt{3}}{2}, \frac{3}{2}\right)$$

34. $\left(\frac{5}{2}, \frac{7\sqrt{3}}{2}\right)$

35. $\left(\frac{x_1 + x_2}{2}, \frac{y_1 + y_2}{2}\right) = \left(\frac{-a + a}{2}, \frac{b + b}{2}\right), \text{ or } \left(\frac{0}{2}, \frac{2b}{2}\right),$
or $(0, b)$

36. $(2, 4\sqrt{2})$

37. $\left(\frac{x_1 + x_2}{2}, \frac{y_1 + y_2}{2}\right) = \left(\frac{6m + (-2m)}{2}, \frac{-7n + n}{2}\right), \text{ or}$
$$\left(\frac{4m}{2}, \frac{-6n}{2}\right), \text{ or } (2m, -3n)$$

38. $\left(\frac{3}{7}, \frac{3}{7}\right)$

39. $\left(\frac{x_1 + x_2}{2}, \frac{y_1 + y_2}{2}\right) =$
$$\left(\frac{-3\sqrt{3} + \sqrt{3}}{2}, \frac{1 - \sqrt{6} + 1 + \sqrt{6}}{2}\right), \text{ or}$$
$$\left(\frac{-2\sqrt{3}}{2}, \frac{2}{2}\right), \text{ or } (-\sqrt{3}, 1)$$

40. $(4.8505, -2.8915)$

41. $\left(\frac{x_1 + x_2}{2}, \frac{y_1 + y_2}{2}\right) = \left(\frac{0 + a}{2}, \frac{0 + b}{2}\right), \text{ or } \left(\frac{a}{2}, \frac{b}{2}\right)$

42. $\left(\frac{\sqrt{2}}{2}, \frac{\sqrt{3}}{2}\right)$

43. $\left(\frac{x_1 + x_2}{2}, \frac{y_1 + y_2}{2}\right) = \left(\frac{\sqrt{a} + (-\sqrt{a})}{2}, \frac{\sqrt{b} + \sqrt{b}}{2}\right), \text{ or}$
$$\left(\frac{0}{2}, \frac{2\sqrt{b}}{2}\right), \text{ or } (0, \sqrt{b})$$

44. (c, d)

45. $2x + 3y = 8,$ (1)
$x - 2y = -3$ (2)

Multiply (2) by -2 and add:
$$2x + 3y = 8$$
$$\underline{-2x + 4y = 6}$$
$$7y = 14 \quad \text{Adding}$$
$$y = 2$$

Substitute 2 for y in (2) and solve for x:
$$x - 2(2) = -3$$
$$x - 4 = -3$$
$$x = 1$$

The solution is $(1,2)$.

46. $\pm 4, \pm 2$

47. $y = x^2 - 4x + 8$
$y = (x^2 - 4x) + 8$
We complete the square inside the parentheses. Take half the x-coefficient and square it, getting 4. Then add 4 inside the parentheses and subtract 4 outside the parentheses.
$y = (x^2 - 4x + 4) + 8 - 4$
$y = (x - 2)^2 + 4$
The graph has the same shape as $y = x^2$ but moved to the right 2 units and up 4 units. The vertex is $(2,4)$.

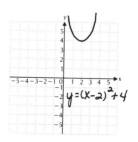

48.

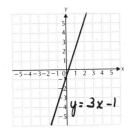

$y = 3x - 1$

49. Find the distance between each pair of points.
Between (9,6) and (-1,2):

$d = \sqrt{(-1 - 9)^2 + (2 - 6)^2} = \sqrt{(-10)^2 + (-4)^2} = \sqrt{116}$

Between (9,6) and (1,-3):

$d = \sqrt{(1 - 9)^2 + (-3 - 6)^2} = \sqrt{(-8)^2 + (-9)^2} = \sqrt{145}$

Between (-1,2) and (1,-3):

$d = \sqrt{[1 - (-1)]^2 + (-3 - 2)^2} = \sqrt{2^2 + (-5)^2} = \sqrt{29}$

Since $(\sqrt{116})^2 + (\sqrt{29})^2 = (\sqrt{145})^2$, the points are the vertices of a right triangle.

50. Yes

51. Find the distance between each pair of points.
Between $(-5, \sqrt{2})$ and $(-3, 2 + \sqrt{2})$:

$d = \sqrt{[-3 - (-5)]^2 + (2 + \sqrt{2} - \sqrt{2})^2} = \sqrt{2^2 + 2^2} = \sqrt{8}$

Between $(-5, \sqrt{2})$ and $(-1, \sqrt{2})$:

$d = \sqrt{[-1 - (-5)]^2 + (\sqrt{2} - \sqrt{2})^2} = \sqrt{4^2 + 0^2} = \sqrt{16}$

Between $(-3, 2 + \sqrt{2})$ and $(-1, \sqrt{2})$:

$d = \sqrt{[-1 - (-3)]^2 + [\sqrt{2} - (2 + \sqrt{2})]^2} = \sqrt{2^2 + 2^2} = \sqrt{8}$

Since $(\sqrt{8})^2 + (\sqrt{8})^2 = (\sqrt{16})^2$, the points are the the vertices of a right triangle.

52. Let (a, y_1) and (a, y_2) be two points on a vertical line. The distance between them is $|y_2 - y_1|$. Using the distance formula, we have

$d = \sqrt{(a - a)^2 + (y_2 - y_1)^2} = \sqrt{(y_2 - y_1)^2} = |y_2 - y_1|.$

Let (x_1, b) and (x_2, b) be two points on a horizontal line. The distance between them is $|x_2 - x_1|$. Using the distance formula, we have

$d = \sqrt{(x_2 - x_1)^2 + (b - b)^2} = \sqrt{(x_2 - x_1)^2} = |x_2 - x_1|.$

53. Let (0,y) be the point on the y-axis that is equidistant from (2,10) and (6,2). Then the distance between (2,10) and (0,y) is the same as the distance between (6,2) and (0,y).

$\sqrt{(0 - 2)^2 + (y - 10)^2} = \sqrt{(0 - 6)^2 + (y - 2)^2}$

$(-2)^2 + (y - 10)^2 = (-6)^2 + (y - 2)^2$

Squaring both sides

$4 + y^2 - 20y + 100 = 36 + y^2 - 4y + 4$

$64 = 16y$

$4 = y$

This number checks. The point is (0,4).

54. (-5,0)

55. Find the distance between each pair of points.
Between A and B:

$d = \sqrt{[0 - (-4)]^2 + (1 - 13)^2} = \sqrt{4^2 + (-12)^2} = \sqrt{160} = 4\sqrt{10}$

Between A and C:

$d = \sqrt{[5 - (-4)]^2 + (-14 - 13)^2} = \sqrt{9^2 + (-27)^2} = \sqrt{810} = 9\sqrt{10}$

Between B and C:

$d = \sqrt{(5 - 0)^2 + (-14 - 1)^2} = \sqrt{5^2 + (-15)^2} = \sqrt{250} = 5\sqrt{10}$

Since $4\sqrt{10} + 5\sqrt{10} = 9\sqrt{10}$, the points lie on the same line.

56. No

57. See the proof in the text answer section.

58. $\sqrt{80} + \sqrt{34} + \sqrt{218} \approx 29.540$

59.

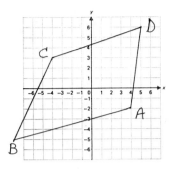

Find the sum of the following distances:

Between A and B:

$d = \sqrt{(-8 - 4)^2 + [-5 - (-2)]^2} = \sqrt{(-12)^2 + (-3)^2} =$

$\sqrt{153} = 3\sqrt{17}$

Between B and C:

$d = \sqrt{[-4 - (-8)]^2 + [3 - (-5)]^2} = \sqrt{4^2 + 8^2} =$

$\sqrt{80} = 4\sqrt{5}$

Between C and D:

$d = \sqrt{[5 - (-4)]^2 + (6 - 3)^2} = \sqrt{9^2 + 3^2} = \sqrt{90} =$

$3\sqrt{10}$

Between D and A:

$d = \sqrt{(5 - 4)^2 + [6 - (-2)]^2} = \sqrt{1^2 + 8^2} = \sqrt{65}$

The perimeter is $3\sqrt{17} + 4\sqrt{5} + 3\sqrt{10} + \sqrt{65} \approx$
38.863.

60. (2,-3), (6,-3)

61. The distance between (4,y) and (5,-2) is $\sqrt{2}$.

$\sqrt{(5 - 4)^2 + (-2 - y)^2} = \sqrt{2}$

$(5 - 4)^2 + (-2 - y)^2 = 2$ Squaring both sides

$1 + 4 + 4y + y^2 = 2$

$y^2 + 4y + 3 = 0$

$(y + 1)(y + 3) = 0$

y = -1 or y = -3

Both numbers check. The points are (4,-1) and
(4,-3).

Exercise Set 10.2

1. $(x + 1)^2 + (y + 3)^2 = 4$

$[x - (-1)]^2 + [y - (-3)]^2 = 2^2$ Standard form

The center is (-1,-3), and the radius is 2.

$(x + 1)^2 + (y + 3)^2 = 4$

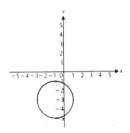

2. Center: (2,-3)

Radius: 1

$(x - 2)^2 + (y + 3)^2 = 1$

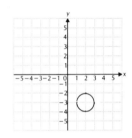

3. $(x - 8)^2 + (y + 3)^2 = 40$

$(x - 8)^2 + [y - (-3)]^2 = (2\sqrt{10})^2$ $(\sqrt{40} = 2\sqrt{10})$

The center is (8,-3), and the radius is $2\sqrt{10}$.

$(x - 8)^2 + (y + 3)^2 = 40$

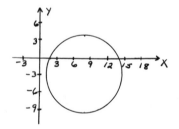

4. Center: (-5,1)

Radius: $5\sqrt{3}$

$(x + 5)^2 + (y - 1)^2 = 75$

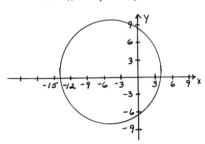

5. $x^2 + y^2 = 2$

   $(x - 0)^2 + (y - 0)^2 = (\sqrt{2})^2$ Standard form

   The center is (0,0), and the radius is $\sqrt{2}$.

   $x^2 + y^2 = 2$

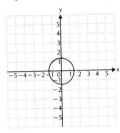

6. Center: (0,0)

   Radius: $\sqrt{3}$

   $x^2 + y^2 = 3$

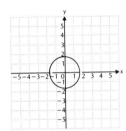

7. $(x - 5)^2 + y^2 = \frac{1}{4}$

   $(x - 5)^2 + (y - 0)^2 = \left[\frac{1}{2}\right]^2$ Standard form

   The center is (5,0), and the radius is $\frac{1}{2}$.

   $(x - 5)^2 + y^2 = \frac{1}{4}$

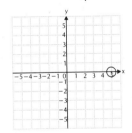

8. Center: (0,1)

   Radius: $\frac{1}{5}$

   $x^2 + (y - 1)^2 = \frac{1}{25}$

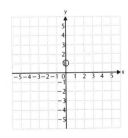

9. $x^2 + y^2 + 8x - 6y - 15 = 0$

              $x^2 + 8x + y^2 - 6y = 15$

   $(x^2 + 8x + 16) + (y^2 - 6y + 9) = 15 + 16 + 9$

                  Completing the square twice

              $(x + 4)^2 + (y - 3)^2 = 40$

   $[x - (-4)]^2 + (y - 3)^2 = (2\sqrt{10})^2$

                      Standard form

   The center is (-4,3), and the radius is $2\sqrt{10}$.

   $x^2 + y^2 + 8x - 6y - 15 = 0$

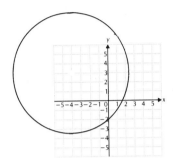

10. Center: (-3,2)

   Radius: $2\sqrt{7}$

   $x^2 + y^2 + 6x - 4y - 15 = 0$

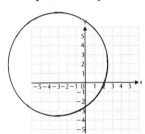

11. $x^2 + y^2 - 8x + 2y + 13 = 0$

              $x^2 - 8x + y^2 + 2y = -13$

   $(x^2 - 8x + 16) + (y^2 + 2y + 1) = -13 + 16 + 1$

                  Completing the square twice

              $(x - 4)^2 + (y + 1)^2 = 4$

              $(x - 4)^2 + [y - (-1)]^2 = 2^2$ Standard form

   The center is (4,-1), and the radius is 2.

   $x^2 + y^2 - 8x + 2y + 13 = 0$

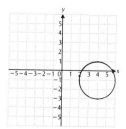

12. Center: (-3,-2)
    Radius: 1
    $x^2 + y^2 + 6x + 4y + 12 = 0$

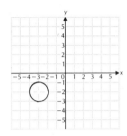

13. $x^2 + y^2 - 4x = 0$
    $x^2 - 4x + y^2 = 0$
    $(x^2 - 4x + 4) + y^2 = 4$ Completing the square
    $(x - 2)^2 + (y - 0)^2 = 2^2$ Standard form
    The center is (2,0), and the radius is 2.
    $x^2 + y^2 - 4x = 0$

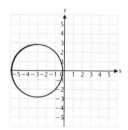

14. Center: (-3,0)
    Radius: 3
    $x^2 + y^2 + 6x = 0$

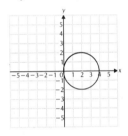

15. $x^2 + y^2 + 10y - 75 = 0$
    $x^2 + y^2 + 10y = 75$
    $x^2 + (y^2 + 10y + 25) = 75 + 25$
    $(x - 0)^2 + (y + 5)^2 = 100$
    $(x - 0)^2 + [y - (-5)]^2 = 10^2$
    The center is (0,-5), and the radius is 10.

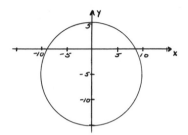

16. Center: (4,0)
    Radius: 10

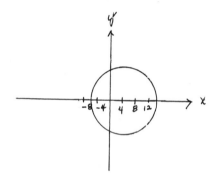

17. $x^2 + y^2 + 7x - 3y - 10 = 0$
    $x^2 + 7x + y^2 - 3y = 10$
    $\left[x^2 + 7x + \dfrac{49}{4}\right] + \left[y^2 - 3y + \dfrac{9}{4}\right] = 10 + \dfrac{49}{4} + \dfrac{9}{4}$
    $\left[x + \dfrac{7}{2}\right]^2 + \left[y - \dfrac{3}{2}\right]^2 = \dfrac{98}{4}$
    $\left[x - \left(-\dfrac{7}{2}\right)\right]^2 + \left[y - \dfrac{3}{2}\right]^2 = \left[\sqrt{\dfrac{98}{4}}\right]^2$
    The center is $\left(-\dfrac{7}{2}, \dfrac{3}{2}\right)$, and the radius is $\sqrt{\dfrac{98}{4}}$, or $\dfrac{\sqrt{98}}{2}$, or $\dfrac{7\sqrt{2}}{2}$.

    $x^2 + y^2 + 7x - 3y - 10 = 0$

18. Center: $\left(\frac{21}{2}, \frac{33}{2}\right)$

   Radius: $\frac{\sqrt{1462}}{2}$

   $x^2 + y^2 - 21x - 33y + 17 = 0$

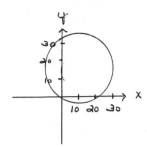

19. $4x^2 + 4y^2 = 1$

             $x^2 + y^2 = \frac{1}{4}$ Multiplying by $\frac{1}{4}$ on both sides

   $(x - 0)^2 + (y - 0)^2 = \left(\frac{1}{2}\right)^2$

   The center is $(0,0)$, and the radius is $\frac{1}{2}$.

   $4x^2 + 4y^2 = 1$

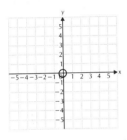

20. Center: $(0,0)$

   Radius: $\frac{1}{5}$

   $25x^2 + 25y^2 = 1$

21. $(x - h)^2 + (y - k)^2 = r^2$ Standard form
   $(x - 0)^2 + (y - 0)^2 = 7^2$ Substituting
                $x^2 + y^2 = 49$ Simplifying

22. $x^2 + y^2 = 16$

23. $(x - h)^2 + (y - k)^2 = r^2$ Standard form
   $[x - (-2)]^2 + (y - 7)^2 = (\sqrt{5})^2$ Substituting
       $(x + 2)^2 + (y - 7)^2 = 5$

24. $(x - 5)^2 + (y - 6)^2 = 12$

25. $(x - h)^2 + (y - k)^2 = r^2$ Standard form
   $[x - (-4)]^2 + (y - 3)^2 = (4\sqrt{3})^2$ Substituting
       $(x + 4)^2 + (y - 3)^2 = 48$
                   $[(4\sqrt{3})^2 = 16\cdot3 = 48]$

26. $(x + 2)^2 + (y - 7)^2 = 20$

27. $(x - h)^2 + (y - k)^2 = r^2$
   $[x - (-7)]^2 + [y - (-2)]^2 = (5\sqrt{2})^2$
       $(x + 7)^2 + (y + 2)^2 = 50$

28. $(x + 5)^2 + (y + 8)^2 = 18$

29. $(x - h)^2 + (y - k)^2 = r^2$
   $[x - (-8)]^2 + (y - t)^2 = (1.3)^2$
       $(x + 8)^2 + (y - t)^2 = 1.69$

30. $(x - 2.7)^2 + (y - k)^2 = 4p^2$

31. **Familiarize.** We make a drawing and label it. Let x represent the width of the border.

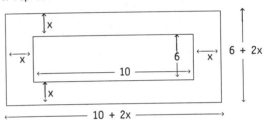

   The perimeter of the larger rectangle is
       $2(10 + 2x) + 2(6 + 2x)$, or $8x + 32$.

   The perimeter of the smaller rectangle is
       $2(10) + 2(6)$, or 32.

   Translate. The perimeter of the larger rectangle is twice the perimeter of the smaller rectangle.

       $8x + 32 = 2\cdot32$

   Carry out. We solve the equation.
       $8x + 32 = 64$
             $8x = 32$
              $x = 4$

   Check. If the width of the border is 4 in., then the length and width of the larger rectangle are 14 in. and 18 in. Thus its perimeter is $2(14) + 2(18)$, or 64 in. The perimeter of the smaller rectangle is 32 in. The perimeter of the larger rectangle is twice the perimeter of the smaller rectangle.

   State. The width of the border is 4 in.

32. 2640 mi

33. $f(x) = 2x^2 - 10x + 7$

   $f(x) = 2(x^2 - 5x) + 7$

   $f(x) = 2\left[x^2 - 5x + \frac{25}{4}\right] + 7 - \frac{25}{2}$ Adding and subtracting $2 \cdot \frac{25}{4}$

   $f(x) = 2\left[x - \frac{5}{2}\right]^2 - \frac{11}{2}$

   The graph has the same shape as $f(x) = 2x^2$ but moved to the right $\frac{5}{2}$ units and down $\frac{11}{2}$ units. The vertex is $\left[\frac{5}{2}, -\frac{11}{2}\right]$.

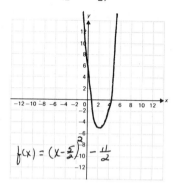

34. $f(x) = -3(x - 4)^2 - 2$

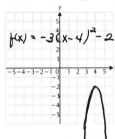

35. We first find the length of the radius which is the distance between $(0,0)$ and $\left[\frac{1}{4}, \frac{\sqrt{31}}{4}\right]$.

   $r = \sqrt{\left[\frac{1}{4} - 0\right]^2 + \left[\frac{\sqrt{31}}{4} - 0\right]^2}$

   $r = \sqrt{\left[\frac{1}{4}\right]^2 + \left[\frac{\sqrt{31}}{4}\right]^2}$

   $r = \sqrt{\frac{1}{16} + \frac{31}{16}}$

   $r = \sqrt{\frac{32}{16}}$

   $r = \sqrt{2}$

   Then we substitute into the standard form for the equation of a circle.

   $(x - h)^2 + (y - k)^2 = r^2$

   $(x - 0)^2 + (y - 0)^2 = (\sqrt{2})^2$

   $x^2 + y^2 = 2$

36. $(x + 4)^2 + (y - 1)^2 = 72$

37. We make a drawing of a circle with center $(3,-5)$ and tangent to the y-axis.

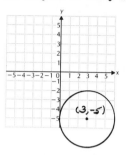

   We see that the circle touches the y-axis at $(0,-5)$. Hence, the radius is the distance between $(0,-5)$ and $(3,-5)$, or $|3 - 0|$, or 3. Now we write the equation of the circle.

   $(x - h)^2 + (y - k)^2 = r^2$

   $(x - 3)^2 + [y - (-5)]^2 = 3^2$

   $(x - 3)^2 + (y + 5)^2 = 9$

38. $(x + 7)^2 + (y + 4)^2 = 16$

39. First we use the midpoint formula to find the center:

   $\left[\frac{7 + (-1)}{2}, \frac{3 + (-3)}{2}\right]$, or $\left[\frac{6}{2}, \frac{0}{2}\right]$, or $(3,0)$

   The length of the radius is the distance between the center $(3,0)$ and either endpoint of a diameter. We will use endpoint $(7,3)$ in the distance formula:

   $r = \sqrt{(7 - 3)^2 + (3 - 0)^2}$

   $r = \sqrt{4^2 + 3^2}$

   $r = \sqrt{25}$

   $r = 5$

   Now we write the equation of the circle:

   $(x - h)^2 + (y - k)^2 = r^2$

   $(x - 3)^2 + (y - 0)^2 = 5^2$

   $(x - 3)^2 + y^2 = 25$

40. $(x + 3)^2 + (y - 5)^2 = 16$

41. The formula for the area of a circle is $A = \pi r^2$, where r is the length of the radius. We use this to find r:

   $\pi r^2 = 64\pi$

   $r^2 = 64$ Multiplying by $\frac{1}{\pi}$

   $r = 8$ Taking the positive square root

   Now we write the equation of the circle:

   $(x - h)^2 + (y - k)^2 = r^2$

   $[x - (-5)]^2 + [y - (-6)]^2 = 8^2$

   $(x + 5)^2 + (y + 6)^2 = 64$

42. Center: (7,8); radius: $\dfrac{5\sqrt{2}}{2}$

43. We write the equation of a circle with center (0,0) and radius 6.4.

$$(x - h)^2 + (y - k)^2 = r^2$$
$$(x - 0)^2 + (y - 0)^2 = (6.4)^2$$
$$x^2 + y^2 = 40.96$$

44. $x^2 + (y - 30.6)^2 = 590.49$

45. Determine whether the pair $(0,-1)$ is a solution of the equation.

$$\frac{x^2 + y^2 = 1}{\begin{array}{c|c} 0^2 + (-1)^2 & 1 \\ 0 + 1 & \\ 1 & \end{array}}$$

The point $(0,-1)$ lies on the unit circle.

46. Yes

47. Determine whether the pair $(\sqrt{2} + \sqrt{3}, 0)$ is a solution of the equation.

$$\frac{x^2 + y^2 = 1}{\begin{array}{c|c} (\sqrt{2} + \sqrt{3})^2 + 0^2 & 1 \\ 2 + 2\sqrt{6} + 3 + 0 & \\ 5 + 2\sqrt{6} & \end{array}}$$

The point $(\sqrt{2} + \sqrt{3}, 0)$ does not lie on the unit circle.

48. No

49. Determine whether the pair $\left[\dfrac{\sqrt{2}}{2}, \dfrac{\sqrt{2}}{2}\right]$ is a solution of the equation.

$$\frac{x^2 + y^2 = 1}{\begin{array}{c|c} \left[\dfrac{\sqrt{2}}{2}\right]^2 + \left[\dfrac{\sqrt{2}}{2}\right]^2 & 1 \\ \dfrac{2}{4} + \dfrac{2}{4} & \\ 1 & \end{array}}$$

The point $\left[\dfrac{\sqrt{2}}{2}, \dfrac{\sqrt{2}}{2}\right]$ lies on the unit circle.

50. Yes

51. See the proof in the answer section in the text.

Exercise Set 10.3

1. $\dfrac{x^2}{4} + \dfrac{y^2}{1} = 1$

$\dfrac{x^2}{2^2} + \dfrac{y^2}{1^2} = 1$

The x-intercepts are $(2,0)$ and $(-2,0)$, and the y-intercepts are $(0,1)$ and $(0,-1)$. We plot these points and connect them with an oval-shaped curve.

$\dfrac{x^2}{4} + \dfrac{y^2}{1} = 1$

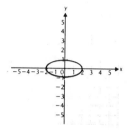

2. $\dfrac{x^2}{1} + \dfrac{y^2}{4} = 1$

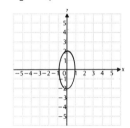

3. $\dfrac{x^2}{16} + \dfrac{y^2}{25} = 1$

$\dfrac{x^2}{4^2} + \dfrac{y^2}{5^2} = 1$

The x-intercepts are $(4,0)$ and $(-4,0)$, and the y-intercepts are $(0,5)$ and $(0,-5)$. We plot these points and connect them with an oval-shaped curve.

$\dfrac{x^2}{16} + \dfrac{y^2}{25} = 1$

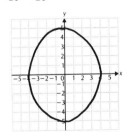

4. $\dfrac{x^2}{9} + \dfrac{y^2}{25} = 1$

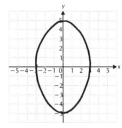

5. $4x^2 + 9y^2 = 36$

$\dfrac{1}{36}(4x^2 + 9y^2) = \dfrac{1}{36}(36)$ Multiplying by $\dfrac{1}{36}$

$\dfrac{x^2}{9} + \dfrac{y^2}{4} = 1$

$\dfrac{x^2}{3^2} + \dfrac{y^2}{2^2} = 1$

The x-intercepts are (3,0) and (-3,0), and the y-intercepts are (0,2) and (0,-2). We plot these points and connect them with an oval-shaped curve.

$4x^2 + 9y^2 = 36$

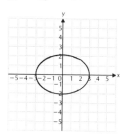

6. $9x^2 + 4y^2 = 36$

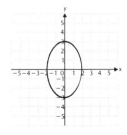

7. $16x^2 + 9y^2 = 144$

$\dfrac{x^2}{9} + \dfrac{y^2}{16} = 1$ Multiplying by $\dfrac{1}{144}$

$\dfrac{x^2}{3^2} + \dfrac{y^2}{4^2} = 1$

The x-intercepts are (3,0) and (-3,0), and the y-intercepts are (0,4) and (0,-4). We plot these points and connect them with an oval-shaped curve.

$16x^2 + 9y^2 = 144$

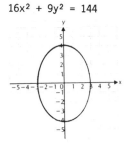

8. $9x^2 + 16y^2 = 144$

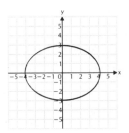

9. $2x^2 + 3y^2 = 6$

$\dfrac{x^2}{3} + \dfrac{y^2}{2} = 1$ Multiplying by $\dfrac{1}{6}$

$\dfrac{x^2}{(\sqrt{3})^2} + \dfrac{y^2}{(\sqrt{2})^2} = 1$

The x-intercepts are $(\sqrt{3},0)$ and $(-\sqrt{3},0)$, and the y-intercepts are $(0,\sqrt{2})$ and $(0,-\sqrt{2})$. We plot these points and connect them with an oval-shaped curve.

$2x^2 + 3y^2 = 6$

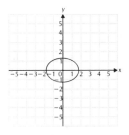

10. $5x^2 + 7y^2 = 35$

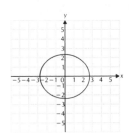

11. $4x^2 + 9y^2 = 1$

$$\frac{x^2}{\frac{1}{4}} + \frac{y^2}{\frac{1}{9}} = 1 \qquad 4x^2 = \frac{x^2}{\frac{1}{4}}, \quad 9y^2 = \frac{y^2}{\frac{1}{9}}$$

$$\frac{x^2}{\left(\frac{1}{2}\right)^2} + \frac{y^2}{\left(\frac{1}{3}\right)^2} = 1$$

The x-intercepts are $\left(\frac{1}{2},0\right)$ and $\left(-\frac{1}{2},0\right)$, and the y-intercepts are $\left(0,\frac{1}{3}\right)$ and $\left(0,-\frac{1}{3}\right)$. We plot these points and connect them with an oval-shaped curve.

$4x^2 + 9y^2 = 1$

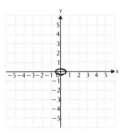

12. $25x^2 + 16y^2 = 1$

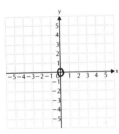

13. $5x^2 + 12y^2 = 60$

$$\frac{x^2}{12} + \frac{y^2}{5} = 1 \qquad \text{Multiplying by } \frac{1}{60}$$

$$\frac{x^2}{(\sqrt{12})^2} + \frac{y^2}{(\sqrt{5})^2} = 1$$

The x-intercepts are $(\sqrt{12},0)$ and $(-\sqrt{12},0)$, or $(2\sqrt{3},0)$ and $(-2\sqrt{3},0)$, and the y-intercepts are $(0,\sqrt{5})$ and $(0,-\sqrt{5})$. We plot these points and connect them with an oval-shaped curve.

$5x^2 + 12y^2 = 60$

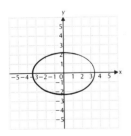

14. $8x^2 + 3y^2 = 24$

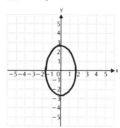

15. $\dfrac{3 - \sqrt{a}}{2 + \sqrt{a}} = \dfrac{3 - \sqrt{a}}{2 + \sqrt{a}} \cdot \dfrac{2 - \sqrt{a}}{2 - \sqrt{a}} = \dfrac{6 - 3\sqrt{a} - 2\sqrt{a} + a}{4 - a} =$

$\dfrac{6 - 5\sqrt{a} + a}{4 - a}$

16. $\dfrac{9 - a}{6 + 5\sqrt{a} + a}$

17. $(1 + x)(1 - x) = 3x^2 - 4$

$$1 - x^2 = 3x^2 - 4$$

$$5 = 4x^2$$

$$\frac{5}{4} = x^2$$

$$\pm \frac{\sqrt{5}}{2} = x$$

18. $\dfrac{-7 \pm 3\sqrt{41}}{8}$

19.   We make a drawing.

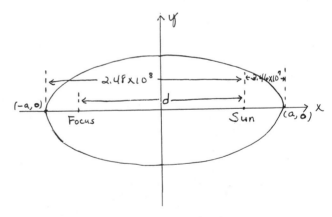

The distance between vertex (a,0) and the sun is the same as the distance between vertex (-a,0) and the other focus. Then

$d = 2.48 \times 10^8 - 3.46 \times 10^7 =$

    $2.48 \times 10^8 - 0.346 \times 10^8 = 2.134 \times 10^8$ mi.

20.   Bank shots originating at one focus (the tiny dot) are deflected to the other focus (the hole).

21. Position the ellipse on a coordinate system as shown below.

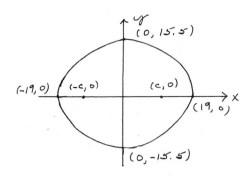

In order to best use the room's acoustics, the President and the advisor should be seated at the foci of the ellipse, or at (-c,0) and (c,0). We use the equation relating the coordinates of the foci and the intercepts to find c:

$$c^2 = a^2 - b^2$$
$$c^2 = (19)^2 - (15.5)^2$$
$$c^2 = 120.75$$
$$c \approx 11$$

We make a sketch.

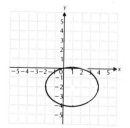

22. $\dfrac{x^2}{81} + \dfrac{y^2}{121} = 1$

23. $4(x - 1)^2 + 9(y + 2)^2 = 36$

$\dfrac{(x - 1)^2}{9} + \dfrac{(y + 2)^2}{4} = 1$ Standard form

$\dfrac{(x - 1)^2}{3^2} + \dfrac{[y - (-2)]^2}{2^2} = 1$

The center is (1,-2) and the vertices are (4,-2), (-2,-2), (1,0), and (1,-4).

$4(x - 1)^2 + 9(y + 2)^2 = 36$

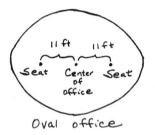

24. Center: (-3,4)

Vertices: (-2,4), (-4,4), (-3,8), (-3,0)

$16x^2 + y^2 + 96x - 8y + 144 = 0$

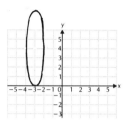

25.

$$4x^2 + 25y^2 - 8x + 50y = 71$$
$$4x^2 - 8x + 25y^2 + 50y = 71$$
$$4(x^2 - 2x) + 25(y^2 + 2y) = 71$$
$$4(x^2 - 2x + 1) + 25(y^2 + 2y + 1) = 71 + 4 + 25$$
$$4(x - 1)^2 + 25(y + 1)^2 = 100$$
$$\dfrac{(x - 1)^2}{25} + \dfrac{(y + 1)^2}{4} = 1 \quad \text{Standard form}$$

The center is (1,-1), and the vertices are (6,-1), (-4,-1), (1,1), and (1,-3).

$4x^2 + 25y^2 - 8x + 50y = 71$

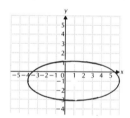

26. Center: (-3,2)

Vertices: $\left(-3\tfrac{1}{5},2\right)$, $\left(-2\tfrac{4}{5},2\right)$, $\left(-3,2\tfrac{1}{3}\right)$, $\left(-3,1\tfrac{2}{3}\right)$

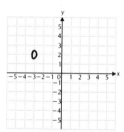

27. The center of the ellipse is (h,k), or (0,0). Then a = 8 and b = 2. We substitute in the standard form for the equation of an ellipse centered at the origin.

$$\dfrac{x^2}{a^2} + \dfrac{y^2}{b^2} = 1$$
$$\dfrac{x^2}{8^2} + \dfrac{y^2}{2^2} = 1$$
$$\dfrac{x^2}{64} + \dfrac{y^2}{4} = 1$$

28. $\dfrac{(x - 5)^2}{25} + \dfrac{(y - 2)^2}{9} = 1$

29. Since two vertices are (2,0) and (-2,0), we know that k = 0. Also

$a + h = 2,$

$-a + h = -2,$

so h = 0 and a = 2.

Substituting, we have

$\dfrac{x^2}{2^2} + \dfrac{y^2}{b^2} = 1,$ or $\dfrac{x^2}{4} + \dfrac{y^2}{b^2} = 1.$

To find b^2, we substitute the coordinates of the point $(1, 2\sqrt{3})$.

$\dfrac{1^2}{4} + \dfrac{(2\sqrt{3})^2}{b^2} = 1$

$\dfrac{1}{4} + \dfrac{12}{b^2} = 1$

$\dfrac{12}{b^2} = \dfrac{3}{4}$

$16 = b^2$

Thus, the equation of the ellipse is

$\dfrac{x^2}{4} + \dfrac{y^2}{16} = 1.$

30. A circle centered at (h,k) with radius a.

31. a), b) See the answer section in the text.

Exercise Set 10.4

1. $y = x^2$

a) This is equivalent to $y = (x - 0)^2 + 0$. The vertex is (0,0).

b) We choose some x-values on both sides of the vertex and compute the corresponding values of y. The graph opens upward, because the coefficient of x^2, 1, is positive.

x	y
0	0
1	1
2	4
-1	1
-2	4

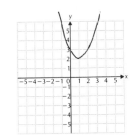

2. $x = y^2$

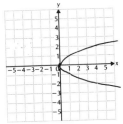

3. $x = y^2 + 4y + 1$

a) We find the vertex by completing the square.

$x = (y^2 + 4y + 4) + 1 - 4$

$x = (y + 2)^2 - 3$

The vertex is (-3,-2).

b) To find ordered pairs, we choose values for y and compute the corresponding values for x. The graph opens to the right, because the coefficient of y^2, 1, is positive.

x	y
-3	-2
-2	-3
-2	-1
1	-4
1	0

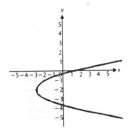

4. $y = x^2 - 2x + 3$

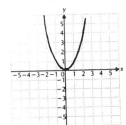

5. $y = -x^2 + 4x - 5$

a) We can find the vertex by computing the first coordinate, x = -b/2a, and then substituting to find the second coordinate:

$x = -\dfrac{b}{2a} = -\dfrac{4}{2(-1)} = 2$

$y = -x^2 + 4x - 5 = -(2)^2 + 4(2) - 5 = -1$

The vertex is (2,-1).

b) We choose some x-values and compute the corresponding values for y. The graph opens downward because the coefficient of x^2, -1, is negative.

x	y
2	-1
3	-2
4	-5
1	-2
0	-5

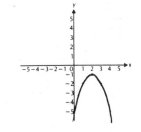

6. x = 4 - 3y - y²

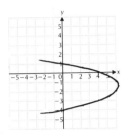

7. x = y² + 1
    a) x = (y - 0)² + 1
        The vertex is (1,0).

    b) To find ordered pairs, we choose y-values and compute the corresponding values for x. The graph opens to the right, because the coefficient of y², 1, is positive.

x	y
1	0
2	1
5	2
2	-1
5	-2

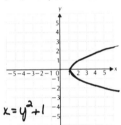

8. x = 2y²

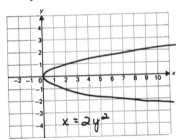

9. x = -1·y²
    a) x = -1·(y - 0)² + 0
        The vertex is (0,0).

    b) We choose y-values and compute the corresponding values for x. The graph opens to the left, because the coefficient of y², -1, is negative.

x	y
0	0
-1	1
-4	2
-1	-1
-4	-2

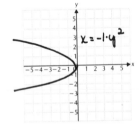

10. x = y² - 1

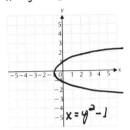

11. x = -y² + 2y
    a) We find the vertex by computing the second coordinate, y = -b/2a and then substituting to find the first coordinate:

$$y = - \frac{b}{2a} = - \frac{2}{2(-1)} = 1$$

$$x = -y^2 + 2y = -(1)^2 + 2(1) = 1$$

   The vertex is (1,1).

    b) We choose y-values and compute the corresponding values for x. The graph opens to the left, because the coefficient of y², -1, is negative.

x	y
1	1
0	0
-3	-1
0	2
-3	3

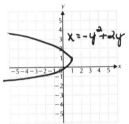

12. x = y² + y - 6

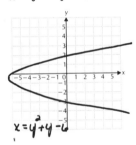

13. x = 8 - y - y²
    a) We find the vertex by completing the square.
        $$x = -(y^2 + y) + 8$$
        $$x = -\left[y^2 + y + \frac{1}{4}\right] + 8 + \frac{1}{4}$$
        $$x = -\left[y + \frac{1}{2}\right]^2 + \frac{33}{4}$$

   The vertex is $\left[\frac{33}{4}, -\frac{1}{2}\right]$.

13. (continued)

b) We choose y-values and compute the corresponding values for x. The graph opens to the left, because the coefficient of y², -1, is negative.

x	y
$\frac{33}{4}$	$-\frac{1}{2}$
8	0
6	1
2	2
8	-1
6	-2
2	-3

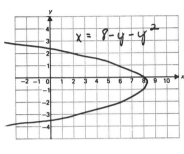

14. y = x² + 2x + 1

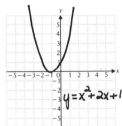

15. y = x² - 2x + 1

a) y = (x - 1)² + 0

The vertex is (1,0).

b) We choose x-values and compute the corresponding values for y. The graph opens upward, because the coefficient of x², 1, is positive.

x	y
1	0
0	1
-1	4
2	1
3	4

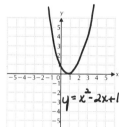

16. y = $-\frac{1}{2}$x²

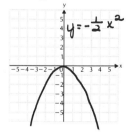

17. x = -y² + 2y + 3

a) We find the vertex by computing the second coordinate, y = -b/2a, and then substituting to find the first coordinate.

$$y = -\frac{b}{2a} = -\frac{2}{2(-1)} = 1$$

$$x = -y^2 + 2y + 3 = -(1)^2 + 2(1) + 3 = 4$$

The vertex is (4,1).

b) We choose y-values and compute the corresponding values for x. The graph opens to the left, because the coefficient of y², -1, is negative.

x	y
4	1
3	0
0	-1
3	2
0	3

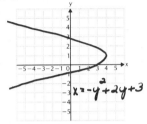

18. x = -y² - 2y + 3

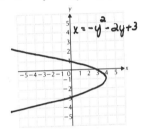

19. x = -2y² - 4y + 1

a) We find the vertex by completing the square.

$$x = -2(y^2 + 2y) + 1$$
$$x = -2(y^2 + 2y + 1) + 1 + 2$$
$$x = -2(y + 1)^2 + 3$$

The vertex is (3,-1).

b) We choose y-values and compute the corresponding values for x. The graph opens to the left, because the coefficient of y², -2, is negative.

x	y
3	-1
1	-2
-5	-3
1	0
-5	1

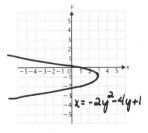

20. $x = 2y^2 + 4y - 1$

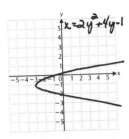

21. <u>Familiarize</u>. 7 oz of ammonia + 4 oz of water = 11 oz of mixture. Let x represent the number of ounces of water to be added to the 11-oz mixture. The new mixture contains 11 + x oz, and the 7 oz of ammonia will be 25% of the new mixture.

<u>Translate</u>.

Ammonia is 25% of new mixture.

$$7 = 0.25(11 + x)$$

<u>Carry out</u>. We solve the equation.

$$7 = 2.75 + 0.25x$$
$$4.25 = 0.25x$$
$$17 = x$$

<u>Check</u>. If she adds 17 oz of water to the 11-oz mixture, the new mixture contains 28 oz. Since it contains 7 oz of ammonia, the new mixture is $\frac{7}{28}$, or 25%, ammonia. The value checks.

<u>State</u>. Julie should add 17 oz of water.

22. $\dfrac{-2 \pm \sqrt{6}}{2}$

23. $x^2 + 2 = 0$
$$x^2 = -2$$
$$x = \pm\sqrt{-2}$$
$$x = \pm i\sqrt{2}$$

24. $-\dfrac{7}{2}, 1$

25. $x = y^2 - y - 6$
$$x = \left[y^2 - y + \frac{1}{4}\right] - 6 - \frac{1}{4}$$
$$x = \left[y - \frac{1}{2}\right]^2 - \frac{25}{4}$$

The vertex is $\left[-\dfrac{25}{4}, \dfrac{1}{2}\right]$.

x	y
$-\dfrac{25}{4}$	$\dfrac{1}{2}$
-6	1
-4	2
0	3
-6	0
-4	-1
0	-2

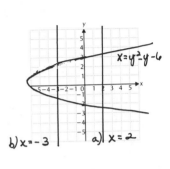

a) Graph x = 2 on the same set of axes as $x = y^2 - y - 6$ and approximate the y-coordinates of the points of intersection. (See the graph above.) The solutions are approximately 3.4 and -2.4.

b) Graph x = -3 on the same set of axes as $x = y^2 - y - 6$ and approximate the y-coordinates of the points of intersection. (See the graph above.) The solutions are approximately 2.3 and -1.3.

26.

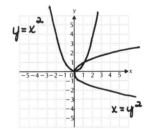

Reflect one graph across the line y = x to obtain the other.

27.

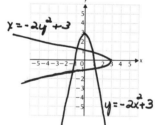

Reflect one graph across the line y = x to obtain the other.

28.

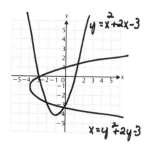

Reflect one graph across the line y = x to obtain the other.

29.

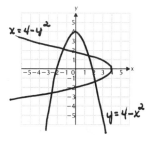

Reflect one graph across the line y = x to obtain the other.

30.

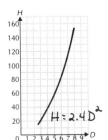

31. Since the line of symmetry is parallel to the y-axis, the equation is of the form $y = ax^2 + bx + c$. We substitute the three given points to obtain a system of equations.

(0,3): $3 = a(0)^2 + b(0) + c$, or

   $3 = c$ (1)

(-1,6): $6 = a(-1)^2 + b(-1) + c$, or

   $6 = a - b + c$ (2)

(2,9): $9 = a(2)^2 + b(2) + c$, or

   $9 = 4a + 2b + c$ (3)

Substituting 3 for c in equations (2) and (3), we have

   $6 = a - b + 3$,

   $9 = 4a + 2b + 3$,   or

   $3 = a - b$,   (4)

   $6 = 4a + 2b$.   (5)

31. (continued)

Multiply equation (4) by 2 and add it to equation (5).

   $6 = 2a - 2b$

   $\underline{6 = 4a + 2b}$

   $12 = 6a$ Adding

   $2 = a$

Substitute 2 for a in equation (4) and solve for b.

   $3 = 2 - b$

   $b = -1$

The equation of the parabola is $y = 2x^2 - x + 3$.

32. $x = 4(y - 1)^2 + 2$

33. See the answer section in the text.

34. $\dfrac{4ac - b^2}{4a}$

Exercise Set 10.5

1. $\dfrac{x^2}{16} - \dfrac{y^2}{16} = 1$

   $\dfrac{x^2}{4^2} - \dfrac{y^2}{4^2} = 1$

a) a = 4 and b = 4, so the asymptotes are $y = \dfrac{4}{4}x$ and $y = -\dfrac{4}{4}x$, or y = x and y = -x. We sketch them.

b) Replacing y with 0 and solving for x, we get $x = \pm 4$, so the intercepts are (4,0) and (-4,0).

c) We plot the intercepts and draw smooth curves through them that approach the asymptotes.

   $\dfrac{x^2}{16} - \dfrac{y^2}{16} = 1$

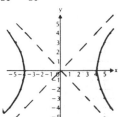

2. $\dfrac{y^2}{9} - \dfrac{x^2}{9} = 1$

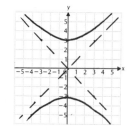

$\underline{3}$. $\dfrac{y^2}{16} - \dfrac{x^2}{9} = 1$

$\dfrac{y^2}{4^2} - \dfrac{x^2}{3^2} = 1$

a) a = 3 and b = 4, so the asymptotes are $y = \dfrac{4}{3}x$ and $y = -\dfrac{4}{3}x$. We sketch them.

b) Replacing x with 0 and solving for y, we get y = ±4, so the intercepts are (0,4) and (0,-4).

c) We plot the intercepts and draw smooth curves through them that approach the asymptotes.

$\dfrac{y^2}{16} - \dfrac{x^2}{9} = 1$

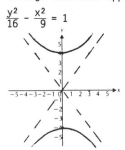

$\underline{4}$. $\dfrac{x^2}{9} - \dfrac{y^2}{4} = 1$

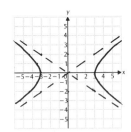

$\underline{5}$. $\dfrac{x^2}{25} - \dfrac{y^2}{36} = 1$

$\dfrac{x^2}{5^2} - \dfrac{y^2}{6^2} = 1$

a) a = 5 and b = 6, so the asymptotes are $y = \dfrac{6}{5}x$ and $y = -\dfrac{6}{5}x$. We sketch them.

b) Replacing y with 0 and solving for x, we get x = ±5, so the intercepts are (5,0) and (-5,0).

c) We plot the intercepts and draw smooth curves through them that approach the asymptotes.

$\dfrac{x^2}{25} - \dfrac{y^2}{36} = 1$

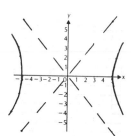

$\underline{6}$. $\dfrac{y^2}{9} - \dfrac{x^2}{25} = 1$

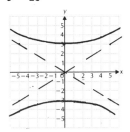

$\underline{7}$. $x^2 - y^2 = 4$

$\dfrac{x^2}{4} - \dfrac{y^2}{4} = 1$ Multiplying by $\dfrac{1}{4}$

$\dfrac{x^2}{2^2} - \dfrac{y^2}{2^2} = 1$

a) a = 2 and b = 2, so the asymptotes are $y = \dfrac{2}{2}x$ and $y = -\dfrac{2}{2}x$, or y = x and y = -x. We sketch them.

b) Replacing y with 0 and solving for x, we get x = ±2, so the intercepts are (2,0) and (-2,0).

c) We plot the intercepts and draw smooth curves through them that approach the asymptotes.

$x^2 - y^2 = 4$

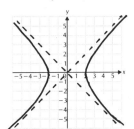

$\underline{8}$. $y^2 - x^2 = 25$

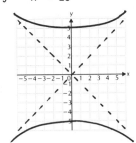

$\underline{9}$. $4y^2 - 9x^2 = 36$

$\dfrac{y^2}{9} - \dfrac{x^2}{4} = 1$ Multiplying by $\dfrac{1}{36}$

$\dfrac{y^2}{3^2} - \dfrac{x^2}{2^2} = 1$

a) a = 2 and b = 3, so the asymptotes are $y = \dfrac{3}{2}x$ and $y = -\dfrac{3}{2}x$. We sketch them.

b) Replacing x with 0 and solving for y, we get y = ±3, so the intercepts are (0,3) and (0,-3).

9. (continued)

    c) We plot the intercepts and draw smooth curves through them that approach the asymptotes.

$4y^2 - 9x^2 = 36$

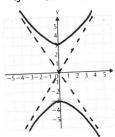

10. $25x^2 - 16y^2 = 400$

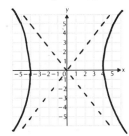

11. $xy = 6$

    $y = \dfrac{6}{x}$ Solving for y

We find some solutions, keeping the results in a table.

x	y
1	6
2	3
3	2
6	1
$\frac{1}{2}$	12
$\frac{1}{3}$	18
-1	-6
-2	-3
-3	-2
-6	-1
$-\frac{1}{2}$	-12
$-\frac{1}{3}$	-18

xy = 6

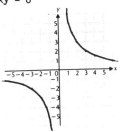

Note that we cannot use 0 for x. The x-axis and the y-axis are the asymptotes.

12. $xy = -4$

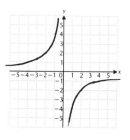

13. $xy = -9$

    $y = -\dfrac{9}{x}$ Solving for y

x	y
1	-9
3	-3
9	-1
$\frac{1}{2}$	-18
-1	9
-3	3
-9	1
$-\frac{1}{2}$	18

xy = -9

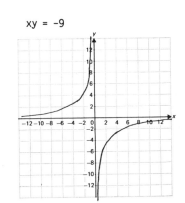

Note that we cannot use 0 for x. The x-axis and the y-axis are the asymptotes.

14. $xy = 3$

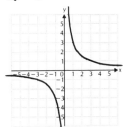

15. $xy = -1$

    $y = -\dfrac{1}{x}$ Solving for y

x	y
1	-1
2	$-\frac{1}{2}$
4	$-\frac{1}{4}$
$\frac{1}{2}$	-2
$\frac{1}{4}$	-4
-1	1
-2	$\frac{1}{2}$
-4	$\frac{1}{4}$
$-\frac{1}{2}$	2
$-\frac{1}{4}$	4

xy = -1

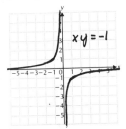

Note that we cannot use 0 for x. The x-axis and the y-axis are the asymptotes.

16. xy = -2

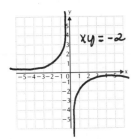

17. xy = 2

$y = \dfrac{2}{x}$ Solving for x

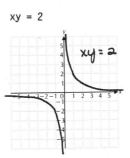

x	y
1	2
2	1
4	$\frac{1}{2}$
$\frac{1}{2}$	4
-1	-2
-2	-1
-4	$-\frac{1}{2}$
$-\frac{1}{2}$	-4

xy = 2

Note that we cannot use 0 for x.
The x-axis and the y-axis are
the asymptotes.

18. xy = 1

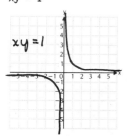

19. $x^2 + y^2 - 10x + 8y - 40 = 0$

Completing the square twice, we obtain an
equivalent equation:

$$(x^2 - 10x) + (y^2 + 8y) = 40$$
$$(x^2 - 10x + 25) + (y^2 + 8y + 16) = 40 + 25 + 16$$
$$(x - 5)^2 + (y + 4)^2 = 81$$

The graph is a circle.

20. parabola

21. $9x^2 - 4y^2 - 36 = 0$
$$9x^2 - 4y^2 = 36$$
$$\frac{x^2}{4} - \frac{y^2}{9} = 1$$

The graph is a hyperbola.

22. parabola

23. $4x^2 + 25y^2 - 100 = 0$
$$4x^2 + 25y^2 = 100$$
$$\frac{x^2}{25} + \frac{y^2}{4} = 1$$

The graph is an ellipse.

24. circle

25. $$x^2 + y^2 = 2x + 4y + 4$$
$$x^2 - 2x + y^2 - 4y = 4$$
$$(x^2 - 2x + 1) + (y^2 - 4y + 4) = 4 + 1 + 4$$
$$(x - 1)^2 + (y - 2)^2 = 9$$

The graph is a circle.

26. circle

27. $$4x^2 = 64 - y^2$$
$$4x^2 + y^2 = 64$$
$$\frac{x^2}{16} + \frac{y^2}{64} = 1$$

The graph is an ellipse.

28. hyperbola

29. $x - \dfrac{3}{y} = 0$
$$x = \frac{3}{y}$$
$$xy = 3$$

The graph is a hyperbola.

30. parabola

31. $y + 6x = x^2 + 6$
$$y = x^2 - 6x + 6$$
The graph is a parabola.

32. hyperbola

33. $$9y^2 = 36 + 4x^2$$
$$9y^2 - 4x^2 = 36$$
$$\frac{y^2}{4} - \frac{x^2}{9} = 1$$

The graph is a hyperbola.

34. circle

35. $\sqrt[3]{125t^{15}} = \sqrt[3]{5^3 \cdot (t^5)^3} = 5t^5$

36. $\pm i\sqrt{5}$

37. $\dfrac{4\sqrt{2} - 5\sqrt{3}}{6\sqrt{3} - 8\sqrt{2}} = \dfrac{4\sqrt{2} - 5\sqrt{3}}{6\sqrt{3} - 8\sqrt{2}} \cdot \dfrac{6\sqrt{3} + 8\sqrt{2}}{6\sqrt{3} + 8\sqrt{2}}$

$= \dfrac{24\sqrt{6} + 32 \cdot 2 - 30 \cdot 3 - 40\sqrt{6}}{36 \cdot 3 - 64 \cdot 2}$

$= \dfrac{-26 - 16\sqrt{6}}{-20}$

$= \dfrac{-2(13 + 8\sqrt{6})}{-2 \cdot 10}$

$= \dfrac{13 + 8\sqrt{6}}{10}$

38. Smaller plane: 400 mph; larger plane: 720 mph

39. Since the intercepts are (0,8) and (0,-8), we know that the hyperbola is of the form $\dfrac{y^2}{b^2} - \dfrac{x^2}{a^2} = 1$ and that b = 8. The equations of the asymptotes tell us that b/a = 4, so

$\dfrac{8}{a} = 4$

$a = 2.$

The equation is $\dfrac{y^2}{8^2} - \dfrac{x^2}{2^2} = 1$, or $\dfrac{y^2}{64} - \dfrac{x^2}{4} = 1.$

40. $\dfrac{x^2}{64} - \dfrac{y^2}{1024} = 1$

41. Center: (2,-1)

Vertices: (-1,-1), (5,-1)

Asymptotes: $y + 1 = \dfrac{4}{3}(x - 2)$, $y + 1 = -\dfrac{4}{3}(x - 2)$

$\dfrac{(x - 2)^2}{9} - \dfrac{(y + 1)^2}{16} = 1$

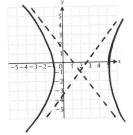

42. $\dfrac{(x + 3)^2}{1} - \dfrac{(y - 2)^2}{4} = 1$

Center: (-3,2)

Vertices: (-4,2), (-2,2)

Asymptotes: $y - 2 = 2(x + 3)$, $y - 2 = -2(x + 3)$

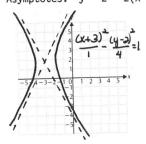

43. $4y^2 - 25x^2 - 8y - 100x - 196 = 0$

$4(y^2 - 2y) - 25(x^2 + 4x) = 196$

$4(y^2 - 2y + 1) - 25(x^2 + 4x + 4) = 196 + 4 - 100$

$4(y - 1)^2 - 25(x + 2)^2 = 100$

$\dfrac{(y - 1)^2}{25} - \dfrac{(x + 2)^2}{4} = 1$ Standard form

Center: (-2,1)

Vertices: (-2,6), (-2,-4)

Asymptotes: $y - 1 = \dfrac{5}{2}(x + 2)$, $y - 1 = -\dfrac{5}{2}(x + 2)$

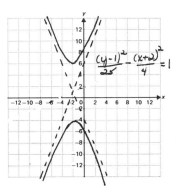

44. $\dfrac{(x - 2)^2}{9} - \dfrac{(y + 1)^2}{9} = 1$

Center: (2,-1)

Vertices: (5,-1), (-1,-1)

Asymptotes: $y + 1 = x - 2$, $y + 1 = -(x - 2)$

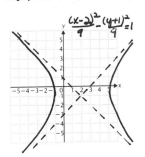

Exercise Set 10.6

<u>1.</u> $x^2 + y^2 = 25,$ (1)
  $y - x = 1$ (2)

First solve Eq. (2) for y.
$y = x + 1$ (3)

Then substitute x + 1 for y in Eq. (1) and solve for x.

$$x^2 + y^2 = 25$$
$$x^2 + (x + 1)^2 = 25$$
$$x^2 + x^2 + 2x + 1 = 25$$
$$2x^2 + 2x - 24 = 0$$
$$x^2 + x - 12 = 0 \quad \text{Multiplying by } \tfrac{1}{2}$$
$$(x + 4)(x - 3) = 0 \quad \text{Factoring}$$

$x + 4 = 0$ or $x - 3 = 0$ Principle of zero
                          products
  $x = -4$ or $x = 3$

Now substitute these numbers into Eq. (3) and solve for y.
$y = -4 + 1 = -3$
$y = 3 + 1 = 4$

The pairs (-4,-3) and (3,4) check, so they are the solutions.

We sketch the graphs to confirm the solutions.

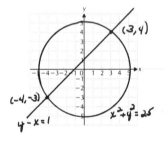

<u>2.</u> (-8,-6), (6,8)

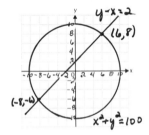

<u>3.</u> $4x^2 + 9y^2 = 36,$ (1)
  $3y + 2x = 6$ (2)

First solve Eq. (2) for y.
$3y = -2x + 6$
$y = -\tfrac{2}{3}x + 2$ (3)

Then substitute $-\tfrac{2}{3}x + 2$ for y in Eq. (1) and solve for x.

$$4x^2 + 9y^2 = 36$$
$$4x^2 + 9\left(-\tfrac{2}{3}x + 2\right)^2 = 36$$
$$4x^2 + 9\left(\tfrac{4}{9}x^2 - \tfrac{8}{3}x + 4\right) = 36$$
$$4x^2 + 4x^2 - 24x + 36 = 36$$
$$8x^2 - 24x = 0$$
$$x^2 - 3x = 0$$
$$x(x - 3) = 0$$

$x = 0$ or $x = 3$

Now substitute these numbers in Eq. (3) and solve for y.

$y = -\tfrac{2}{3} \cdot 0 + 2 = 2$

$y = -\tfrac{2}{3} \cdot 3 + 2 = 0$

The pairs (0,2) and (3,0) check, so they are the solutions.

We sketch the graphs to confirm the solutions.

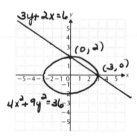

<u>4.</u> (2,0), (0,3)

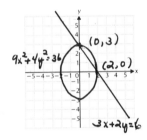

5. $y^2 = x + 3$,   (1)

$2y = x + 4$ (2)

First solve Eq. (2) for x.

$2y - 4 = x$ (3)

Then substitute $2y - 4$ for x in Eq. (1) and solve for y.

$$y^2 = x + 3$$
$$y^2 = (2y - 4) + 3$$
$$y^2 = 2y - 1$$
$$y^2 - 2y + 1 = 0$$
$$(y - 1)(y - 1) = 0$$

$y = 1$ or $y = 1$

Now substitute 1 for y in Eq. (3) and solve for x.

$2 \cdot 1 - 4 = x$

$-2 = x$

The pair $(-2,1)$ checks. It is the solution.

We sketch the graphs to confirm the solution.

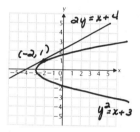

6. $(2,4)$, $(1,1)$

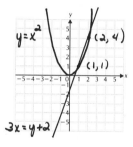

7. $x^2 - xy + 3y^2 = 27$,   (1)

$x - y = 2$ (2)

First solve Eq. (2) for y.

$x - 2 = y$ (3)

Then substitute $x - 2$ for y in Eq. (1) and solve for x.

$$x^2 - xy + 3y^2 = 27$$
$$x^2 - x(x - 2) + 3(x - 2)^2 = 27$$
$$x^2 - x^2 + 2x + 3x^2 - 12x + 12 = 27$$
$$3x^2 - 10x - 15 = 0$$

$$x = \frac{-(-10) \pm \sqrt{(-10)^2 - 4(3)(-15)}}{2 \cdot 3}$$

$$= \frac{10 \pm \sqrt{100 + 180}}{6}$$

$$= \frac{10 \pm \sqrt{280}}{6}$$

$$= \frac{10 \pm 2\sqrt{70}}{6}$$

$$= \frac{5 \pm \sqrt{70}}{3}$$

Now substitute these numbers in Eq. (3) and solve for y.

$$y = \frac{5 + \sqrt{70}}{3} - 2 = \frac{-1 + \sqrt{70}}{3}$$

$$y = \frac{5 - \sqrt{70}}{3} - 2 = \frac{-1 - \sqrt{70}}{3}$$

The pairs $\left(\dfrac{5 + \sqrt{70}}{3}, \dfrac{-1 + \sqrt{70}}{3}\right)$ and $\left(\dfrac{5 - \sqrt{70}}{3}, \dfrac{-1 - \sqrt{70}}{3}\right)$ check, so they are the solutions.

8. $\left(\dfrac{11}{4}, -\dfrac{9}{8}\right)$, $(1,-2)$

9. $x^2 + 4y^2 = 25$,   (1)

$x + 2y = 7$ (2)

First solve Eq. (2) for x.

$x = -2y + 7$ (3)

Then substitute $-2y + 7$ for x in Eq. (1) and solve for y.

$$x^2 + 4y^2 = 25$$
$$(-2y + 7)^2 + 4y^2 = 25$$
$$4y^2 - 28y + 49 + 4y^2 = 25$$
$$8y^2 - 28y + 24 = 0$$
$$2y^2 - 7y + 6 = 0$$
$$(2y - 3)(y - 2) = 0$$

$y = \dfrac{3}{2}$ or $y = 2$

Now substitute these numbers in Eq. (3) and solve for x.

$x = -2 \cdot \dfrac{3}{2} + 7 = 4$

$x = -2 \cdot 2 + 7 = 3$

The pairs $\left(4, \dfrac{3}{2}\right)$ and $(3,2)$ check, so they are the solutions.

9. (continued)

We sketch the graphs to confirm the solutions.

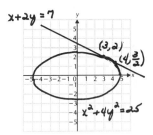

10. $\left(-\frac{5}{3}, -\frac{13}{3}\right)$, (3,5)

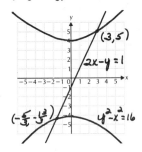

11. $x^2 - xy + 3y^2 = 5$,   (1)

    $x - y = 2$ (2)

First solve Eq. (2) for y.

$x - 2 = y$ (3)

Then substitute $x - 2$ for y in Eq. (1) and solve for x.

$$x^2 - xy + 3y^2 = 5$$
$$x^2 - x(x - 2) + 3(x - 2)^2 = 5$$
$$x^2 - x^2 + 2x + 3x^2 - 12x + 12 = 5$$
$$3x^2 - 10x + 7 = 0$$
$$(3x - 7)(x - 1) = 0$$

$x = \frac{7}{3}$ or $x = 1$

Now substitute these numbers in Eq. (3) and solve for y.

$y = \frac{7}{3} - 2 = \frac{1}{3}$

$y = 1 - 2 = -1$

The pairs $\left(\frac{7}{3}, \frac{1}{3}\right)$ and (1,-1) check, so they are the solutions.

12. $\left(\frac{3 + \sqrt{7}}{2}, \frac{-1 + \sqrt{7}}{2}\right)$, $\left(\frac{3 - \sqrt{7}}{2}, \frac{-1 - \sqrt{7}}{2}\right)$

13. $3x + y = 7$,   (1)

    $4x^2 + 5y = 24$ (2)

First solve Eq. (1) for y.

$y = 7 - 3x$ (3)

Then substitute $7 - 3x$ for y in Eq. (2) and solve for x.

$$4x^2 + 5y = 24$$
$$4x^2 + 5(7 - 3x) = 24$$
$$4x^2 + 35 - 15x = 24$$
$$4x^2 - 15x + 11 = 0$$
$$(4x - 11)(x - 1) = 0$$

$x = \frac{11}{4}$ or $x = 1$

Now substitute these numbers in Eq. (3) and solve for y.

$y = 7 - 3 \cdot \frac{11}{4} = -\frac{5}{4}$

$y = 7 - 3 \cdot 1 = 4$

The pairs $\left(\frac{11}{4}, -\frac{5}{4}\right)$ and (1,4) check, so they are the solutions.

We sketch the graphs to confirm the solutions.

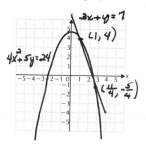

14. $\left(-3, \frac{5}{2}\right)$, (3,1)

15. $a + b = 7$,   (1)

    $ab = 4$ (2)

First solve Eq. (1) for a.

$a = -b + 7$ (3)

Then substitute $-b + 7$ for a in Eq. (2) and solve for b.

$(-b + 7)b = 4$

$-b^2 + 7b = 4$

$0 = b^2 - 7b + 4$

$b = \frac{-(-7) \pm \sqrt{(-7)^2 - 4 \cdot 1 \cdot 4}}{2 \cdot 1}$

$b = \frac{7 \pm \sqrt{33}}{2}$

Now substitute these numbers in Eq. (3) and solve for a.

$a = -\left(\frac{7 + \sqrt{33}}{2}\right) + 7 = \frac{7 - \sqrt{33}}{2}$

$a = -\left(\frac{7 - \sqrt{33}}{2}\right) + 7 = \frac{7 + \sqrt{33}}{2}$

15. (continued)

The pairs $\left(\dfrac{7 - \sqrt{33}}{2}, \dfrac{7 + \sqrt{33}}{2}\right)$ and

$\left(\dfrac{7 + \sqrt{33}}{2}, \dfrac{7 - \sqrt{33}}{2}\right)$ check, so they are the

solutions.

We sketch the graphs to confirm the solutions.

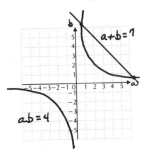

16. (1,-7), (-7,1)

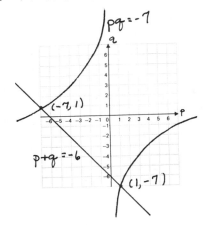

17. $2a + b = 1$,   (1)

$b = 4 - a^2$ (2)

Eq. (2) is already solved for b. Substitute $4 - a^2$ for b in Eq. (1) and solve for a.

$2a + 4 - a^2 = 1$

$\qquad 0 = a^2 - 2a - 3$

$\qquad 0 = (a - 3)(a + 1)$

$a = 3$ or $a = -1$

Substitute these numbers in Eq. (2) and solve for b.

$b = 4 - 3^2 = -5$

$b = 4 - (-1)^2 = 3$

The pairs (3,-5) and (-1,3) check.

We sketch the graphs to
confirm the solutions.

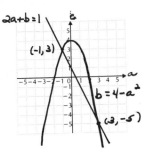

18. (3,0), $\left(-\dfrac{9}{5}, \dfrac{8}{5}\right)$

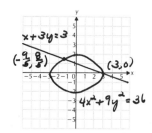

19. $a^2 + b^2 = 89$,   (1)

$a - b = 3$ (2)

First solve Eq. (2) for a.

$a = b + 3$ (3)

Then substitute $b + 3$ for a in Eq. (1) and solve for b.

$\quad (b + 3)^2 + b^2 = 89$

$b^2 + 6b + 9 + b^2 = 89$

$\quad 2b^2 + 6b - 80 = 0$

$\quad\ \ b^2 + 3b - 40 = 0$

$\quad (b + 8)(b - 5) = 0$

$b = -8$ or $b = 5$

Substitute these numbers in Eq. (3) and solve for a.

$a = -8 + 3 = -5$

$a = 5 + 3 = 8$

The pairs (-5,-8) and (8,5) check.

We sketch the graphs to confirm the solutions.

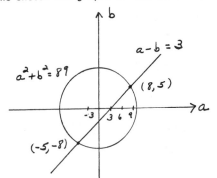

20. (1,4), (4,1)

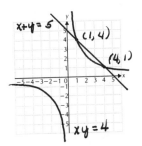

<u>21.</u> $x^2 + y^2 = 5$,   (1)

$x - y = 8$ (2)

First solve Eq. (2) for x.

$x = y + 8$ (3)

Then substitute y + 8 for x in Eq. (1) and solve for y.

$(y + 8)^2 + y^2 = 5$

$y^2 + 16y + 64 + y^2 = 5$

$2y^2 + 16y + 59 = 0$

$y = \dfrac{-16 \pm \sqrt{(16)^2 - 4(2)(59)}}{2 \cdot 2}$

$y = \dfrac{-16 \pm \sqrt{-216}}{4}$

$y = \dfrac{-16 \pm 6i\sqrt{6}}{4}$

$y = -4 \pm \dfrac{3}{2}i\sqrt{6}$

Now substitute these numbers in Eq. (3) and solve for x.

$x = -4 + \dfrac{3}{2}i\sqrt{6} + 8 = 4 + \dfrac{3}{2}i\sqrt{6}$

$x = -4 - \dfrac{3}{2}i\sqrt{6} + 8 = 4 - \dfrac{3}{2}i\sqrt{6}$

The pairs $\left(4 + \dfrac{3}{2}i\sqrt{6},\ -4 + \dfrac{3}{2}i\sqrt{6}\right)$ and $\left(4 - \dfrac{3}{2}i\sqrt{6},\ -4 - \dfrac{3}{2}i\sqrt{6}\right)$ check.

We sketch the graphs to confirm that there are no real-number solutions.

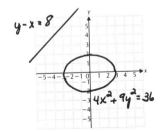

<u>22.</u> $\left(-\dfrac{72}{13} + \dfrac{6}{13}i\sqrt{51},\ \dfrac{32}{13} + \dfrac{6}{13}i\sqrt{51}\right)$,

$\left(-\dfrac{72}{13} - \dfrac{6}{13}i\sqrt{51},\ \dfrac{32}{13} - \dfrac{6}{13}i\sqrt{51}\right)$

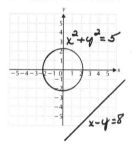

<u>23.</u> <u>Familiarize.</u> Let x = one number and y = the other number.

<u>Translate.</u> We translate to a system of equations.

The sum of two numbers is 12.

$x + y \qquad = 12$

The sum of their squares is 90.

$x^2 + y^2 \qquad = 90$

<u>Carry out.</u> We solve the system:

$x + y = 12$,   (1)

$x^2 + y^2 = 90$ (2)

First solve Eq. (1) for y.

$y = 12 - x$ (3)

Then substitute 12 - x for y in Eq. (2) and solve for x.

$x^2 + y^2 = 90$

$x^2 + (12 - x)^2 = 90$

$x^2 + 144 - 24x + x^2 = 90$

$2x^2 - 24x + 54 = 0$

$x^2 - 12x + 27 = 0$

$(x - 9)(x - 3) = 0$

$x = 9$ or $x = 3$

Now substitute these numbers in Eq. (3) and solve for y.

$y = 12 - 9 = 3$

$y = 12 - 3 = 9$

<u>Check.</u> If the numbers are 9 and 3, the sum is 9 + 3, or 12. The sum of their squares is 81 + 9, or 90. The numbers check. The pair (3,9) does not give us another solution.

<u>State.</u> The numbers are 9 and 3.

<u>24.</u> 8 and 7

<u>25.</u> <u>Familiarize.</u> We first make a drawing. We let ℓ and w represent the length and width, respectively.

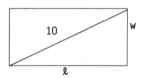

<u>Translate.</u> The perimeter is 28 cm.

$2ℓ + 2w = 28$, or $ℓ + w = 14$

Using the Pythagorean property we have another equation.

$ℓ^2 + w^2 = 10^2$, or $ℓ^2 + w^2 = 100$

25. (continued)

Carry out. We solve the system:

$\ell + w = 14$,   (1)

$\ell^2 + w^2 = 100$ (2)

First solve Eq. (1) for w.

$w = 14 - \ell$ (3)

Then substitute $14 - \ell$ for w in Eq. (2) and solve for ℓ.

$$\ell^2 + w^2 = 100$$
$$\ell^2 + (14 - \ell)^2 = 100$$
$$\ell^2 + 196 - 28\ell + \ell^2 = 100$$
$$2\ell^2 - 28\ell + 96 = 0$$
$$\ell^2 - 14\ell + 48 = 0$$
$$(\ell - 8)(\ell - 6) = 0$$

$\ell = 8$ or $\ell = 6$

If $\ell = 8$, then $w = 14 - 8$, or 6. If $\ell = 6$, then $w = 14 - 6$, or 8. Since the length is usually considered to be longer than the width, we have the solution $\ell = 8$ and $w = 6$, or (8,6).

Check. If $\ell = 8$ and $w = 6$, then the perimeter is $2 \cdot 8 + 2 \cdot 6$, or 28. The length of a diagonal is $\sqrt{8^2 + 6^2}$, or $\sqrt{100}$, or 10. The numbers check.

State. The length is 8 cm, and the width is 6 cm.

26. 1 m by 2 m

27. Familiarize. We first make a drawing. Let ℓ = the length and w = the width of the rectangle.

Translate.

    Area: $\ell w = 20$

    Perimeter: $2\ell + 2w = 18$, or $\ell + w = 9$

Carry out. We solve the system:

Solve the second equation for ℓ: $\ell = 9 - w$

Substitute $9 - w$ for ℓ in the first equation and solve for w.

$$(9 - w)w = 20$$
$$9w - w^2 = 20$$
$$0 = w^2 - 9w + 20$$
$$0 = (w - 5)(w - 4)$$

$w = 5$ or $w = 4$

If $w = 5$, then $\ell = 9 - w$, or 4. If $w = 4$, then $\ell = 9 - 4$, or 5. Since length is usually considered more than width, we have the solution $\ell = 5$ and $w = 4$, or (5,4).

Check. If $\ell = 5$ and $w = 4$, the area is $5 \cdot 4$, or 20. The perimeter is $2 \cdot 5 + 2 \cdot 4$, or 18. The numbers check.

State. The length is 5 in. and the width is 4 in.

28. 1 yd by 2 yd

29. Familiarize. We make a drawing of the field. Let ℓ = the length and w = the width.

Since it takes 210 yd of fencing to enclose the field, we know that the perimeter is 210 yd.

Translate.

    Perimeter: $2\ell + 2w = 210$, or $\ell + w = 105$

    Area: $\ell w = 2250$

Carry out. We solve the system:

Solve the first equation for ℓ: $\ell = 105 - w$

Substitute $105 - w$ for ℓ in the second equation and solve for w.

$$(105 - w)w = 2250$$
$$105w - w^2 = 2250$$
$$0 = w^2 - 105w + 2250$$
$$0 = (w - 30)(w - 75)$$

$w = 30$ or $w = 75$

If $w = 30$, then $\ell = 105 - 30$, or 75. If $w = 75$, then $\ell = 105 - 75$, or 30. Since length is usually considered more than width, we have the solution $\ell = 75$ and $w = 30$, or (75,30).

Check. If $\ell = 75$ and $w = 30$, the perimeter is $2 \cdot 75 + 2 \cdot 30$, or 210. The area is 75(30), or 2250. The numbers check.

State. The length is 75 yd and the width is 30 yd.

30. 5 ft by 12 ft

31. $3x^2 + 6 = 5x$

    $3x^2 - 5x + 6 = 0$

$$x = \frac{-(-5) \pm \sqrt{(-5)^2 - 4 \cdot 3 \cdot 6}}{2 \cdot 3}$$

$$x = \frac{5 \pm \sqrt{-47}}{6}$$

$$x = \frac{5 \pm i\sqrt{47}}{6}$$

32. $\dfrac{5 \pm \sqrt{97}}{6}$

33. $\sqrt{48} = \sqrt{16 \cdot 3} = 4\sqrt{3}$

34. $2a^6 d^2 \sqrt[4]{2d}$

35. $(x - h)^2 + (y - k)^2 = r^2$ Standard form

We substitute each ordered pair for (x,y).

$(4 - h)^2 + (6 - k)^2 = r^2$
$(-6 - h)^2 + (2 - k)^2 = r^2$
$(1 - h)^2 + (-3 - k)^2 = r^2$

$16 - 8h + h^2 + 36 - 12k + k^2 = r^2$
$36 + 12h + h^2 + 4 - 4k + k^2 = r^2$
$1 - 2h + h^2 + 9 + 6k + k^2 = r^2$

$-8h - 12k + 52 + h^2 + k^2 = r^2$ (1)
$12h - 4k + 40 + h^2 + k^2 = r^2$ (2)
$-2h + 6k + 10 + h^2 + k^2 = r^2$ (3)

Multiplying equation (1) by −1 and adding it to equation (2) we get

$20h + 8k - 12 = 0$, or $5h + 2k = 3$

Multiplying equation (1) by −1 and adding it to equation (3) we get

$6h + 18k - 42 = 0$, or $h + 3k = 7$

Solving the system

$5h + 2k = 3$,
$h + 3k = 7$

we get $\left[-\dfrac{5}{13}, \dfrac{32}{13}\right]$.

We now substitute $-\dfrac{5}{13}$ for h and $\dfrac{32}{13}$ for k in one of the original equations and solve for r^2.

$(4 - h)^2 + (6 - k)^2 = r^2$

$\left[4 - \left(-\dfrac{5}{13}\right)\right]^2 + \left[6 - \dfrac{32}{13}\right]^2 = r^2$

$\left(\dfrac{52}{13} + \dfrac{5}{13}\right)^2 + \left(\dfrac{78}{13} - \dfrac{32}{13}\right)^2 = r^2$

$\left(\dfrac{57}{13}\right)^2 + \left(\dfrac{46}{13}\right)^2 = r^2$

$\dfrac{3249}{169} + \dfrac{2116}{169} = r^2$

$\dfrac{5365}{169} = r^2$

The equation of the circle that passes through the points (4,6), (−6,2), and (1,−3) is

$\left[x - \left(-\dfrac{5}{13}\right)\right]^2 + \left[y - \dfrac{32}{13}\right]^2 = \dfrac{5365}{169}$, or

$\left[x + \dfrac{5}{13}\right]^2 + \left[y - \dfrac{32}{13}\right]^2 = \dfrac{5365}{169}$.

36. $\dfrac{1}{4}$ and 8

37. It is helpful to draw a picture.

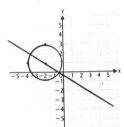

37. (continued)

Let (h,k) represent the point on the line $5x + 8y = -2$ which is the center of a circle that passes through the points (−2,3) and (−4,1). The distance between (h,k) and (−2,3) is the same as the distance between (h,k) and (−4,1). This gives us one equation:

$\sqrt{[h - (-2)]^2 + (k - 3)^2} = \sqrt{[h - (-4)]^2 + (k - 1)^2}$

$(h + 2)^2 + (k - 3)^2 = (h + 4)^2 + (k - 1)^2$

$h^2 + 4h + 4 + k^2 - 6k + 9 = h^2 + 8h + 16 + k^2 - 2k + 1$

$4h - 6k + 13 = 8h - 2k + 17$

$-4h - 4k = 4$

$h + k = -1$

We get a second equation by substituting (h,k) in $5x + 8y = -2$.

$5h + 8k = -2$

We now solve the following system:

$h + k = -1$
$5h + 8k = -2$

The solution, which is the center of the circle, is (−2,1).

Next we find the length of the radius. We can find the distance between either (−2,3) or (−4,1) and the center (−2,1). We use (−2,3).

$r = \sqrt{[-2 - (-2)]^2 + (1 - 3)^2}$

$= \sqrt{0^2 + (-2)^2}$

$= \sqrt{4}$

$= 2$

We now have an equation knowing that the center is (−2,1) and the radius is 2.

$(x - h)^2 + (y - k)^2 = r^2$

$[x - (-2)]^2 + (y - 1)^2 = 2^2$

$(x + 2)^2 + (y - 1)^2 = 4$

38. One piece 50.72 cm long, the other 49.28 cm long.

39. See the answer section in the text.

40. $\dfrac{x^2}{10} + \dfrac{y^2}{5} = 1$

41. $\dfrac{x^2}{a^2} - \dfrac{y^2}{b^2} = 1$

Substitute each ordered pair for (x,y).

$\dfrac{(-3)^2}{a^2} - \dfrac{\left[-\dfrac{3\sqrt{5}}{2}\right]^2}{b^2} = 1$,

$\dfrac{(-3)^2}{a^2} - \dfrac{\left[\dfrac{3\sqrt{5}}{2}\right]^2}{b^2} = 1$,

$\dfrac{\left[-\dfrac{3}{2}\right]^2}{a^2} - \dfrac{0^2}{b^2} = 1$

$\dfrac{9}{a^2} - \dfrac{45}{4b^2} = 1$, (1)

$\dfrac{9}{a^2} - \dfrac{45}{4b^2} = 1$, (2)

$\dfrac{9}{4a^2} = 1$ (3)

41. (continued)

Note that Eq. (1) and Eq. (2) are identical.
Multiply both sides of Eq. (3) by 4:

$$\frac{9}{a^2} = 4$$

Substitute 4 for $\frac{9}{a^2}$ in Eq. (1) and solve for b^2.

$$4 - \frac{45}{4b^2} = 1$$

$$16b^2 - 45 = 4b^2$$

$$12b^2 = 45$$

$$b^2 = \frac{45}{12}, \text{ or } \frac{15}{4}$$

Solve Eq. (3) for a^2.

$$\frac{9}{4a^2} = 1$$

$$\frac{9}{4} = a^2$$

The equation of the hyperbola is $\frac{x^2}{\frac{9}{4}} - \frac{y^2}{\frac{15}{4}} = 1$.

42. $\left(\frac{2a^2 + 6ab + 5b^2}{2a + 4b}, \frac{-2ab - 3b^2}{2a + 4b}\right)$

43. $\frac{x}{a - b} + \frac{y}{a + b} = 1$,   (1)

$x^2 - y^2 = (a - b)^2$ (2)

Solve Eq. (1) for x.

$x(a + b) + y(a - b) = (a - b)(a + b)$

$x(a + b) + ay - by = a^2 - b^2$

$x(a + b) = a^2 - b^2 - ay + by$

$x = \frac{a^2 - b^2 - ay + by}{a + b}$ (3)

Substitute for x in Eq. (2) and solve for y.

$\left[\frac{a^2 - b^2 - ay + by}{a + b}\right]^2 - y^2 = (a - b)^2$

$(a^2 - b^2 - ay + by)^2 - y^2(a + b)^2 =$
    $(a - b)^2(a + b)^2$

$a^4 - 2a^2b^2 + b^4 - 2a^3y + 2a^2by + 2ab^2y - 2b^3y +$
    $a^2y^2 - 2aby^2 + b^2y^2 - a^2y^2 - 2aby^2 - b^2y^2 =$
    $a^4 - 2a^2b^2 + b^4$

$-2a^3y + 2a^2by + 2ab^2y - 2b^3y - 4aby^2 = 0$
                    Simplifying

$-a^3y + a^2by + ab^2y - b^3y - 2aby^2 = 0$
                    Multiplying by $\frac{1}{2}$

$y(-a^3 + a^2b + ab^2 - b^3 - 2aby) = 0$

$y = 0$ or $-a^3 + a^2b + ab^2 - b^3 - 2aby = 0$

$y = 0$ or $y = \frac{-a^3 + a^2b + ab^2 - b^3}{2ab} =$

    $-\frac{(a^2 - b^2)(a - b)}{2ab}$

Substitute in Eq. (3) to find x.

43. (continued)

When $y = 0$, $x = \frac{a^2 - b^2 - a(0) + b(0)}{a + b} = \frac{a^2 - b^2}{a + b} =$

$\frac{(a + b)(a - b)}{a + b} = a - b$.

When $y = -\frac{(a^2 - b^2)(a - b)}{2ab}$,

$x = \frac{a^2 - b^2 - a\left[-\frac{(a^2-b^2)(a-b)}{2ab}\right] + b\left[-\frac{(a^2-b^2)(a-b)}{2ab}\right]}{a + b} =$

$\frac{2ab(a^2 - b^2) + (a^2 - b^2)(a - b)(a - b)}{2ab(a + b)} =$

$\frac{2ab(a + b)(a - b) + (a + b)(a - b)^3}{2ab(a + b)} =$

$\frac{(a+b)[2ab(a-b) + (a-b)^3]}{2ab(a+b)} = \frac{2ab(a-b) + (a-b)^3}{2ab} =$

$\frac{2a^2b - 2ab^2 + a^3 - 3a^2b + 3ab^2 - b^3}{2ab} =$

$\frac{a^3 - a^2b + ab^2 - b^3}{2ab} = \frac{a^2(a - b) + b^2(a - b)}{2ab} =$

$\frac{(a - b)(a^2 + b^2)}{2ab}$

The solutions are (a - b,0) and
$\left(\frac{(a - b)(a^2 + b^2)}{2ab}, -\frac{(a^2 - b^2)(a - b)}{2ab}\right)$.

Exercise Set 10.7

1. $x^2 + y^2 = 25$,   (1)

$y^2 = x + 5$ (2)

We substitute x + 5 for y^2 in Eq. (1) and solve
for x.

$$x^2 + y^2 = 25$$
$$x^2 + (x + 5) = 25$$
$$x^2 + x - 20 = 0$$
$$(x + 5)(x - 4) = 0$$

$x + 5 = 0$ or $x - 4 = 0$
    $x = -5$ or $x = 4$

Next we substitute these numbers for x in either
Eq. (1) or Eq. (2) and solve for y. Here we use
Eq. (2).

$y^2 = -5 + 5 = 0$ and $y = 0$.

$y^2 = 4 + 5 = 9$ and $y = \pm 3$.

The possible solutions are (-5,0), (4,3), and
(4,-3).

Check:
For (-5,0):

$x^2 + y^2 = 25$		$y^2 = x + 5$	
$(-5)^2 + 0^2$	25	0^2	$-5 + 5$
$25 + 0$		0	0
25			

1. (continued)

For (4,3):

$x^2 + y^2 = 25$	
$4^2 + 3^2$	25
$16 + 9$	
25	

$y^2 = x + 5$	
3^2	$4 + 5$
9	9

For (4,-3):

$x^2 + y^2 = 25$	
$4^2 + (-3)^2$	25
$16 + 9$	
25	

$y^2 = x + 5$	
$(-3)^2$	$4 + 5$
9	9

The solutions are (-5,0), (4,3), and (4,-3).

2. (0,0), (1,1)

3. $x^2 + y^2 = 9,$ (1)
$x^2 - y^2 = 9$ (2)

Here we use the addition method.

$$x^2 + y^2 = 9$$
$$\underline{x^2 - y^2 = 9}$$
$$2x^2 \quad\quad = 18 \quad \text{Adding}$$
$$x^2 = 9$$
$$x = \pm3$$

If $x = 3$, $x^2 = 9$, and if $x = -3$, $x^2 = 9$, so substituting 3 or -3 in Eq. (1) give us

$$x^2 + y^2 = 9$$
$$9 + y^2 = 9$$
$$y^2 = 0$$
$$y = 0.$$

The possible solutions are (3,0) and (-3,0).
Check:

$x^2 + y^2 = 9$	
$(\pm3)^2 + (0)^2$	9
$9 + 0$	
9	

$x^2 - y^2 = 9$	
$(\pm3)^2 - (0)^2$	9
$9 - 0$	
9	

The solutions are (3,0) and (-3,0).

4. (0,2), (0,-2)

5. $x^2 + y^2 = 25,$ (1)
$xy = 12$ (2)

First we solve Eq. (2) for y.

$$xy = 12$$
$$y = \frac{12}{x}$$

Then we substitute $\frac{12}{x}$ for y in Eq. (1) and solve for x.

$$x^2 + y^2 = 25$$
$$x^2 + \left(\frac{12}{x}\right)^2 = 25$$
$$x^2 + \frac{144}{x^2} = 25$$
$$x^4 + 144 = 25x^2 \quad \text{Multiplying by } x^2$$
$$x^4 - 25x^2 + 144 = 0$$
$$u^2 - 25u + 144 = 0 \quad \text{Letting } u = x^2$$
$$(u - 9)(u - 16) = 0$$
$$u = 9 \quad \text{or} \quad u = 16$$

We now substitute x^2 for u and solve for x.

$$x^2 = 9 \quad \text{or} \quad x^2 = 16$$
$$x = \pm3 \quad \text{or} \quad x = \pm4$$

Since $y = 12/x$, if $x = 3$, $y = 4$; if $x = -3$, $y = -4$; if $x = 4$, $y = 3$; and if $x = -4$, $y = -3$. The pairs (3,4), (-3,-4), (4,3), (-4,-3) check. They are the solutions.

6. (-5,3), (-5,-3), (4,0)

7. $x^2 + y^2 = 4,$ (1)
$16x^2 + 9y^2 = 144$ (2)

$$-9x^2 - 9y^2 = -36 \quad \text{Multiplying (1) by -9}$$
$$\underline{16x^2 + 9y^2 = 144}$$
$$7x^2 \quad\quad = 108 \quad \text{Adding}$$
$$x^2 = \frac{108}{7}$$
$$x = \pm\sqrt{\frac{108}{7}} = \pm6\sqrt{\frac{3}{7}}$$
$$x = \pm\frac{6\sqrt{21}}{7} \quad \text{Rationalizing the denominator}$$

Substituting $\frac{6\sqrt{21}}{7}$ or $-\frac{6\sqrt{21}}{7}$ for x in Eq. (1) gives us

$$\frac{36 \cdot 21}{49} + y^2 = 4$$
$$y^2 = 4 - \frac{108}{7}$$
$$y^2 = -\frac{80}{7}$$
$$y = \pm\sqrt{-\frac{80}{7}} = \pm4i\sqrt{\frac{5}{7}}$$
$$y = \pm\frac{4i\sqrt{35}}{7}. \quad \text{Rationalizing the denominator}$$

7. (continued)

The pairs $\left(\frac{6\sqrt{21}}{7}, \frac{4i\sqrt{35}}{7}\right)$, $\left(\frac{6\sqrt{21}}{7}, -\frac{4i\sqrt{35}}{7}\right)$,

$\left(-\frac{6\sqrt{21}}{7}, \frac{4i\sqrt{35}}{7}\right)$, and $\left(-\frac{6\sqrt{21}}{7}, -\frac{4i\sqrt{35}}{7}\right)$ check.

They are the solutions.

8. $(0,5)$, $(0,-5)$

9. $x^2 + y^2 = 16$, $x^2 + y^2 = 16$,   (1)
   $\qquad\qquad\qquad$ or
   $y^2 - 2x^2 = 10$ $-2x^2 + y^2 = 10$ (2)

   Here we use the addition method.

   $2x^2 + 2y^2 = 32$ Multiplying (1) by 2
   $\underline{-2x^2 + \;\; y^2 = 10}$
   $\qquad\quad 3y^2 = 42$ Adding
   $\qquad\qquad y^2 = 14$
   $\qquad\qquad\; y = \pm\sqrt{14}$

   Substituting $\sqrt{14}$ or $-\sqrt{14}$ for y in Eq. (1) gives us

   $\qquad x^2 + 14 = 16$
   $\qquad\qquad x^2 = 2$
   $\qquad\qquad\; x = \pm\sqrt{2}$

   The pairs $(-\sqrt{2},-\sqrt{14})$, $(-\sqrt{2},\sqrt{14})$, $(\sqrt{2},-\sqrt{14})$, and $(\sqrt{2},\sqrt{14})$ check. They are the solutions.

10. $(-3,-\sqrt{5})$, $(-3,\sqrt{5})$, $(3,-\sqrt{5})$, $(3,\sqrt{5})$

11. $x^2 + y^2 = 5$,   (1)
    $xy = 2$ (2)

    First we solve Eq. (2) for y.
    $xy = 2$

    $y = \frac{2}{x}$

    Then we substitute $\frac{2}{x}$ for y in Eq. (1) and solve for x.

    $\qquad\quad x^2 + y^2 = 5$

    $\qquad x^2 + \left(\frac{2}{x}\right)^2 = 5$

    $\qquad\quad x^2 + \frac{4}{x^2} = 5$

    $\qquad\quad x^4 + 4 = 5x^2$ Multiplying by x^2
    $x^4 - 5x^2 + 4 = 0$

    $\quad u^2 - 5u + 4 = 0$ Letting $u = x^2$
    $(u - 4)(u - 1) = 0$

    $u = 4$ or $u = 1$
    We now substitute x^2 for u and solve for x.
    $\quad x^2 = 4$ or $x^2 = 1$
    $\quad\; x = \pm 2$ $x = \pm 1$

    Since $y = 2/x$, if $x = 2$, $y = 1$; if $x = -2$, $y = -1$; if $x = 1$, $y = 2$; and if $x = -1$, $y = -2$. The pairs $(2,1)$, $(-2,-1)$, $(1,2)$, $(-1,-2)$ check. They are the solutions.

12. $(4,2)$, $(-4,-2)$, $(2,4)$, $(-2,-4)$

13. $x^2 + y^2 = 13$,   (1)
    $xy = 6$ (2)

    First we solve Eq. (2) for y.
    $xy = 6$

    $y = \frac{6}{x}$

    Then we substitute $\frac{6}{x}$ for y in Eq. (1) and solve for x.

    $\qquad\quad x^2 + y^2 = 13$

    $\qquad x^2 + \left(\frac{6}{x}\right)^2 = 13$

    $\qquad\quad x^2 + \frac{36}{x^2} = 13$

    $\qquad\quad x^4 + 36 = 13x^2$ Multiplying by x^2
    $x^4 - 13x^2 + 36 = 0$

    $\quad u^2 - 13u + 36 = 0$ Letting $u = x^2$
    $(u - 9)(u - 4) = 0$

    $u = 9$ or $u = 4$
    We now substitute x^2 for u and solve for x.
    $\quad x^2 = 9$ or $x^2 = 4$
    $\quad\; x = \pm 3$ $x = \pm 2$

    Since $y = 6/x$, if $x = 3$, $y = 2$; if $x = -3$, $y = -2$; if $x = 2$, $y = 3$; and if $x = -2$, $y = -3$. The pairs $(3,2)$, $(-3,-2)$, $(2,3)$, $(-2,-3)$ check. They are the solutions.

14. $(4,1)$, $(-4,-1)$, $(2,2)$, $(-2,-2)$

15. $3xy + x^2 = 34$,   (1)
    $2xy - 3x^2 = 8$ (2)

    $6xy + 2x^2 = \;\; 68$ Multiplying (1) by 2
    $\underline{-6xy + 9x^2 = -24}$ Multiplying (2) by -3
    $\qquad 11x^2 = \;\; 44$ Adding
    $\qquad\quad x^2 = 4$
    $\qquad\qquad x = \pm 2$

    Substitute for x in Eq. (1) and solve for y.
    When $x = 2$: $3\cdot 2\cdot y + 2^2 = 34$
    $\qquad\qquad\qquad 6y + 4 = 34$
    $\qquad\qquad\qquad\quad 6y = 30$
    $\qquad\qquad\qquad\quad\; y = 5$
    When $x = -2$: $3(-2)(y) + (-2)^2 = 34$
    $\qquad\qquad\qquad\quad -6y + 4 = 34$
    $\qquad\qquad\qquad\qquad -6y = 30$
    $\qquad\qquad\qquad\qquad\quad y = -5$

    The pairs $(2,5)$ and $(-2,-5)$ check. They are the solutions.

16. $(2,1)$, $(-2,-1)$

17. $xy - y^2 = 2$,   (1)
    $2xy - 3y^2 = 0$ (2)

    $\underline{\begin{array}{l} -2xy + 2y^2 = -4 \quad \text{Multiplying (1) by -2} \\ 2xy - 3y^2 = 0 \end{array}}$
    $ -y^2 = -4$
    $ y^2 = 4$
    $ y = \pm 2$

    We substitute for y in Eq. (1) and solve for x.
    When $y = 2$: $x \cdot 2 - 2^2 = 2$
    $ 2x - 4 = 2$
    $ 2x = 6$
    $ x = 3$
    When $y = -2$: $x(-2) - (-2)^2 = 2$
    $ -2x - 4 = 2$
    $ -2x = 6$
    $ x = -3$

    The pairs $(3,2)$ and $(-3,-2)$ check. They are the solutions.

18. $\left(2, -\dfrac{4}{5}\right)$, $\left(-2, -\dfrac{4}{5}\right)$, $(5,2)$, $(-5,2)$

19. $x^2 - y = 5$,   (1)
    $x^2 + y^2 = 25$ (2)

    We solve Eq. (1) for y.
    $x^2 - 5 = y$ (3)

    Substitute $x^2 - 5$ for y in Eq. (2) and solve for x.
    $ x^2 + (x^2 - 5)^2 = 25$
    $x^2 + x^4 - 10x^2 + 25 = 25$
    $ x^4 - 9x^2 = 0$
    $ u^2 - 9u = 0 \quad \text{Letting } u = x^2$
    $ u(u - 9) = 0$

    $u = 0$ or $u = 9$
    $x^2 = 0$ or $x^2 = 9$
    $x = 0$ or $x = \pm 3$

    Substitute in Eq. (3) and solve for y.
    When $x = 0$: $y = 0^2 - 5 = -5$
    When $x = 3$ or -3: $y = 9 - 5 = 4$
    The pairs $(0,-5)$, $(3,4)$, and $(-3,4)$ check. They are the solutions.

    (This exercise could also be solved using the addition method.)

20. $(-\sqrt{2}, \sqrt{2})$, $(\sqrt{2}, -\sqrt{2})$

21. $a^2 + b^2 = 14$,   (1)
    $ab = 3\sqrt{5}$ (2)

    Solve Eq. (1) for b.
    $b = \dfrac{3\sqrt{5}}{a}$

    Substitute $\dfrac{3\sqrt{5}}{a}$ for b in Eq. (1) and solve for a.
    $ a^2 + \left(\dfrac{3\sqrt{5}}{a}\right)^2 = 14$
    $ a^2 + \dfrac{45}{a^2} = 14$
    $ a^4 + 45 = 14a^2$
    $a^4 - 14a^2 + 45 = 0$
    $u^2 - 14u + 45 = 0 \quad \text{Letting } u = a^2$
    $(u - 9)(u - 5) = 0$

    $u = 9$ or $u = 5$
    $a^2 = 9$ or $a^2 = 5$
    $a = \pm 3$ or $a = \pm\sqrt{5}$

    Since $b = 3\sqrt{5}/a$, if $a = 3$, $b = \sqrt{5}$; if $a = -3$, $b = -\sqrt{5}$; if $a = \sqrt{5}$, $b = 3$; and if $a = -\sqrt{5}$, $b = -3$. The pairs $(3, \sqrt{5})$, $(-3, -\sqrt{5})$, $(\sqrt{5}, 3)$, $(-\sqrt{5}, -3)$ check. They are the solutions.

22. $\left(i\sqrt{3}, -\dfrac{81i\sqrt{3}}{3}\right)$, $\left(-i\sqrt{3}, \dfrac{81i\sqrt{3}}{3}\right)$

23. $x^2 + y^2 = 25$,   (1)
    $9x^2 + 4y^2 = 36$ (2)

    $\underline{\begin{array}{l} -4x^2 - 4y^2 = -100 \quad \text{Multiplying (1) by -4} \\ 9x^2 + 4y^2 = 36 \end{array}}$
    $ 5x^2 = -64$
    $ x^2 = -\dfrac{64}{5}$
    $ x = \pm\sqrt{\dfrac{-64}{5}} = \pm\dfrac{8i}{\sqrt{5}}$
    $ x = \pm\dfrac{8i\sqrt{5}}{5} \quad \begin{array}{l}\text{Rationalizing the} \\ \text{denominator}\end{array}$

    Substituting $\dfrac{8i\sqrt{5}}{5}$ or $-\dfrac{8i\sqrt{5}}{5}$ for x in Eq. (1) and solving for y gives us
    $ -\dfrac{64}{5} + y^2 = 25$
    $ y^2 = \dfrac{189}{5}$
    $ y = \pm\sqrt{\dfrac{189}{5}} = \pm 3\sqrt{\dfrac{21}{5}}$
    $ y = \pm\dfrac{3\sqrt{105}}{5}. \quad \begin{array}{l}\text{Rationalizing the} \\ \text{denominator}\end{array}$

    The pairs $\left(\dfrac{8i\sqrt{5}}{5}, \dfrac{3\sqrt{105}}{5}\right)$, $\left(-\dfrac{8i\sqrt{5}}{5}, \dfrac{3\sqrt{105}}{5}\right)$, $\left(\dfrac{8i\sqrt{5}}{5}, -\dfrac{3\sqrt{105}}{5}\right)$, and $\left(-\dfrac{8i\sqrt{5}}{5}, -\dfrac{3\sqrt{105}}{5}\right)$ check. They are the solutions.

24. $\left(\dfrac{4\sqrt{10}}{5}, \dfrac{3i\sqrt{15}}{5}\right)$, $\left(\dfrac{4\sqrt{10}}{5}, -\dfrac{3i\sqrt{15}}{5}\right)$,

    $\left(-\dfrac{4\sqrt{10}}{5}, \dfrac{3i\sqrt{15}}{5}\right)$, $\left(-\dfrac{4\sqrt{10}}{5}, -\dfrac{3i\sqrt{15}}{5}\right)$

25. <u>Familiarize.</u> Let x and y represent the numbers.
    <u>Translate.</u>

    The product of two numbers is 156.

    $$xy = 156 \qquad (1)$$

    The sum of their squares is 313.

    $$x^2 + y^2 = 313 \qquad (2)$$

    <u>Carry out.</u> We solve the system of equations.
    First solve Eq. (1) for y.

    $$xy = 156$$

    $$y = \frac{156}{x}$$

    Then we substitute $\dfrac{156}{x}$ for y in Eq. (2) and solve
    for x.

    $$x^2 + y^2 = 313 \qquad (2)$$

    $$x^2 + \left(\frac{156}{x}\right)^2 = 313$$

    $$x^2 + \frac{24{,}336}{x^2} = 313$$

    $$x^4 + 24{,}336 = 313x^2$$

    $$x^4 - 313x^2 + 24{,}336 = 0$$

    $$u^2 - 313u + 24{,}336 = 0 \qquad \text{Letting } u = x^2$$

    $$(u - 169)(u - 144) = 0$$

    $$u = 169 \quad \text{or} \quad u = 144$$

    We now substitute x^2 for u and solve for x.

    $$x^2 = 169 \quad \text{or} \quad x^2 = 144$$

    $$x = \pm13 \quad \text{or} \quad x = \pm12$$

    Since y = 156/x, if x = 13, y = 12; if x = -13,
    y = -12; if x = 12, y = 13; and if x = -12,
    y = -13. The possible solutions are (13,12),
    (-13,-12), (12,13), and (-12,-13).

    <u>Check.</u> If x = 13 and y = 12, their product is
    156. If x = -13 and y = -12, their product is
    156. The sum of the squares in either case is
    $(\pm13)^2 + (\pm12)^2 = 169 + 144 = 313$. The pairs
    (12,13) and (-12,-13) do not give us any other
    solutions.

    <u>State.</u> The numbers are 13 and 12 or -13 and -12.

26. 6 and 10 or -6 and -10

27. <u>Familiarize.</u> We first make a drawing. Let
    ℓ = the length and w = the width.

    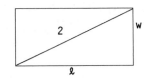

    <u>Translate.</u>
    Area: $\ell w = \sqrt{3}$ (1)
    From the Pythagorean theorem: $\ell^2 + w^2 = 2^2$ (2)

27. (continued)

    <u>Carry out.</u> We solve the system of equations.
    We first solve Eq. (1) for w.

    $$\ell w = \sqrt{3}$$

    $$w = \frac{\sqrt{3}}{\ell}$$

    Then we substitute $\dfrac{\sqrt{3}}{\ell}$ for w in Eq. (2) and solve
    for ℓ.

    $$\ell^2 + \left(\frac{\sqrt{3}}{\ell}\right)^2 = 4$$

    $$\ell^2 + \frac{3}{\ell^2} = 4$$

    $$\ell^4 + 3 = 4\ell^2$$

    $$\ell^4 - 4\ell^2 + 3 = 0$$

    $$u^2 - 4u + 3 = 0 \quad \text{Letting } u = \ell^2$$

    $$(u - 3)(u - 1) = 0$$

    $$u = 3 \quad \text{or} \quad u = 1$$

    We now substitute ℓ^2 for u and solve for ℓ.

    $$\ell^2 = 3 \quad \text{or} \quad \ell^2 = 1$$

    $$\ell = \pm\sqrt{3} \quad \text{or} \quad \ell = \pm1$$

    Length cannot be negative, so we only need to
    consider $\ell = \sqrt{3}$ and $\ell = 1$. Since $w = \sqrt{3}/\ell$, if
    $\ell = \sqrt{3}$, w = 1 and if $\ell = 1$, $w = \sqrt{3}$. Length is
    usually considered to be longer than width, so we
    have the solution $\ell = \sqrt{3}$ and w = 1, or $(\sqrt{3},1)$.

    <u>Check.</u> If $\ell = \sqrt{3}$ and w = 1, the area is
    $\sqrt{3}\cdot1 = \sqrt{3}$. Also $(\sqrt{3})^2 + 1^2 = 3 + 1 = 4 = 2^2$.
    The numbers check.

    <u>State.</u> The length is $\sqrt{3}$ m, and the width is 1 m.

28. $\sqrt{2}$ m by 1

29. <u>Familiarize.</u> We let x = the length of a side of
    one peanut bed and y = the length of a side of
    the other peanut bed. Make a drawing.

    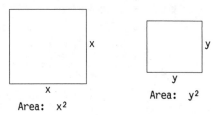

    Area: x^2 Area: y^2

    <u>Translate.</u>
    The sum of the areas is 832 ft².

    $$x^2 + y^2 = 832$$

    The difference of the areas is 320 ft².

    $$x^2 - y^2 = 320$$

29. (continued)

Carry out. We solve the system of equations.

$$x^2 + y^2 = 832$$
$$\underline{x^2 - y^2 = 320}$$
$$2x^2 \qquad = 1152 \qquad \text{Adding}$$
$$x^2 = 576$$
$$x = \pm 24$$

Since length cannot be negative, we consider only x = 24. Substitute 24 for x in the first equation and solve for y.

$$24^2 + y^2 = 832$$
$$576 + y^2 = 832$$
$$y^2 = 256$$
$$y = \pm 16$$

Again, we consider only the positive value, 16. The possible solution is (24,16).

Check. The areas of the peanut beds are 24^2, or 576, and 16^2, or 256. The sum of the areas is 576 + 256, or 832. The difference of the areas is 576 − 256, or 320. The values check.

State. The lengths of the beds are 24 ft and 16 ft.

30. $125, 6%

31. $\dfrac{\sqrt{x} - \sqrt{h}}{\sqrt{x} + \sqrt{h}} = \dfrac{\sqrt{x} - \sqrt{h}}{\sqrt{x} + \sqrt{h}} \cdot \dfrac{\sqrt{x} + \sqrt{h}}{\sqrt{x} + \sqrt{h}} = \dfrac{x - h}{x + 2\sqrt{xh} + h}$

32. $\dfrac{x - 2\sqrt{xh} + h}{x - h}$

33. Familiarize. Let r represent the speed of the boat in still water and t represent the time of the trip upstream. Organize the information in a table.

	Speed	Time	Distance
Upstream	r − 2	t	4
Downstream	r + 2	3 − t	4

Recall that rt = d, or t = d/r.

Translate. From the first line of the table we obtain $t = \dfrac{4}{r - 2}$. From the second line we obtain $3 - t = \dfrac{4}{r + 2}$.

Carry out. Substitute $\dfrac{4}{r - 2}$ for t in the second equation and solve for r.

$$3 - \dfrac{4}{r - 2} = \dfrac{4}{r + 2}$$

$$\text{LCM is } (r - 2)(r + 2)$$

$$(r - 2)(r + 2)\left[3 - \dfrac{4}{r - 2}\right] = (r - 2)(r + 2) \cdot \dfrac{4}{r + 2}$$

$$3(r - 2)(r + 2) - 4(r + 2) = 4(r - 2)$$
$$3(r^2 - 4) - 4r - 8 = 4r - 8$$
$$3r^2 - 12 - 4r - 8 = 4r - 8$$
$$3r^2 - 8r - 12 = 0$$

33. (continued)

$$r = \dfrac{-(-8) \pm \sqrt{(-8)^2 - 4(3)(-12)}}{2 \cdot 3}$$

$$r = \dfrac{8 \pm \sqrt{208}}{6} = \dfrac{8 \pm 4\sqrt{13}}{6}$$

$$r = \dfrac{4 \pm 2\sqrt{13}}{3}$$

Since negative speed has no meaning in this problem, we consider only the positive square root.

$$r = \dfrac{4 + 2\sqrt{13}}{3} \approx 3.7$$

Check. Left to the student.

State. The speed of the boat in still water is approximately 3.7 mph.

34. $\dfrac{1}{2}, \dfrac{1}{4}; \ \dfrac{1}{2}, -\dfrac{1}{4}; \ -\dfrac{1}{2}, \dfrac{1}{4}; \ -\dfrac{1}{2}, -\dfrac{1}{4}$

35. $(x - h)^2 + (y - k)^2 = r^2$ Standard form

Substitute each ordered pair for (x,y).

$$(4 - h)^2 + (-20 - k)^2 = r^2,$$
$$(10 - h)^2 + (-2 - k)^2 = r^2,$$
$$(-4 - h)^2 + (-4 - k)^2 = r^2$$
$$16 - 8h + h^2 + 400 + 40k + k^2 = r^2,$$
$$100 - 20h + h^2 + 4 + 4k + k^2 = r^2,$$
$$16 + 8h + h^2 + 16 + 8k + k^2 = r^2$$

$$-8h + 40k + 416 + h^2 + k^2 = r^2, \qquad (1)$$
$$-20h + 4k + 104 + h^2 + k^2 = r^2 \qquad (2)$$
$$8h + 8k + 32 + h^2 + k^2 = r^2 \qquad (3)$$

Multiplying Eq. (2) by −1 and adding it to Eq. (1), we get

12h + 36k + 312 = 0, or h + 3k = −26.

Multiplying Eq. (2) by −1 and adding it to Eq. (3), we get

28h + 4k − 72 = 0, or 7h + k = 18.

Solving the system

$$h + 3k = -26,$$
$$7h + k = 18,$$

we get (4,−10).

We substitute in one of the original equations to find r^2.

$$(4 - 4)^2 + [-20 - (-10)]^2 = r^2$$
$$0^2 + (-10)^2 = r^2$$
$$100 = r^2$$

The equation of the circle is

$(x - 4)^2 + (y + 10)^2 = 100.$

36. 10 in. by 7 in. by 5 in.

37. $x^2 + xy = a,$ (1)

$y^2 + xy = b$ (2)

Solve Eq. (1) for y.

$xy = a - x^2$

$y = \dfrac{a - x^2}{x}$ (3)

Substitute for y in Eq. (2) and solve for x.

$\left[\dfrac{a - x^2}{x}\right]^2 + x\left[\dfrac{a - x^2}{x}\right] = b$

$\dfrac{a^2 - 2ax^2 + x^4}{x^2} + a - x^2 = b$

$a^2 - 2ax^2 + x^4 + ax^2 - x^4 = bx^2$ Clearing the fraction

$a^2 = ax^2 + bx^2$

$a^2 = x^2(a + b)$

$\dfrac{a^2}{a + b} = x^2$

$\pm\dfrac{a}{\sqrt{a + b}} = x$

$\pm\dfrac{a\sqrt{a + b}}{a + b} = x$ Rationalizing the denominator

Substitute for x in Eq. (3) and solve for y.

When $x = \dfrac{a\sqrt{a + b}}{a + b}$: $y = \dfrac{a - \left[\dfrac{a\sqrt{a + b}}{a + b}\right]^2}{\dfrac{a\sqrt{a + b}}{a + b}} =$

$\dfrac{a - \dfrac{a^2}{a + b}}{\dfrac{a\sqrt{a + b}}{a + b}} = \dfrac{a^2 + ab - a^2}{a\sqrt{a + b}} = \dfrac{ab}{a\sqrt{a + b}} = \dfrac{b}{\sqrt{a + b}} =$

$\dfrac{b\sqrt{a + b}}{a + b}$

When $x = -\dfrac{a\sqrt{a + b}}{a + b}$: $y = \dfrac{a - \left[-\dfrac{a\sqrt{a + b}}{a + b}\right]^2}{-\dfrac{a\sqrt{a + b}}{a + b}} =$

$\dfrac{a - \dfrac{a^2}{a + b}}{-\dfrac{a\sqrt{a + b}}{a + b}} = \dfrac{a^2 + ab - a^2}{-a\sqrt{a + b}} = \dfrac{ab}{-a\sqrt{a + b}} = -\dfrac{b}{\sqrt{a + b}} =$

$-\dfrac{b\sqrt{a + b}}{a + b}$

The pairs $\left[\dfrac{a\sqrt{a + b}}{a + b}, \dfrac{b\sqrt{a + b}}{a + b}\right]$ and

$\left[-\dfrac{a\sqrt{a + b}}{a + b}, -\dfrac{b\sqrt{a + b}}{a + b}\right]$ check. They are the solutions.

38. $(a,-b)$

39. $p^2 + q^2 = 13$ (1)

$\dfrac{1}{pq} = -\dfrac{1}{6}$ (2)

Solve Eq. (2) for p.

$\dfrac{1}{q} = -\dfrac{p}{6}$

$-\dfrac{6}{q} = p$

Substitute -6/q for p in Eq. (1) and solve for q.

$\left[-\dfrac{6}{q}\right]^2 + q^2 = 13$

$\dfrac{36}{q^2} + q^2 = 13$

$36 + q^4 = 13q^2$

$q^4 - 13q^2 + 36 = 0$

$u^2 - 13u + 36 = 0$ Letting $u = q^2$

$(u - 9)(u - 4) = 0$

$u = 9$ or $u = 4$

$x^2 = 9$ or $x^2 = 4$

$x = \pm 3$ or $x = \pm 2$

Since p = -6/q, if q = 3, p = -2; if q = -3, p = 2; if q = 2, p = -3, and if q = -2, p = 3. The pairs (-2,3), (2,-3), (-3,2), and (3,-2) check. They are the solutions.

40. $\left[\dfrac{1}{3},\dfrac{1}{2}\right]$, $\left[\dfrac{1}{2},\dfrac{1}{3}\right]$

Exercise Set 10.8

---

1. $(x - 5)(x + 3) > 0$

The solutions of $(x - 5)(x + 3) = 0$ are 5 and -3. They are not solutions of the inequality, but they divide the real-number line in a natural way. The product $(x - 5)(x + 3)$ is positive or negative, for values other than 5 and -3, depending on the signs of the factors $x - 5$ and $x + 3$.

Sign of x - 5:   - - - - | - - - - | + + + +

Sign of x + 3:   - - - - | + + + + | + + + +

Sign of product: + + + + | - - - - | + + + +

$x - 5 > 0$ when $x > 5$
$x - 5 < 0$ when $x < 5$

$x + 3 > 0$ when $x > -3$
$x + 3 < 0$ when $x < -3$

For the product $(x - 5)(x + 3)$ to be positive, both factors must be positive or both factors must be negative. We see from the diagram that numbers satisfying x < -3 or x > 5 are solutions. The solution set of the inequality is $\{x | x < -3$ or $x > 5\}$, or $(-\infty,-3) \cup (5,\infty)$.

2. $\{x | x < -1$ or $x > 4\}$, or $(-\infty,-1) \cup (4,\infty)$

3. $(x + 1)(x - 2) \leqslant 0$

   The solutions of $(x + 1)(x - 2) = 0$ are -1 and 2. They divide the number line into three intervals as shown:

   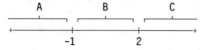

   We try test numbers in each interval.

   A: Test -2, $f(-2) = (-2 + 1)(-2 - 2) = 4;$

   B: Test 0, $f(0) = (0 + 1)(0 - 2) = -2;$

   C: Test 3, $f(3) = (3 + 1)(3 - 2) = 4$

   Since $f(0)$ is negative, the function value will be negative for all numbers in the interval containing 0. The inequality symbol is $\leqslant$, so we need to include the intercepts.

   The solution set is $\{x | -1 \leqslant x \leqslant 2\}$, or $[-1,2]$.

4. $\{x | -3 \leqslant x \leqslant 5\}$, or $[-3,5]$

5. $x^2 - x - 2 < 0$

   $(x + 1)(x - 2) < 0$ Factoring

   See the diagram and test numbers in Exercise 3. The solution set is $\{x | -1 < x < 2\}$, or $(-1,2)$.

6. $\{x | -2 < x < 1\}$, or $(-2,1)$

7. $9 - x^2 \leqslant 0$

   $(3 - x)(3 + x) \leqslant 0$

   The solutions of $(3 - x)(3 + x) = 0$ are 3 and -3. They divide the real-number line in a natural way. The product $(3 - x)(3 + x)$ is positive or negative, for values other than 3 and -3, depending on the signs of the factors $3 - x$ and $3 + x$.

   Sign of $3 - x$:   + + + + | + + + + | - - - -

   Sign of $3 + x$:   - - - - | + + + + | + + + +

   Sign of product: - - - - | + + + + | - - - -

   $\qquad\qquad\qquad$ -3 $\qquad\quad$ 3

   $3 - x > 0$ when $x < 3$
   $3 - x < 0$ when $x > 3$

   $3 + x > 0$ when $x > -3$
   $3 + x < 0$ when $x < -3$

   For the product $(3 - x)(3 + x)$ to be negative, one factor must be positive and the other negative. We see from the diagram that numbers satisfying $x < -3$ or $x > 3$ are solutions. The intercepts are also solutions. The solution set of the inequality is $\{x | x \leqslant -3 \ \text{or} \ x \geqslant 3\}$, or $(-\infty,-3] \cup [3,\infty)$.

8. $\{x | -2 \leqslant x \leqslant 2\}$, or $[-2,2]$

9. $x^2 - 2x + 1 \geqslant 0$

   $(x - 1)^2 \geqslant 0$

   The solution of $(x - 1)^2 = 0$ is 1. For all real-number values of x except 1, $(x - 1)^2$ will be positive. Thus the solution set is $\{x | x \text{ is a real number}\}$, or $(-\infty,\infty)$.

10. $\emptyset$

11. $\qquad\qquad x^2 + 8 < 6x$

    $\qquad\quad x^2 - 6x + 8 < 0$

    $(x - 4)(x - 2) < 0$

    The solutions of $(x - 4)(x - 2) = 0$ are 4 and 2. They are not solutions of the inequality, but they divide the real-number line in a natural way. The product $(x - 4)(x - 2)$ is positive or negative, for values other than 4 and 2, depending on the signs of the factors $x - 4$ and $x - 2$.

    Sign of $x - 4$:   - - - - | - - - - | + + + +

    Sign of $x - 2$:   - - - - | + + + + | + + + +

    Sign of product: + + + + | - - - - | + + + +

    $\qquad\qquad\qquad\qquad$ 2 $\qquad\quad$ 4

    $x - 4 > 0$ when $x > 4$
    $x - 4 < 0$ when $x < 4$

    $x - 2 > 0$ when $x > 2$
    $x - 2 < 0$ when $x < 2$

    For the product $(x - 4)(x - 2)$ to be negative one factor must be positive and the other negative. The only situation in the table for which this happens is when $2 < x < 4$. The solution set of the inequality is $\{x | 2 < x < 4\}$, or $(2,4)$.

12. $\{x | x < -2 \ \text{or} \ x > 6\}$, or $(-\infty,-2) \cup (6,\infty)$

13. $3x(x + 2)(x - 2) < 0$

    The solutions of $3x(x + 2)(x - 2) = 0$ are 0, -2, and 2. They divide the number line into four intervals as shown.

    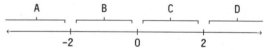

    We try test numbers in each interval.

    A: Test -3, $f(-3) = 3(-3)(-3 + 2)(-3 - 2) = -45$

    B: Test -1, $f(-1) = 3(-1)(-1 + 2)(-1 - 2) = 9$

    C: Test 1, $f(1) = 3(1)(1 + 2)(1 - 2) = -9$

    D: Test 3, $f(3) = 3(3)(3 + 2)(3 - 2) = 45$

    Since $f(-3)$ and $f(1)$ are negative, the function value will be negative for all numbers in the intervals containing -3 and 1. The solution set is $\{x | x < -2 \ \text{or} \ 0 < x < 2\}$, or $(-\infty,-2) \cup (0,2)$.

14. $\{x | -1 < x < 0 \ \text{or} \ x > 1\}$, or $(-1,0) \cup (1,\infty)$

15. $(x + 3)(x - 2)(x + 1) > 0$

The solutions of $(x + 3)(x - 2)(x + 1) = 0$ are
-3, 2, and -1. They are not solutions of the
inequality, but they divide the real number
line in a natural way. The product
$(x + 3)(x - 2)(x + 1)$ is positive or negative,
for values other than -3, 2, and -1, depending
on the signs of the factors $x + 3$, $x - 2$, and
$x + 1$.

Sign of $x + 3$: $- - - -|+ + + +|+ + + +|+ + + +$

Sign of $x - 2$: $- - - -|- - - -|- - - -|+ + + +$

Sign of $x + 1$: $- - - -|- - - -|+ + + +|+ + + +$

Sign of product: $- - - -|+ + + +|- - - -|+ + + +$

```
←——————+————+————+——————→
       -3   -1   2
```

$x + 3 > 0$ when $x > -3$
$x + 3 < 0$ when $x < -3$

$x - 2 > 0$ when $x > 2$
$x - 2 < 0$ when $x < 2$

$x + 1 > 0$ when $x > -1$
$x + 1 < 0$ when $x < -1$

The product of three numbers is positive when all
three are positive or two are negative and one is
positive. We see from the diagram that numbers
satisfying $-3 < x < -1$ or $x > 2$ are solutions.
The solution set of the inequality is
$\{x|-3 < x < -1$ or $x > 2\}$, or $(-3,-1) \cup (2,\infty)$.

16. $\{x|x < -2$ or $1 < x < 4\}$, or $(-\infty,-2) \cup (1,4)$

17. $(x + 3)(x + 2)(x - 1) < 0$

The solutions of $(x + 3)(x + 2)(x - 1) = 0$ are
-3, -2, and 1. They divide the number line into
four intervals as shown:

```
    A      B    C      D
——————+———+——+——+———+——————
     -3  -2      1
```

We try test numbers in each interval.

A: Test -4, $f(-4) = (-4 + 3)(-4 + 2)(-4 - 1) =$
            -10

B: Test $-\frac{5}{2}$, $f\left(-\frac{5}{2}\right) =$

$$\left(-\frac{5}{2} + 3\right)\left(-\frac{5}{2} + 2\right)\left(-\frac{5}{2} - 1\right) = \frac{7}{8}$$

C: Test 0, $f(0) = (0 + 3)(0 + 2)(0 - 1) = -6$

D: Test 2, $f(2) = (2 + 3)(2 + 2)(2 - 1) = 20$

The function value will be negative for all
numbers in intervals A and C. The solution set
is $\{x|x < -3$ or $-2 < x < 1\}$, or
$(-\infty,-3) \cup (-2,1)$.

18. $\{x|x < -1$ or $2 < x < 3\}$, or $(-\infty,-1) \cup (2,3)$

19. $\frac{1}{x - 4} < 0$

We write the related equation by changing the
< symbol to =:

$$\frac{1}{x - 4} = 0$$

We solve the related equation.

$$(x - 4) \cdot \frac{1}{x - 4} = (x - 4) \cdot 0$$
$$1 = 0$$

The related equation has no solution.

Next we find the replacements that are not
meaningful by setting the denominator equal to 0
and solving:

$$x - 4 = 0$$
$$x = 4$$

We use 4 to divide the number line into two
intervals as shown:

```
        A        B
——————+————+————+——————
              4
```

We try test numbers in each interval.

A: Test 0,

$$\begin{array}{c|c} \dfrac{1}{x - 4} < 0 & \\ \hline \dfrac{1}{0 - 4} & 0 \\ -\dfrac{1}{4} & \end{array}$$

The number 0 is a solution of the inequality,
so the interval A is part of the solution set.

B: Test 5,

$$\begin{array}{c|c} \dfrac{1}{x - 4} < 0 & \\ \hline \dfrac{1}{5 - 4} & 0 \\ 1 & \end{array}$$

The number 5 is not a solution of the inequality,
so the interval B is not part of the solution
set. The solution set is $\{x|x < 4\}$, or $(-\infty,4)$.

20. $\{x|x > -5\}$, or $(-5,\infty)$

21. $\frac{x + 1}{x - 3} > 0$

Solve the related equation.

$$\frac{x + 1}{x - 3} = 0$$
$$x + 1 = 0$$
$$x = -1$$

Find replacements that are not meaningful.

$$x - 3 = 0$$
$$x = 3$$

Use the numbers -1 and 3 to divide the number
line into intervals as shown:

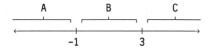

```
        A        B        C
——————+————+————+————+————+——————
             -1        3
```

Try test numbers in each interval.

21. (continued)

A: Test -2,
$$\frac{x + 1}{x - 3} > 0$$

$$\frac{-2 + 1}{-2 - 3} \,\Big|\, 0$$

$$\frac{-1}{-5}$$

$$\frac{1}{5}$$

The number -2 is a solution of the inequality so the interval A is part of the solution set.

B: Test 0,
$$\frac{x + 1}{x - 3} > 0$$

$$\frac{0 + 1}{0 - 3} \,\Big|\, 0$$

$$-\frac{1}{3}$$

The number 0 is not a solution of the inequality, so the interval B is not part of the solution set.

C: Test 4,
$$\frac{x + 1}{x - 3} > 0$$

$$\frac{4 + 1}{4 - 3} \,\Big|\, 0$$

$$\frac{5}{1}$$

$$5$$

The number 4 is a solution of the inequality, so the interval C is part of the solution set. The solution set is $\{x \mid x < -1 \text{ or } x > 3\}$, or $(-\infty, -1) \cup (3, \infty)$.

22. $\{x \mid -5 < x < 2\}$, or $(-5, 2)$

23. $\frac{3x + 2}{x - 3} \leqslant 0$

Solve the related equation.

$$\frac{3x + 2}{x - 3} = 0$$

$$3x + 2 = 0$$

$$3x = -2$$

$$x = -\frac{2}{3}$$

Find replacements that are not meaningful.

$$x - 3 = 0$$

$$x = 3$$

Use the numbers $-\frac{2}{3}$ and 3 to divide the number line into intervals as shown.

```
         A         B         C
  <--+---+----+----+----+---+-->
         -2/3       3
```

Try test numbers in each interval.

23. (continued)

A: Test -1,
$$\frac{3x + 2}{x - 3} \leqslant 0$$

$$\frac{3(-1) + 2}{-1 - 3} \,\Big|\, 0$$

$$\frac{-1}{-4}$$

$$\frac{1}{4}$$

The number -1 is not a solution of the inequality, so the interval A is not part of the solution set.

B: Test 0,
$$\frac{3x + 2}{x - 3} \leqslant 0$$

$$\frac{3 \cdot 0 + 2}{0 - 3} \,\Big|\, 0$$

$$\frac{2}{-3}$$

$$-\frac{2}{3}$$

The number 0 is a solution of the inequality, so the interval B is part of the solution set.

C: Test 4,
$$\frac{3x + 2}{x - 3} \leqslant 0$$

$$\frac{3 \cdot 4 + 2}{4 - 3} \,\Big|\, 0$$

$$14$$

The number 4 is not a solution of the inequality, so the interval C is not part of the solution set. The solution set includes the interval B. The number $-\frac{2}{3}$ is also included since the inequality symbol is $\leqslant$ and $-\frac{2}{3}$ is the solution of the related equation. The number 3 is not included since it not a meaningful replacement. The solution set is $\left\{x \mid -\frac{2}{3} \leqslant x < 3\right\}$, or $\left[-\frac{2}{3}, 3\right)$.

24. $\left\{x \mid x < -\frac{3}{4} \text{ or } x \geqslant \frac{5}{2}\right\}$, or $\left(-\infty, -\frac{3}{4}\right) \cup \left[\frac{5}{2}, \infty\right)$

25. $\dfrac{x - 1}{x - 2} > 3$

Solve the related equation.

$\dfrac{x - 1}{x - 2} = 3$

$x - 1 = 3(x - 2)$

$x - 1 = 3x - 6$

$5 = 2x$

$\dfrac{5}{2} = x$

Find replacements that are not meaningful.

$x - 2 = 0$

$x = 2$

Use the numbers $\dfrac{5}{2}$ and 2 to divide the number line into intervals as shown:

Try test numbers in each interval.

A: Test 0, $\dfrac{x - 1}{x - 2} > 3$

$\left.\dfrac{0 - 1}{0 - 2}\;\right|\;3$

$\left.\dfrac{1}{2}\;\right|$

The number 0 is not a solution of the inequality, so the interval A is not part of the solution set.

B: Test $\dfrac{9}{4}$, $\dfrac{x - 1}{x - 2} > 3$

$\left.\dfrac{\frac{9}{4} - 1}{\frac{9}{4} - 2}\;\right|\;3$

$\left.\dfrac{\frac{5}{4}}{\frac{1}{4}}\;\right|$

$5 \;|$

The number $\dfrac{9}{4}$ is a solution of the inequality, so the interval B is part of the solution set.

C: Test 3, $\dfrac{x - 1}{x - 2} > 3$

$\left.\dfrac{3 - 1}{3 - 2}\;\right|\;3$

$2 \;|$

The number 3 is not a solution of the inequality, so the interval C is not part of the solution set. The solution set is $\left\{x \,\middle|\, 2 < x < \dfrac{5}{2}\right\}$, or $\left(2, \dfrac{5}{2}\right)$.

26. $\left\{x \,\middle|\, x < \dfrac{3}{2} \text{ or } x > 4\right\}$, or $\left(-\infty, \dfrac{3}{2}\right] \cup (4, \infty)$

27. $\dfrac{(x - 2)(x + 1)}{x - 5} < 0$

Solve the related equation.

$\dfrac{(x - 2)(x + 1)}{x - 5} = 0$

$(x - 2)(x + 1) = 0$

$x = 2 \text{ or } x = -1$

Find replacements that are not meaningful.

$x - 5 = 0$

$x = 5$

Use the numbers 2, -1, and 5 to divide the number line into intervals as shown:

Try test numbers in each interval.

A: Test -2, $\dfrac{(x - 2)(x + 1)}{x - 5} < 0$

$\left.\dfrac{(-2 - 2)(-2 + 1)}{-2 - 5}\;\right|\;0$

$\left.\dfrac{-4(-1)}{-7}\;\right|$

$\left.-\dfrac{4}{7}\;\right|$

Interval A is part of the solution set.

B: Test 0, $\dfrac{(x - 2)(x + 1)}{x - 5} < 0$

$\left.\dfrac{(0 - 2)(0 + 1)}{0 - 5}\;\right|\;0$

$\left.\dfrac{-2 \cdot 1}{-5}\;\right|$

$\left.\dfrac{2}{5}\;\right|$

Interval B is not part of the solution set.

C: Test 3, $\dfrac{(x - 2)(x + 1)}{x - 5} < 0$

$\left.\dfrac{(3 - 2)(3 + 1)}{3 - 5}\;\right|\;0$

$\left.\dfrac{1 \cdot 4}{-2}\;\right|$

$-2 \;|$

Interval C is part of the solution set.

D: Test 6, $\dfrac{(x - 2)(x + 1)}{x - 5} < 0$

$\left.\dfrac{(6 - 2)(6 + 1)}{6 - 5}\;\right|\;0$

$\left.\dfrac{4 \cdot 7}{1}\;\right|$

$28 \;|$

Interval D is not part of the solution set.

The solution set is $\{x \mid x < -1 \text{ or } 2 < x < 5\}$, or $(-\infty, -1) \cup (2, 5)$.

28. $\{x \mid -4 < x < -3 \text{ or } x > 1\}$, or $(-4, -3) \cup (1, \infty)$

29. $\dfrac{x}{x-2} \geqslant 0$

Solve the related equation.

$$\dfrac{x}{x-2} = 0$$
$$x = 0$$

Find replacements that are not meaningful.

$$x - 2 = 0$$
$$x = 2$$

Use the numbers 0 and 2 to divide the number line into intervals as shown.

```
        A         B         C
  <----------|--------|--------->
             0        2
```

Try test numbers in each interval.

A: Test -1, $\dfrac{x}{x-2} \geqslant 0$

$$\dfrac{-1}{-1-2} \;\Big|\; 0$$
$$\dfrac{1}{3} \;\Big|$$

Interval A is part of the solution set.

B: Test 1, $\dfrac{x}{x-2} \geqslant 0$

$$\dfrac{1}{1-2} \;\Big|\; 0$$
$$-1 \;\Big|$$

Interval B is not part of the solution set.

C: Test 3, $\dfrac{x}{x-2} \geqslant 0$

$$\dfrac{3}{3-2} \;\Big|\; 0$$
$$3 \;\Big|$$

Interval C is part of the solution set.

The solution set includes intervals A and C. The number 0 is also included since the inequality symbol is $\geqslant$ and 0 is the solution of the related equation. The number 2 is not included since it is not a meaningful replacement. The solution set is $\{x \mid x \leqslant 0 \text{ or } x > 2\}$, or $(-\infty, 0] \cup (2, \infty)$.

30. $\{x \mid -3 \leqslant x < 0\}$, or $[-3, 0)$

31. $\dfrac{x-5}{x} < 1$

Solve the related equation.

$$\dfrac{x-5}{x} = 1$$
$$x - 5 = x$$
$$-5 = 0$$

The related equation has no solution.

Find replacements that are not meaningful.

$$x = 0$$

Use the number 0 to divide the number line into two intervals as shown.

```
        A         B
  <----------|--------->
             0
```

31. (continued)

Try test numbers in each interval.

A: Test -1, $\dfrac{x-5}{x} < 1$

$$\dfrac{-1-5}{-1} \;\Big|\; 1$$
$$6 \;\Big|$$

Interval A is not part of the solution set.

B: Test 1, $\dfrac{x-5}{x} < 1$

$$\dfrac{1-5}{1} \;\Big|\; 1$$
$$-4 \;\Big|$$

Interval B is part of the solution set.
The solution set is $\{x \mid x > 0\}$, or $(0, \infty)$.

32. $\{x \mid 1 < x < 2\}$, or $(1, 2)$

33. $\dfrac{x-1}{(x-3)(x+4)} < 0$

Solve the related equation.

$$\dfrac{x-1}{(x-3)(x+4)} = 0$$
$$x - 1 = 0$$
$$x = 1$$

Find replacements that are not meaningful.

$$(x-3)(x+4) = 0$$
$$x = 3 \text{ or } x = -4$$

Use the numbers 1, 3, and -4 to divide the number line into intervals as shown:

```
       A        B        C        D
  <--------|-------|-------|------->
          -4       1       3
```

Try test numbers in each interval.

A: Test -5, $\dfrac{x-1}{(x-3)(x+4)} < 0$

$$\dfrac{-5-1}{(-5-3)(-5+4)} \;\Big|\; 0$$
$$\dfrac{-6}{-8(-1)} \;\Big|$$
$$-\dfrac{3}{4} \;\Big|$$

Interval A is part of the solution set.

B: Test 0, $\dfrac{x-1}{(x-3)(x+4)} < 0$

$$\dfrac{0-1}{(0-3)(0+4)} \;\Big|\; 0$$
$$\dfrac{-1}{-3\cdot4} \;\Big|$$
$$\dfrac{1}{12} \;\Big|$$

Interval B is not part of the solution set.

33. (continued)

C: Test 2,

$$\frac{\dfrac{x-1}{(x-3)(x+4)} < 0}{\dfrac{2-1}{(2-3)(2+4)} \,\Big|\, 0}$$

$$\frac{1}{-1\cdot 6}$$

$$-\frac{1}{6} \,\Big|$$

Interval C is part of the solution set.

D: Test 4,

$$\frac{\dfrac{x-1}{(x-3)(x+4)} < 0}{\dfrac{4-1}{(4-3)(4+4)} \,\Big|\, 0}$$

$$\frac{3}{1\cdot 8}$$

$$\frac{3}{8} \,\Big|$$

Interval D is not part of the solution set.

The solution set is {x|x < -4 or 1 < x < 3}, or (-∞,-4) ∪ (1,3).

34. {x|-7 < x < -2 or x > 2}, or (-7,-2) ∪ (2,∞)

35. $2 < \dfrac{1}{x}$

Solve the related equation.

$$2 = \frac{1}{x}$$

$$x = \frac{1}{2}$$

Find replacements that are not meaningful.

$$x = 0$$

Use the numbers $\frac{1}{2}$ and 0 to divide the number line into intervals as shown.

Try test numbers in each interval.

A: Test -1,

$$\frac{2 < \dfrac{1}{x}}{2 \,\Big|\, \dfrac{1}{-1}}$$

$$\Big|\, -1$$

Interval A is not part of the solution set.

B: Test $\frac{1}{4}$,

$$\frac{2 < \dfrac{1}{x}}{2 \,\Big|\, \dfrac{1}{\frac{1}{4}}}$$

$$\Big|\, 4$$

Interval B is part of the solution set.

35. (continued)

C: Test 1,

$$\frac{2 < \dfrac{1}{x}}{2 \,\Big|\, \dfrac{1}{1}}$$

$$\Big|\, 1$$

Interval C is not part of the solution set.

The solution set is $\left\{x \,\Big|\, 0 < x < \frac{1}{2}\right\}$, or $\left[0, \frac{1}{2}\right]$.

36. $\left\{x \,\Big|\, x < 0 \text{ or } x \geqslant \frac{1}{3}\right\}$, or $(-\infty, 0) \cup \left[\frac{1}{3}, \infty\right)$

37. $x^2 - 2x \leqslant 2$

$x^2 - 2x - 2 \leqslant 0$

The graph of the quadratic function $f(x) = x^2 - 2x - 2$ opens upward since the coefficient of x^2 is positive. Function values will be negative between the intercepts. We find the intercepts by setting the polynomial equal to 0 and solving.

Here we use the quadratic formula.

$x^2 - 2x - 2 = 0$

$$x = \frac{-(-2) + \sqrt{(-2)^2 - 4(1)(-2)}}{2\cdot 1}$$

$$= \frac{2 \pm \sqrt{12}}{2}$$

$$= \frac{2 \pm 2\sqrt{3}}{2}$$

$$= 1 \pm \sqrt{3}$$

The intercepts are $x = 1 - \sqrt{3}$ and $x = 1 + \sqrt{3}$. The solution set is {x|1 - √3 ⩽ x ⩽ 1 + √3}, or [1 - √3, 1 + √3].

38. {x|x < -1 - √5 or x > -1 + √5}, or (-∞, -1 - √5) ∪ (-1 + √5, ∞)

39. $x^2 + 3 > 0$

The graph of $f(x) = x^2 + 3$ is a parabola with vertex (0,3) that opens upward. Thus function values are positive for any real-number value of x. The solution set is {x|x is a real number}, or (-∞, ∞).

40. ∅

41. $x^4 - 2x^2 > 0$

   $u^2 - 2u > 0$ Letting $u = x^2$

   $u(u - 2) > 0$

   The solutions of $u(u - 2) = 0$ are 0 and 2.
   Replacing u with x^2 we have

   $x^2 = 0$ or $x^2 = 2$

   $x = 0$ or $x = \pm\sqrt{2}$.

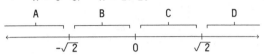

   A: Test -2, $f(-2) = (-2)^4 - 2(-2)^2 = 8$

   B: Test -1, $f(-1) = (-1)^4 - 2(-1)^2 = -1$

   C: Test 1, $f(1) = 1^4 - 2(1)^2 = -1$

   D: Test 2, $f(2) = 2^4 - 2(2)^2 = 8$

   The solution set is $\{x | x < -\sqrt{2}$ or $x > \sqrt{2}\}$, or
   $(-\infty, -\sqrt{2}) \cup (\sqrt{2}, \infty)$

42. $\{x | -\sqrt{2} \leqslant x \leqslant \sqrt{2}\}$, or $[-\sqrt{2}, \sqrt{2}]$

43. a) $-3x^2 + 630x - 6000 > 0$

   $x^2 - 210x + 2000 < 0$ Multiplying by $-\frac{1}{3}$

   $(x - 200)(x - 10) < 0$

   The solutions of $f(x) = (x - 200)(x - 10) = 0$
   are 200 and 10. They divide the number line
   as shown:

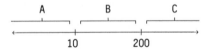

   A: Test 0, $f(0) = 0^2 - 210 \cdot 0 + 2000 = 2000$

   B: Test 20, $f(20) = 20^2 - 210 \cdot 20 + 2000 =$
        -1800

   C: Test 300, $f(300) = 300^2 - 210 \cdot 300 + 2000 =$
        $29,000$

   The company makes a profit for values of x
   such that $10 < x < 200$, or for values of x in
   the interval $(10, 200)$.

   b) See part a). Keep in mind that x must be
   nonnegative.

   The company loses money for values of x such
   that $0 \leqslant x < 10$ or $x > 200$, or for values of
   x in the interval $[0, 10) \cup (200, \infty)$.

44. a) $\{t | 0 < t < 2\}$, or $(0, 2)$

   b) $\{t | t > 10\}$, or $(10, \infty)$

45. We find values of n such that $N \geqslant 66$ <u>and</u> $N \leqslant 300$.

   For $N \geqslant 66$:

   $\dfrac{n(n - 1)}{2} \geqslant 66$

   $n(n - 1) \geqslant 132$

   $n^2 - n - 132 \geqslant 0$

   $(n - 12)(n + 11) \geqslant 0$

   The solutions of $f(n) = (n - 12)(n + 11) = 0$ are
   12 and -11. They divide the number line as shown:

   However, only positive values of n have meaning in
   this exercise so we need only consider the
   intervals shown below:

   A: Test 1, $f(1) = 1^2 - 1 - 132 = -132$

   B: Test 20, $f(20) = 20^2 - 20 - 132 = 248$

   Thus, $N \geqslant 66$ for $\{n | n \geqslant 12\}$.

   For $N \leqslant 300$:

   $\dfrac{n(n - 1)}{2} \leqslant 300$

   $n(n - 1) \leqslant 600$

   $n^2 - n - 600 \leqslant 0$

   $(n - 25)(n + 24) \leqslant 0$

   The solutions of $f(n) = (n - 25)(n + 24) = 0$ are
   25 and -24. They divide the number line as shown.

   However, only positive values of n have meaning
   in this exercise so we need only consider the
   intervals shown below:

   A: Test 1, $f(1) = 1^2 - 1 - 600 = -600$

   B: Test 30, $f(30) = 30^2 - 30 - 600 = 270$

   Thus, $N \leqslant 300$ (and $n > 0$) for $\{n | 0 < n \leqslant 25\}$.

   Then $66 \leqslant N \leqslant 300$ for $\{n | 12 \leqslant n \leqslant 25\}$, or for all
   values of n in the interval $[12, 25]$.

46. $\{n | 9 \leqslant n \leqslant 23\}$, or $[9, 23]$

47. Use a compute or a graphic calculator to graph the function.

$f(x) = x^3 - 2x^2 - 5x + 6$

From the graph we determine the following:

The solutions of $f(x) = 0$ are -2, 1, and 3.

The solution of $f(x) < 0$ is
$\{x \mid x < -2 \text{ or } 1 < x < 3\}$, or $(-\infty, -2) \cup (1,3)$.

The solution of $f(x) > 0$ is
$\{x \mid -2 < x < 1 \text{ or } x > 3\}$, or $(-2,1) \cup (3,\infty)$.

48. $f(x) = \frac{1}{3}x^3 - x + \frac{2}{3}$

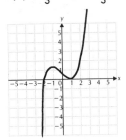

$f(x) = 0$ for $x = -2$ and $x = 1$;

$f(x) < 0$ for $\{x \mid x < -2\}$, or $(-\infty, -2)$;

$f(x) > 0$ for $\{x \mid -2 < x < 1 \text{ or } x > 1\}$, or $(-2,1) \cup (1,\infty)$

49. Use a computer or a graphic calculator to graph the function.

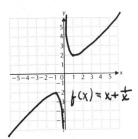

From the graph we determine the following:

$f(x)$ has no zeros.

The solutions $f(x) < 0$ are $\{x \mid x < 0\}$, or $(-\infty, 0)$.

The solutions of $f(x) > 0$ are $\{x \mid x > 0\}$, or $(0,\infty)$.

50. $f(x) = x - \sqrt{x}, \ x \geqslant 0$

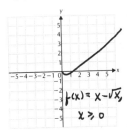

$f(x) = 0$ for $x = 0$ and $x = 1$;

$f(x) < 0$ for $\{x \mid 0 < x < 1\}$, or $(0,1)$;

$f(x) > 0$ for $\{x \mid x > 1\}$, or $(1,\infty)$

51. Use a computer or a graphic calculator to graph the function.

$f(x) = x^4 - 4x^3 - x^2 + 16x - 12$

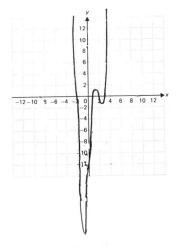

From the graph we determine the following:

The solutions of $f(x) = 0$ are -2, 1, 2, and 3.

The solutions of $f(x) < 0$ are
$\{x \mid -2 < x < 1 \text{ or } 2 < x < 3\}$, or $(-2,1) \cup (2,3)$.

The solutions of $f(x) > 0$ are
$\{x \mid x < -2 \text{ or } 1 < x < 2 \text{ or } x > 3\}$, or
$(-\infty, -2) \cup (1,2) \cup (3,\infty)$.

52. $f(x) = \dfrac{x^3 + x^2 - 2x}{x^2 - x - 6}$

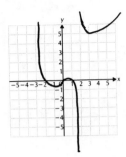

f(x) = 0 for x = -2, x = 0, and x = 1;

f(x) < 0 for
{x│x < -3 or -2 < x < 0 or 1 < x < 2}, or
(-∞,-3) ∪ (-2,0) ∪ (1,2);

f(x) > 0 for
{x│-3 < x < -2 or 0 < x < 1 or x > 2}, or
(-3,-2) ∪ (0,1) ∪ (2,∞)

Exercise Set 11.1

1. $a_n = 3n + 1$
$a_1 = 3 \cdot 1 + 1 = 4$, $\quad a_4 = 3 \cdot 4 + 1 = 13$;
$a_2 = 3 \cdot 2 + 1 = 7$, $\quad a_{10} = 3 \cdot 10 + 1 = 31$;
$a_3 = 3 \cdot 3 + 1 = 10$, $\quad a_{15} = 3 \cdot 15 + 1 = 46$

2. 2, 5, 8, 11; 29; 44

3. $a_n = \dfrac{n}{n + 1}$
$a_1 = \dfrac{1}{1+1} = \dfrac{1}{2}$, $\quad a_4 = \dfrac{4}{4+1} = \dfrac{4}{5}$;
$a_2 = \dfrac{2}{2+1} = \dfrac{2}{3}$, $\quad a_{10} = \dfrac{10}{10+1} = \dfrac{10}{11}$;
$a_3 = \dfrac{3}{3+1} = \dfrac{3}{4}$, $\quad a_{15} = \dfrac{15}{15+1} = \dfrac{15}{16}$

4. 2, 5, 10, 17; 101; 226

5. $a_n = n^2 - 2n$
$a_1 = 1^2 - 2 \cdot 1 = -1$, $\quad a_4 = 4^2 - 2 \cdot 4 = 8$;
$a_2 = 2^2 - 2 \cdot 2 = 0$, $\quad a_{10} = 10^2 - 2 \cdot 10 = 80$;
$a_3 = 3^2 - 2 \cdot 3 = 3$, $\quad a_{15} = 15^2 - 2 \cdot 15 = 195$

6. $0, \dfrac{3}{5}, \dfrac{4}{5}, \dfrac{15}{17}; \dfrac{99}{101}; \dfrac{112}{113}$

7. $a_n = n + \dfrac{1}{n}$
$a_1 = 1 + \dfrac{1}{1} = 2$, $\quad a_4 = 4 + \dfrac{1}{4} = 4\dfrac{1}{4}$;
$a_2 = 2 + \dfrac{1}{2} = 2\dfrac{1}{2}$, $\quad a_{10} = 10 + \dfrac{1}{10} = 10\dfrac{1}{10}$;
$a_3 = 3 + \dfrac{1}{3} = 3\dfrac{1}{3}$, $\quad a_{15} = 15 + \dfrac{1}{15} = 15\dfrac{1}{15}$

8. $1, -\dfrac{1}{2}, \dfrac{1}{4}, -\dfrac{1}{8}; -\dfrac{1}{512}; \dfrac{1}{16,384}$

9. $a_n = (-1)^n n^2$
$a_1 = (-1)^1 1^2 = -1$, $\quad a_4 = (-1)^4 4^2 = 16$;
$a_2 = (-1)^2 2^2 = 4$, $\quad a_{10} = (-1)^{10} 10^2 = 100$;
$a_3 = (-1)^3 3^2 = -9$, $\quad a_{15} = (-1)^{15} 15^2 = -225$

10. $-4, 5, -6, 7; 13; -18$

11. $a_n = (-1)^{n+1}(3n - 5)$
$a_1 = (-1)^{1+1}(3 \cdot 1 - 5) = -2$,
$a_2 = (-1)^{2+1}(3 \cdot 2 - 5) = -1$,
$a_3 = (-1)^{3+1}(3 \cdot 3 - 5) = 4$,
$a_4 = (-1)^{4+1}(3 \cdot 4 - 5) = -7$;
$a_{10} = (-1)^{10+1}(3 \cdot 10 - 5) = -25$;
$a_{15} = (-1)^{15+1}(3 \cdot 15 - 5) = 40$

12. $0, 7, -26, 63; 999; -3374$

13. $a_n = \dfrac{n + 2}{n + 5}$
$a_1 = \dfrac{1+2}{1+5} = \dfrac{3}{6} = \dfrac{1}{2}$, $\quad a_4 = \dfrac{4+2}{4+5} = \dfrac{6}{9} = \dfrac{2}{3}$;
$a_2 = \dfrac{2+2}{2+5} = \dfrac{4}{7}$, $\quad a_{10} = \dfrac{10+2}{10+5} = \dfrac{12}{15} = \dfrac{4}{5}$;
$a_3 = \dfrac{3+2}{3+5} = \dfrac{5}{8}$, $\quad a_{15} = \dfrac{15+2}{15+5} = \dfrac{17}{20}$

14. $-1, \dfrac{3}{2}, 1, \dfrac{7}{8}; \dfrac{19}{26}; \dfrac{29}{41}$

15. $a_n = 4n - 7$
$a_8 = 4 \cdot 8 - 7 = 32 - 7 = 25$

16. 56

17. $a_n = (3n + 4)(2n - 5)$
$a_7 = (3 \cdot 7 + 4)(2 \cdot 7 - 5) = 25 \cdot 9 = 225$

18. 400

19. $a_n = (-1)^{n-1}(3.4n - 17.3)$
$a_{12} = (-1)^{12-1}[3.4(12) - 17.3] = -23.5$

20. $-37,916,508.16$

21. $a_n = 5n^2(4n - 100)$
$a_{11} = 5(11)^2(4 \cdot 11 - 100) = 5(121)(-56) = -33,880$

22. $528,528$

23. $a_n = \left(1 + \dfrac{1}{n}\right)^2$
$a_{20} = \left(1 + \dfrac{1}{20}\right)^2 = \left(\dfrac{21}{20}\right)^2 = \dfrac{441}{400}$

24. $\dfrac{2744}{3375}$

25. $a_n = \log 10^n$
$a_{43} = \log 10^{43} = 43$

26. 67

27. $a_n = 1 + \dfrac{1}{n^2}$
$a_{38} = 1 + \dfrac{1}{38^2} = 1\dfrac{1}{1444}$, or $\dfrac{1445}{1444}$

28. -8

29. $1, 3, 5, 7, 9, \ldots$
These are odd integers, so the general term may be $2n - 1$.

30. 3^n

31. $-2, 6, -18, 54, \ldots$
We can see a pattern if we write the sequence as
$-1 \cdot 2 \cdot 1, 1 \cdot 2 \cdot 3, -1 \cdot 2 \cdot 9, 1 \cdot 2 \cdot 27, \ldots$
The general term may be $(-1)^n 2(3)^{n-1}$.

32. $5n - 7$

33. $\frac{2}{3}, \frac{3}{4}, \frac{4}{5}, \frac{5}{6}, \frac{6}{7}, \ldots$

    These are fractions in which the denominator is 1 greater than the numerator. Also, each numerator is 1 greater than the preceding numerator. The general term may be $\frac{n + 1}{n + 2}$.

34. $\sqrt{2n}$

35. $\sqrt{3}, 3, 3\sqrt{3}, 9, 9\sqrt{3}, \ldots$

    These are powers of $\sqrt{3}$. The general term may be $(\sqrt{3})^n$, or $3^{n/2}$.

36. $n(n + 1)$

37. $-1, -4, -7, -10, -13, \ldots$

    Each term is 3 less than the preceding term. The general term may be $-1 - 3(n - 1)$. After removing parentheses and simplifying, we can express the general term as $-3n + 2$, or $-(3n - 2)$.

38. $\log 10^{n-1}$, or $n - 1$

39. $1, 2, 3, 4, 5, 6, 7, \ldots$
    $S_7 = 1 + 2 + 3 + 4 + 5 + 6 + 7 = 28$

40. -8

41. $2, 4, 6, 8, \ldots$
    $S_5 = 2 + 4 + 6 + 8 + 10 = 30$

42. $\frac{5269}{3600}$

43. $\displaystyle\sum_{k=1}^{5} \frac{1}{2k} = \frac{1}{2\cdot1} + \frac{1}{2\cdot2} + \frac{1}{2\cdot3} + \frac{1}{2\cdot4} + \frac{1}{2\cdot5}$

    $= \frac{1}{2} + \frac{1}{4} + \frac{1}{6} + \frac{1}{8} + \frac{1}{10}$

    $= \frac{60}{120} + \frac{30}{120} + \frac{20}{120} + \frac{15}{120} + \frac{12}{120}$

    $= \frac{137}{120}$

44. $\frac{1}{3} + \frac{1}{5} + \frac{1}{7} + \frac{1}{9} + \frac{1}{11} + \frac{1}{13} = \frac{43,024}{45,045}$

45. $\displaystyle\sum_{k=0}^{5} 2^k = 2^0 + 2^1 + 2^2 + 2^3 + 2^4 + 2^5$

    $= 1 + 2 + 4 + 8 + 16 + 32$

    $= 63$

46. $\sqrt{7} + \sqrt{9} + \sqrt{11} + \sqrt{13} \approx 12.5679$

47. $\displaystyle\sum_{k=7}^{10} \log k = \log 7 + \log 8 + \log 9 + \log 10 \approx 3.7024$

48. $0 + \pi + 2\pi + 3\pi + 4\pi = 10\pi \approx 31.4159$

49. $\displaystyle\sum_{k=1}^{8} \frac{k}{k + 1} = \frac{1}{1 + 1} + \frac{2}{2 + 1} + \frac{3}{3 + 1} + \frac{4}{4 + 1} +$

    $\frac{5}{5 + 1} + \frac{6}{6 + 1} + \frac{7}{7 + 1} + \frac{8}{8 + 1}$

    $= \frac{1}{2} + \frac{2}{3} + \frac{3}{4} + \frac{4}{5} + \frac{5}{6} + \frac{6}{7} + \frac{7}{8} + \frac{8}{9}$

    $= \frac{15,551}{2520}$

50. $-\frac{1}{4} + 0 + \frac{1}{6} + \frac{2}{7} = \frac{17}{84}$

51. $\displaystyle\sum_{k=1}^{5} (-1)^k = (-1)^1 + (-1)^2 + (-1)^3 + (-1)^4 + (-1)^5$

    $= -1 + 1 - 1 + 1 - 1$

    $= -1$

52. $1 - 1 + 1 - 1 + 1 = 1$

53. $\displaystyle\sum_{k=1}^{8} (-1)^{k+1} 3^k = (-1)^2 3^1 + (-1)^3 3^2 + (-1)^4 3^3 +$

    $(-1)^5 3^4 + (-1)^6 3^5 + (-1)^7 3^6 +$

    $(-1)^8 3^7 + (-1)^9 3^8$

    $= 3 - 9 + 27 - 81 + 243 - 729 +$

    $2187 - 6561$

    $= -4920$

54. $-4^2 + 4^3 - 4^4 + 4^5 - 4^6 + 4^7 - 4^8 = -52,432$

55. $\displaystyle\sum_{k=1}^{6} \frac{2}{k^2 + 1} = \frac{2}{1^2 + 1} + \frac{2}{2^2 + 1} + \frac{2}{3^2 + 1} + \frac{2}{4^2 + 1} +$

    $\frac{2}{5^2 + 1} + \frac{2}{6^2 + 1}$

    $= \frac{2}{2} + \frac{2}{5} + \frac{2}{10} + \frac{2}{17} + \frac{2}{26} + \frac{2}{37}$

    $= 1 + \frac{2}{5} + \frac{1}{5} + \frac{2}{17} + \frac{1}{13} + \frac{2}{37}$

    $= \frac{75,581}{40,885}$

56. $1\cdot2 + 2\cdot3 + 3\cdot4 + 4\cdot5 + 5\cdot6 + 6\cdot7 + 7\cdot8 + 8\cdot9 + 9\cdot10 + 10\cdot11 = 440$

57. $\displaystyle\sum_{k=0}^{5} (k^2 - 2k + 3) = (0^2 - 2\cdot0 + 3) +$

    $(1^2 - 2\cdot1 + 3) + (2^2 - 2\cdot2 + 3) +$

    $(3^2 - 2\cdot3 + 3) + (4^2 - 2\cdot4 + 3) +$

    $(5^2 - 2\cdot5 + 3)$

    $= 3 + 2 + 3 + 6 + 11 + 18$

    $= 43$

58. $4 + 2 + 2 + 4 + 8 + 14 = 34$

59. $\displaystyle\sum_{k=1}^{10} \frac{1}{k(k+1)} = \frac{1}{1(1+1)} + \frac{1}{2(2+1)} + \frac{1}{3(3+1)} +$

$$\frac{1}{4(4+1)} + \frac{1}{5(5+1)} + \frac{1}{6(6+1)} +$$

$$\frac{1}{7(7+1)} + \frac{1}{8(8+1)} + \frac{1}{9(9+1)} +$$

$$\frac{1}{10(10+1)}$$

$$= \frac{1}{2} + \frac{1}{6} + \frac{1}{12} + \frac{1}{20} + \frac{1}{30} + \frac{1}{42} + \frac{1}{56} +$$

$$\frac{1}{72} + \frac{1}{90} + \frac{1}{110}$$

$$= \frac{10}{11}$$

60. $\dfrac{2}{3} + \dfrac{4}{5} + \dfrac{8}{9} + \dfrac{16}{17} + \dfrac{32}{33} + \dfrac{64}{65} + \dfrac{128}{129} + \dfrac{256}{257} + \dfrac{512}{513} +$

$\dfrac{1024}{1025} \approx 9.2365$

61. $\dfrac{1}{2} + \dfrac{2}{3} + \dfrac{3}{4} + \dfrac{4}{5} + \dfrac{5}{6} + \dfrac{6}{7}$

This is a sum of fractions in which the denominator is one greater than the numerator. Also, each numerator is 1 greater than the preceding numerator. Sigma notation is

$$\sum_{k=1}^{6} \frac{k}{k+1}.$$

62. $\displaystyle\sum_{k=1}^{5} 3k$

63. $-2 + 4 - 8 + 16 - 32 + 64$

This is a sum of powers of 2 with alternating signs. Sigma notation is

$$\sum_{k=1}^{6} (-1)^k 2^k, \text{ or } \sum_{k=1}^{6} (-2)^k.$$

64. $\displaystyle\sum_{k=1}^{5} \frac{1}{k^2}$

65. $4 - 9 + 16 - 25 + \ldots + (-1)^n n^2$

This is a sum of terms of the form $(-1)^k k^2$, beginning with $k = 2$ and continuing through $k = n$. Sigma notation is

$$\sum_{k=2}^{n} (-1)^k k^2.$$

66. $\displaystyle\sum_{k=3}^{n} (-1)^{k+1} k^2$

67. $5 + 10 + 15 + 20 + 25 + \ldots$

This is a sum of multiples of 5, and it is an infinite series. Sigma notation is

$$\sum_{k=1}^{\infty} 5k.$$

68. $\displaystyle\sum_{k=1}^{\infty} 7k$

69. $\dfrac{1}{1 \cdot 2} + \dfrac{1}{2 \cdot 3} + \dfrac{1}{3 \cdot 4} + \dfrac{1}{4 \cdot 5} + \ldots$

This is a sum of fractions in which the numerator is 1 and the denominator is a product of two consecutive integers. The larger integer in each product is the smaller integer in the succeeding product. It is an infinite series. Sigma notation is

$$\sum_{k=1}^{\infty} \frac{1}{k(k+1)}.$$

70. $\displaystyle\sum_{k=1}^{\infty} \frac{1}{k(k+1)^2}$

71. $\log_3 3 = 1$

1 is the power to which you raise 3 to get 3.

72. 0

73. $\log_3 3^7 = 7$

7 is the power to which you raise 3 to get 3^7.

74. 1

75. $a_n = \dfrac{1}{2^n} \log 1000^n$

$a_1 = \dfrac{1}{2^1} \log 1000^1 = \dfrac{1}{2} \log 10^3 = \dfrac{1}{2} \cdot 3 = \dfrac{3}{2}$

$a_2 = \dfrac{1}{2^2} \log 1000^2 = \dfrac{1}{4} \log (10^3)^2 = \dfrac{1}{4} \log 10^6 = \dfrac{1}{4} \cdot 6 = \dfrac{3}{2}$

$a_3 = \dfrac{1}{2^3} \log 1000^3 = \dfrac{1}{8} \log (10^3)^3 = \dfrac{1}{8} \log 10^9 = \dfrac{1}{8} \cdot 9 = \dfrac{9}{8}$

$a_4 = \dfrac{1}{2^4} \log 1000^4 = \dfrac{1}{16} \log (10^3)^4 = \dfrac{1}{16} \log 10^{12} = \dfrac{1}{16} \cdot 12 = \dfrac{3}{4}$

$a_5 = \dfrac{1}{2^5} \log 1000^5 = \dfrac{1}{32} \log (10^3)^5 = \dfrac{1}{32} \log 10^{15} = \dfrac{1}{32} \cdot 15 = \dfrac{15}{32}$

$S_5 = \dfrac{3}{2} + \dfrac{3}{2} + \dfrac{9}{8} + \dfrac{3}{4} + \dfrac{15}{32} = \dfrac{171}{32}$

76. $i, -1, -i, 1; i; i$

77. $a_n = \ln(1 \cdot 2 \cdot 3 \cdots n)$

$a_1 = \ln 1 = 0$

$a_2 = \ln(1 \cdot 2) = \ln 2 \approx 0.693$

$a_3 = \ln(1 \cdot 2 \cdot 3) = \ln 6 \approx 1.792$

$a_4 = \ln(1 \cdot 2 \cdot 3 \cdot 4) = \ln 24 \approx 3.178$

$a_5 = \ln(1 \cdot 2 \cdot 3 \cdot 4 \cdot 5) = \ln 120 \approx 4.787$

$S_5 \approx 0 + 0.693 + 1.792 + 3.178 + 4.787 = 10.45$

<u>78.</u> a) 41, 43, 47, 53

  b) $41 + n(n - 1)$

  c) 1681, yes

<u>79.</u> $a_n = \left(1 + \frac{1}{n}\right)^n$

  $a_1 = \left(1 + \frac{1}{1}\right)^1 = 2$

  $a_2 = \left(1 + \frac{1}{2}\right)^2 = (1.5)^2 = 2.25$

  $a_3 = \left(1 + \frac{1}{3}\right)^3 = 2.370370$

  $a_4 = \left(1 + \frac{1}{4}\right)^4 = 2.441406$

  $a_5 = \left(1 + \frac{1}{5}\right)^5 = 2.488320$

  $a_6 = \left(1 + \frac{1}{6}\right)^6 = 2.521626$

<u>80.</u> 0.414214, 0.317837, 0.267949, 0.236068, 0.213422, 0.196262

<u>81.</u> $a_n = \frac{1}{n} \cdot \frac{1}{n + 1}$

  $a_1 = \frac{1}{1} \cdot \frac{1}{1 + 1} = \frac{1}{2}$ $\quad a_4 = \frac{1}{4} \cdot \frac{1}{4 + 1} = \frac{1}{20}$

  $a_2 = \frac{1}{2} \cdot \frac{1}{2 + 1} = \frac{1}{6}$ $\quad a_5 = \frac{1}{5} \cdot \frac{1}{5 + 1} = \frac{1}{30}$

  $a_3 = \frac{1}{3} \cdot \frac{1}{3 + 1} = \frac{1}{12}$ $\quad a_6 = \frac{1}{6} \cdot \frac{1}{6 + 1} = \frac{1}{42}$

  $S_1 = \frac{1}{2}$

  $S_2 = \frac{1}{2} + \frac{1}{6} = \frac{2}{3}$

  $S_3 = \frac{1}{2} + \frac{1}{6} + \frac{1}{12} = \frac{3}{4}$

  $S_4 = \frac{1}{2} + \frac{1}{6} + \frac{1}{12} + \frac{1}{20} = \frac{4}{5}$

  Note the following pattern:

  $S_1 = 1 - \frac{1}{1 + 1} = \frac{1}{2}$ $\quad S_3 = 1 - \frac{1}{3 + 1} = \frac{3}{4}$

  $S_2 = 1 - \frac{1}{2 + 1} = \frac{2}{3}$ $\quad S_4 = 1 - \frac{1}{4 + 1} = \frac{4}{5}$

  Thus $S_n = 1 - \frac{1}{n + 1}$.

<u>82.</u> 1, 1, 1, 1, 1, 1

<u>83.</u> $a_1 = 0$, $a_{n+1} = a_n^2 + 4$

  $a_1 = 0$

  $a_2 = a_1^2 + 4 = 0^2 + 4 = 4$

  $a_3 = a_2^2 + 4 = 4^2 + 4 = 20$

  $a_4 = a_3^2 + 4 = 20^2 + 4 = 404$

  $a_5 = a_4^2 + 4 = 404^2 + 4 = 163,220$

  $a_6 = a_5^2 + 4 = 163,220^2 + 4 = 26,640,768,404$

<u>84.</u> 1, 2, 4, 8, 16, 32, 64, 128, 256, 512, 1024, 2048, 4096, 8192, 16,384, 32,768, 65,536

<u>85.</u> Find each term by multiplying the preceding term by 0.75:

  $5200, $3900, $2925, $2193.75, $1645.31, $1233.98, $925.49, $694.12, $520.59, $390.44

<u>86.</u> $4.20, $4.35, $4.50, $4.65, $4.80, $4.95, $5.10, $5.25, $5.40, $5.55

Exercise Set 11.2

<u>1.</u> 2, 7, 12, 17, . . .

  $a_1 = 2$

  $d = 5$ $(7 - 2 = 5, 12 - 7 = 5, 17 - 12 = 5)$

<u>2.</u> $a_1 = 1.06$, $d = 0.06$

<u>3.</u> 7, 3, -1, -5, . . .

  $a_1 = 7$

  $d = -4$ $(3 - 7 = -4, -1 - 3 = -4, -5 - (-1) = -4)$

<u>4.</u> $a_1 = -9$, $d = 3$

<u>5.</u> $\frac{3}{2}$, $\frac{9}{4}$, 3, $\frac{15}{4}$, . . .

  $a_1 = \frac{3}{2}$

  $d = \frac{3}{4}$ $\left(\frac{9}{4} - \frac{3}{2} = \frac{3}{4}, 3 - \frac{9}{4} = \frac{3}{4}\right)$

<u>6.</u> $a_1 = \frac{3}{5}$, $d = -\frac{1}{2}$

<u>7.</u> $2.12, $2.24, $2.36, $2.48, . . .

  $a_1 = 2.12

  $d = 0.12 $($2.24 - $2.12 = $0.12, $2.36 - $2.24 = $0.12, $2.48 - $2.36 = $0.12)$

<u>8.</u> $a_1 = 214, $d = -$3$

<u>9.</u> 2, 6, 10, . . .

  $a_1 = 2$, $d = 4$, and $n = 12$

  $a_n = a_1 + (n - 1)d$

  $a_{12} = 2 + (12 - 1)4 = 2 + 11 \cdot 4 = 2 + 44 = 46$

<u>10.</u> 0.57

<u>11.</u> 7, 4, 1, . . .

  $a_1 = 7$, $d = -3$, and $n = 17$

  $a_n = a_1 + (n - 1)d$

  $a_{17} = 7 + (17 - 1)(-3) = 7 + 16(-3) = 7 - 48 = -41$

<u>12.</u> $-\frac{17}{3}$

13. $1200, $964.32, $728.64, . . .

   $a_1 = \$1200$, $d = \$964.32 - \$1200 = -\$235.68$,
      and $n = 13$
   $a_n = a_1 + (n - 1)d$
   $a_{13} = \$1200 + (13 - 1)(-\$235.68) =$
      $\$1200 + 12(-\$235.68) = \$1200 - \$2828.16 =$
      $-\$1628.16$

14. $7941.62

15. $a_1 = 2$, $d = 4$
   $a_n = a_1 + (n - 1)d$
   Let $a_n = 106$, and solve for n.
   $106 = 2 + (n - 1)(4)$
   $106 = 2 + 4n - 4$
   $108 = 4n$
    $27 = n$
   The 27th term is 106.

16. 33rd

17. $a_1 = 7$, $d = -3$
    $a_n = a_1 + (n - 1)d$
   $-296 = 7 + (n - 1)(-3)$
   $-296 = 7 - 3n + 3$
   $-306 = -3n$
    $102 = n$
   The 102nd term is -296.

18. 46th

19. $a_n = a_1 + (n - 1)d$
   $a_{17} = 5 + (17 - 1)6$ Substituting 17 for n,
             5 for a_1, and 6 for d
      $= 5 + 16 \cdot 6$
      $= 5 + 96$
      $= 101$

20. -43

21. $a_n = a_1 + (n - 1)d$
   $33 = a_1 + (8 - 1)4$ Substituting 33 for a_8, 8 for
                 n, and 4 for d
   $33 = a_1 + 28$
    $5 = a_1$
   (Note that this procedure is equivalent to
   subtracting d from a_8 seven times to get a_1:
   $33 - 7(4) = 33 - 28 = 5$)

22. -54

23. $a_n = a_1 + (n - 1)d$
   $-76 = 5 + (n - 1)(-3)$ Substituting -76 for a_n,
                  5 for a_1 and -3 for d
   $-76 = 5 - 3n + 3$
   $-76 = 8 - 3n$
   $-84 = -3n$
    $28 = n$

24. 39

25. We know that $a_{17} = -40$ and $a_{28} = -73$. We would
   have to add d eleven times to get from a_{17} to a_{28}.
   That is,
      $-40 + 11d = -73$
         $11d = -33$
          $d = -3.$
   Since $a_{17} = -40$, we subtract d sixteen times to
   get to a_1.
      $a_1 = -40 - 16(-3) = -40 + 48 = 8$
   We write the first five terms of the sequence:
      8, 5, 2, -1, -4

26. $a_1 = \frac{1}{3}$; $d = \frac{1}{2}$; $\frac{1}{3}$, $\frac{5}{6}$, $\frac{4}{3}$, $\frac{11}{6}$, $\frac{7}{3}$

27. 5 + 8 + 11 + 14 + . . .
   Note that $a_1 = 5$, $d = 3$, and $n = 20$. We use
   Formula 3.
   $S_n = \frac{n}{2}\left[2a_1 + (n - 1)d\right]$
   $S_{20} = \frac{20}{2}[2 \cdot 5 + (20 - 1)3] = 10[10 + 57] = 10 \cdot 67 =$
      670

28. -210

29. The sum is 1 + 2 + 3 + . . . + 299 + 300. This
   is the sum of the arithmetic sequence for which
   $a_1 = 1$, $a_n = 300$, and $n = 300$. We use Formula 2.
      $S_n = \frac{n}{2}(a_1 + a_n)$
   $S_{300} = \frac{300}{2}(1 + 300) = 150(301) = 45,150$

30. 80,200

31. The sum is 2 + 4 + 6 + . . . + 98 + 100. This is
   the sum of the arithmetic sequence for which
   $a_1 = 2$, $a_n = 100$, and $n = 50$. We use Formula 2.
      $S_n = \frac{n}{2}(a_1 + a_n)$
   $S_{50} = \frac{50}{2}(2 + 100) = 25(102) = 2550$

32. 2500

33. The sum is 7 + 14 + 21 + . . . + 91 + 98. This is the sum of the arithmetic sequence for which $a_1 = 7$, $a_n = 98$, and n = 14. We use Formula 2.

$$S_n = \frac{n}{2}(a_1 + a_n)$$

$$S_{14} = \frac{14}{2}(7 + 98) = 7(105) = 735$$

34. 34,036

35. $S_n = \frac{n}{2}\left[2a_1 + (n - 1)d\right]$

$S_{20} = \frac{20}{2}[2\cdot2 + (20 - 1)5]$ Substituting 20 for n, 2 for a_1, and 5 for d

= 10[4 + 19·5]

= 10[4 + 95]

= 10·99

= 990

36. -1264

37. We first find how many plants will be in the last row.

Familiarize. The sequence is 35, 31, 27, ... It is an arithmetic sequence with $a_1 = 35$ and d = -4. Since each row must contain a positive number of plants, we must determine how many times we can add -4 to 35 and still have a positive result.

Translate. We find the largest integer x for which 35 + x(-4) > 0. Then we evaluate the expression 35 - 4x for that value of x.

Carry out. We solve the inequality.

35 - 4x > 0

35 > 4x

$\frac{35}{4}$ > x

$8\frac{3}{4}$ > x

The integer we are looking for is 8. Thus 35 - 4x = 35 - 4(8) = 3.

Check. If we add -4 to 35 eight times we get 3, a positive number, but if we add -4 to 35 more than eight times we get a negative number.

State. There will be 3 plants in the last row.

Next we find how many plants there are altogether.

Familiarize. We want to find the sum 35 + 31 + 27 + . . . + 3. We know $a_1 = 35$, $a_n = 3$, and, since we add -4 to 35 eight times, n = 9. (There are 8 terms after a_1, for a total of 9 terms.)

Translate. We use Formula 2.

$$S_9 = \frac{9}{2}(35 + 3)$$

Carry out. We calculate to obtain $S_9 = 171$.

Check. We can check the calculations by doing them again. We could also do the entire addition:

35 + 31 + 27 + . . . + 3.

State. There are 171 plants altogether.

38. 62; 950

39. Familiarize. We go from 50 poles in a row, down to one pole in the top row, so there must be 50 rows. We want the sum 50 + 49 + 48 + . . . + 1. Thus we want the sum of an arithmetic sequence. We will use the formula $S_n = \frac{n}{2}(a_1 + a_n)$.

Translate. We want to find the sum of an arithmetic sequence with $a_1 = 50$, $a_n = 1$, and n = 50. Substituting into the formula, we have

$$S_{50} = \frac{50}{2}(50 + 1).$$

Carry out. We calculate to obtain 1275.

Check. We can do the calculation again, or we can do the entire addition:
50 + 49 + 48 + . . . + 1.

State. There will be 1275 poles in the pile.

40. $49.60

41. Familiarize. We want to find the sum of an arithmetic sequence with $a_1 = \$600$, d = $100, and n = 20. We will use the formula

$$S_n = \frac{n}{2}[2a_1 + (n - 1)d].$$

Translate. Substituting into the formula, we have

$$S_{20} = \frac{20}{2}[2\cdot600 + (20 - 1)100].$$

Carry out. We calculate to obtain 31,000.

Check. We can do the calculation again.

State. They save $31,000 (disregarding interest).

42. $10,230

43. Familiarize. We want to find the sum of an arithmetic sequence with $a_1 = 28$, d = 4, and n = 50. We will use the formula

$$S_n = \frac{n}{2}[2a_1 + (n - 1)d].$$

Translate. Substituting into the formula, we have

$$S_{50} = \frac{50}{2}[2\cdot28 + (50 - 1)4].$$

Carry out. We calculate to obtain 6300.

Check. We can do the calculation again.

State. There are 6300 seats.

44. $462,500

45. $\log_a P = k \Longleftrightarrow a^k = P$ The logarithm is the exponent. The base does not change.

46. $e^a = t$

47. $e^t = Q \Longleftrightarrow t = \log_e Q$, or t = ℓn Q   The exponent is the logarithm. The base remains the base.

48. $\log_{49} 7 = \frac{1}{2}$

49. We find the sum of an arithmetic sequence with $a_1 = 1$, $a_n = n$, and $n = n$. We use the formula

$$S_n = \frac{n}{2}(a_1 + a_n).$$

$$S_n = \frac{n}{2}(1 + n), \text{ or } \frac{n(n + 1)}{2}$$

50. n^2

51. Familiarize. Let x represent the first number in the sequence, and let d represent the common difference. Then the three numbers in the sequence are x, x + d, and x + 2d.

Translate.

The sum of the first and third numbers is 10.

$$x + x + 2d \qquad\qquad = 10$$

The product of the first and second numbers is 15.

$$x(x + d) \qquad\qquad = 15$$

Carry out. Solving the system of equations we get x = 3 and d = 2. Thus the numbers are 3, 5, and 7.

Check. The numbers are in an arithmetic sequence. Also 3 + 7 = 10 and 3·5 = 15. The numbers check.

State. The numbers are 3, 5, and 7.

52. $a_1 = p - 5q$; $d = 3p + 2q$

53.
$a_1 = \$8760$
$a_2 = \$8760 + (-\$798.23) = \$7961.67$
$a_3 = \$8760 + 2(-\$798.23) = \$7163.54$
$a_4 = \$8760 + 3(-\$798.23) = \$6365.31$
$a_5 = \$8760 + 4(-\$798.23) = \$5567.08$
$a_6 = \$8760 + 5(-\$798.23) = \$4768.85$
$a_7 = \$8760 + 6(-\$798.23) = \$3970.62$
$a_8 = \$8760 + 7(-\$798.23) = \$3172.39$
$a_9 = \$8760 + 8(-\$798.23) = \$2374.16$
$a_{10} = \$8760 + 9(-\$798.23) = \$1575.93$

54. $51.679.65

55. See the answer section in the text.

56. a) $a_t = \$5200 - \$512.50t$

b) $5200, $4687.50, $4175, $3662.50, $3150, $1612.50, $1100

Exercise Set 11.3

1. 2, 4, 8, 16, . . .

$$\frac{4}{2} = 2, \quad \frac{8}{4} = 2, \quad \frac{16}{8} = 2$$

$r = 2$

2. $-\frac{1}{3}$

3. 1, -1, 1, -1, . . .

$$\frac{-1}{1} = -1, \quad \frac{1}{-1} = -1, \quad \frac{-1}{1} = -1$$

$r = -1$

4. 0.1

5. $\frac{1}{2}, -\frac{1}{4}, \frac{1}{8}, -\frac{1}{16}, \ldots$

$$\frac{-\frac{1}{4}}{\frac{1}{2}} = -\frac{1}{4} \cdot \frac{2}{1} = -\frac{2}{4} = -\frac{1}{2}$$

$$\frac{\frac{1}{8}}{-\frac{1}{4}} = \frac{1}{8} \cdot \left(-\frac{4}{1}\right) = -\frac{4}{8} = -\frac{1}{2}$$

$r = -\frac{1}{2}$

6. -2

7. 75, 15, 3, $\frac{3}{5}$, . . .

$$\frac{15}{75} = \frac{1}{5}, \quad \frac{3}{15} = \frac{1}{5}, \quad \frac{\frac{3}{5}}{3} = \frac{3}{5} \cdot \frac{1}{3} = \frac{1}{5}$$

$r = \frac{1}{5}$

8. 0.1

9. $\frac{1}{x}, \frac{1}{x^2}, \frac{1}{x^3}, \ldots$

$$\frac{\frac{1}{x^2}}{\frac{1}{x}} = \frac{1}{x^2} \cdot \frac{x}{1} = \frac{x}{x^2} = \frac{1}{x}$$

$$\frac{\frac{1}{x^3}}{\frac{1}{x^2}} = \frac{1}{x^3} \cdot \frac{x^2}{1} = \frac{x^2}{x^3} = \frac{1}{x}$$

$r = \frac{1}{x}$

10. $\frac{m}{2}$

11. $780, $858, $943.80, $1038.18, . . .

$$\frac{\$858}{\$780} = 1.1, \quad \frac{\$943.80}{\$858} = 1.1,$$

$$\frac{\$1038.18}{\$943.80} = 1.1$$

$r = 1.1$

12. 0.95

13. 2, 4, 8, 16, . . .

$a_1 = 2$, $n = 6$, and $r = \frac{4}{2}$, or 2.

We use the formula $a_n = a_1 r^{n-1}$.

$$a_6 = 2(2)^{6-1} = 2 \cdot 2^5 = 2 \cdot 32 = 64$$

14. 781,250

15. 2, $2\sqrt{3}$, 6, . . .

   $a_1 = 2$, $n = 9$, and $r = \dfrac{2\sqrt{3}}{2}$, or $\sqrt{3}$

   $a_n = a_1 r^{n-1}$

   $a_9 = 2(\sqrt{3})^{9-1} = 2(\sqrt{3})^8 = 2 \cdot 81 = 162$

16. 1

17. $\dfrac{8}{243}$, $\dfrac{8}{81}$, $\dfrac{8}{27}$, . . .

   $a_1 = \dfrac{8}{243}$, $n = 10$, and $r = \dfrac{\frac{8}{81}}{\frac{8}{243}} = \dfrac{8}{81} \cdot \dfrac{243}{8} = 3$

   $a_n = a_1 r^{n-1}$

   $a_{10} = \dfrac{8}{243}(3)^{10-1} = \dfrac{8}{243}(3)^8 = \dfrac{8}{243} \cdot 19{,}683 = 648$

18. 2,734,375

19. $1000, $1080, $1166.40, . . .

   $a_1 = \$1000$, $n = 12$, and $r = \dfrac{\$1080}{\$1000} = 1.08$

   $a_n = a_1 r^{n-1}$

   $a_{12} = \$1000(1.08)^{12-1} \approx \$1000(2.331638997) \approx$
   $\$2331.64$

20. $1967.15

21. 1, 3, 9, . . .

   $a_1 = 1$ and $r = \dfrac{3}{1}$, or 3

   $a_n = a_1 r^{n-1}$

   $a_n = 1(3)^{n-1} = 3^{n-1}$

22. $a_n = 5^{3-n}$

23. 1, -1, 1, -1, . . .

   $a_1 = 1$ and $r = \dfrac{-1}{1} = -1$

   $a_n = a_1 r^{n-1}$

   $a_n = 1(-1)^{n-1} = (-1)^{n-1}$

24. $a_n = 2^n$

25. $\dfrac{1}{x}$, $\dfrac{1}{x^2}$, $\dfrac{1}{x^3}$, . . .

   $a_1 = \dfrac{1}{x}$ and $r = \dfrac{1}{x}$ (see Exercise 9)

   $a_n = a_1 r^{n-1}$

   $a_n = \dfrac{1}{x}\left(\dfrac{1}{x}\right)^{n-1} = \dfrac{1}{x} \cdot \dfrac{1}{x^{n-1}} = \dfrac{1}{x^{1+n-1}} = \dfrac{1}{x^n}$

26. $a_n = 5\left[\dfrac{m}{2}\right]^{n-1}$

27. 6 + 12 + 24 + . . .

   $a_1 = 6$, $n = 7$, and $r = \dfrac{12}{6}$, or 2

   $S_n = \dfrac{a_1(r^n - 1)}{r - 1}$

   $S_7 = \dfrac{6(2^7 - 1)}{2 - 1} = \dfrac{6(128 - 1)}{1} = 6 \cdot 127 = 762$

28. 10.5

29. $\dfrac{1}{18} - \dfrac{1}{6} + \dfrac{1}{2} - \cdots$

   $a_1 = \dfrac{1}{18}$, $n = 7$, and $r = \dfrac{-\frac{1}{6}}{\frac{1}{18}} = -\dfrac{1}{6} \cdot \dfrac{18}{1} = -3$

   $S_n = \dfrac{a_1(r^n - 1)}{r - 1}$

   $S_7 = \dfrac{\frac{1}{18}[(-3)^7 - 1]}{-3 - 1} = \dfrac{\frac{1}{18}[-2187 - 1]}{-4} = \dfrac{\frac{1}{18}(-2188)}{-4} =$
   $\dfrac{1}{18}(-2188)\left[-\dfrac{1}{4}\right] = \dfrac{547}{18}$

30. 6.6666

31. $1 + x + x^2 + x^3 + \cdots$

   $a_1 = 1$, $n = 8$, and $r = \dfrac{x}{1}$, or x

   $S_n = \dfrac{a_1(r^n - 1)}{r - 1}$

   $S_8 = \dfrac{1(x^8 - 1)}{x - 1} = \dfrac{(x^4 + 1)(x^4 - 1)}{x - 1} =$
   $\dfrac{(x^4 + 1)(x^2 + 1)(x^2 - 1)}{x - 1} =$
   $\dfrac{(x^4 + 1)(x^2 + 1)(x + 1)(x - 1)}{x - 1} =$
   $(x^4 + 1)(x^2 + 1)(x + 1)$

32. $\dfrac{x^{20} - 1}{x^2 - 1}$

33. $200, $200(1.06), $200(1.06)^2$, . . .

   $a_1 = \$200$, $n = 16$, and $r = \dfrac{\$200(1.06)}{\$200} = 1.06$

   $S_n = \dfrac{a_1(r^n - 1)}{r - 1}$

   $S_{16} = \dfrac{\$200(1.06^{16} - 1)}{1.06 - 1} \approx \dfrac{\$200(2.540351685 - 1)}{0.06} \approx$
   $\$5134.51$

34. $60,893.30

35. 4 + 2 + 1 + . . .

   $|r| = \left|\dfrac{2}{4}\right| = \left|\dfrac{1}{2}\right| = \dfrac{1}{2}$, and since $|r| < 1$, the series does have a sum.

   $S_\infty = \dfrac{a_1}{1 - r} = \dfrac{4}{1 - \frac{1}{2}} = \dfrac{4}{\frac{1}{2}} = 4 \cdot \dfrac{2}{1} = 8$

36. $\dfrac{49}{4}$

37. $25 + 20 + 16 + \ldots$

$|r| = \left|\frac{20}{25}\right| = \left|\frac{4}{5}\right| = \frac{4}{5}$, and since $|r| < 1$, the series does have a sum.

$S_\infty = \frac{a_1}{1 - r} = \frac{25}{1 - \frac{4}{5}} = \frac{25}{\frac{1}{5}} = 25 \cdot \frac{5}{1} = 125$

38. 48

39. $100 - 10 + 1 - \frac{1}{10} + \ldots$

$|r| = \left|\frac{-10}{100}\right| = \left|-\frac{1}{10}\right| = \frac{1}{10}$, and since $|r| < 1$, series does have a sum.

$S_\infty = \frac{a_1}{1 - r} = \frac{100}{1 - \left(-\frac{1}{10}\right)} = \frac{100}{\frac{11}{10}} = 100 \cdot \frac{10}{11} = \frac{1000}{11}$

40. No

41. $8 + 40 + 200 + \ldots$

$|r| = \left|\frac{40}{8}\right| = |5| = 5$, and since $|r| \not< 1$ the series does not have a sum.

42. -4

43. $0.3 + 0.03 + 0.003 + \ldots$

$|r| = \left|\frac{0.03}{0.3}\right| = |0.1| = 0.1$, and since $|r| < 1$, the series does have a sum.

$S_\infty = \frac{a_1}{1 - r} = \frac{0.3}{1 - 0.1} = \frac{0.3}{0.9} = \frac{3}{9} = \frac{1}{3}$

44. $\frac{37}{99}$

45. $\$500(1.02)^{-1} + \$500(1.02)^{-2} + \$500(1.02)^{-3} + \ldots$

$|r| = \left|\frac{\$500(1.02)^{-2}}{\$500(1.02)^{-1}}\right| = |(1.02)^{-1}| = (1.02)^{-1}$, or $\frac{1}{1.02}$, and since $|r| < 1$, the series does have a sum.

$S_\infty = \frac{a_1}{1 - r} = \frac{\$500(1.02)^{-1}}{1 - \left(\frac{1}{1.02}\right)} = \frac{\frac{\$500}{1.02}}{\frac{0.02}{1.02}} = $

$\frac{\$500}{1.02} \cdot \frac{1.02}{0.02} = \$25,000$

46. $12,500

47. $0.4444\ldots = 0.4 + 0.04 + 0.004 + 0.0004 + \ldots$

This is an infinite geometric series with $a_1 = 0.4$.

$|r| = \left|\frac{0.04}{0.4}\right| = |0.1| = 0.1 < 1$, so the series has a sum.

$S_\infty = \frac{a_1}{1 - r} = \frac{0.4}{1 - 0.1} = \frac{0.4}{0.9} = \frac{4}{9}$

Fractional notation for $0.4444\ldots$ is $\frac{4}{9}$.

48. 10

49. $0.55555 = 0.5 + 0.05 + 0.005 + 0.0005 + \ldots$

This is an infinite geometric series with $a_1 = 0.5$.

$|r| = \left|\frac{0.05}{0.5}\right| = |0.1| = 0.1 < 1$, so the series has a sum.

$S_\infty = \frac{a_1}{1 - r} = \frac{0.5}{1 - 0.1} = \frac{0.5}{0.9} = \frac{5}{9}$

50. $\frac{2}{3}$

51. $0.15151515\ldots = 0.15 + 0.0015 + 0.000015 + \ldots$

This is an infinite geometric series with $a_1 = 0.15$.

$|r| = \left|\frac{0.0015}{0.15}\right| = |0.01| = 0.01 < 1$, so the series has a sum.

$S_\infty = \frac{a_1}{1 - r} = \frac{0.15}{1 - 0.01} = \frac{0.15}{0.99} = \frac{15}{99} = \frac{5}{33}$

52. $\frac{4}{33}$

53. Familiarize. The rebound distances form a geometric sequence:

$\frac{1}{4} \times 16, \quad \left(\frac{1}{4}\right)^2 \times 16, \quad \left(\frac{1}{4}\right)^3 \times 16, \ldots,$

or $4, \quad \frac{1}{4} \times 4, \quad \left(\frac{1}{4}\right)^2 \times 4, \ldots$

The height of the 6th rebound is the 6th term of the sequence.

Translate. We will use the formula $a_n = a_1 r^{n-1}$, with $a_1 = 4$, $r = \frac{1}{4}$, and $n = 6$:

$a_6 = 4\left(\frac{1}{4}\right)^{6-1}$

Carry out. We calculate to obtain $a_6 = \frac{1}{256}$.

Check. We can do the calculation again.

State. It rebounds $\frac{1}{256}$ ft the 6th time.

54. $5\frac{1}{3}$ ft

55. Familiarize. In one year, the population will be $100,000 + 0.03(100,000)$, or $(1.03)100,000$. In two years, the population will be $(1.03)100,000 + 0.03(1.03)100,000$, or $(1.03)^2 100,000$. Thus the populations form a geometric sequence:

$100,000, \quad (1.03)100,000, \quad (1.03)^2 100,000, \ldots$

The population in 15 years will be the 16th term of the sequence.

Translate. We will use the formula $a_n = a_1 r^{n-1}$ with $a_1 = 100,000$, $r = 1.03$, and $n = 16$:

$a_{16} = 100,000(1.03)^{16-1}$

Carry out. We calculate to obtain $a_{16} \approx 155,797$.

Check. We can do the calculation again.

State. In 15 years the population will be about 155,797.

56. About 24 years

57. Familiarize. The amounts owed at the beginning of successive years form a geometric sequence:

$1200, (1.12)$1200, (1.12)^2$1200,

(1.12)^3$1200, . . .

The amount to be repaid at the end of 13 years is the amount owed at the beginning of the 14th year.

Translate. We use the formula $a_n = a_1 r^{n-1}$ with $a_1 = 1200$, $r = 1.12$, and $n = 14$:

$$a_{14} = 1200(1.12)^{14-1}$$

Carry out. We calculate to obtain $a_{14} \approx 5236.19$.

Check. We can do the calculation again.

State. At the end of 13 years, $5236.19 will be repaid.

58. 10,485.76 in.

59. Familiarize. The lengths of the falls form a geometric sequence:

$556, \left[\frac{3}{4}\right]556, \left[\frac{3}{4}\right]^2556, \left[\frac{3}{4}\right]^3556, \ldots$

The total length of the first 6 falls is the sum of the first six terms of this sequence. The heights of the rebounds also form a geometric sequence:

$\left[\frac{3}{4}\right]556, \left[\frac{3}{4}\right]^2556, \left[\frac{3}{4}\right]^3556, \ldots,$ or

$417, \left[\frac{3}{4}\right]417, \left[\frac{3}{4}\right]^2417, \ldots$

When the ball hits the ground for the 6th time, it will have rebounded 5 times. Thus the total length of the rebounds is the sum of the first five terms of this sequence.

Translate. We use the formula $S_n = \frac{a_1(r^n - 1)}{r - 1}$ twice, once with $a_1 = 556$, $r = \frac{3}{4}$, and $n = 6$ and a second time with $a_1 = 417$, $r = \frac{3}{4}$, and $n = 5$.

D = Length of falls + length of rebounds

$$= \frac{556\left[\left[\frac{3}{4}\right]^6 - 1\right]}{\frac{3}{4} - 1} + \frac{417\left[\left[\frac{3}{4}\right]^5 - 1\right]}{\frac{3}{4} - 1}.$$

Carry out. We use a calculator to obtain $D \approx 3100.35$.

Check. We can do the calculations again.

State. The ball will have traveled about 3100.35 ft.

60. 3892 ft

61. Familiarize. The amounts form a geometric series:

$0.01, $0.01(2), $0.01(2)^2, $0.01(2^3) + \ldots + ($0.01)(2)^{27}$

Translate. We use the formula $S_n = \frac{a_1(r^n - 1)}{r - 1}$ to find the sum of the geometric series with $a_1 = 0.01$, $r = 2$, and $n = 28$:

$$S_{28} = \frac{0.01(2^{28} - 1)}{2 - 1}$$

61. (continued)

Carry out. We use a calculator to obtain S_{28} = $2,684,354.55.

Check. We can do the calculation again.

State. You would earn $2,684,354.55.

62. $645,826.93

63. $\log_a x^2y^3z^5 = \log_a x^2 + \log_a y^3 + \log_a z^5$

$$= 2 \log_a x + 3 \log_a y + 5 \log_a z$$

64. $2 \log_a x + 3 \log_a y - 5 \log_a z$

65. $\log_a \frac{\sqrt{x^2y^4}}{z^{10}} = \log_a \frac{xy^2}{z^{10}}$ Taking the square root

$$= \log_a xy^2 - \log_a z^{10}$$

$$= \log_a x + \log_a y^2 - 10 \log_a z$$

$$= \log_a x + 2 \log_a y - 10 \log_a z$$

66. $\frac{4}{3} \log_a x + \frac{5}{3} \log_a y - 9 \log_a z$

67. $1 + x + x^2 + \ldots$

This is a geometric series with $a_1 = 1$ and $r = x$.

$$S_n = \frac{a_1(r^n - 1)}{r - 1} = \frac{1(x^n - 1)}{x - 1} = \frac{x^n - 1}{x - 1}, \text{ or } \frac{1 - x^n}{1 - x}$$

68. $\frac{x^2[1 - (-x)^n]}{x + 1}$

69. Familiarize. The length of a side of the first square is 16 cm. The length of a side of the next square is the length of the hypotenuse of a right triangle with legs 8 cm and 8 cm, or $8\sqrt{2}$ cm. The length of a side of the next square is the length of the hypotenuse of a right triangle with legs $4\sqrt{2}$ cm and $4\sqrt{2}$ cm, or 8 cm. The areas of the squares form a sequence:

$(16)^2, (8\sqrt{2})^2, (8)^2, \ldots,$ or

256,   128,   64, ...

This is a geometric sequence with $a_1 = 256$ and $r = \frac{1}{2}$.

Translate. We find the sum of the infinite geometric series $256 + 128 + 64 + \ldots$.

$$S_\infty = \frac{a_1}{1 - r}$$

$$S_\infty = \frac{256}{1 - \frac{1}{2}}$$

Carry out. We calculate to obtain $S_\infty = 512$.

Check. We can do the calculation again.

State. The sum of the areas is 512 cm².

70. $S_1 = 2$, $S_2 = 2\frac{1}{2}$, $S_3 = 2\frac{2}{3}$, $S_4 = 2\frac{17}{24}$, $S_5 = 2\frac{43}{60}$, $S_6 = 2\frac{517}{720}$; $S_\infty \approx 2\frac{3}{4}$

Exercise Set 11.4

1. $(m + n)^5$

   Note that a = m, b = n, and n = 5. We will use the 6th row of Pascal's triangle:

   1   5   10   10   5   1

   $(m + n)^5 = m^5 + 5m^4n + 10m^3n^2 + 10m^2n^3 + 5mn^4 + n^5$

2. $a^4 - 4a^3b + 6a^2b^2 - 4ab^3 + b^4$

3. $(x - y)^6$

   We will use the preceding coefficient method. We find each term in sequence. There are 7 terms in the expansion. After we find the 4th coefficient, we can use symmetry to obtain the others.

   The 1st term is x^6.

   The 2nd term is $\frac{1 \cdot 6}{1}x^5(-y)^1 = -6x^5y$.

   (Note that the coefficient is 6. The "−" comes from the y-term.)

   The 3rd term is $\frac{6 \cdot 5}{2}x^4(-y)^2 = 15x^4y^2$.

   The 4th term is $\frac{15 \cdot 4}{3}x^3(-y)^3 = -20x^3y^3$.

   The rest of the coefficients are 15, 6, and 1. The complete expansion is

   $(x - y)^6 = x^6 - 6x^5y + 15x^4y^2 - 20x^3y^3 +$
   $\qquad 15x^2(-y)^4 + 6x(-y)^5 + (-y)^6$
   $\qquad = x^6 - 6x^5y + 15x^4y^2 - 20x^3y^3 + 15x^2y^4 -$
   $\qquad 6xy^5 + y^6$

4. $p^7 + 7p^6q + 21p^5q^2 + 35p^4q^3 + 35p^3q^4 + 21p^2q^5 + 7pq^6 + q^7$

5. $(x^2 - 3y)^5$

   Expand using factorial notation. Note that a = x^2 and b = -3y.

   $(x^2 - 3y)^5 = \binom{5}{0}(x^2)^5 + \binom{5}{1}(x^2)^4(-3y)^1 +$
   $\qquad \binom{5}{2}(x^2)^3(-3y)^2 + \binom{5}{3}(x^2)^2(-3y)^3 +$
   $\qquad \binom{5}{4}(x^2)^1(-3y)^4 + \binom{5}{5}(-3y)^5$
   $\qquad = \frac{5!}{5!0!}x^{10} + \frac{5!}{4!1!}(x^8)(-3y) +$
   $\qquad \frac{5!}{3!2!}(x^6)(9y^2) + \frac{5!}{2!3!}(x^4)(-27y^3) +$
   $\qquad \frac{5!}{1!4!}(x^2)(81y^4) + \frac{5!}{0!5!}(-243y^5)$
   $\qquad = x^{10} - 15x^8y + 90x^6y^2 - 270x^4y^3 +$
   $\qquad 405x^2y^4 - 243y^5$

6. $729c^6 + 1458c^5d + 1215c^4d^2 + 540c^3d^3 + 135c^2d^4 + 18cd^5 + d^6$

7. $(3c - d)^6$

   Note that a = 3c and b = -d. We will use the 7th row of Pascal's triangle.

   1   6   15   20   15   6   1

   $(3c - d)^6 = (3c)^6 + 6(3c)^5(-d)^1 + 15(3c)^4(-d)^2 +$
   $\qquad 20(3c)^3(-d)^3 + 15(3c)^2(-d)^4 +$
   $\qquad 6(3c)^1(-d)^5 + (-d)^6$
   $\qquad = 729c^6 - 1458c^5d + 1215c^4d^2 - 540c^3d^3 +$
   $\qquad 135c^2d^4 - 18cd^5 + d^6$

8. $t^{-12} + 12t^{-10} + 60t^{-8} + 160t^{-6} + 240t^{-4} + 192t^{-2} + 64$

9. $(x - y)^3$

   We will use the preceding coefficient method. There are 4 terms in the expansion. After we find the 2nd coefficient, we can use symmetry to obtain the others.

   The first term is x^3.

   The second term is $\frac{1 \cdot 3}{1}x^2(-y)^1 = -3x^2y$

   (Note that the coefficient is 3. The "−" comes from the y-term.)

   The other two coefficients are 3 and 1. The complete expansion is

   $(x - y)^3 = x^3 - 3x^2y + 3x(-y)^2 + (-y)^3$
   $\qquad = x^3 - 3x^2y + 3xy^2 - y^3$

10. $x^5 - 5x^4y + 10x^3y^2 - 10x^2y^3 + 5xy^4 - y^5$

11. $\left[\frac{1}{x} + y\right]^7$

    Expand using factorial notation. Note that a = $\frac{1}{x}$ and b = y.

    $\left[\frac{1}{x} + y\right]^7 = \binom{7}{0}\left[\frac{1}{x}\right]^7 + \binom{7}{1}\left[\frac{1}{x}\right]^6 y^1 + \binom{7}{2}\left[\frac{1}{x}\right]^5 y^2 +$
    $\qquad \binom{7}{3}\left[\frac{1}{x}\right]^4 y^3 + \binom{7}{4}\left[\frac{1}{x}\right]^3 y^4 + \binom{7}{5}\left[\frac{1}{x}\right]^2 y^5 +$
    $\qquad \binom{7}{6}\left[\frac{1}{x}\right] y^6 + \binom{7}{7} y^7$
    $\qquad = \frac{7!}{7!0!}\left[\frac{1}{x^7}\right] + \frac{7!}{6!1!}\left[\frac{y}{x^6}\right] + \frac{7!}{5!2!}\left[\frac{y^2}{x^5}\right] +$
    $\qquad \frac{7!}{4!3!}\left[\frac{y^3}{x^4}\right] + \frac{7!}{3!4!}\left[\frac{y^4}{x^3}\right] + \frac{7!}{2!5!}\left[\frac{y^5}{x^2}\right] +$
    $\qquad \frac{7!}{1!6!}\left[\frac{y^6}{x}\right] + \frac{7!}{0!7!}(y^7)$
    $\qquad = x^{-7} + 7x^{-6}y + 21x^{-5}y^2 + 35x^{-4}y^3 +$
    $\qquad 35x^{-3}y^4 + 21x^{-2}y^5 + 7x^{-1}y^6 + y^7$

12. $8s^3 - 36s^2t^2 + 54st^4 - 27t^6$

13. $\left(a - \dfrac{2}{a}\right)^9$

Note that $a = a$ and $b = \left(-\dfrac{2}{a}\right)$. We will use the 10th row of Pascal's triangle:

1   9   36   84   126   126   84   36   9   1

$\left(a - \dfrac{2}{a}\right)^9 = a^9 + 9a^8\left(-\dfrac{2}{a}\right)^1 + 36a^7\left(-\dfrac{2}{a}\right)^2 +$

$\qquad 84a^6\left(-\dfrac{2}{a}\right)^3 + 126a^5\left(-\dfrac{2}{a}\right)^4 +$

$\qquad 126a^4\left(-\dfrac{2}{a}\right)^5 + 84a^3\left(-\dfrac{2}{a}\right)^6 + 36a^2\left(-\dfrac{2}{a}\right)^7 +$

$\qquad 9a\left(-\dfrac{2}{a}\right)^8 + \left(-\dfrac{2}{a}\right)^9$

$\qquad = a^9 - 18a^7 + 144a^5 - 672a^3 + 2016a -$

$\qquad 4032a^{-1} + 5376a^{-3} - 4608a^{-5} + 2304a^{-7} -$

$\qquad 512a^{-9}$

14. $512x^9 + 2304x^7 + 4608x^5 + 5367x^3 + 4032x + 2016x^{-1} + 672x^{-3} + 144x^{-5} + 18x^{-7} + x^{-9}$

15. $(a^2 + b^3)^5$

We will use the preceding coefficients method. There are 6 terms in the expansion. After we find the 3rd coefficient, we can use symmetry to obtain the others.

The first term is $(a^2)^5 = a^{10}$.

The second term is $\dfrac{1\cdot 5}{1}(a^2)^4(b^3)^1 = 5a^8b^3$.

The third term is $\dfrac{5\cdot 4}{2}(a^2)^3(b^3)^2 = 10a^6b^6$.

The other coefficients are 10, 5, and 1. The complete expansion is

$(a^2 + b^3)^5 = a^{10} + 5a^8b^3 + 10a^6b^6 + 10(a^2)^2(b^3)^3 +$

$\qquad 5(a^2)(b^3)^4 + (b^3)^5$

$\qquad = a^{10} + 5a^8b^3 + 10a^6b^6 + 10a^4b^9 +$

$\qquad 5a^2b^{12} + b^{15}$

16. $x^{18} + 12x^{15} + 60x^{12} + 160x^9 + 240x^6 + 192x^3 + 64$

17. $(\sqrt{3} - t)^4$

Expand using factorial notation. Note that $a = \sqrt{3}$ and $b = -t$.

$(\sqrt{3} - t)^4 = \begin{pmatrix}4\\0\end{pmatrix}(\sqrt{3})^4 + \begin{pmatrix}4\\1\end{pmatrix}(\sqrt{3})^3(-t)^1 +$

$\qquad \begin{pmatrix}4\\2\end{pmatrix}(\sqrt{3})^2(-t)^2 + \begin{pmatrix}4\\3\end{pmatrix}(\sqrt{3})^1(-t)^3 +$

$\qquad \begin{pmatrix}4\\4\end{pmatrix}(-t)^4$

$\qquad = \dfrac{4!}{4!0!}(9) + \dfrac{4!}{3!1!}(-3\sqrt{3}t) + \dfrac{4!}{2!2!}(3t^2) +$

$\qquad \dfrac{4!}{1!3!}(-\sqrt{3}t^3) + \dfrac{4!}{0!4!}(t^4)$

$\qquad = 9 - 12\sqrt{3}t + 18t^2 - 4\sqrt{3}t^3 + t^4$

18. $125 + 150\sqrt{5}t + 375t^2 + 100\sqrt{5}t^3 + 75t^4 + 6\sqrt{5}t^5 + t^6$

19. $6! = 6\cdot 5\cdot 4\cdot 3\cdot 2\cdot 1 = 720$

20. 24

21. $1! = 1$

22. 1

23. $\begin{pmatrix}5\\0\end{pmatrix} = \dfrac{5!}{(5 - 0)!0!} = \dfrac{5!}{5!\cdot 1} = 1$

24. 7

25. $\begin{pmatrix}8\\4\end{pmatrix} = \dfrac{8!}{(8 - 4)!4!} = \dfrac{8!}{4!4!} = \dfrac{8\cdot 7\cdot 6\cdot 5\cdot 4!}{(4\cdot 3\cdot 2\cdot 1)\cdot 4!} = \dfrac{8\cdot 7\cdot 6\cdot 5}{4\cdot 3\cdot 2\cdot 1} = 2\cdot 7\cdot 5 = 70$

26. 91

27. Find the 3rd term of $(a + b)^6$.

Note that $3 = 2 + 1$, $a = a$, $b = b$, and $n = 6$. Then the 3rd term of the expansion is

$\begin{pmatrix}6\\2\end{pmatrix}a^{6-2}b^2$, or $\dfrac{6!}{4!2!}a^4b^2$, or $15a^4b^2$.

28. $21x^2y^5$

29. Find the 12th term of $(a - 2)^{14}$.

Note that $12 = 11 + 1$, $a = a$, $b = -2$, and $n = 14$. Then the 12th term of the expansion is

$\begin{pmatrix}14\\11\end{pmatrix}a^{14-11}(-2)^{11}$, or $\dfrac{14!}{3!11!}a^3(-2048)$, or $-745,472a^3$.

30. $3,897,234x^2$

31. Find the 5th term of $(2x^3 - \sqrt{y})^8$.

Note that $5 = 4 + 1$, $a = 2x^3$, $b = -\sqrt{y}$, and $n = 8$. Then the 5th term of the expansion is

$\begin{pmatrix}8\\4\end{pmatrix}(2x^3)^{8-4}(-\sqrt{y})^4$, or $\dfrac{8!}{4!4!}(2x^3)^4(-\sqrt{y})^4$, or $1120x^{12}y^2$

32. $\dfrac{35}{27}b^{-5}$

33. $\log_2 x + \log_2(x - 2) = 3$

$\qquad \log_2 x(x - 2) = 3$

$\qquad\quad x(x - 2) = 2^3$

$\qquad\quad x^2 - 2x - 8 = 0$

$\qquad (x - 4)(x + 2) = 0$

$x = 4$ or $x = -2$

Only 4 checks.

34. $\dfrac{5}{2}$

35. $e^t = 280$

      $\ln e^t = \ln 280$

         $t = \ln 280$

         $t \approx 5.6348$

36. ± 5

37. Find the third term of $(0.313 + 0.687)^5$:

$\binom{5}{2}(0.313)^{5-2}(0.687)^2 = \frac{5!}{3!2!}(0.313)^3(0.687)^2 \approx$

0.145

38. $\binom{8}{5}(0.15)^3(0.85)^5 \approx 0.084$

39. Find and add the 3rd through 6th terms of $(0.313 + 0.687)^5$:

$\binom{5}{2}(0.313)^3(0.687)^2 + \binom{5}{3}(0.313)^2(0.687)^3 +$

$\binom{5}{4}(0.313)(0.687)^4 + \binom{5}{5}(0.687)^5 \approx 0.964$

40. $\binom{8}{6}(0.15)^2(0.85)^6 + \binom{8}{7}(0.15)(0.85)^7 +$

$\binom{8}{8}(0.85)^8 \approx 0.89$

41. There are 11 terms in the expansion of $(2u - 3v^2)^{10}$, so the 6th term is the middle term.

$\binom{10}{5}(2u)^5(-3v^2)^5 = \frac{10!}{5!5!}(32u^5)(-243v^{10}) =$

$-1{,}959{,}552u^5v^{10}$

42. $30x\sqrt{x}$, $30\sqrt{3}x$

43. Expand $(x^{-2} + x^2)^4$ using the Binomial Theorem. Note that $a = x^{-2}$, $b = x^2$, and $n = 4$.

$(x^{-2} + x^2)^4 = \binom{4}{0}(x^{-2})^4 + \binom{4}{1}(x^{-2})^3(x^2) +$

$\binom{4}{2}(x^{-2})^2(x^2)^2 + \binom{4}{3}(x^{-2})(x^2)^3 +$

$\binom{4}{4}(x^2)^4$

$= \frac{4!}{4!0!}(x^{-8}) + \frac{4!}{3!1!}(x^{-6})(x^2) +$

$\frac{4!}{2!2!}(x^{-4})(x^4) + \frac{4!}{1!3!}(x^{-2})(x^6) +$

$\frac{4!}{0!4!}(x^8)$

$= x^{-8} + 4x^{-4} + 6x^0 + 4x^4 + x^8$

$= x^{-8} + 4x^{-4} + 6 + 4x^4 + x^8$

44. $x^{-3} - 6x^{-2} + 15x^{-1} - 20 + 15x - 6x^2 + x^3$

45. $(\sqrt{2} + 1)^6 - (\sqrt{2} - 1)^6$

First, expand $(\sqrt{2} + 1)^6$.

$(\sqrt{2} + 1)^6 = \binom{6}{0}(\sqrt{2})^6 + \binom{6}{1}(\sqrt{2})^5(1) +$

$\binom{6}{2}(\sqrt{2})^4(1)^2 + \binom{6}{3}(\sqrt{2})^3(1)^3 +$

$\binom{6}{4}(\sqrt{2})^2(1)^4 + \binom{6}{5}(\sqrt{2})(1)^5 +$

$\binom{6}{6}(1)^6$

$= \frac{6!}{6!0!} \cdot 8 + \frac{6!}{5!1!} \cdot 4\sqrt{2} + \frac{6!}{4!2!} \cdot 4 +$

$\frac{6!}{3!3!} \cdot 2\sqrt{2} + \frac{6!}{2!4!} \cdot 2 + \frac{6!}{1!5!} \cdot \sqrt{2} +$

$\frac{6!}{0!6!}$

$= 8 + 24\sqrt{2} + 60 + 40\sqrt{2} + 30 + 6\sqrt{2} +$
1

$= 99 + 70\sqrt{2}$

Next, expand $(\sqrt{2} - 1)^6$.

$(\sqrt{2} - 1)^6 = \binom{6}{0}(\sqrt{2})^6 + \binom{6}{1}(\sqrt{2})^5(-1) +$

$\binom{6}{2}(\sqrt{2})^4(-1)^2 + \binom{6}{3}(\sqrt{2})^3(-1)^3 +$

$\binom{6}{4}(\sqrt{2})^2(-1)^4 + \binom{6}{5}(\sqrt{2})(-1)^5 +$

$\binom{6}{6}(-1)^6$

$= \frac{6!}{6!0!} \cdot 8 - \frac{6!}{5!1!} \cdot 4\sqrt{2} + \frac{6!}{4!2!} \cdot 4 -$

$\frac{6!}{3!3!} \cdot 2\sqrt{2} + \frac{6!}{2!4!} \cdot 2 - \frac{6!}{1!5!} \cdot \sqrt{2} +$

$\frac{6!}{0!6!}$

$= 8 - 24\sqrt{2} + 60 - 40\sqrt{2} + 30 - 6\sqrt{2} + 1$

$= 99 - 70\sqrt{2}$

$(\sqrt{2} + 1)^6 - (\sqrt{2} - 1)^6 = (99 + 70\sqrt{2}) -$

                                $(99 - 70\sqrt{2})$

                $= 99 + 70\sqrt{2} - 99 + 70\sqrt{2}$

                $= 140\sqrt{2}$

46. 34

47. $\binom{16}{11} = \frac{16!}{5!11!} = \frac{16 \cdot 15 \cdot 14 \cdot 13 \cdot 12 \cdot 11!}{5 \cdot 4 \cdot 3 \cdot 2 \cdot 1 \cdot 11!}$

        $= \frac{16 \cdot 15 \cdot 14 \cdot 13 \cdot 12}{5 \cdot 4 \cdot 3 \cdot 2 \cdot 1}$

        $= 4368$

$\binom{16}{5} = \frac{16!}{11!5!} = \frac{16!}{5!11!} = 4368$

48. Both are $6.350135596 \times 10^{11}$.

49. See the answer section in the text.

50. $\frac{55}{144}$

51. The expansion of $(x^2 - 6y^{3/2})^6$ has 7 terms, so the 4th term is the middle term.

$$\binom{6}{3}(x^2)^3(-6y^{3/2})^3 = \frac{6!}{3!3!}(x^6)(-216y^{9/2}) =$$

$$-4320x^6y^{9/2}$$

52. $-\dfrac{\sqrt[3]{q}}{2p}$

53. The $(r + 1)$st term of $\left(\sqrt[3]{x} - \dfrac{1}{\sqrt{x}}\right)^7$ is

$\binom{7}{r}(\sqrt[3]{x})^{7-r}\left(-\dfrac{1}{\sqrt{x}}\right)^r$. The term containing $\dfrac{1}{x^{1/6}}$

is the term in which the sum of the exponents is $-1/6$. That is,

$$\left(\frac{1}{3}\right)(7 - r) + \left(-\frac{1}{2}\right)(r) = -\frac{1}{6}$$

$$\frac{7}{3} - \frac{r}{3} - \frac{r}{2} = -\frac{1}{6}$$

$$-\frac{5r}{6} = -\frac{15}{6}$$

$$r = 3$$

Find the $(3 + 1)$st, or 4th term.

$$\binom{7}{3}(\sqrt[3]{x})^4\left(-\frac{1}{\sqrt{x}}\right)^3 = \frac{7!}{4!3!}(x^{4/3})(-x^{-3/2}) = -35x^{-1/6},$$

or $-\dfrac{35}{x^{1/6}}$.

54. 8